# Contents

**Part III Questions regarding imitation, "language," and cultural transmission in apes and monkeys**



**Part IV Developmental perspectives on social intelligence and communication in great apes**



**Part V Development of numerical and classificatory abilities in chimpanzees and other vertebrates**



**Part VI Comparative developmental perspectives on ape "language"**

b

**"Language" a[...] es**
Comparative dev[...]

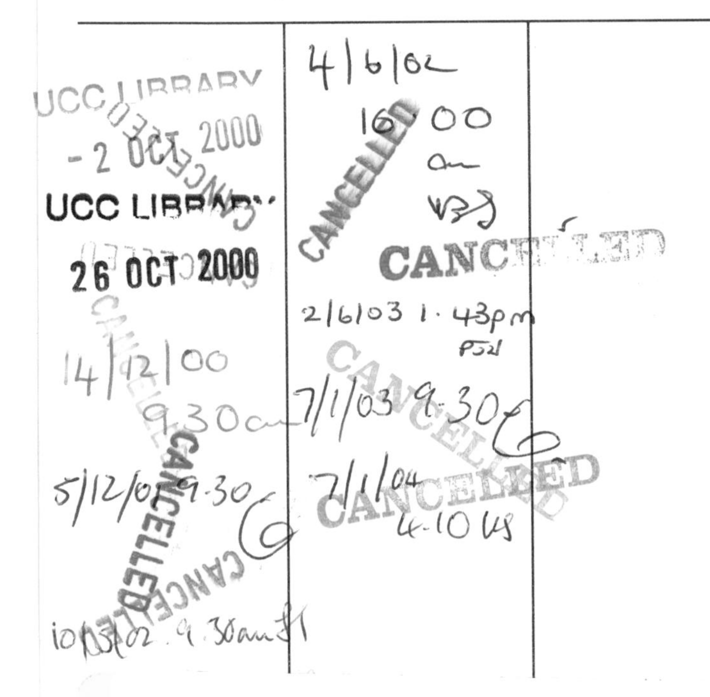

# “Language” and intelligence in monkeys and apes

## Comparative developmental perspectives


*Edited by*

SUE TAYLOR PARKER
*Sonoma State University*

AND

KATHLEEN RITA GIBSON
*University of Texas*

Published by the Press Syndicate of the University of Cambridge
The Pitt Building, Trumpington Street, Cambridge CB2 1RP
40 West 20th Street, New York, NY 10011-4211, USA
10 Stamford Road, Oakleigh, Melbourne 3166, Australia



First published 1990
First paperback edition 1994

Printed in the United States of America

*Library of Congress Cataloging-in-Publication Data*

"Language" and intelligence in monkeys and apes :
comparative development perspectives / edited by
Sue Taylor Parker and Kathleen Rita Gibson.
p. cm.
ISBN 0-521-38028-6
1. Primates – Psychology. 2. Animal intelligence. 3. Cognition in animals.
4. Animal communication. 5. Genetic psychology. 6. Psychology, Comparative.
I. Parker, Sue Taylor. II. Gibson, Kathleen Rita.
QL737.P9L26 1990
599.8′0451-dc20 89-13998

*British Library Cataloguing-in-Publication applied for*

ISBN 0-521-38028-6 hardback
ISBN 0-521-45969-9 paperback

## Contributors

Francesco Antinucci
Istituto di Psicologia
Consiglio Nazionale delle Ricerche
via Ulisse Aldrovandi, 16/B
00197 Roma, Italia

Bernard Baars
Wright Institute
2728 Durant
Berkeley, CA 94704

Kim A. Bard
Yerkes Regional Primate Research Center
Emory University
Atlanta, GA 30322

Gary G. Berntson
Primate Cognition Project
Department of Psychology
Ohio State University
1885 Neil Avenue
Columbus, OH 43210 *and*
Yerkes Regional Primate Research Center

Ben G. Blount
Department of Anthropology and Linguistics
University of Georgia
Athens, GA 30602

Sarah T. Boysen
Primate Cognition Project
Department of Psychology
Ohio State University
1885 Neil Avenue
Columbus, OH 43210 *and*
Yerkes Regional Primate Research Center

Dorothy Munkenbeck Fragaszy
Department of Veterinary and Comparative Anatomy, Pharmacology, and Physiology
College of Veterinary Medicine
Washington State University
Pullman, WA 99164-4830

Kathleen Rita Gibson
Department of Anatomy, Dental Branch
Health Sciences Center
6516 John Freeman Avenue
University of Texas
Houston, TX 77225

Juan Carlos Gómez
Departamento de Psicologia Evolutiva
Universidad Autónoma de Madrid
28049 Madrid, Spain

Patricia Marks Greenfield
Department of Psychology
University of California at Los Angeles
Los Angeles, CA 90024

Howard E. Gruber
Program in Developmental Psychology
Teachers College
Columbia University
525 West 120 Street
New York, NY 10027

Tetsuro Matsuzawa
Primate Research Institute
Kyoto University
Inuyama, Aichi 484, Japan

H. Lyn White Miles
Department of Sociology and Anthropology
University of Tennessee
308 Brock Hall
615 McCallie Avenue
Chattanooga, TN 37403

Sue Taylor Parker
Department of Anthropology
Sonoma State University
Rohnert Park, CA 94928

Irene Maxine Pepperberg
Department of Anthropology
Northwestern University
1810 Hinman Avenue
Evanston, IL 60208

Patricia Poti'
Istituto di Psicologia
Consiglio Nazionale delle Ricerche
via Ulisse Aldrovandi, 16/B
00197 Roma, Italia

Anne E. Russon
Department of Psychology
Glendon College
York University
2275 Bayview Avenue
Toronto, ON M4N 3M6, Canada

E. Sue Savage-Rumbaugh
Yerkes Regional Primate Research Center
Emory University
Atlanta, GA 30322

Michael Tomasello
Department of Psychology
Emory University
Atlanta, GA 30322 *and*
Yerkes Regional Primate Research Center

Jacques Vauclair
Laboratoire de Neurosciences Foncionelles
Unité de Neurosciences Cognitives
31 chemin Joseph-Aiguier
Marseille Cedex 9, France

Elisabetta Visalberghi
Istituto di Psicologia
Consiglio Nazionale delle Ricerche
via Ulisse Aldrovandi, 16/B
00197 Roma, Italia

# Foreword

*Howard E. Gruber*

We live in an era of new disciplines. They emerge for different reasons, often because explosions of knowledge in several related areas permit, indeed demand, reorganization and new syntheses. Such has been the case with cognitive science, ecology, and, in the present instance, "comparative developmental evolutionary psychology," hereinafter CDEP.

But one asks, almost immediately, Didn't Darwin and others do it in the last century? Aren't *The Descent of Man* and *The Expression of Emotions in Man and Animals* prime examples of CDEP? Yes and no. Until one reads this volume, one might advance a tentative yes. But the work reported here shows how far short of the mark the Darwin cohort fell. They simply did not have the densely packed, detailed information necessary to describe the behavior of "closely related species in the same genus" (S. T. Parker's watchword). One obvious consequence of this paucity of knowledge was Darwin's careful refusal to publish a concrete taxonomy of the primates. A branching tree diagram is the only illustration in *On the Origin of Species*, but it is highly abstract, and Darwin used it only to make very general points about the process of evolution. Versions of the tree diagram appear throughout Darwin's notes, beginning in 1837, and one might have expected that by 1871, when he published *Descent*, he would have hazarded a concrete taxonomy. In his private notes there is such a diagram, drawn on April 21, 1868 (reproduced as Figure 1.1a in this volume), but wisely he refrained from publishing it. It remained a complex hypothesis, a private dream of a discipline to come.

In this volume we see the work of those who are making a reality of that dream. Even now, a fleshed-out picture of the evolution of language and intelligence is beyond our grasp. But fragments of that ancient history are falling into place. Parker points out that there have been, in the decade just ending, over 40 scientific papers on primate intelligence, in the neo-Piagetian tradition, aimed at building up our comparative developmental evolutionary image. Contrasted with the dearth of previous decades, this is a great wealth, and it is fascinating to see what the authors of this volume have done with it, as well as with the research stemming from other traditions. On the other hand, that is only four papers per year, or one paper for every four or five extant primate species – so even now the characters of the dream are just assembling.

More important, perhaps, the effort to carry out such a program clarifies the scientific criteria we must satisfy and illuminates the life processes by which behavioral and cognitive evolution must have come about. But here we encounter a grave difficulty. The CDEP program is ambitious and demanding. Theoretical guidance is needed. One of the essential strands of thought must be our theoretical understanding of development. But the dominant theories of development – Freud, Erikson, Piaget – all depict linear, nonbranching series: Stage 1, then Stage 2, then Stage 3, and so forth. Can a linear model of development be mapped onto a branching model of evolution? Or do we need, as I believe, nonlinear models of development?

In principle, Piaget's general theory does not require a linear model. Indeed, in Piaget's borrowing of Waddington's image of the "epigenetic landscape" there is a hint of the greater capaciousness of a branching model. And in this volume – for example, in Antinucci's account of buccal prehension in the macaque – we see the emergence of a novel behavior seemingly as a peculiarity of developmental timing: In a species with very early locomotion a creature can interact with an object before it quite knows what an object is. This departure from CDEP linearity is just the sort of dialectical fact we need in great profusion in order to realize the CDEP dream.

I am struck by the many analogies between the insights of CDEP and my own studies of human creativity. The two fields share the need to combine carefully the diligent search for whatever is general (similarities among related species in one case, among creators in the other) and the sensitive awareness of what is special about this unique organism on the tree of life or that unique person in the pantheon of creators.

The two fields share also a need to combine different research methods – field studies, laboratory studies, and case studies of single individuals such as Gomez's gorilla Muni or Gibson's cebus Andy. Gibson makes a valuable remark about insight: One cannot know whether or not a given moment of behavior is novel unless one knows the particular history of the individual in question. If an insight must be novel, it follows that case studies that track the history of the individual are essential for the scientific investigation of moments of insight. In my own work on the development of Darwin's thinking, I did in fact follow Gibson's precept, but I cannot say that I ever thought of it so clearly, and I believe that no better can be said of other students of creative insight. In this volume, Gibson's remarks on the role of infant canalization in the later high development of manipulative skills are also valuable in suggesting a developmental model of human "giftedness." Another theme unites the two fields: the emphasis, in this volume and in my work on creativity (Wallace & Gruber, 1989), on intentionality. This emphasis is necessarily coupled with a refusal to be intimidated by charges of teleological thinking. Each discipline illuminates the other.

These examples bring out a more general point: The emergence of a new discipline presents new epistemological problems and new opportunities. Returning to the contrast between Darwinian beginnings and contemporary advances, we see a notable difference in strategy. Darwin had a two-pronged strategy: On the one hand, he needed to demonstrate *continuity* in nature. For that purpose he could exploit the rudiments of intelligence to be found in lower organisms (such as his work on the behavior of earthworms).

On the other hand, he needed to illuminate the adaptive significance, within an evolutionary framework, of specialized structures and functions. This he could do with great skill and command of detail in certain organisms, notably in his botanical work. But he simply did not have the material to work with to make the same kind of point about the behavior of primates. Thus, *The Descent of Man* is primarily an exploration of rudiments, homologies, and continuities. The area he left almost untouched, the study of the adaptive significance of the *differences* in the development of intelligence among closely related species, has become the vacuum into which modern research now rushes.

Inevitably, the way in which a new synthesis comes about reflects the often unspoken scientific values of the investigators. In this volume we see choices being made: Specialization or synthesis? Anthropomorphism as methodological trap or as valuable asset? The prolonged study of the individual as a failure of scientific discipline or as a deeper penetration into living reality?

There is a certain irony in the difference between the speed of gorilla development and the speed of human scientific development. It has taken us decades to come as far as we have in treating the human individual not as an object but as a living subject. It appears from Gomez's work that gorillas make that transition in a matter of months.

## References

Darwin, C. (1859). *On the origin of species.* London: Murray.

Darwin, C. (1871). *The descent of man and selection in relation to sex.* London: Murray.

Darwin, C. (1872). *The expression of emotions in man and animals.* London: Murray.

Gruber, H. E. (1978). Darwin's irregularly branching "tree of nature" and other images of wide scope. In J. Wechsler (Ed.), *On aesthetics in science* (pp. 121–140). Cambridge, MA: M.I.T. Press.

Gruber, H. E. (1981). *Darwin on man: A psychological study of scientific creativity.* University of Chicago Press. (Original work published 1974)

Piaget, J. (1971). *Biology and knowledge.* University of Chicago Press. (Original work published 1967)

Waddington, C. H. (1957). *The strategy of the genes.* London: Allen & Unwin.

Wallace, D. B., & Gruber, H. E. (1989). *Creative people at work: Twelve cognitive case studies.* Oxford University Press.

# Preface

The editors conceived this book with the double purpose of introducing to a larger public a new approach to the study of animal abilities and of stimulating additional studies in the new research program described in this volume. This research program, which we call "comparative developmental evolutionary psychology," is based on the application of frameworks from human developmental psychology and evolutionary biology to comparative studies of primate abilities. All of the studies reported in this volume are either directly or indirectly developmental; that is, they either trace the ontogenetic development of specific abilities within developmental frameworks or use such frameworks to compare the nature and level of abilities of their subjects to those of human infants and children at various stages of development. None of the studies in this volume have been published elsewhere. The majority of the chapters grew out of presentations made at a 3-day workshop held as a presession of the Ninth International Congress of Primatology at Göttingen, West Germany, in July 1986. The additional chapters are by colleagues of the original contributors or by researchers whose work we discovered subsequent to the workshop.

This book is the second volume devoted specifically and exclusively to studies of animal intelligence and cognition based on comparative developmental (Piagetian, neo-Piagetian, and developmental psycholinguistic) perspectives. The first volume, which was published last year, reported on long-term studies of intellectual development in cebus, gorillas, and macaques at the Comparative Psychology Laboratory at the Consiglio Nazionale delle Ricerche (CNR) in Rome (Antinucci, 1989). Several members of that laboratory are contributors to this book.

The book is divided into six parts. Part I, "Theoretical frameworks for comparative developmental studies," comprises four chapters. In Chapter 1, Parker examines the history of biological anthropology and developmental and comparative psychology in the United States and traces the strands that contributed to the emergence of *comparative developmental evolutionary psychology*. In chapter 2, Parker and Baars examine the usages of a variety of terms used to describe animal abilities and show how these usages reflect theoretical positions. Gibson, in chapter 3, presents a new information-processing model for the neurological mechanisms underlying intelligent behavior, emphasizing the critical importance of quantitative parameters in differentiating instinct and intelligence. In chapter 4, Parker uses life history strategy theory to examine the role of energy constraints in the rate of maturation of intelligent, large-brained primates.

Part II, "Comparative developmental perspectives on cebus intelligence," comprises the next four chapters. In chapter 5, Antinucci compares rates of development of sensorimotor intelligence, prehension, and locomotion in cebus monkeys, long-tailed macaques, gorillas, and human infants and proposes a causal model to explain the relationship between third- and fifth-stage sensorimotor intelligence and early development of prehension versus locomotion. In chapter 6, Fragaszy introduces the developmental concepts of *dynamic systems theory* to the study of motor development in nonhuman primates. Applying these concepts to comparative study of motor development in hand-reared and colony-reared cebus, she demonstrates complex interactions among variables in the system. In chapter 7, Gibson describes the intellectual abilities and disabilities of her pet cebus monkey, analyzing the complex interplay of smart and dumb behaviors from a neo-Piagetian perspective. In chapter 8, Parker and Poti' describe the use of a stick as a tool by captive cebus. Comparing the development of the scheme in juveniles and in an infant in terms of Piaget's sensorimotor period stages, they also emphasize the complex interplay between innate and intelligent aspects of cebus behavior.

Part III, "Questions regarding imitation, 'language,' and cultural transmission in apes and monkeys," contains three theoretical pieces. In chapter 9, Visalberghi and Fragaszy examine the imitational capacities of monkeys and apes, combining an analysis of the literature on object manipulation and tool use in monkeys with an ingenious series of experiments designed to identify alternative mechanisms of learning. Tomasello, in chapter 10, reviews experimental approaches to imitation, and then examines published reports of cultural transmission of tool-using behaviors and communicatory signals in chimpanzees, emphasizing the need to avoid overinterpretation of behaviors and to consider simpler alternative mechanisms than cultural transmission. In chapter 11, Vauclair reintroduces de Saussure's original meaning of the concept of "arbitrariness" and discusses problems arising from misinterpretations of this and other linguistic concepts in the literature on ape language learning.

Part IV, "Developmental perspectives on social intelligence and communication in great apes," consists of four chapters. In chapter 12, Gómez describes the spontaneous development of prelinguistic signals in a captive gorilla infant interacting with human caretakers, interpreting these instrumental abilities within a Piagetian framework. Bard, in chapter 13, describes a similar process occurring among infant orangutans reared by rehabilitated orangutan mothers in a wild setting, also using a Piagetian framework of analysis. In chapter 14, Russon describes the stages of emergence of peer play in captive chimpanzees using a Piagetian perspective. In chapter 15, Blount describes age related differences in the social use of space among captive pygmy chimpanzees.

Part V, "Development of numerical and classificatory abilities in chimpanzees and other vertebrates," comprises three chapters. Boysen and Berntson, in chapter 16, describe the development of number concepts in captive chimpanzees, comparing their conceptual levels to those of human children. In chapter 17, Matsuzawa makes a related comparison between the classificatory abilities of a chimpanzee and those of human children of various ages. Pepperberg, in chapter 18, compares the numerical and classificatory abilities of an African Grey parrot to those of human children and chimpanzees, demonstrating an interesting case of convergent evolution of specific mental abilities.

There are two chapters in Part VI, "Comparative developmental perspectives on ape 'language.'" In chapter 19, Miles describes the stages of development reference in a signing orangutan in relation to various aspects of cognitive development, including the development of imitation, symbolic play, deception, and self-recognition, using neo-Piagetian frameworks. Her data are particularly significant for their long-term developmental perspective (she presents data spanning 8 years of development). Finally, in chapter 20, Greenfield and Savage-Rumbaugh describe the linguistic abilities of pygmy chimpanzees, focusing particularly on the parallels between grammatical development in this species and in human children.

In addition to their comparative developmental focus, these chapters betray a variety of related preoccupations: questions regarding continuities and discontinuities among human and nonhuman primates, the evolutionary significance of various abilities and the proximal mechanisms generating specific abilities and developmental patterns, the interpretation and analysis of behaviors, and the neurological bases of cognitive abilities. The diversity of approaches to these questions revealed in the chapters in this volume reflects the diversity of disciplinary training and cultural traditions of the contributors. We hope that all of the contributors will continue to pursue these themes and continue to interact to refine their questions and techniques and to broaden their theoretical and methodological perspectives. Working with them has certainly done that for us.

### Reference

Antinucci, F. (Ed.). (1989). *Cognitive structures and development in nonhuman primates.* Hillsdale, NJ: Erlbaum.

# Acknowledgments

We thank the International Primatological Society (IPS) and the University of Göttingen for hosting the three-day Workshop on Comparative Cognition preceding the XIth Congress of the IPS in 1986, which led to this volume. We especially thank the Department of Anatomy of the University of Göttingen for providing the facilities for our workshop, and Dr. Ulrich Zeller of the Anatomy Department, who was the gracious host to our workshop. We also thank the National Science Foundation for providing travel funds for some of the participants.

In a larger frame, we are grateful for many circumstances, institutions, and scholars who have inspired, shaped, supported, and encouraged us, individually and jointly, over the past 30 years. In Gibson's case, these include Marian Diamond, Theodore McCown, and Sherwood Washburn of the University of California, Berkeley. In Parker's case, these include Erving Goffman, Richard Dawkins, Jane Lancaster, Phyllis Dolhinow, and John S. Watson (all of whom taught at the University of California, Berkeley, during the late 1960s). We also thank our fellow students of the Departments of Anatomy, Anthropology, Psychology, Sociology, and Zoology at the University of California, Berkeley.

We are indebted to Sonoma State University and the University of Texas, respectively, for supporting us for the past 20 years and for providing the working conditions that made this volume possible. We are also indebted to the International Society of Primatologists, the American Society of Primatologists, The Animal Behaviour Society, the American Anthropology Association, the American Association of Physical Anthropologists, the British Society for Developmental Psychology, the Jean Piaget Archives, the Consiglio Nazionale delle Ricerche of Rome, the Fulbright Foundation, the California State Colleges and Universities Grants for Research, Scholarship and Creative Activity, and the Social Science Research Council, which have provided encouragement and support over the years. In addition, we are indebted to numerous colleagues who have made it possible for us to publish and to participate in national and international symposia and workshops that have expanded our perspectives and sharpened our thinking. We are particularly grateful to A. B. Chiarelli, former editor of the *Journal of Human Evolution* (current editor of *Human Evolution*), and to S. Harnad, editor of the *Behavioral and Brain Sciences* (BBS), for publishing our initial theoretical articles, and to generous reviewers who encouraged us in our joint venture in BBS. Not least, we are beholden to the many scholars who have worked in Comparative Developmental Evolutionary Psychology (CDEP), and especially to the contributors to our workshop and this volume. We are especially pleased at the complementary research into CDEP done since the early eighties at the Consiglio Nazionale delle Ricerche of Rome under the direction of F. Antinucci.

We are grateful to our parents for nurturing our intellectual interests when we were children, and to our partners and children who have encouraged us and patiently read our manuscripts over the years. Parker particularly thanks her friend Andrew Wilson, her son Aron Branscomb, and his father Elbert Branscomb.

Finally, we thank Michael Gnat and other editors at Cambridge University Press for their assistance.

*Part I*

# Theoretical frameworks for comparative developmental studies

# 1 Origins of comparative developmental evolutionary studies of primate mental abilities

*Sue Taylor Parker*

This chapter traces the origins of a new research program in *comparative developmental evolutionary psychology* (CDEP). This new program integrates frameworks from developmental psychology and evolutionary biology for the purpose of studying the ontogeny of mental abilities in monkeys and apes and reconstructing the evolution of primate mental abilities. Because CDEP addresses issues that were suppressed during the professionalization of psychology and anthropology, it should be seen in relation to the fate of the 19th-century evolutionary psychology and anthropology of Darwin, Wallace, Haeckel, and Spencer.

## Nineteenth-century evolutionary psychology and anthropology

Charles Darwin (1809–1882) was a developmental psychologist, a physical anthropologist, a comparative psychologist, and an ethologist as well as a naturalist and taxonomist. His developmental approach can be seen in his use of comparative embryological data (Darwin, 1859) and in his article "A Biological Sketch of an Infant" (1877, reprinted in Gruber, 1981). His comparative psychological and ethological approach can be seen in his discussion of instinct (Darwin, 1859) and in his discussion of the evolution of facial expressions (Darwin, 1872). His physical anthropological approach can be seen in his discussion of the origins of human mental powers and the origins of racial differences (Darwin, 1871). Darwin conceived of both natural selection and sexual selection, but never resolved the mystery of inheritance. Although he recognized that variation was prerequisite to selection and saw that spontaneous variations arose, he also invoked the Lamarckian mechanism of the inheritance of acquired characteristics. He emphasized the continuities in the mental and emotional life of animals and humans and considered humans to be a species closely allied to the apes. He argued that humans belong to one species whose racial differences arose primarily through sexual selection.

Alfred Russell Wallace (1823–1913) was a physical and cultural anthropologist as well as a naturalist and taxonomist. He independently conceived of natural selection, but rejected sexual selection and other "psychological mechanisms" of mental choice in favor of ecological mechanisms. Unlike

Darwin, he also rejected the inheritance of acquired characteristics. Wallace's anthropological approach can be seen in his article "The Origin of Human Races and the Antiquity of Man Deduced from the Theory of 'Natural Selection'" (1864), wherein he discusses the role of natural selection in human origins and racial origins.

Wallace concluded that the human body stopped evolving after the human intellect began to free the species from selection, but that the human head and brain continued to undergo modification and will continue to do so "till the world is again inhabited by a single homogeneous race, no individual of which will be inferior to the noblest specimens of existing humanity" (Wallace, 1864, p. clxix). He also said that intelligence allows humans to transcend nature: "He is indeed a being apart, since he is not influenced by the great laws which irresistibly modify all other organic beings," and hence humans could reasonably be placed in their own order or class (Wallace, 1864, p. clxviii). Wallace always doubted the continuity between the mental powers of animals and humans, and later in life he carried that belief to its ultimate conclusion – that the brain could not be explained by selection and must have a divine origin. Wallace (1889), who was younger than Darwin and lived longer, had the "last word" in his book *Darwinism*. Unlike Darwin, Wallace was of humble origin and struggled to make a living. Wallace was an idealist and enjoyed the role of rebel and iconoclast, embracing vegetarianism, feminism, spiritualism, and other unpopular causes (Glickman, 1985; Hartman, 1989; S. Glickman & H. Hartman, pers. commun.).

Herbert Spencer (1820–1903) was a philosopher, sociologist, and psychologist who wrote *The Principles of Sociology*, *The Principles of Biology*, and *The Principles of Psychology* (1875) as parts of a grand scheme in "synthetic philosophy." He is best known for his developmental evolutionary scheme, which he first outlined in 1852 in a paper entitled "Development Hypothesis" (7 years before Darwin's *Origin of Species*) and later elaborated in various other books. That hypothesis, which was based on von Baer's ideas about embryo development, stipulated universal evolution in stages from undifferentiated homogeneous to differentiated heterogeneous forms, and it applied across all inorganic and organic phenomena. Spencer developed a scheme of evolution of social emotions, intelligence, language, religion, and human institutions, which he defined as "superorganic" phenomena arising from the coordinated efforts of individuals to adapt to their environment (Kardiner & Preeble, 1961, p. 41). Spencer believed in progressive evolution of the mind through Lamarckian mechanisms of inheritance. Unlike Darwin and Wallace, Spencer lacked experience as a naturalist, and unlike them he relied almost exclusively on deduction. Kardiner and Preeble (1961, p. 38) quote Thomas Henry Huxley's aphorism: "Spencer's definition of a tragedy was the spectacle of a deduction killed by a fact."

Ernst von Haeckel (1834–1919) was a developmental biologist and a physical anthropologist who supported Darwin's theory of descent with modification with a variety of evidence, including his theory of development, which postulated the

> inner causal connection between ontogeny and phylogeny, the parallelism between the individual history of the development of organisms, and the palaeontological history of the development of their ancestors, a connection which is actually established by the laws of Inheritance and Adaptation, and which may be summed up in the words: ontogeny, according to the laws of inheritance and adaptation, repeats in its large features the outlines of phylogeny. (Haeckel, 1876, p. 352)

Like Darwin and Spencer, Haeckel accepted Lamarck's theory of the inheritance of acquired characteristics. He also wrote on the stages of evolution leading to humans and on the nature and polygenic origins of human races.

Whereas Darwin's and Wallace's ideas of natural selection and descent with modification upset some applecarts, Haeckel's and Spencer's ideas generally fit into and reinforced 18th- and 19th-century notions of the "Great Chain of Being," the superiority of European males, the perfection and fixity of species, an essentialist, typological concept of species and races, polygenic origin of races, purposive creation of life forms, vaguely distinguished ontogenetic and phylogenetic "stages" of development, and the inheritance of acquired characteristics, especially mental characteristics. Most people simply assimilated the new ideas to their preconceived notions and ignored the inconsistencies.

Ontogeny and phylogeny were fused in an undifferentiated global concept of development: According to the popular view, even after Darwin,

> each [species] could realize its inner potential, which gradually unfolded. The development of species was conceived on the analogy of the evolution of the embryo. Ontogeny, it was held, recapitulated phylogeny. The term *evolution* itself was generally used in this embryological sense until about 1880, and neither Darwin in *The Origin of Species* nor Morgan in *Systems* or *Ancient Society* used the word *evolution* at all. (Kuper, 1985, p. 4)

The ideas of Darwin, Wallace, Haeckel, and Spencer exerted an enormous if confused influence on the ideas of scholars throughout the world in the period from approximately 1840 to 1890. Indeed, these four men are in the distinctly ironic position of being the fathers or stepfathers of a number of mutually incompatible research programs, including ethology and comparative and developmental psychology, and cultural and physical anthropology. A brief review of the origins of these programs is an appropriate introduction to the discussion of a new research program that turns back to the neglected issues of the evolution of mentality, the ontogeny of mentality, and the evolution of ontogeny.

According to Kroeber, Darwin's publication of the *Origin of Species* had a tremendous effect on anthropology:

> A crop of founding fathers of a science of anthropology sprung up within two years. . . . Here is the list: [1861 Bachofen, *The Matriarchate,* and Maine, *Ancient Law;* 1864 Fustel de Coulanges, *The Ancient City;* 1865 McLennan, *Primitive Marriage*, and Tylor, *Researches into the Early History of Mankind;* 1870 Lubbock, *Origin of Civilization;* 1871 Morgan, *Systems of Consanguinity,* and Tylor, *Primitive Culture*]. The [principal] additions that might be made to this early cluster would be

Morgan's *Ancient Society* in 1877 and Tylor's *Anthropology* of 1881. (Kroeber, 1960, pp. 10–11)

(In reality, these books derive more from the ideas of the disreputable Spencer and Haeckel than from those of Darwin.) Kroeber goes on to say that those early anthropologists failed to achieve Darwin's success because the subject matter of culture is "protean" and because they lacked an appropriate data base of the kind Darwin had:

Their answers were mostly of little worth, because they asked the wrong questions; but they founded a science. They did this by occupying a phenomenal area which rested on the organic but extended beyond it, and which therefore had a degree of autonomy, including processes of its own. A central field became defined in these first dozen years: it was that of institutions and custom. (Kroeber, 1960, p. 12)

What Kroeber fails to note is that the central field of "institutions and custom" (i.e., cultural anthropology) was founded on a massive reaction against the ideas of Morgan and others and, by extension, against the ideas of Darwin, Haeckel, and Spencer. He also fails to note that the reaction cut across psychology as well as anthropology and that it involved the demise of studies of ontogeny and mental evolution and the triumph of Wallace's ideas of discontinuity between the mentality of humans and animals and the human's mental transcendence of nature (without reference to Wallace's ideas that the brain continues to evolve, and "higher" races continue to displace "lower" races [Wallace, 1864]).

In the period following the one Kroeber describes, from about 1890 to 1920, various research programs in the social sciences emerged and became professionalized as academic disciplines in American universities (Spencer, 1982; Stocking, 1968) through reaction to these ideas. During that period, the founders and practitioners of these new disciplines struggled to distinguish themselves from their predecessors and from one another. Psychology protected itself from the tainted recapitulationism of Haeckel and G. Stanley Hall by throwing out the concept of ontogeny and distinguished itself from philosophy by throwing out the unobjective concept of mind. Anthropology protected itself from the tainted idea of cultural evolution by throwing out the concept of evolution, except as it applied to bodily structures, and protected itself from the tainted concept of race by substituting the concept of culture.

## Origins of anthropology in the United States

Interest in exotic peoples and in archeological and fossil remains of earlier peoples long predated the work of Darwin and Wallace, as did the first anthropological society, which was founded in 1799. The Société des Observateurs de l'Homme, which included biologists (e.g., Cuvier and Lamarck), linguists, explorers, physicians, and other scholars, was

dedicated to "the science of man in his physical, moral and intellectual aspects," . . . a proposed but never published volume of the Société's memoirs . . . included suggestions for a "methodical classification of race" on the basis of a complete comparative anatomy of peoples; a *Comparative anthropology* of the customs and usages of peoples; ... and a *Comparative dictionary of all known languages.* (Stocking, 1968, p. 16)

Although the members of that society planned a "grandiose" research and collecting expedition that sailed toward Australia in 1800, its aims were largely unfulfilled because of various problems (Stocking, 1968).

Edward B. Tylor (1832–1917), known as the institutional father of cultural anthropology, lacked a university degree and gained his knowledge of other societies through his personal travels, on the basis of which he began to lecture and write on the origins of religion and language. He was appointed keeper of the University Museum at Oxford and then in 1896 became Oxford University's first professor of anthropology. Tylor was basically an 18th-century rationalist who believed in progress and saw cultural practices as springing from rational bases. He was a reductionist in the sense that he thought that social structure and culture derived from the rational mind. His "comparative method" lacked an ethnographic focus: For Tylor, "the value of the comparative framework is that it allows the anthropologist to discover, through induction, the common processes of thought behind human institutions.... All traits which share the same cause belong together, regardless of where they are found" (Hatch, 1973, pp. 36–37).

Lewis Henry Morgan (1818–1881), an American whose ideas on cultural evolution were quite influential, was also an amateur anthropologist, though a banker by profession. His work on the Iroquois and his magnum opus, *Ancient Society* (1877), made him world-famous. He met and was praised by Darwin, and his work strongly influenced Marx and Engels (Langness, 1987).

Although the subject of his magnum opus was the evolution of social institutions, Morgan's [*Ancient Society*] may also be regarded as the explication of a scheme of mental evolution. Institutions developed out of ideas in the human mind, out of "germs of thought."... But the evolution of these germs was more than cultural; it involved a Lamarckian evolution of the brain itself. (Stocking, 1968, p. 116)

Morgan's idea of the stages of evolution of subsistence, family, government, and religion profoundly shaped American anthropology through antithesis, providing the springboard for Franz Boas's antievolutionary school of historical particularism (Langness, 1987; Stocking, 1968), which in turn profoundly biased American anthropology against the study of cultural and mental evolution.[1]

Franz Boas (1858–1942), the father of American anthropology, was born in Germany and educated in the German idealist tradition of thought as a physicist and geographer (Hatch, 1973; Kuper, 1988). As would be expected because of his German roots, Boas took a diffusionist position in the continuing debate with evolutionists. Despite his honorary title, Boas was not the first institutional anthropologist in the United States.

Anthropology in America had its institutional origins around the turn of the century in local museums and offices of Indian affairs in Philadelphia,

Washington, Boston, and New York. John Wesley Powell set up the Bureau of Ethnology at the Smithsonian Institution in Washington, and Frederic Ward Putnam took over a chair in linguistics and folklore at Harvard that was similar to one Daniel Garrison Brinton held in Philadelphia. Putnam then went on to found anthropological institutes in Boston (the Peabody), Chicago (the Field Museum), New York City (the American Museum), and San Francisco, where he helped found the Department of Anthropology at the University of California at Berkeley. Harvard began a graduate program that emphasized archeology and graduated 15 PhDs between 1894 and 1919 (Stocking, 1968).

Although not the first, Boas was the most influential anthropologist of the period. In 1897, after several years of struggle, Boas obtained a position at Columbia University, where he established a PhD program in cultural anthropology that trained the major figures of the following generation, including Kroeber, Benedict, Mead, Goldenweiser, and White. Boas brought systematic data gathering and sophisticated data analysis to anthropology. He defined cultural anthropology largely in antithesis to physical anthropology and devoted his institutional efforts (as opposed to his research efforts) solely to cultural anthropology. Boas's program

> vocally proclaimed its independence from biology, relegated the study of man as a physical organism to a distinctly secondary position, denied in large part the significance of biological race, and raised to central theoretical importance a concept [culture] which had not yet shed the aura of dilettantish humanism. (Stocking, 1968, p. 189)

Culture, in Boas's terms, and especially in Kroeber's terms, was "super-organic" in the twin sense that it transcended individuals and transcended nature. In many ways that concept of culture harkened back to Spencer. It was also a transformation of Wallace's notion (1864) of the human mind as transcendent over nature. Given that that concept of culture was an antithesis to racialist and evolutionist notions of the mind, Boas's neglect of physical anthropology is understandable. It is also ironic, however, given that his research helped shape American physical anthropology into an empirical science.[2]

Boas pioneered in the application of statistical techniques, investigating processes that produce asymmetries in trait distributions. He also invented longitudinal growth studies and demonstrated the influence of environment on growth (e.g., demonstrating that the children of immigrants had head forms unlike those of their parents). Finally, in his book *The Mind of Primitive Man*, Boas (1911/1938) presented an amazingly modern definition of race as a collection of family lines and argued against the commonly held notion that some races were superior to others and that race, culture, and language were connected, hereditarily determined attributes. Boas's empiricist, positivist research program and his redefinition of culture and subordination of biology to environment were consistent with the themes of environmentalism and behaviorism that were emerging in academic departments in the United States. Indeed, if anything, Boas identified the concept of race as the focus of physical anthropology, and then, by distinguishing culture from race and demonstrating the fallacies of race, he drove physical anthropology out of the field.

Although Stocking credits Broca with the founding of modern physical anthropology, Brace follows Hrdlička in arguing that the distinction belongs to Samuel George Morton, Philadelphia physician who did the first comparative anthropometric study of prehistoric skeletal material.

> If the florescence of anthropometry was largely the result of the energy and application of Broca . . . the foundation on which Broca built had been solidly laid by Morton, even though Morton has never been given the credit he is due. The techniques that Broca elaborated from Morton's beginnings were adopted in England following the translation of Topinard's book . . . and, later, Hrdlička and Hooton saw to it that these returned to American anthropology. (Brace, 1982, p. 19)

(See Parker & Baars, P&G2,[3] for further discussion of Morton and Broca.)

Physical anthropology had its origins in two related lines of study in 18th-century Europe: the study of the brain and the study of racial types. Long before the time of Darwin and Wallace, philosophers and anthropologists fought over the monogenic and polygenic theories of racial origins. Whereas that argument originally pertained to whether or not Adam and Eve were the progenitors of all races, it was later carried over into an evolutionary frame (e.g., Wallace, 1864). One of the hotly debated issues concerned the hybridity of the races (i.e., whether or not they could produce fertile offspring). The polygenists argued that races were created for life in separate geographic provinces, were mutually infertile, and would die out if moved to other regions. They also believed, as a corollary, that the original purity of races had been sullied by race mixture, and they worried about race mixture causing degeneracy.

The polygenists charged the monogenists, who argued that races had a single origin, with explaining problems of acclimation to new regions.

> Indeed, it might be argued that the two sides of the unity controversy corresponded to two basic approaches to the study of man, the monogenists representing cultural and the polygenists physical anthropology. However this may be, there is no denying a very close relationship between late nineteenth-century physical anthropology and the earlier polygenist writers. It was the nineteenth-century polygenist of greatest retrospective stature, Paul Broca, who is generally acknowledged as the founder of modern physical anthropology, and who dominated the field until his death in 1880. (Stocking, 1968, p. 56)

Broca founded the Société d'Anthropologie de Paris in 1859, the Laboratoire d'Anthropologie in 1867, and the Ecole d'Anthropologie in 1876 (Spencer, 1982).

While cultural anthropology was gaining momentum as an academic discipline, physical anthropology in the United States grew slowly. Although Harvard University produced the first PhD in physical anthropology in 1898 under Frederic Ward Putnam, the program languished after Putnam's retirement and was revived by Earnest Albert Hooton (1887–1954) only in the 1920s. After 1926, a small but steady stream of PhDs began to come out of Harvard, but the majority of them worked as anatomists in medical schools or anatomy departments or museums (Spencer, 1982; S. L. Washburn, pers. commun.). Aleš Hrdlička (1869–1943), who worked at the National Museum of Natural History in Washington, DC, during 1903–1913, founded the *American Journal of Physical Anthropology* in 1918. Hrdlička tried but failed to turn his laboratory into an institute modeled after that of Broca (Spencer, 1982).

In contrast to Boas, Hooton worked in the typological anthropometric tradition of racial studies (e.g., *Up from the Ape*). Indeed, in the 1950s and 1960s three of Hooton's students published major typologies of race that diverged in interesting ways from that of Hooton (Hooton, 1946; Coon, 1962; Coon, Garn, & Birdsell, 1950; Garn, 1961). Although those authors struggled to reconcile various concepts from population biology into their schemes, in the end their schemes were typological, and even polygenist (Coon, 1962). The full implications of the modern synthesis of evolutionary biology were slow to dawn on physical anthropologists.

Before, during, and after these typological efforts, other anthropologists and social scientists were struggling to defuse the concept of race. In 1950 the United Nations Educational, Scientific, and Cultural Organization (UNESCO) published a statement on race, coauthored by an international group of social scientists led by Ashley Montagu, affirming that all humans belong to one species, that differences between groups are due to evolutionary forces, and that all races are mentally equal. In 1951, UNESCO published a revised statement, coauthored by a group that included biologists, that omitted the last claim (Haraway, 1988). The eventual repudiation of typological concepts of race came not only through better training of students but also through a massive social and academic reaction against racism that arose in response to Nazism and the genocide of World War II (Montagu, 1964).

Just as the early studies of race were dominated by typological, essentialist doctrines attended by the technique of anthropometry, so were the early studies of fossil humans. Just as the brain and the skull were considered the most important characteristics distinguishing the races, they were considered the most important characteristics distinguishing hominid species. Brain size was seen as the visible expression of the Great Chain of Being extending down from the angels to white European males to females to non-European savages and apes and monkeys. Paleoanthropologists agreed implicitly with Wallace's notions of the great antiquity of humans, the discontinuity between the mentality of animals and humans, and the transcendence of human mentality (Wallace, 1864).

As late as the 1950s the dominant model of hominid evolution, which was promulgated by G. Elliot Smith and Arthur Keith, was that the branching between hominids and nonhominids was ancient, probably occurring in the

Oligocene epoch (35–23 million years ago), that the large human brain evolved early, and that the various primitive fossil forms such as Neanderthal and Pithecanthropus had been side branches that arose and became extinct after the more modern, large-brained "presapiens." This latter view was part and parcel of the "multilinear" model of hominid phylogeny championed by the British and the French, as opposed to the unilinear model championed by the Americans and the Germans. Marcellin Boule of the Museum of Natural History in Paris had reconstructed the first Neanderthal skeleton, from Chapelle-aux-Saints, as a primitive, barely bipedal creature that could be safely relegated to a side branch, whereas Hrdlička, of the National Museum of Natural History in Washington, DC, argued that there was a Neanderthal phase in human evolution (Hammond, 1988).

This model of hominid evolution, which Hammond (1988) calls the "shadowman" paradigm, had a profound influence on the course of fossill human studies that is still felt today. Most notably, it laid the foundation for relegation of Neanderthal and Pithecanthropus to side branches, for the acceptance of the Piltdown forgery as genuine, and for rejection of the diagnosis of australopithecines as hominids. Because the discovery in England in 1911 of an "ancient human" with a modern skull and a primitive jaw exactly fit the shadowman paradigm, the find was immediately embraced as "the first Englishman." The theoretical confirmation it offered the dominant model made paleoanthropologists reluctant to accept the hominid status of the small-brained, bipedal Taung baby that Raymond Dart discovered in South Africa in 1924 (Hammond, 1988).

That the eventual acceptance of the Taung baby came not as a result of discovery of increasing numbers of related specimens in South Africa but rather through the influence of Le Gros Clark suggests the deep influence such paradigms can exert. It is interesting to reflect on the preevolutionary and antievolutionary nature of this paradigm and the protection if offered against addressing the issue of continuity between apes and humans. Ironically, the anthropologist who was to address this issue squarely, and who was to reconstruct physical anthropology in light of the modern synthesis of evolutionary biology, was Hooton's student Sherwood L. Washburn.

Sherwood L. Washburn had an early exposure to interdisciplinary research as a member of an expedition that went to Thailand in 1936 under the direction of Harold Coolidge of the Museum of Comparative Zoology at Harvard. During the course of that expedition Washburn worked with the anatomist Adolf Schultz, who was collecting anatomical specimens, and the psychologist Clarence Ray Carpenter, who was studying primate behavior. The time he spent with Carpenter observing the behavior of gibbons gave him a new appreciation of "anatomy in action" that was to come to full fruition many years later. Washburn's appreciation of form and function had also been shaped by his experiences studying anatomy with W. T. Dempster at the University of Michigan and with Le Gros Clark at Oxford University. Those experiences and his work at the Museum of Comparative Zoology allowed

Washburn to "appreciate the best of Hooton but stay removed from his research methods and some conclusions" (Washburn, 1983, p. 2).

Washburn's independence from Hooton was reinforced by his friendship with Theodosius Dobzhansky, the great evolutionary biologist, whom he met when he began teaching anatomy at the Columbia University Medical School. That friendship resulted in a jointly organized symposium at Cold Spring Harbor, New York, in 1950 that was designed to bring genetics into physical anthropology (Washburn, 1983). Washburn elaborated that theme and his emancipation from Hooton's research program in his declaration of a new paradigm: "The old physical anthropology was primarily a technique.... The new physical anthropology is primarily an area of interest, the desire to understand the process of primate evolution and human variation by the most efficient techniques available.... This is essentially a return to Darwinism, but with a difference" (Washburn, 1951, p. 299). He went on to emphasize that races must be studied in terms of population genetics and processes of evolution. He also emphasized the evolutionary implications of form and function and outlined the idea of factoring out independent functional complexes that evolved at different times in hominid evolution. In subsequent years he continued to discuss evolutionary issues at conferences with such major figures in the evolutionary synthesis as Julian Huxley, Ernst Mayr, and George G. Simpson.

In 1955, during a trip to collect baboon skeletons after the Pan-African Congress, Washburn widened the scope of his interest in behavior to include the social behavior of monkeys and apes. During the 1956–1957 academic year, while he was a fellow at the Center for Advanced Study in the Behavioral Sciences at Stanford University, that experience crystallized into a major shift in research emphasis. He also met his future collaborator, David Hamburg, and was hired by the University of California at Berkeley Anthropology Department, where he supervised many PhD theses on primate behavior in the wild.

In 1958, just as he was leaving the University of Chicago to go to Berkeley, Washburn sent a social anthropology graduate student named Irven DeVore to Kenya to study baboons, and Phyllis Jay to India to study langur monkeys. Washburn also carried out his own study of baboons at the Amboseli Game Reserve with his former student Irven DeVore during 1958–1959 (Washburn, 1983). That was the beginning of a decade of intensive field studies of monkeys and apes by Washburn's students. In 1959, Harold Coolidge, who had led the expedition to Thailand, arranged funding for John Emlen and George Schaller to study gorillas in the wild. At approximately the same time, Louis B. Leakey arranged funding for Jane Goodall to study chimpanzees, then for Dian Fossey to study gorillas, and finally for Birute Galdikas to study orangutans, all in the wild (Ribnick, 1982). Jay and Goodall were among the first to focus specifically on primate development in the wild.

The earliest academic studies of monkeys and apes in the wild had been undertaken by psychologists: Henry Nissen studied chimpanzees in 1929,

Harold Bingham studied gorillas in 1930 (both under the direction of Robert Yerkes), and Clarence Ray Carpenter studied howler monkeys and red spider monkeys in Panama and gibbons in Thailand in the 1930s. Although Carpenter set up a colony of free-ranging (but provisioned) rhesus monkeys on the island of Cayo Santiago off Puerto Rico in 1938, and the Japanese began a long-term study of provisioned Japanese macaques in 1948, no field studies of free-ranging monkeys and apes occurred between 1937 and 1958. When field studies were resumed, they had a profound impact on the anthropological view of human evolution (Ribnick, 1982).

During that period, Washburn and his students were developing a comprehensive behavioral model of "man the hunter" that drew on primate behavior as well as fossil remains, archeological associations, and hunter-gatherer lifeways (Haraway, 1988):

> Human hunting is made possible by tools, but it is far more than a technique or even a variety of techniques. It is a way of life, and the success of this adaptation . . . has dominated the cosurse of human evolution for hundreds of thousands of years. In a very real sense our intellect, interests, emotions, and basic social life – all are evolutionary products of the success of the hunting adaptation. . . . It is important to stress, as noted before, that human hunting is a set of ways of life. It involves divisions of labor between male and female, sharing according to custom, cooperation among males, planning, knowledge of many species and large areas, and technical skills. (Washburn & Lancaster, 1968, p. 213)[4]

Several features of the "man the hunter" model are significant in light of earlier controversies. First, this hominid had split off from its nearest primate relative fairly *recently*, perhaps as recently as 4–5 million years ago, according to Washburn's student Vince Sarich (Sarich & Wilson, 1968). In other words, the split between humans and other primates was recent, not ancient as Smith and the shadowman hypothesis assumed. Second, this hominid had an ape ancestor, as was clearly indicated by the new time scale, derived from molecular evidence concerning living primates and comparative anatomical evidence concerning the brachiation complex of the shoulder and thorax. Third, this hominid was bipedal and had small canine teeth, and it had a *small brain*. In other words, the brain evolved last, as Wallace had proposed, not first as the shadowman hypothesis suggested. Behaviorally, this was a cooperative hunter, a tool user, and a food sharer. The demands of the hunting way of life led to the evolution of language and intelligence and hence brain expansion through natural selection operating on genetic variants, not through inheritance of acquired characteristics.

Here – just after the demise of the shadowman and the exposure of the Piltdown forgery – was an overtly evolutionary model that explicitly treated the evolution of human behavior, brain, and mind and furthermore based its case on the Darwinian notion of continuity between human and nonhuman primate behavior. This time, however, the model was based on the modern synthetic theory of evolution and on a growing data base on human and nonhuman social behavior.

Despite the breathtaking leap that Washburn's model embodies, it sidesteps some of the harsher implications of Darwin's model. Haraway analyzes some presuppositional differences between Washburn and the evolutionary biologists:

There was, however, considerable slippage in what behavior meant to the different interpretive communities. Simpson argued that the [principal] problem of evolutionary biology was the origin of adaptation.... From here it was a short step to redefining the key problem of evolutionary behavioral science (including Washburn's influential physical anthropology) as the origin of adaptive behavior in a functioning social group, with adaptation seen more in terms of the integration of groups than in terms of differential reproductive success.... However, if an adaptationist approach were to be taken without substituting the functionalist integrated group for the Darwinian population, then ... differential reproductive success might lead instead to an emphasis on competition, individualism, antagonistic difference, and [a] game theory view of life as a problem in strategic decision-making. (Haraway, 1988, p. 229)[5]

From its 19th-century preevolutionary, racialist beginnings, anthropology in the United States moved, largely through the influence of Boas and his students, into an antievolutionary period of reaction against racialism and naive models of cultural evolution. Following that, largely because of the work of Hooton and his students, physical anthropology resurrected racialist views and then finally rejected them under the twin pressures of population genetics and the social scientists' reactions against racism following World War II. Later still, largely because of the efforts of Hooton's student Washburn, physical anthropologists introduced the modern synthesis of evolutionary biology into the study of primate and hominid evolution, bringing the question of the evolution of mind back to the forefront of the field.

Whereas anthropologists have focused on describing and classifying manifestations of mind within and among groups (cultures), on inferring mental abilities on the basis of the brains and tools of early hominids (and hence the evolution of the mind), and on understanding their adaptive significance, psychologists have focused on describing and classifying manifestations of mind in individuals and on understanding their causes, environmental and neurophysiological. In the jargon of evolutionary biology, psychologists have concerned themselves with understanding proximate causation (the physiological mechanisms underlying phenomena), whereas anthropologists have concerned themselves with understanding ultimate causation (the adaptive significance of phenomena and the evolutionary forces shaping them) (see Daly & Wilson, 1984, for definitions).

## Origins of American psychology and the rise of behaviorism, developmental psychology, and the cognitive sciences

Because psychology is rooted in various European religious and philosophical traditions, as well as in 18th-century evolutionary concepts, it was shaped by debates over the nature of the mind: monism versus dualism, rationalism versus empiricism, innatism versus associationism, objectivism versus subjectivism, and free will versus determinism.

When Harvard University was founded in the 1600s, its students studied Hebrew, Aramaic, Greek, Latin, metaphysics, logic, and physics, plus theology and ethics: "True to the mediaeval tradition, psychology came under the head of physics, but to set it off from mere inanimate matter and its properties, the soul was treated as a special division of physics called *pneumatics* or *pneumatology*, which included matter of the angels as well" (Roback, 1964, p. 30). Psychology was at that time a series of religiously derived postulates and definitions and had no empirical base or broader philosophical range.

By the time of the American Revolution, universities in the British colonies were offering specializations by subject, and the scholasticism of the early universities had been enlivened by theological debates over determinism versus free will and finally had given way to foreign philosophical influences. Scottish realism came to the colonies through John Witherspoon, who was selected as president of Princeton University (then New Jersey College) just before the American Revolution. The Scottish realists originated the method of introspection for classifying and analyzing experience (though the term "introspection" was first used by Schmucker in his textbook *Psychology* in 1842) (Roback, 1964).

By the 1840s, Harvard, Princeton, Yale, Columbia, and the University of Pennsylvania were teaching psychology in a course called "mental" or "intellectual" philosophy. This emerging discipline drew on competing traditions of Scottish realism, English empiricism and German idealism. German idealism was introduced to American philosophers in 1840 by a German immigrant named Frederick Augustus Rauch, who was the first to use the term "psychology" in place of "intellectual philosophy." In his book *Psychology or a View of the Human Soul Including Anthropology*, Rauch dealt with the relationship between organisms and environment, consciousness, personality, language, and semiotics (Roback, 1964).

It was the domains that Herbert Spencer outlined in his book *The Principles of Psychology* (1855/1875), however, that were to have a major influence on the development of psychology. William James (1842–1910), who studied philosophy at Harvard University and held the first chair in psychology in the United States, drew liberally from Spencer's framework – including most of the topics in Spencer's table of contents – in his own historic book of the same title (James, 1890). James, the putative father of American psychology, defined the "scope of psychology" in the following way: "Psychology is the Science of Mental Life, both of its phenomena and of their conditions: The [phenomena] are such things as we call feelings, desires, cognitions, reasonings, and the like" (James, 1890, p. 1).

James's student G. Stanley Hall (1844–1924) drew heavily on the ideas of Spencer, L. H. Morgan, and Haeckel in his work on the development of play in childhood, postulating that as children develop ontogenetically, they recapitulate the stages of cultural evolution in the games they play. Hall, who received the first PhD in psychology in the United States, founded and chaired the first department of psychology in America (at Clark University in 1887), founded the *American Journal of Psychology*, and trained the first generation of American psychologists, is rarely credited for his important role in founding American psychology (Roback, 1964; Ross, 1972). Failure to credit Hall may reflect a wish to deny the existence of an embarrassing forebear who had the bad taste to embrace the soon discredited doctrine of recapitulation. A variety of other factors, including early repudiation of his anecdotal methods, his failure to work consistently in one field, and the problems of Clark University following his retirement as president, contributed to his subsequent neglect (Kessen, 1965).

Although Hall's student James Mark Baldwin (1844–1924) joined Hugo Munsterberg, William James, and Edward Thorndike (1901) in their attack on Hall's methods, he followed Hall into developmental studies, pursuing the interconnected themes of "the evolution of behavior in the race and the child" and a "genetic theory of thought" (Kessen, 1965). As a man of his time, influenced by Lamarck's idea of the inheritance of acquired characteristics, Baldwin (1896) argued that behavior played an active role in the evolution of mentality, proposing "ontogenic selection," which the evolutionary biologist G. G. Simpson later labeled "the Baldwin Effect" (Simpson, 1953).

Baldwin emphasized the roles of habit and imitation, describing a particular variety of adaptation that he called "circular reactions" whereby the fortuitous effect of an action led to its repetition and hence its reinforcement. This mechanism led the child to discover and remember new associations (Baldwin, 1896). Imitation led the child to benefit from the discoveries of others and hence to expand the repertoire of adaptive behavior. Baldwin's concept of circular reactions and his notions of adaptation, assimilation, and accommodation were picked up and elaborated by Jean Piaget (Case, 1985).

Perhaps because he left North America in the middle of his career, perhaps because his approach was antithetical to behaviorism, perhaps because he explicitly acknowledged Darwin's influence on his work (Baldwin, 1909), Baldwin had little direct impact on American psychologists. He has had indirect and delayed effects, however, through his considerable influence on Piaget, Heinz Werner, and G. H. Mead (Broughton & Freeman-Moir, 1982; Case, 1985; Kessen, 1965).

After 1915, American psychology experienced a violent reaction against the concepts of mind, consciousness, evolution, and development. That reaction took the form of the "behaviorist revolution," which has variously been traced back to the British comparative psychologist C. Lloyd Morgan (Boakes, 1984), to the American biologists Jennings and Loeb (Reed, 1987), and to the Russian physiologists Bekhterev and Pavlov (Roback, 1964). Although Boakes argues that behaviorism had its roots in the reaction of C. L. Morgan and E. Thorndike against the anecdotalism and overinterpretation of Romanes and other evolutionists, Roback classifies these psychologists as forerunners of the "functionalist school," which originated in the 1890s at the University of Chicago under John Dewey and J. R. Angel.

Although John B. Watson (1878–1958) often is referred to as the father of behaviorism, Roback argues that he is more accurately described as the promoter of behaviorism:

> It was the Russian physiologists, most of all, who supplied the basis for American behaviorism, even before Pavlov stirred up the scientific world by his experiments on the conditioned reflex. I. M. Sechenov had been developing a mechanistic psychology, and I. P. Pavlov's contemporary and equally able physiologist, Vl. M. Bekhterev elaborated a reflexology that was practically the same as Watson's behaviorism. (Roback, 1964, pp. 268–269)

In any case, between 1913, when he wrote "Psychology as the Behaviorist Sees It," and 1919, when he wrote *Psychology from the Standpoint of a Behaviorist*, Watson acknowledged the potential of Pavlov's work on conditioning for explaining behavior. Fancher argues that "Watson . . . was seeking a general principle that could account for many different kinds of behavior and he realized that Pavlov's conditioned reflex could be employed as a *model* for many other kinds of responses as well" (Fancher, 1979, p. 326).

Watson received the first U.S. PhD in animal psychology (at the University of Chicago, where he studied evolution under Jennings), studied the behavioral correlates of myelin development in the rat brain, and worked with Jennings and Yerkes as an editor of the *Journal of Comparative Psychology* (Fancher, 1979). After Watson was made a full professor of psychology at Johns Hopkins University in 1909 at the age of 31, he exerted considerable power, which he used to sever ties with the Philosophy Department and strengthen those with the Biology Department.

Although Watson abandoned academic psychology and became an advertising executive, under his influence and that of Pavlov, American learning theorists shifted their attention from broader concepts of learning to the narrower realm of conditioning (Baars, 1986; Parker & Baars, P&G2). B. F. Skinner (1938) later contributed the concepts of reinforcement, shaping, and operant conditioning to learning theory, which before Skinner's work had focused exclusively on associationist or classical conditioning. Mainstream American psychology had thrown out subjective, introspective methods and declared the concepts of mind, consciousness, and representation irrelevant and unscientific. The American infant was born a blank slate whose future was determined solely by conditioning.

The receptivity to behaviorism among American psychologists probably was conditioned by a variety of historical factors: rebellion against the authority of such European intellectuals as Wilhelm Wundt, who had trained the first generation of American psychologists in his laboratories; the difficulties of European languages and philosophical traditions; the anti-German sentiments attending World War I; antiintellectual pragmatism; and above all a desire for a uniquely American science (Baars, 1986; Lerner, 1983). As many people have noted, behaviorism was also a reaction against hereditary determinism; infinite educability promised equal opportunity for all and continuing progress, which were consistent with the new American political ideals. Finally, the prestige of the physical sciences must have played a role in this movement, as it did in the rise of historical particularism under Franz Boas. (Given the anti-Russian sentiment following the Russian Revolution, however, it is odd that Russian physiologists had so much influence.)

Of course, some comparative psychologists resisted the pressures of behaviorism during that period. Edward Tolman (1925), for example, introduced the concepts of purpose and cognition into learning theory. He was careful to argue, however, that these were objective descriptive terms. Tolman's cognitive perspective on learning and especially his concept of cognitive maps (Tolman, 1948) are compatible with ethological, Piagetian, and information-processing models (Vauclair, 1987).

The grip of learning theory loosened in the 1930s as a number of influential European émigrés took positions in U.S. psychology and psychiatry departments. These included Wolfgang Köhler, Kurt Lewin, Heinz Werner, and Erik Erikson (Baars, 1986; Lerner, 1983). The theories of Freud, Piaget, and Lewin were included in the first edition of Murchison's handbook in 1931 (Bronfenbrenner, 1983), and Piaget's early work had been translated into English in the 1920s, but these currents were not widely acknowledged by psychologists at that time. Indeed, learning theory continued its domination of American psychology until the "cognitive revolution" in the late 1950s.

In 1948, Heinz Werner, who was teaching at Clark University, published his book *Comparative Psychology of Mental Development* in English, which reintroduced many of Hall's ideas, as well as those of Piaget, Lewin, and other Gestalt psychologists, into American psychology. Werner, who had been trained in Germany in the Gestalt tradition, rejected mechanistic (associationist) approaches to development in favor of organic (holistic, configurational) approaches. He trained Bernard Kaplan (with whom he later wrote an important book) and the important neo-Piagetian researchers Jonas Langer and John Flavell.

Jean Piaget's work became increasingly well known in the 1960s through summaries published by J. McVicker Hunt (1961) and John Flavell (1963), and through Kohlberg's adaptation of his ideas on moral development. In 1966, Jerome Bruner, who had visited Piaget in Geneva, dedicated his book *The Growth of Cognitive Studies* (Bruner, Oliver, & Greenfield, 1966) to Piaget. Piaget (1896–1980) began to study the development of logic in children after he noticed systematic age related errors in children's answers to the intelligence tests he was standardizing in Binet's laboratory.

Piaget's structuralist model of intellectual development is the antithesis of American learning theory models; throughout his work, Piaget explicitly counterposed a "constructionivist" alternative to "associationist" and "innatist" models of development. His model describes the epigenetic "construction" of mental schemes and operations beginning with the innate "reflex" actions of the newborn and culminating in the hypotheticodeductive reasoning of the adult. Construction occurs through interacting processes of assimilation of objects to action schemes, and accommodations of these schemes to the properties of assimilated objects, and through reflective abstraction of the implications of various actions on objects. Piaget described, for example, the child's realization that no matter how one arranges a series of objects in space, they continue to add up to the same "number." Through this constructive process (whose driving force continued to pose a challenge to Piaget until his death), the child develops ever more powerful, flexible, and stable means for understanding the world (e.g., Piaget, 1970).

Piaget distinguished four sequential periods of intellectual development from infancy to adolescence: sensorimotor intelligence from birth to 2 years (Piaget, 1952, 1954, 1962), preoperational intelligence from 2 years to 6–7 years, operational intelligence from 6–7 years to 11–12 years (e.g., Inhelder & Piaget, 1964), and formal operational intelligence from 11–12 years on. Each period is characterized by a sequence of developmental stages that cut across several series (e.g., space, time, causality, classification, number). These series fall under three major domains: physical, infralogical, and logical mathematical reasoning. Each stage represents the highest level of intellectual functioning available to the child at the time, though lower levels of functioning continue to be available at all stages (see Piaget & Inhelder, 1969, for a summary).

Although Piaget described periods of development marked by series of sequential stages, he also described time lags or *decalages*, both across domains and across stages. Conservation of quantities, for example, occurs in early concrete operations, whereas conservation of volume, which involves an added variable, occurs later in early formal operations (e.g., Piaget & Inhelder, 1974).

Piaget's formulation drew not only on Kantian categories, which had been widely used by other psychologists of the mind, but also on the idea of developmental stages, derived from James Mark Baldwin through the work of G. Stanley Hall and Herbert Spencer, and on Baldwin's idea of "circular reactions" (which Piaget elaborated into three kinds: primary, secondary, and tertiary, each developing out of the earlier one during the sensorimotor period). Finally, Piaget's model drew on biological concepts of adaptation and evolution, especially the notions of assimilation and accommodation, and on Lamarckian ideas on the origins of adaptations.

As might be expected, the reintroduction of concepts of mind and stages of development, and the introduction of a constructionist model of development that demoted the role of learning, met with enormous resistance from American psychologists. Not only were the presuppositions contrary to the American spirit, but the methodology was unstandardized and hence unscientific: Beginning with his own infants, Piaget studied the spontaneous behaviors of children solving problems he posed; later, he and his colleague Barbel Inhelder developed what they called the "clinical method" (because it resembled psychoanalytic methods), in which they pressed the children to explain their solutions and to defend those solutions against countersuggestions (e.g., Inhelder & Piaget, 1964).

Those American psychologists who used Piagetian concepts did so mainly as a foil for presenting traditional learning experiments demonstrating that children could solve a given task earlier than Piaget and Inhelder had claimed they could, or that children showed discontinuities within and across domains of knowledge and hence did not display stages. Finding Piaget's clinical method uncongenial and cumbersome, these psychologists set themselves the task of simplifying and standardizing Piagetian tasks (e.g., Brainerd, 1978, 1979; Gelman & Baillargeon, 1983). In accord with their empirical, antitheoretical bias, they focused primarily on age norms for developmental stages, rather than on the processes of construction. In recent years, both neo-Piagetian and non-Piagetian researchers in child development have been working in the new postwar tradition of cognitive sciences.

Piaget's theory was hardly the pivot that turned American psychologists back to issues of the mind and consciousness. The mind returned to American psychology in the company of a much more prestigious alliance of scholars who founded a new discipline called "cognitive science." Cognitive science was a multihybrid creature born of philosophy, psychology, linguistics, neurosciences, and anthropology that came about because of the support for interdisciplinary research by large foundations and because of the increased interest in the functional neurobiology of aphasia after World War II (Gardner, 1987).

Although in 1948 Karl Lashley had cast a gauntlet before the behaviorists in a paper he presented at the Hixton Conference challenging the reigning assumption that all behavior could be explained in terms of associative stimulus–response chains by focusing on the hierarchically arranged behaviors of piano playing and speaking, cognitive science was officially recognized as a new discipline on September 11, 1956, at the Symposium on Information Theory at the Massachusetts Institute of Technology, where seminal papers were presented by the linguist Noam Chomsky and by psychologist George Miller on various aspects of information processing in complex tasks (Gardner, 1987).

A few years later, in 1960, George Miller and Jerome Bruner started the Center for Cognitive Studies at Harvard University, with funding from the Carnegie Foundation. This interdisciplinary center rejoined faculty from the departments of psychology and social relations, which had been separated in 1948. Fellows at the center included linguists Roman Jacobson and Noam Chomsky and an impressive array of visiting scholars, such as Barbel Inhelder, A. I. Luria, Eric Lenneberg, Roger Brown, Barry Brazelton, and Jerry Fodor. Among the graduate students were Patricia Greenfield and Dan Slobin (Bruner, 1988; P. M. Greenfield & D. Slobin, pers. commun.). As Norman and Levelt described it,

> the concepts of information processing were pervading everything: computation, philosophy, communications, engineering, biology, linguistics and, of course, psy-

chology. In the midst of these happenings, the Center for Cognitive Studies gathered together a vibrant group of people with unconventional knowledge and interests, stuck them together in one place, gave them excellent research, meeting, and support facilities, and then allowed what was to happen to happen. (Norman & Levelt, 1988, p. 101)

Hence, information processing became the dominant force in psychology, and mind returned in the 1970s and 1980s (Barrs, 1986; Lerner, 1983).

The earliest modern study of child language acquisition was initiated by Roger Brown, who moved to the Center for Cognitive Studies in 1962, stimulated by "the technological advances of portable tape recorders and transformational grammar" (Slobin, 1988, p. 9). In his pioneering book *A First Language*, Brown (1973) distinguished five stages of language development, beginning with the first word at 6 months and ending with full sentences at 3–4 years. He showed parallels between linguistic and intellectual development, noting that the first words were spoken during Piaget's sensorimotor period.

Greenfield and Smith (1976) noted that a shift toward the study of development prior to the one-word stage was stimulated by Filmore's semantics-based case grammar, which "opens the way to a theoretical treatment of one-word speech as structurally continuous with later grammatical development" (Greenfield & Smith, 1976, pp. 16–17). In a somewhat parallel analysis, Bates, Camaioni, and Volterra (1976) have argued that linguistic studies of infants before the one-word stage awaited a shift in emphasis to pragmatics.

Two generations of North American developmental psychologists trained in the cognitive sciences tradition – Jerome Bruner, Juan Pascual-Leone (a Spanish-born former student of Jean Piaget), Robert Sternberg, Robert Siegler, David Klar, J. Wallace, and Robbie Case – have brought information-processing paradigms to developmental psychology. Juan Pascual-Leone, now of York University in Toronto, developed a major revision of Piaget's model, formalizing task structures and adding the concept of "schematic activation weight," which includes specification of number of cues and their structural relations, salience of cues, and degree of attentional power needed to solve a given task (Pascual-Leone, 1984; Pascual-Leone & Smith, 1969). This model, which strongly influenced Case's recent reformulation of the earlier developmental model, harks back to some of Baldwin's ideas that Piaget rejected (Case, 1985).

In his major synthesis and reformulation of developmental psychology, Robbie Case (1985) has devised a set of critical experiments in information processing to test key Piagetian, Baldwinian, Vygotskian, Brunerian, and Pascual-Leonian concepts, including stage concepts and the roles of social and instructional processes. In contrast to Brainerd and other conditioning theorists, Case concludes that the stage concept is valid: "(1) . . . children go through the same sequence of substages across a wide variety of content domains, and (2) . . . they do so at the same rate, and during the same age range" (Case, 1985, p. 231). Like Piaget and Baldwin, he identifies four major stages of development, but Case avoids many of Piaget's problems by identifying stage-specific levels of ability rather than stage-specific abilities and concepts: "what varies with age is the level of understanding, or degree of sophistication, which children can attain with regard to the concept or ability in question" (rather than the concept or ability itself) (Case, 1985, p. 240). Case also incorporates the roles of social and instructional processes emphasized by Vygotsky, Bruner, and others that were largely ignored by Piaget.

Some of its postulates have been disproved, and others have been modified or recast, but Piaget's model is important because it helped return mind to psychology and because it has stimulated a vigorous ongoing research program in developmental psychology that has grown and flourished since the cognitive revolution. In recent years, for example, students of infancy have revised and elaborated Piagetian theories of object–concept development (Bower, 1979) and logical development (Langer, 1980, 1985, 1986) using microanalysis of filmed sequences of behavior in standardized settings. Langer has identified prelogical action precursors of the stages of logical mathematical reasoning that Piaget described in older children. Others, such as Liz Bates and her colleagues, have found parallels in the development of the first referential gestures and sounds and the stages of sensorimotor intelligence, imitation, and symbolic play (Bates, Benigni, Bretherton, Camaioni, & Volterra, 1979). Still others, such as Kurt Fischer, Inge Bretherton, and Elliott Turiel, have adapted Piagetian concepts to the developmental study of linguistic, social, and affective schemes. Finally, Hermenia Sinclair (1987) and her colleague in Geneva continue to extend and elaborate Piagetian concepts in the study of social development. Since 1970, Piagetian frames have even been employed in comparative studies of nonhuman primates.

### Origins of comparative psychology and ethology

Ironically, both comparative psychology and ethology had their roots in natural history studies of the 19th century and in the work of Charles Darwin, George Romanes, and C. Lloyd Morgan (Boakes, 1984; Glickman, 1985). The two sciences diverged, however, in their later development – comparative psychologists emphasizing methodological rigor and the ethologists emphasizing evolutionary reconstruction.

Darwin argued for continuity in emotional expression and mental and moral abilities between humans and other animals based on descent with modifications (Gruber, 1981). He used the comparative method to demonstrate taxonomic affinities. Darwin gave his notes on animal behavior to the young George Romanes, who followed his mentor in the rich interpretation of animal minds, arguing that introspection could be applied by analogy in *The Mind of Animals* and *Animal Intelligence.* Romanes, however, lacked Darwin's focus on taxonomy.

Romanes eventually passed the torch on to C. L. Morgan, who took a much more rigorous approach than Romanes to the interpretation of behavior, distinguishing between objective data and interpretation, urging the necessity of studying the history of behaviors and of interpreting behavior in the most parsimonious manner (Morgan's canon), and rejecting the Lamarckian interpretations of Spencer, Darwin, and Romanes (Boakes, 1984; Morgan, 1894). Despite his emphasis on rigor, C. L. Morgan, like H. S. Jennings, was an organicist or functionalist (Roback, 1964) who argued that consciousness was an emergent property of life. (The organicist view contrasted with the mechanist view of Jacques Loeb, who argued that behaviors were merely responses to such external forces as temperature and gravity [Demarest, 1987].) Morgan inspired Edward Thorndike with his lectures at Harvard, though Thorndike failed to acknowledge Morgan's influence in his later work (Glickman, 1985).

As comparative psychology developed, it became increasingly laboratory oriented and decreasingly evolutionary and ecologically oriented in its focus. After Watson and Skinner, many comparative psychologists abandoned the organicist model and adopted the mechanist conditioning paradigm, focusing on the search for universal laws of learning. Many narrowed their research to studies of conditioning of laboratory-bred rats and pigeons, which are cheap, reliable, and easy to manipulate (Beach, 1965). When they did study the behavior of other species, they often compared distantly related forms as if they represented links in the Great Chain of Being, or stages of "evolutionary advancement" (Hodos & Campbell, 1969), rather than comparing closely related species in the same genus or family as Darwin had suggested: "The community of certain expressions in distinct though allied species, as in the movements of the same facial muscles during laughter by man and by various monkeys, is rendered somewhat more intelligible, if we believe in their descent through a common progenitor" (Darwin, 1872, p. 12).

In the end, with a few notable exceptions, comparative psychology was neither comparative nor evolutionary, nor developmental. The limitations of comparative psychology reflect the nature of the conditioning paradigm, which is unconducive to developmental studies because it conflates ontogenetic and experiential transformations (Parker & Poti', P&G8) and tends to treat immature animals as miniature adults (Zolman, 1982).[6]

The lack of interest in development may also reflect the choice of a rapidly maturing laboratory species that failed to confront the psychologist with significant age related differences in behavior: Certainly Yerkes and Harlow, and later William Mason, who could not have failed to notice age related differences in their slowly maturing primate subjects, did study development. As discussed later, widespread studies of captive primates were slow to come.[7]

Like Yerkes and Harlow, Schneirla and Rosenblatt, who worked on mother–kitten interactions, focused on developmental changes in the dyad. Indeed, Schneirla was one of a handful of comparative psychologists who focused on both the development and the evolution of behavior (Kuo, 1970;

Scott, 1962; Zimmermann & Torey, 1965). In fact, he considered comparative developmental studies to be the foundation of comparative psychology (Schneirla, 1966). Significantly, he worked with a variety of species, including ants and cats, and studied the behavior of ants in the wild (Piel, 1970). Unfortunately, his approach was unique.

It is interesting in this regard that Thorpe (1979), in his history of ethology, identifies Schneirla, Carpenter, and Beach, all of whom were editors of the journal *Behaviour* when it began in 1948, as ethologists rather than comparative psychologists. Schneirla might have declined the honor, given his attack on the concept of instincts (Piel, 1970). Although the Americans William Morton Wheeler, Charles Otis Whitman, and Wallace Craig laid the foundations for ethology, the movement failed to take hold in the growing climate of behaviorism (Thorpe, 1979).

Ethology did take hold in Europe, however, and became the European branch of animal behavior. Although the earliest ethological studies were those of Darwin, Romanes, and Morgan, ethology as a discipline flourished in Austria, Germany, and Holland under the influence of Oskar Heinroth and Jakob von Uexkull and became a major force between the two world wars under Karl von Frisch, Niko Tinbergen, Konrad Lorenz, Erich von Holst, Adrian Kortlandt, and J. A. Bierens de Haan. Ethology came back to England after the war when Cambridge University established a program under Thorpe and Oxford did so under Tinbergen (Thorpe, 1979).

Unlike comparative psychology, which was dominated by the conditioning paradigm, ethology maintained its evolutionary identity, developing as a branch of systematics, the evolutionary science of taxonomic classification in line with Darwin's suggestion. Behavior, like structure, was treated as a clue to taxonomic affinities: Heinroth, for example, made it clear that "behaviour patterns (as in ducks) can be reliably used to provide evidence for the systematic relations between species; and as a basis for theorizing about the detailed course of evolution of such characteristics" (Thorpe, 1979, p. 53). Their overtly evolutionary perspective led ethologists to focus on the distribution of species-specific behaviors in closely related species in a given genus. It also led them to focus on behavior in the wild as a basis for constructing models.

As is well known, the antithetical perspectives of comparative psychologists and ethologists generated fierce debates over the concepts of learning and instinct during the 1950s and 1960s (e.g., Lehrman, 1953, 1970; Lorenz, 1950, 1957, 1965). Comparative psychologists contended that innateness was a meaningless concept that deflected researchers from investigation of (proximate) cause-and-effect relationships, and ethologists contended that learning, as comparative psychologists defined it, blinded researchers to species-specific adaptations.

That debate yielded two consensual reformulations, both incorporating evolutionary concepts: the recognition of species-specific "constraints on learning" (e.g., Garcia, McGawon, Ervin, & Koelling, 1968; Hinde, 1973;

Marler & Terrace, 1984) and the notion of the animal as strategist who weighs the costs and benefits of potential courses of action (Kamil, 1984; Pepperberg, P&G18). Even following the revitalization of comparative psychology through the infusion of information-processing and ecological models, however, development continues to be neglected. A recent special review issue of the *Journal of Comparative Psychology*, for example, included no articles focusing on development, and only one article mentioned it (Dore & Kirouac, 1987). It is interesting to speculate that the ghosts of Spencer, Romanes, and Hall haunt comparative psychologists to this day, making them wary of development, especially in an evolutionary context.

**Origins of primate laboratory studies in the United States**

As indicated earlier, developmental studies have most often involved large, slowly developing mammalian species. This would suggest that when studies of primates became widespread, developmental research should have flourished. Unfortunately, that was not the case. Ironically, most of the comparative developmental studies of ape and human intelligence were done many years ago in private homes, such as those by Kellogg and Kellogg (1935) and Hayes and Hayes (1971), both colleagues of Yerkes, and by Kohts (1935) in Russia. Those were all "cross-fostering" studies, that is, studies in which an infant of one species was reared by a foster mother of another species (Gardner & Gardner, 1987). As interesting as they were, those studies lacked both theoretical frameworks and the systematic observation and testing paradigms that are possible in laboratory settings.

The first laboratory studies of primates were carried out by Robert M. Yerkes (1876–1956). Yerkes tried for many years before he succeeded in establishing the colony of apes he longed to have. In his book *The Mental Life of Monkeys and Apes*, and in an article in *Science*, Yerkes (1916a,b) spoke eloquently of the need for primate study centers. Yerkes wanted to work with Wolfgang Köhler at Tenerife, but was prevented from doing so by the outbreak of World War I. He first realized his ambition in 1914 on a private estate in Montecito, California, with the help of his friend Dr. G. V. Hamilton. Yerkes, who had been trained in the behaviorist mode, was also a hereditarian in the Galton tradition and an evolutionist in the Darwinian tradition (Reed, 1987). Before turning his attention to primates, Yerkes had studied intelligence in dancing mice and raccoons; he had also participated in intelligence tests conducted by the army during World War I (Cadwaller, 1987; Gould, 1981; Parker & Baars, P&G2). In 1924, Yerkes, who had recently moved from the Philosophy Department at Harvard to the Psychology Department research unit at Yale, received support from Yale University and from the "Rockefeller interests" to begin a primate laboratory in Orange Park, Florida (Reed, 1987). Later that laboratory was moved to Emory University in Atlanta, Georgia, where it remains today.

The second primate laboratory in the United States was established at the

University of Wisconsin by Harry Harlow in 1932 (where it still stands, adjacent to the Wisconsin Regional Primate Research Center that was built in 1961 as one of seven regional primate centers Harlow helped create). Harlow, who like Carpenter and Mason was a student of Calvin Stone, followed in the footsteps of Thorndike and Yerkes in designing a battery of standardized intelligence tests for monkeys and a special apparatus, the Wisconsin General Test Apparatus (TGTA) for "assembly line" testing (e.g., Harlow, 1943, 1944a,b, 1945a–d; Harlow & Bromer, 1938; Young & Harlow, 1943). Those studies led to Harlow's concept of the "learning set" (e.g., Harlow, 1949; Suomi & LeRoy, 1982; H. A. LeRoy, pers. commun.).

Although Harlow had studied the learning abilities of lemurs, cebus, spider monkeys, baboons, guenons, rhesus macaques, gibbons, and orangutans at the zoo and in his first primate colony (e.g., Harlow, 1951; Harlow & Bromer, 1939; Harlow, Uehling, & Maslow, 1932), after a tuberculosis epidemic killed the animals in 1938 he limited the colony to the hardy, dependable worker monkeys: rhesus macaques (Suomi & LeRoy, 1982; H. A. LeRoy, pers. commun.). Later, when William Mason came to Wisconsin as a postdoctoral fellow, he and Harlow developed a battery of age sensitive intelligence tests to follow the ontogeny of learning (e.g., Mason, 1960; Mason & Harlow, 1958a,b; Mason, Blazek, & Harlow, 1956; Mason, Harlow, & Rueping, 1959).

The taxonomic scope of comparative psychological studies of behavior greatly increased after 1960, when the U.S. Congress established seven regional primate research centers under the National Institutes of Health (U.S. DHEW, 1971). These centers, which were the culmination of generations of lobbying by Yerkes, Harlow, and others (Hahn, 1971; H. A. LeRoy, pers. commun.), made prosimians, monkeys, and apes available on a much wider scale.

Since their establishment, the primate research centers have been the settings for myriad studies of primate abilities. Of the handful of senior psychologists who have led major research programs in primate intelligence in the United States – Allan and Beatrice Gardner, Harry Harlow, William Mason, Emil Menzel, David Premack, and Duane Rumbaugh – the majority have worked at these primate centers.

William Mason, after working at Wisconsin, went on to work at the Yerkes Laboratory in Orange Park, at the Delta Regional Primate Research Center at Tulane University, and then at the National Primate Research Center at the University of California, Davis, where he has continued his research on learning and socioemotional development (e.g., Mason, 1979) and has trained or collaborated with dozens of primatologists, including D. M. Fragaszy and E. Visalberghi, contributors to this volume.

Emil Menzel, Jr., who is best known for his early work on the effects of environmental restriction on chimpanzees (Menzel, 1963) and for his work on spatial cognition in chimpanzees (Menzel, 1971; Menzel, Premack, & Woodruff, 1978; Menzel, Savage-Rumbaugh, & Lawson, 1985), worked at the Delta Regional Primate Research Center with William Mason and Hans Kummer in the 1960s after working for several years at a primate center in Japan.

Duane Rumbaugh, who with his colleagues had done extensive work on primate learning and communication (Rumbaugh, 1971, 1973; Rumbaugh & McCormack, 1967; Rumbaugh & Pate, 1984; Rumbaugh & Rice, 1962), came from San Diego State College and the San Diego Zoo to the Yerkes Primate Center, where he has worked for many years while holding a teaching position at Georgia State University. Several of the contributors to this volume have worked at the Yerkes center with Duane Rumbaugh and E. S. Savage-Rumbaugh.

David Premack, who is best known for his research on language learning, intelligence, and consciousness among apes (Premack, 1976a,b; Woodruff & Premack, 1978; Woodruff, Premack, & Kennel, 1978), worked at the Yerkes Laboratory during 1954–1955 before taking a teaching position at the University of California, Santa Barbara, and then at the University of Pennsylvania. Tetsuro Matsuzawa, one of the contributors to this volume, recently worked with Premack.

Of these major figures, only the Gardners, whose work is discussed in a later section, worked entirely outside the network of primate centers, though their meeting with Rumbaugh in 1966 sparked his interest in studying symbol acquisition by chimpanzees (D. Rumbaugh, pers. commun.).

Although the primate centers have been enormously influential in stimulating and supporting comparative psychological research, especially cognitive research, this and related research has been largely nondevelopmental and noncomparative, focusing on learning abilities of adults rather than age-specific intellectual abilities of developing animals (see reviews by Griffin, 1982; Hoage & Goldman, 1986; Jerison, 1987; Meador, Rumbaugh, Pate, & Bard, 1987; Menzel, 1971; Premack, 1976a,b). Both the equipment and the approved methodology of comparative psychology laboratories favor standardized testing of large numbers of animals of the same age and species. These approaches have the virtue of giving rapid results that are valid and reliable, but they have kept comparative psychologists from fulfilling Darwin's mandate to understand the development and evolution of primate intelligence.

## A prescription for CDEP

Darwin recognized that development is a taxonomically relevant domain (Ghiselin, 1969). In fact, he said that development (particularly embryology) is one of the most important subjects in the study of "descent with modification":

> In two or more groups of animals, however they may differ from each other in structure and habits in their adult condition, if they pass through closely similar embryonic stages, we may feel assured that they are all descended from one parent-

form, and are therefore closely related. Thus, community in embryonic structure reveals community of descent; but dissimilarity in embryonic development does not prove discommunity of descent, for in one or two groups, the developmental stages may have been suppressed, or may have been greatly modified through adaptation to new habits of life, as to be no longer recognizable. (Darwin, 1859, p. 345)

Following Darwin, the ethologists recognized that behavior is a taxonomically significant domain or, more precisely, a domain that can reveal phylogenetic relationships. Because embryogenesis is the generation of morphology and behavior, the study of development perforce includes the study of behavioral development.

Therefore, a program in CDEP must bring development under the rubric of evolution, treating developmental patterns as taxonomically relevant characters and charting their taxonomic distributions. The modern evolutionary framework for comparing developmental patterns comes from the branch of behavioral ecology known as life history strategy theory, which treats the evolution of stages of the life cycle (Parker, P&G4). The study of an evolutionary framework for reconstructing the common ancestor of closely related species is known as cladistics. Only exhaustive comparisons of the descendant species in a "clade" in reference to species in a more distantly related group can provide information for reconstructing the pattern of the common ancestral species (Ridley, 1984).

A program that will do these things presupposes a model capable of systematically describing and classifying similarities and differences among closely related and distantly related species in a manner that will highlight unique developmental patterns while simultaneously revealing commonalities. It must therefore be sensitive to the most complex as well as the least complex levels of behavioral organization. In other words, it must be capable of characterizing the emergent phenomena of human intelligence and language, as well as nonhuman intelligence and communication. In order to do this, it must use a flexible, open-ended and opportunistic strategy.

Before considering two such strategies, we should establish general criteria for evaluating theories. According to some recent philosophers of science, theories can be judged by their scope and comprehensiveness, their heuristic power, and by their capacity to generate a vigorous "research programme" (Lakatos, 1970). Some philosophers of science argue that a theory does not stand or fall primarily in relation to its record of accurate prediction, that is, in relation to its falsifiability (Kuhn, 1970); rather, a theory stands while it supports a vigorous program of research, even when various of its auxiliary hypotheses prove false, and a theory falls when it is replaced by another theory of greater heuristic appeal (Kuhn, 1970; Lakatos, 1970). Specifically, frameworks used in comparative studies should display the following strengths:

1. *comprehensiveness* – capacity to classify a wide range of behaviors
2. *flexibility* – utility in both naturalistic and laboratory settings to describe both spontaneous and elicited behaviors

3. *authenticity* – capacity to describe species-specific (unique) behaviors without distortion
4. *discriminativity* – capacity to identify similarities and dissimilarities, that is, to discriminate among various degrees or levels and kinds of abilities manifested by individuals of differing ages and/or species
5. *heuristic value* – ability to suggest hypotheses concerning the proximate and the ultimate mechanisms of behaviors
6. *accessibility* – practical utility for researchers
7. *replicability* – susceptibility to interobserver agreement
8. *compatibility* – susceptibility to coordination with other approaches

## Evaluation of Piaget's model for comparative developmental evolutionary studies of monkey and ape intelligence

To meet the goal of *comprehensiveness*, we need a system capable of analyzing both the simplest and the most complex unstereotyped and intelligent behaviors of our own and other species throughout the life span. Piagetian models are uniquely suited for comparative developmental studies of the organization of unstereotyped behavior and intelligence because they encompass the simplest schemes of infants and the most complex formal reasoning of adults. More important, they provide a series of links connecting the simplest and most complex behaviors (e.g., see Langer's work [1980, 1985, 1986] on protologic in infancy). This connection is made possible by the hierarchical nature of these models, which describe and correspond to developments on increasingly more comprehensive, highly integrated levels of behavioral organization (Case, 1985; Gibson, 1981, 1983, 1990a,b). Unfortunately, Piagetian models are not comprehensive in the domains they cover. Most notably, they fail to treat the critically important domains of social interaction, except in relation to social imitation and moral development. They also fail to treat motor development and state parameters (Fragaszy, P&G6).

To meet the goals of *flexibility* and *authenticity*, we need a system that will address the temporal, spatial, and goal related organization of motor patterns at an abstract level. Piagetian models are well suited to this goal because "Piaget described parameters such as the nature and the locus of coordinations, their elicitors, their goals, the reinforcers, and their temporal patterning which usually remain implicit in behavioral descriptions" (Parker, 1977, p. 44). This allows investigators to discriminate among more and less complex behaviors. It also allows them to describe and compare spontaneous and elicited behaviors and their development in related species (both inside and outside a testing paradigm) without imposing a rigid set of anthropocentric criteria (Parker, 1977). Piagetian concepts have been used to analyze spontaneous behavior in macaques (Parker, 1977), chimpanzees (Russon, P&G14), orangutans (Chevalier-Skolnikoff, Galdikas, & Skolnikoff, 1982; Bard, P&G13; Miles, 1986, P&G19), and gorillas (Gómez, P&G12).

In order to meet the goal of *discriminativity*, we need a system that can detect small differences in ability among individuals of the same age, different ages, and different species across cognitive domains. Because the developmental stages of Piaget's model are epigenetic – that is, because the higher-stage abilities are constructed from components of lower-stage abilities – they are necessarily sequential. This means that stages provide sequential levels of achievement that can be used as criteria for comparing the abilities within a given series. Piagetian stages can be used, for example, to compare the tool-using abilities of various species (Chevalier-Skolnikoff, 1989; Natale, 1989; Parker & Gibson, 1977; Parker & Poti', P&G8) and imitative abilities (Chevalier-Skolnikoff, 1976; cf. Visalberghi & Fragaszy, P&G9). Because each series is described in terms of its own sequence of stages, rates of development in various series can also be compared: The Piagetian concept of *decalage* or temporal displacement of achievements in various series provides a useful frame for discriminating the patterns of development in various species. This concept was useful, for example, in distinguishing the patterns of sensorimotor development in stump-tailed and Japanese macaques from those of great apes and human infants (Antinucci, Spinozzi, Visalberghi, & Volterra, 1982; Chevalier-Skolnikoff, 1977; Parker, 1973, 1977). At a higher level of analysis, Piaget's distinction between physical and logical mathematical knowledge may provide an important basis for discriminating human and nonhuman developmental patterns (Langer, 1989).

Although the *heuristic value* of a framework can be judged only indirectly, two obvious indicators are how well it distinguishes species differences and how much it stimulates adaptive hypotheses: The Piagetian framework is better at identifying species differences than are traditional learning tasks (Meador et al., 1987). It has already stimulated considerable evolutionary hypothesizing (Chevalier-Skolnikoff et al., 1982; Parker, 1977; Parker & Gibson, 1977, 1979; Antinucci, P&G5; Gibson, P&G3; Greenfield & Savage-Rumbaugh, P&G20; Parker & Poti', P&G8).

To meet the goal of *accessibility*, researchers will seek revisions of Piagetian stages and domains and their criteria into simpler, more accessible forms. The Uzgiris and Hunt (1975) ordinal scale of development in the sensorimotor series is one such entity. As indicated elsewhere, Case (1985) has revised and clarified many of the confusing and inaccurate elements of Piaget's model.

To meet the goal of *replicability*, laboratory researchers in particular will seek simple standardized tests that will control for alternative causes of diagnostic behaviors (Thomas & Walden, 1985). Some comparative investigators have used the Uzgiris and Hunt scale (e.g., Redshaw, 1978; Wood, Moriarty, & Gardner, 1980). Antinucci's group in Rome has devised systematic testing procedures across the sensorimotor series specifically for use in testing monkeys in a laboratory setting (Antinucci, 1989). The group in Rome has also been using Langer's extension (1980, 1985) of Piaget's work on logical-mathematical reasoning into the sensorimotor period to study monkeys. These procedures probably will be widely adopted for laboratory studies. Investigators also need straightforward rigorous models for studying spontaneous rather than elicited behavior. The "productive tests of competence" developed by the Gardners (1978) provide a good model for this approach in the case of cross-fostered signing apes.

If they are to gain wider acceptance for comparative developmental studies, Piagetian models should be augmented by more quantitative approaches to the number of schemes and the frequency and distribution of their performance (Gibson, P&G3). Case's revision (1985) holds great promise in this regard.

To meet the goal of *compatibility*, we need a system capable of incorporating, classifying, and coordinating behaviors and processes identified by other approaches. The Piagetian system serves this purpose admirably in the following domains: The classical Köhlerian tool-using tasks fit into the fifth and sixth stages of the sensorimotor means, ends, and causality series (Natale, 1989; Parker & Gibson, 1977; Parker & Poti', P&G8); observational learning and imitation fit into Piaget's imitation series (cf. Visalberghi & Fragaszy, P&G9); prelanguage development in human infants fits into the imitation, play, and sensorimotor intelligence series (e.g., Bates et al., 1979). The use of signs to study intellectual abilities in language-trained apes demonstrates the mutual utility of the two approaches (Gardner & Gardner, 1978; Premack, 1976a; Miles, P&G19). Piagetian approaches also illuminate nonlinguistic communicatory behaviors of great apes (Bard, P&G13; Gómez, P&G12; Russon, P&G13; Tomasello, P&G10).

Perhaps the greatest challenge to comparative neo-Piagetians is to develop descriptive and analytic frameworks that will allow systematic comparison between physical and social manifestations of intelligence. Such frameworks already exist in at least six areas: social play (Piaget, 1962); observational learning and imitation (Piaget, 1962; Tomasello, P&G10; Visalberghi & Fragaszy, P&G9); the use of other animals as means to an end (Bard, P&G13; Blount, P&G15; Gómez, P&G12; Tomasello, P&G10); deception (Chevalier-Skolnikoff, 1986; Miles, 1986); facial communication (Chevalier-Skolnikoff, 1982); linguistic development (Miles, P&G19; Greenfield & Savage-Rumbaugh, P&G20). Another interesting line of research that cuts across social and physical cognition is that of Gallup (1982) on self-recognition in primates (Chevalier-Skolnikoff, 1986). Work of this sort also has a basis in developmental studies of human infants (e.g., Bertenthal & Fischer, 1978).

Development of a systematic comparative framework that will encompass social behavior is particularly important at a time when some primatologists are claiming that various forms of social manipulation in monkeys and apes constitute exalted intellectual achievements (e.g., Whiten & Byrne, 1988). In the absence of such a framework it is impossible to evaluate such claims and to put the behaviors in question in the context of other achievements (Essock-Vitale & Seyfarth, 1987; Mason, 1979).

Given the history of comparative psychology, it is clear that any comparative framework for describing and classifying the intelligence of nonhuman primate species must counter implicit or explicit claims (1) that it is anthropocentric in its aims and in its conceptions and (2) that it is irrelevant to the adaptive challenges that have shaped the species (Kamil, 1984; Pepperberg, P&G18). Indeed, reluctance to face these chanllenges and to explicitly address the implications of the close genetic relationship between humans and great apes probably is one of the factors that have retarded the development of comparative developmental models.

The adoption of a human-based scheme to study monkeys and apes seems reasonable given the close genetic relationships involved: Humans and chimpanzees and gorillas are 98–99% alike in their proteins and DNAs. We are among the most closely related of living species; in fact, chimpanzees probably are more closely related to humans than they are to gorillas (Weiss, 1987). Use of such a scheme also seems reasonable given the similarities in life histories (i.e., in gestation period, weaning, age of sexual maturity) (Parker, P&G4), as well as similarities in brain, behavior (e.g., Goodall, 1986), and intelligence (e.g., Köhler, 1917/1927). As has been notorious since the time of Huxley and Darwin, we are highly similar in our anatomy, physiology, behavior, and development. For all of these reasons, a developmental framework developed on humans should be appropriate for comparing the highest abilities of humans, apes, and monkeys.

The Piagetian framework is useful for identifying similarities and differences among species because it distinguishes levels and rates of development across domains. Hence, it can provide the data necessary for charting the taxonomic distributions of abilities among related primate species that are necessary for making evolutionary reconstructions.

Clearly, traditional comparative psychological studies of primate learning abilities do not offer these benefits and have not stimulated evolutionary hypotheses (e.g., Meador et al., 1987). Even the more sophisticated studies of animal cognition based on behavioral ecology have addressed the adaptive significance of cognitive behaviors on a piecemeal rather than a comprehensive basis (Kamil, 1984). Neither of these approaches offers the potential benefits of a comprehensive, systematic comparative classification of animal abilities and their development.

### Evaluation of developmental psycholinguistic frameworks in comparative studies of ape communication

As in the case of comparative developmental studies of intelligence, the choice of frameworks for the comparative study of communication must be based on the appropriateness of the theory to the problem at hand and on the comprehensiveness, flexibility, authenticity, discriminability, and heuristic power of the framework, as well as its accessibility, replicability, and compatibility with other frames. In the case of ape communication, however, researchers must study two distinct phenomena: (1) innate species-specific communication and (2) provoked cross-species and intraspecies communication via signs and symbols.

Various ethological concepts provide useful frameworks for the study of innate species-specific communication (e.g., Fossey, 1972; Goodall, 1986; Marler & Tenaza, 1977; Snowdon, Brown, & Peterson, 1982). As useful as they are, however, these concepts are inadequate to reveal the more flexible, instrumental use of gesture that Piagetian frames reveal (Bard, P&G13; Gómez, P&G12; Russon, P&G14). Ethological concepts are also inadequate to support comparisons between the use of signs and symbols by cross-fostered apes and human language. Because they grew out of the study of innate action patterns, it is hardly surprising that ethological concepts are inadequate to study behaviors drawn from the complex behavioral systems of the human species.

Likewise, comparative classification schemes such as that of Hockett (1960) are of limited value in comparative primate studies because they are inadequate to embrace the complexity of human language (Parker, 1985). In contrast, two developmental psychological frameworks are especially relevant to comparative studies of communication: social interactionist frames and language acquisition frames. Their value springs in major part from the fact that they embody more sophisticated conceptions of language.

Language is an arbitrary, culturally transmitted communication system capable of encoding an infinite variety of meanings through the use of flexibility interlocking subsystems of sound and meaning (Vauclair, P&G11). The phonological, lexical, grammatical, and semantic subsystems appear to develop somewhat independently during the earliest months, and subsequently they begin to integrate into true language use over a period of many years. During the earliest prelinguistic stages of development, human infants display prespeech movements of the mouth and tongue, and even of the hands in social contexts (Trevarthen, 1979). They also engage in early rhythmic interchanges with their mothers that prefigure later conversational interchanges (Kaye, 1983; Stearn, 1978). By 7–8 months of age they are engaging in highly ritualized games with their mothers that depend on mutual attention to objects. These and many other developmental strands contribute to the emerging linguistic and nonlinguistic communicative abilities of human children.

Given that in most series the great apes display cognitive abilities comparable to those of human infants at Stage 6 of Piaget's sensorimotor period or beyond, models of human communicative development during the sensorimotor period from birth to 2 years of age should be highly relevant to diagnosing the communicative abilities of great apes. Given that great apes have greater voluntary control over their gestures than over their vocalizations (Hayes & Hayes, 1971; Lieberman, 1975), those models arising out of studies of the development of gestures in deaf children and hearing children should be particularly relevant (Volterra, 1987). But finally, psycholinguistic models for language development in human children are critical for comparing continuities and discontinuities in ape and human language related abilities.

**Origins of ape "language" studies**

In 1917, Kafka wrote a story about a chimpanzee with the gift of language, entitled "A Report to the Academy" (Terrace, 1980). Since the turn of the century, several investigators have tried to teach vocal language to great apes: Richard Garner (the explorer), Lightner Witner (a psychologist), Dr. William Furness III (a philosopher and adventurer), Robert M. Yerkes (the psychologist) and his collaborator Blanche Learned, Winthrop Kellogg (the psychologist and Yerkes collaborator) (Kellogg & Kellogg, 1933), and Cathy and Keith Hayes (Yerkes collaborators) (Hayes & Hayes, 1971) all tried to teach chimpanzees (and in one case orangutans) to pronounce words (Hahn, 1971). In each case, great effort on the part of both species resulted in the chimpanzees producing very few words.

Although a few people had had the idea of teaching gestural language to chimpanzees in the early days, the work was not done: In 1913, Dr. Max Rothmann (a physiologist and neuroanatomist who set up the Anthropoid Field Station of the Prussian Academy of Sciences in the Canary Islands in 1912) and the first director of the field station, Eugen Teuber (who had been a student of W. Wundt and Carl Stumpf) (Teuber, 1987); in 1916, William Furness III (in a paper presented to the American Philosophical Society); in 1925, Robert Yerkes in *Almost Human*; in 1968 Winthrop Kellogg (in a review article) (Hahn, 1971).

The first attempt to teach gestural language to a chimpanzee was made by Beatrice Gardner, an ethologist (trained at Oxford by Niko Tinbergen), and Allen Gardner, a learning theory psychologist. In their first formal research report in 1969 in *Science* (Gardner & Gardner, 1969, 1971) they reported that after 36 months of training, the chimpanzee Washoe could appropriately use 85 signs to a rigorous criterion of repetitions (at least one appropriate and spontaneous use for 15 consecutive days), that during the 10th month of training she began to combine signs in ways that paralleled the combinations used by children reported by Roger Brown, that she invented signs and novel sign combinations, that she correctly identified exemplars of objects whose names she knew, and that she engaged in "conversations" with humans referring to absent objects and using negatives, locatives, prepositions, and "wh ...?" questions. (Overall, Washoe compared favorably with human children at Stage III according to Roger Brown's typology.)

After they had studied Washoe's language abilities for 5 years, the Gardners embarked on a multisubject project that involved studying several chimpanzees from birth (Gardner & Gardner, 1978). They found that the chimpanzees exposed to signs from birth began both signing and using signing phrases at a slightly earlier age than did human infants reared under the same conditions.

The second ape language study was initiated in 1968 by David Premack, a psychologist then at the University of California, Santa Barbara. Premack began his project of teaching language to a chimpanzee, Sarah, using portable magnetic plastic tokens that could be placed on a metal board. Premack, a cognitive psychologist with strong philosophical interests, focused on developing a methodology that could control for cueing and observer bias. He published his first report on that research in 1971 in *Science*. Premack demonstrated Sarah's ability to manipulate such abstract concepts as "same/different" and "color of" and "name of" (Premack, 1971a,b, 1976a,b; Pepperberg, P&G18).

The third ape language study had its inception in 1970 when Duane Rumbaugh went to Yerkes Primate Laboratory in Atlanta as assistant director. Rumbaugh, who had learned from the Gardners about their plans in 1966, began to think about using computers to teach language to apes. He and his colleagues employed a symbol system called "Yerkish," using computer keys to represent words (e.g., Rumbaugh & Gill, 1976; Rumbaugh, Gill, & von Glasersfeld, 1973; Savage-Rumbaugh, Rumbaugh, & Boysen, 1978a,b; Savage-Rumbaugh, Rumbaugh, Smith, & Lawson, 1980). Like Premack, this group focused on design of rigorous controls to avoid problems of cueing and observer bias. In their later work they demonstrated that chimpanzees could classify novel objects according to semantic class, such as fruit, tool, and so forth (Savage-Rumbaugh, 1981). They also focused on communication between animals, devising a method for eliciting symbolically mediated cooperative tool use between two chimpanzees (Savage-Rumbaugh et al., 1980).

In recent years, Savage-Rumbaugh and colleagues, including Patricia Greenfield, have begun a project teaching sign language to pygmy chimpanzees (bonobos) under more naturalistic conditions. That study has yielded information on the differences in language learning abilities of common and pygmy chimpanzees (Greenfield & Savage-Rumbaugh, P&G20).

A fourth ape language project, headed by William Lemon of the University of Oklahoma, began in the 1970s. In 1971, when she was 4 years old, Washoe moved from the Gardners' laboratory with her trainer, the young postdoctoral researcher Roger Fouts, to join the chimpanzee colony run by William Lemon. Roger Fouts, along with others, continued to study sign language acquisition in several new chimpanzees, including Bruno, Booee, and Ally, before he took Washoe with him to Western Washington State University in Ellensberg in 1981 (Linden, 1986). The most significant aspect of Fouts's work has been his demonstration that chimpanzees can learn signs from other chimpanzees (Fouts, Hirsch, & Fouts, 1982).

A fifth ape language study began in 1973 when Herbert Terrace, a cognitive psychologist at Columbia University (a student of B. F. Skinner and the linguist Thomas Bever), acquired an infant chimpanzee, Nim Chimpsky, from William Lemon's colony in Oklahoma. After analyzing his data and the data of other investigators, Terrace concluded that the language abilities of apes were more modest than other investigators had implied. He criticized the methodology of the Gardners, Premack, and Rumbaugh's group, arguing that many of the behaviors could be explained by simple chaining and limitation based on food rewards. Terrace's work focused on the question whether or not chimpanzees could use syntax (i.e., create a sentence); he concluded that they could not (Terrace, Petitto, Saunders, & Bever, 1979; Greenfield & Savage-Rumbaugh, P&G20).

The first language acquisition study using an orangutan began in 1978 when Lyn Miles, a biocultural anthropologist, was invited by Duane Rumbaugh to study one of the orangutans at the Yerkes Primate Laboratory. After 1978, she continued linguistic and cognitive studies of Chantuk at the University of Tennessee using sign language (Miles, 1983, 1986; P&G19).

The first language acquisition study using a gorilla was begun in 1972 by Francine Patterson (Patterson & Linden, 1981). Like the Gardners and Miles, Patterson uses sign language as her training vehicle. Although her corpus of material continues to grow, she has published only a fraction of her data.

Although (or perhaps because) students of ape "language" have sought to use psycholinguistic frameworks, the relationships among ape-language investigators, linguists, and developmental psycholinguists have sometimes been tense. The first published professional response to ape language studies was an article by Jacob Bronowski and Ursula Bellugi (1970) (Bellugi is a developmental psycholinguist who had worked with Roger Brown on an early language acquisition study of children) that compared Washoe's language abilities with those of human children. They argued that although chimpanzees might be able to use reference, they could not break grammatical units down to their components and recombine them (a process they termed "reconstitution"). They also claimed that chimpanzees could not ask questions or use negation.

An early direct contact between ape-language investigators and linguists occurred in the late 1960s or early 1970s when Roger Brown of the Harvard Psychology Department invited the Gardners to speak about their research. That meeting stimulated Brown to write an article comparing the language abilities of Washoe with those of human children, as Bronowski and Bellugi had done. Like them, he argued that a profound gap separated the two species (Brown, 1973). Other child language specialists, including Slobin, also studied the Gardners' data.

Another collaboration between a developmental psycholinguist and ape researchers was sparked in 1973 when Patricia Greenfield, then teaching at the University of California, Santa Cruz, saw the Gardners' film of Washoe using signs and invited them to see her film on the one-word stage of language acquisition in human children. Subsequently, Greenfield (1978) was invited by Rumbaugh and Savage-Rumbaugh to be a consultant on their chimpanzee language project. That led to further collaboration between Greenfield and Savage-Rumabugh (P&G20) in studies of language acquisition by pygmy chimpanzees.

Terrace collaborated with Roger Brown and with linguist Thomas Bever (a student of Noam Chomsky who provided Nim Chimpsky's name) in devising a learning experiment with pigeons to demonstrate that pigeons could perform sequential tasks analogous to those performed by the chimpanzee Lana (Terrace, 1979). His former collaborator Laura Petitto also analyzed Nim's language abilities and compared them with the abilities of human children (using Roger Brown's data).

In comparing chimpanzees and humans, Petitto acknowledges that "both possess a set of natural gestures . . . that are used across multiple contexts for multiple referents. Most importantly, both appear to be intentional beings and are able to use gestures to achieve instrumental ends" (Petitto, 1988, p. 221). She notes, however, that whereas human children learn signs and words with great ease, apes must be laboriously taught. Speaking of Nim, she says that "interestingly, he semed to understand the pragmatic or instrumental function of signing, not the symbolic power of signs themselves. He did not reach out to designate objects, people or events around him. He did not use linguistic symbols to identify referents as belonging to some class or kind" (Petitto, 1988, p. 189).

Petitto also argues that the early gestures of human children up to 16 months of age are prelinguistic: Indexical gestures (e.g., pointing) are used communicatively, referentially, and intentionally by human infants 12–18 months of age, but they are prelinguistic, not linguistic (i.e., not true naming), because they are used with a large class of objects – likewise nonindexical gestures (e.g., open-and-close hand gestures) and instrumental gestures (e.g., arms up) used at the same age. Even such iconic gestures as twisting motions of the wrist, such as enacting opening a jar, which emerge at about 16 months, are not linguistic because they do not designate a class of objects. She concludes:

> The developmental moment when the deaf or hearing child departs from the ape is when he or she begins to refer to and represent his or her world with linguistic symbols rather [than] simply enacting actions with gestures. Moreover, around 18–22 months there appears to be a fundamental reorganization in children's knowledge manifested in strong discontinuities between linguistic and nonlinguistic knowledge. (Petitto, 1988, p. 217)

The disagreement is not so much about what the apes do as it is about the relationship between what apes do and human children do. Frameworks drawn from developmental psycholinguistic studies offer a means for clarifying many of these issues because they represent a series of levels of symbolic communicatory ability. Like the various stages of sensorimotor intelligence, these various levels of prelinguistic and linguistic ability offer a framework for discovering both continuities and discontinuities in ape language abilities (Miles, P&G19; Greenfield & Savage-Rumbaugh, P&G20). Ultimately, because they are the experts on language, linguists and psycholinguists must have the last word in the matter of ape language abilities.

### Origins of comparative (neo-Piagetian) developmental studies of primate abilities

Comparative Piagetian studies of primate intelligence were first done in the 1970s in the United States. During the decade from 1977 to 1988, more than 20 articles analyzing monkey and ape intelligence from a Piagetian perspective were published in the following journals:

*American Journal of Primatology*
*Animal Behaviour*
*Animal Learning and Behavior*
*Annali dell'Istituto Superiore di Sanita*
*Behavioral and Brain Sciences*
*Canadian Psychological Review*
*Developmental Psychobiology*
*Developmental Psychology*
*Human Evolution*
*Journal of Comparative Psychology*
*Journal of Human Evolution*
*Psychological Reports*
*Science*

A. B. Chiarelli, through his editorship of the *Journal of Human Evolution* and *Human Evolution* and his editorship of several conference proceedings, has been influential in disseminating this perspective.

During the same period, several edited books have included more than 20 articles on the same topic:

*Primate Biosocial Development* (Chevalier-Skolnikoff & Poirier, 1977)
*Infancy and Epistemology* (Butterworth, 1981)
*Advanced Views in Primatology* (Chiarelli & Corruccini, 1982)
*Primate Behavior and Sociobiology* (Chiarelli & Corruccini, 1983)
*Current Perspectives in Primate Social Dynamics* (Taub & King, 1986)
*Primate Ontogeny and Social Behaviour* (Else & Lee, 1986)

However, this volume and *Cognitive Structures and Development* (Antinucci, 1989) are the first devoted exclusively to comparative neo-Piagetian studies of intellectual development.

Allison Jolly (1964, 1972, 1974) apparently was the first to suggest in print that Piaget's developmental scheme would be useful for comparative studies of primate intelligence (her first mention of the possibility was a brief note at the end of a paper on discrimination and delayed responses in prosimians suggesting that Piaget's object concept might be relevant [Jolly, 1964]). At virtually the same time that Jolly (1972) reported her first attempt to apply Piaget's broader scheme to published data on monkey and ape intelligence, Etienne (1973) outlined the significance of Piagetian frames for understanding species-specific learning abilities. Behavioral studies within the Piagetian framework began to be published at roughly the same time.

Comparative Piagetian studies of primate intelligence fall into two categories: developmental and nondevelopmental. The first comparative developmental study was of cats, rather than primates (Gruber, Girgus, & Banuazizi, 1971). The first developmental studies of primates began to be published in 1972: Vaughter, Smotherman, and Ordy's study (1972) of the subject concept in squirrel monkeys, Parker's study (1973, 1977) of sensorimotor intelligence in an infant stump-tailed macaque, and Hughes and Redshaw's study (1973) of an infant gorilla. Studies continued through the 1970s and 1980s: Mathieu, Bouchard, Granger, and Herscovitch (1976); Mathieu, Daudelin, Dagenais, and Ducarie (1980); Mathieu and Bergeron (1983); Chevalier-Skolnikoff (1976, 1977, 1981, 1983, 1989); Redshaw (1978); Wood et al. (1980); Vauclair (1982); Vauclair and Bard (1983); Antinucci et al. (1982); Antinucci, Spinozzi, and Natale (1986); Natale, Antinucci, Spinozzi, and Poti' (1986); Antinucci (1989); Hallock and Worobey (1984); Gómez (1988).

Meanwhile, nondevelopmental studies of primate intelligence from a Piagetian perspective were also done in the 1970s, often within a conventional training paradigm: Wise, Wise, and Zimmermann's study (1974) of object permanence in rhesus monkeys; Thomas and Peay's study (1976) of length judgment in squirrel monkeys; the Woodruff et al. (1978) and Muncer (1983) studies of conservation of quantities in chimpanzees; Pasnak's study (1979) of conservation in rhesus macaques; the Brown (1973) and Braggio, Hall, Buchanan, and Nadler (1979) studies of classification in chimpanzees. Most of these nondevelopmental studies and some of the developmental studies were done using training methods devised for traditional learning tasks.

These independent applications of Piaget's framework to studies of primate intelligence by several investigators within the span of a few years in the early 1970s suggest that the idea was in the air. In retrospect, it seems surprising that it came so late. Its tardiness can perhaps be attributed to the dominance of learning theory in comparative psychology and to the concomitant resistance to Piaget's model. Other contributing factors may have been the subtlety and difficulty of the model, the prolixity and opacity of Piaget's writing, and the lack of familiarity with comparative neo-Piagetian developmental frameworks among ethologists. Fortunately, Piagetian concepts have become more accessible with the recent publication of Case's systematic recasting (1985) of Piagetian concepts in light of more recent experimental data and in light of models from information processing.

The early developmental investigators included American comparative psychologists (Vaughter, Smotherman, and Ordy), American biological anthropologists who had been students of Washburn and Jay at the University of California, Berkeley (Parker, Gibson, and Chevalier-Skolnikoff), French-Canadian psychologists from the University of Montreal (Mathieu, Bergeron, Daudelin, Dagenais, Ducarie, Russon), a British psychologist from the Jersey Wildlife Preservation Trust (Redshaw), a French-Swiss psychologist working at Yerkes Primate Center who had been a student of Piaget and of the ethologist Etienne at Geneva (Vauclair), and Italian developmental psycholinguists and psychologists (Antinucci, Volterra, Natale, Spinozzi, and Poti') at the Consiglio Nazionale delle Ricerche in Rome. Researchers in comparative Piagetian development are diverse in training and methods of study. Antinucci and Tomasello are developmental psycholinguists by training: Antinucci took his degree at the University of Rome, and Tomasello took his degree at Georgia State University under von Glasersfeld.

Juan Carlos Gómez, a more recent investigator, is now an instructor at the Universidad Autonoma de Madrid. His former department head studied with Piaget in Geneva, and his dissertation supervisor, J. Linaza, studied with Jerome Bruner and with Robert Boakes at Oxford. His department has strong ties with Bruner, Pascual-Leone, Alan Leslie, and Harry McGurk.

Research strategies tend to correlate with disciplinary training: Comparative psychologists usually work in laboratory and/or colony settings with manipulated cage- or colony-reared animals, relying primarily on formal testing with checklists or scoring of set items, using statistical analysis of scores and frequencies, and using large numbers of subjects in cross-sectional studies. They have tended to use Piaget's object concept series exclusively or a single task such as conservation, because these are most easily quantifiable. Developmental psychologists, including psycholinguists, often work in colony settings with one or a few animals and use either running notes or formal testing. They tend to use a broader set of series (e.g., all of Piaget's sensorimotor series), and whenever possible they do longitudinal studies. Zoologists usually work in a free-ranging colony, field station, or wild setting using checklists and running notes and recording qualitative descriptions. Their use of cross-sectional or longitudinal data depends on the length of the study. Like zoologists, anthropologists usually work in a colony, field station, or wild setting. Like psycholinguists, they tend to do longitudinal studies of one or two individuals (Dore & Dumas, 1987).

Questions and interpretations of data reflect the discipline of the investigator. Predictably, according to their learning theoretic training, traditional comparative psychologists are concerned with proximal questions concerning the nature of such developmental mechanisms as reinforcement, social modeling and facilitation, motivation, and neuromuscular control. Developmental psychologists are concerned with such proximate issues as the animal's conception of the relationships among objects, the nature and definition of tertiary circular reactions, the animal's conception of differences between social and physical objects, and the meaning of signs and symbols. Anthropologists are more concerned with such evolutionary questions as taxonomic distributions of intellectual abilities, their adaptive significance, and possible phylogeny. They are also concerned with related questions of brain evolution.

The major contribution of biological anthropologists has been the introduction of evolutionary approaches: adapting Piaget's sensorimotor period stages to describe and analyze *spontaneous behavior* (Chevalier-Skolnikoff, 1977, 1981, 1983, 1989; Parker, 1973, 1977); seeking the evolutionary significance of Piagetian intelligence using data from primate field studies (Boesch & Boesch, 1984; Chevalier-Skolnikoff, 1976, 1981; Chevalier-Skolnikoff et al., 1982); reconstructing phylogeny using ethology and systematics (Parker,

1973, 1977, 1983, 1985; Parker & Gibson, 1977, 1979, 1982); seeking neuroanatomic correlates of intelligence and language (Chevailer-Skolnikoff, 1976; Gibson, 1970, 1977, 1981, 1983, 1990a,b, P&G3).

These disciplinary approaches are only modal tendencies; in fact, psychologists of all varieties have become increasingly interested in ultimate evolutionary questions (Antinucci, P&G5; Greenfield & Savage-Rumbaugh, P&G20; Tomasello, P&G10; Vauclair, P&G11).

In general, comparative developmental studies of intelligence have been confined to the infant stage of development. Given the long life span and slow development of monkeys and apes, these animals need to be studied from birth to adulthood. Some of the apparent limitations of the great apes, for example, may disappear as the animals approach adulthood (Gardner & Gardner, 1978; Chevalier-Skolnikoff, 1986; Miles, P&G19).

Variations in settings, degrees of intervention and manipulation, and rigor in testing in various developmental studies produce unique and complementary data. Only laboratory settings, which almost always involve hand-reared animals, allow the kinds of controlled testing that can eliminate alternative interpretations. Only descriptive studies in wild settings provide the kinds of naturalistic data that give clues to the normal contexts and frequencies of particular patterns of behavior, which might give clues to their advantages. Only interventive studies in colony and/or field station settings allow the manipulation of social and physical contexts and continuous close-range observations of spontaneous behaviors of the same individuals. Likewise, cross-sectional and longitudinal sampling techniques produce complementary data.

Reconciliation of diverse data should be possible if investigators recognize such sources of differences as the effects of training on the rate of development of performance and the effects of differences in methodology on the difficulty of the task (Dore & Dumas, 1987). Given their frequent exposure to the task, we might expect animals in training experiments, especially (cross-fostered) hand-reared animals, to show an accelerated rate of development as compared with untested animals reared by their mothers (Fragaszy, P&G6). Given the importance of context and shaping, we should expect that animals tested with very specifically focused tasks that isolate specific elements of a configuration should show higher terminal levels of performance and achieve them at earlier ages than should animals tested with complex configurations. Given the variety of foci and settings, the development of a full picture depends on correlations. The greater the number of series studied, the greater the number of settings, the greater the number of species, the better. The introduction of such neo-Piagetian frames as Jonas Langer's stages (1980, 1985, 1986) of prelogic development into comparative developmental studies has expanded the scope of these studies in an important way (Antinucci, 1989, P&G5).

In comparative intellectual studies, as in ape language studies, the Piagetian and neo-Piagetian developmental psychologists must have the last word concerning the methodology and interpretation of Piaget's model, whereas evolutionary biologists and biological anthropologists should have the last word concerning the methodologies of evolutionary reconstruction.

**Evolutionary conceptions in CDEP**

The primate order comprises approximately 180 living species, which are classified into two suborders (Strepsirhini and Haplorhini or, alternatively, Prosimii and Simii). The anthropoid primates include two superfamilies of monkeys (Ceboidea, Cercopithecoidea) and one superfamily of apes and humans (Hominoidea). The superfamily Cercopithecoidea (Old World monkeys) includes two subfamilies (Cercopithecinae and Colobinae), each of which includes several genera. The largest monkey genus, *Cercopithecus*, includes more than 20 species, and the genus *Macaca* includes more than 10 species. Each genus, family, and so forth, represents an independent adaptive radiation of a clade (Figure 1.1, Table 1.1).[8]

The bulk of the comparative Piagetian developmental studies to date have concerned the first few years of life in three taxa: the great apes, cebus monkeys, and macaques. Preliminary results reveal interesting decalages among the various sensorimotor series (i.e., object concept, sensorimotor intelligence [means–ends], causality, space, and time): whereas all these species complete at least the fifth stage of the object concept series, at least two species of macaques (*Macaca arctoides* and *M. fuscata*) fail to display secondary circular reactions of the third stage and tertiary circular reactions of the fifth stage of the sensorimotor intelligence causality series, as well as the imitation series. In contrast, both cebus monkeys and great apes complete elements characteristic of five stages of the sensorimotor period in the means–ends, space, and causality series. Only the great apes also complete all six stages in these series. Great apes also complete the sixth stage in the imitation series, although they do so only in the gestural modality. The great apes also display some abilities characteristic of the preoperations period intelligence.

In general, both monkeys and apes display smaller numbers of schemes in each domain and perform them at a much lower frequency than do human children. Moreover, these species achieve the various stages at markedly different rates (Antinucci, P&G5). (An important limitation of these data is that they are based on studies of infants only.) See Table 1.2 for a summary comparison.[9]

Only when we have exhaustive data on all the species in each genus and family, as well as more distantly related species, will we be able to reconstruct the probable abilities of their common ancestor (Ridley, 1984). Preliminary studies of a few individuals in four species of macaques – stump-tailed macaques (*Macaca arctoides*) (Parker, 1977), Japanese macaques (*M. fuscata*) (Antinucci et al., 1982), crab-eating macaques (*M. fascicularis*) (Antinucci, 1989), and lion-tailed macaques (*M. silenus*) (Westergaard, 1987) – suggest

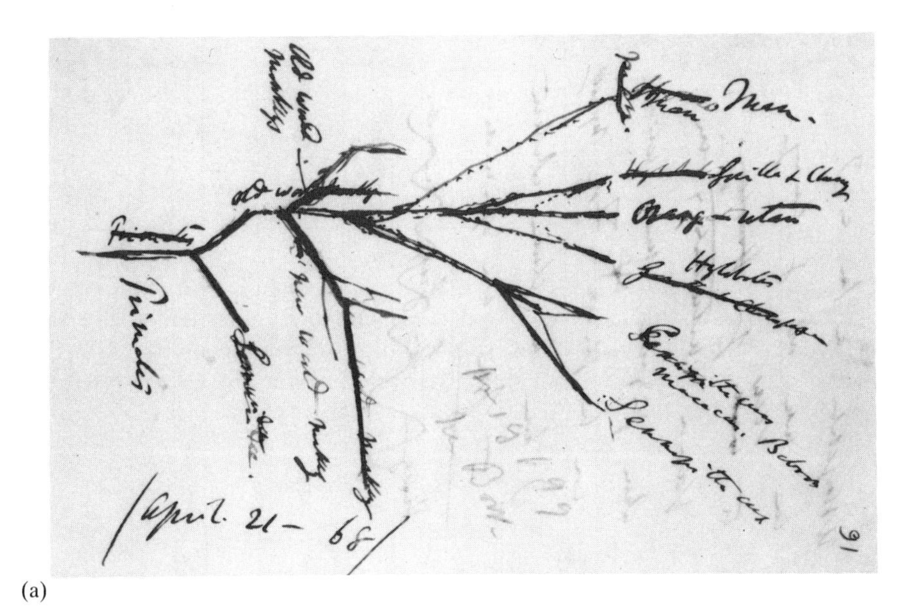

(a)

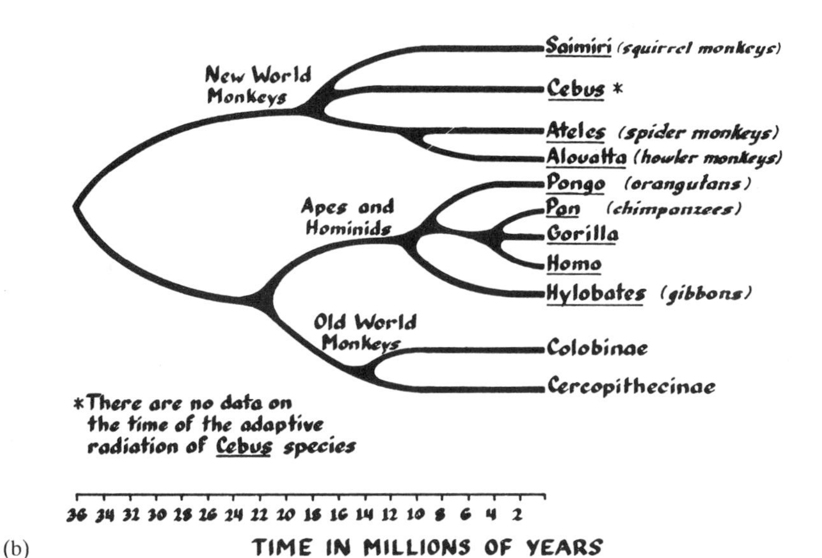


(b)

Figure 1.1. (a) A page from Darwin's notebook (reproduced from Gruber, 1981, by permission of the Syndics of Cambridge University Library).. (b) The phylogeny of anthropoid primates (from Parker & Gibson, 1977).

Table 1.1. *Classification of the primates*

| Suborder | Superfamily | Family | Subfamily | Genus | Common name |
|---|---|---|---|---|---|
| Strepsirhini | Lemuroidea | Cheirogaleidae | | *Allocebus* | Hairy-eared dwarf lemur |
| | | | | *Cheirogaleus* | Dwarf lemur |
| | | | | *Microcebus* | Mouse lemur |
| | | | | *Mirza* | Coquerel's dwarf lemur |
| | | | | *Phaner* | Fork-marked lemur |
| | | Daubentoniidae | | *Daubentonia* | Aye-aye |
| | | Indriidae | | *Avahi* | Woolly lemur |
| | | | | *Indri* | Babakoto |
| | | | | *Propithecus* | Sifaka |
| | | Lemuridae | | *Lemur* | Lemur |
| | | | | *Varecia* | Ruffed lemur |
| | | Lepilemuridae | Hapalemurinae | *Hapalemur* | Gentle lemur |
| | | | Lepilemurinae | *Lepilemur* | Sportive lemur |
| | Lorisoidea | Lorisidae | Galaginae | *Euoticus* | Needle-clawed galago |
| | | | | *Galago* | Bushbaby |
| | | | Lorisinae | *Arctocebus* | Angwantibo |
| | | | | *Loris* | Slender loris |
| | | | | *Nycticebus* | Slow loris |
| | | | | *Perodicticus* | Potto |

| Suborder | Infraorder | Superfamily | Family | Subfamily | Genus | Common name |
|---|---|---|---|---|---|---|
| Haplorhini | Tarsii | Tarsioidea | Tarsiidae | | *Tarsius* | Tarsier |
| | Platyrrhini | Ceboidea | Callimiconidae | | *Callimico* | Goeldi's marmoset |
| | | | Callithricidae | | *Callithrix* | Marmoset |
| | | | | | *Cebuella* | Pygmy marmoset |
| | | | | | *Leontopithecus* | Golden lion tamarin |
| | | | | | *Saguinus* | Tamarin |

| Suborder | Infraorder | Superfamily | Family | Subfamily | Genus | Common name |
|---|---|---|---|---|---|---|
| Haplorhini | Platyrrhini | Ceboidea | Cebidae | Alouattinae | *Alouatta* | Howler monkey |
| | | | | Aotinae | *Aotus* | Owl or night monkey |
| | | | | Atelinae | *Ateles*<br>*Brachyteles*<br>*Lagothrix* | Spider monkey<br>Woolly spider monkey<br>Woolly monkey |
| | | | | Callicebinae | *Callicebus* | Titi monkey |
| | | | | Cebinae | *Cebus* | Capuchin monkey |
| | | | | Pitheciinae | *Cacajao*<br>*Chiropotes*<br>*Pithecia* | Uakari<br>Bearded saki<br>Saki |
| | | | | Saimiriinae | *Saimiri* | Squirrel monkey |
| | Catarrhini | Cercopithecoidea | Cercopithecidae | Cercopithecinae | *Allenopithecus*<br>*Cercocebus*<br>*Cercopithecus*<br>*Erythrocebus*<br>*Macaca*<br>*Miopithecus*<br>*Papio*<br>*Theropithecus* | Swamp monkey<br>Mangabey<br>Guenon<br>Patas<br>Macaque<br>Talapoin<br>Savanna baboon<br>Gelada baboon |
| | | | | Colobinae | *Colobus*<br>*Nasalis*<br>*Presbytis*<br>*Pygathrix*<br>*Rhinopithecus* | Colobus monkey<br>Proboscis monkey<br>Langur<br>Douc langur<br>Golden monkey |
| | | Hominoidea[a] | Hominidae | Gorillinae | *Gorilla*<br>*Pan* | Gorilla<br>Chimpanzee |
| | | | | Homininae | *Homo* | Human |
| | | | Hylobatidae | | *Hylobates* | Gibbon |
| | | | Pongidae | | *Pongo* | Orang-utan |

*Source:* Adapted from Hershkovitz 1977. Classification of the Lemuroidea follows Tattersall 1982; the Lorisoidea, Charles-Dominque 1977; the Cercopithecidae, Thorington and Groves 1970; and the Hominoidea, Andrews and Cronin 1982.
[a]This classification is based on recent molecular evidence. Traditional classifications put *Gorilla* and *Pan* in a separate family from *Homo*.

From Richard (1985).

Table 1.2. *Terminal stage of sensorimotor period stages in monkeys and apes for three series*

| Species | Stage in specified series: Object concept | Means–ends (sensorimotor intelligence) | Imitation |
|---|---|---|---|
| *Cebus* | | | |
| *C. apella* | 5 or 6 | 5 or 6 | 3 |
| *Macaca* | | | |
| *M. arctoides* | 5 | 4 | 3 |
| *M. fuscata* | 5 | 4 | 3 |
| *M. silenus* | 5 | 5 | ? |
| *Pan* | | | |
| *P. troglodytes* | 6 | 6 | 6[a] |
| *Gorilla* | | | |
| *G. gorilla* | 6 | 6[b] | 6[a] |
| *Pongo* | | | |
| *P. pygmeous* | 6 | 6 | 6[a] |
| *Key achievements of included stages* | | | |
| | 5: uncovering visibly displaced object in two or more locations<br>6: reconstructing location of invisibly displaced object | 4: simple means–ends schemes; subordinating one scheme to another, e.g., pulling in object on a string<br>5: trial-and-error discovery of instrumental means<br>6: insightful discovery of instrumental means | 3: imitation schemes from own repertoire<br>5: imitation of novel schemes<br>6: deferred imitation of novel schemes |

[a] Disputed (Antinucci, Spinozzi, & Natale, 1986; Chevalier-Skolnikoff, 1977; Gómez, 1988; Parker, 1976; Redshaw, 1978).
[b] Gestural modality only.

that lion-tailed macaques may show unique convergences in tool-using abilities with the great apes and cebus monkeys (Westergaard & Fragaszy, 1987). Not until we have studied all the species in the genus *Macaca*, the genus *Papio*, the genus *Cercocebus*, and the genus *Cercopithecus*, for example, will we be able to accurately gauge the range of variation in intellectual abilities in the subfamily Cercopithecinae. Likewise for the species in genera of the subfamily Colobinae, which are the closest relatives of the Cercopithecidae. Phylogenetic comparison also demands studies of the lesser ape species in the genera of gibbons (*Hylobates*) and siamangs (*Symphalangus*), which, excluding humans, are the closest relatives of the great apes.

Of greater immediate interest is the range of variation in sensorimotor intelligence among such New World genera as *Cebus*, *Ateles* (spider monkeys), *Brachyteles* (woolly spider monkeys), *Saimiri* (squirrel monkeys), and so forth, which form the family Cebidae, and their relatives, the species and genera in the family Callithricidae, the marmosets (see the chapters on cebus: Antinucci, P&G5; Fragaszy, P&G6; Gibson, P&G7; Parker & Poti', P&G8). The greater time depth of the South American adaptive radiation of monkeys, the lack of competition from apes and prosimians, and the greater morphological variation in this group suggest a great variety of intellectual abilities. See Table 1.1 for taxonomic relationships.

It is already clear from the limited comparative data available that the taxonomic distribution of mental abilities does not fit a simple grade-level model. Cebus monkeys and lion-tailed macaques, for example, display certain abilities lacking in their closest relatives and otherwise found only in great apes and human children. Such cases of convergent or parallel evolution should be expected when similar selection pressures operate on animal populations with similar sensorimotor organizations. Likewise, functionally similar, but structurally dissimilar, abilities should be expected in distantly related species subject to similar selection pressure; therefore, the convergences in classification and numerical abilities of grey parrots and chimpanzees should not dismay us (Boysen & Berntson, P&G16; Matsuzawa, P&G17; Pepperberg, P&G18). Indeed, such exotic similarities may provide important clues to the nature of the selection pressures operating in common (Hamilton, 1973; Parker & Gibson, 1977).

Unlike our 19th-century predecessors, we who seek to reconstruct the evolution of ontogeny in primates today have a rich array of evolutionary models to draw on: behavioral ecology and life history strategy theory, sociobiology, developmental genetics and heterochrony theory, and systematics and cladistic theory.

The study of cladistics provides the evolutionary framework for reconstructing the evolution of development. Cladistic taxonomists use three primary methods for assessing evolutionary relationships: (1) outgroup comparison, (2) embryological similarities, and (3) paleonotological succession (Ridley, 1984). Cope's rule (that the common ancestor of an adaptive radiation of species into new feeding niches generally will be found to have been relatively small in body size and generalist in feeding strategies) suggests caution in reconstructing behavior because it implies that few descendant species live under conditions that resemble those known by the common ancestor from which they evolved. The descendant species, generally the smallest of the adaptive array, that continues the basal adaptation offers the clues most relevant for reconstructing the origins of the character (Parker & Gibson, 1979, 1982).[10]

The study of heterochrony (i.e., differences in the timing of development of various body features and behaviors) is a related science of great relevance to CDEP. Gould's synthesis (1977b) of evolutionary biology and the mechanisms of heterochrony has put the subject in perspective and helped to temper the phobia psychologists suffered after the recapitulation controversy that surrounded G. Stanley Hall's work (e.g., Brainerd, 1979; MacNamara, 1979). Heterochrony theory has in turn stimulated a new controversy on the role of neoteny (the retention of ancestral juvenile or fetal characteristics in the adult stage in descendant species) in the evolution of the human brain (Gibson, 1990, P&G3). Whatever the disputes in interpretation, heterochrony offers a fascinating set of mechanisms for explaining the evolution of development.

The study of life history strategy theory provides the ecological basis for understanding the evolution of development. Systematic data collection on such life history variables as the duration of gestation, lactation, and locomotor dependency, the birth interval, the rate of brain growth, the timing of onset and cessation of reproduction, and the age of death is an important element in an evolutionary research program. Data of this sort can provide important clues to constraints on the evolution and development of intelligence and hence the role of selection in the evolution of ontogeny (Antinucci, P&G5; Fragaszy, P&G6; Gibson, P&G3; Parker, P&G4).

Behavioral ecology and sociobiology provide guidelines for hypotheses concerning the adaptive significance of abilities and developmental patterns. These frameworks require that hypotheses be based on data from field studies of intelligent behaviors in relation to studies of behavioral ecology (Kamil, 1984) and reproductive strategies. Until now, most of the field studies have focused on object manipulation and tool use (e.g., Teleki, 1975), rather than on the development of object manipulation in the wild (e.g., McGrew's study [1977] of the development of object manipulation in wild chimpanzees and Goodall's studies [1986] of the development of tool use in chimpanzee infants), and have not used a Piagetian model. Chevalier-Skolnikoff et al. (1982) published a brief preliminary report on uses of sensorimotor intelligence by wild orangutans based on 2 months of observations at Galdikas's study site (Parker, 1983). Because few field workers know or are interested in the Piagetian model (Christopher and Helwig Boesch provide an important exception [Boesch & Boesch, 1984]), the most likely scenario for such studies is a collaborative plan in which comparative developmental psychologists work at established field sites with a team that is doing studies of behavioral ecology and reproduction.

The roles of selection and the attributions of adaptation are issues of considerable controversy in evolutionary studies, including comparative developmental evolutionary studies. At one extreme of interpretation, investigators invoke selection to explain every behavior, failing to distinguish current advantages of a character or behavior in a particular population from the past advantages that may have operated as selection pressures in the evolution of that character (Williams, 1966). At the other extreme of interpretation, investigators demote selection and invoke constraints on development to explain many characters (Gould & Lewontin, 1979).[11]

Another controversial issue in comparative developmental evolutionary studies is the Lamarckian model that Piaget explicitly reintroduced into the psychological arena in *Biology and Knowledge* (1971), *Behavior and Evolution* (1978), and *Adaptation and Intelligence* (1980). In this way, Piaget has retained a 19th-century global concept of ontogeny and phylogeny. Interestingly, Lamarckian explanations, which led to rejection of the work of Spencer and Hall and ultimately to the rejection of stage theory in the United States, were never rejected by French biologists (Stocking, 1968). Despite the lack of evidence for the operation of Lamarckian mechanisms in the evolution of animals, many Piagetians find Piaget's argument compelling because they believe that the mechanisms of mutation, recombination, and selection are inadequate to explain the evolution of higher intelligence. Unfortunately, Piaget's Lamarckian ideas caused many zoologists and biological anthropologists in the United States to reject Piaget's model (Thomas, 1977).[12]

Although disciplinary differences cause misinterpretations, misunderstandings, and conflicting interpretations, they also generate fruitful controversy and reexamination of basic principles. They are natural features of a growing research program.

**Summary**

Early investigators of animal and human intelligence in the United States and England were influenced by a confusing mixture of 19th-century ideas about mental, cultural, and racial evolution imbued with a fused concept of evolution and development, progressive stages, recapitulationism, and Lamarckism. As social science disciplines became established in U.S. universities around the turn of the century, both anthropologists and psychologists rejected these concepts of mind, development, and evolution that had been imbued with Lamarckian and recapitulationist ideas and thereby turned away from comparative developmental evolutionary studies of mental abilities. Instead, they developed a spartan conditioning model of animal and human abilities.

As a consequence of the enormous power of learning theory in psychology, comparative studies of primate abilities through the 1960s focused almost exclusively on standardized laboratory tests of learning in adult animals of a few easily managed species. Those studies were largely nondevelopmental, noncomparative, and nonevolutionary in their aims. Likewise, as a consequence of the power of historical particularism in anthropology, physical anthropologists focused almost exclusively on morphological evolution, avoiding behavioral evolution.

Only in the 1950s and 1960s did physical anthropologists begin to study primate populations in the wild and begin to focus on the evolution of primate and human behavior and intelligence. Only in the 1970s, following the lead of the newly emerging field of cognitive science, after an expectable time lag, did comparative psychologists readmit the concept of mind and begin to explore "cognitive" processes in animals. At roughly the same time that the rise of cognitive sciences influenced comparative psychologists toward more cognitive approaches to animal abilities, an increasing focus on monkeys and apes as research subjects was stimulated by the establishment of large-scale primate research facilities. Beginning in the 1970s, primatologists began to use developmental (Piagetian and psycholinguistic) models in the comparative study of primate intelligence and communication. Comparative studies of the same species in the wild and in colony and laboratory settings have yet to develop. Likewise, systematic genuswide comparisons have yet to be made.

## Conclusions

This new research program in CDEP gives comparative investigators bases for distinguishing levels of complexity of behavioral organization within series and domains and for making systematic comparisons of the patterns of abilities across behavioral series and domains, thus giving them bases for studying similarities and differences in developmental patterns in closely related and distantly related primate species. Finally, in conjunction with the comparative method and data from field studies, it facilitates reconstruction of the patterns of common ancestors of various primate lineages and ultimately the reconstruction of the evolution of ontogeny. It will thereby be possible to create a truly comparative developmental evolutionary psychology based on the integration of developmental psychology, biological anthropology, and evolutionary biology.[13]

### Acknowledgments

Writing this chapter has been a humbling but enlightening experience for me: humbling because I knew very little about the history of anthropology and less about the history of psychology. Had I more time, I would have done more and better research. As it was, I learned a great deal from the materials I read and, especially, from the following individuals, who provided historical, bibliographical, and biographical data: Francesco Antinucci, Beatrice Gardner, Patricia Greenfield, Howard Gruber, Helen LeRoy, William Mason, Emile Menzel, Lyn Miles, David Premack, Duane Rumbaugh, Daniel Slobin, Steven Suomi, and Sherwood Washburn. I also learned from the numerous references, leads, and helpful suggestions of the following colleagues: Francesco Antinucci, Bernard Baars, Kathleen Gibson, Juan Carlos Gómez, Patricia Greenfield, Helen LeRoy, William Mason, Constance Milbrath, Steven Pulos, Anne Russon, Michael Tomasello, Jacques Vauclair, and Andrew Wilson. I especially thank Kathleen Gibson for pushing me to do more on biological anthropology, and an anonymous reviewer for making me revise the manuscript the 30 millionth time. Most of all I thank Jonas Langer, Hy Hartman, Steve Glickman, and other participants at a seminar in cognitive evolution in the Psychology Department, University of California, Berkeley, spring 1989.

### Notes

1. Evolutionary perspectives on culture and the human psyche also came back into cultural anthropology in the 1950s in the guise of cultural ecology through the work of Julian Stewart, Leslie White, Elwyn Service, and Marshall Sahlins. Leslie White, who taught cultural evolution at the University of Michigan, had been one of Boas's students. White became interested in Lewis H. Morgan's work partly through his Marxist affiliations (Langness, 1987). Beginning in the 1970s, evolutionary perspectives were given a powerful new tool in the form of sociobiological theory, which has been particularly strong among faculty and students from the University of Michigan.

Finally, interest in "the psychic unity of mankind" and in the evolutionary underpinnings of the human psyche has been a continuing strand in the field of personality and culture and psychological anthropology. From the early days, various cultural anthropologists, including Margaret Mead and Bronislaw Malinowski, have used psychoanalytic concepts in their field studies. Freud, of course, was heavily influenced, via James Frazier and G. Stanley Hall, by the 19th-century ideas of cultural evolution and recapitulationism (Langness, 1987).

2. Boas's ambivalent attitude toward physical anthropology probably was based on his knowledge of the racialist views of German physical anthropologists (whose ideas provided the springboard for much of his own work aimed at disproving the idea that temperament and gesture are inherited). The attitude of prewar German physical anthropologists was exemplified by the career of Eugen Fischer, whose professional interests made him so valuable to the Nazis. Fischer's attempt to apply Mendelian principles to the study of race (reported in his 1913 study *Rehobother Bastards*) led to his appointment to the chair in anthropology at the University of Berlin, which in turn determined the focus of German anthropology: "Fischer's appointment as professor at Berlin and director of Germany's most prestigious anthropological institute guaranteed that, for the next two decades, genetics, eugenics, and the study of racial differences would constitute the primary focus of anthropological research" (Proctor, 1988, p. 155). Fischer subsequently presided over the "'cleansing' of the Jews from Germany's most prestigious university" (Proctor, 1988, p. 164) and served as a judge in Berlin's Erbgesundheits-obergericht (Appellate Genetic Health Court), which determined who should be sterilized to prevent the birth of "genetically diseased offspring." Moreover, Fischer's "Kaiser Wilhelm Institut served as a training school for SS physicians in what was generally known as 'genetic and racial care'" (Proctor, 1988, p. 160). Proctor shows that by the 1920s, long before the rise of the Nazis, "race had become the single most important concept in German anthropology, which was well on the way to redenomination as *Rassenkunde*, or 'racial science'" (Proctor, 1988, p. 148). When Hitler came to power in 1933, Boas, then 75 years old, began a fruitless campaign to mobilize leading American and British scientists to speak out against German anti-Semitism (Barkan, 1988).
3. Citations of other chapters in this volume will use an abbreviation for the editors' names, followed by the chapter number (Parker & Gibson, chapter 2, becomes P&G2).
4. Although the "man the hunter" model was perhaps most influential in the rise of the "new physical anthropology," many other models and many other versions of this model have been presented, beginning with Darwin and going on through Engels and Dart and into recent times (Kinzey, 1987). Before the "man the hunter" model, the most influential model in anthropology probably was that of Bartholomew and Birdsell (1953).
5. Haraway's analysis is supported by reference to Washburn's critique of sociobiology (e.g., Washburn, 1980). It also suggests why Washburn has avoided sexual selection theory in his models.
6. Although most investigators point to the paucity of CDEP studies, Dewsbury claims that "the developmental approach, together with the evolutionary emphasis may indeed be regarded as the major emphasis in comparative psychology throughout the century" (Dewsbury, 1984, p. 220). Despite his claim, he provides few examples of such an approach. An examination of the early papers he cites by Wesley Mills (1898) reveals little beyond one-sentence quotations on the desirability of such studies. Likewise, the paper he cites by Lashley and Watson (1913) on the development of a rhesus monkey stands as an isolated effort unaccompanied by a theoretical justification. This is significant given that Dewsbury christens Mills and Watson the "fathers" of "comparative developmental psychology."
7. Arnold Gesell, a pediatrician, founded the Clinic for Child Development at Yale University Medical School. He published comparative data on motor and perceptual development in a rhesus macaque (collected by Lashley and Watson [1913]) and a brief description of the performance of a 5-year-old cebus monkey on a series of manipulative tests. Gesell concluded, however, that "it would be profitless to review the performances of Jocko and to assign to them psychological age values in human terms" (Gesell, 1928, p. 348).

8. Because all the species in a genus or family or higher taxonomic category are descendants of a single common ancestral species, they are more closely related to one another than they are to species in other genera, families, etc. These species groups are clades or adaptive arrays that have diverged in certain ways from one another – usually in diet and body size and reproductive behavior – while maintaining some common features; one species typically remains more like the common ancestor than do the others.
9. It should be noted here that some controversy exists concerning achievement of elements of the third and fifth stages of the means–ends and causality series in gorillas. Two investigators saw fifth-stage achievements in this species (Chevalier-Skolnikoff, 1977; Gómez, 1988; Parker, 1976), but two others failed to see such achievements (Antinucci, Spinozzi, & Natale, 1986; Natale, 1989; Redshaw, 1978). These differences may reflect differences in methodology or differences in the ages of the subjects.
10. The method of outgroup comparison uses a species that is distantly related to the two species being compared as a reference point to decide which of two character states (in the two species being compared) is ancestral. The objective of this exercise is to see which of the two states was present in the common ancestor of the larger unit that embraces both the ingroup and the outgroup. The second method uses embryological similarities as a clue to common ancestry, relying on von Baer's first law, i.e., that more primitive, general features tend to develop before more derived or specialized features. It should be noted, however, that taxonomy is also a highly controversial field within evolutionary biology and that the three competing schools of taxonomy (cladism, phenetism, and evolutionary taxonomy) disagree on many issues (Ridley, 1984). Note that the method of outgroup comparison demands the study of distantly related species of primates or other mammals as a referent.
11. Design constraints are factors that constrain free modification and reorganization of existing structural and behavioral elements. Some of these constraints are imposed by the laws of growth and form (e.g., Gould, 1984; Thompson, 1941), including the cube-square law of volume to surface area, which is responsible for allometric changes in organ size concomitant with evolutionary changes in body size (Gould, 1977b). Other constraints are imposed by the phylogenetic inertia of such complex, integrated organ systems as the vertebrate ground plan or the placental membrane system that will not operate if they change radically.
12. Piaget and many of his students have argued that his formulation is not Lamarckian, because it involves as "active" process of epigenetic adaptation involving indirect "reequilibrations" in response to "disequilibrations" that occur during embryogenesis. According to Piaget, the resulting process of reequilibration, rather than the characteristics themselves, is somehow transmitted to the genotype: "there is a relationship between phenotype and genotype, but [that] the latter copies the former, and not, as is usually believed, the reverse" (Piaget, 1980, p. 54). Piaget denied that acquired characteristics could be inherited, hypothesizing instead that the equilibration processes during embryogenesis led to the substitution of adaptive for maladaptive processes in the genome as well as the embryo. Although Piaget's formulation differs from that of Lamarck in its emphasis on the process of equilibration, both models are Lamarckian from a biological perspective because they involve genetic transmission of adaptive patterns acquired after conception and hence violate Weismann's central dogma of the inaccessibility of the germ plasm in animals.
13. Although I do not address the issue in this chapter, I believe that comparative developmental evolutionary studies will help us understand the mechanisms by which ontogeny evolves, as well as the nature of the constraints operating on the evolution of ontogeny. In other words, I believe that it will help us understand the tantalizing relationships between development and evolution.

## References

Antinucci, F. (Ed.). (1989). *Cognitive structures and development in nonhuman primates.* Hillsdale, NJ: Erlbaum.

Antinucci, F., Spinozzi, G., & Natale, F. (1986). Stage V cognition in an infant gorilla. In D.

Taub & F. A. King (Eds.), *Current perspectives in primate social dynamics* (pp. 403–415). New York: Van Nostrand Reinhold.

Antinucci, F., Spinozzi, G., Visalberghi, E., & Volterra, V. (1982). Cognitive development in a Japanese macaque (*Macaca fuscata*). *Annali dell'Istituto Superiore di Sanita*, *18*(2), 177–184.

Baars, B. (1986). *The cognitive revolution in psychology*. New York: Guilford Press.

Baldwin, J. M. (1896). A new factor in evolution. *American Naturalist*, *30*, 441–451, 536–553.

Baldwin, J. M. (1906). *Mental development in the child and the race* (2nd ed.). New York: Macmillan. (Original work published 1896)

Balwin, J. M. (1909). *Library of genetic science and philosophy. Vol. 2. Darwin and the humanities*. Baltimore: Review Publishing.

Barkan, E. (1988). Mobilizing scientists against Nazi racism, 1933–1939. In G. W. Stocking, Jr. (Ed.), *History of anthropology. Vol. 5. Bones, bodies, behavior: Essays on biological anthropology* (pp. 180–205). Madison: University of Wisconsin Press.

Bartholomew, G., & Birdsell, J. (1953). Ecology and the protohominids. *American Anthropologist*, *55*, 481–98.

Bates, E., Benigni, L., Bretherton, I., Camaioni, L., & Volterra, V. (1979). *The emergence of symbols: Cognition and communication in infancy*. New York: Academic Press.

Bates, E., Camaioni, L., & Volterra, V. (1976). The acquisition of performatives prior to speech. *Merrill-Palmer Quarterly*, *21*(3), 205–226.

Beach, F. (1965). The snark is a boojum. In T. E. McGill (Ed.), *Readings in animal behavior* (pp. 3–14). New York: Holt, Rinehart & Winston.

Bertenthal, B. L., & Fischer, K. W. (1978). Development of self-recognition in the infant. *Developmental Psychology*, *14*, 44–50.

Boakes, R. (1984). *From Darwin to behaviorism: Psychology and minds of animals* (pp. 23–60). Cambridge University Press.

Boas, F. (1938). *The mind of primitive man* (2nd ed.). New York: Free Press. (Original work published 1911)

Boesch, C., & Boesch, H. (1984). Mental maps in wild chimpanzees: An analysis of hammer transports for nut cracking. *Primates*, *25*, 160–170.

Bower, T. G. R. (1979). *Human development*. San Francisco: Freeman.

Brace, C. L. (1982). The roots of the race concept in American physical anthropology. In F. Spencer (Ed.), *A history of American physical anthropology 1930–1980* (pp. 11–30). New York: Academic Press.

Braggio, J. T., Hall, R. D., Buchanan, J. P., & Nadler, R. D. (1979). Cognitive capacities of juvenile chimpanzees on a Piagetian-type multiple classification task. *Psychological Reports*, *44*, 1087–1097.

Brainerd, C. R. (1978). *Piaget's theory of intelligence*. Englewood-Cliffs, NJ: Prentice-Hall.

Brainerd, C. (1979). Recapitulation, Piaget, and the evolution of intelligence: Deja vu. Commentary on Parker and Gibson. *Behavioral and Brain Sciences*, *2*, 381–382.

Bronfenbrenner, U. (1983). The context of development and the development of context. In R. M. Lerner (Ed.), *The development of developmental psychology* (pp. 147–184). Hillsdale, NJ: Erlbaum.

Bronowski, J., & Bellugi, U. (1970). Language, name, and concept. *Science*, *168*, 699.

Broughton, J. M., & Freeman-Moir, D. J. (Eds.). (1982). *The cognitive developmental psychology of James Mark Baldwin: Current theory and research in genetic epistemology*. Chicago: Aldine.

Brown, R. (1973). *A first language: Early stages*. Cambridge, MA: Harvard University Press.

Bruner, J. (1983). *In search of mind: Essays in autobiography*. New York: Harper & Row.

Bruner, J. (1988). Founding the Center for Cognitive Studies. In W. Hirst (Ed.), *The making of cognitive science: Essays in honor of George Miller* (pp. 90–98). Cambridge University Press.

Bruner, J., Oliver, R., & Greenfield, P. (1966). *The growth of cognitive studies*. New York: Wiley.

Butterworth, G. (Ed.). (1981). *Infancy and epistemology*. Brighton, England: Harvester Press.

Cadwaller, T. C. (1987). Early zoological input to comparative animal psychology at the University of Chicago. In E. Tobach (Ed.), *Historial perspectives and the international status of comparative psychology* (pp. 37–80). Hillsdale, NJ: Erlbaum.
Case, R. (1985). *Intellectual development birth to adulthood.* New York: Academic Press.
Chevalier-Skolnikoff, S. (1976). The ontogeny of primate intelligence and its implications for communicative potential: A preliminary report. Origins and Evolution of language and speech. *Annals of the New York Academy of Sciences*, *280*, 173–211.
Chevalier-Skolnikoff, S. (1977). A Piagetian model for describing and comparing socialization in monkey, ape, and human infants. In S. Chevalier-Skolnikoff & F. Poirier (Eds.), *Primate biosocial development* (pp. 159–188). New York: Garland Press.
Chevalier-Skolnikoff, S. (1981). The clever Hans phenomenon, cueing, and ape signing: A Piagetian analysis of methods for instructing animals. In T. A. Sebeok & R. Rosenthal (Eds.), *The clever Hans phenomenon: Communication with horses, whales, apes, and people* (pp. 60–93). New York Academy of Sciences.
Chevalier-Skolnikoff, S. (1982). A cognitive analysis of facial behavior in Old World monkeys, apes, and human beings. In C. Snowdon, C. H. Brown, & M. Petersen (Eds.), *Primate communication* (pp. 303–368). Cambridge University Press.
Chevalier-Skolnikoff, S. (1983). Sensorimotor development in orangutans and other primates. *Journal of Human Evolution*, *12*, 545–561.
Chevalier-Skolnikoff, S. (1986). An exploration of the ontogeny of deception in human beings and nonhuman primates. In R. W. Mitchell & N. S. Thompson (Eds.), *Deception: Perspectives on human and nonhuman deceit* (pp. 205–220). New York: SUNY Press.
Chevalier-Skolnikoff, S. (1989). Spontaneous tool use and sensorimotor intelligence in *Cebus* compared with other monkeys and apes. *Behavioral and Brain Sciences*, *12*(3), 561–627.
Chevalier-Skolnikoff, S., Galdikas, B., & Skolnikoff, A. (1982). The adaptive significance of higher intelligence in wild orangutans: A preliminary report. *Journal of Human Evolution*, *11*, 639–652.
Chevalier-Skolnikoff, S., & Poirier, F. (Eds.). (1977). *Primate biosocial development.* New York: Garland Press.
Chiarelli, A. B., & Corruccini, R. S. (Eds.). (1982). *Advanced views in primatology.* Main lectures of the VIIth Congress of the International Primatological Society. Berlin: Springer-Verlag.
Chiarelli, A. B., & Corruccini, R. S. (Eds.). (1981). *Primate behavior and sociobiology.* Proceedings of the VII Congress of the International Primatological Society. Berlin: Springer-Verlag.
Coon, C. (1962). *Origin of races.* New York: Knopf.
Coon, C., Garn, S., & Birdsell, J. (1950). *Races.* Springfield, IL: Thomas.
Daly, M., & Wilson, M. (1984). *Sex, evolution and behavior* (2nd ed.). Boston: Willard Grant.
Darwin, C. (1859). *On the origin of species.* London: Murray.
Darwin, C. (1871). *The descent of man and selection in relation to sex.* London: Murray.
Darwin, C. (1872). *The expression of emotions in man and animals.* London: Murray.
Demarest, J. (1987). Two comparative psychologies. In E. Tobach (Ed.), *Historical perspectives and the international status of comparative psychology* (pp. 127–158). Hillsdale, NJ: Erlbaum.
Dewsbury, D. A. (1984). *Comparative psychology in the twentieth century.* Stroudsburg, PA: Hutchinson Ross Publishing.
Dore, F. Y., & Dumas, C. (1987). Psychology of animal cognition: Piagetian studies. *Psychological Bulletin*, *102*(2), 219–233.
Dore, F. Y., & Kirouac, G. (1987). What comparative psychology is about: Back to the future. *Journal of Comparative Psychology*, *101*, 242–248.
Else, J. C., & Lee, P. C. (Eds.). (1986). *Primate ontogeny and social behaviour.* Cambridge University Press.
Essock-Vitale, S., & Seyfarth, R. (1987). Intelligence and social cognition. In B. B. Smuts, D. L. Cheney, R. M. Seyfarth, R. W. Wrangham, & T. T. Struhsaker (Eds.), *Primate societies* (pp. 452–461). University of Chicago Press.

Etienne, A. (1973). Developmental stages and cognitive structures as determinants of what is learned. In R. Hinde & J. Stevenson (Eds.), *Constraints on learning: Limitations and predispositions* (pp. 371–395). New York: Academic Press.
Fancher, R. E. (1979). *Pioneers in pyschology.* New York: Norton.
Feinstein, H. M. (1984). *Becoming William James.* Ithaca: Cornell University Press.
Fischer, K. W. (1980). A theory of cognitive development: The control of hierarchies of skill. *Psychological Review*, *87*, 477–531.
Flavell, J. (1963). *The developmental psychology of Jean Piaget.* New York: Van Nostrand.
Fossey, D. (1972). Vocalizations of the Mountain gorilla (*Gorilla gorilla beringei*). *Animal Behavior*, *20*, 568–581.
Fouts, R., Hirsch, A., & Fouts, D. (1982). Cultural transmission of a human language in a chimpanzee mother and infant. In H. Fitzgerald, J. Mullens, & P. Gage (Eds.), *Child nurturance* (Vol. 3, pp. 159–193). New York: Plenum.
Gallup, G. G. (1982). Self-awareness and the emergence of mind in primates. *American Journal of Primatology*, *2*, 237–248.
Garcia, J., McGawon, B. K., Ervin, F. R., & Koelling, R. A. (1968). Cues: Their effectiveness as a function of reinforcer. *Science*, *160*, 794–795.
Gardner, B., & Gardner, A. (1969). Teaching sign language to a chimpanzee. *Science*, *165*, 664–672.
Gardner, B., & Gardner, A. (1971). Two-way communication with an infant chimpanzee. In A. Schrier & F. Stolnitz (Eds.), *Behavior of nonhuman primates* (Vol. 4, pp. 117–183). New York: Academic Press.
Gardner, H. (1987). *The mind's new science.* New York: Basic Books.
Gardner, R. A., & Gardner, B. T. (1978). Comparative psychology and language. *Annals of the New York Academy of Sciences*, *309*, 37–76.
Gardner, R. A., & Gardner, B. T. (1987). The role of cross-fostering in sign language studies of chimpanzees. *Human Evolution*, *2*(6), 65–79.
Garn, S. (1961). *Human races.* Springfield, IL: Thomas.
Gelman, R., & Baillargeon, R. (1983). A review of Piagetian concepts. In J. H. Flavell & E. M. Markman (Eds.), *Handbook of child psychology. Vol. 3. Cognitive development* (pp. 167–230). New York: Wiley.
Gesell, A. (1928). *Infancy and human growth.* New York: Macmillan.
Ghiselin, M. (1969). *The triumph of the Darwinian method.* Berkeley: University of California Press.
Ghiselin, M. (1986). The assimilation of Darwinism in developmental psychology. *Developmental Psychology*, *29*, 12–21.
Gibson, K. R. (1970). Sequence of myelinization in the brain of *Macaca mulatta.* PhD dissertation in anthropology, University of California, Berkeley.
Gibson, K. R. (1977). Brain structure and intelligence in macaques and human infants from a Piagetian perspective. In S. Chevalier-Skolnikoff & F. Poirier (Eds.), *Primate biosocial development* (pp. 113–158). New York: Garland Press.
Gibson, K. R. (1981). Comparative neuro-ontogeny: Its implications for the development of human intelligence. In G. Butterworth (Ed.), *Infancy and epistemology* (pp. 52–82). Brighton, England: Harvester Press.
Gibson, K. R. (1983). Comparative neurobehavioral ontogeny: The constructionist perspective in the evolution of language, object manipulation and the brain. In E. de Grolier (Ed.), *Glossogenetics* (pp. 41–61). New York: Academic Press.
Gibson, K. R. (1986). Cognition, brain size and extraction of embedded foods. In J. C. Else & P. C. Lee (Eds.), *Primate ontogeny and social behaviour* (pp. 93–105). Cambridge University Press.
Gibson, K. R. (1990a). Brain size and the evolution of language. In A. Lock & C. Peters (Eds.), *The genesis of language: A different judgement of evidence. Handbook of symbolic intelligence.* Oxford University Press.
Gibson, K. R. (1990b). Brain and cognition in the growing child: A comparative perspective on the questions of neoteny, altriciality and intelligence. In K. R. Gibson & A. Peterson

(Eds.), *Brain maturation and behavioral development: Biosocial dimensions.* Chicago: Aldine.

Glickman, S. (1985). Some thoughts on the evolution of comparative psychology. In S. Koch & D. Vevy (Eds.), *A century of psychology as a science* (pp. 738–781). New York: McGraw-Hill.

Gómez, J. C. (1988). *The development of intentional manipulations and communication in the gorilla.* Paper presented at the Workshop on Comparative Cognitive Development in Ape and Human Infants, Madrid, April.

Goodall, J. (1986) *Chimpanzees of the Gombe Reserve.* Cambridge, MA: Harvard University Press.

Gould, S. J. (1977a). Size and shape. In S. J. Gould (Ed.), *Ever since Darwin* (pp. 171–178). New York: Norton.

Gould, S. J. (1977b). *Ontogeny and phylogeny.* Cambridge, MA: Harvard University Press.

Gould, S. J. (1981). *The mismeasure of man.* New York: Norton.

Gould, S. J. (1984). How a zebra gets its stripes. In S. J. Gould (Ed.), *Hens' teeth and horse's toes* (pp. 366–375). New York: Norton.

Gould, S. J., & Lewontin, R. (1979). The spandrels of San Marcos and the Panglossian paradigm: A critique of the adaptationist programme. *Proceedings of the Royal Society, London, 205B*, 581–598.

Greenfield, P. M. (1978). Commentary on Savage-Rumbaugh, S., Rumbaugh, D., & Boysen, S. (1978). Linguistically mediated tool use and exchange by chimpanzees (*Pan troglodytes*). *Behavioral and Brain Sciences, 4*, 539–554.

Greenfield, P. M., & Smith, J. H. (1976). *The structure of communication in early language development.* New York: Academic Press.

Griffin, D. R. (Ed.). (1982). *Animal mind – human mind.* New York: Springer-Verlag.

Gruber, H. E. (1981). *Darwin on man: A psychological study of scientific creativity* (2nd ed.). University of Chicago Press.

Gruber, H. E., Girgus, J. S., & Banuazizi, A. (1971). The development of object permanence in the cat. *Developmental Psychology, 4*, 9–15.

Haeckel, E. (1876). *The history of creation: or the development of the earth and its inhabitants by the action of natural causes* (2 vols.). New York: Appleton.

Hahn, E. (1971). *On the side of the apes.* New York: Arena Books.

Hallock, M. B., & Worobey, J. (1984). Cognitive development in chimpanzee infants (*Pan troglodytes*). *Journal of Human Evolution, 13*, 441–447.

Hamilton, W. (1973). *Life's color code.* New York: McGraw-Hill.

Hammond, M. (1988). The shadow man paradigm in paleoanthropology: 1911–1945. In G. W. Stocking, Jr. (Ed.), *History of anthropology. Vol. 5. Bones, bodies, behavior: Essays on biological anthropology* (pp. 117–137). Madison: University of Wisconsin Press.

Haraway, D. (1988). Remodelling the human way of life: Sherwood Washburn and the new physical anthropology: 1950–1980. In G. W. Stocking, Jr. (Ed.), *History of anthropology. Vol. 5. Bones, bodies, behavior: Essays on biological anthropology* (pp. 206–259). Madison: University of Wisconsin Press.

Harlow, H. (1943). Solution by rhesus monkeys of a problem involving the Weigl principle using the matching-from-sample method. *Journal of Comparative Psychology, 36*, 217–227.

Harlow, H. (1944a). Studies in discrimination learning by monkeys: I. Learning of discrimination series and the reversal of discrimination series. *Journal of Comparative Psychology, 30*, 3–12.

Harlow, H. (1944b). Studies in discrimination learning by monkeys: II. Discrimination learning without primary reinforcement. *Journal of Comparative Psychology, 30*, 13–21.

Harlow, H. (1945a). Studies in discrimination learning in monkeys: III. Factors influencing the facility of solution of discrimination problems by rhesus monkeys. *Journal of Genetic Psychology, 32*, 213–227.

Harlow, H. (1945b). Studies in discrimination learning by monkeys: IV. Relative difficulty of discriminations between stimulus objects and between comparable patterns with homogenous and with heterogenous grounds. *Journal of Genetic Psychology*, *32*, 317–321.

Harlow, H. (1945c). Studies in discrimination learning by monkeys: V. Initial performance by experimentally naive monkeys on stimulus–object and pattern discrimination. *Journal of Genetic Psychology*, *33*, 3–10.

Harlow, H. (1945d). Studies in discrimination learning by monkeys: VI: Discriminations between stimuli differing in both color and form, only in color, and only in form. *Journal of Genetic Psychology*, *33*, 225–235.

Harlow, H. (1949). The formation of learning sets. *Psychological Review*, *56*, 51–65.

Harlow, H. (1951). Primate learning. In C. S. Stone (Ed.), *Comparative psychology* (pp. 183–238). Englewood Cliffs, NJ: Prentice-Hall.

Harlow, H., & Bromer, J. (1938). A test-apparatus for monkeys. *Psychological Record*, *2*, 434–436.

Harlow, H., & Bromer, J. (1939). Comparative behavior of primates. VIII. The capacity of platyrrhine monkeys to solve delayed reaction tests. *Journal of Comparative Psychology*, *28*, 299–304.

Harlow, H., Uehling, H., & Maslow, A. (1932). Comparative behavior of primates: I. Delayed reaction tests on primates from the lemur to the orangutan. *Journal of Comparative Psychology*, *13*, 133–143.

Hartman, H. (1989). *The evolution of natural selection, Darwin versus Wallace.* Manuscript distributed in Psychology 240, "Cognitive Evolution," Department of Psychology, Spring 1989, University of California, Berkeley.

Hatch, E. (1973). *Theories of man and culture.* New York: Columbia University Press.

Hayes, K., & Hayes, C. (1971). The intellectual development of a home raised chimpanzee. *Proceedings of the American Philosophical Society*, *95*, 105–109.

Hinde, R. A. (1973). Constraints on learning: An introduction to the problems. In R. Hinde & J. Stevenson-Hinde (Eds.), *Constraints on learning* (pp. 1–19). New York: Academic Press.

Hoage, R. J., & Goldman, L. (Eds.). (1986). *Animal intelligence.* Washington, DC: Smithsonian Institution Press.

Hockett, C. (1960). The origin of speech. *Scientific American*, *203*, 88–96.

Hodos, W., & Campbell, C. B. G. (1969). Scala naturae: Why there is no theory of comparative psychology. *Psychological Review*, *76*, 337–350.

Hooton, E. O. (1946). *Up from the ape* (rev. ed.). New York: Macmillan.

Hothersall, D. (1984). *History of psychology.* Philadelphia: Temple University Press.

Hughes, J., & Redshaw, M. (1973). Cognitive, manipulative and social skills in gorillas. *Jersey Wildlife Preservation Trust Annual Report of 1974*, pp. 53–61.

Hunt, J. M. (1961). *Intelligence and experience.* New York: Ronald Press.

Inhelder, B., & Piaget, J. (1964). *The early growth of logic in the child.* New York: Norton.

James, W. (1875). *The principles of psychology* (2 vols.). New York: Holt. (Original work published 1855)

Jerison, H. (Ed.). (1987). *Intelligence and evolutionary biology.* New York: Springer-Verlag.

Jolly, A. (1964). Choice of cue in prosimian learning. *Animal Behavior*, *12*(4), 571–577.

Jolly, A. (1972). *The evolution of primate behavior.* New York: Macmillan.

Jolly, A. (1974). The study of primate infancy. In K. Connolly & J. Bruner (Eds.), *The growth of competence* (pp. 49–74). New York: Academic Press.

Kamil, A. (1984). Adaptation and cognition: Knowing what comes naturally. In H. L. Roitblat, T. G. Bever, & H. S. Terrace (Eds.), *Animal cognition* (pp. 533–544). Hillsdale, NJ: Erlbaum.

Kardiner, A., & Preeble, E. (1961). *The studied man.* New York: Mentor.

Kaye, K. (1983). *The mental and social life of babies.* University of Chicago Press.

Kellogg, W. (1968). Communication and language in the home-reared chimpanzee. *Science*, *162*, 423–427.

Kellogg, W., & Kellogg, L. (1933). *Development of ape and child.* New York: McGraw-Hill.

Kessen, W. (1965). *The child.* New York: Wiley.

Kinzey, W. (Ed.). (1987). *The evolution of human behavior: Primate models.* New York: SUNY Press.
Köhler, W. (1927). *The mentality of apes.* New York: Vintage Press. (Original work published 1917 as Intelligenzprufungen an Anthropoiden. *Abhandlungen der Preussische Akademie der Wissenschaften Physikalische-Mathematische Klasse*).
Kohts, N. (1935). *Infant ape and human child.* Moscow: Scientific Memoirs of the Museum Darwinium.
Kroeber, A. F. (1960). Evolution, history, and culture. In S. Tax (Ed.), *The evolution of man. Vol. 2. Evolution after Darwin* (pp. 1–16). University of Chicago Press.
Kuhn, T. S. (1970). *The structure of scientific revolutions* (2nd ed.). University of Chicago Press.
Kuo, Z. (1970). The need for coordinated efforts in developmental studies. In L. R. Aronson, E. Tobach, D. S. Lehrman, & J. S. Rosenblatt (Eds.), *Development and the evolution of behavior: Essays in memory of T. C. Schneirla* (pp. 181–193). San Francisco: Freeman.
Kuper, A. (1985). The development of Henry Lewis Morgan's evolutionism. *Journal of the History of the Behavioral Sciences, 21*, 3–22.
Kuper, A. (1988). *The invention of primitive society.* London: Routledge & Kegan Paul.
Lakatos, E. (1970). Falsification and the methodology of scientific research programmes. In E. Lakatos & A. Musgrave (Eds.), *Criticism and the growth of knowledge* (pp. 91–196). Cambridge University Press.
Langer, J. (1980). *The origins of logic: Six to twelve months.* New York: Academic Press.
Langer, J. (1985). Necessity and possibility during infancy. *Archives de Psychologie, 53*, 61–75.
Langer, J. (1986). *The origins of logic: One to two years.* New York: Academic Press.
Langer, J. (1989). Comparison with the human child. In F. Antinucci (Ed.), *Cognitive structures and development in nonhuman primates* (pp. 229–242). Hillsdale, NJ: Erlbaum.
Langness, L. L. (1987). *The study of culture* (rev. ed.). Navoto, CA: Chandler & Sharp Publishers.
Lashley, K. S., & Watson, J. B. (1913). Notes on the development of a young monkey. *Journal of Animal Behavior, 3*, 114–139.
Lehrman, D. S. (1953). A critique of Konrad Lorenz's theory of instinctive behavior. *Quarterly Review of Biology, 28*, 337–363.
Lehrman, D. S. (1970). Semantics and conceptual issues in the nature–nurture problem. In L. R. Aronson, E. Toback, & D. S. Lehrman (Eds.), *Development and evolution of behavior* (pp. 17–52). San Francisco: Freeman.
Lerner, R. (1983). The history of philosophy and the philosophy of history in development. In R. Lerner (Ed.), *On the development of developmental psychology* (pp. 3–28). Hillsdale, NJ: Erlbaum.
Lieberman, P. (1975). *On the origins of language.* New York: Macmillan.
Linden, E. (1986). *Silent partners.* New York: Ballantine.
Lorenz, K. (1950). The comparative method in studying innate behavior patterns. *Symposium on Social and Experiential Biology, 4*, 221–268.
Lorenz, K. (1957). Companionship in bird life. In C. H. Schiller (Ed.), *Instinctive behavior: The development of a modern concept* (pp. 83–128). New York: International Universities Press.
Lorenz, K. (1965). *The evolution and modification of behavior.* University of Chicago Press.
McGrew, W. (1977). Socialization and object manipulation of wild chimpanzees. In S. Chevalier-Skolnikoff & F. Poirier (Eds.), *Primate biosocial development* (pp. 261–288). New York: Garland Press.
MacNamara, J. (1979). Doubts about the form of development. Commentary on Parker and Gibson. *Behavioral and Brain Sciences, 2*, 393–394.
Marler, P., & Tenaza, R. (1977). Signaling behavior of apes with special reference to vocalization. In T. Sebeok (Ed.), *How animals communicate* (pp. 965–1033). Bloomington: Indiana University Press.
Marler, P., & Terrace, H. (Eds.). (1984). *The biology of learning.* New York: Springer-Verlag.
Mason, W. (1960). The effects of social restriction on the behavior of the rhesus monkeys: I. *Journal of Comparative and Physiological Psychology, 53*, 582–589.

Mason, W. (1979). Ontogeny of social behavior. In P. Marler & J. Vandenberg (Ed.), *Handbook of behavioral neurobiology. Vol. 3. Social behavior and communication* (pp. 1–28). New York: Plenum.
Mason, W., Blazek, N., & Harlow, H. (1956). Learning capacities of the infant rhesus monkey. *Journal of Comparative and Physiological Psychology*, *51*, 71–74.
Mason, W., & Harlow, H. (1958a). Formation of conditioned responses in infant monkeys. *Journal of Comparative and Physiological Psychology*, *51*, 68–70.
Mason, W., & Harlow, H. (1958b). Performance of infant rhesus monkeys on a spatial discrimination problem. *Journal of Comparative and Physiological Psychology*, *51*, 71–74.
Mason, W., Harlow, H., & Rueping, R. (1959). The development of manipulatory responsiveness in the infant rhesus monkey. *Journal of Comparative and Physiological Psychology*, *52*, 555–558.
Mathieu, M., & Bergeron, G. (1983). Piagetian assessment on cognitive development in chimpanzee (*Pan troglodytes*). In A. B. Chiarelli & R. Corruccini (Eds.), *Primate behavior and sociobiology* (pp. 142–147). New York: Springer-Verlag.
Mathieu, M., Bouchard, M., Granger, L., & Herscovitch, J. (1976). Piagetian object permanence in *Cebus capucinus*, *Lagothrica flavicauda* and *Pan troglodytes*. *Animal Behaviour*, *24*, 575–588.
Mathieu, M., Daudelin, N., Dagenais, Y., & Ducarie, T. (1980). Piagetian object causality in two house-reared chimpanzees *Pan troglodytes*. *Canadian Journal of Psychology*, *34*, 179–186.
Meador, D. M., Rumbaugh, D. M., Pate, J. L., & Bard, K. (1987). Learning, problem-solving, cognition, and intelligence. In G. Mitchell & J. Erwin (Eds.), *Comparative primate biology* (Vol. 2, Part B, pp. 17–83). New York: Alan R. Liss.
Menzel, E. W., Jr. (1963). Patterns of responsiveness in chimpanzees reared through infancy under conditions of environmental restriction. *Psychologische Forschung*, *27*, 337–365.
Menzel, E. W., Jr. (1971). Communication about the environment in a group of young chimpanzees. *Folia Primatologica*, *15*, 220–232.
Menzel, E. W., Jr., Premack, D., & Woodruff, G. (1978). Map reading by chimpanzees. *Folia Primatologica*, *29*, 241–249.
Menzel, E. W., Jr., Savage-Rumbaugh, S., & Lawson, J. (1985). Chimpanzee (*Pan troglodytes*) spatial problem-solving with the use of mirrors and televised equivalents of mirrors. *Journal of Comparative Psychology*, *99*, 211–217.
Miles, L. (1983). Apes and language: The search for communicative competence. In J. DeLuce & H. T. Wilder (Eds.), *Language in apes* (pp. 43–61). New York: Springer-Verlag.
Miles, L. (1986). How can I tell a lie? Apes, language, and the problem of deception. In R. W. Mitchell & N. S. Thompson (Eds.), *Deception: Perspectives on human and nonhuman deceit* (pp. 245–266). New York: SUNY Press.
Mills, W. (1898). The nature of animal intelligence and the methods of investigating it. *Psychological Review*, *6*, 262–272.
Montagu, A. (Ed.). (1964). *The concept of race*. New York: Collier Books.
Morgan, C. L. (1894). *An introduction to comparative psychology*. London: Walter Scott.
Morgan, L. H. (1877). *Ancient society*. New York: World Publishing.
Muncer, S. (1983). "Conservations" with a chimpanzee. *Developmental Psychobiology*, *16*(1), 1–11.
Natale, F. (1989). The stick problem: Causality II. In F. Antinucci (Ed.), *Cognitive structure and development in nonhuman primates* (pp. 121–133). Hillsdale, NJ: Erlbaum.
Natale, F., Antinucci, F., Spinozzi, G., & Poti', P. (1986). Stage 6 object concept in nonhuman primate cognition: A comparison between gorilla (*Gorilla gorilla gorilla*) and Japanese macaque (*Macaca fuscata*). *Journal of Comparative Psychology*, *100*, 335–339.
Norman, D. A., & Levelt, W. J. M. (1988). Life at the center. In W. Hirst (Ed.), *The making of cognitive science: Essays in honor of George A. Miller* (pp. 100–110). Cambridge University Press.
Parker, S. T. (1973). *Piaget's sensorimotor series in an infant macaque: The organization of non-*

*stereotyped behavior in the evolution of intelligence*. PhD dissertation, Department of Anthropology, University of California, Berkeley.
Parker, S. T. (1976). *A comparative longitudinal study of sensorimotor development in a macaque, a gorilla, and a human infant from a Piagetian perspective*. Paper presented at an Animal Behavior Society meeting, Boulder.
Parker, S. T. (1977). Piaget's sensorimotor series in an infant macaque: A model for comparing unstereotyped behavior and intelligence in human and nonhuman primates. In S. Chevalier-Skolnikoff & F. Poirier (Eds.), *Primate bio-social development* (pp. 43–112). New York: Garland Press.
Parker, S. T. (1983). Intelligent locomotion? Leaping to conclusions about orang-utan intelligence. (Letter to the editor) *Journal of Human Evolution*, *12*, 495–497.
Parker, S. T. (1985). A social technological model for the evolution of language. *Current Anthropology, 26*(5), 617–639.
Parker, S. T., & Gibson, K. R. (1977). Object manipulation, tool use, and sensorimotor intelligence as feeding adaptations in cebus monkeys and great apes. *Journal of Human Evolution, 6*, 623–641.
Parker, S. T., & Gibson, K. R. (1979). A developmental model for the evolution of language and intelligence in early hominids. *Behavioral and Brain Sciences*, *2*, 367–408.
Parker, S. T., & Gibson, K. R. (1982). The importance of theory for reconstructing the evolution of language and intelligence. In A. B. Chiarelli & R. S. Corruccini (Eds.), *Advanced views in primate biology* (pp. 42–64). New York: Springer-Verlag.
Pascual-Leone, J. (1984). Attentional, dialectic and mental effort. In M. L. Commons, F. A. Richards, & C. Armon (Eds.), *Beyond formal operations* (pp. 182–215). New York: Plenum.
Pascual-Leone, J., & Smith, J. (1969). The encoding and decoding of symbols by children: A new experimental paradigm and a neo-Piagetian model. *Journal of Experimental Child Psychology*, *8*, 328–355.
Pasnak, R. (1979). Acquisition of prerequisites to conservation by macaques. *Journal of Experimental Psychology: Animal Behavior Processes*, *5*(2), 194–210.
Patterson, F., & Linden, E. (1981). *The education of Koko*. New York: Holt, Rinehart & Winston.
Pavlov, I. (1961). Conditioned reflexes: An investigation of the physiological activity of the cerebral cortex. In T. Shipley (Ed.), *Classics in psychology* (pp. 756–797). New York: Philosophical Library. (Original work published 1929).
Petitto, L. (1988). "Language" in the prelinguistic child. In F. S. Kessel (Ed.), *The development of language and language researchers: Essays in honor of Roger Brown* (pp. 187–222). Hillsdale, NJ: Erlbaum.
Piaget, J. (1952). *The origins of intelligence in children*. New York: Norton.
Piaget, J. (1954). *The construction of reality in children*. New York: Basic Books.
Piaget, J. (1962). *Play, dreams and imitation in childhood*. New York: Norton.
Piaget, J. (1968). *Structuralism*. New York: Harper.
Piaget, J. (1971). *Biology and knowledge*. University of Chicago Press.
Piaget, J. (1978). *Behavior and evolution*. New York: Pantheon.
Piaget, J. (1980). *Adaptation and intelligence: Organic selection and phenocopy*. University of Chicago Press.
Piaget, J., & Inhelder, B. (1969). *The psychology of the child*. New York: Basic Books.
Piaget, J., & Inhelder, B. (1974). *The child's conception of quantities*. London: Routledge & Kegan Paul.
Piel, G. (1970). The comparative psychology of T. C. Schneirla. In L. R. Aronson, E. Tobach, D. Lehrman, & J. Rosenblatt (Eds.), *Development and evolution of behavior: Essays in memory of T. C. Schneirla* (pp. 1–13). San Francisco: Freeman.
Premack, D. (1971a). On the assessment of language in chimpanzees. In A. Schrier & F. Stollnitz (Eds.), *Nonhuman primate behavior* (Vol. 3, pp. 185–228). New York: Academic Press.
Premack, D. (1971b). Language in chimpanzee? *Science*, *172*, 808–822.
Premack, D. (1976a). *Intelligence in ape and man*. Hillsdale, NJ: Erlbaum.
Premack, D. (1976b). On the study of intelligence in chimpanzees. *Current Anthropology*, *28*, 516–521.

Proctor, R. (1988). From *Anthropologie* to *Rassenkunde* in the German anthropological tradition. In G. W. Stocking, Jr. (Ed.), *History of Anthropology. Vol. 5. Bones, bodies, behavior: Essays on biological anthropology* (pp. 138–179). Madison: University of Wisconsin Press.
Redshaw, M. (1978). Cognitive development in human and gorilla infants. *Journal of Human Evolution*, *7*, 113–141.
Reed, J. (1987). Robert M. Yerkes and the comparative method. In E. Tobach (Ed.), *Historical perspectives and the international status of comparative psychology* (pp. 91–102). Hillsdale, NJ: Erlbaum.
Ribnick, R. (1982). A short history of primate field studies: Old World monkeys and apes. In F. Spencer (Ed.), *A history of American physical anthropology 1930–1980* (pp. 49–73). New York: Academic Press.
Richard, A. F. (1985). *Primates in nature*. San Francisco: Freeman.
Ridley, M. (1984). *Classification and evolution*. New York: Longman.
Roback, A. A. (1964). *A history of American psychology* (rev. ed.). New York: Collier Books.
Romanes, G. J. (1882). *Animal intelligence*. London: Kegan Paul & French.
Ross, D. (1972). *G. Stanley Hall*. University of Chicago Press.
Rumbaugh, D. (1971). Evidence of qualitative differences in learning among primates. *Journal of Comparative and Physiological Psychology*, *76*, 250–255.
Rumbaugh, D. (1973). The importance of nonhuman primate studies of learning and related phenomena for understanding cognitive development. In G. H. Bourne (Ed.), *Nonhuman primates and medical research* (pp. 415–429). New York: Academic Press.
Rumbaugh, D., & Gill, T. V. (1976). Language and the acquisition of language-type skills by a chimpanzee (*Pan*). *Annals of the New York Academy of Sciences*, *270*, 90–123.
Rumbaugh, D., Gill, T. V., & von Glasersfeld, E. C. (1973). Reading and sentence completion by a chimpanzee (*Pan*). *Science*, *9*, 343–347.
Rumbaugh, D., & McCormack, D. (1967). The learning skills of primates: A comparative study of monkeys and apes. In D. Stark, R. Schneider, & H. J. Kuhn (Eds.), *Progress in primatology* (pp. 289–306). Stuttgart: Fischer.
Rumbaugh, D., & Pate, J. L. (1984). The evolution of primate cognition: A comparative perspective. In T. Roitblat, T. G. Bever, & H. S. Terrace (Eds.), *Animal cognition* (pp. 569–587). Hillsdale, NJ: Erlbaum.
Rumbaugh, D., & Rice, C. P. (1962). Learning set formation in young great apes. *Journal of Comparative and Physiological Psychology*, *55*, 866–868.
Sarich, V., & Wilson, A. (1968). Immunological time scale for hominid evolution. *Science*, *158*, 1200.
Savage, A., & Snowdon, C. T. (1982). Mental retardation and neurological deficits in a twin orangutan. *American Journal of Primatology*, *3*, 239–251.
Savage-Rumbaugh, S. (1981). Can apes use symbols to represent their world? In T. A. Sebeok & R. Rosenthal (Eds.), *The clever Hans phenomenon: Communication with horses, whales, apes, and people* (pp. 35–59). New York Academy of Sciences.
Savage-Rumbaugh, S., Rumbaugh, D. M., & Boysen, S. (1978a). Symbolic communication between two chimpanzees (*Pan troglodytes*). *Science*, *201*, 641–644.
Savage-Rumbaugh, S., Rumbaugh, D., & Boysen, S. (1978b). Linguistically mediated tool use and exchange by chimpanzees (*Pan troglodytes*). *Behavioral and Brain Sciences*, *4*, 539–554.
Savage-Rumbaugh, S., Rumbaugh, D., Smith, S. T., & Lawson, J. (1980). Reference: The linguistic essential. *Science*, *210*, 922–925.
Schneirla, T. C. (1966). Behavioral development and comparative psychology. *Quarterly Review of Biology*, *41*(3), 283–303.
Schneirla, T. C., & Rosenblatt, J. S. (1961). Behavioral organization and genesis of the social bond in insects and mammals. *American Journal of Orthopsychiatry*, *31*, 223–253. Reprinted 1974 in T. E. McGill (Ed.), *Readings in animal behavior* (pp. 287–290). New York: McGraw-Hill.
Scott, J. P. (1962). Critical periods in behavioral development. *Science*, *138*, 949–958. Re-

printed 1974 in T. E. McGill (Ed.), *Readings in animal behavior* (pp. 271–286). New York: McGraw-Hill.
Simpson, G. G. (1953). The Baldwin effect. *Evolution*, *7*, 110–117.
Sinclair, H. (1987). Symbolic systems and interpersonal relations in infancy. In J. Montangero, A. Trython, & S. Dionnet (Eds.), *Symbolism and knowledge* (pp. 129–141). Geneva: Cahiers de la Fondation Archives Jean Piaget (No. 8).
Skinner, B. F. (1980). *The behavior of organisms*. New York: Appleton-Century-Crofts.
Skinner, B. F. (1957). *Verbal behavior*. New York: Appleton-Century-Crofts.
Slobin, D. (1988). From the Garden of Eden to the Tower of Babel. In F. S. Kessel (Ed.), *The development of language and language researchers: Essays in honor of Roger Brown* (pp. 9–22). Hillsdale, NJ: Erlbaum.
Snowden, C., Brown, C. H., & Peterson, M. R. (Eds.), (1982). *Primate communication*. Cambridge University Press.
Spencer, F. (Ed.). (1982). *A history of American physical anthropology: 1930–1980*. New York: Academic Press.
Spencer, H. (1875). *The principles of psychology* (2 vols.). New York: Appleton. (Original work published 1855)
Stearn, D. (1978). *The first relationship*. Cambridge, MA: Harvard University Press.
Stocking, G. W., Jr. (1968). *Race, culture, and evolution: Essays in the history of anthropology*. University of Chicago Press.
Suomi, S. J., & LeRoy, H. A. (1982). In memorium: Harry F. Harlow (1905–1981). *American Journal of Primatology*, *2*, 319–342.
Taub, D., & King, F. A. (Eds.). (1986). *Current perspective in primate social dynamics*. New York: Van Nostrand Reinhold.
Teleki, G. (1975). Chimpanzee subsistence technology: Materials and skills. *Journal of Human Evolution*, *3*, 575–594.
Terrace, H. (1979). *Nim. A chimpanzee who learned sign language*. New York: Knopf.
Terrace, H. (1980). A report to the academy, 1980. In T. Sebeok & R. Rosenthal (Eds.), *The clever Hans phenomenon: Communication with horses, whales, apes, and people* (pp. 94–114). New York Academy of Sciences.
Terrace, H. (1984). Animal cognition. In H. L. Roitblat, T. G. Bever, & H. S. Terrace (Eds.), *Animal cognition* (pp. 7–25). Hillsdale, NJ: Erlbaum.
Terrace, H., Petitto, L. A., Saunders, R. J., & Bever, T. (1979). Can an ape create a sentence? *Science*, *206*, 891–902.
Teuber, M. (1987). *Founding of the primate station, Tenerife, Canary Islands* (pp. 1–11). Paper presented to the International Society for the History of Behavioral and Social Sciences 19th annual meeting, Bowdoin College, Brunswick, ME.
Thomas, F. J. (1977). Piaget and Lamarck. *Science Education*, *61*(3), 279–286.
Thomas, R. K., & Peay, L. (1976). Length judgments by squirrel monkeys: Evidence for conservation? *Developmental Psychology*, *12*, 349–352.
Thomas, R. K., & Walden, E. L. (1985). The assessment of cognitive development in human and nonhuman primates. In T. C. Arnand Kumar (Ed.), *Nonhuman primate models for human growth and development* (pp. 187–215). New York: Alan R. Liss.
Thompson, D. W. (1941). *On growth and form*. Cambridge University Press.
Thorndike, E. L. (1901). *Notes on child study*. New York: Macmillan.
Thorndike, E. L. (1911). *Animal intelligence, experimental studies*. New York: Macmillan.
Thorpe, W. H. (1979). *Origins and rise of ethology*. London: Praeger.
Tinbergen, N. (1951). *The study of instinct*. Oxford University Press.
Tolman, E. D. (1925). Purpose and cognition: The determiners of animal learning. *Psychological Review*, *32*, 285–297.
Tolman, E. D. (1932). *Purposive behavior in men and animals*. New York: Century.
Tolman, E. D. (1948). Cognitive maps in rats and man. *Psychological Review*, *55*, 189–208.
Trevarthen, C. (1979). Communication and cooperation in early infancy. A description of primary intersubjectivity. In M. Bullowa (Ed.), *Before speech* (pp. 321–347). Cambridge University Press.

U.S. DHEW (1971). *NIH Primate Research Centers: A major scientific resource*. U.S. Department of Health, Education, and Welfare, Public Health Service, National Institutes of Health (Stock no. 1740-0320). Washington, DC: U.S. Government Printing Office.

Uzgiris, I. C., & Hunt, J. M. (1975). *Assessment in infancy: Ordinal scales of psychological development*. Urbana: University of Illinois Press.

van Lawick-Goodall, J. (1970). Tool using in primates and other vertebrates. In D. Lehrman, R. Hinde, & E. Shaw (Eds.), *Advances in the study of behavior* (Vol. 3, pp. 195–249). New York: Academic Press.

Vauclair, J. (1982). Sensorimotor intelligence in human and nonhuman primates. *Journal of Human Evolution*, *11*, 257–264.

Vauclair, J. (1987). Representation et intentionnalité dans la cognition animale. In M. Siguan (Ed.), *Comportement, cognition et conscience* (pp. 59–87). Paris: Presses Unversitaires de France.

Vauclair, J., & Bard, K. (1983). Development of manipulations with objects in ape and human infants. *Journal of Human Evolution*, *12*, 631–645.

Vaughter, R. M., Smotherman, W., & Ordy, J. M. (1972). Development of object permanence in the infant squirrel monkey. *Developmental Psychology*, *7*, 34–38.

Volterra, V. (1987). From single communicative signal to linguistic combination in hearing and deaf children. In *Symbolism and knowledge* (shorter English ed.). Geneva: Cahiers de la Fondation Archives Jean Piaget (No. 8).

Wallace, A. R. (1864). The origin of human races and the antiquity of man deduced from the theory of "natural selection." *Anthropological Review*, *2*, clviii–clxx.

Washburn, A. R. (1889). *Darwinism*. London: Macmillan.

Washburn, S. L. (1951) The new physical anthropology. *Transactions of the New York Academy of Sciences Series II*, *13*, 298–304. Reprinted 1966 in T. W. McKern (Ed.), *Readings in physical anthropology* (pp. 2–6). Englewood Cliffs, NJ: Prentice-Hall.

Washburn, S. L. (1980). Human biology and social science. In E. A. Hoebel, R. Currier, & S. Kaiser (Eds.), *Crisis in anthropology: The view from Spring Hill* (pp. 321–332). New York: Garland Press.

Washburn, S. L. (1983). Evolution of a teacher. *Annual Review of Anthropology*, *12*, 1–24.

Washburn, S. L., & Lancaster, C. S. (1968). The evolution of hunting. In I. DeVore & R. Lee (Eds.), *Man the hunter* (pp. 293–303). Chicago: Aldine. Reprinted 1968 in S. L. Washburn & P. C. Jay (Eds.), *Perspectives on human evolution* (pp. 213–229). New York: Holt, Rinehart & Winston.

Watson, J. B. (1913). Psychology as a behaviorist sees it. *Psychological Review*, *20*, 158–177.

Weiss, M. (1987). Nucleic acid evidence bearing on hominoid relationships. *Yearbook of the American Journal of Physical Anthropology*, *30*, 41–74.

Werner, H. (1948). *Comparative psychology of mental development*. New York: International Universities Press.

Westergaard, C. (1987). Lion-tailed macaques manufacture and use tools. *American Journal of Primatology*, *12*, 376. (Abstract).

Westergaard, C., & Fragaszy, D. (1987). The manufacture and use of tools by capuchin monkeys (*Cebus apella*). *Journal of Comparative Psychology*, *101*(2), 159–168.

White, S. H. (1983). The idea of developmental psychology. In R. Lerner (Ed.), *On the development of developmental psychology* (pp. 55–78). Hillsdale, NJ: Erlbaum.

Whiten, A., & Byrne, R. W. (1988). Tactical deception in primates. *Behavioral and Brain Sciences*, *11*, 233–273.

Wiliams, G. (1966). *Adaptation and natural selection*. Princeton University Press.

Wise, K. L., Wise, L. A., & Zimmermann, R. R. (1974). Piagetian object permanence in the infant rhesus monkeys. *Developmental Psychology*, *10*, 429–437.

Wood, S., Moriarty, K., Gardner, B., & Gardner, A. (1980). Object permanence in child and chimpanzee. *Animal Learning and Behavior*, *8*, 3–9.

Woodruff, G., & Premack, D. (1978). Does the chimpanzee have a theory of mind? *Behavioral and Brain Sciences*, *4*, 515–526.

Woodruff, G., Premack, D., & Kennel, K. (1978). Conservation of liquid and solid quantity. *Science*, *202*, 991–994.

Yerkes, R. (1916a). *The mental life of monkeys and apes*. New York: Holt.

Yerkes, R. (1916b). Provision for the study of monkeys and apes. *Science*, *43*, 231–234.

Young, M. L., & Harlow, H. (1943). Generalization by rhesus monkeys of a problem involving the Weigl principle using the oddity method. *Journal of Comparative Psychology*, *36*, 201–216.

Zimmermann, R. R., & Torey, C. C. (1965). Ontogeny of learning. In A. Schrier, H. Harlow, & F. Stollnitz (Eds.), *Behavior of nonhuman primates* (Vol. 2, pp. 405–448). New York: Academic Press.

Zolman, J. F. (1982). Ontogeny of learning. In P. Bateson & P. Klopfer (Eds.), *Perspectives in ethology. Vol. 5. Ontogeny* (pp. 275–323). New York: Plenum.

# 2 How scientific usages reflect implicit theories: *Adaptation*, *development*, *instinct*, *learning*, *cognition*, and *intelligence*

*Sue Taylor Parker and Bernard Baars*

As the anthropological linguist Edward Sapir noted, language is largely unconscious. In everyday life, we all use words without knowing exactly what we mean by them. Examination of scholarly works suggests that we extend this practice into our professional lives as well. The words we use and the way we use them reflect our implicit, unconscious knowledge, presuppositions, and theories (and, equally, our implicit, unconscious ignorance). The nature of one's implicit knowledge (and ignorance) is determined by one's disciplinary training and one's personal intellectual history.

Specialists in philosophy, zoology (especially ethology), cognitive sciences, and various branches of psychology, for example, often use the terms *learning*, *intelligence*, and *cognition* in subtly different ways. Scholars in different subdisciplines describe similar phenomena in different terms: Comparative psychologists, especially learning and conditioning theorists, for example, use the term *learning* to describe the acquisition of instrumental behaviors, whereas comparative developmental psychologists use *intelligence* or *cognition* to describe such acquisitions. Through the years, various investigators have used a variety of terms to characterize animal abilities. A brief survey of books on mental abilities of animals, for example, reveals varying usages of the related terms *mind/mentality*, *learning*, *cognition*, *intelligence*, *thought*, and *consciousness* (Table 2.1).

*Instinct*, *learning*, *intelligence*, and *development* are used in different ways by embryologists, learning and conditioning theorists, ethologists, and developmental psychologists. Although William James and John B. Watson spoke of the relationship between instinct and habit, in the 1950s many comparative psychologists defined *learning* and *instinct* in opposition to each other, whereas ethologists defined them in relation to each other. Learning and conditioning theorists and modern cognitive psychologists see *learning* and *intelligence* as similar, basically continuous phenomena, whereas developmental psychologists see these as quite distinct phenomena. Likewise, learning and conditioning theorists and cognitive psychologists have hardly distinguished the terms *learning* and *development*, focusing primarily on modifications of behavior through experience rather than during ontogeny.

Table 2.1. *Classification of books dealing with the mental abilities of animals*

| | |
|---|---|
| Intelligence | Romanes (1886) *Animal Intelligence*<br>Thorndike (1911) *Animal Intelligence*<br>Premack (1976) *Intelligence in Ape and Man*<br>Hoage & Goldman (Eds.) (1986) *Animal Intelligence*<br>Macphail (1982) *Brain and Intelligence in Vertebrates* |
| Mind/mentality | Yerkes (1916) *The Mental Life of Monkeys and Apes*<br>Köhler (1927) *The Mentality of Apes*<br>Griffin (Ed.) (1982) *Animal Mind; Human Mind* |
| Thought | Walker (1983) *Animal Thought* |
| Awareness | Griffin (1976) *The Question of Animal Awareness* |
| Cognition | Hulse & Fowler (Eds.) (1978) *Cognitive Processes in Animal Behavior*<br>Roitblat, Bever, & Terrace (Eds.) (1984) *Animal Cognition*<br>Else & Lee (1986) *Primate Ontogeny, Cognition and Social Behaviour*<br>Antinucci (1989) *Cognitive Structures and Development in Nonhuman Primates* |
| Learning | Hinde & Stevenson-Hinde (Eds.) (1973) *Constraints on Learning*<br>Marler & Terrace (Eds.) (1984) *The Biology of Learning* |
| Behavior | Klüver (1933) *Behavior Mechanisms in Monkeys*<br>Masterton, Hodos, & Jerison (Eds.) (1976) *Evolution, Brain, and Behavior* |
| Miscellaneous | Kellogg & Kellogg (1933) *Development of Ape and Child*<br>Kohts (1935) *Infant Ape and Human Child*<br>Hayes (1951) *Ape in Our House*<br>Patterson & Linden (1981) *The Education of Koko*. New York: Holt, Rinehart & Winston |

Both denotations and connotations of words vary from discipline to discipline, and even from subdiscipline to subdiscipline. Certain words, such as *intelligence* and *recapitulation*, become unpopular because they have negative political connotations. Meanwhile, other words with related meanings, such as *cognition* and *learning*, become popular because they lack negative connotations and/or because they are associated with such prestigious models as information processing.

We focus on the history of discipline-specific use of certain terms because it reveals a good deal about the implicit knowledge and presuppositions of various disciplines. New usages and changing relationships among the domains of terms signal theoretical shifts. Although shifts in usage can be discovered in the professional literature, they are rarely discussed by practitioners. A brief history of the terms *development*, *instinct*, *learning*, *intelligence*, and *cognition* may shed some light on the significance of current usages of these and related terms and provide the background for a discussion of differing views of the relationships among them. Although the definitions we present here are few among many, we have tried to sample the perspectives of influential thinkers and authors of standard reference books on the grounds that they represent some sort of "linguistic authority" (Nunberg, 1984).

## Definitions

### *Development*

*Webster's Third New International Dictionary* (1981) defines *development* as follows:

> **1**: the act, process, or result of developing: the state of being developed: a gradual unfolding by which something (as a plan or method, an image upon a photographic plate, a living body) is developed. . . : gradual advance or growth through progressive changes: EVOLUTION (a stage of ~): a making usable or available . . . **2 a**: the whole process of growth and differentiation by which the potentialities of a zygote, spore, or embryo are realized: *broadly*: ONTOGENY. **b**: the gradual differentiation of an ecological community or a natural group; *sometimes*, PHYLOGENY. (p. 618)

*Development* is then a more general concept than *ontogeny*, which, according to *Webster's* (1981), is "the biological development or course of development of an individual organism – distinguished from *phylogeny*" (p. 1577).

In his *Dictionary of Philosophy and Psychology*, James Mark Baldwin defines development as

> the entire series of vital changes normal to the individual organism, from its origin from the parent cell or cells until death. Development refers to the individual (ONTOGENESIS), evolution to the race (PHYLOGENESIS), a distinction of terms (q.v.) made by Haeckel. In most organisms there are well-marked stages in development, characterized by structure changes. (Baldwin, 1901/1960, Vol. 1, p. 274)

*Embryologists.* Embryologists are specialists in development at its inception. For further elaboration of the concept of development, as specified in *Webster's* definition 2a, we find the following definitions of development:

> The cardinal criterion of development: progressive and cumulative change – evident at all levels of biological organization, molecules, cells, tissues, and organs – during an organism's life history. (Ebert & Sussex, 1972, p. 6)

> A unified process, which includes the arising of orderly recognizable patterns as a consequence of (or at least accompanied by) the formation of new constituents, the fabrication of these constituents into larger units, and their rearrangement in space. . . . It is meaningful to consider at least four . . . component processes: determination, differentiation, growth, and morphogenesis. (Ebert & Sussex, 1972, p. 8)[1]

Not only do developing organic systems undergo progressive and cumulative change, but their constituents are arranged hierarchically: "The embryologist is confronted not simply by chemical substances, but by a whole hierarchy of more complex organized entities, such as subcellular organelles, cells, tissues and organs, in each of which the material substance has some definite spatial arrangement" (Waddington, 1962, p. 2).

Embryogenesis had been described classically in terms of the competence of tissues to be transformed into specific entities and of the induction of such changes through chemical action on these tissues. With the discovery of the genetic code and protein synthesis, some embryologists shifted their attention to the roles of genes and gene products in development.

Waddington was perhaps the first biologist to envision these standard embryological concepts in terms of the actions of interacting "gene action systems" that define and maintain stabilized pathways of change (Waddington, 1962). As Trevarthen said, "what is inherited is not gene information in any strict sense, but a highly specific instability of prefabricated building blocks, a strategy or program of how to assimilate components from the environment and how to develop further and thereby automatically gain structure" (Trevarthen, 1973, p. 91). These gene action systems continue to operate during development after birth.

*Systems theorists.* Recently, under the influence of ecological and systems theory, some investigators have emphasized historical processes of information construction:

> In the same way that it is unreasonable to say that information regarding when to emerge is "in" the fly's body or "in" the environment . . . it is unreasonable to say that information for developing a body with just these sequential sensitivities and reactivities is "in" the constituent cells or anywhere else. It is true not only that "information" on emergence schedule is a joint construction of organism and surround but also that *information for that state of sensory and behavioral readiness which makes emergence possible is assembled and defined in ontogeny.* (Oyama, 1985, p. 136, italics added)

According to Oyama, "fate is constructed, amended and reconstructed, partly by the emerging organism itself. It is known to no one, not even the genes" (1985, p. 121). Only the means are inherited; phenotypes are constructed during ontogeny, first through the interplay of initial states of information systems in the fertilized egg, including nuclear DNA, regionalized cytoplasmic macromolecules (particularly messenger RNA from the mother), and the cytoskeletal matrix, and later through signals from other cells and from outside the embryo: "Since all aspects of phenotype are products of ontogenesis, they are all in some sense acquired" (Oyama, 1985, p. 125). Viewed from this perspective, "canalization, or developmental homeostasis, is . . . a result of a system of relationships and their consequences, not of some regulating force emanating from the genome" (Oyama, 1985, p. 130).

The shift in paradigm from a teleonomic readout of preformed information to the contingent construction of developmental information will inevitably influence models of behavioral development and hence comparative developmental psychology.

*Developmental psychologists.* Because later development is continuous with embryogenesis, embryogenesis provides models for other kinds of developmental processes. Visions of development follow models of embryogenesis. Trevarthen's view of development, for example, is based on his view of embryogenesis:

> The principles of embryogenesis apply in all growth including psychological growth, and not just to the morphogenesis of the body of the embryo. . . . Birth is the begin-

ning of a process which may be called *psychogenesis* – the maturation of a system prefabricated to become an intelligent, conscious individual who will assimilate knowledge of objects and beings in the outside world according to inherent rules. Seen in this light, even learning is a part of the embryogenesis of a psychological mechanism, depending upon the formulations of growing brain circuits and upon selection of prepared alternatives of structure in them. Every aspect of psychological development is conditioned by the growth and differentiations of nerve cells. (Trevarthen, 1973, pp. 89–91)

This view goes back at least to the 19th century:

When Huxley defined the scope of human embryology, he made it coextensive with the entire life cycle. The organism has a wholeness in time as well as space. Under the concept of hierarchical continuity, all the morphogenetic events, somatic and functional, which occur in the entire life cycle must have a close kinship in their determination. (Gesell, 1945, p. 185)

Even today, embryogenesis is a model for brain development, and hence for psychological development after birth: "Embryogenesis concerns the development of the body, but it concerns as well the development of the nervous system and the development of mental functions. In the case of the development of knowledge in children, embryogenesis ends only in adulthood" (Piaget, 1978, p. 228).

Rejecting two formulations that he variously labels empiricist and rationalist, and associationist and innatist, Piaget describes development as an epigenetic process of *construction* (Piaget, 1968). The central and unsolved problem at the heart of Piaget's work was the question of what constructive mechanisms drive the developmental transitions from stage to stage.

In his middle period, Piaget (1970) specified the following processes in the development of knowledge, that is, of "operations" or reversible interiorized actions:

. . . first of all, *maturation*, in the sense of Gesell, since this development is a continuation of embryogenesis; second, the role of *experience* of the effects of the physical environment on the structures of intelligence; third, *social transmission* in the broad sense (linguistic transmission, education, etc.); and fourth, a factor which is too often neglected but one which seems to me fundamental and even the principal factor. I shall call this the factor of *equilibration* or, if you prefer it, of self-regulation. (Piaget, 1978, p. 230)

In his later discussion of constructive processes, Piaget (1985) develops a convoluted model involving observables and coordinations, regulations, compensations, and reflective abstraction.

In his recent revision of Piagetian theory, Case (1985) explains intellectual development in terms of two processes: hierarchical integration, which involves "the assembly of superordinate structure from previously existing component structures" (Case, 1985, p. 281), and maturation, by which he refers specifically to myelinization of the nervous system, which increases operating speed (Gibson, P&G3). According to Case, hierarchical integration accounts for the universal order of emergence of abilities within a domain, as well as individual and cultural differences in rates and terminal levels of development, whereas maturation accounts for the long course of intellectual development, changes in developmental rate, cessation of development, synchronies in development across domains, and resistance to instruction.

### *Learning*

According to *Webster's* (1981), *learning* has the following definition:

> **1 a** (1): the act or experience of one that learns . . . (2) the process of acquisition and extinction of modifications in existing knowledge, skills, habits, or action tendencies in a motivated organism through experience, practice, or exercise – compare MATURATION **b** (1): something that is learned or taught . . . ACQUIREMENT **2 b**: knowledge accumulated and handed down by generations of scholars: CULTURE. (p. 1286)

Perhaps significantly, Baldwin's *Dictionary of Philosophy and Psychology* (1901/1960) has no listing for "learning."

Thorndike (1911) argued that all learning falls into the following categories:

> (1) *learning by trial and accidental success*, by the strengthening of the connections between the sense-impressions representing the situation and the acts – impulses and acts – representing our successful response to it and by the inhibition of similar connections with unsuccessful responses; (2) *learning by imitation*, where the mere performance by another of a certain act in a certain situation leads us to do the same; and (3) *learning by ideas*, where the situation calls up some idea (or ideas) which then arouse the act or in some way modifies it. (Thorndike, 1911, p. 174)

As *Webster's* definition and Thorndike's definition imply, "learning," in general parlance, refers to a broad range of phenomena, from acquisition of motor and perceptual skills to intellectual skills involved in algebra and English syntax to imitation. None of these kinds of learning, within a modern perspective, resembles the stimulus–response conditioning that was considered the prototype and essence of learning during much of the behavioristic era. For cultural reasons, "conditioning" came to occupy a central place in the psychology of learning.

Americans were well versed in British associationism, but they had linguistic and cultural problems understanding such German theorists as Kant and Wilhelm Wundt (Blumenthal, 1977; Fancher, 1979). American behavioristic associationism, however, was more extreme than British associationism, which dealt with connections between ideas, rather than between stimuli and responses. Under the influence of Watson, American behavioristic associationists reduced perceptual learning, gestalt learning, abstract rule learning, and the like to "associations," which in turn were reduced to conditioned connections between physical stimuli and physical responses (Baars, 1986; Fancher, 1979).

American psychologists, in reaction to European thought, were drawn to the "reflex-reduction program" of such Russian physiologists as Pavlov and

Bekhterev. Many physiologists and physicalistic psychologists sought to demonstrate the fallacy of such popular and quasi-religious ideas as free will and teleology. Thus, for many years before his famous experiment in classical conditioning, Pavlov wanted to show that all "mental" phenomena could be reduced to normal causal chains of neural events (hence falsifying the notion of free will or an élan vital). When he was a young man, Pavlov's scientific hero I. M. Sechenov, a Russian physiologist, argued that all mental events could be understood as a mere chain of reflexes:

If a conscious psychical act is not accompanied by any external manifestations, it still remains a reflex. . . . A thought is two-thirds of a psychical reflex . . . only the end of the reflex, i.e., movement, is completely absent. . . . Association is effected through a continuous series of reflexes in which the end of a preceding reflex coincides in time with the beginning of a subsequent [one]. (Quoted in Razran, 1965, p. 354)[2]

*Physiological and comparative psychologists.* Pavlov defined *reflex* as follows:

An external or internal stimulus falls on some one or other nervous receptor and gives rise to a nervous impulse; this nervous impulse is transmitted along nerve fibres to the central nervous system, and here, on account of existing nervous connections, it gives rise to a fresh impulse which passes along outgoing nerve fibres to the active organ, where it excites a special activity of the cellular structures. Thus a stimulus appears to be connected of necessity with a definite response, as cause with effect. (Pavlov, 1904/1961, pp. 764–765)

In some contexts, Pavlov spoke as if any demonstrable causal relationship was a reflex. When one of Pavlov's experimental dogs suddenly began to struggle against its restraining harness, for example, Pavlov spoke of the "freedom reflex" (Pavlov, 1904/1961, p. 770). On the basis of his experiments he distinguished two kinds of reflexes: inborn or "unconditional" reflexes and signal or "conditional" reflexes (later mistranslated as "conditioned"). He argued that there is little difference between instincts and reflexes, but that the term *reflex* has priority.

The importance of Pavlov's famous experiment was that it explained the acquisition of new "conditional" connections between novel stimuli and reflexive responses. (Because animals are able to acquire *new* connections between stimuli and responses, the reflex-reduction program therefore had to demonstrate that old reflexes could be connected to new stimuli.)

Even so, Skinner argued that classical conditioning explains only a small part of behavior:

The conditioned reflex is a simple principle of limited scope. . . . A very different process, through which a person comes to deal effectively with the environment is operant conditioning: Many things in the environment, such as food and water, sexual contact and escape from harm, are crucial to the survival of the individual and the species, and any behavior which produces them therefore has survival value. Through the process of operant conditioning, behavior having this consequence becomes more likely to occur. The behavior is said to be *strengthened* by its consequences, and for that reason the consequences themselves are called "reinforcers." (Skinner, 1974, pp. 43–44)

Before Skinner, Tolman had changed and expanded the concept of learning beyond the restrictions of stimulus–response psychology. Tolman is perhaps best known for the concepts of cognitive maps, latent learning, and vicarious trial-and-error learning (Tolman, 1948/1951). He described his position as follows:

> First, learning consists not in stimulus–response connections but in the building up in the nervous system of sets which function like cognitive maps, and second, that such cognitive maps may be usefully characterized as varying from a narrow strip variety to a broader comprehensive variety. (p. 246)

Tolman's concepts laid the groundwork for modern studies of spatial cognition in animals (Vauclair, 1987).

*Ethologists.* Hinde (1973) reviewed research on animal learning and its implications for revision of "the traditional learning theory approach." Focusing on "Limitations Within and Differences Between Species in Learning in Particular Contexts," he listed the following items:

> (i) *Constraints on what is learned within a species*. . . . [a] failure to learn in one context by animals which possess the appropriate perceptual and motor apparatus, and whose learning ability in other contexts seems well developed. . . . (ii) *Preparedness to learn.* The converse . . . involves cases in which learning occurs with exceptional rapidity. . . . (iii) *Species differences in what is learned.* Of particular interest are cases in which closely related species differ in the extent to which learning occurs in similar functional contexts. (a) *Egg recognition* . . . (b) *Species recognition* . . . (c) *Song development* . . . (d) "*Tool-using*" . . . (e) *Orientation in fish* . . . (f) *Orientation in bees* . . . (g) *Language development.* (pp. 7–10)

Focusing on "Some Mechanisms Whereby Learning Is Constrained," Hinde gave the following list:

> (i) *Constraints on the stimuli which become associated* . . . (ii) *Constraints on the response* . . . (iii) *Species and sex differences in reinforcing effects* . . . (iv) *Response/reinforced interactions* . . . (v) *Diversity of reinforcing effects* . . . (vi) *Interference by behaviour elicited by aspects of the situation not relevant to learning task or not specifically reinforced* . . . (vii) *Age changes in what is learned* . . . (a) song learning . . . (b) imprinting. (Hinde, 1973, pp. 10–15)

In a recent article aimed at integrating field (ethological) and laboratory learning, Gould and Marler (1984) affirm the utility of the following traditional functional categories of learning: habituation, sensitization, associative learning (classical conditioning), trial-and-error learning (operant conditioning), and higher-order (i.e., cognitive trial-and-error) learning. In conclusion, they argue that learning consists primarily

> of specialized, dedicated but well integrated subroutines, based on a small number of general learning strategies which have been "customized" as appropriate for each context and each species . . . a relatively straightforward consideration of how natural selection has wired up these various components to create more complex, species-typical behavior is likely to be more fruitful than a search for a General Theory. (Gould & Marler, 1984, p. 70)

*Cognitive psychologists.* When we look at research in learning in the last 10 years, we find the world once more turned upside down. Learning theory has been transformed into cognitive psychology. Rescorla (1988), for example, recently published an article entitled "Classical Conditioning – It's Not What You Think It Is," in which he describes evidence indicating that even classical conditioning, the kingpin of learning theory, is a sophisticated representational process in which animals actively explore and establish knowledge about their world:

> Pavlovian conditioning is not a stupid process by which the organism willy-nilly forms associations between any two stimuli that happen to co-occur. Rather, the organism is better seen as an information seeker using logical and perceptual relations among events, along with its own preconceptions, to form a sophisticated representation of its world. (p. 154)

Newer typologies of learning include representation, imitation, problem solving, and reasoning under the rubric of cognition. Cognitively oriented comparative psychologists now speak of animal cognition (Terrace, 1984).

Thomas (1980), for example, in a revision of a system proposed by Gagne, proposed the following eight levels of "intellective" learning across animal species:

> 1) habituation; 2) signal learning (classical conditioning); 3) stimulus–response learning (operant conditioning); 4) chaining (of operants); 5) concurrent discriminations; 6) affirmative concepts (relative and absolute); 7) disjunctive concepts, conjunctive concepts, conditional concepts; and, 8) biconditional concepts. (p. 460)

In an exhaustive review of laboratory tests on primate abilities, Meador, Rumbaugh, Pate, and Bard (1987) classify learning into the following seven categories:

1. simple conditioning and discrimination learning;
2. complex discrimination learning (learning set, reversal learning, concurrent discrimination);
3. time-based, response-patterned and match-to-sample learning (delayed responses, alternation problems, match-to-sample problems);
4. probability learning and cross-modal transfer (probability learning, cross-modal transfer and recognition);
5. complex learning (problem solving, tool use, cognitive maps, oddity problems, same-difference learning, higher-order sign problems, use of number and quantity concepts, reasoning, and self-awareness);
6. observational learning; and
7. language training.

As we might expect, various degrees of assimilation of the new cognitive framework coexist, so that some investigators continue to slip into more traditional conditioning stances, while others emphasize information processing and representation, which is the hallmark of the new cognitive psychology. Recently, psychologists who study animal cognition have adopted an information-processing model of animals as strategists who optimize their foraging through decision making based on cost–benefit analysis (Kamil, 1984).

*Instinct*

We begin again with *Webster's* (1981):

> **2**: a natural or inherent aptitude, tendency, impulse, or capacity ... **3 a**: complex and specific response on the part of an organism to environmental stimuli that is largely hereditary and unalterable though the pattern of behavior through which it is expressed may be modified by learning, that does not involve reason, and that has as its goal the removal of a somatic tension or excitation **b**: behavior that is mediated by reactions (as reflex arcs) below the conscious level – usu. not used technically (p. 1171)

*Comparative and physiological psychologists.* Romanes (1886) defined *instinct*, in contrast to *reflex*, as

> reflex action into which there is imported the element of consciousness. The term is therefore a generic one, comprising all those faculties of mind which are concerned in conscious and adaptive action, antecedent to individual experience, without necessary knowledge of the relation between means employed and ends attained, but similarly performed under similar and frequently recurring circumstances by all individuals of the same species. (p. 17)

William James devoted an entire chapter in his historic book *The Principles of Psychology* (1890/1983) to instinct. He begins with the following statement:

> *Instinct is usually defined as the faculty of acting in such a way as to produce certain ends, without foresight of the ends, and without previous education in the performance*.... They are functional correlates of structure. With the presence of a certain organ goes, one may say, almost always a native aptitude for its use. (p. 1004)

In a later section he lists the following catalogue of instincts in infants: sucking, biting, clasping, carrying in the mouth, crying, turning the head to the side, holding the head erect, sitting up and standing, vocalization, gestural and vocal imitation, rivalry, anger, sympathy, fear (of black things, dark places, high places, and strangers), curiosity and play, jealousy, sexuality, secretiveness, and parental love. He also speaks of a hunting instinct.

Pavlov (1904/1961) credits Spencer with the idea that instinctive reactions are reflexes, and he argues that "between the simplest reflex and the instinct we can find numerous stages of transition, and among these we are puzzled to find any line of demarcation" (p. 766). After discussing several claims concerning the distinctiveness of instincts (e.g., their complexity, their length, their internal origins, their holistic nature), Pavlov concludes that

> instincts and reflexes are alike the inevitable responses of the organism to internal and external stimuli, and therefore we have no need to call them by two different terms. Reflex has the better claim of the two, in that it has been used from the very beginning with a strictly scientific connotation. (Pavlov, 1904/1961, pp. 768–769)

Baldwin (1901/1960) defines *instinct* this way:

> (1) An inherited reaction of the sensorimotor type, relatively complex and markedly adaptive in character, and common to a group of individuals.... This definition makes instinct a definitely biological not a psychological conception.... The line of

difficulty with this definition lies in the distinction of instinct from reflex action, but the facts that instinct is definitely associated with stimulation through higher centres (sensorimotor), and that it is highly adaptive and relatively complex, while reflexes are relatively simple and not always evidently adaptive, serve to differentiate them. On the other hand, the distinction of instinct from action of the secondary-automatic type is in their origin respectively, the former being congenital, the latter acquired. (Vol. 2, p. 55)

The comparative psychologist Schneirla (1971), in contrast, suggests the following definition of instinct: "*species-typical behavior, studied from the standpoint of phylogenesis and of individual development*" (p. 35).

*Psychoanalysts.* Freud's English translators used the word "instinct" for the German word *Trieb*, which is more accurately rendered as "drive" or "impulse":

This is hardly a "slight complication." Freud used the word *Instinkt* when it seemed appropriate to him – to refer to the inborn instincts of animals – and he shunned it when he was speaking of human beings. Since Freud made a clear distinction between what he had in mind when he spoke of instincts and what he had in mind when he spoke of *Trieb*, the importance of retaining the distinction seems obvious. (Bettelheim, 1984, p. 104)

The immediate effect of the Freud (1915) English translation of *Trieb* into "instinct" (the word "drive" had not yet been used in English in a psychological sense) was immense confusion. When used (allegedly by McDougall) first to list the behaviors of animals, and then to explain them, "instinct" had been attacked by behaviorists as a circular concept. Among psychologists, the word "instinct" was in such poor repute that for some decades any reference to species-specific behavior was banned in behavioristic circles (e.g., Hilgard & Marquis, 1940). When Freud's work appeared (mistakenly) to raise the tarnished flag of instinct just when psychologists thought they had banished it forever from scientific discourse, his theory fell on critical ears (Bettelheim, 1984).

*Ethologists.* Darwin (1859/1960) wrote a chapter on instinct in the *Origin of Species* in which he argued that instincts evolve through natural selection. Although he declined to define "instinct," he claimed that instincts are adaptive: "It will be universally admitted that instincts are as important as corporeal structures for the welfare of each species" (Darwin, 1859/1960, p. 185). He also noted the parallels between instincts and habits:

This comparison gives, I think, an accurate notion of the frame of mind under which an instinctive action is performed but not necessarily of its origin. How unconsciously many habitual actions are performed, indeed not rarely in direct opposition to our conscious will! yet they may be modified by the will or reason.... Once acquired, they often remain constant throughout life. Several other points of resemblance between instincts and habits could be pointed out. As in repeating a well-known song, so in instincts, one action follows another by a sort of rhythm; if a person be interrupted in a song, or in repeating anything by rote, he is generally forced to go back to recover the habitual train of thought. (pp. 184–185)

Lorenz and Tinbergen argue as follows:

However one might wish to interpret the word "instinct," the existence of certain motor patterns rigidly innate in the individual will always have to be recognized. These movements may have great taxonomic value for a species, a genus, or even a whole phylum. Since they constitute the very core of what earlier students of animal behavior regarded as the manifestations of instinct, we have used, or rather retained, the term "instinctive action" for them. Instinctive actions are remarkably independent in form from all receptor processes; not only from "experience" in the broadest sense of the word, but also from the stimuli that affect the organism during their operation.... Once the movement is released, it seems quite independent of the animal's receptors as well as of further external stimulation. Unlike taxes, which are innate reaction patterns, instinctive actions are innate motor patterns.... the two most important and distinctive aspects of instinctive action, which present insurmountable difficulties to any other explanation, can be interpreted quite simply by adopting a hypothesis of automatic rhythmical generation and coordination of impulses. These are [a] decrease in threshold to releasing stimuli, and the discharge *in vacuo* of instinctive acts, which may as a result occur in situations where the act has no survival value. (Lorenz & Tinbergen, 1957, pp. 176–179)

On the other hand, instinctive actions occur in response to preordained stimuli operating on "innate releasing mechanisms" in the animal's nervous system. The animal's probability of encountering these stimuli is increased by its adaptive and variable searching or "appetitive behaviors": "Psychologically the most important and striking quality of instinctive action ... is that its purely automatic performance, while not directed toward a goal after which the subject strives, represents such a goal as a whole: the organism has an 'appetite' for discharging its own instinctive activities" (Lorenz & Tinbergen, 1957, p. 183).

Other investigators have emphasized the complexity, variability, and plasticity of reflexes. Barlow (1968, 1977), for example, shows that "fixed action patterns" vary in duration and in the sequence and number of components from one occassion to the next. He also shows that they are subject to modification during performance on the basis of feedback from their target. For these reasons he proposes calling them "modal action patterns." More recently, Berkinblit, Feldmans, and Fukson (1986) have discussed neurophysiological models for the variability and plasticity of "innate motor patterns."

### *Cognition*

*Webster's* (1981) has this to say about *cognition*:

**1 a**: the act or process of knowing in the broadest sense; *specif*: an intellectual process by which knowledge is gained about perceptions or ideas – distinguished from *affection* and *conation* **b**: a product of this act, process, faculty, or capacity: KNOWLEDGE, PERCEPTION. (p. 440)

According to Baldwin, cognition is

> the being aware of an OBJECT.... As above defined, cognition is an ultimate mode of consciousness co-ordinate with conation and affection. Cf. CLASSIFICATION (of mental functions). It may well be questioned, however, whether in current usage cognition does not imply judgment, at least in a rudimentary form.... Knowledge is practically synonymous, but lacks an adjective form. (Baldwin, 1901/1960, Vol. 1, p. 102)

*Physiological psychologists.* Wundt was one of the earliest psychologists to use the term *cognition.* According to Wundt's model of mental processing,

> When the subject was concentrating on the stimulus, the stimulus first had to be *perceived*, that is, simply registered in consciousness, and then *apperceived*, or consciously "interpreted" in light of the response that was associated with it. With attention focused on the response, the added step of apperception or conscious interpretation of the stimulus was not necessary.... Still other complications were added to the experiment to measure the time of mental processes more complicated than apperception. Reaction times were longer here than when the subject was concentrating on a single kind of expected stimulus. The difference was said by Wundt to represent the time for *cognition* to occur; that is, the stimulus had not only to be perceived and apperceived, but also recognized and differentiated from the other stimuli that were not supposed to elicit responses.... reaction times were longer still if different stimuli were presented, each calling for a different response.... The time this required in excess of the cognition reaction was assumed to be the time for a process of *association.* (Fancher, 1979, pp. 138–139)

Although related to modern usage, this usage is highly specialized.

*Cognitive psychologists.* In their famous book *Plans and the Structure of Behavior*, Miller, Galanter, and Pribram (1960) characterized the conflicting views of conditioning theorists and cognitive theorists:

> What an organism does depends on what happens around it. As to the way in which this dependency should be described, however, there are, as in most matters of modern psychology, two schools of thought. On the one hand are the optimists, who claim to find the dependency simple and straightforward. They model the stimulus response relation after the classical, physiological pattern of the reflex arc and use Pavlov's discoveries to explain how new reflexes can be formed through experience.... Arrayed against the reflex theorists are the pessimists, who think that living organisms are complicated, devious, poorly-designed for research purposes, etc. They maintain that the effect an event will have upon behavior depends on how the event is represented in the organism's picture of itself and its universe. They are quite sure that any correlations between stimulation and response must be mediated by an organized representation of the environment, a system of concepts and relations within which the organism is located ... a schema, a simulacrum, a cognitive map, an Image. (pp. 8–9)

According to Ulric Neisser, "'cognition' refers to all processes by which the sensory input is transformed, reduced, elaborated, stored, recovered, and used. It is concerned with these processes even when they operate in the absence of relevant stimulation, as in images and hallucinations" (quoted in Flanagan, 1984, p. 178).

In his brief history and review of cognitive psychology and artificial intelligence, Flanagan concludes that

(1) Cognitive psychology is committed to the reasonable view that the mind is a representational system, that is, an intentional system that transforms, processes, stores, and retrieves information about the world.
(2) The cognitive psychologist follows Kant in viewing this representational system as consisting of a rich system of a priori structures, processors, and categories which we use to create an orderly "picture" of the world. . . .
(3) Any psychology, therefore, that fails to talk about mental events and processes will not be remotely adequate. . . .
(5) Cognitive psychologists pitch their explanations at the higher functional levels of analysis (at what Dennett calls the intentional and design stances), rather than at the level of brain physiology (the physical stance). . . .
(6) Cognitive psychology has several important philosophical implications, of which three stand out. . . . People often lack knowledge of underlying mental processes; we are not adept at identifying the causes of our behavior and mental states and we are even sometimes mistaken about the contents of our minds. . . . Many psychologists and philosophers . . . believe that the data support a more modular view of the mind, that is, a view which sees the mind as a system of many different special-purpose processors, most of which have no idea what the others are doing. (Flanagan, 1984, pp. 243–244)

In other words, cognitive sciences have brought "mind" and representation back into psychology. On the subject of representation, Baars has this comment:

In the mid-1950's, it was still common to ridicule the notion of mental representation. Twenty years later, the issue became the core of the new psychology. Over several decades, psychologists were forced, by clear empirical results, to adopt a highly abstract theoretical language. Instead of chains of stimuli and responses, they found it necessary to speak of abstract semantic representations that encode not just the meaning of individual words, but an inferential, deeply interpreted representation of the world. (Baars, 1986, p. 170)

*Developmental psychologists.* Flavell (1977), while declining to define "cognition," says that a traditional image of cognition

includes such Higher-Mental-Processes types of psychological entities as knowledge, consciousness, intelligence, thinking, imagining, creating, generating plans and strategies, reasoning, inferring, problem solving, conceptualizing, classifying and relating, symbolizing and perhaps fantasizing and dreaming. . . . Certain components would have a somewhat humble, less purely cerebral-intellectual cast to them. Organized motor movements . . . perception, imagery, memory, attention and learning . . . all varieties of social cognition . . . and the social communicative versus private-cognitive uses of language. (Flavell, 1977, p. 2)

### *Intelligence*

*Intelligence* is defined as follows by *Webster's* (1965):

**1 a** (1): the faculty of understanding: capacity to know or apprehend: INTELLECT, REASON . . . **b**: the available ability as measured by intelligence tests or by other social criteria to use one's existing knowledge to meet new situations and to solve new problems, to learn, to foresee problems, to use symbols or relationships, to create new relationships, to think abstractly: ability to perceive one's environment, to deal with it symbolically, to deal with it effectively, to adjust to it, to work toward a goal: the degree of one's alertness, awareness, or acuity: ability to use with awareness the mechanism of reasoning whether conceived as a unified intellectual factor or as the aggregate of many intellectual factors or abilities, as intuitive or as analytic, as organismic, biological, physiological, psychological, or social in origin and nature **c**: mental acuteness: SAGACITY, SHREWDNESS. (p. 1174)

Baldwin defines "intellect" (or "intelligence") as

the faculty or capacity of knowing; intellection or, better, COGNITION (q.v.) denotes the process. . . . The earlier English psychologists used the word understanding rather than intellect. . . . There is a tendency to apply the term intellect more especially to the capacity for conceptual thinking. This does not hold in the same degree of the connected word intelligence. We speak freely of "animal intelligence"; but the phrase "animal intellect" is unusual. However, the restriction to "conceptual process" is by no means so fixed and definite as to justify us in including it in the definition. (Baldwin, 1901/1960, Vol. 2, p. 559)

Intelligence is the province of psychometricians, behavioral geneticists, neuroscientists, cognitive psychologists, cognitive scientists, developmental psychologists, comparative psychologists, biologists, and anthropologists, each group focusing on differing aspects of the phenomenon. As their name suggests, psychometricians are interested in classifying, measuring, and comparing degrees of intelligence within the human species; behavioral geneticists are interested in the inheritance of intelligence within families and within species; neuroscientists, the physiological bases of intelligence; cognitive psychologists and cognitive scientists, modeling information processing; developmental psychologists, the ontogenetic development of intelligence; comparative psychologists, species differences in intelligence; early physical anthropologists, craniometry; and, finally, biologists and anthropologists, the evolution of intelligence.

*Comparative psychologists.* Romanes (1886), in his book *Animal Intelligence*, defined reason or intelligence as "the faculty which is concerned in the intentional adaptation of means to ends. It therefore implies the conscious knowledge of the relation between means employed and ends attained, and may be exercised in adaptation to circumstances novel alike to the experience of the individual and to that of the species" (p. 16).

Tolman's definition reflects the shift away from the concept of consciousness: "What we mean by intelligence is probability of success in reaching goals; by motivation, probability of persistence in striving towards goals; and by emotional stability, probable tendencies not to exhibit unacceptable divagation in the pursuance of such goals" (Tolman, 1948/1951, p. 232).

*Psychometricians and behavioral geneticists.* According to Sternberg (1982),

traditionally, psychometricians . . . have sought to discover the nature of intelligence by searching for common sources of individual-differences variation in performance on large collections of tests consensually believed to measure intelligence. Reasoning and problem solving have played important parts in virtually every theory of intelligence that has been factor-analytically derived. (p. 226)

Historically, three movements that preceded and led to psychometry or IQ psychology were craniometry, localizationism, and phrenology. All three can be traced back to the German physician Franz Joseph Gall (1758–1828). Gall was a brilliant comparative anatomist who discovered the cerebral commissures, lateralization of control of movement, and the distinction between gray and white matter. He also demonstrated the relationship between brain size and mental abilities in various species, and conceived localization:

Not content to stop with the bare assertion that the mind was localized somewhere and somehow in the brain [Gall] held that discrete psychological "faculties" were localized in specific small parts of the brain. Furthermore, Gall believed that bumps and indentations on the surface of the skull provided accurate measures of the underlying brain parts, and hence of the different faculties. (Fancher, 1979, pp. 45–46)

In their early period, all three were heavily associated with hereditarian, racist views. Samuel Morton, for example, published three major works on the sizes of human skulls: "*Crania Americana* of 1839 . . . the *Crania Aegyptiaca* of 1844; and the epitome of his entire collection in 1849. Each contained a table summarizing his results on average skull volumes arranged by race" (Gould, 1981, p. 53). Paul Broca believed

that the higher mental functions were localized in anterior regions of the cortex, and that posterior areas busied themselves with the more mundane, though crucial, roles of involuntary movement, sensation, and emotion. Superior people should have more in front, less behind. . . . He accepted Gratiolet's classification of human groups into *races frontales*: (whites with anterior and frontal lobes most highly developed), "*races parietales*" (Mongolians with parietal or mid lobes most prominent), and "*races occipitales*" (blacks with most in the back). (Gould, 1981, p. 97)

Paul Broca was also, of course, the discoverer of Broca's area in the frontal lobe, whose destruction causes speech loss (Fancher, 1979).

It was Darwin's cousin Francis Galton who, shortly after Broca's time, invented statistical techniques for comparing degrees of intelligence and for demonstrating the inheritance or intelligence: "The arguments by which I endeavour to prove that genius is hereditary, consist in showing how large is the number of instances in which men who are more or less illustrious have eminent kinsfolk" (Galton, 1892/1962, p. 49).

Galton approached the question of the relative frequency of genius in a variety of ways, including comparing the frequency of achievements of the ranges of marks among the honors students in mathematics at Cambridge University taking the wrangler's exam (from 74 scoring 500–1,000 points to 1 scoring 7,500–8,000 points). He then demonstrated that "the deviations from average – upwards towards genius, and downward towards stupidity, must follow the law that governs deviations from all true averages" (Galton, 1892/1962, p. 72). After examining 300 families of 1,000 eminent and 415 illustrious men in various fields (literature, science, art, music, poetry, the law, the church, and the military), Galton concluded that "the general uniformity in the distribution of ability among the kinsmen in different groups, is strikingly manifest" (p. 374). (By "different groups" he meant grades of kinship.) Galton also compared the intelligence of the "negro race" unfavorably with that of Europeans, and he argued for selective breeding to increase the average level of intelligence (Galton, 1892/1962; Gould, 1981).

Alfred Binet, director of the psychology laboratory at the Sorbonne, best known as the father of modern intelligence tests, followed in the footsteps of Galton and Broca. After publishing nine papers on craniometry and becoming discouraged with the technique, in 1904 Binet was commissioned by the French minister of public education to devise techniques for identifying public school children in need of special education:

> Binet decided to assign an age level to each task, defined as the youngest age at which [a] child of normal intelligence should be able to complete the task successfully. . . . The age associated with the last task he could perform became his "mental age," and his general intellectual age was calculated by subtracting his mental age from his true chronological age. . . . In 1912 the German psychologist W. Stern argued that mental age should be divided by chronological age, not subtracted from it, and the intelligence *quotient*, or IQ, was born. (Gould, 1981, pp. 149–150)

Although Binet assigned little meaning to the score, believing that intelligence was too multifaceted and complex to be captured in one score, others interpreted it differently.

Goddard was the first to use Binet's intelligence tests in the United States, in 1916; but Louis Terman, a professor at Stanford University, revised and popularized them in the so-called Stanford–Binet tests. It was Terman who introduced the concept of general intelligence and set the average IQ at 100. Terman was a hereditarian who advocated universal testing of children to establish their vocational potentials (Gould, 1981).

In 1917, Terman, Yerkes, Goddard, and others met at Yerkes's invitation to develop IQ tests for universal testing of army recruits. Yerkes supervised testing of 1.76 million recruits, with the goal of determining their fitness for various levels of service, and with the hope of gathering uniform data on a large pool of subjects. Despite the fact that the tests were culturally biased and unevenly administered, they were interpreted by Yerkes (1921) and others as accurate indicators of hereditary, racial, and class differences in intelligence (Gould, 1981).

Meanwhile, other psychometricians were arguing over how many discrete factors are involved in intelligence. In 1904, Charles Spearman, professor of psychology at University College, London, developed factor analysis "as a procedure for deciding between the two- vs. many-factor theory by determining whether the common variance in a matrix of correlation coefficients could be reduced to a single 'general' factor, or only to several independent 'group' factors" (Gould, 1977, p. 257). In 1904, Spearman calculated a single general factor: "He called it *g*, or general intelligence, and imagined that he had identified a unitary quality underlying all cognitive mental activity – a quality that could be expressed as a single number and used to rank people on a unilinear scale of intellectual worth" (Gould, 1977, p. 251). (Cyril Burt argued that Spearman's *g* factor proved the innateness of intelligence.) In 1934, L. L. Thurstone, professor of psychology at the University of Chicago, published *The Vectors of Mind*, the first of several books in which he argued against a single *g* factor model of intelligence, and in favor of a multifactor model: seven (or more) *primary mental abilities* (PMAs). Like Spearman, his interpretation was based on the development of a new mathematical technique: vector rotation.

Although Thurstone attacked Spearman for Spearman's reification of his principal component factor into a real biological entity, he made the same logical error in his own interpretation of the PMAs (Gould, 1977). The argument continued: Guilford (1956), for example, taking the multifactorial approach, Jensen (1969), taking the single-factor approach.

*Information processing: Cognitive science psychologists.* Sternberg argues that information-processing analysis is

> like the psychometric approach in its application to quantitative indexes of intelligent behavior (rather than to quantitative indexes of concepts of intelligent behavior), but differs from the psychometric approach in its use of stimulus variation rather than individual differences variation as the means of isolating elementary units of intelligence. The motivating idea in information-processing analysis is to decompose performance on tasks into elementary information-processing components, and then to show the interrelations among the components used to solve various tasks requiring intelligent performance. In this approach, too, reasoning and problem solving have been found to be critical ingredients of intelligence. (Sternberg, 1982, p. 226)

*Developmental psychologists.* According to Piaget, intelligence is

> the most plastic and at the same time the most durable structural equilibrium of behavior.... It is the most highly developed form of mental adaptation, that is to say, the indispensable instrument for interaction between the subject and the universe when the scope of the interaction goes beyond immediate and momentary contacts to achieve far reaching and stable relations.... life brings with it indirect interaction between subject and object, which takes effect at ever increasing spatio-temporal distances and along ever more complex paths.... Only intelligence, capable of all its detours and reversal by action and by thought, tends towards an all-embracing equilibrium by aiming at the assimilation of the whole of reality and the accommodation to it of action. (Piaget, 1966, pp. 8–9)

### *Adaptation*

*Webster's* (1981) has this to say about *adaptation*:

> **1 a**: the act or process or adapting, fitting, or modifying ... **b**: the state or condition of being adapted or adjusted or of adapting or adjusting oneself ... **2**: adjustment to environmental conditions: as **a**: adjustment of a sense organ (as the eye) to the intensity or quality of stimulation (as light) prevailing at the moment effected by

changes in sensitivity and occurring as a heightened sensitivity . . . or as a physical adjustment to meet changed conditions . . . or as decline or loss of sensitivity to a constant stimulus **b**: modification of an organism or of its parts or organs fitting it more perfectly for existence under the conditions of its environment and resulting from the action of natural selection upon variation – compare NATURAL SELECTION, VARIATION 6 **c**: the continuing process through which the organization of groups is modified to meet the requirements of their social and physical environment **3**: something that is adapted: **a**: a modification for a new use: an alteration or change in form or structure . . . **b**: a composition rewritten into a new form. (p. 23)

*Ethologists and evolutionary biologists.* Although Darwin apparently did not use the word, the concept denoted by *adaptation* is central to the theory of natural selection. As such, it has been subject to considerable scrutiny under the rubric of "adaptationism" by some biologists in recent years.

Ronald Fisher (1940) made the following comments on adaptation:

In order to consider in outline the consequences to the organic world of the progressive increase of fitness of each species of organism, it is necessary to consider the abstract nature of the relationship which we term "adaptation." . . . An organism is regarded as adapted to a particular situation, or to the totality of situations which constitute its environment, only in so far as we can imagine an assemblage of slightly different organic forms, which would be less well adapted to that environment. (p. 41)

George Williams (1966), in his treatise on fallacious concepts of evolutionary progress and group selection, circumscribed the boundaries of adaptation: "the mere presence of adaptation is no argument for its necessity, either for the individual or the population. It is evidence only that during the evolutionary development of the adaptation [genes that augmented] its development survived *at a greater rate* than those that did not" (Williams, 1966, p. 29). (He also lays out criteria for distinguishing adaptations from beneficial effects.)

*Developmental psychologists.* According to Baldwin (1901/1960), adaptation has two distinct meanings requiring two separate listings. The first listing defines adaptation this way:

(1) A word signifying adjustment or fitness; as of means to ends, organ to function, &c. (2) In biology, adaptation is a general term used to signify the adjustment of the organism or its organs to the environment, with special reference to other organisms. . . . In view of modern biological theory and discussion, two modes of adaptation should be distinguished: (1) adaptation through variation (heredity); (2) adaptation through modification (acquired). For the functional adjustment of the individual to its environment (2, above), J. Mark Baldwin suggested the term ACCOMMODATION (q.v.) recommending that adaptation be confined to the structural adjustments which are congenital and hereditary (1, above). On this distinction adaptations are phylogenic . . . and accommodations are ontogenic. (p. 15)

Jean Piaget borrowed the biological concept of adaptation and used it to describe the adjustments to the environment achieved through the interplay of assimilation and accommodation:

> Adaptation must be described as an equilibrium between the actions of the organism on the environment and vice versa. Taking the term in its broadest sense, "assimilation" may be used to describe the action of the organism on surrounding objects, in so far as this action depends on previous behavior involving the same or similar objects.... Conversely the environment acts on the organism and, following the practice of biologists, we can describe this converse action by the term "accommodation." ... This being so, we can then define adaptation as an equilibrium between assimilation and accommodation. (Piaget, 1966, pp. 7–8)

*Cognitive psychologists.* Turning on the other meaning of the term, many cognitive psychologists are now turning to the study of "adaptive systems" (Grossberg, 1980; Rumelhart, McClelland, & PDP Group, 1986). Adaptive systems are variants of neural networks, whose nodes and connections operate in parallel (i.e., parallel processors). Such systems, equipped with a simple, neurally plausible set of properties, are capable of modeling a variety of perceptual, motoric, and learning phenomena.

These "connectionist" networks usually are discussed in a mixture of "representational" and "adaptational" vocabularies. Indeed, the case can be made that these two vocabularies are functionally equivalent. Thus, we can say that a connectionist model *represents* the phonemes of English as well as some lexical items, and also that the network *adapts to* new, previously unknown phonemes and words. Somehow, the two vocabularies manage to co-exist with little confusion.

## Various conceptions about relationships and domains

### *Intelligence and cognition*

Although the dictionary definitions suggest a distinction between cognition and intelligence, the Baldwin (1901/1960) and Flavell (1977) definitions and current usage of the term reveal that *cognition* is often employed as a synonym for intelligence, especially by comparative psychologists, ethologists, and cognitive psychologists; indeed, writers often use the two terms interchangeably (e.g., Terman, Piaget, and Flavell). The term *cognition*, perhaps because of its associations with information processing and artificial intelligence, seems to have a slightly more rigorous or scientific connotation. Indeed, "animal cognition" may be the name on the treaty that settles the 30 years' war between ethologists and comparative psychologists. Ironically, *animal cognition* embraces concepts of representation and adaptation that were banished by the behaviorists, while incorporating the rigor and prestige of mathematics through information-processing models. *Intelligence*, perhaps because of its historical association with intelligence tests and their various political uses, is currently disfavored in many quarters.

### *Intelligence, cognition, and learning*

*Comparative psychologists.* The terminology used in psychological studies of animal abilities seems to reflect the disciplinary training and philosophy of the investigators: Reports of animal abilities that use standard tests tend to use the term *learning*; reports that use seminaturalistic observations, and those focusing on development, tend to use the term *intelligence* or *cognition*. This can be true even when the same behaviors are under investigation, for example, in the case of studies of tool use and other forms of instrumentation.

On the other hand, some comparative psychologists have assimilated "cognition" to "learning" (e.g., Fobes & King, 1986; Meador et al. 1987), treating it as a category of learning. This perspective is explicitly stated by Thomas (1986) in a review paper on vertebrate intelligence: "Intelligence will be treated as being equivalent to *learning ability*, and differences in intelligence will be regarded as differences in learning ability" (p. 38). In an earlier paper on the evolution of intelligence, the same author says that

> students of intelligence in nonhuman animals who have explicitly related learning abilities to intelligence (or some synonymous terms such as "higher mental functions" or "higher nervous activity") include: *Bitterman* (1965); *Corning et al* (1976); *Harlow* (1958); *Hayes and Nissen* (1971); *Jolly* (1972); *Masterson and Skeen*(1972); *Passingham* (1975); *Razran* (1971); *Rumbaugh and Gill* (1974); *Viaud* (1960). These and other references to be cited will attest that the measurement of intelligence in animals has usually been regarded to be synonymous with the measurement of learning ability. (Thomas, 1980, p. 455)

*Developmental psychologists.* Although educational psychologists did much to popularize Piaget's ideas in the United States, they were primarily concerned with problems of learning, and Piaget's model subordinates learning to development. According to Piaget's staunch critic Charles Brainerd, for example,

> Piaget . . . views the study of intellectual development as a branch of embryology. . . . This reliance on biological metaphors is the principal difference between Piaget's theory and those approaches to children's intelligence that have traditionally been popular in North America. These latter approaches are characterized by their emphasis on learning as a source of intellectual development . . . and by their reliance on metaphors borrowed from the simple conditioning paradigms of animal psychology. . . . By comparison, learning does not play a central role in Piaget's theory. In fact, it plays essentially no role at all. (Brainerd, 1979, pp. 168–169)

In contrast, according to Gallagher and Reid (1981), Piaget's theory is a learning theory. They say that "the contribution of genetic epistemology to our understanding of learning has just begun to be recognized," and they list the following six contributions:

> 1) learning is an internal process of construction. . . . 2) learning is subordinated to development, that is, competence is a precondition for learning; 3) children learn not only by observing objects but also by reorganizing . . . what they learn from coordinating their activities; 4) growth in knowledge is often sparked by a feedback process that proceeds from questions, contradictions, and consequent mental reorganization; 5) questions, contradictions, and the consequent reorganization of thought are often stimulated by social interaction; 6) since awareness . . . is a process of reconstruction . . . understanding lags behind action. (pp. 10–11)

*Instinct and learning*

William James discussed the relationship between instinct and experience as follows:

> It is obvious that every instinctive act, in an animal with memory, must cease to be "blind" after being once repeated, and must be accompanied with foresight of its "end" just so as that end may have fallen under the animal's cognizance.... It is plain, then, that no matter how well endowed an animal may originally be in the way of instincts, his resultant actions will be most modified if the instincts combine with experience. (James, 1890/1983, pp. 1010–1011)

John B. Watson (1914/1967) contrasted instinct and habit as follows:

> Instinct and habit differ so far as concerns the origin of the *pattern* (number and localization of simple reflex arcs involved) and the *order* (temporal relations) of the unfolding of the elements composing that pattern. In instinct both pattern and order are inherited: in habit both are acquired.... After habits are perfected they function in all particulars as do instincts. (pp. 184–185)

*Comparative psychologists.* Many comparative psychologists, following Hebb, argued that instinct was a meaningless concept, defined only in opposition to learning. Others, following Lehrman (1953), argued that an unspecifiably diffuse relationship exists between instinct and learning.

Schneirla (1971) expressed this viewpoint in his discussion of behavioral development and comparative psychology. He argued for a new developmental paradigm based on redefinitions of maturation and experience:

> Accordingly, I suggest that *maturation* be redefined as the contributions to development of growth and of tissue differentiation, together with their organic and functional trace effects surviving from earlier development; and that *experience* be defined as the contributions to development of the effects of stimulation from all available sources (external and internal), including their functional trace effects surviving from earlier development.... The developmental contributions of the two complexes, *maturation* and *experience*, must be viewed as "fused" (i.e., as inseparably coalesced) at all stages in the ontogenesis of any organism.... *Interaction*, the conventional term used to denote the interrelationship of "genetic" and "environmental" factors, assumes the existence of major developmental agencies that remain separate entities and that are supposed to engage in distinct functional interchanges. (p. 37)

According to Schneirla (1971),

> the term *experience* is frequently used in a restricted way to denote those processes akin to learning which are based on external events that affect the organism, and which begin at the neonatal stages.... In contrast, the concept *experience* is defined broadly here to denote classes of stimulative effects that result in functional changes ranging all the way from stimulus-involved biochemical and physiological processes to conditioning and learning. *Experience* includes any class of stimulative effect on the organism and is clearly not equivalent to learning. (p. 38)

Lorenz (1965), in sharp contrast, argued that innate and learned elements are distinguishable and are intercalated during development. He defined the innate not only as that which is not learned, "but what must be in existence before all learning is possible" (p. 44). Specifically, according to Lorenz, phylogenetically patterned motor and receptor organization act as an "innate schoolmarm":

The teaching function of the consummatory act is based partly on phylogenetically acquired information contained in the motor pattern itself and partly on that contained in the receptor organization of the releasing mechanism but mainly on the interaction of both, that is to say, on the reafference which the organism produces for itself by performing the consummatory act in the adequate consummatory situation. (Lorenz, 1965, p. 88)

Lorenz argued that phylogenetic adaptation and adaptive modification through learning were the two exclusive channels by which information about the environment could be fed into the organism. He went on to say that

practically every functional unit of behavior contains individually acquired information in the form of a stimulus-input which releases and directs reflexes, taxes, etc.... and thus determines the time and the place at and in which the behavior is performed. It is by no means a logical necessity that adaptive modification of behavioral mechanisms is invariably concerned in the survival function of these short-notice responses [which] in their very essence, are the opposite of "what we formerly called learned," being, indeed as already mentioned what Pavlov has termed "unconditioned reflexes." (Lorenz, 1965, p. 18)

Gould and Marler (1987) headline a recent article, entitled "Learning by instinct," with the following summary: "Usually seen as diametric opposites, learning and instinct are partners: the process of learning, in creatures at all levels of mental complexity, is often initiated and controlled by instinct" (p. 74).

### *Instinct and intelligence (or reason)*

Romanes (1886) starts by defining *instinct*:

mental action ... directed towards the accomplishing of adaptive movement, antecedent to individual experience, without necessary knowledge of the relation between the means employed and the ends attained, but similarly performed under the same appropriate circumstances by all the individuals of the same species. (p. 16)

He continues:

Now in every one of these respects, with the exception of containing a mental constituent and in being concerned in adaptive action, instinct differs from reason. For reason, besides being concerned in adaptive action is always subsequent to individual experience, never acts but upon a definite and often laboriously acquired knowledge of the relation between means and ends, and is very far from being always similarly performed under the same appropriate circumstances. (p. 16)

He notes, however, that "instinct passes into reason by imperceptible degrees" (p. 16).

Although instinct and intelligence customarily are placed at opposite ends of a spectrum, cognitive psychological and ethological theories both imply that intelligence, like learning, is made possible by various elements of innate organization. In particular, Schiller's work (1952, 1957) suggests that innately determined motor patterns are the necessary raw materials for the development of intelligent problem solving (e.g., Gould & Marler, 1987; Lorenz, 1965; Schiller, 1952, 1957).

From his study of tool use in chimpanzees, Schiller concluded that

> as the ultimate units of which adaptive patterns are composed, individual or elementary grasping, pointing, shaking, carrying, placing, pulling and so forth can be conceived as being at the chimpanzee's disposal as his sensorimotor capacities develop in the course of maturation and connecting general experience. In this sense these motor elements, rather than stimulus–response connections or unconditioned reflexes, are the basis of adaptive behavior. Their organization is serialization, or rather condensation, not selection. They can be connected with very different perceptions. Since no external stimulus is definitely correlated to the response, it is fair to replace the term "response" with that of "emittance" of patterns determined more by the internal state of the organism than by the external stimulus. (Schiller, 1957, p. 286)

Clearly, Schiller is implying that intelligent problem solving – for example, tool use – is made possible by innate organization.

In a recent discussion of "innate motor patterns," Berkinblit et al. (1986) make an even more explicit connection between instinctive and voluntary behavior: "We will hypothesize that *voluntary movements are constructed by the central nervous system using the same mechanisms that underlie the variability of innate motor patterns*" (p. 586).

The predictable hierarchical construction of intelligent schemes from simple motor acts during infancy clearly implies innate organization of intelligence (Parker & Poti', P&G8; Gibson, P&G3). Likewise, many lines of evidence converge to suggest that language acquisition by human children is innately programmed in every element (Chomsky, 1957; Lenneberg, 1967; Siegler, 1987; Gibson, P&G3; see Parker, 1985, for a review of the evidence).

### *Adaptation, instinct, and learning*

According to ethologists, innate organizations, including instinctual behaviors, are taxon-specific adaptations that have arisen through natural selection, partly as adaptations for species-specific learning. Many modern ethologists and comparative psychologists are pursuing research that tests specific hypotheses concerning the organization and adaptive significance of animal learning (e.g., Gould & Marler, 1987; Marler & Terrace, 1984).

### *Adaptation and intelligence*

Piaget defines intelligence as an adaptation and argues that it is the most complex form of mental adaptation, though interestingly he doubts that the highest forms of intelligence in humans (i.e., formal operations) could have a direct selective advantage (Piaget, 1971).

Arguments over whether human intelligence is an adaptation that arose through natural selection or a product of divine creation go back at least to Darwin and Wallace (Gould, 1980). Arguments over whether human intelligence is specifically an adaptation or a by-product of other adaptations continue today (e.g., Faitkowski, 1986; Gould, 1981; Parker & Gibson, 1979).

The claim that intelligence is an adaptation has sometimes been confounded with the claim that more intelligent creatures are better adapted than less intelligent creatures, or that all creatures would benefit from higher intelligence. Neither of these claims is implied by the claim that intelligence is an adaptation in specific creatures, anymore than the claim that sonar is an adaptation in bats and dolphins implies that creatures without sonar are less well adapted. Intelligence is only one of many successful adaptive strategies – one that has been relatively rare in earth history, perhaps because of its high costs (Parker, P&G4).

### *Development and learning*

A brief survey of articles and books suggests that comparative psychologists rarely treat the relationship between *development* and *learning*. When development and learning are discussed, two opposing perspectives emerge: the view that they are basically the same phenomenon and the view that they are basically different phenomena.

*Comparative psychologists.* Few comparative psychologists have compared the learning of immature and mature animals of the same species (though Mason & Harlow, have provided an interesting counterexample) (Parker, P&G1), and when they have done so they have focused primarily on the age at which a particular learning ability has emerged (Zolman, 1982). One reason for this lay in the assumption that the young animal was an incompletely formed adult, and hence differences between adults and young were strictly quantitative. Even now, "the two most popular psychological approaches are to either specify the general principles of learning applicable to all species and all developmental ages or to differentiate species and developing animals on behavioral constructs" (Zolman, 1982, p. 293).

*Developmental psychologists.* Piaget, in an essay on development and learning, made a sharp distinction between the two processes:

> For some psychologists development is reduced to a series of specific learned items, and development is thus the sum, the accumulation of this series of specific items. I think this is an atomistic view which deforms the real state of things. In reality, development is the essential process and each element of learning occurs as a function of total development. (Piaget, 1982, p. 228)

Piaget objects to the classic stimulus–response model of learning, arguing that "a stimulus is a stimulus only to the extent that it is significant and it becomes significant only to the extent that there is a structure which permits its assimilation, a structure which can integrate this stimulus but which at the same time sets off the response" (Piaget, 1982, p. 234). According to his view, development is primary because assimilatory schema are constructed through the various processes of development. Until they are constructed they are unavailable for assimilation. This explains why children cannot be trained to understand certain relationships such as transitivity before achieving a certain stage of development of understanding of logicomathematical structures through equilibration.

Furth (1969) interprets Piaget's theory as postulating two aspects of ontogenetic experience: universal, species-specific *ontogenetic development* and environmentally particular *ontogenetic learning*:

> Ontogenetic development is the acquisition of novel behavior during ontogeny, that (a) is a joint function of the species-specific possibilities within the genome for acquiring new behavioral structures and the individual's experience in the species environment, (b) derives its information primarily from feedback from the subject's actions on the environment, (c) is chiefly motivated by internal regulative mechanisms, (d) leads to a restructuring on a higher plane of species behavioral structure, and (e) is normally irreversible.... Ontogenetic learning is the acquisition of novel behavior during ontogeny that (a) is a joint function of species-specific available behavioral structures ... and the individual's experience in a particular environment ... (b) derives its information primarily from ... the environment, (c) is chiefly motivated by external reinforcing contingencies, (d) leads to special application or to a cumulative increase in the range of application of available behavioral structures, (e) is subject to forgetting. (p. 66)

*Information-processing theorists.* Oyama (1985), in contrast, takes the view that development and learning are the same phenomenon from an information-processing perspective:

> In describing types of feedback, Wiener observes that if information fed back to the system changes the system's "general method and pattern of performance," learning may be considered to have occurred.... It is not accidental that this description fits development in general, since I think the traditional ways of distinguishing learning from other kinds of development are often difficult to defend. In any case, what is important here is that the machine's "taping" (programming) is itself altered, and with it the manner of processing data. (p. 120)

Both adults and children think; both learn things. Thinking and learning both entail information processing. Several information-processing theories address the issue of how children's thinking and learning differ from those of adults. In fact, according to Siegler (1987), "information-processing theories assume that our understanding of how children think can be greatly enriched by knowledge of how adults think" (p. 63). After reviewing several such theories (i.e., his own and those of Sternberg, Case, Newell & Simon, and Klahr & Wallace), Siegler identifies the following four mechanisms that operate in both learning and development: strategy construction, automatization, generalization, and encoding.

Case (1985), in his revision of Piaget's model that is also based on information-processing concepts, states the apparent paradox concerning learning and development very nicely:

> While the data on adults' functioning appear to implicate practice as the factor that produces the change in children's basic operative capacities, the data on children's functioning appear to implicate maturation. . . . The most obvious two-factor explanation would run as follows (1) whenever anyone receives practice in a novel operation, operational efficiency and speed increase up to some asymptotal value which is set by the efficiency of their psychological system, and (2) as children mature, these physiological limits change, at a rate and in a fashion which is set by maturation. (pp. 367–368)

By "maturation," Case refers specifically to maturation of the short-term storage space (STSS) in the memory, probably mediated by myelinization of the nervous system: The STSS "sets limits on the new executive control structures children can assemble, by limiting the complexity of the structures they can detect in the course of applying their existing executive structures" (Case, 1985, p. 307).

### *Development and intelligence*

*Psychometric approaches.* In that IQ represents a *relationship* between "mental age" and "chronological age," it implies development. Binet began his work on intelligence tests in order to test schoolchildren. Aside from devising age-specific questions, however, psychometricians have evinced little interest in intellectual development.

*Developmental psychologists.* Piaget, who began his career in psychology working on Binet's intelligence tests, soon shifted his attention to the study of age- and stage-specific intellectual abilities (i.e., to developmental questions). According to Piaget, intelligence develops epigenetically from birth through adolescence through four major periods: the sensorimotor period (from birth to 2 years), the preoperations period (from 2 to 5 years), the concrete operations period (from 5 to 11 years), and the formal operations period (from 11 years on). Each period is characterized by a more complex form of reasoning. The highest form of intelligence, "formal operational" reasoning, which develops when children are about 11 years old, involves hypotheticodeductive reasoning or scientific methodology of systematic comparisons of all variables, holding all but one variable constant (Inhelder & Piaget, 1958).

Some neo-Piagetian developmental psychologists and some information-processing developmental psychologists argue that differences in ability between adults and children are based not on differences in reasoning but on differences in processing capacity, differences in strategy construction, and/or differences in encoding mechanisms (Case, 1985; Siegler, 1987; Sternberg, 1982). Although the content of Piaget's stage model has been heavily criticized and widely revised, the existence of four periods or stages of intellectual development is widely accepted by developmental psychologists with an information theory bent (e.g., Case, 1985; Siegler, 1987). Indeed, the idea of four developmental stages goes back at least to Spencer and Baldwin (Parker, P&G1).

*Adaptation and development*

Although some ethologists have studied the ontogeny of behavior (Hailman, 1982), development has not been a major focus in ethology (Parker, P&G1). In recent years, however, behavioral ecologists have focused on the adaptive significance of the overall shape of the life cycle for various species. The theoretical formulation they have developed to explain the evolution of differing developmental patterns is known as *life history strategy theory* (e.g., Cole, 1954; Stearns, 1980). This relatively new theory provides a framework for comparative developmental psychology, which has the potential to explain species differences in developmental rate and pattern and in intelligence and brain size (Parker, P&G4).

## Conclusion

When we examine the various definitions and positions touched on in this chapter, we see that there are many ways to cut a given cake: Different questions are posed; different answers are given. Even so, we can see several quests cross-cutting the various positions: to describe and classify abilities, to identify factors underlying abilities in adult animals and factors underlying abilities in immature animals, to discover mechanisms of ontogenetic and experiential development of abilities, and to understand how natural selection has acted on various abilities.

The first major issue implicit (and, sometimes, explicit) in these formulations concerns the nature and complexity of mental organization. After a long period of banishment, mental representation has reemerged as a key concept in cognition and intelligence. Differences between ethologists and comparative psychologists are in the process of reconciliation through the concepts of animal cognition (Kamil, 1984; Terrace, 1984) and constraints on learning (Hinde, 1973).

The second major issue concerns the mechanisms involved in modifications of behavior during ontogeny and after experience – most specifically their similarities and differences. Differences in the positions of developmental psychologists and cognitive psychologists are in the process of reconciliation through the influence of information-processing theorists (e.g., Case, 1985; Siegler, 1987).

The third major issue concerns the relationships among cognition, learning, and development. Information-processing concepts may also provide a bridge between animal cognition and developmental psychology and hence stimulate the growth of comparative developmental psychology.

Finally, the fourth major issue concerns the relationships among evolutionary processes and learning, development, and cognition. Concepts of optimum foraging theory (Kamil, 1984) and life history strategy theory (Parker, P&G1), both drawn from behavioral ecology, look most promising in this domain.

## Notes

1. To elaborate on this definition: "*Determination* is the process by which a cell or a part of an embryo becomes restricted to a given pathway: *differentiation* is the actual appearance of new properties. . . . Growth means permanent enlargement – that is, developmental increase in total mass. . . . The term *morphogenesis* means simply the generation of form, the assumption of a new shape" (Ebert & Sussex, 1972, pp. 8–10).
2. The ideological nature of the reflex-reduction program is easy to demonstrate (Baars, 1986). In midcentury, the four most influential young physiologists were Herman von Helmholtz, Emil Du Bois-Reymond, Ernst Brucke, and Carl Ludwig. Pavlov studied under Ludwig, Freud worked for 6 years in Brucke's laboratory, and Wilhelm Wundt, who founded experimental psychology under the rubric of physiological psychology, studied with Du Bois-Reymond and Helmholtz. The four men formed a private club in Berlin that was pledged to destroy the last remnants of religious thinking in biological matters, declaring that "no other forces than the common physical-chemical ones are active within the organism" (Jones, quoted by Baars, 1986, p. 23). And when Pavlov published his famous demonstration, the results were hailed not only in scientific circles but also among the lay proscientific public. Thus, H. G. Wells described Pavlov as "a star which lights the world, shining above a vista hitherto unexplored" (Baars, 1986, p. 50).

## References

Antinucci, F. (Ed.). (1989). *Cogntive structures and development in nonhuman primates.* Hillsdale, NJ: Erlbaum.

Baars, B. (1986). *The cognitive revolution in psychology*. New York: Guilford Press.

Baldwin, J. M. (1960). *Dictionary of Philosophy and Psychology*, Vols. 1–3. London: Peter K. Smith. (Original work published 1901)

Barlow, G. (1968). Ethological units of behavior. In David Ingle (Ed.), *The central nervous system and fish behavior* (p. 217). University of Chicago Press.

Barlow, G. (1977). Modal action patterns. In T. Seboek (Ed.), *How animals communicate* (pp. 98–134). Bloomington: Indiana University Press.

Berkinblit, M.B., Feldman, A. G., & Fukson, O. I. (1986). Adaptibility of innate motor patterns and motor control mechanisms. *Behavioral and Brain Sciences*, *9*, 585–638.

Bettelheim, B. (1984). *Freud and man's soul.* New York: Vintage Books.

Blumenthal, A. L. (1977). *The process of cognition*. Englewood Cliffs, NJ: Prentice-Hall.

Brainerd, C. (1979). *Piaget's theory of intelligence.* Englewood Cliffs, NJ: Prentice-Hall.

Case, R. (1985). *Intellectual development: Birth to adulthood.* New York: Academic Press.

Chomsky, N. (1957). *Syntactic structures.* The Hague: Mouton.

Cole, L. C. (1954). The population consequences of life history phenomena. *Quarterly Review of Biology*, *29*(2), 103–137.

Darwin, C. (1960). *On the origin of species* (2nd ed.). New York: Modern Library. (Original work published 1859)

Ebert, J. D., & Sussex, I. M. (1972). *Interacting systems of development* (2nd ed.). New York: Holt, Rinehart & Winston.

Else, J. G., & Lee, P. C. (Eds.). (1986). *Primate ontogeny, cognition and social behavior.* New York: Cambridge University Press.

Faitkowski, K. R. (1986). A mechanism for the origin of the human brain: A hypothesis. *Current Anthropology*, *27*(3), 288–306.

Fancher, R. E. (1979). *Pioneers in psychology*. New York: Norton.

Fisher, R. A. (1940). *The genetical theory of natural selection*. New York: Dover.

Flanagan, O. J., Jr. (1984). *The science of the mind*. Cambridge, MA: M.I.T. Press.

Flavell, J. H. (1977). *Cognitive development*. Englewood Cliffs, NJ: Prentice-Hall.

Fobes, J. L., & King, J. E. (Eds.). (1986). *Primate behavior*. New York: Academic Press.

Freud, S. (1915). Instincts and their vicissitudes. In *The complete psychological works of Sigmund Freud* (Vol. 14, pp. 117–140). London: Hogarth.

Freud, S. (1954). Project for a scientific psychology. In S. Freud (Ed.), *The origins of psychoanalysis*. New York: Basic Books.

Furth, H. (1969). *Piaget and knowledge*. Englewood Cliffs, NJ: Prentice-Hall.

Gallagher, J. M., & Reid, D. K. (1981). *The learning theory of Piaget and Inhelder*. Monterey: Brooks/Cole.

Galton, F. (1962). *Hereditary genius* (reprint of 2nd ed.). New York: World Publishing. (Original work published 1892)

Gardner, H. (1987). *The mind's new science*. New York: Basic Books.

Gesell, A. (1945). *The embryology of behavior*. New Haven, CT: Greenwood Press.

Gould, J. L., & Marler, P. (1984). Ethology and the natural history of learning. In P. Marler & H. S. Terrace (Eds.), *The biology of learning* (pp. 47–74). New York: Springer-Verlag.

Gould, J., & Marler, P. (1987). Learning by instinct. *Scientific American*, *256*(1), 74–85.

Gould, S. J. (1977). *Ontogeny and phylogeny*. Cambridge, MA: Harvard University Press.

Gould, S. J. (1980) Natural selection and the human brain: Darwin vs. Wallace. In S. J. Gould (Ed.), *The panda's thumb* (pp. 47–57). New York: Norton.

Gould, S. J. (1981). *The mismeasure of man*. New York: Norton.

Griffin, D. R. (1976). *The question of animal awareness: Evolutionary continuity of mental experience*. New York: Rockefeller University Press.

Griffin, D. R. (Ed.). (1982). *Animal mind – human mind*. New York: Springer-Verlag.

Grossberg, S. (1980). How does a brain build a cognitive code? *Psychological Review*, *87*, 1–51.

Guilford, J. P. (1956). The structure of intellect. *Psychological Bulletin*, *63*, 267–293.

Hailman, J. P. (1982). Ontogeny: Toward a general theoretical framework for ethology. In P. Bateson & P. Klopfer (Eds.), *Perspectives in ethology. Vol. 5. Ontogeny* (pp. 133–189). New York: Plenum.

Hayes, C. (1951). *The ape in our house*. New York: Harper.

Hilgard, E. R., & Marquis, D. G. (1940). *Conditioning and learning*. New York: Appleton-Century-Crofts.

Hinde, R. A. (1973). Constraints on learning: An introduction to the problems. In R. Hinde & J. Stevenson-Hinde (Eds.), *Constraints on learning* (pp. 1–19). New York: Academic Press.

Hoage, R. J., & Goldman, L. (Eds.). (1986). *Animal intelligence*. Washington, DC: Smithsonian Institution Press.

Hulse, S. H., & Fowler, H. (Eds.). (1978). *Cognitive processes in animal behavior*. Hillsdale, NJ: Erlbaum.

Inhelder, B., & Piaget, J. (1958). *The growth of logical thinking from childhood to adolescence*. New York: Basic Books.

James, W. (1983). *The principles of psychology*. Cambridge, MA: Harvard University Press. (Original work published 1890)

Jensen, A. D. (1969). How much can we boost I.Q. and scholastic achievement? *Harvard Educational Review*, *33*, 1–123.

Kamil, A. C. (1984). Adaptation and cognition: Knowing what comes naturally. In H. Roitblat, T. Bever, & H. Terrace (Eds.), *Animal cognition* (pp. 533–544). Hillsdale, NJ: Erlbaum.

Karmilov-Smith, A., & Inhelder, B. (1975). If you want to get ahead, get a theory. *Cognition*, *3*: 195–212.

Kellogg, W., & Kellogg, L. (1933). *Development of ape and child.* New York: McGraw-Hill.
Klüver, H. (1933). *Behavior mechanisms in monkeys.* University of Chicago Press.
Köhler, W. (1927). *The mentality of apes.* New York: Random House.
Kohts, N. (1935). *Infant ape and human child.* Moscow: Scientific.
Langer, J. (1969). *Theories of development.* New York: Holt, Rinehart.
Lehrman, D. (1953). A critique of Konrad Lorenz's theory of instinctive behavior. *Quarterly Review of Biology, 28,* 337–363.
Lenneberg, E. (1967). *Biological foundations of language.* New York: Wiley.
Lorenz, K. (1965). *The evolution and modification of behavior.* University of Chicago Press.
Lorenz, K., & Tinbergen, N. (1957). Taxis and instinct. In C. Schiller (Ed.), *Instinctive behavior* (pp. 176–208). New York: International Universities Press.
Marler, P., & Terrace, H. (Eds.). (1984). *The biology of learning.* New York: Springer-Verlag.
Masterton, R. B., Hodos, W., & Jerison, H. (Eds.). (1976) *Evolution, brain, and behavior* (2 vols.). New York: Halsted.
Meador, D. M., Rumbaugh, D. M., Pate, J. L., & Bard, K. (1987). Learning, problem-solving, cognition, and intelligence. In G. Mitchell & J. Erwin (Eds.), *Comparative primate biology* (Vol. 2, Part B, pp. 17–83). New York: Alan R. Liss.
Miller, G., Galanter, E., & Pribram, K. (1960). *Plans and the structure of behavior.* New York: Holt, Rinehart & Winston.
Nunberg, G. (1984). *Linguistic authority.* Presented at the monthly colloquium of the Linguistics Department, University of California, Berkeley.
Oyama, S. (1985). *The ontogeny of information.* Cambridge University Press.
Parker, S. T. (1985). A social technological model for the evolution of language. *Current Anthropology, 26*(5), 617–639.
Parker, S. T., & Gibson, K. R. (1979). A developmental model for the evolution of language and intelligence. *Behavioral and Brain Sciences, 2,* 367–408.
Pavlov, I. (1961). Conditioned reflexes, an investigation of the physiological activity of the cerebral cortex. In T. Shipley (Ed.), *Classics in psychology* (pp. 756–797). New York: Philosophical Library. (Original work published 1904)
Piaget, J. (1966). *Psychology of intelligence.* New York: Littlefield, Adams & Company.
Piaget, J. (1970). *Structuralism.* New York: Harper Colophon.
Piaget, J. (1971). *Biology and knowledge.* University of Chicago Press.
Piaget, J. (1982). Development and learning. In R. Ripple & V. Rockcastle (Eds.), *Piaget rediscovered* (pp. 7–20). Ithaca, NY: Cornell School of Education Press.
Piaget, J. (1985). *The equilibration of structures.* University of Chicago Press.
Piaget, J., & Inhelder, B. (1969). *The psychology of the child.* New York: Basic Books.
Premack, D. (1976). *Intelligence in ape and man.* Hillsdale, NJ: Erlbaum.
Razran, G. (1965). Russian physiologist's psychology and American experimental psychology: A historical and a systematic collation and a look into the future. *Psychological Bulletin, 63,* 42–64.
Rescorla, R. A. (1988). "Classical conditioning" – it's not what you think it is. *American Psychologist, 43,* 151–160.
Roitblat, T., Bever, T., & Terrace, H. (Eds.) (1984). *Animal cognition: Proceedings of the theory Frank Guggenheim conference.* Hillsdale, NJ: Erlbaum.
Romanes, G. (1886). *Animal intelligence.* (4th ed.). London: Kegan Paul, Trench, & Co.
Rumelhart, D. E., McClelland, J. R., & the PDP Research Group. (1986). *Parallel distributed processing: Explorations in the microstructure of cognition. Vol. 1. Foundations. Vol. 2. Biological and psychological models.* Cambridge, MA: M.I.T. Press.
Schiller, P. (1952). Innate constituents of complex responses in primates. *Psychological Review, 59*(3), 177–191.
Schiller, P. (1957). Innate motor patterns as a basis of learning: Manipulation in a chimpanzee. In C. H. Schiller (Ed.), *Instinctive behavior* (pp. 264–287). New York: International Universities Press.
Schneirla, T. C. (1971). Behavioral development and comparative psychology. In C. L. Kutscher

(Ed.), *Readings in comparative studies of animal behavior* (pp. 21–45). Waltham: Xerox College Publishing.
Siegler, R. (1987). *Children's thinking*. Englewood Cliffs, NJ: Prentice-Hall.
Skinner, B. F. (1974). *About behaviorism*. New York: Random House.
Stearns, S. (1980). A new view of life history evolution. *Oikos*, *35*, 266–281.
Sternberg, R. J. (1982). Reasoning, problem solving, and intelligence. In R. Sternberg (Ed.), *Handbook of human intelligence* (pp. 227–307). Cambridge University Press.
Terrace, H. S. (1984). Animal cognition. In H. Roitblat, T. Bever, & H. Terrace (Eds.), *Animal cognition* (pp. 7–25). Hillsdale, NJ: Erlbaum.
Thomas, R. K. (1980). Evolution of intelligence: An approach to its assessment. *Brain, Behavior and Evolution*, *17*, 454–472.
Thomas, R. K. (1986). Vertebrate intelligence: a review of laboratory research. In R. J. Hoage & L. Goldman (Eds.), *Animal intelligence: Insights into the animal mind* (pp. 37–56). Washington, DC: Smithsonian Institution Press.
Thorndike, E. (1911). *Animal intelligence, experimental studies*. New York: Macmillan.
Tolman, E. C. (1951). Cognitive maps in rats and men. In E. C. Tolman (Ed.), *Collected papers in psychology* (pp. 241–264). Berkeley: University of California Press. (Original work published 1948)
Trevarthen, C. (1973). Behavioral embryology. In E. Carterette & M. Friedman (Eds.), *Handbook of perception* (Vol. 3, pp. 90–117). New York: Academic Press.
Vauclair, J. (1987). Representation et intentionalité dans la cognition animale. In M. Siguan (Ed.), *Comportement, cognition et conscience* (pp. 59–87). Presses Universitaires de France.
Waddington, C. H. (1962). *New patterns in genetics and development*. New York: Columbia University Press.
Walker, S. (1983). *Animal thought*. New York: Methuen.
Watson, J. B. (1967). *Behavior: An introduction to comparative psychology*. New York: Holt, Rinehart & Winston. (Original work published 1914)
*Webster's Third New International Dictionary*. (1981). Springfield, MA: Merriam-Webster.
Williams, G. (1966). *Adaptation and natural selection*. Princeton University Press.
Yerkes, R. (1916). *The mental life of monkeys and apes*. New York: Holt.
Yerkes, R. (Ed.). (1921). Psychological examining in the United States Army. *Memoirs of the National Academy of Sciences*, *15*, 890.
Zolman, J. F. (1982). Ontogeny of learning. In P. Bateson & P. Klopfer (Eds.), *Perspectives in ethology. Vol. 5. Ontogeny* (pp. 275–323). New York: Plenum.

# 3 New perspectives on instincts and intelligence: Brain size and the emergence of hierarchical mental constructional skills

*Kathleen Rita Gibson*

Two great religious traditions, Christian and Hindu, provide sharply dichotomous views of humanity. For the Christian, unbridgeable spiritual gaps separate humans from animals. Humans have souls; animals do not. For the Hindu, animal and human minds represent different points on a continuous progression toward Nirvana. These two philosophical perspectives permeate scientific as well as religious concepts of the human mind.

Twenty-five years ago, the dominant anthropological and psychological paradigms followed the Christian tradition in considering humans and animals to be separated by qualitative, seemingly unbridgeable, behavioral gaps: Humans were the only animals who could symbolize, make a tool, or recognize their own self-images. Within that theoretical climate, Ralph Holloway proposed his now classic theory of the evolution of the human brain. Brain size alone cannot account for the behavioral distinctions of the human species. Rather, the human brain has been reorganized, thereby permitting language and other mental skills (Holloway, 1966, 1968). Holloway's theory met with wide acclaim. It seemed intuitively correct.

Nearly simultaneously, Harry Jerison set forth a quite different theory, more in keeping with the Eastern tradition of a continuum between animal and human minds: Human intelligence reflects changes in brain size and in total information-processing capacity (Jerison, 1973). Unlike Holloway, Jerison received a cool reception. Those who considered human behavior qualitatively distinct from that of the apes also expected qualitative, rather than quantitative, differences in brain structure.

In the intervening years, our knowledge of primate behavior has expanded. All of the human–ape dichotomies so cherished by the anthropologists and psychologists of the early 1960s have fallen (Ettlinger, 1984; Gibson, 1988a). Those who hold steadfast to the belief that unbridgeable, qualitative behavioral gaps separate humans and animals find themselves in retreat, continually forced to pose increasingly less elegant definitions of humanity or of language.

Similarly, 25 years of valiant search for neural reorganization have proved distinctly unsatisfying. It seems that when pressed for evidence of neural reorganization, Holloway and his followers speak primarily in quantitative terms (Holloway, 1979; Holloway & Post, 1982). As the human brain has enlarged, certain portions, such as the neocortical association areas, have expanded more than others, and the ratio of neuronal connections to neuronal cell bodies has increased. Others note that expansion of the neocortical association areas and increased neuronal connectivity regularly accompany brain enlargement in higher primates (Jerison, 1979, 1982; Passingham, 1975). Thus, neural reorganization reflects increased brain size.

In this changing climate, some have begun to take Jerison's views more seriously. One recent reformulation proposes that the intellectual differences between human and ape reflect brain-size-mediated increases in a particular form of information-processing capacity. Specifically, expanded brain size results in the capacity to hold more perceptual, motor, and conceptual units in mind simultaneously. As a result, humans possess more discrete motor and sensory units than other animals and can combine and recombine these discrete units into more varied and more hierarchically constructed behavioral patterns (Gibson, 1988a, 1990a). Thus, humans possess more finely differentiated oral and manual sensorimotor patterns. They also construct tool-using, linguistic, and social schemes consisting of several hierarchical levels and possessing constructional mobility (i.e., the ability to combine and recombine elements in a varied manner) at each level in the hierarchy. Apes, for instance, can stack blocks and use hammers and probes (Beck, 1980; Boesch & Boesch, 1981, 1983, 1984; Goodall, 1986; McGrew, 1974), but only humans combine hammering, piercing, and stacking actions into goal-directed hierarchical actions such as nailing two blocks of wood together (Gibson, 1990a).[1] Similarly, apes use single gestural symbols and short two- and three-gesture combinations (Fouts, 1973; Gardner & Gardner, 1969, 1980; Miles, 1983; Terrace, 1979, 1980). They also possess rudimentary grammar (Greenfield & Savage-Rumbaugh, P&G20). Humans, however, combine multiple gestures and words into highly varied, hierarchically organized sentences containing embedded clauses and phrases. Apes and monkeys engage in social and political interactions involving up to three animals (De Waal, 1983; Kummer, 1967), but humans construct hierarchical political systems.

Among nonhuman vertebrates, the range of variation in brain size is vast, ranging from less than a gram to several pounds (Jerison, 1973). Consequently, if human intelligence results from brain-size-mediated expansions of mental constructional skills, then animal species should vary among themselves in intellectual levels (cf. Macphail, 1987). Some evidence suggests that this is the case. Piagetian tasks, for instance, are inherently constructional in nature (Case, 1985). Primates exhibit varying levels of performance on these tasks, with the large-brained cebus monkeys and great apes outperforming other groups (Chevalier-Skolnikoff, 1989; Gibson, 1986; Parker & Gibson, 1977, 1979).

The concept that intelligence varies among animal species is, of course, not new, as evidenced by classic dichotomies between instinct and learning. According to classic theories, instincts are inherited behavioral patterns of stereotyped form, generally consisting of two major components: species-typical fixed action patterns and innate releasing mechanisms. Described in this manner, instincts appear qualitatively different from intelligence.

We now know, however, that instincts are not so stereotyped as once believed. Even seemingly simple behaviors, such as the wiping reflex of frogs, are highly variable, constructed acts (Berkinblit, Feldman, & Fukson, 1986). For this reason, Barlow (1977) suggests replacing the term *fixed action pattern* with *modal action pattern*.

Similarly, decades of deprivation experiments have demonstrated that species-typical "instinctive" behaviors require suitable conditions for their development, and behavioral geneticists have long since documented the inherited nature of intelligence. Thus, the issue of instincts versus learning as determinants of behavior should be dead. In fact, the controversy should have died in the late 1960s following three major publications: an article, "How an Instinct Is Learned" (Hailman, 1969), and two textbooks (Hinde, 1970; Marler & Hamilton, 1966) demonstrating the interactions of inherited and environmental factors in the development of behavior.

Nonetheless, the instinct concept lives. Scientists and laypeople alike experience continuing difficulty describing the behaviors not only of many lower vertebrates but also of primate and human infants without resorting to instincts or allied concepts (e.g., Parker & Poti', P&G8). Some textbooks on animal behavior still unabashedly use the term *instinct* (Alcock, 1975). Chomsky (1980) continues to maintain that language is innate, and Gould and Marler have recently published the paper "Learning by Instinct" (1987).

Perhaps, despite multiple declarations of its death, the instinct concept lives because the concept is only partially flawed. Scientists may err in drawing sharp qualitative dichotomies between genes and learning and between stereotypy versus variability of behavior, just as they have erred in sharply dichotomizing human behavior and animal behavior. Rather, instinct and intelligence may represent two ends of a continuum of mental constructional abilities. This chapter explores this concept and suggests that instincts and intelligence differ in the degree of mental constructional ability involved, particularly in the degree of hierarchicalization of behavior and in the degree of neural combination and recombination occurring within each hierarchical layer.

Heuristic neural models based on modern parallel distributive processing theories of brain function have been presented (McClelland & Rumelhart, 1986; Rumelhart & McClelland, 1986). These models suggest that brain size (or the size of individual neural processing units) and neural interconnectivity are the critical variables determining whether a behavior appears instinctive or intelligent. Specifically, increasing numbers of neurons and increasing numbers of connections per neuron automatically result in increased differentiation of sensory and motor units and hence in increases in the numbers of discrete sensory and motor behaviors possessed. Similarly, duplication of neural processing areas automatically produces an increased ability to process varied information in parallel (i.e., in synchrony) (Hinton, McClelland, & Rumelhart, 1986) and hence an increased mental constructional capacity of the sort described by Case (1985), Gibson (1983, 1988a), and Langer (1986).

Ontogenetically, the infantile behaviors of humans and other animals derive from small, subcortical neural processing structures. Hence, they appear instinctive. With neocortical maturation, larger-brained species develop the capacity to combine and recombine simpler behavioral units in a hierarchical fashion. Thus, their behavior appears more intelligent and more varied.

## Neural processing models of visual perception

This section considers hypothetical models of the potential effects of variations in quantitative neural parameters on visual perception. The postulated constructional levels are the author's own. To keep the discussion relatively brief and simple, as well as to emphasize quantitative neural variations of potential relevance to species differences in intelligence, the models are highly simplified. Complexities of neural function or general aspects of function common to all brains are ignored.

### *No constructional networks – feature detectors are also object detectors*

Many individual neurons within the vertebrate visual nervous system function as feature detectors. Feature detectors respond preferentially to single items, such as color, line orientation, or movement direction (Hubel & Wiesel, 1968). Variations in the numbers of feature detectors and in the degree of interconnectivity among detectors can dramatically affect perceptual outcome.

Integration of the outputs of varied feature detectors requires direct or indirect connections between them. In other words, it requires a neuronal network. A system lacking interconnectivity would also lack the capacity to construct diverse object images from separate features. Rather, feature detectors would function as object detectors. Although no vertebrate nervous system is composed entirely of object detector neurons, some species do possess feature detectors that double as object detectors. Thus, the retina of the leopard frog contains neurons preferentially responsive to small, moving dark images with convex edges – in other words, to moving bugs within the frog's visual field (Maturana, Lettvin, McCulloch, & Pitts, 1960). Frogs, rabbits, and possibly many other vertebrates possess cells that respond to a moving edge that suddenly stops. In Alcock's view, these cells function as predator detectors. Similarly, frogs possess cells that may function as sky detectors, as they are preferentially responsive to the color blue (Alcock, 1975).

Object detector neurons may confer significant advantages to species that live in predictable environments and need to distinguish limited numbers of regularly recurring objects. Not only do such neurons respond to salient objects, but they may actually bias the nervous system to preferentially detect them. A system composed entirely of object detector neurons, however, would have serious limitations. An animal possessing such a system would perceive only objects for which it possessed feature detectors. An expanded brain size in terms of increased numbers of object detector neurons would provide the ability to detect a greater number of objects. Nonetheless, such a system could not provide for all perceptual eventualities. Even with its modified object detector system, for instance, the frog lacks the ability to perceive stationary objects and may actually starve to death in the presence of edible, but stationary, food (Sarnat & Netsky, 1981).

### *Unilevel constructional network – constructed object images based on key stimuli*

A unilevel constructional network would provide constructed object images, thereby permitting the detection of greater numbers of objects than can be detected by an equal number of object detector neurons. The numbers of feature detectors and the degree of interconnectivity determine the actual number of detectable objects.

Envision, for instance, a small network consisting of 14 feature detectors, grouped together into six categories: color, shape, movement, size, distance, and vertical versus horizontal orientation. Assume that the system contains feature detectors for three colors (red, dark, light), two shapes (round, long), two conditions (moving, stationary), three sizes (large, medium, small), two distances (within reach, out of reach), and two positions (vertical, horizontal). Neural connections exist between neurons of different categories, but not between neurons of the same category. This system is considered a unilevel constructional network because combination and recombination of neuronal output can occur at only one level – between categories. No construction is possible within feature categories.

This network distinguishes as a separate object any structure that contains a unique combination of features. As such, it can potentially distinguish 154 objects.[2] In contrast, a simple feature detector–object detector system composed of 14 neurons could distinguish only 14 objects. Even greater versatility and precision of object recognition could be added to such an object constructor network by increasing the numbers of distinct feature detectors in each category. Thus, a system containing 60 interconnected neurons including 10 colors, 10 shapes, 10 degrees of movement, 10 sizes, 10 distances, and 10 positions could potentially detect up to 1 million objects, as opposed to 60 potentially distinguishable objects in a feature detector–object detector system.

Thus, neuronal networks and hence neuronal interconnectivity provide major benefits in terms of increased perceptual capabilities by permitting the integration of varied features into constructed object images. Still, a system in which interconnections and constructional mobility exist only at one level – between major feature categories – possesses distinct limitations. For one, the range of discrimination within each category remains slim. The 14-neuron system described earlier provides the ability to distinguish only three colors and two orientations; the 10-neuron system provides the ability to distinguish only 10 colors. Thus, such unilevel networks can respond to key stimuli only. They cannot distinguish objects possessing finely graded perceptual features. Some items recognizable by the 14-neuron network and items with which they might be confused are shown in Table 3.1.

Many of the innate releasing mechanisms described by ethologists share properties of this unilevel object constructor network. The neonatal laughing gull, for instance, will peck preferentially at the maternal bill, but it can be fooled by a cardboard model pointing downward and painted red (Tinbergen, 1960). Similarly, adult male stickleback fish attack other fish with red bellies, but they can be fooled by cardboard models painted red (Tinbergen, 1952). Human infants smile at a human face, but they can be fooled by a cardboard model containing two dark spots in the position of the eyes (Kagan, 1970).

### *Bilevel constructional networks – hierarchical construction of precise and highly varied object images*

Distinguishing a wide range of potential objects based on fine gradations of individual features requires a large, bilevel constructional system containing separate neuronal networks for each feature category (color, brightness, size, shape, movement, position, orientation), as well as an object constructor network capable of constructing images based on the combined output of separate feature constructor networks. Parallel or synchronous processing among the feature constructor networks will provide the essential speed needed for efficient and timely behavior (Feldman & Ballard, 1982). This system is termed a bilevel constructional network because neuronal interconnections and interactions occur at both the feature constructor and the object constructor levels.

The potential implications of neuronal numbers and neuronal interconnectivity for the precise discrimination of individual features have previously been discussed with respect to color (Gibson, 1988b). Envision, for instance, a system of color feature detectors. Each color detector connects with each of the others. Whenever one feature detector or any combination of feature detectors fires, a specific color is perceived. Such a system can perceive $2^n - 1$ colors.[3] Thus, a system of 5 feature detectors could perceive 31 colors; a system of 10 feature detectors, 1,023 colors. This contrasts with the 5 or 10 colors perceptible by individual color detector neurons working in isolation. Thus, increased neuronal numbers and increased neuronal interconnectivity permit increased perceptual discrimination, but increased neuronal connectivity has by far the greatest effect. See McClelland (1986) for more exten-

Table 3.1. *Object recognition in the 14-neuron network*

| Object features | Potential corresponding objects |
|---|---|
| Round, red, small, stationary | Cherry, cranberry, strawberry, marble |
| Round, red, medium size, stationary | Apple or ball |
| Long, red, pointed down, moving | Herring gull's bill, mammalian tongue |
| Long, red, moving, horizontal | Belly of male stickleback fish, toy fire engine |
| Dark, round, moving | Bug, gravel, or dirt blowing in the wind |
| Dark, large, oblong or round, small or big, stationary | Hole or stone |
| Dark, long, moving | Snake or worm |

sive discussion of the roles of neuronal numbers and connectivity in feature discrimination.

A bilevel feature constructor/object constructor network could distinguish virtually the entire range of visually perceptible objects provided the range and numbers of feature detectors are great enough. In fact, human perceptual behavior evidences constructional mechanisms of this nature (Bower, 1974). Thus, bilevel constructional networks may exist within both visual and other sensory systems.

Despite their great versatility, bilevel constructional systems lack one major property of highly intelligent systems: As long as a neural system contains only one object constructor network, then only one object can be perceived at a time, because such a network can possess only one firing state at a time. In other words, a single-object constructor network could perceive either a chimpanzee or an onion, but it could not perceive a chimpanzee eating an onion (Hinton et al., 1986).

### *Multilevel constructional networks – hierarchical constructions of relationships between objects*

In order to perceive two or more objects simultaneously, two or more object constructor neuronal networks must be present and functioning in parallel (Hinton et al., 1986). In addition, separate object constructor networks must interconnect, directly or indirectly, in order to construct higher-order concepts of object–object relationships. Thus, the perception of object–object relationships requires a multilevel constructional system containing multiple interconnected object constructor networks. Presumably, the greater the numbers of such networks, the larger the numbers of objects that can be perceived (or conceived) simultaneously, and the greater the hierarchical or multilevel processing capacity. The greater the degree of interconnectivity between object constructor networks, the greater the constructional mobility (ability to combine and recombine object images) at higher levels in the hierarchy.

*Summary of visual models*

In summary, brain size potentially impacts upon perceptual systems in a manner highly consistent with the behavioral distinctions actually observed. Small brains with few or poorly interconnected feature detectors will recognize objects based on key stimuli only. Large brains possessing greater numbers of individual feature detectors and more feature constructor and object constructor neuronal networks will provide the capacities for fine perceptual discrimination and for the construction of object images; very large brains, for the construction of object–object relationships.

## Neural processing models of motor function

### *No constructional network – coordination of movement by single command neurons*

Any purposive action requires the simultaneous firing of many motor neurons. Interneurons coordinate these firings. At the simplest level, single interneurons termed *command neurons* may function to coordinate entire actions (Wiersma & Ikeda, 1978). By definition, "such 'command fibers' can thus release the whole coordinated pattern without assistance from sensory feedback" (Wiersma, 1978). Though the concept of command neurons has occasioned much debate (Kupfermann & Weiss, 1978), some neurons do seem to meet these criteria, most notably the Mauthner cell, which mediates propulsive escape movements in fish and amphibians (Sarnat & Netsky, 1981).

Command neurons effectively engage stereotyped species-typical motor patterns. Nonetheless, they possess inherent disadvantages. For one, "these command fibers are not able to vary their command to suit the changing conditions in the periphery, they merely have the ability to fire or not to fire" (Wiersma, 1978). Even very simple adjustments to changes in spatial orientation of movements are beyond their capabilities. For another, command neurons provide little motor flexibility. Expansions in the numbers of command neurons can produce an expanded motor repertoire, but by themselves command neurons provide no ability for complex motor combinatory and recombinatory actions. Combinatory ability requires functional connections between command systems.

### *Unilevel constructional networks – aiming and other environmental accommodations*

Just as object constructor networks provide more versatility than do separate object detector neurons, so constructional (command) networks provide more motor versatility than do individual command neurons. For instance, even very simple motor acts, such as the tongue flicking responses of frogs and many lizards, must be aimed. Aiming requires that an animal take into account both the position of its own body and that of the stimulus (Berkinblit et al., 1986). Recent research indicates that aiming trajectories reflect the vectorial product of numerous interacting neurons (Georgopoulos, 1988). Similarly, many other seemingly stereotyped motor actions possess the ability to accommodate to varying environmental conditions (Hinde, 1970). Thus, they more likely reflect the output of interacting neuronal networks than the output of single command neurons.

### *Hierarchically organized unilevel constructional networks – modal action patterns*

Some motor actions involving only a few muscle groups may require only one system of interacting neurons (e.g., only one command neuronal net). Most modal action patterns, however, are more complex. They involve the sequential performance of several motor actions, each constructed by its own neuronal network.

Thus, most modal action patterns are inherently hierarchical in nature. Separate motor patterns on one level must be organized together by interneurons or neuronal linkages at a higher level. The basic nature of modal action patterns, however, suggests that the second level of organization involves minimal constructional mobility, because subcomponents of modal action patterns usually are performed in the same sequence (Alcock, 1975; Hinde, 1970). Even such variability as may exist at higher levels consists of components of the action being dropped under certain environmental circumstances, rather than of subcomponents being combined and recombined in many diverse ways.

Frogs, for instance, possess a wiping reflex (Berkinblit et al., 1986). If a foreign stimulus is applied to the skin of a frog's forearm, the frog will wipe it off with the hind limb. Frogs can do this no matter the position of the foreign substance and no matter the position of the hind limb at the time the substance is applied. This wiping reflex is composed of five subcomponents (flexion of the leg, placement, aiming, wiping, and extension of the leg). Each of these actions involves its own motor neuronal net. Some components of this reflex may be dropped if the frog is jumping at the time the wiping reflex occurs. The actions, however, do not occur in alternate sequences, such as first extending the leg, then placing it.

Similarly, the human neonate possesses a modal action pattern – the neonatal visual reach and grasp. When properly supported and presented with a visual stimulus of the appropriate size and position, the infant will reach and grasp (Bower, 1974). The grasping action always consists of all fingers grasping in unison. As reaching and grasping actions are controlled by separate neuronal networks (Alstermark, Lundberg, Norrsell, & Sybirska, 1981), this neonatal pattern requires that separate actions be coordinated together at a second (hierarchical) level. No constructional mobility exists at this second level, however, because the infant always performs the movement in the same sequence, first reaching and then grasping, with all of the fingers acting in unison.

*Multilevel constructional networks – intelligent hierarchical construction of motor acts*

Multilevel constructional networks provide greater motor flexibility. For instance, unlike the human baby, whose hand actions are basically limited to grasping with all fingers acting in synchrony, the human adult possesses a large repertoire of discrete hand movements. The adult can move each finger separately and possesses special precision and power grips as well as pointing actions. Adults can use these actions in varied sequential combinations. Thus, they can first reach and then grasp with precision or power grips or first reach and then touch or point. Alternately, humans can first grasp or point and then reach. Each of these actions can occur in synchrony with actions of the hand, the foot, or the mouth.

In other words, the adult human must possess many separate neuronal command networks for individual movements of the digits and hands, as well as higher-level networks that permit these separate units to be flexibly combined and recombined with each other. Thus, constructional mobility occurs at several levels in the motor hierarchy. Many primates have similar skills.

Logically, sequential actions could reflect successive actions of the neuronal nets coding for each movement. Neural transmission, however, is so slow that sequential processing would result in slow, choppy, inefficient movements (McClelland & Rumelhart, 1986; Rumelhart & McClelland, 1986). Actual analyses of complex actions suggest that they are processed in parallel. In other words, groups of actions are planned in advance and activated simultaneously (Hinton, 1984; McClelland, Rumelhart, & Hinton, 1986; Rumelhart & Norman, 1982). For instance, skilled typists activate motor units for whole sequences of letters at once. This permits each action in the sequence to vary according to the actions that precede and follow it. Thus, when about to type the letter *e*, one will find one's hand in different positions depending on the letters preceding and following the *e*. Similar advanced planning may well underlie speech, dance, and other skilled movements.

Parallel processing of motor acts is subject to the same neural conditions as other parallel processing activities. For one, parallel processing requires the duplication of motor areas. In fact, higher primates and perhaps other mammals possess duplicate cortical motor areas (Kaas, 1987). Presumably, the larger the numbers of motor processing areas, the greater the parallel processing abilities and the larger the number of motor units that can be activated at once, in other words, the greater the degree of advanced planning of motor actions. Also, the greater the degree of interconnections between motor processing areas, the greater the ability to combine and recombine motor actions.

*Summary of motor models*

Thus, brain size does influence the motor system. Small-brained species with limited numbers of neurons and limited neuronal interconnectivity will have limited motor repertoires. Large-brained species with many highly interconnected motor neuronal networks will possess large repertoires of discrete movements. In addition, increased parallel processing ability will provide them with more hierarchically organized motor behaviors and more constructional mobility at each level in the hierarchy. Hence, large-brained species will construct longer and more varied sequential actions.

## Combining stimulus and response: Automatic linkages or volition

To observers, instinctive behaviors are automatic or semiautomatic responses to specific stimuli. In contrast, intelligent behaviors appear volitional. That is, large-brained animals vary their responses in accord with so many internal and external states that observers have difficulty predicting them.

Distinctions between automatic and voluntary responses must relate the patterns of input to motor units. At one extreme, for instance, a feature detector or object detector neuron could link directly to one motor command unit. The motor unit, in turn, would have no other inputs. Thus, each firing of the object detector neuron would elicit the same motor response. The behavior would appear instinctive or reflexive.

At another level, a "pattern associator" network (McClelland et al., 1986) could determine stimulus–response relationships. In pattern associator systems, one neuronal network links directly to another. The firing of specific combinations of neurons in the first network automatically results in the firing of specific combinations of neurons in the second. Behaviors mediated by such systems would also appear instinctive.

At a third level of complexity, numerous object or pattern detector systems could connect with a motor decision network. Thus, visual tactile, auditory, olfactory, visceral, and emotional impulses would all converge on a single motor decision center. This would permit information from varied perceptual and motivational systems to interact. Responses to individual stimuli would vary in accordance with numerous additional factors. The resulting behaviors would appear volitional.

Thus, both brain size and brain organization impact on the degree of automaticity versus volitional control of stimulus–response relationships. The larger the system, the greater the range of perceptual/conceptual constructs and the greater the range of potential responses. The greater the number of perceptual, conceptual, and motivational units determining responses, the more likely the resulting behavior will appear volitional.

## Neural models versus neural reality

The foregoing models suggest that brain size and connectivity patterns have profound influences on behavior. Small, poorly connected perceptual systems permit object recognition by means of key stimuli or stereotyped object detector neurons. Larger systems permit constructed object images and constructed images of object–object relationships. Small motor systems permit hierarchically organized motor behaviors exhibiting little variability in actual motor sequences, that is, modal action patterns. Large motor systems permit hierarchical organization of long but highly varied motor sequences. Similarly, small, minimally connected neural systems result in automatic linkages between stimulus and response; large, maximally interconnected systems permit volitional behaviors. In other words, the smaller the neuronal system controlling a particular behavior, the more likely the behavior is to appear instinctive; the larger the system, the more likely behavior will appear intelligent.

Like all models, these are oversimplifications. In an effort to highlight salient aspects of hierarchical processing skills, other critical neural processing mechanisms have been ignored. Thus, actual brain function requires many more neurons (McClelland, 1986) and additional hierarchical processing levels. Nonetheless, neural variations potentially capable of producing divergent mental constructional skills do exist both within and between brains.

The human brain, for instance, contains both neocortical and subcortical sensory and motor processing centers. The necortex is extremely large, overshadowing all subcortical regions combined. Thus, it possesses the quantitative parameters essential for complex parallel processing and hierarchical constructional functions (Mountcastle, 1978). In fact, the neocortex contains varied processing regions. Some, such as the primary senory areas, have all of the functional characteristics expected of complex feature constructor networks. Others, such as the secondary sensory areas, may function as object constructor networks. Still others, the association areas, contain duplicate sensory and motor regions and thus permit complex hierarchical constructional capacities.[4] In contrast, small subcortical sensory and motor nuclei are structurally more suited for perceiving key stimuli and for executing modal action patterns.

The human brain also contains processing areas capable of providing relatively automatic stimulus–response linkages as well as complexly determined volitional behaviors. The structure and function of the optic tectum (or superior colliculus), for instance, suggest that it functions as a pattern associator mediating automatic, but precise, spatial relationships between visual and somatosensory perceptions (Gaither & Stein, 1979; Trevarthen, 1968). In other animals, for instance, the tectum mediates reflexlike functions such as the tongue flicking responses of frogs, the lunging and grasping reactions of lizards, and the tendency of frogs and other animals to dash for cover or cower in response to looming objects (Casagrande, Harting, Hall, Diamond, & Martin, 1972; Ingle, 1970, 1972; Sprague & Meickle, 1965).

In contrast, the basal ganglia and amygdala receive varied visual, somatosensory, auditory, olfactory, and visceral inputs. Thus, they function in the choice of actions based on multiple inputs and complex decision-making processes (Passingham, 1987).

**Mental constructional skills and animal behavior**

Brain sizes differ markedly among vertebrate species. In general, the brains of fish and amphibians are very small; those of reptiles, small; and those of birds and mammals, large. Even within the mammalian class, however, brain size varies from less than a gram to more than 5,000 g. The human brain, at 1,300 g, is extremely large, but is nonetheless exceeded in size by the brains of elephants, porpoises, and white and blue whales (Jerison, 1973; Sarnat & Netsky, 1981).

Brain size also varies allometrically with body size, and for reasons that remain poorly understood, intelligence may correlate more closely with relative brain size than with absolute brain size. Methods that adjust for allometric brain–body size relationships restore humans to their preeminent position as the "brainiest" species. Allometric adjustments also yield other relative brain size differentials of potential behavioral significance. In general, cartilaginous fish have larger relative brain sizes than bony fish; varanid lizards have larger brain sizes than other lizards; and carnivores, elephants, primates, and dolphins have larger brain sizes than other mammals (Jerison, 1973; Sarnat & Netsky, 1981).

The major vertebrate classes also manifest differential enlargement of specific portions of the brain (Sarnat & Netsky, 1981). Among fish and amphibians, the optic tectum is proportionately the largest neural processing center. Reptiles and birds exhibit enlarged basal ganglia; mammals have enlarged basal ganglia, limbic systems, and neocortices. Mammals also differ from each other in the sizes of specific neocortical sensory, motor, and association areas and in the degrees of duplication of sensory and motor processing areas.

If the mental constructional model presented here is correct, differences in size of the brain and of its component parts will produce observable behavioral effects. The smaller the brain, the more likely the animal will be to display minimally constructed behaviors (instincts). The larger the brain, the more likely it is that the animal's behavior will appear highly intelligent. Thus, small-brained species will possess limited motor repertoires organized into modal action patterns, and they will tend to identify food items or mates on the basis of key stimuli only. Large-brained species will have large motor repertoires and will construct complex, variable, goal directed motor sequences. They will recognize foods, mates, and other objects by means of object images constructed from varied perceptual features. Large-brained species will also exhibit greater ability to accommodate motor actions to environmental stimuli, and their behavior will appear more volitional. The greatest mental constructional skills will be found among those animals with the largest neocortices and neocortical association areas.

Unfortunately, little research has been directed specifically to these questions of mental constructional ability. Much can be learned, however, from descriptions of natural behavior. In analyzing zoological and ethological descriptions, it is helpful to ask the following questions:

1. How many discrete movements of the oral, hand, foot, and tail regions does the species possess? How many such actions can the species organize into one goal directed behavior? How variable are the goal directed sequences manifested by the species?

2. Do species members swallow food whole? Do they practice external fertilization? Alternately, can they accommodate motor responses to sensory information sufficiently well to manipulate prey, sexual partners, or other objects? Thus, can they break food into small pieces prior to ingestion? Can they engage in internal fertilization and other direct contact sexual and social behaviors? If so, what manipulators do they use for these tasks?

3. Do species members identify food, mates, competitors, or predators on the basis of a few key stimuli, or do they construct actual object images? Do they, for instance, always feed on foods that share certain perceptual properties, such as a particular size, movement pattern, or odor? Alternately, do they feed on foods possessing varied properties? Can they recognize a food item whether it is stationary or moving? Can they recognize individual foods by their varied properties, such as olfactory, visual, touch, and so forth? Can they recognize competitors or mates as individuals, or do they respond to species members simply as members of specific age–sex classes based on a few key stimuli, such as color or odor?

4. Do species members simply sit and wait for food items to move within view, or do they actively search for food? If they search for food, does their search appear random, or do they show signs of constructed cognitive maps of the environment? Such mapping requires the storage and organization of large amounts of environmentally relevant information.

5. Do species members mentally construct images of relationships between two or more objects? Can they, for instance, find food that is hidden under, behind, or inside another object? Can they use tools? If so, how many different tool-using schema do they coordinate together to meet a single end?

6. Do species members mentally construct and predict the actions of living prey or of other species members? Do they, for instance, actively pursue, chase, or herd prey, as opposed to sitting and waiting for it to appear? Do they engage in social relationships involving three or more individuals? Do they intentionally deceive others?

In general, these predictions coincide with spontaneous animal behaviors in the wild. Small-brained fish and amphibians manifest one extreme of these behavioral patterns; large-brained mammals, the other. Thus, most small-brained fish and amphibians recognize both food and conspecifics by key stimuli. Many of them, for instance, identify food primarily by movement and size. They either sit and wait for it to appear or randomly cruise the environment in search of it, showing no signs of constructed cognitive maps (Hamilton, 1973; Regal, 1978). Most fish and amphibians also possess limited motor repertoires. In particular, they lack manipulative skills. Thus, they must swallow food whole and practice external fertilization. In keeping with the relative predominance of the optic tectum, most fish and amphibians capture prey by simple, possibly reflexive, lunging and grasping actions.

In contrast, the larger-brained sharks and rays do possess sufficient manipulative skills to manipulate foods and practice internal fertilization. Sharks, for instance, can break prey into smaller pieces prior to ingestion, and rays can crack mollusk shells and engage in some manipulative actions of the tail and wings (Doubilet, 1988).

In conjunction with their larger relative brain size (Jerison, 1973; Sarnat & Netsky, 1981), reptiles display a broader behavioral repertoire than do fish or amphibians, particularly with respect to the sensory and motor processes involved in social behavior. Thus, most reptiles possess motorically complex aggressive and sexual displays (Greenberg, 1978), and they practice internal fertilization. The largest-brained reptiles also exhibit more complex feeding practices than do most fish. Thus, the monitor lizards and the Komodo dragon have sufficient manipulative and perceptual skills to dismember prey and may construct rudimentary cognitive maps of the environment (Auffenberg, 1978; Regal, 1978).

In comparison with reptiles, birds possess relatively large brains for their body sizes. Similarly, avian behavior is complex. Avian perceptual skills are sufficient to permit the recognition of both stationary and cryptic prey. Some birds find hidden foods, such as worms or embedded insects. This suggests that they construct at least rudimentary images of object–object relationships. Many possess sufficient manipulative skill to crack nuts, to manipulate tiny seeds, or even to use tools to extract embedded insects or crack eggshells (Beck, 1980). They also construct complex nests and engage in elaborate courtship and parental routines.

In general, however, ethological descriptions of avian behaviors suggest that avian motor behavior is dominated by modal action patterns, rather than by hierarchically varied motor constructs (Alcock, 1975; Collias, 1964; Hinde, 1970; Lorenz, 1971; Tinbergen, 1960). In addition, even seemingly highly constructed behaviors such as tool use occur in limited contexts (Parker & Gibson, 1977). These factors may account for the smaller relative brain sizes of birds in comparison with mammals.

The largest neocortices are found among members of the mammalian class. Mammalian neocortical enlargement correlates with major advances in feeding strategies (Gibson, 1986). Nearly all mammals manifest advanced manipulative capacity in comparison with reptiles. Mammals chew food and may use other manipulators such as prehensile tongues or lips, hands, feet, tails, or trunks to manipulate it. Many find hidden or cryptic prey by diverse sensory means, such as touch, vision, or hearing. Many extract foods from other matrices. Some possess complex hunting abilities that involve actual stalking or chasing of prey. Some may even herd prey. In addition, all mammals care for their young, and many mammalian groups possess complex social structures.

The largest-brained mammals tend to fall into one of two groups. The first consists of dolphins, sea lions, otters, and some carnivores that actively pursue prey (Jerison, 1973). Dolphins, wolves, and hunting dogs also hunt cooperatively and may herd prey. The pursuit and, most especially, the cooperative hunting of prey place heavy demands on information-processing capacity and mental constructional skills. To pursue prey, an animal must precisely discriminate prey movements and construct mental maps of future prey movements. Cooperative hunting requires even greater information-processing capacity, because animals must predict future actions not only of prey but also of conspecifics. In this sense, cooperative hunting may place similar demands on information-processing systems as do complex social relationships involving three or more animals. The demands of these latter systems have been elaborated by Harcourt (1988).

The mental constructional skills of dolphins are likely to be especially strong (Gibson, 1988b). Dolphins construct object images by means of echolocation. Thus, they construct images of prey and conspecifics from many separate sounds and must also construct "sound images" of future positions of prey and conspecifics.

The second group of mammals to exhibit enlarged brains comprises omnivorous or frugivorous extractive foragers (Gibson, 1986). These species, which include many monkeys, great apes, frugivorous bats, and perhaps elephants,[5] construct seemingly complex cognitive maps of the environment in order to locate widely dispersed foods (Clutton-Brock & Harvey, 1980; Eisenberg & Wilson, 1978; Milton, 1988). In order to extract foods from other matrices in which they are encased, they must also recognize relationships between objects. Thus, they must recognize that one object may lie inside another object. In addition, they must possess the manipulative skills to crack shells, peel fruits, unravel leaves, and so forth. The more omnivorous the extractor, the greater its required range of perceptual and manipulative skills. Some nonhuman primate extractive foragers with extremely large brains for their body sizes, notably cebus and chimpanzees, possess constructional and manipulative abilities sufficient to permit proficient tool use in varied contexts to meet varied ends (Parker & Gibson, 1977).

Primate extractive foragers derive their skills from two aspects of neocortical structure. For one, the sensory and motor cortices controlling the primate hand are particularly enlarged, as is the visual cortex (Clark, 1960). For another, the largest-brained primates exhibit differential enlargement of the neocortical association areas (Passingham, 1973, 1975). Such enlargement reflects duplication or multiplication of sensory and motor processing areas (Kaas, 1987). According to the analyses presented here and elsewhere, the greater the duplication of sensory and motor areas, the more the parallel processing ability, and in particular the greater the ability to construct relationships between multiple objects and multiple actions (Hinton, 1984). Thus, differential enlargement of neocortical association areas provides precisely the skills needed for constructing complex object relationships and complex manipulative schemes.

The human brain represents the culmination of the primate trend toward enlarged neocortical association areas (Passingham, 1973, 1987). Some other neural structures are also enlarged, particularly Broca's area, the pulvinar, and certain limbic nuclei (Armstrong, 1982, 1986). No unique structures, however, are known for the human brain.

The major expansion of the human neocortical association areas suggests that humans possess extensive duplication or multiplication of sensory and motor processing units (Kaas, 1987) and hence extensive parallel processing and hierarchical constructional capacities. Certainly, humans possess extensive constructional capacities. Humans are, for instance, omnivorous extractive foragers par excellence (Gibson, 1983; Gibson & Parker, 1986; Parker & Gibson, 1979). Thus, many human dietary staples are foods that must be extracted from the ground or from other matrices, such as tubers, nuts, shellfish, grains, or meat encased in thick hides. Humans gain access to these foods via complex cognitive mapping and tool-using strategies. Like dolphins and some other carnivores, humans also pursue game and hunt cooperatively. Human social skills are such that they construct complex social hierarchies.

Human mental constructional skills are perhaps the most obvious within the realms of tool use and language. Other papers have detailed human–ape differences in technological and linguistic constructional skills (Gibson, 1983, 1985, 1988a, 1990a; Reynolds, 1981, 1982, 1983). To briefly review, humans construct precise, highly differentiated and highly variable oral and manual motor acts. Human tool use exhibits constructional advances over those of the apes in other realms as well. Humans make tools composed of varied subcomponents (e.g., spearhead and shaft, bow and arrow), as well as those that require advanced visualization of complex three-dimensional shapes and of the sequential steps needed to produce these shapes (Wynn, 1979, 1981). Similarly, they use several tools simultaneously or in succession to meet a single end.

Apes possess only the rudiments of these skills. They use several types of tools, but they do not use them together as part of one tool-using sequence. They make tools by modifying the structure of existing branches, but they do not construct tools of varied components or complex shapes. Examination of fossil hominid brains and tool-making techniques indicates that as the hominid brain expanded and reached its present form, tools increasingly manifested evidence of these mental constructional skills (Gibson, 1986; cf. Wynn, 1988). Similarly, apes possess the rudiments of symbolic language, but human language is more highly constructed at all levels. Human speech demands a wide motor repertoire of discrete oral movements and the ability to rapidly combine and recombine specific movements into highly complex vocal sequences. Humans construct words from separate phonemes, phrases and clauses from separate words, and hierarchical sentences from separate clauses and phrases. Also, unlike apes, humans construct stories or recount events using constructed oral or written paragraphs.

## The ontogenetic connection

### *Cross-species similarities in behavioral and neural development*

According to Piagetian theory, human intelligence matures through a series of successive stages demarcated by increasing cognitive skills. Each successive stage builds on the accomplishments of its predecessor (Piaget, 1952, 1954, 1955, 1969). During the early infantile period, the behavioral repertoire of monkeys, apes, and humans consists of reflexes and other behaviors that exhibit minimal intelligence (Chevalier-Skolnikoff, 1977; Parker, 1977). Thus, many infantile motor actions are holistic behaviors that can neither be combined with other actions nor be broken into discrete units (Piaget, 1952). Human and nonhuman primate neonates, for instance, possess grasping abilities of a holistic nature. All of the fingers act at once. Small infants can neither execute individual actions of the digits nor control the strength of the grasping action (Connelly & Elliot, 1972; Halverson, 1943; Hines, 1942; Jensen, 1961).

Similarly, infants perceive key perceptual stimuli, rather than construct finely differentiated object images. Thus, the smiling responses of the human infant suggest an inability to distinguish crude cardboard models and facial masks from actual faces (Kagan, 1970). Young infants also recognize objects on the basis of single qualities alone, such as movement or position, rather than on the basis of entire constellations of features, such as size, color, shape, position, and movement (Bower, 1974).

Finally, to observers, neonates seem lacking in volitional behaviors. Piaget (1952) considered neonatal behaviors to be entirely reflexive in nature. Although we now know that infantile behavioral repertoires and learning capacities are much greater than Piaget believed, observers continue to describe neonatal behaviors as elicited rather than voluntary (Bower, 1974).

With maturation, the repertoire of discrete motor actions increases, perceptual skills expand, and young humans and other primates begin to combine discrete perceptual and motor units into new and varied hierarchical constructs. Human ontogeny is characterized by increasing mental constructional skills, at least to the time of adolescence. In particular, as children mature, their information-processing capacities expand, and they begin to hold greater numbers of concepts in mind simultaneously. As a result, they construct mental schemes containing greater amounts of information and greater numbers of hierarchical levels (Case, 1985). So strong are these developing mental skills that mental construction is considered the essence of human cognitive development (Langer, 1986). In particular, hierarchical or multilevel mental construction underlies the development of language, mathematics, logic, morality, and social interactions (Case, 1985; Kohlberg, 1984; Langer, 1986; Piaget, 1952, 1954, 1955, 1969).

Comparative studies emphasizing the Piagetian framework indicate that primate intellectual development is also a constructional event (Parker, 1977; Bard, P&G13; Blount, P&G15; Gómez, P&G12; Russon, P&G14). In fact, primate and human intellectual developments exhibit striking parallels (A. Diamond, 1990; Parker & Gibson, 1979; Russon, P&G14). In a very real sense, then, the ontogeny of intelligence parallels its phylogeny. Expansions in constructional abilities mediate both processes.

These ontogenetic/phylogenetic parallels reflect neural maturational parameters (Bronson, 1981; Gibson, 1981, 1983; Parker & Gibson, 1979). Specifically, the neocortex is both the last structure to expand phylogenetically and the last to mature ontogenetically.

Judging by the timing of myelination, synaptogenesis, and electrophysiological development, brain stem structures mature in advance of cortical structures (Gibson, 1977, 1981). Thus, neonatal behavior reflects brain stem control rather than cortical control. As brain stem structures are extremely small in comparison with cortical structures, primate and human neonates are, in a functional sense, small-brained animals. Their reflexive, instinctive, and other minimally constructed behaviors mirror their neural status.

In both monkeys and humans, the neocortex matures gradually. Synaptogenesis, which confers functional potential, occurs early and may occur synchronously across all cortical areas (Rakic, Bourgeois, Eckenhoff, Zeceivc, & Goldman-Rakic, 1986). In contrast, myelination, which confers the functional speed and specificity so critical for parallel processing mechanisms, matures slowly and regionally (Conel, 1939–1967; Flechsig, 1920; Gibson, 1970), as does dendritic branching, which confers complex decision-making processes upon individual neurons (Conel, 1939–1967; Scheibel, 1990). Not only do both myelination and dendritic branching exhibit regional maturation patterns, but they both exhibit the same regional patterns. Areas that myelinate early also achieve dendritic maturity early. Thus, the most thoroughly studied maturational parameter, myelination, serves as a reasonable guide to cortical maturational sequences, as well as to the development of efficient long-distance neural transmission mechanisms.

Regional myelination patterns correlate with behavioral maturation, as reported by Piagetian scholars. The primary sensory and motor areas that provide fine within-modality discriminatory capacity myelinate in advance of the association areas, which provide higher-level hierarchical constructional skills (Gibson, 1977, 1981, 1990b). Similarly, lower cortical layers, which interconnect and interact primarily with brain stem structures, mature in advance of higher cortical layers, which provide for corticocortical integration. Like intellectual development, the entire myelination process takes many years and is not completed until adolescence or later (Yakovlev & LeCours, 1967).

*Species-typical infantile behaviors and the channeling of intellectual development*

Although monkeys, apes, and humans share common developmental patterns, some differences occur. For one, large-brained species, notably humans, great apes, and possibly cebus monkeys, eventually develop more cortical constructional capacity than others; that is, in adulthood they manifest greater intelligence (Chevalier-Skolnikoff, 1989; Parker & Gibson, 1977, 1979). For another, primate species manifest differing species-typical infantile behavioral patterns (Vauclair, 1984; Vauclair & Bard, 1983; Antinucci, P&G5; Parker & Poti', P&G8).

The human infant, in particular, possesses species-typical behavioral patterns lacking in most monkeys and apes. For instance, human neonates imitate simple facial expressions, such as tongue thrusting (Meltzoff & Moore, 1977). Beginning at 6 weeks of age, the human baby will spontaneously smile at the mother's face, and at about 6 months of age babies will babble.

Primate infants also differ in their object manipulation habits. Cebus infants, like humans, manipulate objects manually prior to developing mature locomotor patterns (Antinucci, P&G5). In contrast, locomotion develops prior to object manipulation in macaques, whereas locomotion and object manipulation develop at approximately the same time in chimpanzees. Human infants are also more likely than chimpanzees to manipulate objects bimanually (Vauclair, 1984; Vauclair & Bard, 1983).

Between 4 and 8 months of age, human infants repeatedly shake rattles, kick mobiles, and engage in other repetitive actions designed to make interesting spectacles last – the secondary circular reactions of Piaget (1952). Stump-tailed macaques do not exhibit these behaviors (Parker, 1977). Chimpanzees may do so (Chevalier-Skolnikoff, 1977), but in my experience with home-reared chimpanzees, they do so infrequently.[6]

Beginning about 12 months of age, human infants experiment with object–object and object–force interactions, such as repeatedly dropping objects from highchairs or banging objects together while varying force, speed, and other parameters – the tertiary circular reactions of Piaget (1952). The ability to use tools is a natural outgrowth of these actions and also emerges at about 1 year of age in human children. Similarly, during the second year of life, human children stack blocks and engage in other object–object constructional tasks. Again, infant stump-tailed macaques fail to manifest these behaviors (Parker, 1977), but cebus infants may manifest some of them (Parker & Poti', P&G8). All of these behaviors are surprisingly rare in chimpanzee and other great ape infants, given their tool-using skills in adulthood. Great apes are 3 to 4 years of age before they engage in significant object–object related behaviors (Chevalier-Skolnikoff, 1983, 1989; Mignault, 1985; Morell, 1985; Russon, P&G14; Vauclair, 1984; Miles, P&G19; K. R. Gibson, pers. obs.).

No differences in neural maturation patterns are known to account for differential timings in the development of object manipulation skills. Thus, these differences may reflect differences in size or organization of neural structures. Judging by the age of emergence, the neural control of the human social smile and early infantile imitation skills resides with subcortical structures, whereas the control of babbling and of the secondary and tertiary circular reactions lies either with brain stem structures or with the lowest, and earliest maturing, cortical layers (Gibson, 1970; Konner, 1990). Thus, the evolutionary emergence of infantile species-specific behaviors may reflect expanded subcortical structures or very rudimentary functioning of expanded cortical regions.

The early emergence of species-typical behaviors appears critical to human intellectual development. These infantile behaviors generate other inputs essential for mental development, such as maternal interactions, baby talk, and information about object properties. They also provide the infant with opportunities to practice oral and manual skills. Thus, these behaviors channel intellectual development in particular directions by assuring access to appropriate environmental inputs. In particular, human species-typical infantile behaviors channel development in linguistic and tool-using domains. The first words of human infants, for instance, represent a major mental constructional milestone involving the integration of a number of prior achievements in verbal, social, and object oriented domains (Case, 1985).

Perhaps the presence of channeling mechanisms in human, but not ape, infants explains the earlier and more consistent development of linguistic and tool-using skills in the human species. Apes clearly manifest some symbolic and tool-using capabilities, but they begin using tools at a later age, and they do so less consistently. Their language skills emerge only subsequent to massive human instruction.

These considerations suggest that in addition to possessing greater mental constructional skills, highly intelligent species may also possess greater numbers of species-typical, minimally constructed infantile behaviors that channel adult development in specific directions. Unfortunately, although others have discussed the potential significance of "innate schoolmarms" (Gould & Marler, 1987; Lorenz, 1965; see also Gibson, 1981; Gibson & Parker, 1982; Parker & Poti', P&G8), comparative infantile behavior remains a largely untapped research area. In the quest for delineating genetic versus environmental causes of infantile behaviors, the potential influence of minimally constructed, species-typical behaviors on subsequent cognitive development has been ignored. This major research task lies ahead.

### *Learning by instinct versus learning by intelligence; fallacies of the altriciality and neoteny theories of intellectual development*

This view that high intelligence reflects developmentally channeled, brain-size-mediated increases in mental constructional capacities constrasts with other views that high intelligence reflects changes in ontogenetic rates and/or neoteny (Gould, 1977).

Although all vertebrates exhibit similar patterns of brain maturation, the timing of brain maturation varies significantly among taxa (Gibson, 1990b). Among mammals, for instance, major differences exist in the state of brain maturity at birth (Portmann, 1967; Sacher & Staffeldt, 1974). Thus, bears acquire as little as 2% of their final brain size by birth; rats, dogs, and cats, about 10%; ungulates, as much as 75%.

In comparison with most mammals, higher primates are neurologically precocial at birth (Gibson, 1990b). Portmann (1967), for instance, considers any mammals whose neonatal brain size exceeds 20% of its adult brain size to be neurologically precocial. The neonatal rhesus monkey brain is about 60–70% as large as the adult brain (Sacher & Staffeldt, 1974). Other neonatal monkey brains are at least 40% of their final size; chimpanzee brains, about 35–40%. By the time of birth, the human brain reaches about 25% of its adult size. Thus, the state of human brain maturation at birth is by no means unique. In fact, it is totally nondistinctive, being in the middle range among mammals as a whole and only somewhat less mature than that of the neonatal chimpanzee.

The nondistinctiveness of human neonatal brain maturity throws into question the popular view that human intelligence derives from neonatal altriciality. Further, if neural altriciality causes intelligence, then altricial mammals in general should have large brains, and opossums, white mice, and laboratory rats should be intellectual giants. In actuality, most altricial mammals have small brains, and most large-brained mammals give birth to precocial young (Portmann, 1967).

Postnatally, it takes the human brain surprisingly little time to "catch up" to monkey brains in terms of myelination and other maturational parameters (Gibson, 1990b). By 3 to 4 months of age, the human brain is as mature, judging by myelination standards, as the brain of the highly precocial rhesus monkey is at birth.

Although myelination data are unavailable for the chimpanzee, in terms of percentage brain size achieved by birth, chimpanzees are intermediate between rhesus monkeys and humans. This suggests that the human brain at 6 weeks to 2 months of age is as mature as the chimpanzee brain at birth (see Gibson, 1990b, for a detailed discussion of comparative brain maturation). This estimate coincides with other data indicating that chimpanzee behavior at birth is only slightly advanced in comparison with that of the human neonate and may actually lie within the range of variation of human neonatal behavior (Bard, Platzman, & Suomi, 1988). Consequently, if human neonatal immaturity plays a causative role in the development of human intelligence, this role must relate to intellectual development in the first weeks of postnatal life, a period when human infants primarily sleep, when the neocortex is barely functional, and when most species-typical channeling behaviors, such as babbling, have yet to appear.

Even if neonatal immaturity is not a cause of high intelligence, other maturational parameters may play roles. Most large-brained mammals, for instance, have a long period of postnatal growth (Rensch, 1959). Great apes and humans, in particular, do not reach full physical maturity until their teenage years or later (Galdikas, 1981, 1985; Goodall, 1986; Tuttle, 1987; van Lawick-Goodall, 1973). In humans and rhesus monkeys, neural myelination occurs at least to puberty and may extend beyond it (Gibson, 1970, 1990b; Yakovlev & LeCours, 1967). Thus, a long period of neural maturation may typify all higher primates.

Prolonged maturation provides at least two intellectual advantages (Gibson, 1990b; Lancaster, 1985; Lancaster & Lancaster, 1983). First, it allows young animals time to play and learn prior to foraging on their own. In apes, this learning period extends to 5 or 6 years of life; in humans it sometimes extends for 2 or 3 decades. Second, a long period of brain growth may serve as a developmental buffer for large-brained species. The brain is an extremely expensive organ from metabolic and nutritive standpoints, especially during the period of brain growth, and serious malnutrition can cause permanent brain damage (Morgan, 1990). Prolonged growth permits essential nutrients to be acquired over a longer time frame.

Some theorists suggest that prolonged maturation confers additional benefits: that immature brains possess greater plasticity than mature brains and hence superior learning abilities (Gould, 1977). The relationship among neural plasticity, learning abilities, and intelligence, however, is not clear (Gibson, 1990b). Much neural plasticity occurs in relationship to environmental insult. Thus, infant monkeys may lose previously developed visual functions if they fail to receive appropriate environmental input (Hubel & Wiesel, 1970; Hubel, Wiesel, & LeVay, 1977). Similar deprivation at later ages produces no visual defects. Alternately, damage to the left cerebral hemisphere in adulthood produces permanent language deficits. Following similar damage in childhood, the right hemisphere assumes language functions (Bub & Whitaker, 1980). This neural lability in response to environmental or physical trauma has no obvious relationship to intelligence.

Moreover, it is fallacious to equate learning and intelligence. Much learning, such as conditioned reflexes, is fundamentally unintelligent because it requires minimal mental constructional skills. As noted by Hailman (1969), even instinctive behaviors are learned. Laughing gulls, for instance, develop increasingly precise pecking actions with practice and also become increasingly capable of distinguishing maternal from nonmaternal stimuli. Such learning, however, exhibits none of the multilevel combination and recombination of separate behavioral and mental concepts considered by Piagetian and neo-Piagetian scholars to be the essence of intelligence (Case, 1985; Langer, 1986; Parker & Poti', P&G8). This hierarchical constructional capacity requires fully functioning neocortical association areas. As such, it requires a mature, rather than an immature, brain. For this reason, school systems delay the teaching of higher mathematics and other subjects requiring abstract thought to puberty, rather than attempting to teach them to preschoolers, whose brains simply cannot comprehend them.

These considerations point out serious fallacies with a popular theory of prolonged immaturity as a determinant of intelligence. According to Gould, human brain and behavior not only experience prolonged development but also are neotenous; that is, in adulthood, they resemble the brains and behaviors of immature animals, specifically with respect to learning capacity (Gould, 1977). If adult human behavior were really neotenous, it would consist primarily of reflexes and modal action patterns rather than of highly intelligent, constructed acts. Similarly, if the adult brain were neotenous, it would be dominated by brain stem rather than cortical structures.

The view that the human brain is neotenous is one that is held by scholars such as Gould who have never studied brain development. Developmental neurobiologists often have claimed just the opposite, that neural ontogeny parallels neural phylogeny. In other words, young human brains resemble the adult brains of other animals (Ebbesson, 1984; Jacobson, 1978). Certainly, human children resemble the adults of other species in their degree of cortical constructional capacity (Parker & Gibson, 1979). Older children and humans, however, exceed other animals in this critical intellectual facility.

Moreover, no need exists to postulate special developmental mechanisms to permit learning in adulthood as Gould (1977) and others (Lerner, 1984) have done. Even rats continue to learn and to experience brain growth in response to environmental stimulation to very old age (M. C. Diamond, 1990). Thus, some learning related neural plasticity occurs throughout the life span even in animals whose postnatal maturation period is quite short.

## Conclusion

Maturing human intelligence reflects increasing hierarchical (or multilevel) mental constructional skills (Case, 1985; Langer, 1986). Thus, as children grow, they develop increasing information-processing capacities. This permits them to hold numerous perceptual, motor, and conceptual schemes in mind simultaneously and to combine and recombine these schemes into new mental constructs. Newly formed mental constructs can also be combined and recombined with each other. This leads to highly complex, multilayer mental constructions. Comparative data indicate that the evolution of intelligence, like its ontogeny, involved increases in hierarchical mental constructional skills. Thus, reflexes and instincts represent the lower end of a quantitative continuum of mental constructional levels, and human intelligence, language, and technological skills represent the upper end.

Neural models suggest that species differences in intelligence reflect quantitative variations in numbers of neurons, numbers of connections per neuron, and degree of duplication of neural processing functions. Maturational differences within a species reflect the functional maturity of the neocortex.

Primate and human newborns have only minimal cortical functions, and hence only minimal intelligence. Human and primate infants do, however, possess a broad repertoire of infantile species-typical behaviors. Although these behaviors represent the lower end of a continuum of mental constructional capacity, they nevertheless play critical roles in intellectual development by channeling mental constructional skills in species-typical directions. Thus, species differences in intelligence reflect both brain-size-mediated adult mental constructional skills and less intelligent species-typical infantile behaviors that channel intellectual development. They do not reflect degree of maturity at birth or neoteny.

**Acknowledgment**

The concepts expressed in this chapter were developed over many years of continual collaboration and discussions with Sue Taylor Parker.

**Notes**

1. This action is considered hierarchical because it involves several layers of construction. Humans must construct individual images of each object (wood, hammer, and nails) from multiple perceptual information (Bower, 1974). They must also construct several relationships between two objects (e.g., hammering, stacking, and piercing). Finally, they must be able to construct advanced mental images of the interrelationships between multiple objects and multiple actions (hammering, stacking, and piercing). The chimpanzee reaches only the second level of this constructional hierarchy – constructing relationships between two objects.
2. 3 colors × 2 shapes × 2 conditions × 3 sizes × 2 distances × 2 positions.
3. A standard algebraic formula.
4. The neocortex contains primary auditory, visual, tactile, and motor areas, as well as secondary sensory and motor regions and major association areas. Primary sensory areas possess the properties expected of complex feature detecting networks, including large numbers of highly interconnected neurons, each responsive to one discrete feature, such as line orientation, tone, or point of skin stimulated (Hubel & Wiesel, 1968; Mountcastle, 1978).

   Secondary sensory areas interact with primary areas and respond to larger, more holistic object images, such as faces or hands. Patients with lesions in these regions maintain perceptual capacity in the sense that they can hear, see, and feel, but they cannot recognize constructed object images. In other words, secondary areas seem to function as object constructor networks. In fact, some two decades ago the great Russian neurologist Alexander Luria suggested that these areas synthesize separate perceptions into holistic object images (Luria, 1966).

   The great parietotemporal and frontal association areas contain multiple representations of sensory and motor processing centers (Kaas, 1987). Thus, they provide the duplication of function essential for parallel processing and hierarchical construction of complex sensory and motor acts. Similarly, according to Luria (1966), these regions take separate sensory and motor constructs and organize them into higher-level simultaneous and sequential wholes. Lesions in the parietotemporal regions result in cognitive defects characterized by an inability to keep multiple concepts in mind simultaneously. Patients exhibit particular problems with tasks involving cross-modal integration and higher intellectual processes such as object naming, grammatical constructs, map reading, dressing, tool use, gestural abilities, and drawing.

   The frontal association areas provide behavioral flexibility and possess complex functions related to setting goals, planning sequential behavioral acts, and determining whether or not behavioral results match behavioral plans (Nauta, 1971; Warren & Ackert, 1964). Thus, the frontal areas possess functions that Parker and Poti' (P&G8)

consider essential for hierarchical integration. Damage to the frontal lobes leads to childish behavior, characterized by an inability to plan actions in advance. These planning difficulties are evident across all behavioral domains, including tool use, language, and social behaivor.

5. Elephants, strictly speaking, are not extractive foragers. They do, however, strip bark and eat it. Bark stripping is very similar to the peeling and stripping actions of many primates who eat fruit or pith. Thus, it demands similar manipulative skills.
6. For several years I followed the development of three chimpanzees raised in the home of Jim and June Cook of Conroe, Texas (Morell, 1985). The Cooks tried to raise these animals very much like human children. I visited the Cook home on a weekly basis and also took the chimpanzee Leah to my home for a period of 24 hours per week for 6 months when she was between 3 and 9 months of age. My primary interest was the development of object manipulation skills. Thus, in my visits, I particularly tried to encourage the animals to stack blocks and engage in other constructive tasks, with little success until they were about 3 years of age.

## References


Alcock, J. (1975). *Animal behavior: An evolutionary approach*. Sunderland, MA: Sinauer Associates.

Alstermar, B. A., Lundberg, U., Norrsell, U., & Sybirska, E. (1981). Integration in descending motor pathways controlling the forelimb of the cat. 9. Differential behavioral defects after spinal cord lesions interrupting defined pathways from higher centres to motor neurons. *Experimental Brain Research*, *42*, 299–318.

Armstrong, E. (1982). Mosaic evolution in the primate brain: Differences and similarities in the hominid thalamus. In E. Armstrong & D. Falk (Eds.), *Primate brain evolution* (pp. 131–162). New York: Plenum.

Armstrong, E. (1983). Relative brain size in mammals. *Science*, *220*, 1302–1304.

Armstrong, E. (1986). Enlarged limbic structures in the human brain: The anterior thalamus and medial mammillary body. *Brain Research*, *362*, 394–397.

Auffenberg, W. (1978). Social and feeding behavior in *Varanus komodoensis*. In N. Greenberg & P. D. MacLean (Eds.), *Behavior and neurology of lizards* (pp. 301–331). Bethesda, MD: National Institute of Mental Health.

Bard, K., Platzman, K. A., & Suomi, S. J. (1988). *Neurobehavioral responsiveness in neonatal chimpanzees: Orientation to animate and inanimate stimuli*. Paper presented at the 12th Congress of the International Primatological Society, Brasilia, Brasil, July 24–29.

Barlow, G. (1977). Modal action patterns. In T. A. Sebeok (Ed.), *How animals communicate* (pp. 98–134). Bloomington: Indiana University Press.

Beck, B. (1980). *Animal tool behavior: The use and manufacture of tools by animals*. New York: Garland Press.

Berkinblit, M. B., Feldman, A. G., & Fukson, O. I. (1986). Adaptability of innate motor patterns and motor control mechanisms. *Behavioral and Brain Sciences*, *9*, 585–638.

Boesch, C., & Boesch, H. (1981). Sex differences in the use of natural hammers by wild chimpanzees: A preliminary report. *Journal of Human Evolution*, *10*, 585–593.

Boesch, C., & Boesch, H. (1983). Optimisation of nut cracking with natural hammers in chimpanzees. *Behavior*, *83*, 265–286.

Boesch, C., & Boesch, H. (1984). Possible causes of sex differences in the use of natural hammers by wild chimpanzees. *Journal of Human Evolution*, *13*, 415–440.

Bower, T. G. R. (1974). *Development in infancy*. San Francisco: Freeman.

Bronson, G. (1981). Structure, status and characteristics of the nervous system at birth. In P. M. Stratton (Ed.), *Psychobiology of the newborn* (pp. 99–118). New York: Wiley.

Bub, D., & Whitaker, H. A. (1980). Language and verbal processes. In M. C. Whittrock (Ed.), *The brain and psychology* (pp. 211–244). New York: Academic Press.

Casagrande, V. A., Harting, J. K., Hall, W. C., Diamond, I. T., & Martin, G. F. (1972). Su-

perior colliculus in the tree shrew: A structural and functional subdivision into superficial and deep layers. *Science*, *177*, 444–447.
Case, R. (1985). *Intellectual development: Birth to adulthood*. New York: Academic Press.
Changeaux, J. P. (1985). *Neuronal man*. New York: Pantheon Books.
Chevalier-Skolnikoff, S. (1977). A Piagetian model for describing and comparing socialization in monkey, ape and human infants. In S. Chevalier-Skolnikoff & F. E. Poirer (Eds.), *Primate biosocial development* (pp. 159–187). New York: Garland Press.
Chevalier-Skolnikoff, S. (1983). Sensorimotor development in orang-utans and other primates. *Journal of Human Evolution*, *12*, 545–561.
Chevalier-Skolnikoff, S. (1989). Spontaneous tool use in *Cebus* compared with other monkeys and apes. *Behavioral and Brain Sciences*, *12*, 561–627.
Chomsky, N. (1980). Human language and other semiotic systems. In T. A. Sebeok & J. Umiker-Sebeok (Eds.), *Speaking of apes* (pp. 429–440). New York: Plenum.
Clark, W. E. Le Gros (1960). *The antecedents of man*. Chicago: Quadrangle Books.
Clutton-Brock, T. H., & Harvey, P. H. (1980). Primates, brains and ecology. *Journal of Zoology*, *190*, 309–323.
Collias, N. E. (1964). The evolution of nests and nest-building in birds. *American Zoologist*, *4*, 175–190.
Conel, J. L. (1939–1967). *The postnatal development of the human cerebral cortex*, Vols. 1–8. Cambridge, MA: Harvard University Press.
Connelly, K., & Elliot, J. (1972). The evolution and ontogeny of hand function. In N. B. Jones (Ed.), *Ethological studies of child behavior* (pp. 329–383). Cambridge University Press.
Count, E. W. (1947). Brain and body weight in man : Their antecedents in growth and evolution. *Annals of the New York Academy of Sciences*, *46*, 993–1122.
De Waal, F. (1983). *Chimpanzee politics*. New York: Harper & Row.
Diamond, A. (1990). Frontal lobe involvement in cognitive changes during the first year of life. In K. R. Gibson & A. Petersen (Eds.), *Brain and behavioral maturation: Biosocial perspectives*. Hawthorne, NY: Aldine.
Diamond, M. C. (1990). Environmental influences on the young brain. In K. R. Gibson & A. Petersen (Eds.), *Brain and behavioral maturation: Biosocial perspectives*. Hawthorne, NY: Aldine.
Doubilet, D. (1988). Ballet with stingrays. *National Geographic*, *175*(1), 84–94.
Ebbesson, S. O. E. (1984). Evolution and ontogeny of neural circuits. *Behavioral and Brain Sciences*, *7*, 321–366.
Eisenberg, J. F., & Wilson, D. E. (1978). Relative brain size and feeding strategies in Chiroptera. *Evolution*, *32*, 740–751.
Ellingson, R. J., & Rose, G. H. (1970). Ontogenesis of the electroencephalogram. In W. A. Himwich (Ed.), *Developmental neurobiology* (pp. 441–474). Springfield, IL: Thomas.
Ettlinger, G. (1984). Humans, apes and monkeys: The changing neuropsychological viewpoint. *Neuropsychologia*, *22*, 685–696.
Ewer, R. F. (1968). *Ethology of mammals*. London: Elek Science.
Feldman, J. A., & Ballard, D. H. (1982). Connectionist models and their properties. *Cognitive Science*, *6*, 205–254.
Flechsig, P. (1920). *Anatomie des menschlichen Gehirns und Rückenmarks auf myelogenetischer Grundlage*. Leipzig : Georg Thomas.
Fouts, R. S. (1973). Acquisition and testing of gestural signs in four young chimpanzees. *Science*, *180*, 973–980.
Gaither, N. S. & Stein, B. E. (1979). Reptiles and mammals use similar sensory organizations in the midbrain. *Science*, *205*, 595–597.
Galdikas, B. M. F. (1981). Orangutan reproduction in the wild. In C. E. Graham (Ed.), *Reproductive behavior of the great apes* (pp. 281–300). New York: Academic Press.
Galdikas, B. M. F. (1985). Subadult male orangutan sociality: Reproductive behavior at Tanjung Puting. *American Journal of Primatology*, *8*, 87–99.

Gardner, R. A., & Gardner, B. T. (1969). Teaching sign language to a chimpanzee. *Science, 165*, 664–672.
Gardner, R. A., & Gardner, B. T. (1980). Comparative psychology and language acquisition. In T. A. Sebeok & J. Umiker-Sebeok (Eds.), *Speaking of apes* (pp. 287–330). New York: Plenum.
Georgopoulos, A. P. (1988). Neural integration of movement: Role of motor cortex in reaching. *FASEB Journal, 2*, 2849–2857.
Gibson, K. R. (1970). *Sequence of myelinization in the brain of Macaca mulatta*. Unpublished PhD dissertation, University of California, Berkeley.
Gibson, K. R. (1977). Brain structure and intelligence in macaques and human infants from a Piagetian perspective. In S. Chevalier-Skolnikoff and F. E. Poirer (Eds.), *Primate biosocial development* (pp. 113–157). New York: Garland Press.
Gibson, K. R. (1981). Comparative neuroontogeny, its implications for the development of human intelligence. In G. Butterworth (Ed.), *Infancy and epistemology* (pp. 52–82). Brighton, England: Harvester Press.
Gibson, K. R. (1983). Comparative neurobehavioral ontogeny: The constructionist perspective in the evolution of language, object manipulation and the brain. In E. DeGrolier (Ed.), *Glossogenetics* (pp. 41–66). New York: Harwood Academic Publishers.
Gibson, K. R. (1985). Has the evolution of intelligence stagnated since Neanderthal man? In G. Butterworth, J. Rutkouska, & M. Scaife (Eds.), *Evolution and development* (pp. 102–114). Brighton, England: Harvester Press.
Gibson, K. R. (1986). Cognition, brain size and the extraction of embedded food resources. In J. G. Else & P. C. Lee (Eds.), *Primate ontogeny, cognitive and social behavior* (pp. 93–105). Cambridge University Press.
Gibson, K. R. (1988a). Brain size and the evolution of language. In M. Landsberg (Ed.), *The genesis of language: A different judgement of evidence* (pp. 149–172). Berlin: Mouton de Gruyter.
Gibson, K. R. (1988b). Fish, sea snakes, dolphins, teeth, and brains – some evolutionary paradoxes. *Behavioral and Brain Sciences, 11*, 93–94.
Gibson, K. R. (1990a). The ontogeny and evolution of the brain, cognition and language. In A. Lock & C. Peters (Eds.), *Handbook of symbolic intelligence*. Oxford University Press.
Gibson, K. R. (1990b). Brain and cognition in the growing child: A comparative perspective on the questions of neoteny, altriciality and intelligence. In K. R. Gibson & A. Petersen (Eds.), *Brain maturation and behavioral development: Biosocial dimensions*. Hawthorne, NY: Aldine.
Gibson, K. R., & Parker, S. T. (1982). Brain structure, Piaget and adaptation: "No, I think, therefore I eat." *Behavioral and Brain Sciences, 5*, 288–294.
Gibson, K. R., & Parker, S. T. (1986). Extraction: A non-sexist interpretation of early hominid origins. *American Journal of Physical Anthropology, 69*, 204.
Goodall, J. (1986). *The chimpanzees of Gombe*. Cambridge, MA: Harvard University Press.
Gould, J. L., & Marler, P. M. (1987). Learning by instinct. *Scientific American, 256*(1), 74–85.
Gould, S. J. (1977). *Ontogeny and phylogeny*. Cambridge, MA: Harvard University Press.
Greenberg, N. (1978). Ethological considerations in the experimental study of lizard behavior. In N. Greenberg & P. D. McLean (Eds.), *Behavior and neurology of lizards* (pp. 203–224). Bethesda, MD: National Institute of Mental Health.
Hailman, J. (1969). How an instinct is learned. *Scientific American, 221*(6), 96–108.
Halverson, H. M. (1943). The development of prehension in infants. In R. G. Barker, J. S. Kounin, & H. F. Wright (Eds.), *Child development and behaviour* (pp. 49–65). New York: McGraw-Hill.
Hamilton, W. J. (1973). *Life's color code*. New York: McGraw-Hill.
Harcourt, A. H. (1988). Alliances in contests and social intelligence. In R. W. Byrne & A. Whiten (Eds.), *Machiavellian intelligence* (pp. 132–152). Oxford: Clarendon Press.
Hinde, R. (1970). *Animal behavior*. New York: McGraw-Hill.

Hines, M. (1942). The development and regression of reflexes and progression in the young macaque. *Carnegie Institute of Washington, Contributions to Embryology, 28*, 309–451.
Hinton, G. (1984). Parallel computations for controlling an arm. *Journal of Motor Behavior, 16*, 171–194.
Hinton, G., McClelland, J. L., & Rumelhart, D. E. (1986). Distributed representations. In D. Rumelhart & J. McClelland (Eds.), *Parallel distributed processing. Vol. 1. Foundations* (pp. 77–109). Cambridge, MA: M.I.T. Press.
Holloway, R. L. (1966). Cranial capacity, neural reorganization, and hominid evolution: A search for more suitable parameters. *American Anthropologist, 68*, 103–121.
Holloway, R. L. (1968). The evolution of the primate brain: Some aspects of quantitative relationships. *Brain Research, 7*, 121–172.
Holloway, R. L. (1969). Culture, a human domain. *Current Anthropology, 10*, 395–412.
Holloway, R. L. (1979). Brain size, allometry and reorganization: A synthesis. In M. E. Hahn, B. C. Dudek, & C. Jensen (Eds.), *Development and evolution of brain size* (pp. 59–88). New York: Academic Press.
Holloway, R. L., & Post, D. (1982). The relativity of relative brain size measures and hominid evolution. In E. Armstrong & D. Falk (Eds.), *Primate brain evolution, methods and concepts* (pp. 57–76). New York: Plenum.
Hubel, D. H., & Wiesel, T. N. (1968). Receptive fields and functional architecture of monkey striate cortex. *Journal of Physiology, 195*, 215–243.
Hubel, D. H., & Wiesel, T. N. (1970). The period of susceptibility to the physiological effects of unilateral eye closure in kittens. *Journal of Physiology (London), 206*, 419–436.
Hubel, D. H., Wiesel, T. N. & LeVay, S. (1977). Plasticity of ocular dominance columns in the monkey striate cortex. *Philosophical Transactions of the Royal Society of London (Biology), 278*, 377–409.
Humphrey, N. K. (1970). What the frog's eye tells the monkey's brain. *Brain, Behavior and Evolution, 3*, 324–327.
Huttenlocher, P. R. (1979). Synaptic density in human frontal cortex – developmental changes and effects of aging. *Brain Research, 163*, 195–205.
Huttenlocher, P. R., Courten, C., Garey, L., & Van Der Loos, D. (1982). Synaptogenesis in human visual cortex – evidence for synapse elimination during normal development. *Neuroscience Letters, 33*, 247–252.
Ingle, D. (1970). Visuomotor functions of the frog optic tectum. *Brain, Behavior and Evolution, 3*, 57–71.
Ingle, D. (1972). Two visual systems in the frog. *Science, 181*, 1053–1055.
Inhelder, B., & Piaget, J. (1958). *The growth of logical thinking from childhood to adolescence.* New York: Basic Books.
Jacobsen, S. (1963). Sequence of myelinization in the brain of the albino rat: A. Cerebral cortex, thalamus and related structures. *Journal of Comparative Neurology, 121*, 5–29.
Jacobson, M. (1978). *Developmental neurobiology*. New York: Plenum.
Jensen, G. D. (1961). The development of prehension in the macaque. *Journal of Comparative Physiology and Psychology, 54*, 11–12.
Jerison, H. J. (1973). *Evolution of the brain and intelligence.* New York: Academic Press.
Jerison, H. J. (1979). The evolution of diversity in brain size. In M. Hahn, C. Jensen, & B. Dudek (Eds.), *Development and evolution of brain size: Behavioral implications* (pp. 29–57). New York: Academic Press.
Jerison, H. J. (1982). Allometry, brain size, cortical surface, and convolutedness. In E. Armstrong & D. Falk (Eds.), *Primate brain evolution* (pp. 77–84). New York: Plenum.
Kaas, J. H. (1987). The organization of the neocortex in mammals: Implications for theories of brain function. *Annual Review of Psychology, 38*, 129–151.
Kagan, J. (1970). The determinants of attention in the infant. *American Scientist, 56*, 298–306.
Kohlberg, L. (1984). *The moral development of the child. Vol. 2. The psychology of moral development.* New York: Harper & Row.

Konner, M. (1990) Universals of behavioral development in relation to myelination. In K. R. Gibson & A. Petersen (Eds.), *Brain and behavioral development*. Hawthorne, NY: Aldine.

Kretschman, H. J. (1967). Die Myelinogenese eines Nestflüchters (*Acomys cahirnes minous*, Bate, 1906) im Vergleich zu der Nesthockers (Albinomaus). *Journal für Hirnforschung*, *9*, 373–396.

Kummer, H. (1967). Tripartite relations in hamadryas baboons. In S. Altmann (Ed.), *Social communication among primates* (pp. 63–71). University of Chicago Press.

Kupfermann, I., & Weiss, K. (1978). The command neuron concept. *Behavioral and Brain Sciences*, *1*, 3–39.

Lancaster , J. B. (1985). Evolutionary perspectives on sex differences in the higher primates. In A. Rossi (Ed.), *Gender and the life course* (pp. 3–27). Chicago: Aldine.

Lancaster, J. B., & Lancaster, C. (1983). Parental investment and the hominid adaptation. In D. Ortner (Ed.), *How humans adapt: A biocultural odyssey* (pp. 33–56). Washington, DC: Smithsonian Institution Press.

Langer, J. (1986). *The origins of logic*. New York: Academic Press.

Lerner, R. M. (1984). *On the nature of human plasticity*. Cambridge University Press.

Lorenz, K. (1965). *The evolution and modification of behavior*. University of Chicago Press.

Lorenz, K. (1971). *Studies in animal and human behavior*, Vols. 1–2. Cambridge, MA: Harvard University Press.

Luria, A. (1966). *Higher cortical functions in man*. New York: Basic Books.

McClelland, J. (1986). Resource requirements of standard and programmable nets. In D. Rumelhart & J. McClelland (Eds.), *Parallel distributed processing. Vol. 1. Foundations* (pp. 460–487). Cambridge, MA: M.I.T. Press.

McClelland, J., & Rumelhart, D. (1986). *Parallel distributed processing. Vol. 2. Psychological and biological models*. Cambridge, MA: M.I.T. Press.

McClelland, J. L., Rumelhart, D. E., & Hinton, G. E. (1986). The appeal of parallel distributed processing. In D. E. Rumelhart & J. L. McClelland (Eds.), *Parallel distributed processing. Vol. 1. Foundations* (pp. 3–44). Cambridge, MA: M.I.T. Press.

McGrew, W. C. (1974). Tool use by wild chimpanzees in feeding upon driver ants. *Journal of Human Evolution*, *3*, 501–508.

MacLean, P. D. (1978). Why brain research on lizards? In N. Greenberg & P. D. McLean (Eds.), *Behavior and neurology of lizards* (pp. 1–10). Bethesda, MD: National Institute of Mental Health.

Macphail, E. (1987). The comparative psychology of intelligence. *Behavioral and Brain Sciences*, *10*, 645–696.

Marler, P., & Hamilton, W. J. (1966). *Mechanisms of animal behavior*. New York: Wiley.

Martin, R. P. (1962). Entwicklungszeiten des Zentralnervensystems von Nagern mit Nesthocker and nestflüchterontogenese (Cavia cobaya Schreb. und *Rattus norvegicus* Erxleben). *Revue Suisse de Zoologie*, *69*, 617–727.

Martin, R. P. (1982). Allometric approaches to the primate nervous system. In E. Armstrong & D. Falk (Eds.), *Primate brain evolution* (pp. 39–56). New York: Plenum.

Maturana, H. R., Lettvin, J. Y., McCulloch, W. S., & Pitts, W. H. (1960). Anatomy and physiology of vision in the frog (*Rana pipiens*). *Journal of General Physiology*, *43*, 129–175.

Meltzoff, A. N., & Moore, M. K. (1977). Imitation of facial and manual gestures by human infants. *Science*, *205*, 217–219.

Mignault, C. (1985). Transition between sensorimotor and symbolic activities in nursery-reared chimpanzees (*Pan troglodytes*). *Journal of Human Evolution*, *14*, 747–758.

Miles, H. L. (1983). Two-way communication with apes and the evolution of language. In E. de Grolier (Ed.), *Glossogenetics: the origin and evolution of language*. Paris: Harwood Academic Publishers.

Milton K. (1988). Foraging behavior and the evolution of primate intelligence. In R. W. Byrne & A. Whiten (Eds.), *Machiavellian intelligence* (pp. 285–306). Oxford: Clarendon Press.

Morell, V. (1985). Challenging chimpanzees. *Equinox: The Magazine of Canadian Discovery*, *22*, 17–18.

Morgan, B. (1990). Nutrition and brain development. In K. Gibson & A. Petersen (Eds.), *Brain and behavioral maturation: Biosocial perspectives*. Hawthorne, NY: Aldine.
Mountcastle, V. (1978). An organizing principle for cerebral function: The unit module and the distributed system. In G. M. Edelmann & V. B. Mountcastle (Eds.), *The mindful brain* (pp. 7–50). Cambridge, MA: M.I.T. Press.
Nauta, W. J. H. (1971). The problem of the frontal lobe: A reinterpretation. *Journal of Psychiatric Research*, *8*, 167–187.
Parker, S. T. (1977). Piaget's sensorimotor period series in an infant macaque: A model for comparing unstereotyped behavior and intelligence in human and nonhuman primates. In S. Chevalier-Skolnikoff & F. Poirer (Eds.), *Primate biosocial development* (pp. 43–112). New York: Garland Press.
Parker, S. T., & Gibson, K. R. (1977). Object manipulation, tool use and sensorimotor intelligence as feeding adaptations in cebus monkeys and great apes. *Journal of Human Evolution*, *6*, 623–641.
Parker, S. T., & Gibson, K. R. (1979). A model of the evolution of language and intelligence in early hominids. *Behavioral and Brain Sciences*, *2*, 367–407.
Passingham, R. E. (1973). Anatomical differences between the neocortex of man and other primates. *Brain, Behavior, and Evolution*, *7*, 337–359.
Passingham, R. E. (1975). Changes in the size and organization of the brain in man and his ancestors. *Brain, Behavior, and Evolution*, *11*, 73–90.
Passingham, R. E. (1987). Two cortical systems for directing movement in motor areas of the cerebral cortex. *Ciba Foundation Symposium*, *132*, 151–164.
Percheron, G., Yelnick, J., & Francois, C. (1984). A Golgi analysis of the primate globus pallidus. III. Spatial organization of the striatopallidal complex. *Journal of Comparative Neurology*, *227*, 214–227.
Piaget, J. (1952). *The origins of intelligence in children*. New York: Norton.
Piaget, J. (1954). *The construction of reality in the child*. New York: Basic Books.
Piaget, J. (1955). *The language and thought of the child*. New York: World Publishing.
Piaget, J. (1969). Genetic epistemology. *Columbia Forum*, *12*, 4–11.
Portman, A. (1967). *Zoologie aus vier Jahrzehnten*. Munich: R. Piper Verlag.
Prideaux, T. (1973). *CroMagnon man*. New York: Time-Life Books.
Purves, D., & Lichtman, J. V. (1985). *Principles of neural development*. Sunderland, MA: Sinauer Associates.
Rakic, P., Bourgeois, J. P., Eckenhoff, M. F., Zecevic, N., & Goldman-Rakic, P. S. (1986). Concurrent overproduction of synapses in diverse regions of the primate cerebral cortex. *Science*, *232*, 232–235.
Regal, P. J. (1978). Behavioral differences between reptiles and mammals: An analysis of activity and mental capabilities. In N. Greenberg & P. D. McLean (Eds.), *Behavior and neurology of lizards* (pp. 183–203). Bethesda, MD: National Institute of Mental Health.
Rensch, B. (1959). *Evolution above the species level*. New York: Columbia University Press.
Reynolds, P. C. (1981). *On the evolution of human behavior*. Berkeley: University of California Press.
Reynolds, P. C. (1982). The primate constructional system: The theory and description of instrumental object use in humans and chimpanzees. In M. von Cranach & R. Harré (Eds.), *The analysis of action* (pp. 343–383). Cambridge University Press.
Reynolds, P. C. (1983). Ape constructional ability and the origin of linguistic structure. In E. De Grolier (Ed.), *Glossogenetics: The origin and evolution of language* (pp. 185–200). New York: Harwood Academic Publishers.
Rumelhart, D., & McClelland, J. (1986). *Parallel distributed processing. Vol. 1. Foundations*. Cambridge, MA: M.I.T. Press.
Rumelhart, D. E., & Norman, D. A. (1982). Simulating a skilled typist: A study of skilled cognitive-motor performance. *Cognitive Science*, *6*, 1–36.
Rutkowska, J. (1985). Does the phylogeny of conceptual development increase our understanding of concepts or of development? In G. Butterworth, J. Rutkowska, & M. Scaife (Eds.),

*Evolution and developmental psychology* (pp. 115–129). Brighton, England: Harvester Press.

Sacher, G. A., & Staffeldt, E. F. (1974). Relation of gestation time to brain weight for placental mammals: Implications for the theory of vertebrate growth. *American Naturalist*, *108*, 593–615.

Sarnat, H. B., & Netsky, M. G. (1981). *Evolution of the nervous system*. Oxford University Press.

Scheibel, A. (1990). Some structural and developmental correlates of speech. In K. R. Gibson & A. Petersen (Eds.), *Brain and behavioral maturation: Biosocial perspectives*. Hawthorne, NY: Aldine.

Short, R. V. (1976). The evolution of human reproduction. *Proceedings of the Royal Society, Series B*, *195*, 3–24.

Sprague, J. M., & Meickle, T. H. (1965). The role of the superior colliculus in visually guided behavior. *Experimental Neurology*, *11*, 115–146.

Terrace, H. S. (1979). *Nim, a chimpanzee who learned sign language*. New York: Knopf.

Terrace, H. S. (1980). Is problem solving language? In T. A. Sebeok & J. Umiker-Sebeok (Eds.), *Speaking of apes* (pp. 385–405). New York: Plenum.

Tinbergen, N. (1952). The curious behavior of the stickleback. *Scientific American*, *187*, 22–26.

Tinbergen, N. (1960). *The herring gull's world*. Garden City, NY: Doubleday.

Trevarthen, C. (1968). Two mechanisms of vision in primates. *Psychologische Forschung*, *31*, 299–337.

Tuttle, R. (1987). *Apes of the world; their social behavior, communication, mentality and ecology*. Park Ridge, NJ: Noyes.

van Lawick-Goodall, J. (1973). The behavior of chimpanzees in their natural habitat. *American Journal of Psychiatry*, *130*, 1–12.

Vauclair, J. (1984). Phylogenetic approach to object manipulation in human and ape infants. *Human Development*, *27*, 321–328.

Vauclair, J., & Bard, K. A. (1983). Development of manipulations with objects in ape and human infants. *Journal of Human Evolution*, *12*, 631–645.

Warren, J. M., & Ackert, K. (1964). *The frontal granular cortex and behavior*. New York: McGraw-Hill.

Wiersma, G. A. G. (1978). The original definition of command neuron. *Behavioral and Brain Sciences*, *1*, 34–35.

Wiersma, G. A. G., & Ikeda, K. (1978). Interneurons commanding swimmeret movements in the crayfish, *Procambarus clarki* (Girard). *Comparative Biochemistry and Physiology*, *12*, 509–525.

Winick, M. (1981). Food and the fetus. *Natural History*, *90*, 76–81.

Wynn, T. G. (1979). The intelligence of later Acheulean hominids. *Man*, *14*, 371–391.

Wynn, T. G. (1981). The intelligence of Oldowan hominids. *Journal of Human Evolution*, *10*, 529–541.

Wynn, T. (1988). Tools and the evolution of human intelligence. In R. W. Byrne & A. Whiten (Eds.), *Machiavellian intelligence* (pp. 271–284). Oxford: Clarendon Press.

Yakovlev, P. I., & LeCours, A. R. (1967). The myelinogenetic cycles of regional maturation of the brain. In A. Minkowski (Ed.), *Regional development of the brain in early life* (pp. 3–70). Oxford: Blackwell Scientific Publications.

# 4 Why big brains are so rare: Energy costs of intelligence and brain size in anthropoid primates

*Sue Taylor Parker*

## Introduction

Little attention has been devoted to the question why so few species are large-brained and highly intelligent. Neglect of this question may have followed from a general resistance to the idea that intelligence arose as an adaptation through natural selection. Controversy about the evolution of human brain size and intelligence has been historic: Darwin (1871/1938) argued that intelligence is the product of selection, whereas Wallace, his codiscoverer of natural selection, argued that it is God-given (Gould, 1980; Parker, P&G1).

Although few living scholars would agree with Wallace, some biologists and anthropologists argue that the large cortex and higher intellectual abilities of the human are fortuitous by-products of other adaptations (Fiatkowski, 1986; Williams, 1966). Others avoid the issue by arguing that intelligence is a meaningless concept, or they define it in such a general way as to preclude investigation of specific adaptive values. On the other hand, those who do think that intelligence is adaptive sometimes overgeneralize the adaptive advantages of intelligence and consequently wonder why it is so rare among animal species.

In this chapter I argue that occurrences of large cortices and higher intelligence in animals are limited by the high energy costs of this functional complex, as well as by phylogenetic inertia. This argument implies, of course, that intelligence is an adaptation (or a series of adaptations), not an epiphenomenon. It also implies that intelligence is a meaningful concept that is susceptible to comparative study, that it is the product of selection and not a fortuitous by-product, and that certain phylogenetic and energy constraints explain its rarity as an adaptation.

Intelligence can be defined as the ability to understand cause-and-effect and logical relationships and to use that understanding to manipulate the physical and social world. In more specific terms, intelligence involves the ability to identify many variables, to construct flexible relationships among them, and to solve problems through trial and error and insightful manipulation of these variables (Parker & Baars, P&G2). It depends on the existence of specific schemes or algorithms for manipulating data and on sufficient information-processing capacities to store and compare these data. It can be compared along the following dimensions:

1. the numbers and kinds of schemes or algorithms available for manipulating data,
2. the kinds of data on which they operate,
3. the numbers of combinations and coordinations in which they can be combined, and
4. the speed with which these can be accomplished (Case, 1985).

The fact that all of these factors depend upon neurological organization explains the close association between brain size and intelligence (Gibson, 1986, P&G3). As discussed throughout this volume, species-specific patterns of intellectual development in monkeys and apes have been described and compared using Piagetian and neo-Piagetian concepts (Antinucci, 1989; Antinucci, Spinozzi, & Natale, 1986; Antinucci, Spinozzi, Visalberghi, & Volterra, 1982; Chevalier-Skolnikoff, 1977, 1983; Mathieu & Bergeron, 1983; Parker, 1973; Piaget, 1952, 1962; Redshaw, 1978; Spinozzi & Natale, 1986; Vauclair, 1982). (See Parker, P&G1, for a discussion of the history of such efforts.)

Large cortices and higher intelligence constitute an expensive adaptive complex from the perspective of energy consumption. In adult humans, brain tissue, especially cortical tissue, is second only to liver tissue in its energy cost per unit (Armstrong, 1985). In the neonate, brain tissue uses the lion's share of metabolic energy. Brain development is therefore apparently a rate-limiting factor in the achievement of onset of reproduction in large-brained species (Lancaster, 1986). This suggests that the evolution of intelligence and large brains will occur only when the benefits exceed the high energy costs.

### Energy expenditures and life history strategy theory

Life history strategy (LHS) theory offers an appropriate framework for understanding the costs and benefits of intelligence as an adaptation because it is concerned with the evolution of the life cycle, that is, with "decisions" about when and how often to breed within the life span (Kenagy, 1987), or how to allocate energy resources throughout the life span. LHS theory, which arose as a branch of behavioral ecology (Cole, 1954), is an optimization theory (Stearns, 1983). It is based on the notion that selection has shaped the timing and sequence of development, that is, that life history parameters are solutions to the problem of partitioning energy optimally among the functions of growth, maintenance, and reproduction at any given age (Gadgil & Bossert, 1970). Investigators disagree concerning relationships among life history variables, particularly the degree to which various features can vary independently (Horn, 1978; Stearns, 1976, 1983).

Although the concept of energy allocation is central to the concept of life history strategies, actual measurement of energy expenditure in reproductive and other activities has lagged behind theoretical development (Kenagy, 1987). Basal metabolic (or resting) rate (BMR) and daily energy expenditure (DEE) are the most common measures of energy expenditure (both of these measures use standard physical units [kilocalories or kilojoules] per time unit). DEE offers the advantage of directly measuring energy expenditure in specific activities (Kenagy, 1987).[1] BMR offers the advantage of context standardization, but it is not the best measure of energy expenditure (McNab, 1988, p. 26).[2] Very few data exist on either BMR or DEE in animals.

Because production processes (growth and reproduction) are the most energy expensive aspects of the life cycle for most mammals, and because relative reproductive success determines the configuration of future generations, selection operates on both the timing and the magnitude of growth and reproduction. These parameters vary within species in relation to age, sex, and body size and in relation to such environmental variables as climate and plant growth. The timing of the breeding and birth seasons, for example, is generally set to minimize the total energy expenditure on reproduction. In the golden-mantled ground squirrel, for example, "while total energy demand rises to an annual peak [in June] because of reproduction, the energy costs for thermostatic heat production are greatly reduced because of increased environmental temperature, and the ease of obtaining food is increased by an abundance of plant foods" (Kenagy, 1987, p. 259). This is an example of the operation of "optimality principles," including the tendency to optimize "net return" on total investment and to seek "joint utility" according to the balance among conflicting selection pressures and within the constraints of fixed costs in the evolution of life history (Ghiselin, 1987).

According to the classic model of *K* and *r* life history strategies, such features as body size, onset of reproduction, length, number and timing of reproductive efforts, and developmental state at birth are coadapted features that change together (MacArthur & Wilson, 1967): *K* strategists are large, long-lived animals with a low reproductive rate (i.e., late maturity and small litters) and a long life span that display considerable parental investment; *r* strategists, in contrast, are small, short-lived animals with a high reproductive rate that display little parental investment.

Also, according to the classical model, *K* and *r* strategies correlate with density dependent and density independent mortality, respectively. Although widely used, and heuristically stimulating, the *K*–*r* dichotomy is inadequate to classify the variety of life history strategies even within the order primates. The classic model suffers from its attempt to explain all life history variables in terms of one selective parameter, density dependent versus density independent selection, ignoring such other parameters as trophic level and successional stage of the habitat (see Parker, 1988, for a summary).

Although many behavioral ecologists argue that life history features are directly shaped by natural selection operating under various constraints (Cole, 1954; Gadgil & Bossert, 1970), others argue that these features are indirectly shaped by allometrically determined consequences of body size

(Western, 1979) and/or metabolic rate (Rasmussen & Izard, 1988), or by brain size (Sacher & Staffeldt, 1974).[3] The allometric relationship between body mass and "standard" metabolic rate has been widely investigated (McNab, 1988). The decreasing metabolic rate that occurs with increasing body size results from greater heat retention attendant on changing surface-to-volume ratios (Gould, 1977b).

One complication in this picture lies in the possible variables intervening between body mass and BMR:

> The influence of body mass on the basal rate of metabolism is simple if, and only if, body mass affects no variable associated with basal rate. Many secondary physiological and ecological relations, however, vary with body mass *and* with the basal rate of metabolism, so these relations are included when basal rate is scaled to body mass, whether such inclusion is intended or not. (McNab, 1988, p. 43)

McNab mentions, for example, the correlation between low BMR and torpor in small marsupial mammals, and conversely the relationship between high BMR and small body size in eutherian mammals, which follows from continuous endothermy. He also mentions the relationship between low BMR and large body size in folivorous mammals that eat plants that contain toxic secondary compounds (McNab, 1978) and animals that eat ants and termites that contain toxic compounds (Rasmussen & Izard, 1988). Although phylogenetic factors also influence BMR, grouping mammals according to diet "food habit" gives a better fit between BMR and body size than grouping according to order (McNab, 1988). Other complicating variables intervening between body mass and BMR include the higher metabolic costs associated with placental births, lactation, and large brains, as discussed later.

## Life history variables in nonhuman primates

Because they are statistical measures of population parameters, life history data on wild populations of primates can come only from long-term field studies. Because they rely on close observation of physiological features, data on sexual maturation and gestation can come only from long-term colony and laboratory studies. Given that these parameters vary among wild populations of the same species and between wild and captive populations (Altmann, 1979), averages require multiple studies and populations, and sample sizes should be specified.

Although studies of primate life history have been increasing in number over the past decade (e.g., Altmann, Altmann, Hausfater, & McCusky, 1977; Drickamer, 1974; Dunbar, 1980; Fairbanks & McGuire, 1986; Harcourt, Fossey, Stewart, & Watts, 1980; Kurland, 1977; Robinson, 1987a,b; Rowell; 1975, 1977a,b; Rowell & Richards, 1979; Sigg, Stolba, Abegglen, & Dasser, 1982; Teleki, Hunt, & Pfiffeling, 1976), relatively few populations and species have been studied from this perspective. Several secondary sources cite life history data on a large number of primate species (e.g., Ardito, 1976; Cutler, 1976; Eisenberg, 1981; Napier & Napier, 1967; Sacher & Staffeldt 1974;

Shultz, 1969), but only one secondary source (Harvey & Clutton-Brock, 1985) provides statistical analyses of the relationships among variables. See Table 4.1 for a reprint of their primary data.

Harvey and Clutton-Brock (1985) have analyzed the relationships among the following life history variables in primates: male and female body weights (kilograms), neonatal weight (grams), adult and neonatal brain weights (grams), litter size, gestation length (days), weaning age (days), female and male ages at maturity (months), female age at first breeding (months), estrus length (days), and maximum recorded life span (years). Although all these variables except cycle length and number of offspring are highly correlated, some are much more highly correlated than others. The highest positive correlations are between neonatal and adult brain sizes (0.99), neonatal and adult brain sizes and ages at sexual maturity (0.97), neonatal and adult brain sizes and age at first breeding in females (0.96), and neonatal weight and adult weight (0.97).

They have also analyzed the taxonomic distribution of variance in these features. Their analysis revealed the following patterns:

> Within species, the greatest variance occurred in the interbirth interval and the age at first breeding, and the least variance occurred in adult brain weight.
> Within genera, the greatest variance occurred in life span, and the least in first breeding age and neonatal brain weight.
> Within subfamilies, the greatest variance occurred in estrus cycle length and gestation length, and the least in neonatal brain weight.
> Within families, the greatest variance occurred in the age at maturity and age at first breeding, and the least in estrus cycle length.

Overall, they show that "around 15% of the total variation in their sample is attributable to differences between species within genera plus between genera within subfamilies; the remaining 85% is attributable to differences among subfamilies and families" (Harvey & Clutton-Brock, 1985, p. 567). A more extended discussion of the significance of life history variables and their taxonomic distribution is given in the following section.

Because of this pattern of variation, and because subfamilies often represent separate radiations, Harvey and Clutton-Brock argue that the subfamily is the most appropriate unit for comparison. Given the unequal temporal, adaptive, and genetic distances among species in various primate genera, subfamilies, and families, however, it seems likely to me that the optimum unit for comparison varies from group to group and from problem to problem. As indicated earlier, for example, feeding habits may sometimes give better correlations with body weight and BMR than will phyletic relationships.

In a carefully controlled study of the relationships among body weight, brain size, and metabolic rate in four species in the family Lorisidae, for example, Rasmussen and Izard (1988) found that BMR correlated highly with gestation length, lactation period, and age at asymptotic approach to adult weight, and somewhat less highly with litter size and growth rate constant.

Table 4.1. *Life history variables in primates*

| | (1) | (2) | (3) | (4) | (5) | (6) | (7) | (8) | (9) | (10) | (11) | (12) | (13) | (14) |
|---|---|---|---|---|---|---|---|---|---|---|---|---|---|---|
| LEMUROIDEA | | | | | | | | | | | | | | |
| LEMURIDAE | | | | | | | | | | | | | | |
| Lemur catta | 2.50 | 2.90 | 135 | 88.2 | 1.2 | 105 | 39 | 30.0 | — | 27.1 | 511 | 30.0 | 8.8 | 25.6 |
| Lemur fulvus | 1.90 | 2.50 | 118 | 81.4 | 1.0 | 135 | — | 27.8 | 10.0 | 30.8 | 547 | 23.3 | 10.7 | 25.2 |
| Lemur macaco | 2.50 | 2.50 | 128 | 100.0 | 1.0 | — | 33 | — | 24.0 | 27.1 | — | — | — | 25.6 |
| Lemur mongoz | 1.80 | 1.80 | 128 | — | 1.1 | — | 37 | — | — | — | — | — | — | 21.8 |
| Lemur rubriventer | — | — | — | — | — | — | — | — | — | — | — | — | — | 27.2 |
| Varecia variegatus | 3.10 | 3.60 | 102 | 107.5 | 2.0 | 90 | 30 | 23.5 | 5.2 | — | 365 | — | 10.6 | 34.2 |
| Hapalemur griseus | 2.00 | 2.00 | 140 | 48.0 | 1.0 | — | — | 28.6 | — | 12.1 | — | 24.0 | — | 14.7 |
| Lepilemur m. mustelinus | 0.64 | 0.61 | 135 | 34.5 | 1.0 | 75 | — | — | 21.0 | — | — | 21.0 | 2.9 | 9.5 |
| Lepilemur m. ruficaudatus | 0.75 | 0.75 | — | — | — | — | — | — | — | — | — | — | — | — |
| CHEIROGALEIDAE | | | | | | | | | | | | | | |
| Cheirogaleus major | 0.40 | 0.40 | 70 | 18.0 | 2.0 | — | 30 | — | — | 8.8 | — | — | — | 5.9 |
| Cheirogaleus medius | 0.18 | 0.18 | — | 19.0 | 1.0 | — | — | — | — | 9.0 | — | — | — | 2.9 |
| Microcebus coquereli | 0.30 | 0.30 | 89 | 12.0 | 1.5 | — | — | — | — | — | — | — | — | — |
| Microcebus murinus | 0.08 | 0.08 | 62 | 6.5 | 1.9 | 40 | 50 | 11.50 | 9.5 | 15.5 | 312 | — | — | 1.8 |
| Phaner furcifer | 0.40 | 0.44 | — | — | — | — | — | — | — | — | — | — | — | 7.3 |
| INDRIIDAE | | | | | | | | | | | | | | |
| Indri indri | 10.50 | 10.50 | 160 | 300.0 | 1.0 | 365 | — | — | — | — | 912 | — | — | 34.5 |
| Avahi laniger | 1.30 | 1.30 | — | — | 1.0 | 150 | — | — | — | — | 365 | — | — | 10.0 |
| Propithecus diadema | 7.50 | 7.50 | — | — | — | — | — | — | — | — | — | — | — | 37.0 |
| Propithecus verreauxi | 3.50 | 3.70 | 140 | 107.0 | 1.0 | 180 | — | — | 30.0 | — | 360 | 30.0 | — | 27.5 |
| DAUBENTONIIDAE | | | | | | | | | | | | | | |
| Daubentonia madagascariensis | 2.80 | 2.80 | — | — | 1.0 | — | — | — | 29.0 | — | 912 | — | — | 45.2 |
| LORISIDAE | | | | | | | | | | | | | | |
| LORISINAE | | | | | | | | | | | | | | |
| Loris tardigradus | 0.26 | 0.29 | 163 | 12.7 | 1.6 | — | 40 | — | 13.0 | 12.0 | — | — | 2.7 | 6.7 |
| Nycticebus coucang | 1.20 | 1.30 | 193 | 49.3 | 1.0 | 90 | 40 | — | — | 14.5 | — | — | 4.0 | 10.0 |
| Arctocebus calabarensis | 0.31 | 0.32 | 134 | 25.2 | 1.0 | 115 | 39 | — | 9.5 | 9.5 | — | — | 2.3 | 7.7 |
| Perodicticus potto | 1.08 | 1.02 | 193 | 46.5 | 1.1 | 150 | 39 | 25.0 | 18.0 | 10.0 | 349 | 18.0 | — | 14.3 |
| GALAGINAE | | | | | | | | | | | | | | |
| Galago alleni | 0.27 | 0.23 | — | 24.0 | 1.3 | — | — | — | 8.0 | — | — | — | — | 6.1 |
| Galago crassicaudatus | 1.26 | 1.42 | 135 | 47.4 | 1.1 | 90 | — | 16.5 | 12.0 | 15.0 | 360 | — | 4.0 | 11.8 |
| Galago demidovii | 0.62 | 0.63 | 111 | 7.5 | 1.2 | 45 | — | 30.8 | 8.0 | 14.0 | 330 | 9.0 | 1.2 | 2.7 |
| Galago elegantulus | 0.28 | 0.29 | 135 | — | — | — | — | — | — | — | — | — | — | 5.8 |
| Galago senegalensis | 0.21 | 0.24 | 124 | 11.5 | 1.6 | 75 | — | 10.8 | 6.7 | 16.0 | 200 | — | 2.3 | 4.8 |
| TARSIIDAE | | | | | | | | | | | | | | |
| Tarsius spectrum | 0.20 | 0.20 | 157 | 30.0 | 1.0 | 68 | 24 | 17.0 | 14.0 | 12.0 | 152 | — | — | 3.8 |
| Tarsius syrichta | 0.12 | 0.13 | 180 | 26.2 | 1.0 | — | 28 | — | — | — | — | — | — | 4.0 |
| CALLITRICHIDAE | | | | | | | | | | | | | | |
| Callithrix humeralifer | 0.30 | 0.30 | — | — | — | — | — | — | — | — | — | — | — | — |
| Callithrix jacchus | 0.29 | 0.31 | 148 | 28.0 | 2.1 | 63 | 16 | 17.0 | 12.0 | 12.0 | 157 | 16.7 | 4.4 | 7.9 |
| Saguinus fuscicollis | 0.37 | 0.42 | 149 | 40.0 | 1.5 | 90 | — | 24.1 | — | — | 242 | — | — | 9.3 |
| Saguinus o. geoffroyi | 0.51 | 0.50 | — | 50.0 | 1.9 | 55 | — | — | — | — | 243 | — | — | 10.5 |
| Saguinus midas midas | 0.53 | 0.60 | 127 | 36.0 | 2.0 | 70 | 16 | 24.0 | 20.0 | 13.0 | 240 | — | — | 10.4 |
| Saguinus nigricollis | 0.46 | 0.47 | — | 43.5 | 1.9 | 80 | — | — | — | — | — | — | — | 8.9 |
| Saguinus oedipus oedipus | 0.51 | 0.45 | 145 | 43.2 | 1.9 | — | 16 | — | 18.0 | 13.0 | 280 | — | 4.9 | 9.0 |
| Saguinus midas tamarin | 0.30 | 0.26 | — | — | — | — | — | — | — | — | — | — | — | 10.6 |
| Leontopithecus rosalia | 0.55 | 0.56 | 129 | 53.6 | 1.8 | 90 | — | 35.6 | 18.0 | 14.0 | 304 | 28.7 | — | 12.9 |
| Cebuella pygmaea | 0.14 | 0.16 | 136 | 16.0 | 2.1 | 90 | — | 24.0 | 24.0 | 10.0 | 154 | 24.0 | — | 4.2 |
| CALLIMICONIDAE | | | | | | | | | | | | | | |
| Callimico goeldii | 0.53 | 0.65 | 154 | 48.6 | 1.0 | 65 | 27 | 15.8 | 8.5 | 9.0 | 167 | 16.5 | 5.8 | 10.8 |
| CEBIDAE | | | | | | | | | | | | | | |
| CEBINAE | | | | | | | | | | | | | | |
| Cebus albifrons | 2.60 | 2.60 | — | 234.0 | — | 270 | — | — | 43.1 | — | — | — | — | 82.0 |
| Cebus apella | 2.10 | 2.86 | 160 | 248.0 | 1.0 | — | 18 | 42.0 | — | 40.0 | — | 56.0 | — | 71.0 |
| Cebus capucinus | 2.70 | 3.80 | — | 230.0 | 1.0 | — | — | — | — | — | — | — | 29.0 | 79.2 |
| Cebus nigrivittatus | 2.30 | 2.90 | — | — | 1.0 | — | — | — | — | — | — | — | — | 80.8 |
| Saimiri oerstedii | 0.58 | 0.75 | — | — | 1.0 | — | — | — | — | — | — | — | — | 25.7 |
| Saimiri sciureus | 0.58 | 0.75 | 170 | 195.0 | 1.0 | — | 18 | 46.3 | — | 21.0 | 414 | — | — | 24.4 |

Table 4.1. (*cont.*)

| | (1) | (2) | (3) | (4) | (5) | (6) | (7) | (8) | (9) | (10) | (11) | (12) | (13) | (14) |
|---|---|---|---|---|---|---|---|---|---|---|---|---|---|---|
| ALOUATTINAE | | | | | | | | | | | | | | |
| Alouatta caraya | 5.70 | 6.70 | — | — | 1.0 | — | — | — | — | — | — | 60.0 | — | 56.7 |
| Alouatta palliata | 5.70 | 7.40 | 187 | 480.0 | 1.0 | 630 | 16 | 45.0 | 45.0 | 13.0 | 675 | — | 30.8 | 55.1 |
| Alouatta seniculus | 6.40 | 8.10 | — | — | 1.0 | — | — | — | — | — | — | — | — | 57.9 |
| ATELINAE | | | | | | | | | | | | | | |
| Ateles belzebuth | 5.80 | 6.20 | — | — | 1.0 | — | — | — | — | — | 760 | — | — | 106.6 |
| Ateles fusciceps | 9.10 | 8.90 | 226 | — | 1.0 | 365 | 26 | 58.5 | 51.0 | 20.0 | 1095 | 57.0 | — | 114.7 |
| Ateles geoffroyi | 5.80 | 6.20 | 229 | 426.0 | 1.0 | — | 26 | — | 48.0 | 20.0 | 870 | — | 64.0 | 110.9 |
| Ateles paniscus | 5.80 | 6.60 | — | 480.0 | 1.0 | — | — | — | — | — | 880 | — | — | 109.9 |
| Lagothrix lagothricha | 5.80 | 6.80 | 225 | 450.0 | 1.0 | 315 | 25 | — | 98.0 | 12.0 | 720 | — | — | 96.4 |
| Brachyteles arachnoides | 9.50 | 9.50 | — | — | — | — | — | — | — | — | — | — | — | 120.1 |
| AOTINAE | | | | | | | | | | | | | | |
| Aotus trivirgatus | 1.00 | 0.92 | 133 | 98.0 | 1.0 | 75 | — | — | — | 12.6 | 220 | — | 10.1 | 18.2 |
| Callicebus moloch | 1.05 | 1.10 | — | — | 1.0 | — | — | — | — | — | 365 | — | — | 19.0 |
| Callicebus torquatus | 1.10 | 1.10 | — | — | 1.0 | 140 | — | 48.0 | — | — | 365 | — | — | 22.4 |
| PITHECIINAE | | | | | | | | | | | | | | |
| Pithecia monachus | — | — | — | — | — | — | — | — | — | — | — | — | — | 38.1 |
| Pithecia pithecia | 1.40 | 1.60 | 163 | — | 1.0 | — | — | — | — | 13.7 | 365 | — | — | 31.7 |
| Chiropotes chiropotes | 3.00 | 3.00 | — | — | — | — | — | — | — | 15.0 | — | — | — | 58.2 |
| Chiropotes satanas | — | — | — | — | — | — | — | — | — | — | — | — | — | 53.0 |
| Cacajao calvus | — | — | — | — | — | — | — | — | — | — | — | — | — | 73.3 |
| Cacajao c. rubicundus | — | — | — | — | 1.0 | — | — | — | — | — | 1095 | — | — | 75.2 |
| CERCOPITHECIDAE | | | | | | | | | | | | | | |
| CERCOPITHECINAE | | | | | | | | | | | | | | |
| Macaca fascicularis | 4.10 | 5.90 | 162 | 346.0 | 1.0 | 420 | 28 | 46.3 | — | — | 390 | — | — | 69.2 |
| Macaca fuscata | 9.10 | 11.70 | 173 | 503.0 | 1.0 | — | 28 | 60.0 | — | — | — | — | — | 109.1 |
| Macaca maurus | 5.10 | 9.50 | 163 | — | 1.0 | — | — | — | — | — | — | — | — | — |
| Macaca mulatta | 3.00 | 6.20 | 167 | 481.0 | 1.0 | — | 29 | 43.3 | 34.0 | 21.6 | 360 | 38.0 | 54.5 | 95.1 |
| Macaca nemestrina | 7.80 | 10.40 | 167 | 473.0 | 1.0 | 365 | — | 47.3 | 35.0 | 26.3 | 405 | — | 66.0 | 106.0 |
| Macaca radiata | 3.70 | 6.60 | 162 | 404.0 | 1.0 | — | 28 | — | — | — | — | — | — | 76.8 |
| Macaca silenus | 5.00 | 6.80 | — | — | 1.0 | — | — | — | — | — | — | — | — | 85.0 |
| Macaca sinica | 3.40 | 6.50 | — | — | 1.0 | — | — | — | — | — | — | — | — | 69.9 |
| Macaca arctoides | 8.00 | 9.20 | 175 | 485.0 | 1.0 | — | 29 | — | — | 30.0 | 525 | — | — | 104.1 |
| Macaca sylvanus | 10.00 | 11.20 | — | — | 1.0 | — | — | 46.0 | 46.0 | — | 945 | — | — | 93.2 |
| Macaca nigra | 6.60 | 10.40 | 176 | 455.0 | 1.0 | — | 36 | 66.0 | 49.0 | 18.0 | 540 | — | — | 94.9 |
| Cercocebus albigena | 6.40 | 9.00 | 177 | 425.0 | 1.0 | 210 | 28 | 72.0 | 48.0 | 21.0 | 510 | — | — | 99.1 |
| Cercocebus atys | 5.50 | 10.20 | 167 | — | 1.0 | — | — | — | — | 18.0 | — | — | — | — |
| Cercocebus galeritus | 5.50 | 10.20 | 171 | — | 1.0 | — | 30 | 78.0 | — | 19.0 | — | — | — | 114.7 |
| Cercocebus torquatus | 5.50 | 8.00 | 171 | — | 1.0 | — | 33 | 78.0 | 32.0 | 20.5 | 390 | — | — | 109.6 |
| Papio c. cynocephalus anubis | 12.00 | 21.00 | 180 | 1068.0 | 1.0 | 420 | 31 | — | — | — | 420 | — | — | 175.1 |
| Papio c. cynocephalus | 15.00 | 20.00 | 175 | 854.0 | 1.0 | — | 31 | 73.0 | 51.0 | — | 630 | 73.8 | 73.5 | 169.1 |
| Papio c. hamadryas | 9.40 | 21.50 | 172 | — | 1.0 | — | — | — | — | 35.6 | — | — | — | 142.5 |
| Papio c. papio | 13.00 | 26.00 | 184 | — | 1.0 | — | — | — | — | — | 423 | — | — | 165.3 |
| Papio c. ursinus | 16.80 | 20.40 | 187 | — | 1.0 | — | — | — | 38.0 | — | — | 60.0 | — | 214.4 |
| Papio leucophaeus | 10.00 | 17.00 | 176 | — | 1.0 | — | 33 | 60.0 | 42.0 | 28.6 | 450 | — | — | 152.7 |
| Papio sphinx | 11.50 | 25.00 | 173 | 613.0 | 1.0 | — | 35 | 60.5 | — | 29.1 | 523 | — | — | 159.4 |
| Theropithecus gelada | 13.60 | 20.50 | 170 | 464.0 | 1.0 | 450 | 34 | 54.0 | 49.5 | — | 525 | — | — | 131.9 |
| Cercopithecus aethiops | 3.56 | 4.75 | 163 | 314.0 | 1.0 | — | 33 | 47.7 | 30.0 | 31.0 | 365 | — | — | 59.8 |
| Cercopithecus ascanius | 2.90 | 4.20 | — | — | 1.0 | 180 | — | — | — | — | — | — | — | 66.5 |
| Cercopithecus campbelli | 3.60 | 3.60 | — | — | 1.0 | — | — | — | 40.0 | — | — | — | — | 65.8 |
| Cercopithecus cephus | 2.90 | 4.10 | — | — | 1.0 | — | — | — | — | — | — | — | — | 63.6 |
| Cercopithecus diana | — | — | — | 450.0 | 1.0 | — | — | — | — | 34.8 | — | — | — | 77.3 |
| Cercopithecus erythrotis | — | — | — | — | 1.0 | 180 | — | — | — | — | — | — | — | 65.2 |
| Cercopithecus l'hoesti | 4.70 | 8.50 | — | — | 1.0 | — | — | — | — | — | — | — | — | 76.0 |
| Cercopithecus mitis | 4.40 | 7.60 | 140 | 402.0 | 1.0 | — | 30 | 55.5 | 62.0 | — | 413 | — | — | 75.0 |
| Cercopithecus mona | 2.50 | 4.40 | — | 284.0 | 1.0 | — | — | — | — | — | — | — | — | 66.0 |
| Cercopithecus neglectus | 3.96 | 7.00 | 182 | 260.0 | 1.0 | — | — | 53.5 | 48.0 | 20.0 | 600 | 72.0 | — | 70.8 |
| Cercopithecus nictitans | 4.20 | 6.60 | — | — | 1.0 | — | 28 | — | — | — | — | — | — | 78.6 |
| Cercopithecus pogonias | 3.00 | 4.50 | — | — | 1.0 | — | — | — | — | — | — | — | — | 71.1 |
| Cercopithecus aethiops pygerythrus | 3.00 | 5.40 | — | 325.0 | 1.0 | — | — | — | — | — | — | — | — | 62.7 |

Table 4.1. (*cont.*)

| | (1) | (2) | (3) | (4) | (5) | (6) | (7) | (8) | (9) | (10) | (11) | (12) | (13) | (14) |
|---|---|---|---|---|---|---|---|---|---|---|---|---|---|---|
| Cercopithecus talapoin | 1.10 | 1.40 | 162 | 180.0 | 1.0 | 180 | 36 | 53.0 | 48.0 | 22.3 | 365 | 114.0 | — | 37.7 |
| Erythrocebus patas | 5.60 | 10.00 | 163 | — | 1.0 | — | — | 36.0 | 33.0 | 20.2 | 420 | 42.0 | — | 106.6 |
| Allenópithecus nigroviridis | — | — | — | — | — | — | — | — | — | — | — | — | — | 62.5 |
| COLOBINAE | | | | | | | | | | | | | | |
| Presbytis aygula | 6.20 | 6.30 | — | — | 1.0 | — | — | — | — | — | — | — | — | 80.3 |
| Presbytis cristatus | 8.10 | 8.60 | — | — | 1.0 | — | — | — | — | — | — | — | — | 64.0 |
| Presbytis entellus | 11.40 | 18.40 | 168 | — | 1.0 | — | 22 | 51.0 | 42.0 | 20.0 | — | — | — | 135.2 |
| Presbytis geei | 8.10 | 8.60 | — | — | 1.0 | — | — | — | — | — | — | — | — | 81.3 |
| Presbytis johnii | 12.00 | 14.80 | — | — | 1.0 | — | — | — | — | — | — | — | — | 84.6 |
| Presbytis melalophos | 6.60 | 6.70 | — | — | 1.0 | — | — | — | — | — | — | — | — | 80.0 |
| Presbytis obscura | 6.50 | 8.30 | 150 | 485.0 | 1.0 | — | — | — | — | — | — | — | — | 67.6 |
| Presbytis potenziani | 6.40 | 6.50 | — | — | 1.0 | — | — | — | — | — | — | — | — | — |
| Presbytis rubicundus | 6.30 | 6.30 | — | — | 1.0 | — | — | — | — | — | — | — | — | 92.7 |
| Presbytis senex | 7.80 | 8.50 | — | 360.0 | 1.0 | 225 | — | — | — | — | 569 | — | — | 64.9 |
| Pygathrix nemaeus | — | — | 165 | — | 1.0 | — | — | — | — | — | 495 | — | — | 108.5 |
| Rhinopithecus roxellanae | — | — | — | — | — | — | — | — | — | — | — | — | — | 121.7 |
| Nasalis larvatus | 9.90 | 20.30 | 166 | 450.0 | 1.0 | — | — | — | — | — | — | — | — | 94.2 |
| Colobus angolensis | 9.00 | 10.70 | — | — | 1.0 | — | — | — | — | — | — | — | — | 73.5 |
| Colobus badius | 5.80 | 10.50 | — | — | 1.0 | — | — | — | — | — | 520 | — | — | 73.8 |
| Colobus guereza | 9.25 | 11.80 | — | 445.0 | 1.0 | 390 | — | 55.0 | — | — | 365 | — | — | 82.3 |
| Colobus polykomos | 8.40 | 10.40 | 170 | 597.0 | 1.0 | — | — | 102.0 | — | 26.0 | 380 | — | 38.5 | 76.7 |
| Colobus satanas | 9.50 | 12.00 | 195 | — | 1.0 | 480 | — | — | — | — | — | — | — | 80.2 |
| Colobus verus | 3.60 | 3.80 | — | — | 1.0 | — | — | — | — | — | — | — | — | 57.8 |
| HYLOBATIDAE | | | | | | | | | | | | | | |
| Hylobates agilis | 5.70 | 6.00 | — | — | 1.0 | — | — | — | — | — | — | — | — | 110.0 |
| Hylobates concolor | 5.80 | 5.60 | — | — | 1.0 | — | — | — | — | — | — | — | — | 131.7 |
| Hylobates hoolock | 6.50 | 6.90 | — | — | 1.0 | 700 | 28 | — | 84.0 | — | — | — | 55.0 | 108.5 |
| Hylobates klossii | 5.90 | 5.70 | 210 | — | 1.0 | 330 | — | — | — | — | — | — | — | 91.1 |
| Hylobates lar | 5.30 | 5.70 | 205 | 410.5 | 1.0 | 730 | 27 | 111.7 | 108.0 | 31.5 | 969 | 78.0 | 50.1 | 107.7 |
| Hylobates moloch | 5.70 | 6.00 | — | — | 1.0 | — | — | — | — | — | — | — | — | 113.7 |
| Hylobates pileatus | — | — | — | — | — | — | — | — | — | — | — | — | — | 114.2 |
| Hylobates syndactylus | 10.60 | 10.90 | 231 | 517.0 | 1.0 | — | — | — | — | — | — | — | — | 121.7 |
| PONGIDAE | | | | | | | | | | | | | | |
| Pongo pygmaeus | 37.00 | 69.00 | 260 | 1728.0 | 1.0 | 1095 | 30 | 128.0 | 84.0 | 50.0 | 1025 | 115.5 | 170.3 | 413.3 |
| Pan troglodytes | 31.10 | 41.60 | 228 | 1756.0 | 1.0 | 1460 | 36 | 138.0 | 118.0 | 44.5 | 1825 | 156.0 | 128.0 | 410.3 |
| Gorilla gorilla | 93.00 | 160.0 | 256 | 2110.0 | 1.0 | 1583 | 28 | 118.2 | 78.0 | 39.3 | 1460 | 120.0 | 227.0 | 505.9 |
| HOMINIDAE | | | | | | | | | | | | | | |
| Homo sapiens | 40.10 | 47.90 | 267 | 3300.0 | 1.0 | 720 | 28 | 232.0 | 198.0 | 60.0 | 1440 | — | 384.0 | 1250.0 |

NOTES: Column headings refer to (1) female weight in kg; (2) male weight in kg; (3) gestation length in days; (4) weight of individual neonates in g; (5) number of offspring per litter; (6) weaning age in days; (7) length of estrous cycle in days; (8) age at first breeding for females in months; (9) age at sexual maturity for females in months; (10) maximum recorded life span in years; (11) interbirth interval in days; (12) age at sexual maturity for males in months; (13) neonatal brain weight in grams; (14) adult brain weight in grams.

*Source:* From Harvey, Martin, & Clutton-Brock (1986), in Smuts et al. (Eds), *Primate societies;* © 1987 by The University of Chicago Press.

They discovered this pattern by holding body and brain size constant across two galago and two loris species (they were also, of course, holding the phyletic relationship constant by examining members of one family). They concluded that the more *K* selected life history of the lorises may reflect a low metabolic rate adapted to a diet of toxic insects.

From a taxonomic perspective, the family level is the only feasible level for comparative analysis among the great apes, whereas among such widely radiated Old World monkeys as macaques, baboons, and guenons, the genus may be the most appropriate unit for analysis. The genus level is also preferable in cases such as that of the cebus and spider and squirrel monkeys, in which genera in the same family have been separated longer than have families in such other groups as the great apes (Cronin & Sarich, 1975).

An interesting pattern of variation in life history features has been described, for example, within the guenons: the vervets (*Cercopithecus aethiops*), blue monkeys (*Cercopithecus mitis*), de Brazzas (*Cercopithecus neglectus*), patas (*Erythrocebus patas*), and talapoin (*Miopithecus talapoin*) (Fairbanks & McGuire, 1986; Rowell, 1975, 1977a,b; Rowell & Richards, 1979). These species, which vary by a factor of 5 in female body weight, also display significant differences in age at first reproduction, gestation length, birth interval, and longevity. The *r* strategists, vervets and patas females, display their first reproductive efforts at 36–47 months and 36 months of age, respectively, whereas the *K* strategists, talapoin and de Brazzas females, give birth first at 60 months of age, and the Sykes monkey females at 66 months. Vervets and patas display 12-month birth intervals, whereas Sykes and de Brazzas display 16- and 20-month intervals, respectively (Rowell & Richards, 1979). Interestingly, the smallest species, the talapoin, is (relatively) a *K* strategist, whereas the largest, the patas, is an *r* strategist. Perhaps dietary specialization on noxious insects may be involved in the talapoin also.

A similar pattern of variation has been described for great apes. These species, which vary by a factor of 3 in female body weight, also display differences in onset of reproduction, birth interval, and longevity. Common chimpanzee females first give birth at 13 years of age, whereas gorilla females give birth at 10 years. The birth interval in chimpanzees is 5.75 years, whereas that in gorillas is 4.12 years (Harcourt et al., 1980; Lancaster & Lancaster, 1983; Tutin, 1979). Gorillas live as long as 40 years, and chimpanzees as long as 50 years.

Life history data on *Macaca* (Drickamer, 1974; Kurland, 1977), *Papio* (Altmann et al., 1977; Rowell, 1969), and the marmoset genus *Saguinus* (Epple, 1970; French, 1983; Terborgh & Goldizen, 1985) are also available for intrageneric analysis. Data on marmosets reveals high reproductive potential associated with early sexual maturity, twinning and annual births.

Newly published data on the life history and demography of *Cebus olivaceous* (formerly called *C. nigrivittatus*) in Venezuela reveal that this species is a *K* strategist: Females experience onset of reproduction at 6–9 years of age, and males become sexually mature at 12–15 years of age. Their maximum longevity is 47 years, and the females display menopause. Interestingly, although these small animals are comparable to great apes in these features, their birth interval is only 2 years (Robinson, 1987a,b). This pattern, which contrasts sharply with the more *r* selected pattern of squirrel monkeys, may be due to the demands of brain growth.

## Relationships among LHS variables in mammals

### *Body size*

Although it has been proposed as a primary determinant of life history, this must be an oversimplification given the opportunistic nature of size: Body sizes typically vary within an array of closely related species that have diverged from a common ancestor (Jolly, 1970; Pilbeam & Gould, 1974). The larger species in the array can be viewed as scaled-up versions of a relatively small common ancestor. Among primates, more than 80% of the variance in male and female body weight occurs within genera, subfamilies, and families (Harvey & Clutton-Brock, 1985). (The phenomenon of the relatively small common ancestor is known as "Cope's law" [Stanley, 1973].) Such arrays are found, for example, among galagos, colobus, macaques, baboons, great apes, and the australopithecines. Variation in body size is less pronounced among such small genera as *Saguinus*, whose foraging and feeding strategies depend on their small size (Terborgh, 1983).

Body size among primates sets some of the parameters for locomotor and predatory adaptations (e.g., the possibility of stealth for small animals), as well as setting the general level of energy requirements (larger animals can subsist on lower-grade, less nutritious foods than smaller animals, which tend to have a higher heat loss and hence metabolic rate). Both the amounts and kinds of food required in turn set the parameters for food species and hence for ranging patterns (i.e., wide ranging for scattered, seasonably variable foods, or narrow ranging for abundant leaves or small, continuously ripening fruits) (Terborgh, 1983).

Because dietary niche correlates with body size (larger species specializing in bulkier, less nutritious foods), related species may reduce competition by diverging in size. Cope's law has been explained by the fact that smaller species are less specialized than larger species and hence have greater evolutionary and ecological access to new niches (through their descendants) (Stanley, 1973).

As body size increases in a series of closely related species in the same clade, some organs increase in size proportionally (or isometrically), whereas other organs or physiological variables increase or decrease in size nonproportionally (or allometrically) (according to some scaling factor). Surface-area-functioning organs that service the entire volume of the body (e.g., cross sections of weight-bearing bones, molar crowns, and intestines) must increase more than proportionally to compensate for the fact that the volume increases more, proportionally, than the surface area when a solid is enlarged. As a consequence, surface-area-functioning organs increase in size more than their proportion to the whole, and volume-related organs increase less.

Larger-bodied species tend to exhibit greater sexual dimorphism than do smaller-bodied species (Table 4.1). This is probably because the energy budgets of smaller species are so tight as to preclude investment in male organs of combat, except perhaps on a seasonal basis, as is seen in the male fattening phenomenon in male squirrel monkeys, talapoin monkeys, and crabeating macaques (Hrdy, 1981).

### *Gestation period and maturity at birth*

Length of gestation correlates with neonatal brain size (0.84) (Harvey & Clutton-Brock, 1985) and with the percentage of adult brain size achieved at

Table 4.2. *Representative postnatal brain size increases*

| Precocial | | Intermediate | Altricial | |
|---|---|---|---|---|
| < 2-fold | 2–3-fold | (3–5-fold) | 5–12-fold | > 12-fold |
| Guinea pig | Noctule bats | Porcupine | Lion (5) | Brown bear (58) |
| Llama | Long-nosed bats | Wild boar | House cat (5.8) | Polar bear (47) |
| Zebra | Nutria | Some deer | Tiger (10) | |
| Fur seal | Chinchilla | Lynx | Red fox (10) | |
| Rhesus monkey | Beaver | Gray fox | Tree shrew (6) | |
| Colobine monkey | European hare | Human | Hedgehog (11) | |
| Howler monkey | Elephant | | European rabbit (7.6) | |
| Gibbon | Most deer | | Mouse (9–10) | |
| Gorilla | Antelope | | Rat (9–10) | |
| | Dolphin | | Squirrel (10) | |
| | Galago | | | |
| | Ring-tailed lemur | | | |
| | Some macaques | | | |
| | Baboons | | | |
| | Cebus monkey | | | |
| | Chimpanzee | | | |
| | Orangutan | | | |

*Source:* From Gibson (in press), after Portmann (1967) and Sacher & Staffeldt (1974).

birth (the index of brain advancement) such that the greater the percentage of adult brain weight achieved at birth, the longer the gestation period (Sacher & Staffeldt, 1974). Hence, contrary to popular opinion, larger-brained species undergo less brain size increase than do smaller-brained species (Portmann, cited in Gibson, in press, P&G3) (Table 4.2).

Increased brain size in anthropoid primates (a doubling of brain weight to body weight), as compared with strepsirhine primates, was made possible by reducing the amount of nonneural somatic tissue associated with neural tissue in fetal brains (Sacher, 1982). Metabolic and aerobic support for large brains became feasible with the evolution of fully invasive (hemochorial) placentas (Eisenberg, 1981; Luckett, 1975).

Length of gestation is one of the critical factors differentiating so-called altricial and precocial species: "In mammals, altricial young usually have shorter gestation periods than precocial species, all else being equal" (Case, 1978, p. 254). The fetal state of altricial neonates correlates with large litter size and low prenatal parental investment per infant. The more advanced state of brain development typical of precocial neonates correlates with smaller litter size and higher prenatal parental investment. Maximum parental investment in altricial mammals occurs during lactation. Altricial neonates are most common among nest- or den-building species; precocial neonates are most common among seminomadic grazing species.

The degree of variation in neonatal condition among species within the same genus or family has not been widely studied. In Lorisidae, galagos are more precocial, and lorises more altricial (Izard & Rasmussen, 1985; Rasmussen & Izard, 1988). Two species in the rodent genus *Acomys* also display contrasting precocial and altricial patterns (B. D'Udine, pers. commun., 1986). Rabbits and hares, members of the family Leporidae, likewise display contrasting developmental patterns (Eisenberg, 1981).[4]

Although human neonates are born in a motorically helpless state similar to that of altricial species, they share more characteristics with precocial species than with altricial species (Gibson, 1981, in press, P&G3). Although the altricial–precocial distinction is widely cited, it has limited value and might better be subsumed under the categories of types and amounts of parental and kin investment per stage of development (Parker, 1988).

Gestation length appears to be a fairly conservative factor within primate genera (only 2.3% of the variance in this trait among primates occurs within genera) (Harvey & Clutton-Brock, 1985). But note that a comparison within the family Lorisidae yields a significant difference between galagos and lorises independent of brain and body size (Rasmussen & Izard, 1988) (Table 4.3).

Clearly, gestation length can vary significantly among closely related species of primates. It seems likely that the existence of such protective mechanisms as nests and shelters can allow the evolution of shorter gestation periods in species pushed by constraints on fetal size. The short gestation length of human infants relative to their motoric development, for example, may have been enforced by the constraints of the bipedal pelvis (Leutenegger, 1982; cf. Lindberg, 1982).

### *Growth rates and milk composition*

Milk composition correlates with neonatal pattern: "The milk of altricial mammals typically has relatively high fat and protein content associated with low sugar content, while the milk of precocial mammals has relatively high sugar content but low fat content" (Martin, 1984, p. 102). Generally, mothers bearing altricial infants have a greater peak of milk energy output as well as richer milk than do mothers of the same body size bearing precocial infants. This is one reason that altricial young exhibit more rapid growth than precocial young.

Growth rates in mammals do not correlate with level of maturity at birth, but do correlate (allometrically) with differences in adult body size (Case, 1978). As we shall see, they also correlate negatively with brain size.

The high-sugar, low-fat composition of human milk seems to be designed to optimize brain growth rather than body growth. It seems to put lactation at the energetic limit for human females (Lancaster, 1986).

Table 4.3. *Life history strategy of Lorisidae*

| | *Loris* | *Nycticebus* | *G. seneg.* | *G. crass.* |
|---|---|---|---|---|
| **Adult male weight (g)** | 197<br>7.8<br>2 | 1281<br>252.4<br>6 | 184<br>14.9<br>10 | 1918<br>197.0<br>10 |
| **Adult female weight (g)** | 190<br>16.3<br>2 | 1249<br>74.2<br>4 | 164<br>9.0<br>10 | 1563<br>69.0<br>10 |
| **Cranial capacity ($cm^3$)** | 5.7<br>0.4<br>2 | 10.1<br>0.6<br>3 | 5.3<br>1.6<br>5 | 12.9<br>1.7<br>6 |
| **Gestation length (days)** | 167<br>1.5<br>4 | 192<br>1.0<br>3 | 124<br>2.4<br>28 | 136<br>3.7<br>21 |
| **Lactation length (days)** | 169<br>32.0<br>5 | 180<br>11.3<br>2 | 101<br>11.1<br>28 | 134<br>10.5<br>9 |
| **First estrus (days)** | 340<br>56.5<br>4 | — | 265<br>95.6<br>17 | 445<br>89.8<br>8 |
| **Litter size** | 1.00<br>—<br>9 | 1.00<br>—<br>9 | 1.55<br>0.50<br>42 | 1.64<br>0.63<br>77 |
| ***K* ($days^{-1}$)** | .013<br>.006<br>6 | .014<br>.003<br>5 | .020<br>.004<br>9 | .019<br>.005<br>10 |
| ***I* (days)** | 68.6<br>18.9<br>6 | 89.5<br>23.5<br>5 | 25.4<br>9.3<br>9 | 48.1<br>7.2<br>10 |
| **BMR (ml $O_2$/g · hr)[1]** | 0.38 | 0.24 | 0.72 | 0.43 |

[1]Müller et al., 1985; Müller, 1979; Dobler, 1982; Müller and Jaksche, 1980.

*Source:* From Rasmussen & Izard (1988).

## *Weaning and birth intervals*

Next to age at first reproductive effort and litter size, the birth interval is the most significant determinant of lifetime reproductive rate. The birth interval is conditioned in large part by the energetic demands of lactation and gestation. In fact, lactation, as the most energetically demanding component of parental investment in mammals, is a rate limiting factor in reproductive rate (Eisenberg, 1981; Kenagy, 1987; Martin, 1984).

Variations in weaning and birth interval can be understood as short-term, facultative responses to variations in nutritional elements. "If there is only one part of the year that is favorable, . . . then each species must make an evolutionary choice, so to speak, as to which parts of the reproductive cycle – conception, gestation, birth, lactation, or weaning – must be protected" (Lancaster & Lee, 1965, p. 504). Because lactation and transport, especially in the later months of the infant's suckling period, are more costly than gestation, this stage of the reproductive cycle may be a pacesetter.

The timing of weaning should be based on an assessment of the cost–benefit ratio of continued lactation versus reproduction of other offspring. This may explain the prevalence of parent–offspring conflict over weaning, "the phase of parental care during which the rate of parental investment drops most sharply" (Martin, 1984, p. 1258). Among monkeys and apes, weaning apparently occurs after the peak velocity of brain growth and after the development of prehensive and locomotor competence sufficient to support independent feeding.

According to Harvey and Clutton-Brock (1985), both birth interval and weaning seem to vary within species and among closely related primate species. With regard to birth interval,

> 15.8% of the variance within primates occurs within species,
> 8.8% within genera,
> 19% within subfamilies, and
> 56.4% within families.

With regard to weaning,

> 7.6% of the variance occurs within species,
> 5.2% within genera,
> 23.9% within subfamilies, and
> 63.3% within families.

In their comparison of reproductive parameters in various species of guenons, mangabeys, and colobus monkeys, for example, Rowell and Richards found considerable variation in birth intervals within and among species (Rowell & Richards, 1979).

Life history theory may also help shed light on the proximate and ultimate factors involved in the timing of various aspects of motor development in monkeys and apes that are closely related to weaning and independent feeding. These should be especially pertinent to the interactions among motor (particularly prehensive vs. locomotor) development (Antinucci, P&G5) and intellectual development and various environmental factors (Antinucci, P&G5; Fragaszy, P&G6; Gibson, P&G3).

### *Juvenile care, practicing, and playing*

Higher primates, like most social mammals, have an extended period of development following weaning and preceding the onset of sexual reproduction. This period, during which they are protected and tolerated by adults, is devoted to play and other forms of practice. The high cost of the juvenile period and of play, as measured in terms of delayed reproduction and increased risk of injury and predation, suggests an even higher payoff in terms of increased reproductive success (Fagen, 1977). Among rhesus monkeys and humans, the payoff comes in relation to increased skill in fighting and other forms of intrasexual competition (Parker, 1984; Symons, 1978). The length of this period may also be determined by the energy demands of reproduction. The timing, length, and intensity of parental investment in male and female offspring will vary with the relative reproductive potential of the two sexes (Clutton-Brock & Albon, 1982; Parker, 1988).

### *Puberty and the onset of reproduction*

Like the birth interval, the timing of the onset of reproduction is a key variable in life history and in reproductive rate (Stearns, 1983). In fact, in late-maturing species, the age of onset of reproduction is more significant to lifetime reproductive rate than is the number of offspring born; this is a consequence of the compound interest effect of offspring reproduction (Cole, 1954).

Mammals in general, higher primates in particular, and humans most notably display adolescent sterility (i.e., a delay between puberty and onset of reproduction); the magnitude of this delay is a significant parameter in life history (Lancaster, 1986; Lancaster & Lancaster, 1983). As in the case of birth interval, the delay may be determined by the energy demands of fetal and neonatal brain development, which require a period of energy storage in the form of body fat. In more general terms, the delay may be determined by the maximum rate of development of the most energy expensive organ. (Among certain mammals, gestation length may be increased by prolonged elaboration of the trophoblast [Eisenberg, 1981].)

The high reproductive payoff of early sexual maturity suggests that it will occur as soon as the benefits of continued growth and development are outweighed by their costs in terms of future reproductive success. In many primate species (e.g., rhesus monkeys), the costs of development after puberty do not preclude the onset of reproduction, and hence the period of adolescent sterility is brief. In humans, on the other hand, the costs of growth and development sufficient to sustain gestation and lactation are so great as to preclude reproduction for several years after puberty (Lancaster, 1986).

Age at first reproduction varies among closely related species: 11% of the variance among primates occurs within species, 0.9% within genera, 7.1% within subfamilies, and 81.1% within families (Harvey & Clutton-Brock, 1985). As indicated earlier, data on various guenons, mangabeys, and colobus monkeys revealed considerable variations among species in age at first birth: Patas females gave birth as early as 2.5 years of age, and vervets as early as 3.5 years, whereas Sykes and blue monkeys and colobus and gray mangabeys gave birth at 4.5 years (Rowell & Richards, 1979, p. 67).

### *Reproductive success*

Because reproductive success relative to other members of the population is both the currency of cost–benefit calculations and the selective mechanism that shapes life histories, it is important to consider this parameter explicitly. Net relative reproductive success of individuals and lineages of course combine to yield reproductive rates of populations. In both cases the net figure depends on the survival rate of offspring, as well as the fecundity of parent and offspring and the timing of their reproductive efforts. As Lack (1966) long ago demonstrated, bearing more offspring may be a less successful strategy than bearing fewer offspring. Fecundity and/or the timing of births and/or the survival of offspring may vary among individuals according to rank, among populations according to resource availability, and among species according to parental strategies. Among rhesus monkeys, for example, "higher ranking females had daughters that matured earlier, they had a higher percent reproduction each year and their infants had a higher probability of surviving to the age of one year" (Drickamer, 1974, p. 77).

### *Cessation of reproduction and longevity*

Cessation of reproduction occurs among a number of large, long-lived species, including great apes and elephants, as well as humans, and in at least one small, long-lived species: *Cebus olivaceous* (Robinson, 1987a,b). It seems to occur when the payoff from investment in juvenile and adult offspring would be greater than the payoff from investment in new offspring (Dawkins, 1978). In human societies in which parental investment extends throughout the life cycle, adult individuals whose parents are still living have higher reproductive success than do those whose parents are dead (Gaulin, 1980). Although data are scanty, maximum life span seems to be species specific and to correlate with *K* or *r* strategies and degree of encephalization (Cutler, 1976; Sacher, 1959).

### *Life history packages*

As indicated earlier, certain life history traits cluster together in primates: Strikingly, both neonatal and adult brain weights correlate positively with age at onset of reproduction for females (0.93 and 0.96) and with age at sexual maturity for males (0.97) (Harvey & Clutton-Brock, 1985).

## The brain as pacemaker of the human life cycle

Slow developmental rates and hence late maturity apparently result from a variety of growth limiting factors, including high metabolic investment in such antipredator defenses as armor, poison, or large body size (Wilbur, Tinkle, & Collins, 1974), as well as from lowered metabolic rates as defenses against plant and insect toxins (McNab, 1978; Rasmussen & Izard, 1988).

In the case of humans, slow developmental rates apparently result from high metabolic investment in a large brain, specifically in a large cortex. The human brain is an energy expensive organ: Among neonatal humans, the brain uses 87% of the BMR; among 2-year-olds, 64%; among 5-year-olds, 44%; among adults, 23%. Only the liver exceeds the brain in its energy

Table 4.4. *Brain size and energy use and basal metabolic rate at various stages of development in humans*

| Approximate age (yr) | Body weight (g) | Brain weight (g) | Brain metabolic rate/body BMR (%) | Glucose needed (g/day) |
|---|---|---|---|---|
| Birth | 3,500 | 475 | 87 | 38 |
| 0.5 | 5,500 | 650 | 64 | 52 |
| 1.0 | 11,000 | 1045 | 53 | 84 |
| 5.0 | 19,000 | 1235 | 44 | 99 |
| 10.0 | 31,000 | 1350 | 34 | 108 |
| Adolescent | 50,000 | 1360 | 27 | 109 |
| Adult | 70,000 | 1400 | 23 | 112 |

*Source:* Adapted from Holliday (1978, p. 124).

Table 4.5. *Cerebral metabolic rate (CMR) in various mammals*

| Species | CMR (cc $O_2$ $[100\ g]^{-1}$ $min^{-1}$) | Brain wt/ body wt | CMR × brain wt/ body wt × BMR |
|---|---|---|---|
| Rat | 7.6 | 0.83 | 4.76 |
| Cat | 4.5 | 1.03 | 6.15 |
| Dog | 3.4 | 0.72 | 4.12 |
| Monkey | 3.7 | 2.57 | 12.42 |
| Human | 3.3 | 2.34 | 20.20 |

*Source:* From Armstrong (1985).

demands; "During the first year of life, up to 65% of the total metabolic rate is devoted to the brain and only 8% to muscle tissue" (Lancaster, 1986, p. 32) (Table 4.4).

Not only is the human brain proportionally larger compared with the brains of other mammals, but its metabolic rate is higher (Mink, Blumenschine, & Adams, 1981); the respiration rate of the cortex is 40% greater than that of the whole brain (Armstrong, 1985). Whereas most vertebrates use 2–8% of their resting metabolism for the central nervous system, primates devote 9–20% of theirs (Table 4.5):

> Primate brains, as represented by *Macaca mulatta* and *Homo sapiens*, use a relatively higher proportion of their body metabolism (9 to 20%, respectively) . . . than do the nonprimate brains of rat, cat, and dog (4 to 6%) . . . a major primate adaptation appears to have been the allocation of a larger proportion of the body's energy supply to the brain. (Armstrong, 1983, p. 1304)

This explains why large-brained primate species must have later onset of reproduction (including a longer period of adolescent sterility), longer gestation periods, longer birth intervals, menopause, and long life spans: At each stage of the life cycle, more energy must be allocated to brain growth and/or maintenance, and for females more energy must be allocated to energy stores for feeding the brains of fetuses and lactating infants.

Given that the brain is the organ of intelligent adaptations, and given that it is apparently the pacesetter of life histories among humans and great apes, we need comparative measures of brain parameters that reflect brain capacities and functions. The cortical surface area, including surfaces buried in sulci and fissures and including the paleocortex and the neocortex, apparently provides such an index because it reflects the number of standard-size cortical modules (25 $cm^2$ each), which in turn reflects processing capacity (Jerison, 1982).

Fissurization, which is a means for increasing surface area while conserving volume, correlates well with cortical area. Larger cortices also differ from smaller cortices in their dendritic arborization, which increases with brain size by an exponent of 0.33 (Jerison, 1982). Brain size does not scale with body surface area, but does with maternal metabolic rate (Martin, 1981). Because "basal (standard) metabolic rate (BMR) estimates the amount of available oxygen and energy per unit time" (Armstrong, 1983, p. 1302), it is a good index of the energy limits on brain functioning.

Surface area of cortex and BMR of cortex should provide consistent measures of brain function, because increased cortical surface area and increased dendritic surface area should demand increased energy expenditure. These measures are superior to such conventional measures of brain function as brain-to-body ratio and encephalization index, both of which discount the brains of large animals and inflate the brains of small animals and hence distort the picture by magnifying differences among closely related species in a clade with the same basic adaptation (Shea, 1985).

### Constraints on the evolution of large brains and intelligence and their implications for ontogeny

The high metabolic costs of large cortices and hence of intelligent adaptations suggest the cause of their rarity. This big brain complex can evolve only under a particular concatenation of events: (1) when its benefits exceed its high costs and (2) when ancestral species have already evolved metabolic systems adequate to support it.

The foregoing model of neurological energy constraints on development in intelligent species suggests that comparative studies of primate intellectual development should be pursued in a life history framework. Such an approach focuses on the relationships among motoric and intellectual development and such life history variables as brain metabolism, reproductive development, particularly sexual maturation and onset of reproduction, and longevity. Given the obvious complexity of the systems involved, such an approach

Table 4.6. *Sensorimotor period intellectual achievement in relation to life history variables in selected primate species*

| Species | Piaget's sensorimotor period intelligence, highest stage[a] | | | Age at highest achievement (yr) | Antinucci's ratio of locomotion to prehension[b] | Age at first birth (yr) | Encephalization index[c] |
|---|---|---|---|---|---|---|---|
| | O.C. | S-M | I | | | | |
| Macaca | 5 | 4 | 3 | 0.5 | 0.25 : 1 | 4 | 8–10 |
| Cebus | 5 | 5 | 3 | 3.0 (?) | 0.75 : 1 | 6 | 10–12 |
| Gorilla | 6 | 6 | 6[d] | 5.0 (?) | 0.50 : 1 | 10 | 6–8 |
| Chimpanzee | 6 | 6 | 6[d] | 5.0 (?) | 0.50 : 1 | 13 | 10–12 |
| Human | 6 | 6 | 6 | 2.0 | 1.50 : 1 | 18 | 28.8 |

[a]O.C. = object concept series; S-M = sensorimotor intelligence series and causality series; I = imitation series.
[b]From Antinucci, P&G5 (see Table 4.4).
[c]From Gibson (1986).
[d]In gestural but not vocal mode.

might benefit from the dynamic systems approach described by Fragaszy (P&G6). Such a combined approach would allow us to see both the proximal developmental relationships among these elements and the broader context of energy constraints and cost–benefit ratios of various life history strategies (Parker, 1988).

On the basis of the currently available evidence, a few species of nonhuman primates can be compared on a few of these variables. They can be ranked according to their highest levels of intelligence, as compared with human children, using a Piagetian framework: great apes, cebus monkeys, lion-tailed macaques (Westergaard, 1987), and stump-tailed and Japanese macaques (Antinucci, 1989; Antinucci et al., 1982, 1986; Chevalier-Skolnikoff, 1977, 1983; Mathien & Bergeron, 1983; Parker, 1973, 1977; Parker & Gibson, 1977; Redshaw, 1978; Spinozzi & Natale, 1986; Vauclair, 1982; Vauclair & Bard, 1983; Visalberghi & Antinucci, 1986). These ranks are consistent with expected differences in age at onset of reproduction and longevity. Data on BMR of the cortex of fissurization are not available for all species.

Comparison of the life histories of these few primate species reveals the following patterns in highly intelligent species, as compared with less intelligent species in the same family, superfamily, or suborder:

- greater percentage of life span devoted to brain growth;
- greater percent of life span devoted to intellectual development;
- slower motor development, specifically later locomotor relative to prehensive development (Antinucci, P&G5);
- later onset of reproduction (Table 4.6).

All of these patterns are consistent with the energy constraint model for slow development in more intelligent primate species, except perhaps for Antinucci's "locomotor/prehensive ratio." Whether this developmental ratio is fortuitous (Antinucci, P&G5), is constrained by cortical maturation schedules (Gibson, P&G3), or is a special adaptation for object manipulation (Parker, 1973; Vauclair & Bard, 1983) is impossible to say at this point. Whatever its cause, its cooccurrence with higher sensorimotor intelligence is significant because it serves to canalize the attention of infants to their own hands and to the properties of objects they manipulate (Parker, 1973; Vauclair & Bard, 1983; Antinucci, P&G5; Gibson, P&G3).

Although consideration of the selection pressures favoring higher sensorimotor intelligence is beyond the scope of this chapter, I favor the hypothesis that certain aspects of fifth and sixth stages of sensorimotor intelligence are adaptive for extractive foraging with tools on a variety of seasonally variable, hidden, embedded foods (Parker & Gibson, 1977, 1979; Gibson, 1986). Likewise, I think that higher, and later developing, periods of intellectual development in humans evolved in conjunction with more advanced forms of extractive foraging on scavenged bones and brains (Parker, 1987) and from social manipulation and social cooperation involved in these and other subsistence activities (Parker, 1984, 1985).

## Conclusions

The evolution and development of intelligence in primates are fully intelligible only in the context of life history strategy theory. Comparative evidence suggests that developmental rates can be influenced by a variety of energy expensive adaptations or, conversely, by reduced metabolic rate. Among humans and great apes, and perhaps cebus, high metabolic investment in a large cortex seems to be the rate limiting factor in development. To test this model we need physiological life history data on a wide variety of primate species. Data on intrageneric and intrafamily variation are particularly important for elucidating questions of evolutionary conservatism and opportunism of specific life history parameters. Comparable physiological life history data on other such large-brained mammals as elephants and dolphins would also be relevant.

### Acknowledgments

I want to thank the various researchers at the Consiglio Nazionale delle Ricerche (CNR) in Rome, Italy, especially the two Francescos, Antinucci and Natale, for numerous discussions and valuable criticisms concerning various ideas in this chapter. I also thank Bruno D'Udine of the CNR for his generous contribution to this chapter. Likewise, I want to thank Kathleen Gibson for her continuing intellectual collaboration with me. I also thank the following readers of this manuscript in its various stages: Aron Branscomb, Thelma Rowell, and Steven Pulos.

## Notes

1. According to Kenagy, "DEE consists of four components of maintenance – resting ($r$), alertness ($A$), locomotion ($L$), and thermoregulation ($T$) – and two components of production – growth ($G$) and reproduction ($R$). Thereby, DEE, in kJ/day is described as

   $$\text{DEE} = r + A + L + T + G + R$$

   These categories are defined in terms of time intervals spent in any condition or activity (measured in field behavioural studies) and corresponding rates of energy expenditure (measured in the laboratory as oxygen consumption and converted to energy equivalents). The $r$ includes basal metabolic rate (BMR) plus the metabolism of maintenance-level digestion.... The production term $G$ (growth) includes storage of energy.... $R$ describes only the metabolic expenditure of energy for production of the testes, gestation of embryos and synthesis of milk; it does not include the intrinsic energy content of sperm ... the newborn young and the milk produced during lactation, which is exported from the mother without being oxidized" (Kenagy, 1987, pp. 265–266).
2. McNab continues: "Mean daily rates of energy expenditure, which may be more appropriate, scale proportionally to approximately $m^{0.50}$, at least at masses less than 100 g" (McNab, 1988, p. 26). He mentions that in other studies the scaling of BMR relative to body mass has varied from $m^{0.67}$ to $m^{0.75}$. He also points out problems inherent in using body mass rather than other measures of body size (e.g., seasonal shifts in fat).
3. Allometry is a special class of nonproportional scaling of organ size relative to total body size when animals grow larger (developmentally or phyletically). Allometry contrasts with isometry, which is proportional scaling of body organs relative to body mass, as discussed later.
4. Spiny desert mice (*Acomys chomus*) and their cogeneric mice (*A. auratis*) differ in the following way in their life histories (D'Udine, pers. commun., 1986):

| | *A. chomus* | *A. auratis* |
|---|---|---|
| Gestation | 42 days | 21 days |
| Litter size | 4 | 12 |
| Age at sexual maturity | 60–80 days | 60–80 days |
| Neonatal state | Eyes open, ears open, body hair | Eyes closed, ears closed, hairless |

## References


Altmann, J. (1979). Age cohorts as paternal sibships. *Behavioral Ecology and Sociobiology*, *6*, 161–164.

Altmann, J., Altmann, S., & Hausfater, G. (1981). Physical maturation and age estimates of yellow baboons, *Papio cynocephalus*, in Amboseli National Park, Kenya. *American Journal of Primatology*, *1*, 389–400.

Altmann, J., Altmann, S. A., Hausfater, G., & McCuskey, S. A. (1977). Life history of yellow baboons: Physical development, reproductive parameters, and infant mortality. *Primates*, *18*(2), 315–330.

Antinucci, F. (Ed.). (1989). *Cognitive structures and development in nonhuman primates*. Hillsdale, NJ: Erlbaum.

Antinucci, F., Spinozzi, G., & Natale, F. (1986). Stage V cognition in an infant gorilla. In D. Taub & F. A. King (Eds.), *Current perspectives in primate social dynamics* (pp. 403–415). New York: Van Nostrand Reinhold.

Antinucci, F., Spinozzi, G., Visalberghi, E., & Volterra V. (1982). Cognitive development in Japanese macaque (*Macaca fuscata*). *Annali dell'Istitute Superiore di Sanita*, *18*(2), 177–184.

Ardito, G. (1976). Check-list of the data on the gestation length of primates. *Journal of Human Evolution*, *5*, 213–221.

Armstrong, E. (1983). Relative brain size and metabolism in mammals. *Science*, *2*, 1302–1304.

Armstrong, E. (1985). Allometric considerations of the adult mammalian brain, with special emphasis on primates. In W. Jungers (Ed.), *Size and scaling in primate biology*. New York: Academic Press.

Case, R. (1985). *Intellectual development: Birth to adulthood.* New York: Academic Press.

Case, T. J. (1978). On the evolution and adaptive significance of postnatal growth rates in the terrestrial vertebrates. *Quarterly Review of Biology*, *53*, 243–280.

Chevalier-Skolnikoff, S. (1977). A Piagetian model for describing and comparing socialization in monkey, ape, and human infants. In S. Chevalier-Skolnikoff & F. Poirier (Eds.), *Primate biosocial development* (pp. 159–188). New York: Garland Press.

Chevalier-Skolnikoff, S. (1983). Sensorimotor development in orangutans and other primates. *Journal of Human Evolution*, *12*, 545–561.

Chism, J., & Rowell, T. E. (1986). Mating and residence patterns of male patas monkeys. *Ethology*, *72*, 31–39.

Chism, J., Rowell, T., & Olsen, D. (1984). Life history patterns of female patas monkeys. In M. Small (Ed.), *Female primates: Studies by women primatologists* (pp. 179–190). New York: Alan R. Liss.

Clutton-Brock, T., & Albon, S. (1982). Parental investment in male and female offspring in mammals. In King's College Biology Group (Eds.), *Current problems in sociobiology* (pp. 223–247). Cambridge University Press.

Cole, L. C. (1954). The population consequences of life history phenomena. *Quarterly Review of Biology*, *29*(2), 103–137.

Cronin, J., & Sarich, V. (1975). Molecular systematics of New World monkeys. *Journal of Human Evolution*, *4*, 357–375.

Cutler, R. G. (1976). Evolution of longevity in primates. *Journal of Human Evolution*, *5*, 169–201.

Darwin, C. (1938). *The descent of man and selection with respect to sex*. New York: Appleton, Century. (Original work published 1871)

Dawkins, R. (1978). *The selfish gene*. Oxford University Press.

Drickamer, L. C. (1974). A ten-year summary of reproductive data for free-ranging *Macaca mulatta. Folia Primatologica*, *21*, 61–80.

Dunbar, R. I. M. (1980). Demographic and life history variables of a population of gelada baboons (*Theropithecus gelada*). *Journal of Animal Ecology*, *49*, 485–506.

Eisenberg, J. F. (1981). *The mammalian radiations*. University of Chicago Press.

Epple, G. (1970). Maintenance, breeding, and development of marmoset monkeys (Callithricidae) in captivity. *Folia Primatologica*, *12*, 56–76.

Fagen, R. M. (1977). Selection for optimal age-dependent schedules of play behavior. *American Naturalist*, *111*, 395–414.

Fairbanks, I. A., & McGuire, M. T. (1986). Age, reproductive value, and dominance-related behaviour in vervet monkey females: Cross-generational influences. *Animal Behavior*, *34*, 1710–1721.

Fiatkowski, K. R. (1986). A mechanism for the origin of the human brain. *Current Anthropology*, *27*, 288–290.

French, J. A. (1983). Lactation and fertility: An examination of nursing and interbirth intervals in cotton-top tamarins (*Saguinus o. oedipus*). *Folia Primatologica*, *40*, 276–282.

Gadgil, M., & Bossert, W. (1970). Life historical consequences of natural selection. *American Naturalist*, *104*, 1–24.

Gaulin, S. J. C. (1980). Sexual dimorphism in the human post-reproductive life-span: Possible causes. *Journal of Human Evolution*, *9*, 277–232.

Ghiselin, M. T. (1987). Evolutionary aspects of marine invertebrate reproduction. In A. C.

Giese, J. S. Pearse, & V. B. Pearse (Eds.), *Reproduction of marine invertebrates*, Vol. 9 (pp. 609–665). Oxford: Blackwell Scientific Publications.
Gibson, K. R. (1981). Comparative neuro-ontogeny: Its implications for the development of human intelligence. In G. Butterworth (Ed.), *Infancy and epistemology* (pp. 52–82). Brighton, England: Harvester Press.
Gibson, K. R. (1986). Cognition, brain size and extraction of embedded foods. In J. C. Else & P. C. Lee (Eds.), *Primate ontogeny and social behaviour* (pp. 93–105). Cambridge University Press.
Gibson, K. R. (in press). Brain and cognition in the growing child: A comparative perspective on the questions of neoteny, altriciality and intelligence. In K. R. Gibson & A. Peterson (Eds.), *Brain maturation and behavioral development: Biosocial dimensions*. Chicago: Aldine.
Gould, S. J. (1977a). *Ontogeny and phylogeny*. Cambridge, MA: Harvard University Press.
Gould, S. J. (1977b). Size and shape. In S. J. Gould (Ed.), *Ever since Darwin* (pp. 171–178). New York: Norton.
Gould, S. J. (1980). Natural selection and the human brain: Darwin vs. Wallace. In S. J. Gould (Ed.), *The panda's thumb*. New York: Norton.
Harcourt, A., Fossey, H. D., Stewart, K. J., & Watts, D. P. (1980). Reproduction in wild gorillas and some comparisons with chimpanzees. *Journal of Reproduction and Fertility* [*Supplement*], *28*, 59–70.
Harvey, P. H., & Clutton-Brock, T. H. (1985). Life history variation in primates. *Evolution*, *39*(3), 559–581.
Harvey, P. H., Martin, R. D., & Clutton-Brock, T. H. (1986). Life histories in comparative perspective. In B. B. Smuts, D. L. Cheney, R. M. Seyfath, R. W. Wangham, & T. T. Struhsaker (Eds.), *Primate societies* (pp. 181–196). University of Chicago Press.
Holliday, M. A. (1978). Body composition and energy needs during growth. In F. Faulkner & N. Tanner (Eds.), *Human growth. Vol. 2. Postnatal growth* (pp. 117–139). New York: Plenum.
Horn, H. S. (1978). Optimal tactics of reproduction and life history. In J. R. Krebs & N. B. Davies (Eds.), *Behavioral ecology, an evolutionary approach* (pp. 411–429). Oxford: Blackwell Scientific Publications.
Hrdy, S. B. (1981). *The woman that never evolved*. Cambridge, MA: Harvard University Press.
Izard, M. K., & Rasmussen, P. T. (1985). Reproduction in the slender loris (*Loris tardignadus malabaricus*). *American Journal of Primatology*, *8*, 153–165.
Jerison, H. (1982). Allometry, brain size and convolutedness. In E. Armstrong & D. Falk (Eds.), *Primate brain evolution: Methods and concepts* (pp. 77–84). New York: Plenum.
Jolly, C. (1970). The large African monkeys as an adaptive array. In J. Napier & P. Napier (Eds.), *Old World monkeys: Evolution, systematics and behavior* (pp. 139–174). New York: Academic Press.
Kenagy, G. J. (1987). Energy allocation for reproduction in the golden-mantled ground squirrel. *Symposia of the Zoological Society* (*London*), *57*, 259–273.
Kurland, J. (1977). Kin selection in the Japanese macaque. *Contributions to Primatology*. Basel, Switzerland: Karger.
Lack, D. (1966). *Population studies of birds*. Oxford University Press.
Lancaster, J. B. (1985). *Commentary on birth in biocultural perspective*. Paper presented to the American Anthropology Association, December 6, Washington, DC.
Lancaster, J. B. (1986). Human adolescence and reproduction. In J. B. Lancaster & B. A. Hamberg (Eds.), *School-age pregnancy and parenthood: Biosocial dimensions* (pp. 17–37). Chicago: Aldine.
Lancaster, J. B., & Lancaster, C. S. (1983). Parental investment: The hominid adaptation. In D. J. Ortner (Ed.), *How humans adapt* (pp. 33–56). Washington, DC: Smithsonian Institution Press.
Lancaster, J. B., & Lee R. (1965). The annual reproductive cycle in monkeys and apes. In I. DeVore (Ed.), *Primate behavior: Field studies of monkeys and apes* (pp. 486–573). New York: Holt, Rinehart & Winston.

Leutenegger, W. (1979). Evolution of litter size in primates. *American Naturalist, 114,* 525–531.
Leutenegger, W. (1982). Encephalization and obstetrics in primates. In E. Armstrong & D. Falk (Eds.), *Primate brain evolution: Methods and concepts*. New York: Plenum.
Lindberg, D. (1982). Primate obstetrics: The biology of birth. *American Journal of Primatology [supplement]*, *1*, 193–199.
Luckett, W. P. (1975). Ontogeny of the fetal membranes and placenta: Their bearing on primate phylogeny. In F. Szalay (Ed.), *Phylogeny of the primates* (pp. 157–182). New York: Plenum.
MacArthur, R. H., & Wilson, E. O. (1967). *Island biogeography*. Princeton University Press.
McNab, B. K. (1978). Energetics of arboreal folivores: Physiological problems and ecological consequences of feeding on an ubiquitous food supply. In G. G. Montgomery (Ed.), *The ecology of arboreal folivores*. Washington, DC: Smithsonian Institution Press.
McNab, B. K. (1988). Complications inherent in scaling the basal rate of metabolism in mammals. *Quarterly Review of Biology*, *63*, 25–54.
Martin, P. (1984). The meaning of weaning. *Animal Behavior*, *32*(4), 1257–1259.
Martin, R. D. (1981). Relative brain size and basal metabolic rate in terrestrial vertebrates. *Nature*, *293*, 57–60.
Martin, R. D. (1975). The bearing of reproductive behavior and ontogeny on strepsirhine phylogeny. In F. Szalay (Ed.), *Phylogeny of the primates* (pp. 265–297). New York: Plenum.
Mathieu, M., & Bergeron, G. (1983). Piagetian assessment of cognitive development in chimpanzee (*Pan troglodytes*). In A. B. Chiarelli & R. Corruccini (Eds.), *Primate behavior and sociobiology* (pp. 142–147). New York: Springer-Verlag.
Mayer, P. J. (1982). Evolutionary advantage of menopause. *Human Ecology*, *10*(4), 477–494.
Mink, J., Blumenschine, R., & Adams, D. B. (1981). Ratio of CNS to body metabolism in vertebrates: Its constancy and functional basis. *American Journal of Physiology*, *241*(3–4), R203.
Napier, J., & Napier, P. (1967). *A handbook of the living primates*. New York: Academic Press.
Nicholson, N. (1986). Infants, mothers and other females. In B. Smuts, D. Cheney, R. Seyfarth, R. Wrangham, & T. Struhsaker (Eds.), *Primate societies* (pp. 330–342). University of Chicago Press.
Parker, S. T. (1973). *Piaget's sensorimotor series in an infant macaque: The organization of nonstereotyped behavior in the evolution of intelligence*. PhD dissertation, Department of Anthropology, University of California, Berkeley.
Parker, S. T. (1977). Piaget's sensorimotor period series in an infant macaque: a model for comparing unstereotyped behavior and intelligence in nonhuman primates. In S. Chevalier-Skolnikoff & F. Poirier (Eds.), *Primate biosocial development* (pp. 43–112). New York: Garland.
Parker, S. T. (1984). Playing for keeps: An evolutionary perspective on human games. In P. K. Smith (Ed.), *Play in animals and humans* (pp. 271–293). Oxford: Blackwell Scientific Publications.
Parker, S. T. (1985). A social technological model for the evolution of language. *Current Anthropology*, *26*(5), 617–639.
Parker, S. T. (1987). A sexual selection model for hominid evolution. *Human Evolution*, *2*, 235–253.
Parker, S. T. (1988). *A life history perspective on human development: Beyond the altricial/precocial distinction*. Unpublished manuscript.
Parker, S. T., & Gibson, K. R. (1977). Object manipulation, tool use, and sensorimotor intelligence as feeding adaptations in cebus monkeys and great apes. *Journal of Human Evolution*, *6*, 623–641.
Parker, S. T., & Gibson, K. R. (1979). A developmental model for the evolution of language and intelligence in early hominids. *Behavioral and Brain Sciences*, *2*, 367–408.
Passingham, R. E. (1975). Changes in the size and organisation of the brain in man and his ancestors. *Brain and Behavioral Evolution*, *11*, 73–90.
Piaget, J. (1952). *The origins of intelligence in children*. New York: Norton.

Piaget, J. (1962). *Play, dreams and imitation*. New York: Norton.
Pianka, E. R. (1970). On *r*- and *K*-selection. *American Naturalist*, *104*, 592–597.
Pilbeam, D., & Gould, S. J. (1974). Size and scaling in human evolution. *Science*, *186*, 892–901.
Portmann, A. (1965). Ueber die Evolution der Tragzeit bei Sangetieren. *Revue Suisse de Zoologie*, *72*, 658–666.
Portmann, A. (1967). *Zoologie aus vier Jahrzeiten*. Munich: Piper & Co.
Pusey, A. E., & Packer, C. (1986). Dispersal and philopatry. In B. Smuts, D. Cheney, R. Seyfath, R. Wrangham, & T. Struhsaker (Eds.), *Primate societies* (pp. 250–266). University of Chicago Press.
Rasmussen, D. T., & Izard, M. K. (1988). Scaling of growth and life history traits relative to body size, brain size, and metabolic rate in lorises and galagos. *American Journal of Physical Anthropology*, *75*, 357–367.
Redshaw, M. (1978). Cognitive development in human and gorilla infants. *Journal of Human Evolution*, *7*, 113–141.
Robinson, J. (1987a). *What can long-term field studies tell us about sociality in primates?* Paper presented at a meeting of the American Society of Primatologists, University of Wisconsin, June.
Robinson, J. (1987b). *Demography and group structure in wedge-capped capuchin monkeys, Cebus olivaceus*. Unpublished manuscript.
Rowell, T. E. (1969). Long-term changes in a population of Uganda baboons. *Folia Primatologica*, *11*, 241–254.
Rowell, T. E. (1975). Reproduction in captive vervet and Sykes' monkeys. *Journal of Mammalogy*, *56*(4), 940–946.
Rowell, T. E. (1977a). Variation in age at puberty in monkeys. *Folia Primatologica*, *27*, 284–296.
Rowell, T. E. (1977b). Reproductive cycles of the talapoin monkey (*Miopithecus talapoin*). *Folia Primatologica*, *28*, 188–202.
Rowell, T. E., & Richards, S. M. (1979). Reproductive strategies of some African monkeys. *Journal of Mammalogy*, *60*, 58–69.
Sacher, G. A. (1959). Relation of lifespan to brain weight and body weight in mammals. In G. Wolstenholme & M. O'Connor (Eds.), *CIBA Foundation colloquia an aging. Vol. 5: The lifespan of animals* (pp. 115–133). London: Churchill.
Sacher, G. A. (1982). Relation of lifespan to brain weight and body weight in mammals. In E. Armstrong & D. Falk (Eds.), *Primate brain evolution: Methods and concepts*. New York: Plenum.
Sacher, G. A., & Staffeldt, E. F. (1974). Relation of gestation time to brain weight for placental mammals: Implications for the theory of vertebrate growth. *American Naturalist*, *108*, 593–615.
Sarich, V., & Wilson, A. (1967). Immunological time scale for hominid evolution. *Science*, *158*, 1002–1003.
Shea, B. (1985). In W. Jungers (Ed.), *Size and scaling* (pp. 175–205). New York: Academic Press.
Shea, B. (1987). Reproductive strategies, body size, and encephalization in primate evolution. *International Journal of Primatology*, *8*(2), 139–156.
Shultz, A. (1969). *Lives of primates*. New York: International Universities Press.
Sigg, H., Stolba, A., Abegglen, J., & Dasser, V. (1982). Life history of hamadryas baboons: Physical development, infant mortality, reproductive parameters and family relationships. *Primates, 23*, 473–487.
Spinozzi, G., & Natale, F. (1986). The interaction between prehension and locomotion in macaque, gorilla and child cognition. In G. Else & P. C. Else (Eds.), *Primate ontogeny, cognition and social behaviour* (pp. 155–160). Cambridge University Press.
Stanley, S. M. (1973). An explanation for Cope's rule. *Evolution*, *27*, 1–26.
Stearns, S. C. (1976). Life-history tactics: A review of the ideas. *Quarterly Review of Biology*, *51*, 3–47.
Stearns, S. C. (1983). A new view of life-history evolution. *Oikos*, *35*, 266–281.

Symons, D. (1978). *Play and aggression in rhesus monkeys*. New York: Columbia University Press.

Teleki, G., Hunt, E. E., Jr., & Pfiffeling, J. H. (1976). Demographic observations (1963–1973) on the chimpanzees of Gombe National Park, Tanzania. *Journal of Human Evolution*, *5*, 559–598.

Terborgh, J. (1983). *Five New World primates*. Washington, DC: Smithsonian Institution Press.

Terborgh, J., & Goldizen, A. W. (1985). On the mating system of the cooperatively breeding saddle-backed tamarin (*Saguinus fuscicollis*). *Behavioral Ecology and Sociobiology*, *16*, 293–299.

Trivers, R. (1972). Parental investment and sexual selection. In B. Campbell (Ed.), *Sexual selection and the descent of man*. Chicago: Aldine.

Trivers, R., & Willard, D. C. (1973). Natural selection of parental ability to vary the sex ratio of offspring. *Science*, *179*, 90–92.

Tutin, C. E. G. (1979). Mating patterns of reproductive strategies in a community of wild chimpanzees (*Pan troglodytes schweinfurthii*). *Behavioral Ecology and Sociobiology*, *6*, 29–38.

Vauclair, J. (1982). Sensorimotor intelligence in human and nonhuman primates. *Journal of Human Evolution*, *11*, 257–264.

Vauclair, J., & Bard, K. (1983) Development of manipulations with objects in ape and human infants. *Journal of Human Evolution*, *12*, 631–645.

Visalberghi, E., & Antinucci, F. (1986). Tool use in the exploitation of food resources in *Cebus apella*. In J. C. Else & P. C. Lee (Eds.), *Primate ecology and conservation* (Vol. 2, pp. 54–62). Cambridge University Press.

Watts, E. S., & Gavan, M. A. (1982). Postnatal growth of nonhuman primates: The problem of the adolescent spurt. *Human Biology*, *54*(1), 53–70.

Westergaard, C. (1987). Lion-tailed macaques manufacture and use tools. *American Journal of Primatology*, *12*, 376. (Abstract)

Western, D. (1979). Size, life history and ecology in mammals. *African Journal of Ecology*, *17*, 185–204.

Wilbur, H., Tinkle D. W., & Collins, J. P. (1974). Environmental certainty, trophic level, and resource availability in life history evolution. *American Naturalist*, *108*, 805–817.

Williams, G. (1966). *Adaptation and natural selection*. Princeton University Press.

*Part II*

# Comparative developmental perspectives on cebus intelligence

# 5 The comparative study of cognitive ontogeny in four primate species

*Francesco Antinucci*

## Theoretical perspectives

The comparative study of cognitive ontogeny in primates is a fundamental key to an understanding of both the structure of human cognition and its evolutionary formation. There are two main reasons why the study of development is a privileged avenue to a theory of cognitive capacity. First, it provides a nonarbitrary basis for the identification of cognitive structures. What we commonly call cognition is a highly complex and integrated set of closely interacting systems that is not easy to pull apart. For example, is memory an independent system? Or language? Or attention? Does the capacity to use tools or that to draw inferences and deductions correspond to a specific cognitive mechanism? Judging from the way research practice in this field is divided up (as shown, for example, by the titles of many monographs), one should think so.

On the other hand, considerable evidence indicates that this is *not* the way cognition is structured. A deeper analysis of those capacities reveals the interaction of a complex set of underlying mechanisms that are called upon and employed in a variety of other functions (e.g., Bever, 1974). In other words, there seems not to be a one-to-one correspondence between overt cognitive capacities and the underlying mechanisms responsible for them. By showing cognitive mechanisms in the process of formation, development helps to bring to light these underlying mechanisms, either because they are not yet integrated in such a complex way or because they are still in the process of differentiation. Systems that are gradually and independently formed are good candidates to be "real" units of cognitive structure.

The second, and more fundamental, reason to emphasize the study of ontogeny is that the very process of development poses fundamental constraints on the form that fully grown structures will have. In a system whose parts are tightly interrelated, even a local modification is likely to reverberate elsewhere, generating widespread reorganizations. If, as is the case in the development of cognition (or in the development of an organism, for that matter, or, to go even further, in the transformation of a social structure), successive transformations must also generate stages that are functionally viable at all times, when powerful constraints are posed by the transformational process itself. Given the initial system, there can be few ways in which it can be transformed, the fewer the higher the integration of the system. Understanding the mechanisms of transformation thus translates into an understanding of the transformed structures. It is for these reasons, I believe, that one of the most powerful, comprehensive, and coherent theories of human cognition put forth, that of Jean Piaget, derives entirely from the study of development.

The *comparative* study of cognitive development adds a further dimension to this approach. In fact, the strongest case for comparative studies relies on the two reasons just discussed. In the ontogeny of the individual and the phylogeny of the species, the same basic issue is at stake: the transformation across time of a highly integrated system. In this case, too, given the initial system (species that undergo evolutionary transformation) and the constraint of functional viability at all stages of transformation, avenues of evolutionary change are severely limited and impose powerful constraints on the resulting structures.

This is the point that the evolutionary biologist S. J. Gould has constantly emphasized, speaking of "design constraints" on evolutionary transformations that form the basis for the important distinction between adaptations and "exaptations" (Gould & Vrba, 1982). In conclusion, taking seriously the constraints posed by both ontogenetic and phylogenetic transformational processes can give us substantial aid in coping with the crucial problem faced by the scientific study of this field: the vast underdetermination of theory by data.

Both sets of constraints have been largely ignored by most investigators theorizing on comparative cognition. The crucial role of transformational processes is ignored when the integrated nature of the cognitive system is not taken into account, both by traditional and by more recent approaches. In behavioristic theories, such as the "learning theory," cognition is seen as a set of independent abilities, each of which can be independently shaped ad libitum by the appropriate reinforcement contingencies. Mechanisms that transform and interrelate these abilities are supposed to be quite primitive and simple. They reduce to generalization and discrimination.

Consequently, within this theoretical paradigm there is no specificity of ontogeny: Ontogenetic transformations are accounted for by the same learning mechanisms that account for adult abilities. Differences are only of a quantitative nature; more complex (i.e., more numerous and longer) "associative chains" can be formed in the second case. Of the same nature are species differences; from rat to human it is merely a question of how much of the same "stuff" (e.g., Harlow, 1949; Hull, 1943; Skinner, 1956). Although popularly thought to be at the opposite extreme, much theorizing about cognition within the ethological and sociobiological traditions shares a number of underlying assumptions with the behavioristic paradigm. Barash, for example, deals with a typical cognitive problem in the following way:

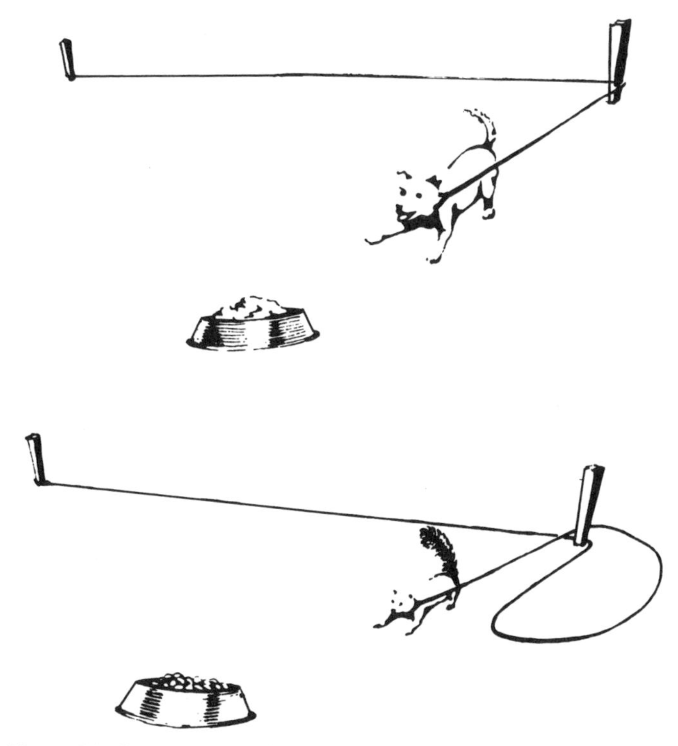

Figure 5.1. Dogs versus squirrels attempting a detour problem (from Barash, 1977).

Consider a simple detour experiment where an animal must go away from a goal in order to reach it eventually [Figure 5.1]. Dogs are no good at such tasks: a dog strains at its leash; whines pitifully; looks at you with its sad brown eyes; runs about wildly; and finally may fall asleep. Then it starts again. Eventually, by luck alone, it might find itself on the far side of the intervening post, whereupon it rushes triumphantly to the food. By contrast, common tree squirrels are uncommonly good at solving this problem. After surveying the situation, the animal proceeds surely to circumnavigate the post and get the food.

Why the difference? This type of result is puzzling to learning theorists, especially since squirrels are universally recognized as less intelligent than dogs. Even their brains are obviously simpler, with fewer convolutions in the cerebral cortex. The answer is actually straightforward if we consider the natural history of squirrels and dogs. Dogs live in a two-dimensional world: If they want something, they go and get it. Squirrels, on the other hand, live in trees. In their three-dimensional world, a squirrel wanting to go from tree to tree has a choice: it can descend the trunk; go along the ground and then climb the tree; or it can remain in the treetops, seeking a place where branches of two adjacent trees are in contact. The former strategy would expose the squirrel to its ground-dwelling predators; the latter would therefore be safer, but it would often require the ability to go initially away from the goal so as to achieve it eventually. In other words, the ancestors of present-day squirrels were good at solving detour problems. Rephrasing this more accurately in the language of evolutionary biology: ancestral squirrels that were relatively more adept at solving such problems left more offspring than did those who were less well endowed. Thus, the ability to conduct successful detours was favored by natural selection, such that each population of squirrels came to be composed of individuals who were good at going away from goals in order to reach them. (Barash, 1977, pp. 3–5)

The obvious difference in the standing on the innate versus acquired issue boils down to whether the "detour behavior" has been shaped by environmental or evolutionary "contingencies" (call the first *reinforcement* and the second *selection*), and thus to whether "learning" is more conveniently applied to the level of the individual or that of the species. The explanatory logic of the two approaches, however, is really the same. Both of them take for granted that there is indeed a specific ability, "operating a detour," that can be isolated and shaped or selected ad libitum. There is a one-to-one correspondence between bits of behavior and their causal determinants.

It is precisely this assumption that is questioned in structural views of the cognitive system. "Operating a detour," for example, is a performance dependent upon several different systems: the way space is cognitively organized, the structure of means–end coordination, and so on. According to these views, one *cannot* modify (whether ontogenetically or phylogenetically) this performance without modifying the underlying cognitive structures it depends upon. If such a modification occurs, then its consequences are not likely to be confined to the single performance in question, but probably will involve a number of behaviors and abilities based on the spatial organization.

A theory that explicitly rejects such atomistic views in favor of a decidedly structural one would be immune from this fallacy. In fact, in recent years a number of studies on comparative cognition and development have been framed within the theory of Jean Piaget, a structural theory par excellence. Positive results have not failed to emerge. To take just one example, consider the Parker and Gibson (1977) analysis of tool use. That attempt to relate some of these performances to more general structures of the cognitive system enabled those authors to distinguish appropriately relevant theoretical categories (like the distinction between tool vs. proto–tool use, or that between context-specific vs. intelligent tool use) and to establish meaningful relations in this rather confused and confusing matter.

Yet even in these cases one must regret that the deeper implications of a structural approach are not fully developed. In comparing the cognitive capacities of macaques, great apes, and humans (Parker & Gibson, 1979), for example, structural constraints are set aside, and a simple additive mechanism is postulated in order to account for the interrelations of the different primate species' cognitive systems: The progression of stages identified in the development of the human cognitive system is mapped onto the phyletic "progression" of the primate species. Phylogenetic transformation means here adding one step to the ontogenetic sequence. Besides the strong general critiques that can be (and have been) raised against "recapitulation" models (Gould, 1977, 1979), this view forces one to overlook or underestimate important empirical data, as I shall try to show later, and the subtle role they

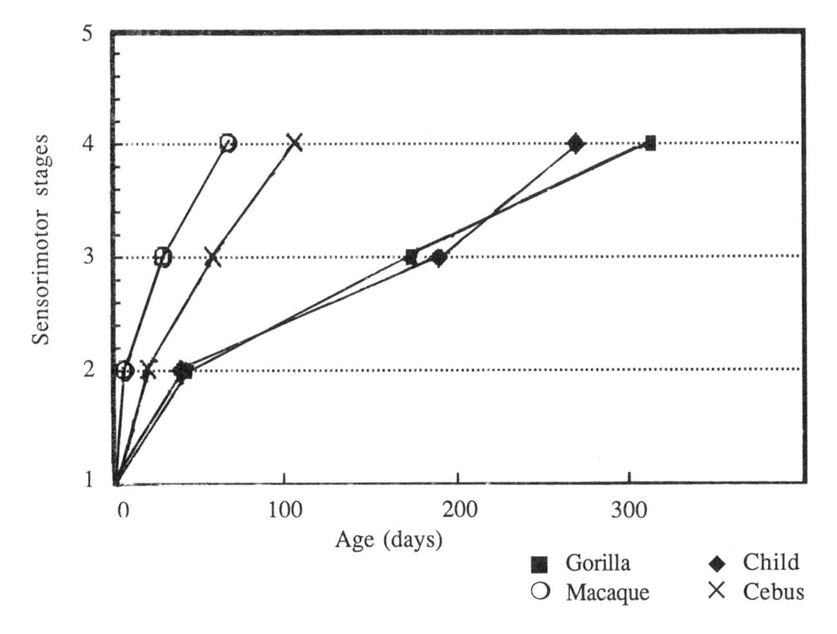


Figure 5.2. Cognitive development.

play in differentiating courses of cognitive development. To restate a fundamental point, species comparison should decrease rather than increase the degrees of freedom allowed in the analysis of the most complex and terminal system.

**Comparative analysis**

Using the theoretical perspective outlined earlier, we examine and compare the early cognitive development of four primate species: macaque (*Macaca fascicularis*), cebus (*Cebus apella*), gorilla (*Gorilla gorilla gorilla*) and human (*Homo sapiens sapiens*). Work on these species has been carried out in our laboratory in the past 7 years and has been reported elsewhere (Antinucci, Spinozzi, Visalberghi, & Volterra, 1982; Poti', 1985; Spinozzi & Antinucci, 1983); thus, I shall not go into any methodological details. Our data on macaque, cebus, and human infants come from several subjects that have been pooled together. Our data on gorillas come from a single subject and therefore need to be replicated. All subjects are followed longitudinally beginning at birth and were tested in the classical Piagetian "clinical" fashion (supplemented by careful screening of videotape recordings) and in more experimentally structured settings (Antinucci, Spinozzi, & Natale, 1986; Natale, Antinucci, Spinozzi, & Poti', 1986).

Figures 5.2 and 5.3 show achievements of the first four stages of sensorimotor intellectual development plotted against time for each of the four

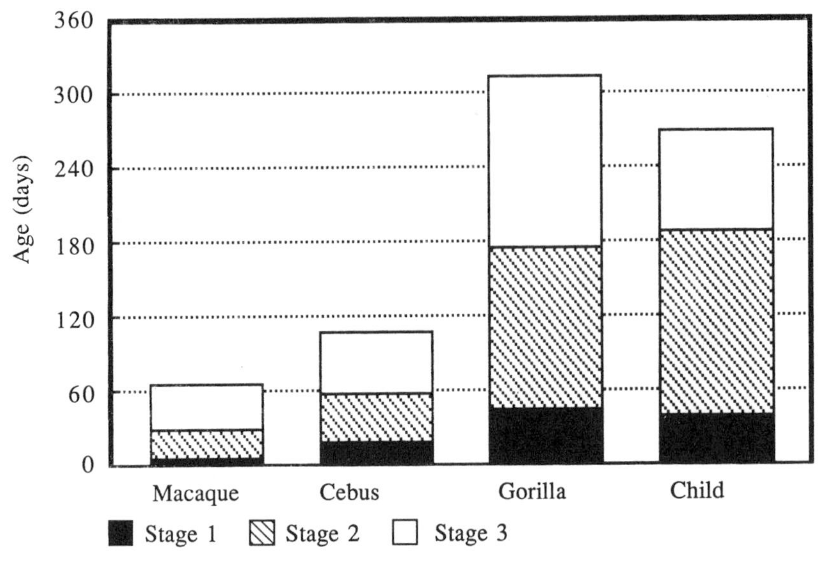


Figure 5.3. Sensorimotor development.

species. Identification of the four stages was as in the work of Piaget (1963; 1971), except for the (very important) qualifications that will be discussed later. Notice first of all the differences in rates of development: Gorilla and child show the slowest rate, followed by the cebus, whose rate is about three times faster, and the macaque, which is between four and five times faster. Let us now examine, stage by stage, each species' development.

### *Stage 1*

In Stage 1, capacities are least differentiated; consequently, the highest degree of uniformity among species occurs at this stage. Exercise of the primary reflexes of grasping, sucking, and visual fixation evolves into primary schemata for tactile, oral, visual, and presumably, auditory assimilation of external stimuli. These schemata are uncoordinated: The visual, oral, and tactile worlds are separate and unconnected realities.

### *Stage 2*

Stage 2 is marked by the beginning of reciprocal coordinations between primary schemata. In the human infant these coordinations develop in a characteristic sequence. The most primitive is the coordination of hand and mouth, usually manifested in thumb sucking. Then coordination between prehension and suction is established: Objects that are grasped are taken to the mouth, and vice versa. Finally, the coordination between vision and prehension is established.

Already at this point, the first important differences in the developmental courses of the four species appear. In the macaque, hand–mouth coordination is well established at 5 days; however, no prehension–suction coordination follows. A completely different coordination, one that is never seen in the human infant, appears, beginning at 7 days: On seeing an object, the macaque grasps it directly with its mouth by targeting its head onto it (we shall call this pattern *buccal prehension*). This is possible because the macaque quickly develops a capacity for locomotion: at 7 days. It can confidently move itself in space. Visually guided grasping by mouth requires a much more elementary coordination than does visually guided hand grasping, because eyes and mouth are both carried on the head; hence, simple visual targeting by means of locomotion brings the mouth onto the object automatically. During the second week of life, buccal prehension becomes the dominant pattern of interaction with objects, whereas hand prehension guided by vision is still unsuccessful.

The first partially successful attempts at visually guided hand prehension begin at 15 days, and the full coordination of vision and prehension is achieved at 27 days. Here the standard sequence is followed: First, prehension is successful only when both the hand and the object are in the same visual field; then the hand grasps without requiring any visual adjustment. Buccal prehension, however, does not disappear even when visually guided hand prehension is perfectly developed; it continues to occur side by side with hand prehension and is as frequent.

In the gorilla, primitive hand–mouth coordination is well established at 56 days. At 66 days, the first successful coordinations of prehension and mouthing occur, thus following the same sequence as in the human infant. Between 66 and 115 days, all attempts to grasp objects through visually guided prehension are unsuccessful. In this period, however, a pattern of buccal prehension identical with that seen in the macaque develops. In fact, at 88 days the gorilla becomes capable of independent locomotion in space, and buccal prehension starts to occur systematically. This is the only pattern through which objects are picked up until 110 days. At 110 days, the first partially successful attempts to grasp objects through visually guided hand prehension occur, and subsequently develop through the usual sequence. At first the hand must be in the same visual field as the object; finally, at 168 days, objects are grasped with a single hand movement without visual adjustment. During this phase, buccal prehension continues to occur, but it significantly decreases in frequency as hand prehension progresses. It appears that the two patterns are in a typical developmental competition: Buccal prehension tends more and more to occur when hand prehension is unsuccessful because of either physical or cognitive difficulties, such as in grasping suspended objects or partially hidden objects.

Cebus development in Stage 2 follows yet another course. Hand–mouth coordination is established at 19 days. First attempts at prehension–mouthing coordination occur at 27 days. These attempts are quite unsystematic, how-

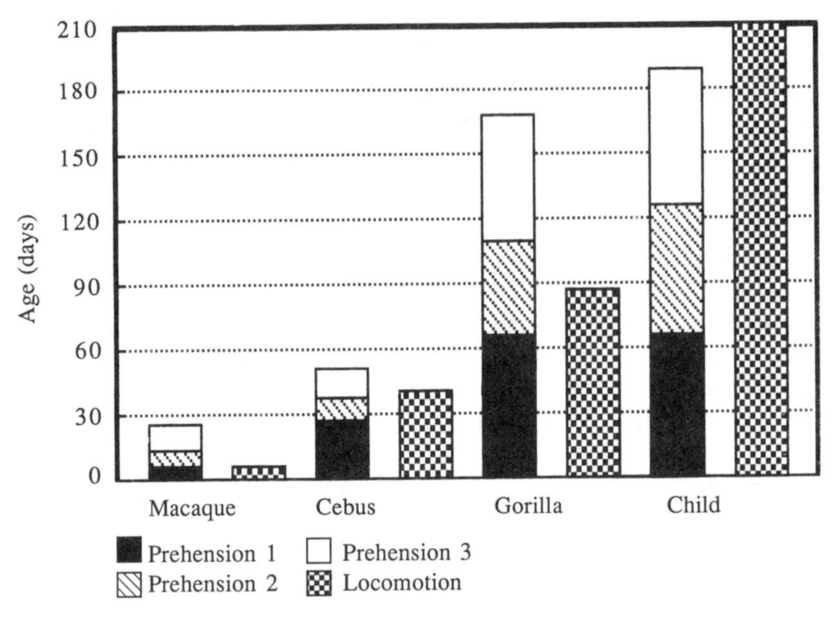


Figure 5.4. Prehension and locomotion.

ever, and are accompanied by attempts to grasp at sight. Neither kind of attempt is successful until 38 days; both developments appear to be very slow.

Contrary to the macaque and gorilla, however, the cebus fails to develop any buccal prehension schema during this period. In fact, at this age the cebus still is not capable of independent locomotion. Only limited head movements are possible; accordingly, some buccal grasping guided by sight is found, but only in the case of the nursing bottle when this comes close enough to its head. At 38 days, successful coordination of vision and prehension begins, which, as for the other species, occurs only when both the hand and the object are in the same visual field. Only 3 days later, at 41 days, the cebus becomes capable of moving freely in space. Prehension development is finally completed at 52 days.

*Prehension and locomotion.* Figure 5.4 summarizes the data presented up to now. To make comparisons convenient, the development of prehension has been subdivided into three phases:

*Phase 1* corresponds to hand–mouth coordination for all species.

*Phase 2* begins with attempts to establish coordinations relative to object grasping. As we have seen, the four species differ from one another.

The macaque begins this phase with a visual–locomotive coordination and a consequent buccal prehension pattern. No prehension–suction coordination ever appears, and the animal is capable of independent movements in space from the very beginning of this phase.

The gorilla, on the other hand, develops the prehension–suction coordination, but as soon as independent locomotion is established, the visual–locomotive coordination appears, and buccal prehension becomes the almost exclusive grasping pattern until the end of this phase.

The cebus attempts both prehension–suction and vision–prehension coordination, but is unsuccessful until the end of this phase. Independent locomotion is absent through the whole phase; hence, no visual–locomotive coordination appears. Buccal prehension is limited to the single case of the nursing bottle.

The human infant develops only the prehension–suction coordination.

*Phase 3* corresponds to the development of visually guided hand prehension, which shows the same sequence for all species: Successful grasping requires initially the visual adjustment of the hand to the object. Differences among the four species are present here too.

The macaque continues to use buccal prehension side by side with hand prehension. In the gorilla, the two prehension patterns compete with each other: Buccal prehension becomes less and less frequent as hand prehension becomes more and more precise.

The cebus acquires independent locomotion during this phase, but does not develop any subsequent pattern of buccal prehension.

The human infant is still limited to head turning and head raising.

### *Stage 3*

In Stage 3 of sensorimotor development, differences among the four species increase. In the human infant, this stage is marked by the appearance and development of "secondary circular reactions." (When an action schema applied to an object produces an "interesting" result on the external environment, the infant tries to conserve and reproduce by repeating the schema.)

This crucial behavior pattern is completely absent in the macaque: Objects are always assimilated through primary schemata (mouthing, touching, lifting, smelling, etc.), even when actions on them produce salient effects that are clearly noticed by the animal (making noise, flashing light, swinging, rotating, etc.). Secondary schemata are never encountered (this is confirmed by the Parker [1977] macaque study), even much later in development.

Secondary circular reactions are also completely absent in the third stage of the gorilla's development. Some limited (albeit controversial) instances of this behavior pattern (like rolling a stick) seem to show up during the second half of the second year, but they are and remain extremely primitive.

The cebus, on the other hand, manifests during this stage at least one clear instance of secondary circular reaction: The repetitive schema consists in lifting an object with both hands and banging it onto a substrate. The action varies in intensity and number of consecutive repetitions; the resulting effect is attentively observed, and the object is sometimes visually explored after the action. In the next stage, this schema differentiates, generating a number of secondary circular reactions, like banging, rubbing, and sliding the object on a series of different substrates. These schemata become more and more frequent and numerous during the first year of development.

Because of these differences, the occurrence and temporal duration of Stage 3 for the three nonhuman primate species are compared on the basis of the object concept developmental sequence. In the object concept domain there is complete uniformity in the four species. The same steps are followed, from anticipatory adjustment to rapid movements, which marks the beginning of the stage, to the reconstruction of an invisible whole from a visible fraction, to the successful recovery of a completely hidden object, which marks the passage to Stage 4.

### *Stage 4*

In view of the increasing divergence among the four species from Stage 1 through Stage 3, the characterization and delimitation of Stage 4 opens up a number of problems that we shall not discuss here. Only a few salient points will be stated. The hallmark of Stage 4, the intercoordination of independent schemata into a means–end relation, can be easily identified in all four species. In the macaque, however, the schemata coordinated are of one kind only: they all involve subject–object relations, such as setting aside an obstacle and reaching an object, raising a screen and picking up an object, or pulling a string and grasping an object attached to it. They never involve object–object relations, such as pushing one object against another, putting one object on top of or inside another, or, in general, combining one object with another (Parker, 1977).

The same limitation is found in the gorilla. In this animal, however, some elementary schemata establishing object–object relations (mainly hitting) are found much later, toward the end of the second year. They do not seem to increase or progress.

In the cebus, too, schemata relating object–object are initially absent. They eventually make their appearance, after quite a temporal lag, at 230 days. Two objects are combined together (e.g., side by side or on top of one another). Contrary to the macaque and the gorilla, however, the cebus during the first months of the second year shows an explosion of object–object and object–substrate composing (Natale et al., 1986).

Development of the object concept, on the other hand, remains constant across the four species: the macaque at 100 days, the gorilla at 360 days, and the cebus at 125 days. All show instances of "residual reaction" in the object displacement task. All four species eventually achieve Stage 5 of the object concept development. Only the cebus, on the other hand, shows clear and definite instances of "tertiary circular reactions," appearing at about the same time of its systematic object composing. Tertiary circular reactions were never seen either in the macaque or in the gorilla.

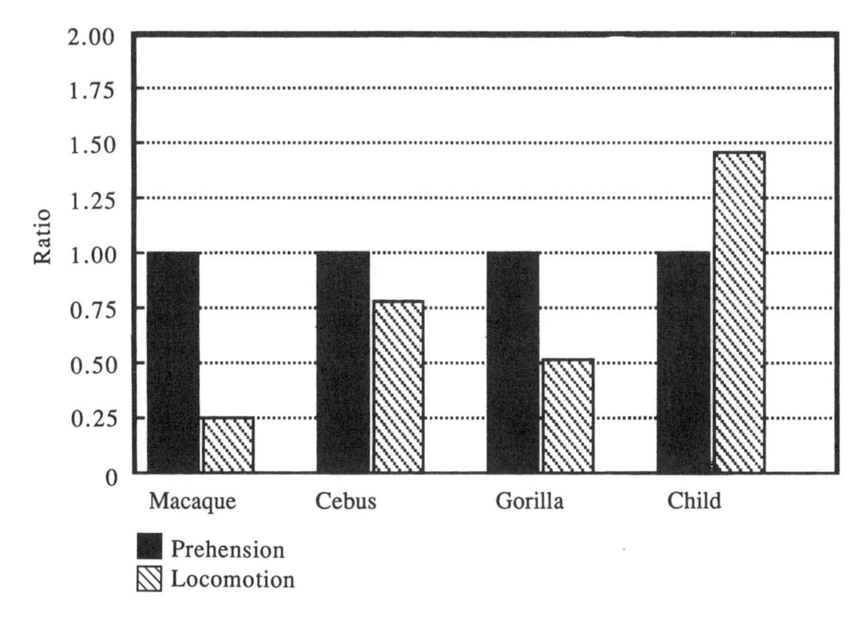


Figure 5.5. Locomotion/prehension ratio.

**Discussion**

Species differences found in Stage 2 have an obvious coorelate in the relative time of onset of independent locomotion. The word "relative" needs to be stressed: It is the time of onset of locomotion in relation to the time of prehension development in each species that is crucial. Figure 5.5 shows this relation as a locomotion/prehension ratio for each species.

Taking as a reference for comparative purposes the final point of development of prehension at the end of the third phase, one can see that the macaque becomes capable of moving in space at about a quarter of the way through the course of its prehensive development (0.26), the gorilla at about half (0.52), and the cebus toward the end (0.79). The human infant, on the other hand, becomes capable of locomoting long after its prehensive development is complete (1.47).

I hypothesize that these differences explain the different developmental patterns of the four species. The macaque can move toward objects *before* any hand coordination develops; hence, in order to mouth them, it can take itself to the objects. When hand coordination develops, this pattern of object exploration is already well established and functional and is not replaced. The gorilla develops locomotion a bit later, when its first hand coordination is already established, but *before* it is capable of grasping at sight. Hence, the gorilla also approaches objects directly by the mouth, not through the hand.

However, because visually guided hand prehension begins to develop immediately after locomotion, buccal prehension enters into competition with hand prehension and comes to be used more as a substitute pattern. Finally, the cebus develops locomotion when visually guided hand prehension is already developing, and consequently no previous pattern of direct mouth exploration has been established: Object exploration is channeled through hand grasping, and buccal prehension never develops. In this respect, the cebus comes closest to the human infant.

To understand how and why these early differences have consequences for the course of subsequent development, consider first what happens in child development. Human infants are practically immobilized for the first 9 months of life. Throughout this long period of their cognitive development, all their physical interactions with external objects are mediated by the hand. Both the mouth and the visual exploration of objects depend on hand grasping. Because neither the infant's body nor head can move, it is the hand that must take objects to the mouth, and it is the hand that can vary the position and perspective of objects (through movement and rotation) in order to allow visual exploration. In addition to that, and again because of their immobility, children have no way of interacting with distant objects (i.e., with objects outside of the range covered by the arm length) except through the use of other intermediary objects.

Consider that the hand is, in an elementary sense, the first "detached" instrument (and, in fact, it takes months to learn to move it under control and to direct it appropriately onto sighted objects) (Piaget, 1963). Given this, we can see that the child's exploration of and interaction with the environment, from the beginning of the first effective hand coordinations to the attainment of independent locomotion, develop almost entirely through the use of intermediaries: the hand at the more primitive levels, the hand plus other objects at the more advanced levels. Systematic use of intermediaries carries with it a concentration on the causal effects produced by the action on the objects, and on the causal and spatial relations of objects among one another.

Consider now, at the opposite extreme, the macaque. At the very beginning of Stage 2, it can move freely in space. It still is not capable of grasping objects with its hands, but it can approach them easily, can mouth them, and can explore them visually by moving its body around them. Furthermore, it can also reach and explore distant objects without having to find a way of acting on them indirectly. In other words, its exploration of and interaction with objects can occur without the use of intermediaries (i.e., through a much simpler route). The explosion of mouth prehension, its dominance and concurrence with respect to hand prehension, and the continued use of only primary schemata in exploring objects all testify that this route is indeed followed. The relative difference in the time of onset of locomotion very soon *channels* the cognitive interaction with the external world into two different modes: action through intermediaries in the child, direct action through body movements in the macaque. Now, it is precisely those action patterns that depend on the first mode (i.e., secondary circular reactions) and actions relating object to object (tertiary circular reactions) that we do not find in the macaque's later development when compared with that of the child.

Gorilla and cebus fall in between these two extremes in terms of their relative times of onset of locomotion, with the gorilla leaning more in the direction of the macaque. Consequently, the gorilla's later developmental stages show essentially the same character as the macaque's, though to a less extreme degree: Limited and primitive secondary circular reactions and some object–object interactions seem to appear much later in development.

The cebus, on the other hand, comes closer to the human infant mode, and we do find secondary circular reactions, object–object interactions, and even tertiary circular reactions, though somewhat retarded in development and certainly less elaborated, less differentiated, and less numerous than those developed by the human infant.

On the other hand, cognitive structures that do not depend on the mode of interaction with objects, such as construction of the object concept, are not affected by this early difference, and the same developmental course is followed in this case by the four species. Only the effect of the different developmental rates is here seen.

The second major difference found in the development of these four primate species, that of developmental rates, might be important in accounting for more "quantitative" differences, rather than the "qualitative" differences in the structures built.

Cognitive constructions develop through the subject's assimilative–accommodative interaction with the environment. The rate of development directly affects the duration of this interaction at each successive developmental stage. One should consider, in fact, that the duration of each relevant environmental event (e.g., instances of specific spatial configurations, or object movements) or action schema acting on the environment is practically the same for any developing subject. Faster developmental rates mean that the subject will experience fewer of these events and will generate fewer of these action schemata at each developmental stage.

The correlate of this longer or shorter duration of the subject's interaction might well be a greater or smaller variety and richness of the constructions at each stage. Within a given style of interaction, richer elaborations can be achieved depending on the amount of time spent in the interaction.

We have seen that two different interacting modes characterize the development of the macaque and gorilla, on the one hand, and the cebus and child, on the other, but we have also noticed differences between the two species within each mode. These differences, which concern more the amount and differentiation of the patterns seen, rather than their type, might be the effect of the longer or shorter "practice" allowed by the two species' vastly different developmental rates.

## Conclusion

The comparative analysis of cognitive ontogeny seems to fulfill the expectations outlined in the initial theoretical discussion. The first of these expectations, the possibility of providing a nonarbitrary basis for identification of cognitive systems, seems to be borne out (Parker, P&G1). Our analysis shows, for example, that the various series across the sensorimotor stage, contrary to what Piaget himself seems to imply, do not compose a unit of cognitive structure in the classical Piagetian sense. There is no necessary relation between secondary circular reactions and "intermediate" schema detaching, as seen in the object concept, or between means–end coordination and the construction of object–object relations (Paker, 1977).

In the second place, our analysis suggests that transformational constraints are responsible for the form of later structures, as we have seen, for example, in the case of physical knowledge about objects (e.g., their modes of action). Its resulting form appears to be shaped back in ontogeny by the temporal interaction of prehension and locomotion. As predicted by a structural view of transformational constraints, a slight difference in early development "snowballs" into major differences in later structures.

Finally, again as predicted, this analysis sheds some light on possible paths of evolutionary transformation. Two "time regulators" appear to be involved. On one side, rates of development (possibly tied to general rates of maturation) seem to have progressively slowed down, and in this case a near perfect concordance exists (at the slowest level) between ape and human. On the other side, the temporal onset of locomotor capacity seems to have varied (possibly in relation to the different postural and locomotive adaptations of the four species) independently of the first factor. In this case the maximal difference occurs between the human child and the three other primate species. Interestingly enough, however, it is the South American cebus that is closer to humans on this second dimension, and, at least in the domains we have considered, this species' cognition appears to be more similar to that of the human child. The proper understanding of this (quite unexpected) result will have to wait for study and comparison of the ontogenetic development of more primate species.

### References

Antinucci, F., Spinozzi, G., & Natale, F. (1986). Stage V cognition in an infant gorilla. In D. M. Taub & F. A. King (Eds.), *Current perspectives in primate social dynamics* (pp. 403–415). New York: Van Nostrand Reinhold.

Antinucci, F., Spinozzi, G., Visalberghi, E., & Volterra, V. (1982). Cognitive development in an infant Japanese macaque. *Annali dell'Istituto Superiore di Sanita, 18*, 177–184.

Barash, D. P. (1977). *Sociobiology and behavior*. Amsterdam: Elsevier.

Bever, T. G. (1974). The interaction of perception and linguistic structures. In T. Sebeok (Ed.), *Current trends in linguistics* (Vol. 9, pp. 1158–1233). The Hague: Mouton.

Gould, S. J. (1977). *Ontogeny and phylogeny*. Cambridge, MA: Belknap Press.

Gould, S. J. (1979). Commentary on Parker & Gibson (1979, pp. 385–387).

Gould, S. J., & Vrba, E. S. (1982). Ex-aptation. *Paleobiology*, *8*, 4.
Harlow, H. F. (1949). The formation of learning sets. *Psychological Review*, *56*, 51–65.
Hull, C. L. (1943). *Principles of behavior*. New York: Appleton-Century-Crofts.
Natale, F., Antinucci, F., Spinozzi, G., & Poti', P. (1986). Stage VI object-concept in nonhuman primate cognition. *Journal of Comparative Psychology*, *4*, 335–339.
Parker, S. T. (1977). Piaget's sensorimotor period series in an infant macaque. In S. Chevalier-Skolnikoff & F. Poirier (Eds.), *Primate biosocial development* (pp. 43–112). New York: Garland Press.
Parker, S. T., & Gibson, K. R. (1977). Object manipulation, tool use, and sensorimotor intelligence as feeding adaptations in cebus monkeys and great apes. *Journal of Human Evolution*, *6*, 623–641.
Parker, S. T., & Gibson, K. R. (1979). A developmental model for the evolution of language and intelligence in early hominids. *Behavioral and Brain Sciences*, *2*, 367–408.
Piaget, J. (1963). *The origins of intelligence in children*. New York: Norton.
Piaget, J. (1971). *The construction of reality in child*. New York: Ballantine.
Poti', P. (1985). Ricerche piagettiane sullo sviluppo cognitivo del genere *Macaca*. *Eta' Evolutiva*, *21*, 19–35.
Skinner, B. F. (1956). A case history in scientific method. *American Psychologist*, *11*, 221–233.
Spinozzi, G., & Antinucci, F. (1983). Lo sviluppo cognitivo di un gorilla nel primo anno di vita. *Eta' Evolutiva*, *15*, 46–55.

# 6 Sensorimotor development in hand-reared and mother-reared tufted capuchins: A systems perspective on the contrasts

*Dorothy Munkenbeck Fragaszy*

## Introduction

Sensorimotor development occupies a special place in cognitive developmental theory. Infants' experiences through their motor activity in the first year or so of life are believed to contribute to their initial understanding of the physical world (e.g., Case, 1985). For this reason, sensorimotor development and cognitive development are inextricably related. Sensorimotor development holds an even greater significance for those using a cognitive approach derived from Piagetian theory in developmental studies of nonhumans, because only the first, sensorimotor period of cognitive development in the Piagetian scheme can be examined across species (e.g., Antinucci, 1989). In this chapter, I identify some limitations of a cognitively oriented approach to the comparative study of sensorimotor development. I also describe an alternative theoretical framework that I anticipate will better enable us to focus on the *process* of development, which many authors have urged (e.g., Fentress, 1984; Hinde & Bateson, 1984). The alternative framework, drawn from dynamical systems theory, is in a nascent stage, but I believe it holds great promise for comparative developmental studies.

A dynamical systems perspective allows us to examine the changing relations among elements in a system (say, walking) during development and the synchronic and diachronic links between the focal system and other systems (e.g., Fentress, 1986; Fentress & McLeod, 1986; Thelen, in press; Thelen, Kelso, & Fogel, 1987; Trevarthen, 1984a,b; Wolff, 1987). This perspective provides a new way of conceptualizing discontinuities, reversals, stability and instability, and vulnerability – phenomena that challenge any closely bounded view of development (e.g., Bateson, 1984; Emde, Gaensbauer, & Harmon, 1976; Golani & Fentress, 1985; Oppenheim, 1981).

The dynamical systems approach emphasizes the pervasive influence of *context* on behavior. I argue that attributing a causal role to any particular factor (e.g., cognitive capacity) for behavior exhibited in a particular circumstance, or for differences between species in behavior in a given circumstance, is premature until we know the constraints on behavior imposed by the setting and task, as well as the organism. I also argue that dynamical systems theory can guide the identification and characterization of said constraints. I hope that this first attempt at fitting systems theory to comparative data (or, more accurately, attempting to restate the data in systems language), which previously have received primarily cognitive interpretations (e.g., Antinucci, P&G5), will spark interest in this approach among comparative developmentalists. Obviously, interpreting data post hoc does not constitute the best forum to evaluate any theory, because circularity is implicit in the exercise. My aims, however, are to see where the exercise takes us and to see if new questions arise.

In the early sections of this chapter I present an overview of dynamical systems theory in general terms, point out the advantages of the theory over a cognitive orientation for the study of sensorimotor development, and review selected data on locomotor and prehensile abilities of human infants using a systems framework. In later sections I compare these aspects of development in mother-reared and hand-reared infant tufted capuchins (*Cebus apella*). Finally, I interpret the findings of the comparative studies in a systems perspective. Studies of infant capuchins observed with their mothers in social groups in captivity (Fragaszy, 1989, 1990; Fragaszy & Adams-Curtis, in press; Fragaszy, Scollay, Baer, & Adams-Curtis, unpublished data) and observations of capuchin infants reared by humans and tested in asocial circumstances (Antinucci, P&G5; D. M. Fragaszy, unpublished data; G. Spinozzi, unpublished data) are used in the exercise.

The focus on capuchins for this exercise is particularly apt. Capuchins have been singled out by people from diverse theoretical backgrounds (e.g., Harlow, 1951; Klüver, 1937; Parker & Gibson, 1977) as especially interesting in comparative considerations of the evolution of problem-solving and tool-using abilities. We know that capuchin monkeys are behaviorally altricial at birth relative to many other primates, perhaps more so than indices of brain development would predict (Fragaszy, 1990; Harvey, Martin, & Clutton-Brock, 1987). Many aspects of life history and behavior are unusual in the genus (Fragaszy, Visalberghi, & Robinson, in press; Martin, in press). The altriciality of capuchins, in combination with their eventual performance of sophisticated instrumental behavior and their unusual status among primates as slow-growing and large-brained monkeys makes them a natural target for studies of sensorimotor development.

## Dynamical systems views of sensorimotor development

A useful theory of sensorimotor development must

1. be consistent with the emergent nature of behavior,
2. encompass stability, discontinuity, reversal, and transformation in development,
3. illuminate variability in behavior across context as well as time, and
4. guide empirical research on development.

Dynamical systems theory, developed around problems in the physical sciences and mathematics, provides the potential for a new framework that will encompass these principles (Haken, 1983; Prigogine & Stengers, 1984).

The key principles of dynamical systems theory have all been in the behavioral literature for two decades, since Bertalanffy's seminal work (1968): the integrity of organization, of relational order, of self-stabilizing and self-organizing capacities. These principles are hardly new to developmental thinking. They are compatible with the strengths of Piagetian theory, which was to some degree an early precursor of systems thinking. The new element in current dynamical systems work is the discovery that rather simple non-linear systems can exhibit previously unsuspected complexity (e.g., May, 1976; see Gleick, 1987, for an entertaining, nontechnical introduction to the topic). The nonlinearity of many real-world physical systems, the appearance of order where previously we saw only disorder, the existence of stable states at far-from-equilibrium conditions, and the potential for a new universal law of physical order have rich implications for the conduct of science and our philosophy of nature (Prigogine & Stengers, 1984).

Several key points of dynamical systems theory (summarized largely from Thelen, in press) are presented next. Rather than present the theory in detail, I refer the interested reader to several recent works that develop the subject in greater detail with reference to behavioral development (Fentress & McLeod, 1986; Thelen, in press; Thelen et al., 1987; Wolff, 1987).

1. A system is composed of elements. Because these elements can combine with each other in many ways, the system possesses a large number of degrees of freedom. The body can be thought of as a system with many (skeletal) elements. Movement of the body potentially involves many degrees of freedom because of the number of elements in the skeletal and muscular systems (the "Bernstein problem" – Bernstein, 1967).

2. The ordered relationship among elements in a system is termed a *state*. Under appropriate conditions of energy flow into the system, the elements of a system interact, and spatial order and temporal order emerge in far-from-equilibrium conditions. This spatiotemporal order serves to reduce the number of degrees of freedom from the many that would exist if each element behaved independently of others. "Chemical clocks," in which millions of molecules behave in synchrony, were among the first examples of emergent order discovered in chemical systems, in apparent contradiction to the laws of equilibrium chemistry (Prigogine & Stenger, 1984). All biological systems exhibit stable, far-from-equilibrium states requiring energy input; this principle is virtually synonymous with the statement that metabolic activity supports life, which is a far-from-equilibrium phenomenon. The order in natural systems is not imposed from the outside nor contained within any element of the system; it "falls out" of the interactions among the elements. It is *emergent*. States are expressions of emergent order.

3. States resulting from energy-dependent interactions can split into two or more new states, producing multiple stable states in one system. In a

dynamical system, all of these states occur far from the nominal equilibrium of the system, that is, where the system would revert without further input of energy. Behavioral states (e.g., waking, sleeping) exemplify the plurality of possible stable states existing over time in one system (an organism, in this case). States represent the range of possible varieties of organization in the system; the organism alternates among them.

4. A dynamical system exhibits stability if it reaches a particular state (called an *attractor state*) from varying initial conditions and if it returns to that state after perturbation (unless the perturbation reaches some critical level, as discussed later). This idea recalls Waddington's notion (1957) of canalization in embryological development – that once embarked on a particular path, the organism develops along that path unless extreme perturbations shift development to an alternative (discontinuous) path. Rhythmic motor actions (e.g., sucking, rocking) exemplify stable behavioral attractor states.

5. Various types of attractor states have been identified and illustrated graphically (many examples are given by Gleick, 1987). *Point* attractors draw the system to a single converging state. *Limit-cycle* attractors, like those in chemical clocks, contain intrinsic characteristics (such as autocatalytic processes) that produce stable periodicity: The system shifts back and forth between alternate states. *Chaotic* attractors, although characterized by a very few degrees of freedom, present "restless, but bounded, activity" (Skarda & Freeman, 1987) that is never redundant. In fact, a chaotic attractor produces behavior with arresting visual order (e.g., fractals, which make lovely graphic art) (Peitgen & Richter, 1986).

The concept of attractor states has particular pertinence for behavioral and developmental applications (Thelen, in press). A limited number of behavioral states can occur at any point in development. As the components of the system change developmentally, so do the attractor states. Attractor states may become more stable (as in the case of increasing stability of sleeping or waking) or may become less stable and eventually disappear (as in various positional behaviors).

6. Shifts between attractor states are discrete (i.e., discontinuous). Sometimes a single state shifts into two or more stable modes (bifurcates). These processes result in new states, and an increasing number of states (change and differentiation). Skarda and Freeman (1987) provide an illustration of shifts between attractor states in the olfactory bulb as a function of stimulation with odorants, and they describe how a new attractor state appears (following a bifurcation from the background chaotic state) that corresponds to the detection of a newly learned odor. A behavioral example can be found in the ontogeny of song in swamp sparrows, in which a single subsong undergoes differentiation to produce many varieties of adult song produced by one individual (Thorpe, 1961).

7. A shift to a different state follows when a change of sufficient scale occurs in the value of an element or elements in the system. Usually a single

element or a small number of elements is involved. Change in the element(s) moves some critical relationship past a threshold, which results in reorganization of the entire system. The critical element or relationship involved in a state shift is called a *controlling parameter*. When the value of the controlling parameter is scaled beyond a critical value, amplification of instabilities disrupts the existing stability, and the system reorganizes into a new state. The elements of the system are largely the same after reorganization, but the relations among the elements are different. In research on motor control, discontinuous coordinative patterns of symmetrical movements of the fingers are neatly described by this model (e.g., Haken, Kelso, & Bunz, 1985). Gait transitions (walk to trot to gallop, in many quadrupeds) are also easy to conceptualize in this way.

8. In developmental systems, in which organismic components mature at different rates, the slowest-developing endogenous component at any particular time serves as the endogenous controlling parameter affecting the transition to a particular state. For example, if the muscles of the leg are too weak to support the weight of the body on one leg, walking will not appear, regardless of the status of other muscles, the degree of coordination present between the limbs, and so forth. Control parameters may be exogenous as well as endogenous. The plane of the substrate in relation to gravity could be a controlling parameter under the right conditions. At some degree of elevation from horizontal, for example, a human walking up an incline switches to a quadrupedal position and crawls or climbs. Location of control (endogenous or exogenous) is irrelevant to the functioning of the system (i.e., whether one walks or crawls). Location of control may change over time even under the same external circumstances, because of changes in the system.

In the language of systems theory, our attention in developmental work should be directed toward identifying controlling parameters through time, probing the stability of the system, and charting the way reorganizations take place at points of transition. Perhaps the simplest message to be gained from systems theory vis-à-vis development is that no single domain possesses the controlling parameter over long periods of development. Models of development that focus on any single domain (whether cognitive features, physical characteristics, or the task) as the determinant of change over development are doomed to fail because they do not recognize that alterations occur in the *system* as well as in its elements as development proceeds.

Dynamical systems theory can be developed into a useful theory of sensorimotor development, in accord with the requirements laid out at the beginning of this section. The emergent nature of behavior (out of relations among elements) and the appearance of new modes of organization across context and time are central features of development that fit comfortably within dynamical systems theory. The last requirement for a useful theory of sensorimotor development – that it guide empirical research – can be achieved in stronger ways than we can point to now. At this stage in the development of the theory, the potential to meet this requirement must be sufficient to support further investigation. The potential is clearly present.

Before reviewing studies of human sensorimotor development from a dynamical systems perspective, I briefly contrast some of the preceding points with those of contemporary neo-Piagetian cognitive developmental theory. Piagetian theory is an appropriate choice for this exercise because of its significant role in developmental work in humans (e.g., Case, 1985, or scan the table of contents of *Developmental Psychology*), because it has played an influential role in interpretation of behavior in nonhumans, especially primates (e.g., Dore & Dumas, 1987), and because it has been identified as an appropriate framework for comparative study of sensorimotor *development* in nonhuman primates (e.g., Parker, 1977; Parker & Gibson, 1977, 1979; Parker, P&G1). Obviously the emphasis in cognitive developmental theory differs from the emphasis in the astructural, acognitive dynamical systems approach, even though the two perspectives start with some similar assumptions. I highlight some of the key differences next.

## Contemporary Piagetian view of sensorimotor development

In many ways, dynamical systems theory and neo-Piagetian cognitive theory share the same starting assumptions. Contemporary cognitive researchers clearly appreciate the dynamic nature of development. Three aspects of the Piagetian approach to development are in accord with assumptions of systems theory:

1. the self-initiated, self-organized nature of activity, at every age,
2. the stability of behavioral organization over periods of time, and
3. the qualitative and occasionally rapid nature of behavioral change.

These aspects of development are explained in Piagetian theory by reference to processes of assimilation and accommodation, which result in reorganization of the cognitive structures that mediate behavior (Piaget, 1952, 1954). Although for many years Piaget's theory dominated the organismic approach to development (i.e., an approach that posited the foregoing points 1–3), the contemporary cognitive program, especially as it applies to sensorimotor development, suffers from serious limitations.

One limitation concerns the conceptualization of the organization of behavior present at birth. Piaget's view that reflexive activity in the human neonate formed the basis for later voluntary behavior has been discredited (e.g., Gibson, 1981). This idea still persists in the literature and in clinical wisdom, however (see Prechtl, 1984, for criticism of the idea's use in recent neonatological practice). And after acknowledging this shortcoming, Piagetian researchers have retreated from the issue of origins of behavior in the infant. Theoretical developments that can explain the organization of coordinated movement and the coordination of sensation and movement in early life are needed. Cognitive theory is not of much use here. Neurogenetic approaches (e.g., Gibson, 1981; Goldman-Rakic, 1987) represent one means of explaining the basis of early behavior and its transition to later forms and a means of relating these to cognitive interpretations of behavioral change. Systems perspectives offer another (e.g., Golani & Fentress, 1985).

Grasping in human neonates is an example of a behavior concerning the origins of which Piagetian theory is now silent. We know now that the early reflexive capacity of grasping upon palmar pressure is functionally unrelated to the organization of reaching and voluntary grasping. They can coexist or not; the appearance of one is unrelated to the appearance of the other (Beek, 1986). As Beek points out, this is not a trivial matter for our conceptions of action systems. They are not merely elaborations of reflexive activities. The integrity of action systems and self-organization is quite distinct from the integrity of what are regarded as reflexive capacities (Prechtl & O'Brien, 1982).

Another problem in the Piagetian program, especially troublesome in comparative studies, lies in its vision of a universal pattern of development toward an adult norm, which is in some sense anticipated by the developing system. There are two assumptions included in the idea of universal patterns in development leading to an adult endpoint that are problematic: the linear direction of change, and the linkage among behavioral capacities in different tasks, leading to the assumption that changes in performance on one task are accompanied by changes in performance on related tasks. Both of these assumptions limit our conceptions of early development.

Linear changes from immature to mature forms cannot be a general rule applying equally to all aspects of development. As many developmentalists (Bateson, 1987; Oppenheim, 1981; Prechtl, 1984) have noted, at each age the organism is a viable concern and perforce possesses adequate means to survive in its environment as it is currently constituted. Behaviorally, the developing organism is never simply an undeveloped version of the adult form. Indeed, infants often possess specializations that serve obvious adaptive functions in their species-typical rearing environments, such as the organization of suckling in rats, which is outside the satiation system governing the cessation of ordinary oral feeding even in rats of the same age (Hall & Williams, 1983). Some behaviors are present only during one part of the life cycle; others appear, disappear, and reappear, such as walking in neonatal rats and badgers (Golani, Broncht, Moualern, & Teitelbaum, 1981; Oppenheim, 1981).

It is difficult to conceptualize these appearances and disappearances in terms of a progressive transformation toward an adult form. On the other hand, occurrences of reversals from later forms to earlier forms and instability of form at particular times (e.g., across settings) pose no conceptual problems in systems theory. Of course, one cannot assume that every pattern exhibited by an immature organism represents an adaptive specialization, just as one cannot assume that all behaviors in adults represent adaptations. Adaptation must be demonstrated, not assumed. Nevertheless, acknowledging that specializations *can* be present at each phase of the life cycle widens the range of possible questions.

The second problematical assumption of universal pattern, linkage among the elements, has occupied a great deal of attention in the guise of cognitive "stages" and decalage (e.g., Brainerd, 1978; Uzgiris & Hunt, 1975). It is now recognized that the frequently observed temporal associations among certain behaviors do not reflect *necessary* synchronies of development. This is a critical point for comparative developmentalists. Although Piagetian theory allows us to identify asynchronies of behavioral change during development (and indeed encourages their discovery, by specifying synchronies to look for) (Parker, P&G1), it does not predict them, nor does it help us to understand why particular temporal relationships among abilities (different from those observed in humans) should exist. Systems theory could help to identify which elements serve as controlling parameters in particular situations and at particular times (e.g., context dependence in behavior).

A third limitation of cognitive theory for the study of sensorimotor development follows directly from its emphasis on cognition. The search for cognitive processes constraining (shaping, organizing) developmental changes in behavior has deflected attention from other, equally potent influences on the same behavior, especially in early life. Physical capacities and environmental constraints are often overlooked. I take as an example the treatment of the onset of walking in human infants as presented by Thelen (1983) and Zelazo (1983). Zelazo argued that the onset of unaided walking in humans at about 9 months reflected (in part) cognitive reorganization, specifically, improved access to memory and improved ability to generate specific associations quickly. Zelazo's argument was framed in a contemporary amalgam of information-processing and neo-Piagetian cognitive frameworks. Thelen, however, argued convincingly that biomechanical considerations *alone* are adequate to explain the timing of onset of unaided walking in human infants.

By overlooking components in the behavioral system other than the inferred cognitive abilities or maturation, and by accepting synchronous change as evidence of functional linkage, investigators have missed potential explanations of developmental transformation. Of course, physical maturation is recognized by everyone as a critical concomitant of the appearance of motor skills. But the interplay between motor capacity, task, setting, and performance is much more subtle than is suggested by the concept of "physical capacity" conceived separately from the task description. Cognitive models of sensorimotor development provide little impetus for investigation of context dependent performance, because in general the physical setting is not regarded as an important source of intrasubject variance in performance. Internal cognitive characteristics are assumed to be more important than context. Because systems theory specifies no particular element as a controlling parameter for behavioral change, it provides an independent framework for identifying sources of control in different settings and at different ages.

### Three aspects of human sensorimotor development seen from systems theory

Results from three studies of human sensorimotor development inspired by systems principles are presented next. These concern stepping behaviors in infants before the onset of walking, the organization of waking and sleeping, and voluntary reaching and grasping, all of which have been treated from other perspectives. Presenting them in concert here emphasizes the coherence provided by a systems approach. It also provides the human comparison for the aspects of sensorimotor development in capuchins that are the focus of a later section of this chapter.

Thelen's work on the stepping behavior of human infants and its eventual culmination in walking stands as one of the best examples of the application of systems theory in motor development (Thelen, 1985, 1986; Thelen & Fisher, 1982; Thelen, Fisher, & Ridley-Johnson, 1984). Thelen and her coworkers first identified an essential kinematic feature of adult bipedal walking: the 180° alternation in the movement of the legs. This pattern is evident in human infants when they begin to walk. In fact, even in newborns it is weakly evident, in the form of spontaneous kicks in the supine position (Thelen, 1985). The stability of alternation is greatly dependent, however, on contextual factors. When 7-month-old infants who are not yet walking are simply held upright with their feet on the floor, they rarely exhibit alternate stepping movements. When they are supported under their arms while their feet contact a slowly moving treadmill, however, alternating movements of the legs are reliably elicited (Thelen, 1986). The appearance of the alternation pattern is sensitive to contextual demands: When a split-belt treadmill is employed, so that one foot is contacting a faster-moving surface than the other, the speed of movement of the two legs is adjusted to maintain the alternation (Thelen, Ulrich, & Niles, 1987). Thus, alternation of stepping movements functions as an *attractor state*, appearing in stable configuration even in the face of perturbation. Postnatal changes in neurological maturation and cognitive processes are not part of the explanation of when or why infants step alternately. Rather, the biomechanical constraints, including the ability of one leg to support the weight of the whole body, and contextual demands, such as the angle of the body and the pull of gravity on the limbs, are all-important for young infants.

Wolff's work (1987) on the organization of behavioral states in infants stands as a second example of systems theory informing study of sensorimotor development. Like Prechtl and O'Brien (1982) and others, Wolff observed that behavior in infants (as in individuals of all ages) is rhythmic, spontaneous, and self-organized into discrete states. (The term *state* as employed to describe stable configurations of behavioral and physiological activity is entirely compatible with its meaning in the formulation of systems theory.) Wolff was particularly concerned with the changing stability of state organization through infancy and the significance of state organization for the infant's interaction with the environment.

The organization of behavioral states is important to sensorimotor development for many reasons, but I focus here on the exclusive occurrence of motor activity involving the coordination of hands and vision during one particular state identified by Wolff, called the alert active state (Wolff, 1987). The alert active state is characterized by the occurrence of "controlled motor actions of limited intensity and frequency that may sometimes have a rhythmical character, or may be goal-directed, while the eyes are open ... and may scan the environment during movement" (Wolff, 1987, p. 25). The alert active state is thus more specific than being awake. Because the infant initiates exploratory motor and instrumental activities only or primarily when in this state, the organization of behavioral states constrains when and for how long an infant can experience contingent control of the environment or practice motor skills.

Wolff observed that states exhibit changing degrees of attraction during development. That is, at one age, an infant will readily shift from one state to another after a perturbation, whereas at a later age the infant will fail to shift states after an identical perturbation. Sleep, for example, is a powerful attractor state in the young infant. Even vigorous sensory stimulation may fail to arouse a young infant from sleep, whereas the same stimulation will more reliably wake an older infant. Another example is the ability of older infants to postpone sleep when engaged in an activity that holds their interest, an ability that is not as well developed in younger infants. Alert activity becomes a more powerful attractor state in older infants than it is in younger infants. Thus, older infants have more opportunities to engage in exploratory motor behavior, because they remain in the appropriate state for such activity for longer periods. They also have more control over transitions out of the state in which such activity is possible. The infant's changing ability to maintain an alert active state for longer periods and to accommodate to larger or more frequent perturbations without shifting states provide changing opportunities for initiating interaction with the environment. In this way, state organization and cognitive development are intimately related.

The activities that take place within the alert active stage change with age, in concert with increasing stability of this state. Consider the capacity for performing two related acts at the same time. At about 8 weeks of age, the human infants Wolff observed were able to continue the performance of one sensorimotor pattern, such as kicking their legs rhythmically, while concurrently fiddling with a toy with their hands. At the same age, infants could use the two hands in independent actions, such as scratching a foot with one hand while reaching for a toy with the other. According to Wolff, the onset of the ability to combine two actions marks an important change in behavioral organization, "the onset of a behavioral state that is probably essential for the discovery of new combinations among ... actions, the linking of means to ends behavior, and the construction of new ends realized by the novel combination of action patterns" (Wolff, 1987, p. 78). These aspects of behavior figure centrally in cognitive development.

In the state of alert activity characterized earlier, the infant's behavior becomes less dependent on specific stimulus conditions, and the infant exhibits choice in its motor activity. The ability to act "spontaneously" is associated with greater control over shifts from the active state. Wolff hypothesizes that maintenance of the state of waking alertness is dependent at first on the actual performance of goal directed actions; later, the *possibility* of goal directed actions is sufficient.

Wolff's work suggests the contribution of a dynamical systems perspective to studies of behavioral development. Focus on the changing *organization* of behavior, rather than on the occurrence of specific elements of behavior, supports an integrated view of cognitive, affective, and motor development. It can be argued that the same could be achieved from other perspectives, that systems principles supply nothing "new" conceptually beyond what is already provided by contemporary epigenetic-constructivist approaches to development (G. Sackett, pers. commun.). Wolff specifically addresses the contribution of systems thinking in relation to traditional epigenetic-constructivist models of the induction of novel behaviors during human development:

> When considering, and then rejecting, the various alternative solutions to the problem of developmental induction, I am assuming that any appeal to a priori autonomous central executive agencies, including those implicit to the epigenetic-constructivist perspective [as exemplified in the writings of Werner, Piaget, and Freud] is ultimately unsatisfactory as a basis for "explaining" development. Further, I am assuming that it may be timely and appropriate to propose a very different perspective on developmental transformations and induction. (Wolff, 1987, p. 248)

One can argue that contemporary epigenetic views, such as expressed by Oppenheim (1981), *are* compatible with Wolff's systems approach. Nevertheless, systems thinking does lead to testable hypotheses about the induction of novel behavior in Wolff's domains of interest that would be unlikely to appear from other sources.

Wolff provides several testable hypotheses about the organization of behavior in infants generated by dynamical systems theory that, although they may be compatible with those perspectives, have failed to emerge from neurological or cognitive perpsectives. He further suggests that by considering affective behaviors as patterns of motor coordination, we can see conceptual links between behavioral states and affective behavior, and hence a new way to study their integration. This last idea could be useful for comparative studies of early development, which generally overlook affect for want of a good way to relate it to other aspects of development.

Reaching and grasping in human infants constitute a third area of behavioral development in which systems principles provide insight. Von Hofsten (1986) has shown that human infants can reach toward a moving object at very young ages (5 days after birth) provided that adequate postural support is given and the task is designed to match the visual capacities of the infant (see Bower, 1979, for similar studies). Vision and movement, von Hofsten argues, are coordinated at birth. Reaching movements are linked with vision, in that movements while the object is visually fixated are better aimed than those occurring without visual fixation (von Hofsten, 1986). Infants can even "catch" objects as soon as they can reach for stationary objects (von Hofsten & Lindhagen, 1979). At no age are vision and movement independent of each other.

Von Hofsten (1986) views his work as revealing innate knowledge linking action and perception in human infants (i.e., "knowing" the link between vision and space, and organizing movement on the basis of visuospatial knowledge). Gibson (1981) suggests a neurological basis for early visuomotor coordination involving midbrain structures, thus enriching our sense of how infants "know" the link, and she further suggests how the integration between sensation and action changes during the course of neurological maturation. Others view visuomotor competence of young human infants as an example of emergent coordination and an opportunity to look for system scaling factors responsible for transitions to new modes of organization (new states) (Clark, 1986). These two perspectives are compatible and should be complementary. Surely visuomotor coordination undergoes changes contemporaneously with maturation, both musculoskeletal and neurological. Systems descriptions of relevant parameters should indicate the manner in which transitions occur and, with further probing, could indicate the relative significance of changes in physical maturity, context, and experience to behavior at specific moments in development.

These three examples of developmental studies of human infants – concerning stepping, the alert active state and dual motor activity, and reaching – highlight aspects of behavioral development that I believe are fundamental and that must be incorporated into any useful theory of sensorimotor development:

1. Behavior is *emergent*. New forms of behavior arise from new organizations of the system. New forms are not built up from reflexive "building blocks," nor inherited, to be expressed only when maturation allows, nor acquired, in the sense of being imposed on the organism from an agent outside itself.
2. Behavior is constrained by context as well as organismic variables.
3. Behavioral organization changes discontinuously (even when underlying organismic changes are gradual) (Hinde & Bateson, 1984).

Systemic analyses of early motor skills challenge cognitive formulations of early behavioral change, especially those involving the idea of cognitive constraints (cognitive assimilation and the construction of knowledge about space and objects) on the appearance of voluntary movement. They are in accord with some other aspects of cognitive formulations, such as temporal stability of organization. They complement the study of the integration of cognitive change with neural ontogeny represented by Gibson's writings (1981, P&G3) and by the work of Goldman-Rakic (1987). More important than their particular points of agreement or disagreement with extant views of motor development, systems theory approaches to behavioral development

1. are completely in accord with the fundamental principles of behavioral development listed earlier, as well as some others omitted for brevity's sake (see Thelen, in press, for a more comprehensive treatment),
2. allow integration and organization of data from a wide variety of sources, providing great explanatory power, and
3. provide a new vocabulary for, and a novel way of thinking about, organization and change.

All that is lacking from our earlier list of requirements for dynamical systems theory to be useful in studies of sensorimotor development (see the section "Dynamical systems views . . . ") is to use it to guide empirical research. This state of affairs is being rectified even now, and we can expect the effort to continue.

Now I turn to studies with capuchins (*Cebus apella*) on the same aspects of early development (behavioral states, reaching and prehension, and locomotion) just reviewed in humans. Although our data are less extensive than the human data, I think they are sufficient to illustrate the value of a systems perspective in the comparative study of sensorimotor development.

## Behavioral states, manipulation, and locomotion in mother-reared tufted capuchins

Our research addresses the development and expression of motor skills and adaptive manipulative behavior in capuchins. Our studies of infant development are aimed at characterizing species-typical development of motor and social skills, particularly in relation to manipulation of objects.

### *Methods*

*Housing.* Several studies of infant capuchins reared in species-typical social groups in our laboratory are summarized here. The groups, composed of an adult male, 5–6 adult females, and their offspring (total group size averaging 15) are housed in double-room indoor cages (total volume 1,800 cubic feet) furnished with bedding, perches, visual barriers, and various manipulanda, including natural browse on an intermittent basis. Food (principally commercial chow, supplemented with smaller items such as seeds) is strewn on the floor (see Fragaszy & Adams-Curtis, in press, for further details). Group composition roughly approximates that seen in the wild (e.g., Terborgh, 1983). The groups are undisturbed by human activity, except for necessary maintenance. Activity budgets in the groups approximate those seen in wild capuchins, with object manipulation and feeding occupying about half of the daylight hours, and movement occupying another 20–30%.

*Procedure.* Formal and informal observations have been made of about two dozen infants born into our laboratory groups. Formal observations have

Table 6.1. *Categories of behavioral activity scored from birth through 11 weeks for an infant capuchin monkey reared by its mother*

| Category[a] | Definition |
|---|---|
| Sleep | Eyes closed for >1 min. Body generally still, although twitching of limbs and tail may occur. |
| Drowsy | Eyes may be closed, opened intermittently, or open and glazed. Some slow body movements or postural adjustments. Differentiated from sleep by the presence of slow movements. After 2 min of no movements, if eyes remained closed, transition to sleep was scored. |
| Alert/inactive | Eyes open and clear; body generally still and relaxed; head turning and postural adjustments could occur. |
| Alert/active | Infant was clearly alert. Gross motor activity resulting in the infant moving on the mother's body or in extrapersonal space sufficient to change the orientation of the core of the body and/or standing with torso unsupported. |
| Nurse | On mother's ventrum, with head oriented toward nipple. Scored in favor of sleep when ambiguous. |

[a] All categories had minimum durations of 3 sec.
*Source*: Adapted from Fragaszy (1989).

included longitudinal scoring of the frequency of certain motor activities and the duration of contact with other animals from birth through 7 months (Visalberghi & Fragaszy, 1990). Behavioral events, including social interaction with specified others, manipulation of the environment, and play (solitary and social) were noted as they occurred during 10-min sampling periods. Five-minute samples of social spacing (contact, with whom, or time spent alone) followed frequency sampling.

More detailed data on manipulation have been collected using semicontinuous scoring (5 sec on, 5 sec off) of manipulative acts and targets of focal animals during 10-min sampling sessions (Fragaszy & Adams-Curtis, 1989). Additionally, videotapes were made of four infants ages 6 weeks to 24 weeks, for 1 hour per week during weeks 6–12, and for half an hour per week thereafter (D. M. Fragaszy & S. Friedman, unpublished data). To date, the tapes of one infant have been scored for the occurrence of manipulation and contact with the mouth to an object or surface, with or without the use of the hands. Targets and hand(s) used were specified, and durations of manipulation were noted. (Scoring is proceeding for the other infants.)

Time budgets (of activity states) were obtained in two studies. In one, an infant was observed 6–8 hours consecutively at 7-day intervals from birth through 77 days of age (Fragaszy, 1989). The timing and nature of transitions in behavioral state and nursing were noted, and occurrences of exploratory manipulative behaviors were marked according to act and target. The scoring vocabulary for behavioral states and nursing used in this study is presented in Table 6.1. Since this study was completed, we have observed two other

Figure 6.1. A healthy, full-term neonatal infant tufted capuchin within hours of birth. Note the limp tone of the body, excepting the flexed extremities. Drawn from a photograph by Wei-zhi Su.

infants with the same protocol during their first 3 months for shorter periods per week.

In the second study of time budgets, an infant in its eighth week was observed around the clock for 5 days, using a scoring scheme nearly identical with that in the study just mentioned (D. M. Fragaszy et al., unpublished data). The study was undertaken to compare a hand-reared infant with a normally reared infant (as described later), and only its normative aspects are cited in this section.

### *Results*

*Status of the neonate.* A capuchin infant is more helpless at birth than would be expected from its size (relative to that of the mother) and the size of its brain (relative to the adult brain and relative to its own body weight) (Figure 6.1) (Fragaszy, 1990). Although able to cling without support, infants are not able to move in directed fashion, to lift their heads, or to track objects visually (Fragaszy, 1989; Spinozzi, 1989). Mothers usually cuddle newborn infants, and an infant may remain on the mother's ventrum for hours even in the daytime if she is still. The altriciality of capuchin infants stands out sharply in relation to the markedly different pattern of motor competence presented by infant cercopithecines and squirrel monkeys (Fragaszy, in press).

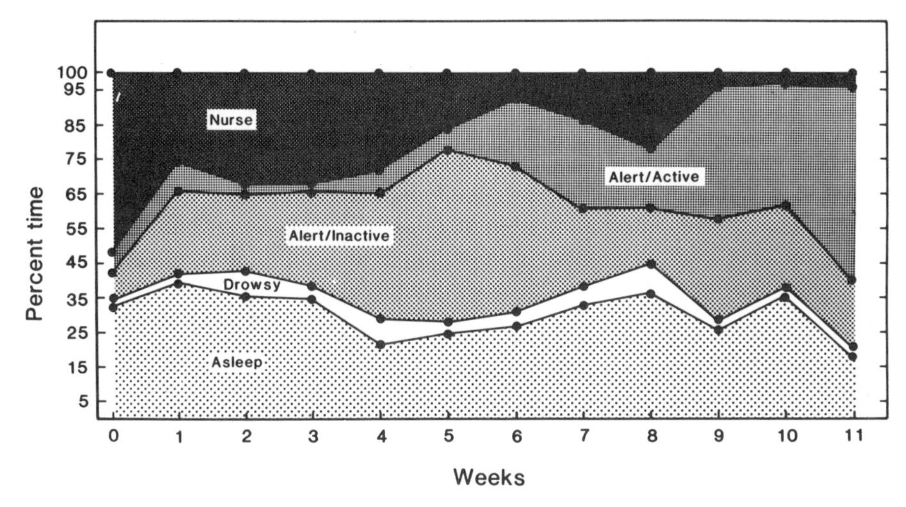


Figure 6.2. The distribution of time among various behavioral states and nursing in a capuchin infant reared by its mother. The infant was observed on the day of birth and weekly thereafter through 11 weeks for 6–8 hours consecutively. Note the increase in time spent in the alert/inactive state at the end of the first month, at the expense of nursing and sleeping, and the increase in time spent in the alert/active state at the start of the third month, at the expense of nursing and (later) alter/inactive. (From Fragaszy, 1989, in *Developmental Psychobiology*, copyright © 1989 by John Wiley and Sons, Inc.; reprinted by permission.)

*Behavioral states.* At birth, the infant nurses and sleeps; movement is rarely apparent, and alert states account for only a small proportion of observed time. Over the next 3 months, major changes in the infant's behavior occur. Data on the proportion of the daytime spent in various states by a representative infant are presented in Figure 6.2. Nursing declined quite sharply from birth to week 1. Changes in the amounts of time in the alert/inactive state (which increased) and nursing (which decreased) occurred between week 3 and weeks 4 and 5. Time spent in the alert/active state began to increase in week 6, but increased sharply at the beginning of the third month, with corresponding reduction in the proportion of time in an alert/inactive state or nursing. Time spent sleeping remained fairly constant through the first 10 weeks, with an evident drop in week 11. Whether this drop indicated the start of a trend or simply fluctuation around a mean value is not known.

The pattern of changes in the proportions of time in various states suggests that, following stabilization after birth, major reorganization of behavior occurs at about 8 weeks of age, with one-third or more of the infant's time in an alert/active state after that week. In conjunction with changes in motor coordination (noted later), the changes occurring around the third month look like a "biobehavioral shift," as identified by Emde et al. (1976). This age corresponds to the shift in growth rate from curvilinear to linear in capuchins (Jungers & Fleagle, 1980), an indication that widespread changes in physiological functioning could be occurring at this time. Such a shift in growth rates

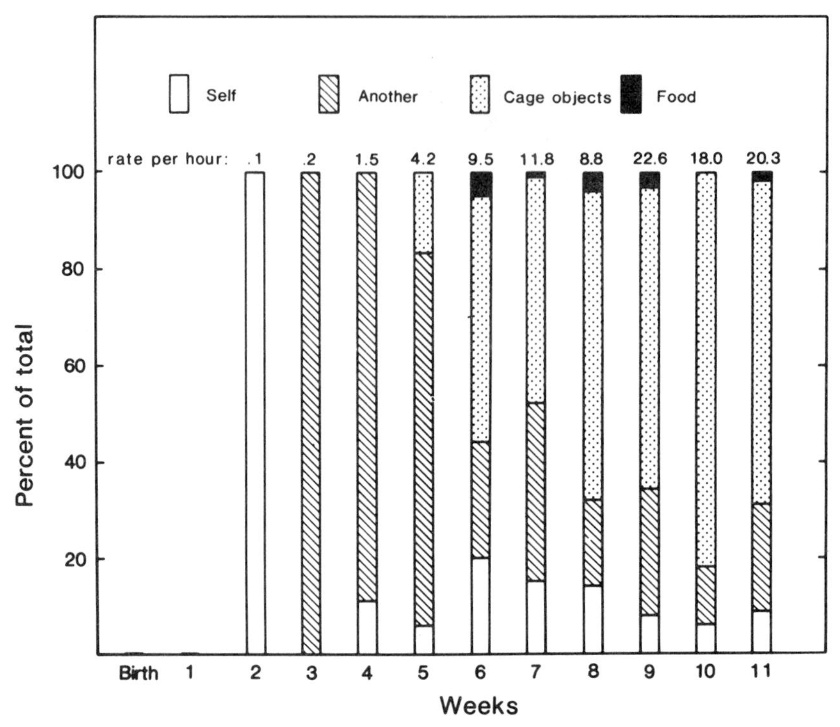


Figure 6.3. The targets of manipulation per week shown as the proportion of all targets for a mother-reared infant capuchin from birth through 11 weeks. The rate of manipulation is shown at the top of each bar. Note that social targets predominate in weeks 3–5, a time when the young infant is not able to reach out into space if such reaching requires that it support its own torso. (From Fragaszy, 1989, in *Developmental Psychobiology*, copyright © 1989 by John Wiley and Sons, Inc.; reprinted by permission.)

in connection with behavioral change seems relevant to sensorimotor organization and cognitive function. Goldman-Rakic (1987) has found associations between certain features of brain growth (density of synapses in the cortex) and shifts in performance on memory tasks between 2 and 4 months after birth in infant rhesus macaques. Although there are large differences between macaques and capuchins in motor function at birth and in early infancy (Fragaszy, 1989, in press), we do not know if these are associated with differences in maturation of the brain. In any case, the data suggest that major changes in the infant's experience with the environment occurred at about 8–9 weeks after birth. In the language of systems theory, the stability of state organization was altered, with alert/active states becoming more stable, and other states less stable.

*Manipulation.* Prehension (including reaching toward a target) is absent in the first month in mother-reared infants (see Figure 6.3 for representative data from one infant). In the second month, an infant will reach out toward

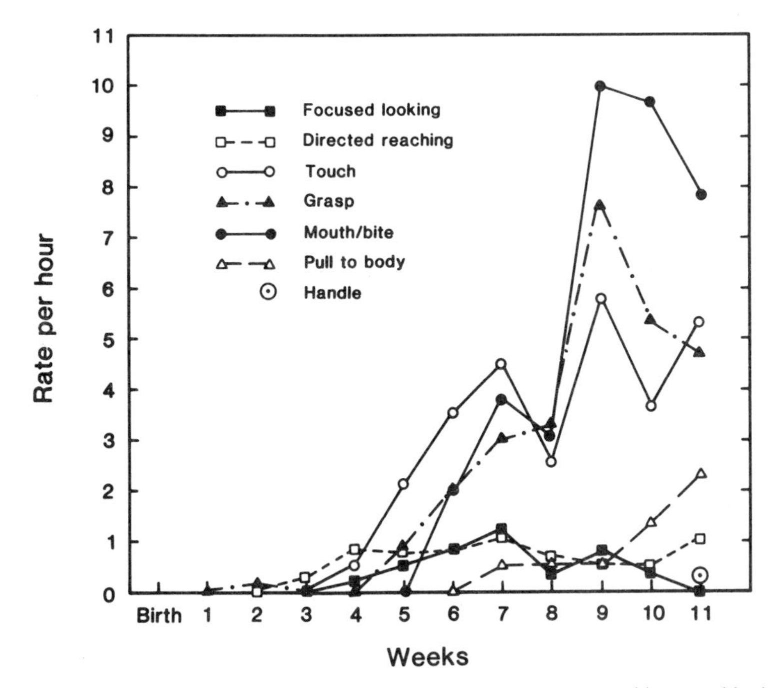


Figure 6.4. The form of manipulation in a young capuchin reared by its mother from birth through 11 weeks. Note that the infant showed large increases in touching in weeks 5–7, and grasping and mouthing (which required holding the object) in week 9. (From Fragaszy, 1989, in *Developmental Psychobiology*, copyright © 1989 by John Wiley and Sons, Inc.; reprinted by permission.)

other animals that come to it, and sometimes will touch them. The fingers are splayed, and no grasping is apparent. In the third month, when the time in alert/active states increases, infants increase their frequency of activity, and the sophistication of their motor control increases tremendously (Figure 6.4). Precision grips first appear at the beginning of the fourth month. Many manipulative patterns characteristic of juvenile and adult capuchins, particularly combinatorial actions such as striking an object on a substrate, first appear months later. The first instances of combinatorial behaviors have been seen at the start of the sixth month in mother-reared infants. Activities involving placement of an object or the hand in specific relation to another object, such as inserting a stick into a hole, first appear near the end of the first year. Table 6.2 shows major events in the development of prehension in infant capuchins, along with locomotor abilities.

*Posture and locomotion.* Neonatal infant capuchins are unable to lift their heads or other parts of their bodies off a supporting surface. Control of the head develops by 2 weeks, and at the same age infants can crawl in directed fashion on the mother (e.g., from back to ventrum, or vice versa) without assistance. Mother-reared infants can support the torso with arms extended

Table 6.2. *Summary of changes with age[a] (in weeks) in posture and prehension in capuchin infants reared by their mothers or by humans, and observed in their home cages*

| Activity | Mother reared[b] | Hand reared[c] |
|---|---|---|
| Lifts head, tracks visually | 2 | 2 |
| Reaches out, unimanually | 3 | 2 |
| Reaches out, bimanually | 7–8 | 2 |
| Grasps | 5 | 5 |
| Extends arms, supports torso | 5 | 3–4 |
| Stands quadrupedally (briefly) | 5 | 3 |
| Walks quadrupedally | 8 | 4 |

[a] Earliest ages when the activity was observed are given.
[b] From Fragaszy (1989) and Fragaszy et al. (unpublished data).
[c] From Spinozzi (1989) and Fragaszy (1989, unpublished data).

(Figure 6.5) and push to stand *on the mother* at the end of 5 weeks, but they dismount from a carrier only several weeks after that, and they first locomote while off the mother or other carrier at about 9 or 10 weeks. Quadrupedal locomotion is competent in the fifth month. Locomotor play (solitary and social) becomes frequent in the sixth month.

**Comparison of mother-reared and hand-reared capuchin infants**

We had the opportunity to study two hand-reared infants when they were orphaned or rejected. We pursued two themes in the study of these infants, echoing those studied in the normally reared infants: (1) spontaneous activity of the infant, measured in terms of locomotion and manipulation in familiar settings and in test situations, and (2) organization of behavioral states. Both infants were healthy at birth, and while being tested they appeared entirely normal in their growth and development. The first infant studied was removed from her mother at 2 weeks of age and observed systematically in the eighth week after birth in the study described next. The second infant was orphaned at 1 week and studied longitudinally thereafter. The data presented here from the second infant concern development in the first 4 months after birth. Our observations of the second infant were exploratory in nature, and we included as many different types of testing as possible on a sometimes serendipitous basis. The infant was observed on several days per week in the home setting using the states protocol described earlier. We also observed the second hand-reared infant with a direct assessment procedure that had been developed by Schneider (1987) for study of rhesus monkeys. Schneider's procedure incorporates elements of several tests used with human infants. The elements of the assessment relevant to this chapter include control of the head while lifted or held, visual tracking in the horizontal and vertical planes, muscle tonus, resistance to passive extension of the limbs, response to palmar

Figure 6.5. A 5-week-old hand-reared infant capuchin supporting its torso with extended arms. The behavior first appears in the same form in mother-reared infants at this age.

pressure, righting from a supine position, and response to a sudden auditory stimulus. Last, we tested this infant's dexterity in a variety of manipulative tasks.

The data from these two infants are supplemented by data on three other hand-reared infants of the same species tested regularly on sensorimotor coordination (e.g., visual tracking, reaching and grasping, locomotion) and in various problem-solving situations (e.g., recovery of an object moved out of sight) by Giovanna Spinozzi (1989).

*Methods*

A hand-reared infant and a mother-reared infant, matched for age and sex, were observed for 5 consecutive days (120 hours) in the eighth week (days 50–57) (D. M. Fragaszy & P. Scollay, unpublished data). Following the same protocol used for the infant observed weekly, we scored behavioral states and prehensile activity. A keyboard method was used in this study, however,

Table 6.3. *Proportions of time spent[a] in various activity states over a 5-day period (120 hours) in the eighth week after birth by a hand-reared infant capuchin and a mother-reared counterpart of the same sex*

| | Activity state | | | | |
|---|---|---|---|---|---|
| | Sleep | Drowsy | Alert/inactive | Alert/active | Nurse[b] |
| Hand reared | 41 | 9 | 42 | 8 | 2 |
| Mother reared | 46 | 6 | 44 | 4 | 2 |

[a] Values are percentages of total observed time.
[b] Nursing occurred concurrently with other states; other categories are mutually exclusive.

rather than noting transition times manually as was done in the longitudinal study of the mother-reared infant discussed earlier. The keyboard provided more accurate notations of elapsed time.

To tap a significant environmental dimension expected to vary between mother rearing and hand rearing, we also scored the mother's locomotor movement. In the case of the hand-reared infant, we scored movement by the carrier when the human holder moved the infant, which clung to an arm. A deliberate attempt was made to provide as much movement stimulation as possible to the hand-reared infant.

Finally, we also scored the occurrences of head cocking, a behavior exhibited especially by young primates of smallish genera in which the head is tipped, typically at right angles to the body axis (Menzel, 1980). We scored this behavior because it is a clear indicator of visual fixation, and in that sense reflects visual exploration of the environment. It also produces vestibular stimulation in the absence of other bodily movement and thus is an energetically cheap way to produce vestibular stimulation.

The hand-reared infant was fed a milk formula on a 3-hour schedule around the clock. The light cycle was the natural light cycle for both infants (about 14 hours of light, 10 hours of dark; midsummer), with the addition of dim lighting at night to feed the hand-reared infant, and red-light heat lamps at night in the outside area of the mother-reared infant's home cage. Except when riding on a human carrier, the hand-reared infant was housed with a paint-roller "surrogate mother" mounted at a 30° angle from horizontal in an incubator held at 85–90° F.

### *Results*

*Behavioral states.* The mother-reared and hand-reared infants spent about the same amount of time asleep or drowsy (50–52%) and alert (48–50%) (Table 6.3). These values are close to those for the hand-reared infant observed longitudinally at the end of week 8 (compare with Figure 6.2). Major

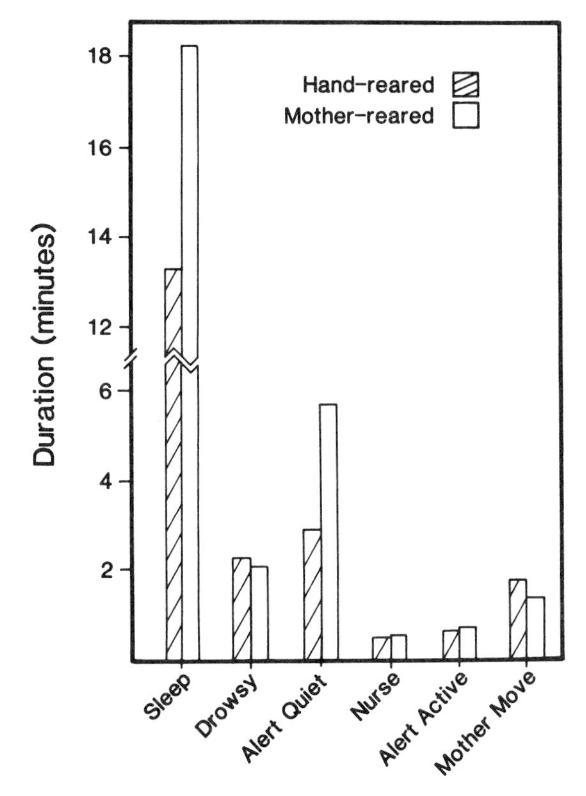


Figure 6.6. The mean duration per bout for various states of the infant and its mother or human carrier in a hand-reared infant and a mother-reared infant capuchin, both in the eighth week after birth. Both infants were observed over 5 days (120 hours). Note the large difference in the durations of bouts of sleeping and alert/quiet periods.

differences, however, were observed between the hand-reared and mother-reared infants in the proportion of time allotted to the alert/active state (8% in the hand-reared infant, 4% in the mother-reared infant) and in the duration of alert/active and sleep states, both of which were longer in the mother-reared infant (Figure 6.6; note that alert/inactive is synonymous with quiet/alert). Although the lengths of bouts of motor activity were similar in the two infants, more bouts of motor activity occurred in the hand-reared infant (Figure 6.7). Also, head cocking and manipulative activity were much more frequent in the hand-reared infant (Figure 6.7). The pattern suggests that the hand-reared infant was not merely more active than the mother-reared infant while awake, but was active in a different cycle, and distributed its exploratory actions differently.

Differences in the cycles of activity between hand-reared and mother-reared infants appear equally pronounced when we examine the distribution of activity across a 24-hour day (Figure 6.8). For example, a strong diurnal cycle is evident for sleep in the mother-reared infant, with sleeping occupying

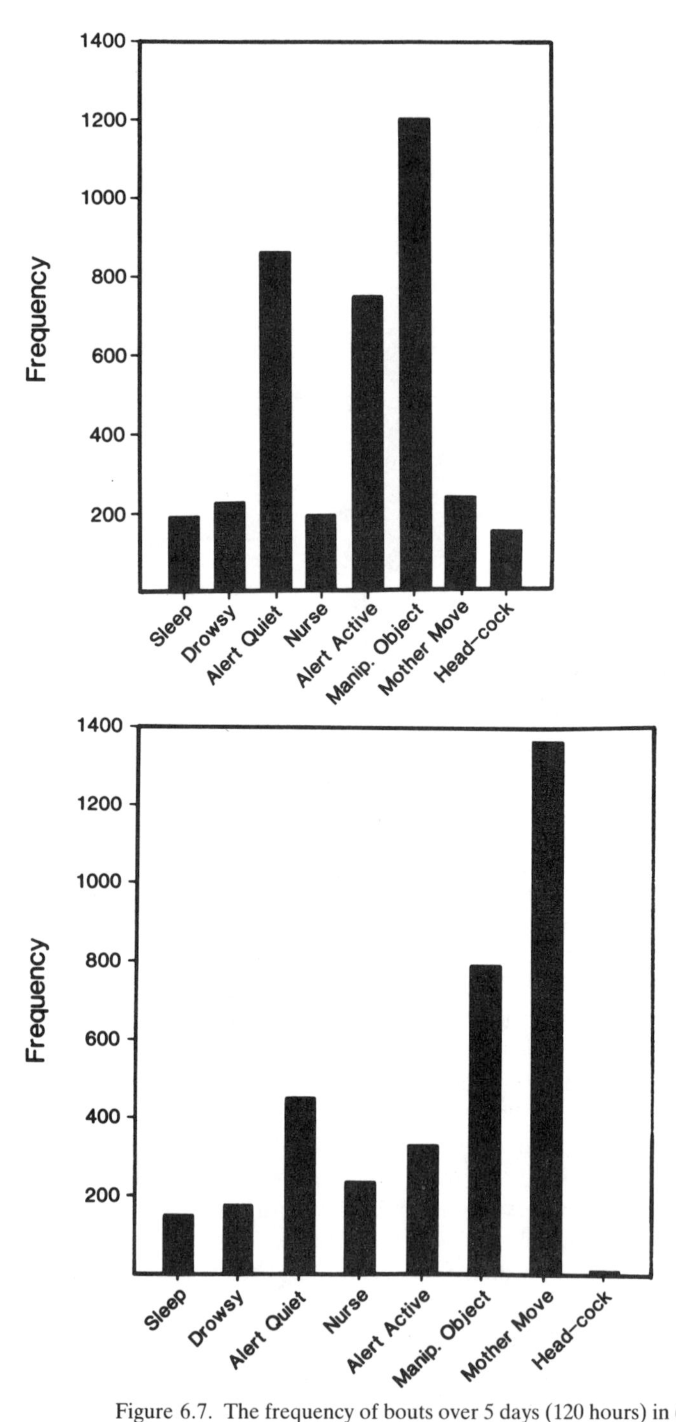


Figure 6.7. The frequency of bouts over 5 days (120 hours) in (a) a hand-reared infant and (b) a mother-reared infant capuchin. Note the greater overall frequency in the hand-reared infant, especially noticeable for alert/quiet, alert/active, and head cock.

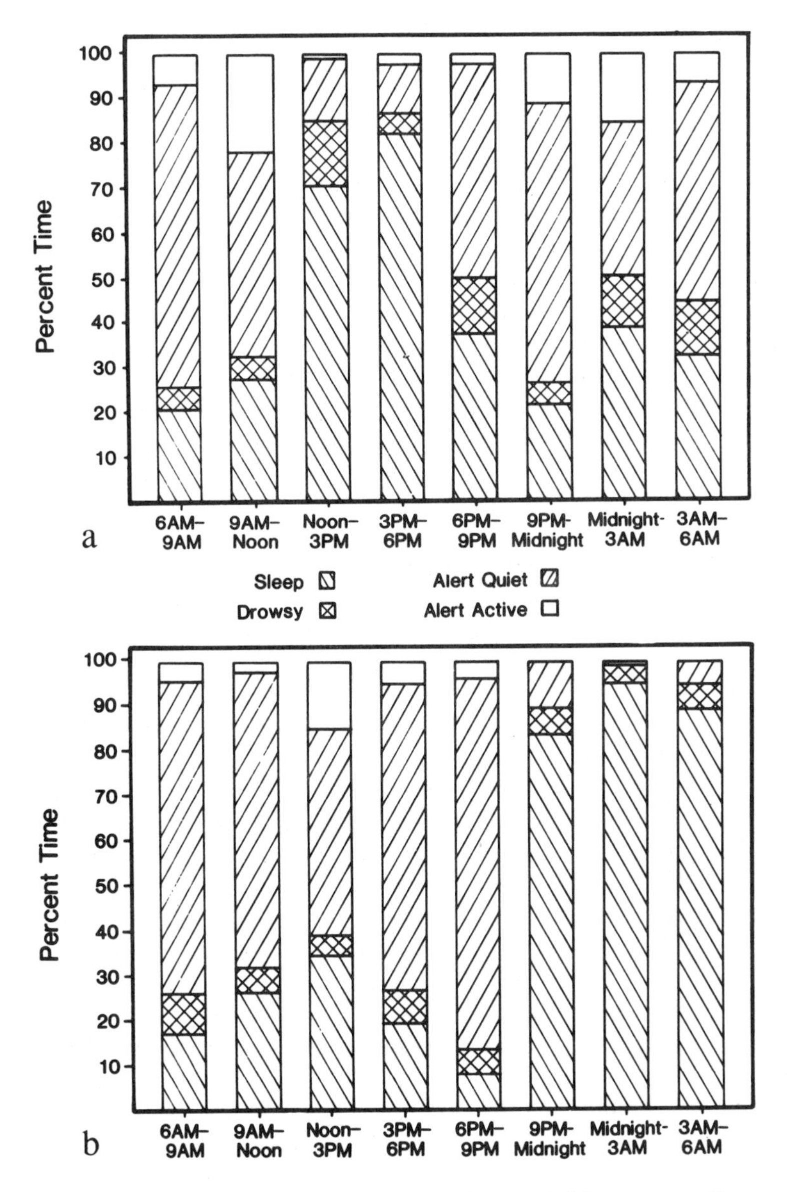


Figure 6.8. Activity states over a 24-hour period (average over 5 consecutive days) in (a) a hand-reared and (b) a mother-reared infant tufted capuchin. Note the clear diurnal pattern of sleeping and waking in the mother-reared infant, and its absence in the hand-reared infant.

almost all of the dark hours. In the hand-reared infant, a diurnal cycle for sleeping is difficult to see. However, a plot from pooled time periods is not a sensitive way to examine periodicities in behavior. We used time series analyses to identify periodicities in activity at finer resolution and to evaluate the strength of the periodicities (Visaberghi & Fragaszy, in press). These analyses showed that 24-hour periodicity was present in both infants for the duration and frequency of most states, but that the periodicity was more pronounced in the mother-reared infant, and the mother-reared infant had fewer short cycles (less than 4 hours) than did the hand-reared infant.

Longitudinal observations of state and motor activity were made for the second hand-reared infant from week 4 through week 15 using the same protocol as for the 5-day study described earlier, but for shorter consecutive periods (2 hours at a time). The data confirm that the amounts of time spent asleep and awake were similar to those for mother-reared infants (through 11 weeks), whereas motor activity occupied more of the hand-reared infant's awake time. That was true before as well as after the burst in motor activity in mother-reared infants occurring at about 9 weeks. It seems that hand-reared infants exhibit a different distribution of activity, different cycles of activity, and earlier increases in motor activity compared with their mother-reared counterparts.

*Manipulation.* The hand-reared infant, when supported so that the head and torso were resting on a stationary surface, reached toward a visually fixated object in its *second week*, both unimanually and bimanually. That was strikingly early in comparison with mother-reared infants, which do not begin to reach out for weeks after that age. Flexion of the fingers was still crude at 2 weeks, and while grasping upon contact with a fixated object was not evident, reaching toward a seen object clearly was (Figure 6.9). The rate of manipulative activity by the hand-reared infant followed weekly was far greater (at least twice as frequent) than the average rate for its mother-reared counterparts through 15 weeks, when collection of quantitative data for the hand-reared infant ended. Observations of the hand-reared infant ended in part because the infant was so busy that the observer had difficulty keeping accurate records of its manipulations. At this age, the infant manipulated objects with great rapidity during a large portion of its alert/active time.

*Locomotion.* The hand-reared infant stood in its home cage in the third and fourth weeks and locomoted in its home cage with its torso off the floor (a kind of creeping locomotion) at the same age. Locomotion by the hand-reared infant outside of its home cage was still absent at this age, but appeared by 8 weeks. Its absence outside the home cage is in line with behavior of mother-reared infants, but its appearance in the home cage during the same period is surprising. When a mother-reared infant is separated from its mother in the first few months, it does not attempt to crawl to her in a strange place, even if she is inches away, and even if it has begun to locomote independently in the home cage. Yet at home, the hand-reared infant was capable of standing and walking at 4 weeks! Locomotion is apparently an unstable state in the infant capuchin when it first appears, evident only under

(a)
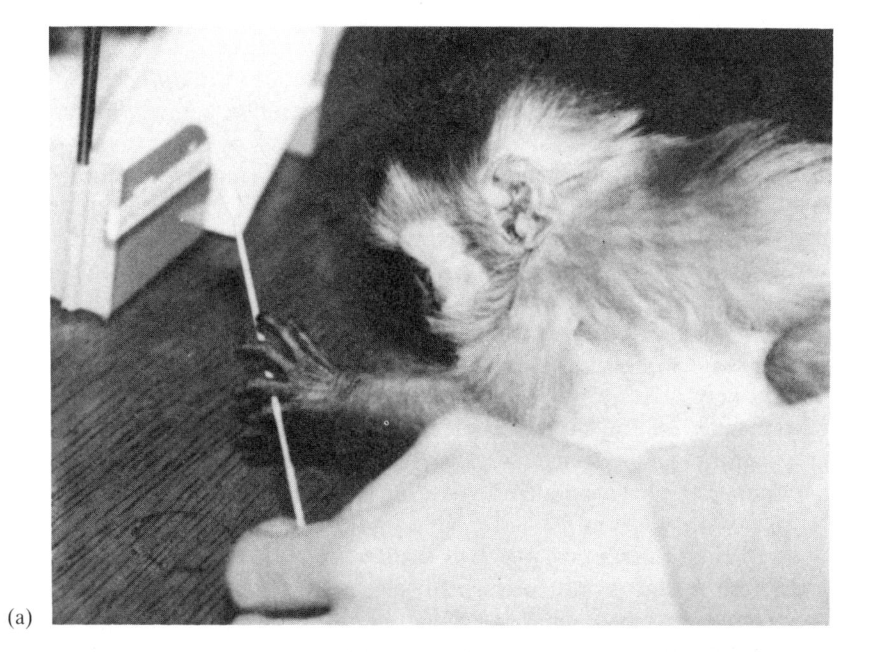

(b)
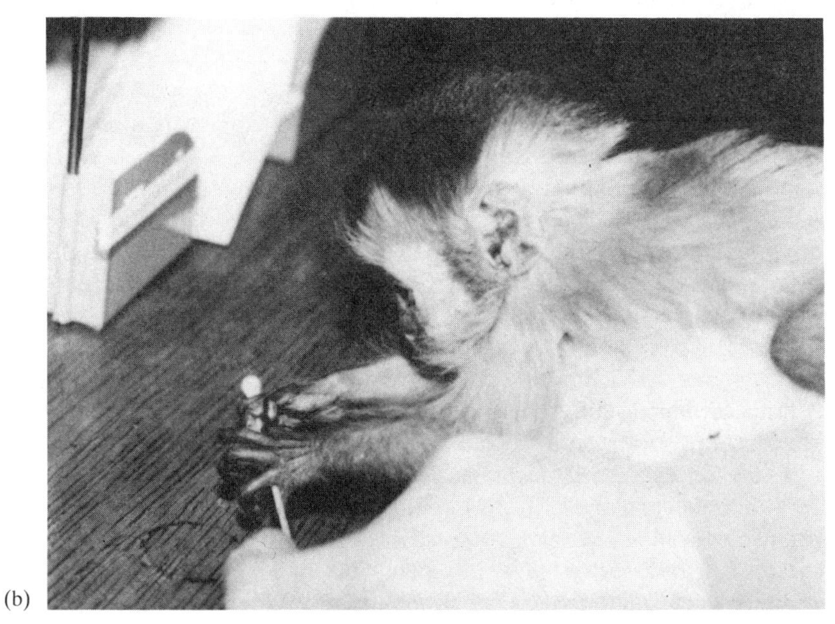

Figure 6.9. A 2-week-old hand-reared infant capuchin reaching with (a) one hand for a visually fixated object within arm's reach and (b) both hands for a visually fixated object within arm's reach.

the most favorable conditions of a familiar setting and a stationary, flat surface.

A telling episode in connection with the onset of locomotion occurred with the hand-reared infant at 8 weeks, when it was fully capable of locomoting all over its cage and climbing the wire sides. When a sudden noise startled the infant, it immediately dropped to the surface, assuming the prone, clinging position characteristic of every frightened infant. The next week, a similar startle caused the infant to run *up* the mesh, a complete reversal in geotaxis. Once again, behavioral organization seems to have undergone a major change at the beginning of the third month.

The differences between hand-reared and mother-reared infants in locomotion and prehension clearly reflect the postural constraint present in the species-normal setting of nearly continuous carriage by the mother or other carrier until the fourth month of life. Restricted prehension in this time period is not cognitively mediated, although the nature of prehension (ballistic vs. visually guided, or controlled during movement) may vary as a function of brain growth (Gibson, 1981, P&G3). Obviously, a hand-reared infant has less need to cling for support when its support is stationary. It can relax its grip, even if its balance is not very steady or its limbs very strong, and reach into space or venture to stand. The mother-reared infant clinging to an unpredictable and moving support cannot afford to let go with either hand until it has sufficient strength and postural control to be secure with a three-point cling. Similarly, the mother-reared infant cannot use both hands until its grip is secure with legs alone.

In the language of systems theory, postural support is a *controlling parameter* constraining the appearance of prehension and locomotion in both hand-reared and mother-reared infants. The *attractor states* of reaching and locomotion involve the synchronous and sequential coordination of many body parts to move the body while counteracting the effects of gravity and inertia. They are quite unstable at first, and equally so in infants reared by the mother or by hand. Slight perturbations are enough to throw the system back to a more stable point state: clinging. Differences in the timing of onset and the frequency of locomotion and manipulation in capuchin infants in different settings are due to differences in postural support for the infant that affect the requirement for postural coordination.

These findings on capuchin infants parallel Thelen's findings on the postural/biomechanical constraints on stepping patterns in prewalking human infants, as well as the work by von Hofsten and others on reaching in human infants. A wide range of motor behaviors can be seen "precocially" in a permissive environment that frees the infant from postural and biomechanical constraints.

Differences between hand-reared and mother-reared infants in the amount of time devoted to motor activity and the frequency of manipulation also reflect a different organization of activity through time. Probably more is involved here than postural support.

The proportion of time and number of bouts in an alert/active state were quite different for the hand-reared and mother-reared infants. Once begun, physical activity endured the same length of time in each infant, suggesting a time-limiting process (such as fatigue) in the stability of that state (a limit-cycle state). The differences between the infants can be described in terms of the relative stability of the resting (alert/inactive) state. In the hand-reared infant, the regulation of states is rarely perturbed from an external source. Its cycles of motor activity and rest are closer to a free-running system generating its own changes in state than are the mother-reared infant's cycles. In the mother-reared infant, the demands of its physical setting (clinging to a moving mother) probably serve as a significant source of regulation for the infant. One effect of regulation is strengthened stability of the sleeping state, and another effect is strengthened stability of the alert/quiet state. Many other factors affecting the stability of sleep and inactive states can be hypothesized, including thermoregulatory processes, the metabolic effects of milk, and so on (Hofer, 1987).

The frequency of head cocking in the hand-reared infant also suggests more free-running cycles of state changes. Head cocking is a means of varying visual input, but it also provides vestibular stimulation. In the hand-reared infant, it may be generated to produce vestibular stimulation that is not provided through the movements of the mother or surrogate, just as body rocking is performed by nursery-reared rhesus monkeys (Mason & Berkson, 1975). Vestibular stimulation may also be a controlling parameter affecting the stability of resting alert states and sleep. We commonly observe that young infants sleep in the daytime if the mother is locomoting steadily, just as human infants enter a sleep state and stay asleep more reliably when rhythmically moved. The significance of vestibular function for cognitive capacity is hinted at by Schneider's work (1987) demonstrating that this single index at birth was the best predictor of individual performance 6 months later on cognitive tasks in rhesus macaques. The contribution of movement, passive or active, to the development of vestibular function is a topic that deserves further work.

In short, a mother-reared infant experiences a different set of contextual constraints on mother activity, and a different set of contextual regulators of behavioral state, than does a hand-reared infant. Jointly, these produce some remarkable differences in the appearance of motor skills in capuchin monkeys. The different timing of appearance of motor skills in the two circumstances reflects different responses to the constraints imposed by the physical setting interacting with the postural status and physical stamina of the infants. This suggests that the appearance of these motor skills is not tied to cognitive maturation. In more extensive studies comparing the development of rhesus monkeys reared by the mother or in a nursery, Schneider (1987) documented a wide range of other differences between infants reared in the two situations. Although differences were not evident on the day of birth, responsivity, muscle strength, and many other characteristics reflecting the organization of behavior differed in infants in the two settings from Day 1 on (Schneider, 1987, pers. commun.).

### Reprise: Coordination, context, and activity

In what sense can a systems approach to sensorimotor development provide insights that are lacking in a cognitive developmental approach? One product of systems thinking (and the one particularly relevant to the comparative data cited earlier) is greater sensitivity to the external context of behavior and to the possibility that context may affect behavior differently at different periods in development.

Clearly, the differences in the physical settings in which the capuchin infants acted affected their behavior in the first months of life. The contrast between hand-reared and mother-reared infants in some aspects of behavior is probably greatest during early development, when, for example, differing postural constraints serve as controlling parameters in the two environments. For older capuchin infants, as for the more precocial macaques at all ages, postural context is less frequently a controlling parameter.

A second product of a systems perspective is the focus of attention on the organization of behavioral states and the cyclic physical processes affecting behavior. In the data reported earlier, clear differences between hand-reared and mother-reared infants were evident in basic cycles of behavior and time budgets. In systems terms, these come about because of differences in attractor states, meaning that different relations among the same elements occur in the two conditions. Having identified differences in state organization in the two settings, it is now possible to probe the parameters responsible for the differences, such as the occurrence of movement by the carrier.

A third product of a systems perspective is attention to the means by which age-specific behaviors in species-normal environments are produced (by the interaction among behavioral capacities and species-typical environmental contexts), and how they are outgrown as the infant matures. This approach does not, by itself, answer the question whether a behavior at a particular age is an age-specific adaptation or is an immature (nonfunctional or inefficiently functional) form of an adult behavior (these are not mutually incompatible possibilities). Answering that question will require converging evidence from studies of function and physical maturation as well as of behavioral organization. The systems approach focuses on the organization of elements, not on the functional consequences of the organization. But a neutral stance allows us to see that age-specific behaviors may appear for reasons independent of cognitive maturation.

Certainly a systems approach is implicit in ethological/psychobiological views of development, and no doubt this contributes to my affinity for this approach. But I believe that making the approach explicit will be useful to others concerned with determining how environmental setting and behavioral organization are related in young animals. Understanding the link between behavior and context is a task of practical importance, as, for example, in the handling of human infants in intensive care nurseries. These infants exhibit a variety of behavioral disturbances that are likely to be related to the abnormality of their environment (Gottfried, Allace-Lande, Sherman-Brown, King, & Coen, 1981). It is also a task of enormous theoretical significance, for cognitive development as much as for mother development.

My discussion has centered on comparisons among members of the same species observed in different settings and on the processes involved in behavioral regulation at the proximate level. Can systems theory also be used productively to guide research on phylogenetic comparisons of development? I see no reason why not. Its use in the domain of sensorimotor development will bring us to many new questions.

## Acknowledgments

The work reported in this chapter has been supported by grant 8503603 from the National Science Foundation and grant R01-MH41543 from the National Institute of Mental Health. The animals were supported during some of the research by the College of Sciences, San Diego State University, and by the College of Veterinary Medicine, Washington State University. The author was supported by a Research Scientist Development Award from the National Institute of Mental Health during the preparation of this manuscript. The chapter was prepared in final form while I was a Visiting Scholar at the Subdepartment of Animal Behavior, University of Cambridge, Great Britain. I thank my host, Professor Hinde, and the many colleagues at the Subdepartment for their hospitality. I also wish to thank the many collaborators who have contributed to the research program, including P. Scollay, L. Adams-Curtis, S. Friedmann, and R. Lammers-Carlson, and a large number of undergraduate research assistants. Thanks also to L. Adams-Curtis. E. Visalberghi, S. Datta, and A. Jolly for commenting on an earlier version of the manuscript. G. Spinozzi graciously shared unpublished data with me. Lastly, thanks to S. Parker and K. Gibson for their judicious editorial advice and patience.

## References

Antinucci, F. (Ed.). (1989). *Cognitive structure and development in nonhuman primates*. Hillsdale, NJ: Erlbaum.

Bateson, P. (1984). Sudden changes in ontogeny and phylogeny. In G. Greenberg & E. Tobach (Eds.), *Behavioral evolution and integrative levels* (pp. 156–166). Hillsdale, NJ: Erlbaum.

Bateson, P. P. G. (1986). Functional approaches to behavioral development. In J. Else & P. Lee (Eds.), *Primate ontogeny, cognition and social behavior* (pp. 183–192). Cambridge University Press.

Bateson, P. (1987). Biological approaches to the study of behavioral development. *International Journal of Behavioral Development*, *10*, 1–22.

Bateson, P. (in press). Additive models may mislead. *International Journal of Behavioral Development*.

Beek, P. J. (1986). Perception–action coupling in the young infant: An appraisal of von Hofsten's research program. In M. G. Wade & H. T. A. Whiting (Eds.), *Motor development in children: Aspects of coordination and control* (pp. 187–196). Dordrecht: Martinus Nijhoff.

Bernstein, N. (1967). *The coordination and regulation of movements*. London: Pergamon Press.

Bertalanffy, L. von. (1968). *General system theory*. New York: Braziller.

Bower, T. G. R. (1979). *Human development*. San Francisco: Freeman.

Brainerd, C. J. (1978). The stage question in cognitive developmental theory. *Behavioral and Brain Sciences*, *2*, 173–213.

Case, R. (1985). *Intellectual development: Birth to adulthood*. New York: Academic Press.

Clark, J. E. (1986). The perception–action system: A commentary on von Hofsten. In M. G. Wade & H. T. A. Whiting (Eds.), *Motor development in children: Aspects of coordination and control* (pp. 197–206). Dordrecht: Martinus Nijhoff.

Dore, F., & Dumas, C. (1987). Psychology of animal cognition. *Psychological Bulletin*, *102*, 219–233.

Ehrlich, A. (1984). Infant development in two prosimian species: Greater galago and slow loris. *Developmental Psychobiology*, *7*, 439–454.

Emde, R., Gaensbauer, T., & Harmon, R. (1976). Emotional expression in infancy. *Psychological Issues*, *10*, 519–528.

Fentress, J. C. (1984). The development of coordination. *Journal of Motor Behavior*, *16*, 99–134.

Fentress, J. C. (1986). Development of coordinated movement: Dynamic, relational and multileveled perspectives. In M. G. Wade & H. T. A. Whiting (Eds.), *Motor control in children: Aspects of control and coordination* (pp. 77–106). Dordrecht: Martinus Nijhoff.

Fentress, J. C., & McLeod, P. J. (1986). Motor patterns in development. In E. Blass (Ed.), *Handbook of behavioral neurobiology* (pp. 35–97). New York: Plenum.

Fragaszy, D. M. (1989). Activity states and motor activity in an infant capuchin monkey (*Cebus apella*) from birth through eleven weeks. *Developmental Psychobiology*, *22*, 141–157.

Fragaszy, D. M. (in press). Early behavioral development in capuchins. *Folia Primatologica*, *54*(3–4).

Fragaszy, D. M., & Adams-Curtis, L. E. (1989). *Interactions with objects in tufted capuchins (Cebus apella) I: Form, frequency, and age differences*. Unpublished manuscript.

Fragaszy, D. M., & Adams-Curtis, L. E. (in press). Challenge in captivity. In H. O. Box (Ed.), *Primate responsiveness to environmental change*. London: Chapman & Hall.

Fragaszy, D. M., Visalberghi, E., and Robinson, J. G. (in press). Variability and adaptability in the genus *Cebus*. *Folia primatologica*, *54*(3–4).

Gibson, K. R. (1981). Comparative neuro-ontogeny, its implications for the development of human intelligence. In G. Butterworth (Ed.), *Infancy and epistemology: An evaluation of Piaget's theory* (pp. 52–84). New York: St. Martin's Press.

Gleick, J. (1987). *Chaos. Making a new science*. New York: Penguin Books.

Golani, I., Broncht, G., Moualern, D., & Teitelbaum, P. (1981). "Warm-up" along dimensions of movement in the ontogeny of exploratory behavior in the infant rat and other infant mammals. *Proceedings of the National Academy of Sciences, USA*, *78*, 7226–7229.

Golani, I., & Fentress, J. C. (1985). Early ontogeny of face grooming in mice. *Developmental Psychobiology*, *18*, 529–544.

Goldman-Rakic, P. S. (1987). Development of cortical circuitry and cognitive function. *Child Development*, *58*, 601–622.

Gottfried, A. W., Allace-Lande, P., Sherman-Brown, S., King, J., & Coen, C. (1981). Physical and social environments of newborn infants in special care units. *Science*, *214*, 673–675.

Haken, H. (1983). *Synergetics: An introduction* (3rd ed.). Heidelberg: Springer-Verlag.

Haken, H., Kelso, J. A. S., & Bunz, H. (1985). A theoretical model of phase transitions in human hand movements. *Biological Cybernetics*, *51*, 347–356.

Hall, W. G., & Williams, C. L. (1983). Suckling isn't feeding, or is it? A search for developmental continuities. *Advances in the Study of Behavior*, *13*, 219–254.

Harlow, H. F. (1951). Primate learning. In C. P. Stone (Ed.), *Comparative psychology* (pp. 183–238). Englewood Cliffs, NJ: Prentice-Hall.

Harvey, P., Martin, R., & Clutton-Brock, T. (1987). Life history strategies in comparative perspective. In B. Smuts, R. Seyfarth, R. Wrangham, & T. Struhsaker (Eds.), *Primate societies* (pp. 181–196). University of Chicago Press.

Hinde, R. A., & Bateson, P. (1984). Discontinuities versus continuities in behavioral development and the neglect of process. *International Journal of Behavioral Development, 7*, 129–143.

Hofer, M. A. (1987). Early social relationships: A psychobiologist's view. *Child Development, 58*, 633–647.

Hofsten, C. von. (1986). The emergence of manual skills. In M. G. Wade & H. T. A. Whiting (Eds.), *Motor development in children* (pp. 167–185). Dordrecht: Martinus Nijhoff.

Hofsten, C. von, & Lindhagen, K. (1979). Observations of the development of reaching for moving objects. *Journal of Experimental Child Psychology, 38*, 158–173.

Jungers, W. L., & Fleagle, J. G. (1980). Postnatal growth allometry of the extremities in *Cebus albifrons* and *Cebus apella*: A longitudinal and comparative study. *American Journal of Physical Anthropology, 53*, 471–478.

Klüver, H. (1937). Re-examination of implement using behavior in a cebus monkey after an interval of three years. *Acta Psychologica, 2*, 347–395.

Martin, R. D. (in press). Evolution of the brain in early hominids. *Ossa.*

Mason, W. A., & Berkson, G. (1975). Effects of maternal mobility on the development of rocking and other behaviors in rhesus monkeys: A study with artificial mothers. *Developmental Psychobiology, 8*, 197–211.

May, R. M. (1976). Simple mathematical models with very complicated dynamics. *Nature, 261*, 459–467.

Menzel, C. (1980). Head-cocking and visual perception in primates. *Animal Behaviour, 28*, 555–558.

Oppenheim, R. W. (1981). Ontogenetic adaptations and retrogressive processes in the development of the nervous system and behavior: A neurobiological perspective. In K. J. Connolly & H. F. R. Prechtl (Eds.), *Maturation and development: Biological and psychological perspectives. Clinics in developmental medicine*, No. 77/78 (pp. 73–109). Philadelphia: Lippincott.

Oppenheim, R. W. (1984). Ontogenetic adaptations in neural and behavioral development: Toward a more ecological developmental psychobiology. In H. F. R. Prechtl (Ed.), *Continuity of neural functions from prenatal to postnatal life. Clinics in developmental medicine*, No. 94 (pp. 16–30). London: Spastics International Medical Publications.

Parker, S. T. (1977). Piaget's sensorimotor period series in an infant macaque: A model for comparing unstereotyped behavior and intelligence in human and nonhuman primates. In S. Chevalier-Skolnikoff & F. Poirier (Eds.), *Primate biosocial development* (pp. 43–112). New York: Garland Press.

Parker, S. T., & Gibson, K. R. (1977). Object manipulation, tool use and sensorimotor intelligence as feeding adaptations in cebus monkeys and great apes. *Journal of Human Evolution, 6*, 623–641.

Parker, S. T., & Gibson, K. R. (1979). A developmental model for the evolution of language and intelligence in early hominids. *Behavioral and Brain Sciences, 2*, 367–408.

Peitgen, H. O., & Richter, P. H. (1986). *The beauty of fractals*. Berlin: Springer-Verlag.

Piaget, J. (1952). *The language and thought of a child.* London: Routledge & Kegan Paul.

Piaget, J. (1954). *The origins of intelligence in children.* New York: International Universities Press.

Prechtl, H. F. R. (1984). Continuity and change in early neural development. In H. F. R. Prechtl (Ed.), *Continuity of neural functions from prenatal to postnatal life. Clinics in developmental medicine*, No. 94 (pp. 1–15). London: Spastics International Medical Publications.

Prechtl, H. F. R., & O'Brien, M. J. (1982). Behavioral states of the full term newborn. Emergence of a concept. In P. Stratton (Ed.), *Pychobiology of the human newborn* (pp. 53–73). New York: Wiley.

Prigogine, I., & Stengers, I. (1984). *Order out of chaos: Man's new dialogue with nature.* New York: Bantam.

Riete, M., & Short, R. (1976). Nocturnal sleep in isolation-reared monkeys: Evidence for environmental independence. *Developmental Psychobiology, 10*, 555–561.

Schneider, M. (1987). *A rhesus monkey model of human infant individual differences*. Unpublished doctoral dissertation, University of Wisconsin.
Skarda, A., & Freeman, W. J. (1987). How brains make chaos in order to make sense of the world. *Behavioral and Brain Sciences*, *10*, 161–195.
Spinozzi, G. (1989). Early sensorimotor development in cebus. In F. Antinucci (Ed.), *Cognitive structure and development of nonhuman primates* (pp. 55–66). Hillsdale, NJ: Erlbaum.
Terborgh, J. (1983). *Five New World primates*. Princeton University Press.
Thelen, E. (1983). Learning to walk is still an "old" problem: A reply to Zelazo. *Journal of Motor Behavior*, *15*, 139–161.
Thelen, E. (1985). Developmental origins of motor coordination: Leg movements in human infants. *Developmental Psychobiology*, *18*, 1–22.
Thelen, E. (1986). Development of coordinated movement: Implications for early human development. In M. G. Wade & T. A. Whiting (Eds.), *Motor development in children: Aspects of coordination and control* (pp. 107–124). Dordrecht: Martinus Nijhoff.
Thelen, E. (in press). Self-organization in developmental processes: Can systems approaches work? In M. Gunnar (Ed.), *Systems in development: The Minnesota Symposium in Child Psychology*, Vol. 22.
Thelen, E., & Fisher, D. M. (1982). Newborn stepping: An explanation for a "disappearing reflex." *Developmental Psychobiology*, *18*, 760–775.
Thelen, E., Fisher, D. M., & Ridley-Johnson, R. (1984). The relationship between physical growth and a newborn reflex. *Infant Behavior and Development*, *7*, 479–493.
Thelen, E., Kelso, J. A., & Fogel, A. (1987). Self-organizing systems and infant motor development. *Developmental Review*, *7*, 39–65.
Thelen, E., Ulrich, B. D., & Niles, D. (1987). Bilateral coordination in human infants: Stepping on a split-belt treadmill. *Journal of Experimental Psychology: Human Perception and Performance*, *13*, 405–410.
Thorpe, W. H. (1961). *Bird song. The biology of vocal communication and expression in birds*. Cambridge University Press.
Trevarthen, C. (1984a). How control of movement develops. In H. Whiting (Ed.), *Human motor actions – Bernstein revisited* (pp. 223–261). New York: Elsevier.
Trevarthen, C. (1984b). Biodynamic structures, cognitive correlates of motive sets, and the development of motives in infants. In W. Prinz & A. Sanders (Eds.), *Cognition and motor processes* (pp. 327–350). New York: Springer-Verlag.
Uzgiris, I. C., & Hunt, J. M. (1975). *Ordinal scales of psychological development*. University of Chicago Press.
Visalberghi, E., & Fragaszy, D. M. (in press). Food using behavior in tufted capuchin monkeys (*Cebus apella*) and crab-eating macaques (*Macaca fascicularis*). *Animal Behavior*.
Waddington, C. H. (1957). *The strategy of the genes*. London: Allen & Unwin.
Wolff, P. H. (1987). *The development of behavioral states and the expression of emotions in early infancy. New proposals for investigation*. University of Chicago Press.
Zelazo, P. (1983). The development of walking: New findings and old assumptions. *Journal of Motor Behavior*, *15*, 99–137.

# 7 Tool use, imitation, and deception in a captive cebus monkey

*Kathleen Rita Gibson*

A recent article in *Behavioral and Brain Sciences* by the British psychologist Euan Macphail (1987) argues that all nonhuman vertebrates are equally intelligent. All animal intelligence derives from one and only one intellectual mechanism: the ability to form associations. No qualitative or quantitative differences can exist among nonhuman vertebrates because none possesses any additional intellectual mechanism.

The view that a uniform level of intelligence characterizes all nonhuman vertebrates is difficult to reconcile with the anatomy of the structure that mediates intelligence: the brain (Gibson, P&G3). Vertebrate brain sizes range from less than a gram to several pounds (Jerison, 1973; Sacher & Staffeldt, 1974). Structurally, the brains of some species are dominated by the brain stem, others by enlarged basal ganglia, and still others by enlarged neocortices (Sarnat & Netsky, 1981). Similarly, species differ in their performances on standard Piagetian tests of human intelligence (Chevalier-Skolnikoff, 1977, 1983, 1989; Dore & Dumas, 1987; Parker, 1977; Parker & Gibson, 1977, 1979; Redshaw, 1978; Wise, Wise, & Zimmerman, 1974; Miles, P&G19; Russon, P&G14).

These considerations suggest caution in interpreting the results of laboratory tests that imply uniformity of intelligence across the vertebrate order. Even if Macphail's assessment that all animal learning and problem solving reflect the ability to form associations should prove correct, his conclusion that all animals are equally intelligent does not follow. Animals may still differ in the complexity and variety of the associations they can make, the speed with which they can make them, and the numbers of associations they can hold in mind simultaneously.

That intelligence may vary rather dramatically in accordance with quantitative parameters of this nature is, in fact, the thrust of much recent work. Modern theories indicate that intelligence is a hierarchically additive process in which two or more separate motor, sensory, or conceptual schemes are added together sequentially or simultaneously to form one new whole (Case, 1985; Piaget, 1952, 1954). Constructional capacity varies with the number of sensory and motor schemes possessed, with the numbers of schemes or objects that can be used simultaneously or in succession to meet a single end, and with the numbers of contexts in which specific motor or object manipulation schemes can be expressed. Animals appear to differ in mental constructional ability (Parker & Gibson, 1977). In particular, large-brained species may have more mental constructional capacity than others (Gibson, 1986, P&G3), and some analyses suggest that human ape intellectual differences primarily reflect quantitative differences in constructional capacities (Case, 1985; Gibson, 1988, 1990).

Much recent research has focused on the intellectual skills of cebus monkeys. Cebus possess very large brains for their body size (Jerison, 1973). In addition, their tool-using capacities and foraging strategies have long suggested that they are more intelligent than most animals (Antinucci & Visalberghi, 1986; Bierens de Haan, 1931; Cooper & Harlow, 1961; Gibson, 1986; Klüver, 1933, 1937; Mathieu, Bouchard, Granger, & Herscovitch, 1976; Parker & Gibson, 1977, 1979; Romanes, 1882; Visalberghi, 1988; Visalberghi & Antinucci, 1986; Warden, Koch, & Fjeld, 1940; Parker, P&G4). The actual level of intelligence reached by cebus monkeys, however, still remains unclear. In particular, it is uncertain whether cebus manipulative skills reflect Stage 5 or 6 of Piaget's sensorimotor series (Piaget, 1952) or whether they may reflect even higher skills (Chevalier-Skolnikoff, 1989; Gibson, 1989). Nor is it clear whether or not cebus intellectual skills extend beyond tool use into imitative and communicative domains (Chevalier-Skolnikoff, 1989; Parker, 1989; Visalberghi & Fragaszy, P&G9).

This chapter provides data of relevance to these latter issues by presenting an in-depth account of the spontaneous behaviors of a captive cebus monkey, Andy, who has been living in my home since 1978. Andy's behavior helps elucidate the behavioral and motivational range that can lead to tool use in cebus monkeys and demonstrates a single monkey's ingenuity in devising his own tool-using tasks, using whatever materials might be available.

Andy possesses the mental constructional skills necessary to use varied objects in varied contexts. In particular, he uses objects for self-grooming and as tools to effect physical changes in other objects and behavioral changes in living animals. His manipulative skills thus appear equivalent to Stage 5 or 6 of Piaget's sensorimotor period (Piaget, 1952). He also possesses deceptive abilities of approximately the same level. In this respect, his mental constructional skills exceed those reported for most vertebrates. Andy's imitative skills, however, are very limited, equivalent only to about Stage 2 or 3 of the sensorimotor period. These and other data suggest that animals can possess schema- or domain-specific intelligence (Gibson, 1989). High intellectual performance in one domain does not guarantee high performance in others.

## Observational conditions

At the time these observations began, Andy was a young adult male $5\frac{1}{2}$ years of age; they continued until the age of 16 years. For the first 5 years of the observational period, he possessed the full complement of adult male canine teeth; later, because of repeated biting episodes, the canines were filed to the level of the other teeth. Andy's origin was uncertain, although he was known to have had at least two previous human owners. Most likely his history was that of many pets who are purchased when they are young and cute and then are discarded when they reach adulthood and become aggressive. Andy's species is also uncertain, but he is probably a *Cebus nigrivittatus*.

Andy's previous owner kept him in a cage 3 × 3 × 3 ft. For his first 5 years in my home, Andy was housed in a cage 4 × 4 × 5 ft, with visibility of yard, kitchen, TV, and human conversational areas. A litter floor lined with disposable polyethylene plastic sheeting sat 2 ft below the cage floor just out of reach of his prehensile tail. Andy was taken out of his cage several times a week for several hours at a time. Outings usually consisted of walks in the yard, periods of being leashed (15-ft leash) in one area of the yard or garage, or "play" in the sink. In 1983, Andy's cage was discarded. Since then he has been kept permanently on a leash. This permits easier handling and play with Andy, and he seems happier leashed than caged. For the last 5 years, Andy has spent the weekends outside in a large lakeside area, except during very cold weather.

On some occasions, Andy has stayed at the home of Jim and June Cook in Conroe, Texas, where he has been caged in a large enclosure 15 × 30 ft adjacent to an equally large enclosure housing two young chimpanzees. Andy has also had frequent opportunities to interact with cats, dogs, and ducks.

Andy is given monkey chow, hard-boiled eggs, meat, nuts, onions, garlic, water, ice cubes, and a wide variety of raw fruits and vegetables on a regular basis. At intervals he is given raw eggs, mealworms, june bugs, crickets, melons, coconuts, sunflower seeds, raw rice and wheat, cheese, fish, milk, and shellfish. Fruits and nuts are given unpeeled and unshelled. On occasion, food is wrapped in plastic or tinfoil and placed in a box prior to being given to him.

During his weekend outings, Andy procures some food on his own. He has been observed eating a wide variety of fresh young vegetation and flowers. He regularly catches flying insects. On a few occasions he has eaten duck's eggs, lizards, turtles, frogs, and snakes. On one occasion he ate poisonous mushrooms.

Nonfood items given to Andy on a regular basis include two terry cloth towels and several paper towels. He is regularly given newspaper, junk mail, rubber bands, tennis balls, Tinkertoys, plastic cups, coins, keys, cloth, stuffed animals, puppets, and other toys. In recent years he has been given a 1-in. segment of an unlighted cigar on a daily basis.[1] When outside, he has access to sticks, stones, garden hoses, and tubs of water.

Most observational periods are informal and occur during cleaning, feeding, playing, and grooming sessions. No attempt has been made to train Andy to use objects by offering artificial rewards for such behaviors, although tool use and other object manipulations have frequently been demonstrated to him.

Many of Andy's manipulative schemes are used in social encounters. The nature of the encounter is recorded as agonistic or playful, depending on the facial gestures, vocalizations, and postures that accompany it. Behaviors accompanied by the open-mouth, bared-teeth threat face or the bared-teeth scream face are judged to be agonistic. Those accompanied by the relaxed, open-mouth play face are judged to be playful (Oppenheimer, 1973; Weigel, 1979).

## Observations

### *General perceptual and motor capacities*

One predictable result of enlarged brain size is increased perceptual acuity and manipulative dexterity (Gibson, P&G3). Another is the ability to combine varied motor actions. Andy possesses these qualities. He has a very fine precision grip between the second and third digits of both hands. This enables him to pick up individual cantaloupe seeds or individual grains of rice. He can also combine actions of all four extremities and tail when exerting banging forces against cages or other items. He is ambidextrous, but seems to prefer the left hand for fine manipulative actions, and the right for banging actions.

Andy's visual skills and attentiveness to the fine details of his surroundings exceed my own (even with contact lens). For instance, the poisonous mushrooms that he once ate were tiny, only a few millimeters wide, and I had checked the area specifically looking for mushrooms just minutes before he found them. On one occasion he was sitting on my lap when he suddenly shrieked, jumped into the garden, and jumped back on my lap holding a decapitated snake about 3–4 in. long, which he then proceeded to devour. All of this occurred in a matter of seconds. I was unaware of the snake's presence until Andy jumped into may lap holding it in his hands. Andy also recognizes human friends at a distance of several blocks, and his hearing is very acute. He always alerts me to the presence of strange noises outside or to pots boiling over on the stove.

Andy is also extremely quick, agile, and strong for his size (11 lb). He is sufficiently strong to have broken the welds of his cages several times. Once a box containing 60 lb of weights was placed on some flooring materials within Andy's reach. He moved the box, weights and all. In fact, Andy's strength, agility, and temperament are such that adult men and nearly grown male chimpanzees have expressed fear of him.

### *Tool use and other object oriented behaviors*

Andy exhibits both curiosity and possessiveness with regard to objects. He accepts all objects offered to him and regularly rattles cages, bangs his leash against the wall, or otherwise exhibits excitement when a novel object is held near him. He grabs for everything within reach. Once he obtains an object, he usually bangs it to pieces or tears it apart. He once opened 100 tea bags and emptied their contents on the kitchen floor. On another occasion he tore apart a large phone book, ripping out each individual page. His favorite activity, however, is shredding newspapers.

Andy is also an accomplished extractor (Parker & Gibson, 1977) who behaves as if extraction is, in itself, a rewarding activity. He readily extracts nut meat, seeds, melons, coconuts, eggs, and shellfish from their coverings. He opens vitamin pills and swallows the contents. Any food or other item encased in a box or wrapping is quickly removed. He exhibits the primate play face when opening boxes and junk mail, unzipping zippers, and placing his hands into paper bags, up sleeves, into pockets, and into crevices between furniture cushions.

Andy frequently manipulates objects with respect to other objects, with respect to people or animals, and with respect to gravity. Andy possesses six object–object manipulation schemes that he manifests in tool-using endeavors. These include banging, raking, rubbing, throwing, placing objects inside each other, and placing objects on top of each other. He can use each scheme in varied contexts:

1. *Banging*: Banging is used to inflict physical damage and to make noise.
   a. *Hammering inanimate objects*: Andy opens nuts, coconuts, and tough fruits by hammering them with or against hard objects.
   b. *Hammering animate or "semianimate" things*: Andy also hammers hard objects against the backs of puppets, stuffed animals, and humans. Hammering against stuffed animals or people is accompanied by primate threat faces and occurs in situations in which he is obviously angry or frightened.
   c. *Noise making*: Andy makes loud noises by banging padlocks or chains against walls. This most often serves to call me from a separate room and to indicate that he is hungry, thirsty, desirous of attention, in need of towels, or disturbed by strange animals in the backyard. The banging ceases when his desires are satisfied.
2. *Raking*: Andy's primary raking tools are the towels with which he is supplied daily. Raking occurs in two variants: raking in out-of-reach objects and sweeping objects away.
   a. *Raking in*: During the years he was caged, Andy frequently raked up the polyethylene lining of his cage and its contained litter. On one occasion he pulled the window drapes into his cage by the use of towels. He has been observed attempting to rake up adhesive tape, scissors, newspapers, books, shoes, tape recorders, and cameras that lie immediately outside his reach. Recently, a kitten, Dixie, has come to live in our home. Andy attempts to rake the kitten in by tossing the towel to the kitten. The kitten grabs the towel with its claws. Andy then pulls on the towel with the attached kitten. Andy and the kitten "play" in this manner for 10–15 min at a time.
   b. *Sweeping*: When caged, Andy would sweep debris or objects that disturbed him from the cage or cage floor with towels.
3. *Throwing*: Andy throws anything and everything within reach: towels, newspaper, feces, balls, sticks, stones, ice cubes, and food. Throwing primarily occurs in social contexts.

a. *Agonism*: Andy regularly throws objects at people, chimpanzees, or other animals when angry or frightened. Events that provoke agonistic throwing toward people including taking objects away from him, making too much noise, two people touching each other, one person removing an object from the hands of another, and telling Andy he is a "bad" monkey. He also throws objects at noisy fighting cats, at dogs and ducks, and at flies, roaches, and human or animal intruders in the backyard. On occasion he has thrown objects at the TV. This has occurred during Muppet shows (which elicit bared-teeth scream and threat faces) and during fist fights, arguments, fencing episodes, and kissing scenes.
b. *Attention attracting*: Andy regularly throws objects at humans, cats, or dogs whose backs are turned to him in order to attract attention. This attention attracting behavior differs from agonistic throwing in that it is usually accompanied by the primate play face.
c. *Play*: Andy throws balls in the air and catches them and plays catch with humans using balls, cloth, monkey chow, or other food.

4. *Placing objects inside each other*: This scheme is usually used in sponging or soaking contexts.
   a. *Sponging*: Andy places paper towels in water, runny eggs, or other semiliquid substances in order to sponge them up and transport them to his mouth.
   b. *Soaking*: Andy soaks monkey chow, seeds, and nuts in two ways. He places monkey chow in cups of water, and he overturns cups of water onto his food tray and then soaks the food items.
   c. *Use of a cup*: Andy is adept at using cups to carry water and other liquids. He can also drink from a cup without spilling water.
5. *Placing objects on top of each other*: Andy places objects on top of each other primarily in contexts that function to protect Andy or other objects from physical or behavioral forces.
   a. *Object support*: Andy uses objects as trays. When caged, Andy would place newspapers or towels on the bottom of his cage and then place food or other small objects on top of them; this prevented them from falling through the cage floor. He also uses paper towels to transport seeds or other small items from one place to another.
   b. *Concealment of inanimate objects*: Andy may use towels to hide his possessions during short-term storage. He sometimes covers food with a towel and saves it for a midnight snack or early morning feeding. Favored objects such as coins may also be concealed in this manner.
   c. *Self-protection from physical forces*:
      (1) "*Pot holders*": Andy places newspapers, paper towels, or terry cloth towels on top of hot hard-boiled eggs before picking them up or cracking them. He may also pick up other hot foods by use of "pot holders."
      (2) *Bodily comfort*: Andy places terry cloth towels or bathroom rugs on the cage bottom or on the floor. He then uses them as "mats" for sleeping or resting. When sleeping or cold, he covers his body with a towel. In the mornings, his eyes are usually covered with a towel.
   d. *Self-protection from behavioral forces*: Andy uses towels to protect himself when frightened.
      (1) *Covering a frightening object*: Andy places towels over the faces of stuffed animals and puppets and over "slinky toys." These items elicit bared-teeth threat and scream faces. Covering, in this con-

text, appears to function as a protective shield. After covering these items, he may then hit and pound them.

(2) *Self-covering*: Andy routinely wraps himself in a towel when sleeping. On several occasions when frightened by loud workmen or by fire scenes or war movies on TV he has wrapped his whole body in a towel, including his eyes.

(3) *Use of a "shield" and "lasso"*: On one occasion when "attacked" by a plastic snapping dragon (manipulated by me) he raised a towel vertically, forming a shield between him and the dragon for approximately 30 sec. Then he swiftly changed tactics and threw the towel over the dragon's head and captured him.

6. *Rubbing*: Rubbing primarily occurs in grooming contexts.
   a. Andy rubs onions, tobacco, smoke, garlic, and shampoo into his skin.
   b. He also rubs towels against the floor and walls of his living area.
7. *Combinations*: Andy has several behavior patterns that require the use of two object–object or object–body manipulations simultaneously or in succession to meet the same end. These behaviors occur in several contexts: self-grooming, agonism, and fluid ingestion.
   a. He wraps himself in a towel, rubs onions and garlic into his skin, and then rubs his skin against the towel.
   b. He sometimes covers the heads of puppets or stuffed animals and then pounds them with a hard object.
   c. He facilitates the melting of ice cubes by banging them together, then sponges up the liquid with a towel.

## *Deception*

Andy occasionally engages in deceptive behaviors. He knows, for instance, that I always come in response to his alarm bark. He frequently uses the alarm bark when there is nothing alarming in the vicinity – he simply wants my attention.

Andy becomes very angry when I am late getting home, and frequently he will bite in those circumstances. When extremely angry, he will inhibit his normal agonistic facial expressions and vocalizations and begin to groom my hand, until he has it within reach; then he will grab and pull with all his strength and bite. Similarly, on one occasion Andy began to groom Leah, a chimpanzee, but then suddenly bit her so hard that not only Leah but also her brothers have been afraid of Andy ever since.

Andy will also turn his head and pretend to be looking in the opposite direction when he wishes to steal my glasses or items that I or others have within our pockets. Then he slowly inches his hands in the direction of the desired object.

## *Imitation*

Andy's imitation skills are quite minimal, seemingly limited to social facilitation. Andy will look at any human, animal, or inanimate object that I look at or point to. When he sees me scrubbing the floor, he sometimes rubs the floor with a towel or paper towel. He will eat the foods that he sees me eating, and he wants any object that he sees in my hands. He has, however, never imitated a novel scheme.

*Representation and symbolism*

Andy has never exhibited any behaviors suggesting that he has representational or symbolic thought.

**Discussion**

Piaget described the development of human intelligence as proceeding through four major periods: sensorimotor from birth to 24 months, preoperations from 24 months to 7 years, concrete operations from 7 years to 11 years, and formal operations from 11 years on. Each of these periods is, in turn, divided into stages.

Andy's tool-using endeavors appear to correspond to Stage 5 or 6 of the sensorimotor period or to the level of manipulative intelligence exhibited by human children in the second year of life. In Piagetian terms, Stage 5 is "The Tertiary Circular Reaction or the Discovery of New Means by Active Experimentation" (Piaget, 1952). Characteristic behaviors of Stage 5 include repeated trial-and-error manipulation of object–object relationships, the use of object–object manipulation schemes in varied contexts to meet varied ends, and the elaboration of a variety of means and a variety of ends. Andy clearly exhibits each of these behaviors to a strong degree.

He has also mastered several of the classic Stage 5 Piagetian tasks, including the ability to draw sticks and other objects through the bars of his cage and the behavior of the support and of the rake (Piaget, 1952). Thus, if an object rests on a support, such as cloth or a box, Andy can obtain it by drawing in the cloth. Although Andy generally prefers to pound sticks until they break, rather than use them as rakes, he is very proficient at using towels as rakes, a somewhat more difficult task. Andy can also drink water from a cup, a rather complicated task requiring the understanding of various physical forces (Bruner, 1968). Drinking from a cup usually develops during Stage 5. Andy's behavior of playing by throwing objects to humans also resembles the Stage 5 human behavior of giving objects to mothers (Bates, 1976).

Stage 6 of the sensorimotor period is "The Invention of New Means through Mental Combinations" (Piaget, 1952). Stage 5 and Stage 6 tool-using behaviors are distinguished primarily on the basis of the origin of the behavior. During Stage 5, infants discover tool-using schemes by trial and error. During Stage 6, they discover them by insight (Piaget, 1952). Piaget considered insight to be a form of internalized trial and error. In other words, an insightful person or animal mentally recombines schemes in the mind in order to derive solutions. Diagnosing insight is extremely difficult, unless an animal's entire history is known. Chevalier-Skolnikoff, for instance, claims that cebus monkeys possess insight (Chevalier-Skolnikoff, 1989). Her conclusions are based on one behavior observed at the San Diego Wild Animal Park. A monkey, when given some nuts, immediately picked up a stick, ran to a rock, placed the nut on the rock, and banged it with the stick. Chevalier-Skolnikoff considered the behavior insightful because she had never given the monkeys nuts before. Thus, she assumed that the animal could not have learned that behavior previously. That conclusion was unwarranted under the circumstances. That was Chevalier-Skolnikoff's first encounter with these monkeys. The animals were born in the wild and had lived in a wild animal park for variable periods of time. Surely they had previously had access to sticks, stones, and rocks.

Similarly, Andy was $5\frac{1}{2}$ years old when I acquired him. In addition, his first manifestations of certain tool-using tasks, such as using towels as rakes, occurred when I was absent. It is thus very difficult to know whether these skills developed by means of trial-and-error behaviors or by insight. Some of these behaviors, such as using towels as potholders or obtaining plastic dragons by lasso techniques, do appear to me to be insightful. Andy, however, has never exhibited tool-using behaviors that I am certain are insightful.

Nor has he ever exhibited any tool-using behavior that classifies as preoperational. Chevalier-Skolnikoff suggested that cebus monkeys at the San Diego Wild Animal Park might possess preoperational intelligence because they behave in ways she considers similar to human behavior during the preoperational stage. For instance, she considers rolling cloth into a ball to be equivalent to drawing a circle (Chevalier-Skolnikoff, 1989). These conclusions demonstrate major misunderstandings of the distinctions between preoperational and sensorimotor intelligence (Gibson, 1989).

Sensorimotor intelligence involves the construction of relationships between objects. In contrast, preoperational intelligence involves the ability to construct relationships between relationships (Case, 1985). One reflection of such an ability would be the construction of relationships between two different tool-using schemes or between two different spatial or physical relationships. Chevalier-Skolnikoff reported no such behaviors. Nor has Andy ever clearly demonstrated the ability to construct relationships between two or more separate tool-using schemes. Some of his behaviors, such as rubbing his back against a towel while self-anointing with onions or tobacco, may come close to the construction of relationships between relationships, but only in an extremely rudimentary sense. Certainly, Andy has never exhibited any of the classic preoperational behaviors, such as drawing circles or lines. Nor does he construct objects from Tinkertoys or blocks.

Consequently, I feel confident in stating that Andy possesses elaborated Stage 5 tool-using skills and possibly Stage 6, but I have seen nothing that qualifies as preoperational intelligence.

Andy's level of intelligence in other respects is more questionable. He does possess elements of intentional deception. Few animals appear to have such skills, and cogent arguments have been presented suggesting that deception is an intelligent behavior that requires hierarchically embedded concepts (Byrne & Whiten, 1988; Dennett, 1988; Mitchell & Thompson, 1986). In Dennett's terms, some of Andy's behaviors appear to correspond to second-order embeddedness of "*x* wants *y* to believe that. . . ." Baboons also seem to possess second-order deceptive techniques (Byrne & Whiten, 1988). In Mitchell's classificatory system, some of Andy's behaviors correspond to Level 3 (repeating an action that had desired results in the past), and others to Level 4 (lying) (Mitchell, 1986). At the present time, however, it is difficult to place Andy's deceptive skills within the Piagetian framework. According to Chevalier-Skolnikoff, human children begin to deceive others at about 18 months of age (Chevalier-Skolnikoff, 1986). This suggests that Andy's deceptive skills may correspond to Stage 5 or 6 of the sensorimotor period.

Andy's imitation skills, however, are poor. During Stage 5, human infants begin to imitate novel actions on objects (Piaget, 1962). In 11 years, Andy has never imitated a novel action. Specifically, he has never imitated any tool-using technique or new object manipulation technique that I have demonstrated. His skills in this regard are so poor that he has never even imitated my actions of unfastening his leash. Rather, his imitation appears to fall within Stage 3 of the sensorimotor period: the imitation of actions already present in the infant's or animal's behavioral repertoire. Andy's lack of imitational skills accords with the findings presented by Visalberghi and Fragaszy (P&G9). In contrast, Chevalier-Skolnikoff claims that the monkeys at the San Diego Wild Animal Park do imitate novel acts (Chevalier-Skolnikoff, 1989). Unfortunately, her conclusions in this regard reflect the same lack of scientific rigor as her conclusions pertaining to insight. She lacked sufficient familiarity with the animals to know if the "imitated" schemes were novel or not. Most of the actions she reported, such as nut cracking, have been reported in many cebus monkeys and may well have developed independently in these animals (Parker, 1989).

The finding of poor imitative skills in Andy and other cebus monkeys is of interest in terms of the genesis of animal tool-using skills. Chevalier-Skolnikoff argues that tool use in cebus and great apes is learned through imitation (Chevalier-Skolnikoff, 1977, 1989). In contrast, Beck (1975) claims that most primate tool use derives from trial-and-error behavior. Certainly, Andy's tool-using behaviors are more likely to reflect trial-and-error discovery than imitation. Thus, his behavior lends credence to Beck's hypothesis. This is not to demean Andy's level of intelligence. Although trial-and-error behavior is sometimes considered unintelligent, Piaget has presented cogent arguments that trial-and-error object–object or object–force imitation is one manifestation of intelligence (Piaget, 1952).

That cebus or other animals might be more intelligent in some domains than in others is neither surprising nor unexpected. After all, cebus also fail to exhibit Stage 5 behaviors in linguistic domains. That is, they do not use verbal or gestural symbols.

The relative advancement of cebus intelligence in tool using, as compared with the imitational domain, raises other issues. What mechanisms might account for such advancement? One possibility suggested elsewhere in this volume is that genetically canalized infantile behaviors help to channel intellectual development in certain directions (Gibson, P&G3; Parker, P&G1). Cebus infants engage in considerable manual manipulation of objects, more so than other monkeys (Antinucci, P&G5), and they also experiment with raking actions (Parker & Poti', P&G8). No similar channeling behaviors exist in imitative or vocal domains. Possibly the differential development of tool using as compared with other domains reflects this differential presence of channeling behaviors. The presence of such channeling behaviors in cebus may also account for the fact that tool use has been reported much more frequently in this taxon than in other monkey species that evidence tool-using abilities in laboratory settings, such as baboons and macaques (Beck, 1980).

To return to Macphail's conclusions (1987) that all species are equally intelligent, it is clear that Andy exhibits the ability to construct object–object relationships of varied kinds and to use them for varied ends. Only a small number of animal species exhibit tool use in laboratory or wild settings (see Beck, 1980, for a review of animal tool use). Even many monkeys commonly raised in human homes, such as squirrel monkeys and spider monkeys, rarely manifest tool use. Further, many animals who do use tools do so in only a single context to meet a single end (Parker & Gibson, 1977). Similarly, very few animals exhibit intentional deception or imitation. Thus, it is clear that if the Piagetian framework is taken into account, species are not equally intelligent. Some, notably cebus monkeys and great apes, possess more mental constructional capacity than do the majority of animal species, particularly with respect to tool-using skills (Chevalier-Skolnikoff, 1989; Parker & Gibson, 1977, 1979).

**Summary**

The spontaneous behaviors of a captive adult male cebus monkey were observed over a period of $10\frac{1}{2}$ years. This monkey proved a very proficient tool user at the level of Stages 5 and 6 of the sensorimotor period. Purposeful tool-using manipulations exhibited by this monkey corresponded roughly to the human use of rakes, hammers, brooms, sponges, balls, aimed missiles, pot holders, mattresses, blankets, trays, shields, and containers. This monkey also possessed rudiments of deceptive behavior similar to those reported for Old World monkeys, such as baboons. These deceptive skills may also represent Stage 5 or 6 of the sensorimotor series. In contrast, his imitation skills were minimal.

These data suggest that decalage has occurred in the evolution of intelligence. A single species or animal may reach higher levels on some Piagetian scales than on others. One possible reason for this decalage is the existence of genetically canalized, species-typical infantile behaviors that help to channel intellectual development in particular directions.

**Acknowledgments**

Dr. Crystal Baker found this monkey in need of a home and kept him for several months before giving him to me. John Kopecki, Jim Cook, and Keath Gibson have helped me care for Andy over the years.

My interest in cebus monkeys, as well as in general problems of intellectual development and evolution, arose over a period of years in conjunction with a long-term collaboration with Sue Taylor Parker.

Ms. Mary Ellen Thames helped in the typing of the manuscript.

**Note**

1. Many cebus monkeys self-anoint with tobacco, orange rind, onion, or garlic. The adaptive significance of this behavior remains unknown. My experiences with Andy, however, suggest that tobacco, onion, and garlic have a curative function with respect to skin infections. Andy is particularly prone to necrotic infections. For a period of several years he developed infections subsequent to every minor scratch. Usually within hours of self-anointing with tobacco or onion these infections showed signs of healing. Since I began giving him tobacco on a prophylactic basis, he has had no skin infections. In Andy's case, tobacco has a stronger curative function than onion or garlic, because he is more prone to use it for self-anointing.

**References**

Antinucci, F., & Visalberghi, E. (1986). Tool use in *Cebus apella*: A case study. *International Journal of Primatology*, 7(4), 349–362.

Bates, E. (1976). *Language and context*. New York: Academic Press.

Beck, B. (1975). Primate tool behavior. In R. Tuttle (Ed.), *Socioecology and psychology of primates* (pp. 413–447). The Hague: Mouton.

Beck, B. (1980). *Animal tool behavior*. New York: Garland Press.

Bierens de Haan, J. A. (1931). Werkzeuggebrauch und Werkzeugherstellung bei einum Niediren Affen (*Cebus hypoleucus humb.*) *Zeitschrift für Vergleichende Physiologie*, *13*, 639–695.

Braggio, J. T., Hall, A. D., Buchanan, J. P., & Nadler, R. D. (1979). Cognitive capacities of juvenile chimpanzees on a Piagetian-type multiple classification task. *Psychological Reports*, *44*, 1087.

Bruner, J. (1968). *Processes of cognitive growth: Infancy*. Worcester, MA: Clark University Press.

Byrne, R., & Whiten, A. (1988). *Machiavellian intelligence*. Oxford: Clarendon Press.

Carpenter, C. R., & Locke, N. M. (1937). Notes on symbolic behavior in a cebus monkey *Cebus apella*. *Journal of Genetic Psychology*, *51*, 267–278.

Case, R. (1985). *Intellectual development: Birth to adulthood*. New York: Academic Press.

Chevalier-Skolnikoff, S. (1976). The ontogeny of primate intelligence and its implication for communicative potential: A preliminary report. *Annals of the New York Academy of Sciences*, *280*, 173–216.

Chevalier-Skolnikoff, S. (1977). A Piagetian model of describing and comparing socialization in monkey, ape, and human infants. In S. Chevalier-Skolnikoff & F. E. Poirier (Eds.), *Primate biosocial development* (pp. 159–187). New York: Garland Press.

Chevalier-Skolnikoff, S. (1983). Sensorimotor development in orang-utans and other primates. *Journal of Human Evolution*, *12*, 545–561.

Chevalier-Skolnikoff, S. (1986). An exploration of the ontogeny of deception in human beings and other non-human primates. In R. W. Mitchell & N. S. Thompson (Eds.), *Deception: Perspectives on human and nonhuman deceit* (pp. 205–220). New York: SUNY Press.

Chevalier-Skolnikoff, S. (1989). Spontaneous tool use and sensorimotor intelligence in *Cebus* compared with other monkeys and apes. *Behavioral and Brain Sciences*, *12*, 561–627.

Cooper, L., & Harlow, H. (1961). Notes on a cebus monkey's use of a stick as a weapon. *Psychological Reports*, *8*, 418.
Dennett, C. (1988). The intentional stance in theory and practice. In R. Byrne & A. Whiten (Eds.), *Machiavellian intelligence* (pp. 180–202). Oxford: Clarendon Press.
Dore, F. Y., & Dumas, C. (1987). Psychology of animal cognition: Piagetian studies. *Psychological Bulletin*, *2*, 219–233.
Gibson, K. R. (1986). Cognition, brain size and the extraction of embedded food resources. In J. G. Else & P. C. Lee (Eds.), *Primate ontogeny, cognition and social behavior* (pp. 93–105). Cambridge University Press.
Gibson, K. R. (1988). Brain size and the evolution of language. In M. Landsberg (Ed.), *The genesis of language: A different judgement of evidence* (pp. 149–172). Berlin: Mouton de Gruyer.
Gibson, K. R. (1989). Tool use in Cebus monkeys: Moving from orthodox to neo-Piagetian approaches. *Behavioral and Brain Sciences*, *12*, 598–599.
Gibson, K. R. (1990). The ontogeny and evolution of the brain, cognition and language. In A. Lock & C. Peters (Eds.), *Handbook of symbolic intelligence*. Oxford University Press.
Jerison, H. (1973). *Evolution of the brain and intelligence*. New York: Academic Press.
Klüver, H. (1933). *Behaviour mechanisms in monkeys*. University of Chicago Press.
Klüver, H. (1937). Re-examination of implement using behavior in a cebus monkey after an interval of three years. *Acta Psychologica*, *2*, 347–397.
Macphail, E. (1987). The comparative psychology of intelligence. *Behavioral and Brain Sciences*, *10*, 645–695.
Mathieu, M., Bouchard, M., Granger, L., & Herscovitch, J. (1976). Piagetian object permanence in *Cebus capuchinus*, *Lagothrica flavicauda*, and *Pan troglodytes*. *Animal Behavior*, *24*, 585–588.
Mathieu, M., Daudelin, N., Dagenais, Y., & Decarie, T. (1980). Piagetian causality in two house-reared chimpanzees (*Pan troglodytes*). *Canadian Journal of Psychology*, *34*, 179–186.
Mitchell, R. W. (1986). A framework for discussing deception. In R. W. Mitchell & N. S. Thompson (Eds.), *Deception: Perspectives on human and non-human deceit* (pp. 3–40). New York: SUNY Press.
Mitchell, R. W., & Thompson N. S. (Eds.) (1986). *Deception: Perspectives on human and non-human deceit*. New York: SUNY Press.
Oppenheimer, J. R. (1973). Social and communicatory behavior in the cebus monkey. In C. R. Carpenter (Ed.), *Behavioral regulators of behavior in primates* (pp. 251–271). Lewisburg, PA: Bucknell University Press.
Parker, S. T. (1977). Piaget's sensorimotor period series in an infant macaque: A model for comparing unstereotyped behavior and intelligence in human and nonhuman primates. In S. Chevalier-Skolnikoff & F. E. Poirier (Eds.), *Primate biosocial development* (pp. 43–112). New York: Garland Press.
Parker, S. T. (1989). Imitation and derivative actions. Commentary on S. Chevalier-Skolnikoff, Spontaneous tool use and sensorimotor intelligence in *Cebus* compared with other monkeys and apes. *Behavioral and Brain Sciences*, *12*, 604.
Parker, S. T., & Gibson, K. R. (1977). Object manipulation, tool use, and sensorimotor intelligence as feeding adaptations in cebus monkeys and great apes. *Journal of Human Evolution*, *6*, 623–641.
Parker, S. T., & Gibson, K. R. (1979). A developmental model for the evolution of language and intelligence in early hominids. *Behavioral and Brain Sciences*, *3*, 367–408.
Piaget, J. (1952). *The origins of intelligence in children*. New York: Norton.
Piaget, J. (1954). *The construction of reality in the child*. New York: Ballantine Books.
Piaget, J. (1962). *Play, dreams and imitation in childhood*. New York: Norton.
Redshaw, M. (1978). Cognitive development in human and gorilla infants. *Journal of Human Evolution*, *7*, 133–141.
Romanes, G. J. (1882). *Animal intelligence*. London: Kegan Paul, Trench & Co.

Sacher, G. A., & Staffeldt, E. F. (1974). Relation of gestation time to brain weight for placental mammals: Implications for the theory of vertebrate growth. *American Naturalist*, *108*, 593–615.
Sarnat, H. B., & Netsky, M. G. (1981). *Evolution of the nervous system*. Oxford University Press.
Ververs, G., & Weiner, J. (1963). Use of a tool by a captive capuchin monkey. *Primates: Symposium of the Zoological Society of London*, *10*, 115–118.
Visalberghi, E. (1988). Responsiveness to objects in two social groups of tufted capuchin monkeys (*Cebus apella*). *American Journal of Primatology*, *15*, 349–360.
Visalberghi, E., & Antinucci, F. (1986). Tool use in the exploitation of food resources in *Cebus apella*. In J. C. Else & P. C. Lee (Eds.), *Primate ecology and conservation* (Vol. 2, pp. 57–62). Cambridge University Press.
Warden, C. J., Koch, A. M., & Fjeld, H. A. (1940). Instrumentation in *Cebus* and *Rhesus* monkeys. *Journal of General Psychology*, *56*, 297–310.
Weigel, R. M. (1979). The facial expression of the brown capuchin monkey (*Cebus apella*). *Behaviour*, *68*(3–4), 250–276.
Wise, K. L., Wise, L. A., & Zimmerman, R. R. (1974). Piagetian object permanence in the infant rhesus monkey. *Developmental Psychology*, *10*, 429–437.

# 8 The role of innate motor patterns in ontogenetic and experiential development of intelligent use of sticks in cebus monkeys

*Sue Taylor Parker and Patricia Poti'*

> In particular cases there is as much difficulty in classifying certain actions as instinctive or rational, as there is in cases where the question lies between instinct and reflex action. And the explanation of this is, as already observed, that instinct passes into reason by imperceptible degrees; so that actions in the main instinctive are very commonly tempered with what Pierre Huber calls "a little dose of judgment or reason," and vice versa. But here, again, the difficulty which attaches to the classification of particular actions has no reference to the validity of the distinctions between the two classes of actions; these are definite and precise. – Romanes (1886, p. 16)

Given that recent investigators have labeled a variety of behaviors in various birds and mammals as tool use (e.g., Alcock, 1972; Beck, 1980; Hall, 1963; Kortlandt & Kooij, 1962; van Lawick-Goodall, 1970), it is appropriate to establish our definition of "tool use." We require that to qualify as a tool user, the animal must move a detached object for the purpose of changing the state and/or position of another object or a behavior (Beck, 1980; Parker & Gibson, 1977; van Lawick-Goodall, 1970). Furthermore, we require that to qualify as an intelligent tool user, the animal must understand that the detached object acts as a detached intermediary capable of displacing the goal object, and the animal must understand how to manipulate the object in relation to the physical constraints of the situation, as described later.

Detailed studies of the tool-using abilities of cebus monkeys have been available at least since the time of Klüver's study (1933), in which, among other things, he replicated many of the Köhler (1927) chimpanzee experiments on brown-capped cebus and Java monkeys (*Cebus apella* and *Macaca fascicularis*). Although Klüver found the cebus clearly superior to the macaque and comparable to the chimpanzee, he was able to elicit systematic use of the stick as a tool in only one cebus monkey, and he expressed the need to study the behavior in a number of animals, especially in light of marked individual variations in performance. Warden, Koch, and Fjeld (1940) and Harlow (1951) repeated the Köhler and Klüver experiments with cebus, with parallel results. Klüver expressed uncertainty about the meaning of the performances, the definition of "intelligence," and the difficulty of distinguishing between insight and trial and error. Harlow also expressed uncertainty concerning the role of intelligence in cebus tool use: "It is possible that the differences in instrumentation abilities relate more to innate propensities toward manipulation than to intellectual factors" (Harlow, 1951, p. 221). More recently, Izawa and Mizuno (1977) have expressed similar doubts about cebus intelligence.

Given that Klüver's and Harlow's uncertainty about the relationship between tool use and intelligence has persisted among later investigators (e.g., Hall, 1963; King, 1986), it is important to address this difficult issue. Clearly, without a general theory of intelligence it is quite arbitrary to assign degrees of intellectual difficulty to given behaviors. Piaget's model, however, sheds new light on the problem by defining stages and levels of intellectual ability across a variety of domains, specifically by recognizing that repeated efforts to use the stick as a detached intermediary and insightful use of a stick as a tool are behaviors characteristic of Stages 5 and 6 of the sensorimotor intelligence series of the sensorimotor period of development: Stage 5 involving directed groping with the stick (and other objects) as means to an end, and Stage 6 involving insightful use of tools.

Use of the stick as a tool, and an understanding of means–ends relations in general, is only one aspect of sensorimotor intelligence; other aspects or series include development of an understanding of simple causal relations (Piaget, 1954, 1962). These series fall primarily under the larger domain of physical knowledge, that is, knowledge concerning the behavior of objects in relation to other objects, and in relation to forces and fields. This kind of knowledge is constructed through reflective abstraction on the results of natural events and induced events. Other abilities, such as classification and seriation and construction of numbers, fall under the domain of logical-mathematical knowledge, that is, knowledge about possible abstract relations among objects. This kind of knowledge is constructed primarily through reflective abstraction on systematic arrangements of objects relative to other objects, as, for example, construction of one-to-one correspondences (Piaget, 1970; Antinucci, P&G5).

One of the advantages of the Piagetian framework is that it gives us a detailed description of the abilities of our own species that can be used as a standard against which to measure the complexity of various abilities and the rates at which they develop in various other species.

In this chapter we report on two related studies:

1. a study of the experiential development of behavior with a stick, based on an attempt to replicate Köhler's and Klüver's experiments with the stick, on three young and two adult brown-capped cebus monkeys (*Cebus apella*);
2. a study of the ontogenetic development of this behavior in a single infant of this species using a Piagetian framework.

## Stick experiments with juvenile and subadult cebus monkeys

### *Subjects and setting*

All the subjects were housed at the Institute of Psychology Laboratory of the Consiglio Nazionale delle Ricerche in the Rome Zoo. The three young animals were born in the Rome Zoo to the same mother: a 2½-year-old female, Carlotta (born February 1984); a 4-year-old female, Brahms (born May 1982); and a 5-year-old nulliparous female, Pippi (born August 1981). All three young animals were hand reared until they were several months old. Carlotta was studied developmentally before being introduced into the cage with the other two young animals and the adult male Cammello at the age of 7 months (Visalberghi & Riviello, 1987). At the time of the study (October–December 1986), the three animals were housed together in a single experimental cage with an outdoor section (1.7 × 3.0 × 2.6 m), an indoor section (1.7 × 1.9 × 2.6 m), and an adjoining experimental area (1.7 × 3.0 × 2.6 m). Habituation and testing were done in the latter two areas. The two adult males, Cammello and Rosso, were housed in two separate cages, each with two adult females. Each cage had an outdoor–indoor area of 6.0 × 7.8 × 6.0 m. Habituation and testing were done in the inner cages.

The following materials were used: a three-tier metal cart with wheels, 42 cm wide × 80 cm across × 82 cm high; round, solid bamboo sticks 20 cm long (short sticks) or 30 cm long (long sticks); round, hollow tubes of opaque black and translucent Plexiglas 15 cm long, with an inside diameter slightly exceeding the outside diameter of the solid sticks (about 2 cm); two plastic cups, the smaller 5 cm in diameter, the larger 10 cm square; shelled peanuts and small lozenge candies as objectives.

### *Experimental design*

The experiment was designed to replicate three tasks designed by Köhler:

1. the use of a stick to rake in an inaccessible object,
2. the use of a shorter stick to rake in a longer stick to get an inaccessible object, and
3. the construction of a compound tool from two short sticks, neither of which was sufficient to reach an object, and the use of that compound tool to achieve the object.

An additional task of

4. raking in two nested cups covering an object

was added at the end of the other experiment.

The main experiment had three phases: During Phase I the animals were presented with the first task; in Phase II the animals were presented with the second task; in Phase III the animals were presented with the third task.

Prior to the initiation of the formal testing in Phase I, the animals were given several days to habituate to the tester, the materials, and the task. For a period of 10 days the young animals were given sticks to play with for 15–30 min per day for a total of approximately 3 hours. During this period, use of the stick was informally demonstrated several times, and the animals were periodically tested to see if they would perform the task. Videotapes were made during this period. The habituation period for the two adult males was cut short because they showed little inclination to manipulate sticks or to use a stick as a tool. Formal testing was initiated when an animal had successfully performed the task several times.

Preceding formal testing for Phase II, the use of a stick to rake in a second longer stick to use to get the objective was demonstrated several times, and the young animals were pretested for the second task for 4 days for a total of 1 hour. In anticipation of testing for Phase III, the young animals were given short sticks and hollow tubes of opaque plastic and of clear plastic, with peanuts inside to encourage insertion of the sticks into the tubes, for approximately 4 hours over a 16-day period. Formal testing for this phase was never initiated, because none of the animals produced a compound tool.

During Phase I the juvenile animals (Carlotta and Brahms) were tested eight times, and the young adult (Pippi) was tested seven times, in the following four conditions (Figure 8.1) of the single-stick task (yielding 32 and 28 trials, respectively):

(a) with the stick vertically arrayed to the right side of the objective relative to the animal,
(b) with the stick vertically arrayed to the left side of the objective,
(c) with the stick horizontally arrayed between the animal and the objective, and
(d) with the stick vertically arrayed and the objective above its distal end.

During Phase II the only animal that succeeded in the pretest (Carlotta) was tested five times in the following three conditions of the double-stick task (yielding 15 trials):

(a) with the shorter stick perpendicular and to the right of the longer stick,
(b) with the shorter stick perpendicular and to the left of the longer stick, and
(c) with the shorter stick parallel to the longer stick, and both sticks horizontal relative to the animal.

The objective was on the distal side of the longer stick in all conditions (Figure 8.2).

During Phase III the only animal who succeeded in Phase II, Carlotta, was pretested in two conditions of the construction task:

(a) with the short stick and the tube lying vertically end-to-end to the right of the objective, and
(b) with the stick in the same configuration to the left of the objective.

The additional task, which involved presentation of an object hidden under two nested cups arrayed to the right and to the left of a single stick, was also presented to Carlotta four times (a total of eight trials).

Data on success rates were collected on check sheets using the focal animal sampling technique (Altmann, 1974). Trials are terminated by one of the following events: achievement of the objective, knocking the tool or the objective off the table, taking the tool away, or leaving the scene without returning within 1 min.

Data on hand use, duration of performance, and techniques of object manipulation were taken from videotapes of pretest performances.

The experimental design for the stick experiments differed from that of Klüver in the following respects: The tasks were more difficult because the

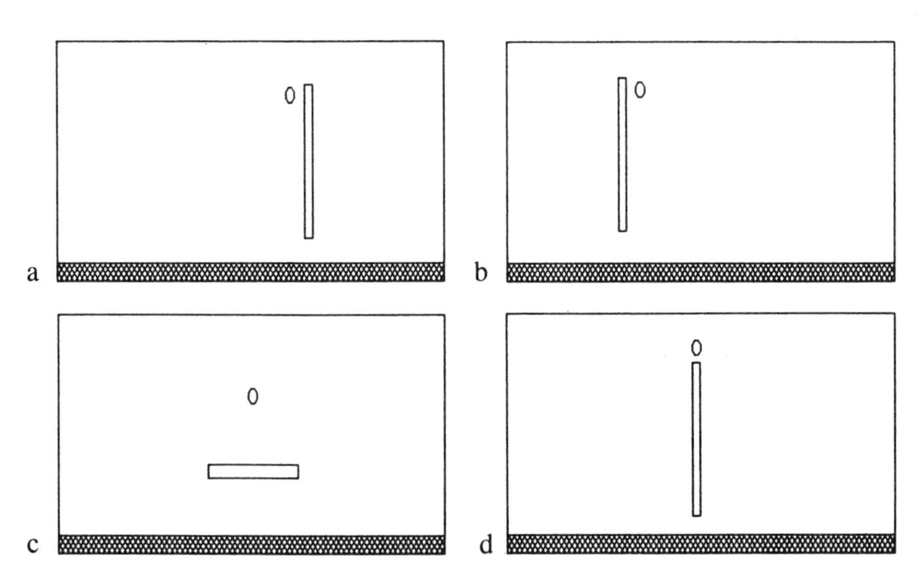


Figure 8.1. Single-stick problem: (a) condition 1, stick to the right of the objective; (b) condition 1, stick to the left of the objective; (c) condition 3, stick horizontal below objective; (d) condition 4, stick vertical below objective.

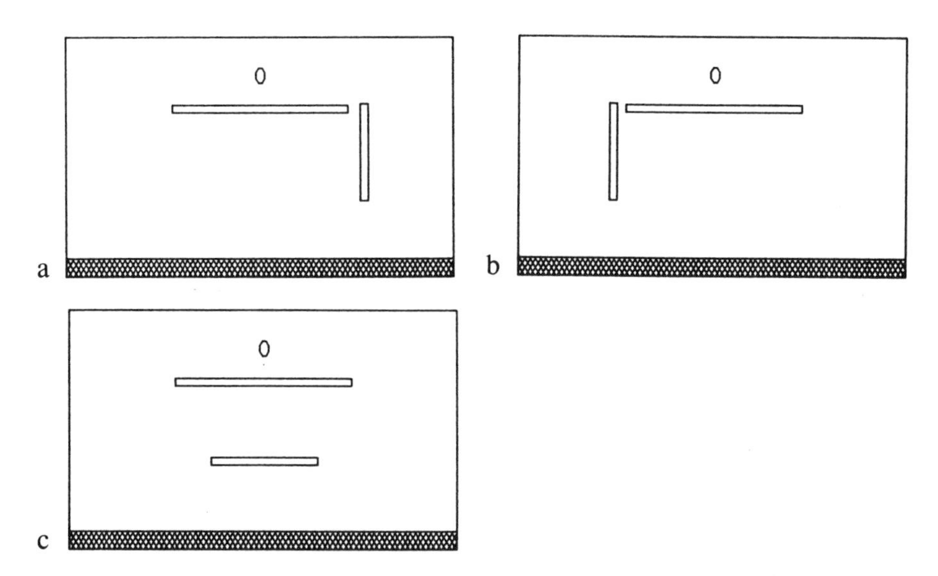


Figure 8.2. Double-stick problem: (a) condition 1, short stick perpendicular to the right of the long stick; (b) condition 2, short stick perpendicular to the left of the long stick; (c) condition 3, short stick parallel below the long stick.

Table 8.1. *Success rate in single-stick problem for Brahms*

| Position of stick | Success | Failure | No trial | Success (%) |
|---|---|---|---|---|
| To right of objective | 1 | 7 | 0 | 13 |
| To left of objective | 7 | 1 | 0 | 87 |
| Horizontal below | 2 | 5 | 1 | 25 |
| Vertical below | 1 | 5 | 2 | 13 |

*Note:* Total number of trials for each task = 8.

animals were given simple straight sticks rather than the L- and T-shaped sticks he used, the animals were confined behind a mesh rather than being tethered, and they performed on a table rather than on the pavement. Both of these latter conditions produced more mechanical difficulties for the animals. Another difference was that the animals were tested in the presence of other animals. The results differ only in the number of animals who systematically performed the task.

*Results*

Of the five experimental animals, only the juveniles and the subadult achieved the criterion behavior of using a stick as a tool in the single-stick task, and only one of the three young animals, Carlotta, achieved the criterion behavior in the double-stick task. None of the animals achieved the criterion behaviors in the construction of a compound-tool task. Neither of the two adult males performed the single-stick task systematically or regularly. These results are consistent with those of Visalberghi (1988), who found in a large-scale study of object manipulation in the same species that juveniles of both sexes showed a higher frequency of contact with novel objects than did adults.

Although all three young animals performed the single-stick task, they differed in their levels of understanding of the task. Whereas Carlotta and Pippi displayed a mature (Period 3) Stage 5 level of understanding of the stick as a detached intermediary, moving the stick consistently in the right direction and adjusting their movements successfully (see the "Discussion" section that follows), Brahms did not achieve this level. She engaged in stereotyped actions, sometimes moving the stick away from the object, suggesting magical thinking. She also displayed a "residual reaction" when after repeatedly failing to get an object she placed the stick in a location (on the lower shelf of the cart) where she had previously succeeded. Her lower level of understanding is reflected in the fact that she succeeded only when the stick was on the left side of the objective: In that condition, a stereotyped sweep of the stick in her left hand often succeeded. See Tables 8.1–8.3 for individual success rates, Table 8.4 for overall success rate, Table 8.5 for hand preferences, and Table 8.6 for time devoted to trials in the single-stick task.[1]

Table 8.2. *Success rate in single-stick problem for Carlotta*

| Position of stick | Success | Failure | No trial | Success (%) |
|---|---|---|---|---|
| To right of objective | 8 | 0 | 0 | 100 |
| To left of objective | 8 | 0 | 0 | 100 |
| Horizontal below | 3 | 5 | 0 | 62 |
| Vertical below | 4 | 4 | 0 | 50 |

*Note:* Total number of trials for each task = 8.

Table 8.3. *Success rate in single-stick problem for Pippi*

| Position of stick | Success | Failure | No trial | Success (%) |
|---|---|---|---|---|
| To right of objective | 6 | 1 | 0 | 85 |
| To left of objective | 2 | 3 | 2 | 28 |
| Horizontal below | 2 | 3 | 2 | 28 |
| Vertical below | — | — | — | — |

*Note:* Total number of trials for each task = 7.

Table 8.4. *Overall success rate in three juvenile subjects on single-stick problem*

| Position of stick | Success (%) | Failure (%) | No trial (%) |
|---|---|---|---|
| To right of objective | 65 | 35 | 0 |
| To left of objective | 74 | 17 | 9 |
| Horizontal below | 30 | 57 | 13 |
| Vertical below | 31 | 56 | 13 |

Table 8.5. *Handedness in the one-stick problem*

| Subject | Right | Left | Alternate R/L |
|---|---|---|---|
| Brahms | 2 | 12 | 2 |
| Carlotta | 12 | 2 | 2 |
| Pippi | — | — | — |

Table 8.6. *Time devoted to trials*

| Subject | Range | Average |
|---|---|---|
| Brahms | 0.03–2.48 | 0.31 |
| Carlotta | 0.01–0.84 | 0.23 |
| Pippi | — | — |

Table 8.7. *Success rate in two-stick problem for Carlotta*

| Position of short stick relative to long stick | Success | Failure | No trial | Success (%) |
|---|---|---|---|---|
| Right side perpendicular | 4 | 2 | 0 | 66 |
| Left side perpendicular | 2 | 4 | 0 | 33 |
| Horizontal below | 4 | 2 | 0 | 33 |

Only one animal, the juvenile Carlotta, performed the double-stick task. In condition (a) (with the short stick to the right of the long stick), her success rate was 66%, in condition (b) (with the short stick to the left), her success rate was 33%, as it was in condition (c) (with the short stick horizontal below the long stick) (Table 8.7). It is interesting to note that in this more difficult double-stick task she showed an effect of hand preference that she did not display in the single-stick task. This suggests that she had not completely mastered the task, as discussed later.

Although none of the animals performed the double-stick construction task, Carlotta did insert small twigs into the hollow tubes of clear plastic that were given the animals in free play (in preparation for the last experiment). In a subsequent elaboration of this task, Visalberghi and Trinka (in press) affixed hollow tubes of clear plastic to a substrate. Under this simplified condition, animals were able to use a stick to push a food reward out of the tube. Klüver (1933) and Harlow (1951) elicited similar performances from cebus using large hollow tubes.[2]

Only Carlotta was tested with the two-cup task. Her overall success rate was 62%: 100% success with the tool to the right of the objective, 25% success with the tool to the left of the objective, in a total of eight trials for two conditions. Again, this pattern suggests that she had not fully mastered the task.

### *Discussion*

Several striking patterns emerged from this study. The first pattern was greater interest and superior performance by the juveniles and subadult as compared with the adult.

The second pattern was the association between an extended period of free manipulation and play with the stick and its systematic use as a tool. Only the animals that engaged in extensive manipulation, that is, two of the three young animals, used the stick effectively and systematically as a tool. Moreover, their systematic use of the stick as a tool followed a more or less extensive period of free manipulation that included repetitive stereotyped activities of bimanual banging and rubbing, as well as scratching, poking, levering, and throwing, and even more primitive schemes of holding and biting and manual and pedal carrying. Neither of the adult males engaged in extensive free object manipulation. One of them, however, used the stick as a lever several times.

The failure of the adult males to perform even the single-stick task seemed to be a failure of motivation. They showed relatively little interest in the stick or the food objective. The failure of the second male, Cammello, was particularly striking because he was the same animal who used stones and wood as tools to crack open nuts (Antinucci & Visalberghi, 1986; Visalberghi & Antinucci, 1986). Both males showed a strong tendency toward vigilance behavior, patrolling the boundaries of the cages more or less continuously. This behavior may be part of an antipredator strategy related to the fact that male cebus spend more time feeding on the ground than do female cebus, at least in some locations (Fragaszy, 1987).

The third pattern was the lightning speed of the movements of the animals, which made it difficult to appreciate, let alone record, their movement patterns without slow-motion film analysis.

The fourth pattern was the elaborated and often extremely delicate trial-and-error manipulation of the stick relative to the objective revealed by slow-motion analysis of films taken during pretest performances. These included the following behaviors:

1. repositioning the hand along the length of the stick,
2. repositioning the extension of the stick (one occasion only),
3. changing hands in order to have a better position for moving the stick relative to the objective,
4. shifting the position of the stick to the other side of the objective consonant with the optimum direction of movement for the preferred hand,
5. sliding the stick on the table to reposition it optimally relative to the objective, and
6. repeated changing of body positions relative to the stick and the objective, often involving letting go of the stick, withdrawing the hand through the mesh, and reinserting it through another hole closer to the stick.

In all of these behaviors the young cebus showed greater flexibility in tool use than did either the macaques or the gorilla (Natale, Poti', & Spinozzi, 1984).

The following samples of behavioral sequences from the videotape give a better sense of the situation:

1. Carlotta sits perched on the mesh, watching as the stick is placed horizontally between her and a piece of candy. She gets the candy and eats it, then jumps away, but returns as the candy is replaced farther from the stick. She grasps the stick in the middle with her right hand, the knuckles facing me, and repositions her hand so that her thumb is on the left side. Watching the candy, she swings the stick in an arc on the surface of the table backward to her right until its distal end is near the candy; then she repositions her hand closer to the distal end and sweeps the stick along in an arc until the left side of the stick contacts the right side of the candy, and then continues to push the candy along in an arc toward her for about 5 cm. As she sees it approaching

her, she lets go of the stick and reaches for the candy with the same hand, extending her arm through the hole and turning her head to her left; she misses the candy, but pushes the stick away. She turns her head to look and reaches again and misses; she withdraws her arm and sticks it through an adjacent hole and gets the candy.

2. Carlotta looks at the double-stick array, with the short stick to the right of the long stick. She reaches through the mesh with her right hand, grasps the middle of the stick, and rolls it gently across the table below the longer stick; then she lets go and regrasps it, turning it backward to reposition it to the right of the right end of the longer stick, and rolls it in an arc toward the right, missing the longer stick. She lets go and regrasps the shorter stick at its proximate end and pushes the distal end into the middle of the longer stick, so that the right end moves away from her and the left end moves closer to her. She then lets go of the shorter stick, pulls her arm through the mesh, moves a few squares, reinserts her right arm, and grasps the longer stick at the proximate left end and rotates the distal end in a smooth arc on the table, bringing the nut closer to the mesh. She lets go of the stick and grasps the nut with the same hand.

These descriptions reveal several features of the behavior that Klüver also noticed: It was highly motivated, highly goal directed, and highly skilled. The errors were virtually always "good errors," in Köhler's terminology; that is, they always revealed the animal's understanding of the task and the situation. Knocking the objective away and knocking it off the table were the most common sources of failure. The animals often touched the objective with the stick in the course of their efforts. These behaviors were incompatible with the notion that the success was inadvertent or fortuitous, though the behavior clearly began in that way. Success required not only an understanding of causality but also an ability to control and adjust actions in accord with the physical constraints of the situation.

It was no accident that Carlotta was the most successful in all the single-stick conditions and also the only one of the three to perform the double-stick task. Carlotta invented an alternative scheme to achieve the objective that she preferred to the use of the stick: alternately pushing the table away and jerking it back, to displace the nut or the candy. Other investigators have reported that cebus use cloths and ropes as whips to rake in objects (Klüver, 1933; Gibson, 1989b, P&G7). It is clear that motivation correlates with success, and vice versa. Carlotta was always the most eager to perform and the most tireless in her performances.

One interesting sidelight has to do with the social conditions of testing. Through trial and error it became clear early on that Brahms and Pippi would perform only when another animal was present. Brahms was always highly excitable and easily distracted by noises. Pippi, who was very shy and subordinate, was difficult to test because Carlotta always displaced her. After Carlotta had performed the tasks many times, I noticed that Pippi not only sat next to her watching but also sometimes tried to do the test. At the end of the test period, I succeeded in getting Pippi to perform alone. Carlotta may have acted as a model rival for Pippi (Pepperberg, P&G18).

This effect cannot be attributed to imitation of novel behaviors (a behavior characteristic of Stage 5 in human infants), because it did not involve obvious attempts to copy specific action patterns (Piaget, 1962). Most investigators agree that cebus do not imitate novel actions (Klüver, 1933; Visalberghi & Fragaszy, P&G9). It may, however, indicate an ability to learn cause-and-effect relationships between two objects through observation, a process Beck (1975) calls stimulus enhancement. Izawa inferred such a process when he saw juvenile cebus observing adult cebus hunting for frogs (Izawa, 1978). If so, this represents a higher-level ability than observational learning concerning a single object (Beck, 1975).

One additional comment concerning the failure to construct a compound tool: That seemed to have been a failure of motivation and of mechanical skill. One animal, Carlotta, occasionally placed small objects in the tubes – peanuts and small splinters of wood – and she several times placed two tubes together side by side, and twice she placed two tubes end-to-end. These were infrequent behaviors, however, especially as compared with their counterparts in chimpanzees and human infants. Unlike the subjects of Klüver and Harlow, she did not use a stick to push the nut out of the tube.

These and other behaviors, such as poking sticks into slots and crevices, suggest that the "putting inside" scheme is more easily elicited when there is an obvious size difference between the container and the contained object. That this failure does not seem to reflect an inability to coordinate several schemes is suggested by Carlotta's success in the single-stick/double-cup task. In that task she typically used the stick in two separate operations, first to rake in the larger cup, and then, after lifting the larger cup, to rake in the smaller cup, before lifting it and taking the objective. That operation involved five sequential coordinations. By the same token, many of the single- and double-stick performances involved long chains of sequential coordinations.

These cebus seem to show an interesting curvilinear relationship between age and performance on the stick task: Although the juveniles and subadult performed better (Visalberghi, 1988) than the adults, as in chimpanzees (Schiller, 1952), the ability to perform the stick task develops late in infancy. Our infant subject did not use the stick as a tool effectively until she was 18 months old, and she had not yet performed the double-stick task at the end of this study, when she was 24 months old.

## Development of the use of the stick by an infant cebus monkey

The second part of the project involved a longitudinal study of the ontogenetic development of stick usage in a hand-reared female, Roberta, born

(July 7, 1986) to the mother of the other three young subjects. These animals were also the subjects of a comprehensive study of sensorimotor intellectual development (Spinozzi & Poti', 1989; Spinozzi, 1989).

*Setting*

Between the ages of 5 and 8 months Roberta was given sessions with the stick, together with other objects, both inside and outside the cage. Between the ages of 9 and 24 months, formal testing was administered. The single-stick problem was presented using the same four conditions as for the other subjects (Figure 8.1). Conditions (a), (b), and (c) were simultaneously introduced, and condition (d) was presented from the age of 10 months on. Roberta was tested alone in an experimental cage 500 × 80 cm. A wooden platform 70 × 70 cm and sticks 25 cm long were used as materials. Conditions were balanced across trials. Roberta received 21 sessions of 28 trials each, on the average. Sessions were videotaped and successively transcribed. Reward was constituted by a food item. From 20 to 24 months of age Roberta was presented with the double-stick problem. The conditions used are illustrated in Figure 8.2.

*Results*

Between 5 and 8 months, no specific behaviors are performed on the stick as an instrument. Roberta pulled in the stick or slapped the proximal end of the stick while looking at the reward. Such behaviors were preceded or followed by a direct reach. In such a way a connection between the two objects was established that consisted in a by-product of a direct reach. Schemes inadequate to put the stick in contact with the reward were performed as if they were effective. These behaviors had a magical character in that they were performed independently of the relations between the two objects, and they were repeated despite their uselessness.

At 8 months, Roberta had already reached Stage 5 in the domain of object concept and had solved the support task in the causality domain. She was in fact able to establish some spatial and objective relations between two detached objects: The hidden object was conceived as existing under the screen independently of the subject's actually perceiving it, and a causal link was recognized to exist between the movements of the support and those of the reward placed onto it. After that time, three developmental periods followed in sequential order:

*First period: 9–14 months.* This was a transitional period during which effective behaviors first emerged, but coexisted with primitive and magic behaviors. Active search for contact was revealed by new, effective schemes applied to the stick: Roberta started sliding or rotating the stick on the platform

Table 8.8. *Contact and success rates in first phase*

| | Contact (%) | | | Success (%) | | | |
|---|---|---|---|---|---|---|---|
| Position of stick | Overall | Direct | Reposition | Overall | Direct | Reposition | No trials |
| To right of objective | 70.6 | 70.6 | — | 53.9 | 53.9 | — | 128 |
| All other positions | 46.3 | 29.4 | 16.7 | 20.8 | 7.9 | 12.9 | 302 |

or launching it toward the reward. Direct reaching completely disappeared.

That Roberta did not yet conceive the stick as a detached intermediary is shown by other behaviors that had a magical character and were totally ineffective in putting the stick in contact with the reward. First of all, when sliding or rotating the stick, Roberta sometimes oriented the movement away from the reward. Second, Roberta also displayed ineffective primitive schemes like slapping the proximal end of the stick, pulling the stick, sliding it back and forth on the platform below the reward, and hitting it on the platform. During this phase, 53% of the trials resulted in contacts between the stick and the reward, and 30% of the trials also led to actual retrieval of the reward.

Wrong directions of movements and ineffective schemes were performed in all of the experimental conditions. However, because Roberta had come to use her right hand predominantly, contacts and successes were more likely to occur in the condition with the stick at the right of the objective. As shown in Table 8.8, in this condition Roberta displayed higher percentages of contacts and of successes. Also, the ratio of successes to contacts was in that condition much higher than in each of the others. Roberta did not, however, just rely on the spontaneous scheme of moving the stick to the left, as is indicated by the number of contacts in the other three conditions.

It must be noted that in the other three conditions two strategies were used to direct the stick toward the reward. Besides moving the stick directly toward the reward, Roberta sometimes repositioned the stick at the right of the objective before trying to retrieve the reward. She used the direct move strategy more often, however, and in fact obtained more contacts through a direct move, though successes were much more numerous after repositioning.

Repositioning was thus an important new scheme that emerged during this period. It took two forms: Sometimes it consisted in a quiet and single move; at other times it was achieved through hitting or launching the stick on the platform toward the reward. At first, launching and hitting led to repositioning the stick by accident, but after a few sessions Roberta used these actions on purpose: She went on hitting or launching the stick for minutes, monitoring the effect until a favorable position was obtained. Then she immediately shifted to recovering the reward.

Table 8.9. *Success rate in second phase*

| Position of stick | Success (%) | Contact (%) | No trial |
|---|---|---|---|
| To right of objective | 64.7 | 94.1 | 17 |
| All other positions | 56.1 | 82.4 | 57 |

Table 8.10. *Success rates in conditions other than with the stick at the right of the objective, according to the different strategy used*

| Strategy | Contact (%) | Success (%) | No trial |
|---|---|---|---|
| Direct move | 66.6 | 22.2 | 27 |
| Repositioning | 96.6 | 86.6 | 30 |

*Second period: 15–17 months.* Three consistent phenomena indicated definitive overcoming of magical thinking, as well as comprehension of the stick as a detached intermediary, despite persisting difficulties in actually retrieving the reward:

1. Roberta never moved the stick in the wrong direction, that is, away from the reward, during this period.
2. Primitive ineffective schemes disappeared.
3. Contact between the stick and the reward was systematically obtained in all conditions (Table 8.9).

Failures to contact the reward were due to movements that were too short or too violent. The percentage of success was lower than that for contacts in all other conditions (Table 8.9). Seemingly, Roberta had problems controlling the effects of the stick movements on the reward even after the contact had been realized. Mistakes consisting of displacing the reward far away were consequences of too violent a movement or of a wrong angle of impact.

Success was more frequent in the condition with the stick at the right of the objective (Table 8.9). When the stick was not to the right of the objective, Roberta directly moved it toward the reward as often as when it was repositioned at the right. But again success was more often achieved by means of repositioning the stick at the right than by moving the stick directly toward the reward (Table 8.10).

*Third period: 18 months on.* Roberta was tested on the single-stick problem until the age of 20 months. During that period Roberta realized the contact between the stick and the reward in 97% of the trials. She actually recovered the reward in 87% of the trials. In the condition with the stick at the right, the percentage of successes was 100%. The strategy Roberta finally used to solve the stick problem was very neat. When the stick was placed at the right, she carefully slid and rotated the stick toward the reward and monitored the movement of the reward. But in the other three conditions she shifted to a strategy of systematically repositioning the stick at the right before starting the retrieval of the reward. That is, she did not try to realize the contact directly any more. She still used both forms of repositioning, thus launching or hitting at times, and sometimes she needed many attempts to relocate the stick in an optimal position.

After solving the single-stick problem, up to 24 months of age Roberta was unable to solve the double-stick problem. She did not discriminate between the two sticks. She happened to draw the longer stick by the shorter, but then she tried to gain the reward using both sticks alternately.

### *Discussion*

Roberta developed the capacity to use a stick as a tool over a period of several months following the beginning of the test. Different developmental periods could be distinguished that represented different levels of understanding of the properties of the stick as a detached instrument. These periods corresponded to the various stages Piaget described for the development of the behavior pattern of the stick in children (Piaget, 1952, 1954). During a first stage, children do not understand the relation between the stick and an out-of-reach object; they rarely connect the handling of the stick with the movement of the objective. In a subsequent stage they begin to systematically pursue contact between the stick and the objective. At this stage, the causal relation between the stick and the objective is not yet fully objectified nor spatialized: Children do not always succeed in contacting the objective with the stick, and often the action itself assumes a magical character; that is, it seems to be effective per se. Subsequently children reach a new stage during which they submit their own action to the contact condition, but without being able to actually obtain the objective. They have difficulties in anticipating and controlling the physical effects of the movements of the stick on the objective, that is, the way in which the strength and rotation they impress to the stick are transmitted to the other object.

Children gradually improve in practical understanding of the rules of the transmission of movement until they reach a stage at which they have no difficulties in retrieving an out-of-reach object by means of a stick. Analogously to children, Roberta passed through three different periods: After a period when she only occasionally established a link between the movement of the stick and that of the reward, she started systematically establishing such a link, but without clear understanding of the spatial and physical aspects of the relation between the stick and the reward. It was this lack of understanding that prevented her from systematically putting the two objects in contact despite her progress in accommodating her action to the different conditions.

In the second phase, Roberta reached a partial understanding of the independent properties of the stick as an instrument that involved only the spatial contact. In any condition she correctly moved the stick toward the reward. In all conditions, however, she had difficulties in mastering the physical aspects of the transmission of movement from one object to the other.

In the last phase, Roberta was able to successfully adjust her action to the physical and dynamic constraints involved in the task. Having already reached the notion of the stick as a detached instrument during the second phase, Roberta learned to cope with the difficulties deriving from the amplified effects of the strength and orientation of the movement from the stick to the reward. In the stick-at-right condition she was successful in 100% of the trials. In the other conditions she devised a successful strategy consisting of systematically repositioning the stick in the favorable location before trying to retrieve the reward.

The double-stick problem involves two embedded means–ends relations. Seemingly it was this further difficulty that prevented Roberta from mastering the double-stick problem up to the age of 24 months. As discussed earlier, another subject, Carlotta, was able to solve the same problem at the age of 31 months. Apparently the problem requires a higher level of understanding about the use of the stick proper and is therefore a later acquisition. This interpretation is consistent with the fact that Schiller (1957) was unable to elicit successful performance of this in chimpanzees less than 5 or 6 years old.[3]

### Parallels between ontogenetic development and experiential development of tool use

When they were first presented with sticks laid next to food items, the two young cebus whose behaviors were described in the first part of this chapter engaged in a sequence of activities that paralleled the ontogenetic stages or phases of development of tool use in the infant Roberta, as described in the preceding section. First, they tried to reach the food object by reaching for it directly through the bars with their hands; second, they hit or threw the stick toward the food object; third, they grasped the stick and moved it toward the food, sometimes contacting the food; finally, they adjusted the position and movement of the stick to make contact with the food in such a way that they could move it toward them. (The third young cebus failed to achieve the final stage, but did go through the earlier stages.) Parallels between ontogenetic and experiential development of problem-solving skills have been reported previously in human children faced with a complex task (Karmilov-Smith & Inhelder, 1975).

## Implications of developmental stages for distinguishing levels of intelligent tool use in various species and individuals within the same species

The developmental stages discussed in the preceding sections suggest a framework for comparing the levels of intelligence in the tool-using performances of animal species. Specifically, these stages suggest that the behavior of obtaining an out-of-reach object with a stick per se does not constitute mature (Period 3) Stage 5 sensorimotor period understanding of causality, and the "use" of the stick as an intermediary between the actor and the objective does not constitute an understanding of its role in transmitting force to the object. Attribution of such an understanding requires evidence that the animal views the stick as a detached intermediary that can, through its actions on another object, change that object's position and that the animal can systematically control the actions of the stick to achieve that end.

This is important, because a variety of behaviors involving acquisition of objects with sticks, including the use of a tethered object that can be accidentally maneuvered without an understanding of its status as a detached intermediary, have been lumped under the general category of tool use. The Piagetian framework of stages of development of tool use provides a basis for making distinctions among these various performances according to the level of intelligence involved.

## Relationships between innate motor patterns and intelligent tool use in cebus: The innate hierarchical integrator model

Is intelligent behavior at the opposite end of the spectrum from, and entirely different from, instinctive behavior, or are the two in some way related (Parker, 1973)? Data on experiential development and ontogenetic development of the use of a stick as a tool in cebus and chimpanzees offer some valuable clues to the mystery. Cebus monkeys display combinations of innate motor patterns and apparently intelligent problem-solving behaviors that are particularly tantalizing in regard to this question. Are they smart or dumb, or perhaps idiot savants? As we shall see in the following section, some investigators dismiss cebus tool use as mere habit, whereas others argue that cebus tool use can involve preoperational intelligence (Chevalier-Skolnikoff, 1989). In order to resolve this question, we must analyze the components of intelligent behaviors.

To start with the numbers of manipulative schemes or action patterns at the service of a species, it is clear that the numbers correlate with brain size and intelligence (Gibson, P&G3): Torigoe (1985) showed, for example, that among 74 nonhuman primate species, cebus monkeys and great apes show the largest numbers of such schemes. In cebus, these schemes include the following: grasping, biting, carrying, pushing, pulling, rubbing, scratching, banging, poking, levering, rolling, throwing, inserting, and hitting. A similar repertoire of object manipulation schemes is available to chimpanzees: "Licking, chewing, stroking, and splitting the stick, banging, poking, hammering with it, and thrusting the end into any available openings are the responses that occur frequently and constitute the basis of complex motor patterns of utilizing the sticks as tools" (Schiller, 1952, p. 184).

Not only large numbers of manipulative schemes but also high frequencies of their performance (perhaps based on self-reinforcing properties of the actions and/or their contingent effects) are necessary to channel the animals' attention to, and allow representation of, contingent relationships and other situationally relevant parameters (Parker, 1987; Siegler, 1986).

In addition to a large number of schemes frequently performed, mobility or free combinability (sequential, simultaneous, reversible) and hierarchical organization of schemes seem to be necessary to the development of intelligent behavior (Case, 1985; Parker, 1973, 1977; Gibson, 1977, 1983, P&G3). Recursive hierarchical organization wherein sequences become subordinated subunits of larger units is particularly important.

Given the proper conditions (e.g., the presence of an out-of-reach food object), these factors conspire to generate efforts to use the stick as a rake, which through practice may become proficient. In other words, they operate as an "innate schoolmarm" (Lorenz, 1965), teaching the animal about physical forces and causality.

Several observers have remarked on the high salience of the banging scheme in cebus monkeys (Antinucci & Visalberghi, 1986; Izawa & Mizuno, 1977; Visalberghi, 1988; Gibson, 1989b). Objects of various sizes and shapes, including blocks of wood, stones, and sticks, are all assimilated to this scheme. Under proper circumstances, this scheme directs the animal to the discovery of the use of a hammer to open encased objects (Izawa & Mizuno, 1977). Longer, thinner objects, like sticks, are assimilated to poking, rolling, and hitting schemes, which under proper conditions can lead to the use of sticks as tools to rake in out-of-reach objects. Because the banging scheme has higher priority than these schemes, it may interfere with the development of tool use of this sort.

In replication of Köhler's study of the development of tool use in captive chimpanzees, Schiller (1952, 1957) tested animals from 1 to 15 years of age. He found that although older animals solved the problems more rapidly than did younger animals, all the subjects required both free play and specific training with a stick to use it efficiently as a tool. (He also found age differences in ability, with only the animals 5 years and older being able to solve the compound-stick problem.) He concluded:

Motor patterns at the disposal of the animal, whether learnt or unlearnt, enter complexes of response sequence that are, as they appear, more or less adaptive and become, by provoking repetition with incentives, solidified, smoothly running units of behavior. These adaptive complexes of generalized routines are conducive to problem solution. A compound operant is conditioned by the intrinsic consequences of the behavior proper. It is suggested that behavioral adjustment to environmental entities is a composite result of innate response patterns. (Schiller, 1952, p. 187)

Lorenz spoke of the "innate schoolmarm," and Schiller provided an instructive example of the role that instinctive action patterns play in learning to use sticks as tools (see Gould & Marler, 1984, 1987, for examples of the role instinct plays in learning in other species). More recent ethological and neurophysiological formulations stress the variability of performance in "innate motor patterns," which is, of course, a prerequisite to selection (Barlow, 1968; Berkinblit, Feldman, & Fukson, 1986; Gibson, P&G3). Before inquiring into the relationship between innate organization and intelligence, we must ask how intelligence differs from learning.

**Intelligence versus learning**

Most investigators agree that intelligence embodies an ability to learn to solve complex tasks through directed trial-and-error groping and insight, especially those tasks requiring indirect means for their solution (Köhler, 1927; Piaget, 1952; Parker & Baars, P&G2; Gibson, P&G3; Antinucci, P&G5). Intelligence, however, is taxonomically a much rarer, more restricted phenomenon than learning. Learning occurs in all animal species and encompasses a variety of levels of adaptation, including such low levels as habituation and associative learning (Thomas, 1985), and is often highly specialized, inflexible, and limited in scope, as well as species-specific (Gould & Marler, 1987; Hinde & Stevenson-Hinde, 1983). Intelligence, which occurs only in a few long-lived, large-brained species, is, in contrast, quite generalized, flexible, and broad in scope, as well as species-specific.

Intellectual development apparently differs from learning in the following features: the number of mobile schemes available for recombination, the flexibility of encoding of novel internal and external situational features, the levels and magnitude of "hierarchical integration," and the number of recursive integrative cycles it entails. Hierarchical integration, that is, integration of two or more schemes (one or more as subordinate schemes), involves the four subprocesses of schematic search, schematic evaluation, schematic retagging, and schematic consolidation (Case, 1985, p. 382). In humans, Case argues that hierarchical integration occurs in four situations: exploration, problem solving, imitation, and mutual regulation (teaching). Although these processes seem to have parallels with Schiller's processes, they are more comprehensive and more elaborated.

Although intellectual development differs in some respects from learning, we can infer the operation of an entity analogous to Lorenz's innate schoolmarm in the course of intellectual development (Gibson, 1985, 1989a, P&G3). Using Case's conception, we might call this entity the "innate hierarchical integrator." The innate hierarchical integrator, like the innate schoolmarm, presupposes, on the motor side, the existence of a pool of self-reinforcing innate motor patterns spontaneously emitted under certain circumstances (though the pool is larger, the individual units are smaller, and their mobility is greater). On the receptor side, it assumes the existence of a set of attractive stimulus configurations that elicit the pool of motor patterns, as well as an interest in outcomes contingent on those motor patterns. It also assumes an abstract central representation of common features of the outcome of classes of actions on classes of objects.

Unlike the innate schoolmarm, the innate hierarchical integrator also presupposes search, evaluation, retagging, and consolidation of schemes (Case, 1985). It may also involve a recursive cyclic repetition of these processes up to some processing limit, which varies from species to species. Like the innate schoolmarm, the innate hierarchical integrator generates normal species-specific achievements only in the appropriate stimulus situation. These processes depend on neocortical constructional processes of internal representation, differentiation, mobility, and hierarchical coordination (Gibson, 1983).

According to this model, it is the existence of higher-level hierarchical transformational processes of retagging and consolidation, not the absence of innate programming, that distinguishes intellectual development from learning. In other words, according to this model, selection has favored a new kind of information-processing paradigm in intelligent species.[4]

The patterns of experiential and ontogenetic development of tool use in captive cebus suggest how these processes may work: A series of innate actions will be tried out on the stick. Then, in the presence of the objective, a smaller subset of such actions (e.g., throwing, hitting, rolling) will be tried on the stick vis-à-vis the objective (search). Then a few of these patterns will be combined and selectively reapplied (e.g., rolling, sliding) (evaluation). Then these actions will be hierarchically organized and coordinated and oriented with respect to the relationship between the stick and the objective (placing, repositioning, moving toward the objective, adjusting force and direction of movement) (retagging). Finally, the entire ensemble will become integrated into a strategy (consolidation).

## Cebus tool use in relation to feeding in the wild

Why selection should have favored innate actions and intelligent tool use in a small Central and South American monkey is an interesting question. Parker and Gibson (1977) suggested that intelligent tool-using abilities evolved independently in great apes and cebus monkeys as an adaptation for extractive foraging on a variety of seasonally variable embedded foods.

Recent field studies have revealed that indeed cebus monkeys use a variety of complex object manipulations, especially banging open hard-shelled fruits against tree trunks (a form of object manipulation classified as "proto–tool use" by Parker and Gibson, i.e., actions involving transformation of an object through object–substrate interactions).

At the Cosha Cashu study site in Peru, the heavier *Cebus apella* crack healthy *Astrocaryum* palm nuts against tree trunks, whereas the lighter *Cebus albifrons* are limited to carefully selected palm nuts that have been rendered easier to open by the action of bruchid beetles. The *C. apella* at Cosha Cashu forage for hidden beetle and other larvae through "destructive foraging" (i.e., smashing and banging open dead branches), whereas the lightweight *C. albifrons* search through leaves and break open wasp nests (Terborgh, 1983).

Terborgh (1983) has contrasted the foraging strategies of five sympatric monkeys species, commenting that because the larger of the two cebus species eats hidden insects, it relies on a random searching strategy rather than on the evoked searching strategy of species that can see their prey. This observation suggests that the high frequency of spontaneous, innate pounding behavior in *C. apella* arose as an adaptation for random searching for hidden insect food.

At the La Macarena study site in Colombia, *C. apella* eat a wide variety of embedded or hidden plant, animal, and insect foods, including ripe Cumare palm nuts (which they open by banging the eyes against a hard node on nearby Guadua trees), large snails (whose shells they apparently crack against trees), termites, ants, beetle grubs (which they extract by banging sticks against tree trunks), and tree frogs (which they extract from tree bowls by tearing off the covering) (Izawa, 1978, 1979; Izawa & Mizuno, 1977). The nut-cracking technique of the cebus varies, but, in general, "making a fulcrum on his hind legs or tail, he balances himself, kicks the air, accelerates and then strikes only the top of the largest part of the husk against the joint of the guadua" (Izawa & Mizuno, 1977, p. 783).

These reports support the hypothesis that the repetitive innate banging scheme in cebus (the "secondary circular reaction," Antinucci, P&G5) is an adaptation for extracting embedded foods. Likewise, these reports suggest that an understanding of orientation and aim in the banging scheme is an adaptation for extracting embedded foods. Both of these schemes contribute to the intelligent use of tools to open embedded foods.

Indeed, Izawa and Mizuno have described several factors that suggest the intelligent nature of this technology: The condition of the nuts is carefully checked, and different techniques are applied, depending on the ripeness; various techniques are employed specifically in banging; only the older animals are successful in their efforts; cebus in other areas do not engage in this practice (Izawa & Mizuno, 1977; Terborgh, 1983).

Despite these features, these authors suggest that this pattern, which is absent in other sympatric primate species (i.e., woolly monkeys and spider monkeys), can be attributed to "a behavioral trait, generally called a habit, in which they tend to strike things against trees or a wall with their hands," not to "higher mentality" (Izawa & Mizuno, 1977, p. 790).

As indicated in the preceding section, it is not necessary to oppose these two explanations: The innate hierarchical integrator model is consistent with the coexistence of innate and intelligent behavior. The remaining question concerns the contexts favoring a recursive hierarchical integrator that can search among the available pool of motor patterns, evaluate their utility for the task at hand, retag them as a functional unit, and consolidate them into a new, higher-order scheme, taking into account dynamic physical constraints on object behavior. Such a program should be a more efficient device for generating instrumental behaviors on a variety of objects with differing properties than should learning a specific chain of behaviors appropriate to each object. Such a system requires self-reinforcing behaviors, a generalized interest in certain stimulus properties, an interest in contingent outcomes, an ability to represent a generalized cause-and-effect relationship among actions and their outcomes, and an ability to differentiate and hierarchically reorganize and integrate behaviors.

## Summary and conclusions

1. We elaborate the definition of tool use, specifying criteria for intelligent tool use.
2. We present a detailed analysis of both the experiential development and the ontogenetic development of the use of a stick as a tool for raking in an out-of-reach object.
3. We classify stages of ontogenetic development of tool use and relate them to Piaget's stages of sensorimotor intelligence manifest in the same task.
4. We suggest that these stages can be used comparatively as a framework for diagnosing levels of intelligence in the use of a stick as a tool.
5. Finally, by analogy with Lorenz's innate schoolmarm, we suggest an innate hierarchical integration model for relating intelligent tool use to innate organization.

## Notes

1. In a subsequent restudy and elaboration of the study of the use of sticks as tools, Natale (1989) found that Brahms used an even more stereotyped one-handed sweep toward the reward.
2. Harlow produced a silent film of his cebus monkey subject that shows the animal, which is tethered by a chain, performing the following problem-solving tasks drawn from Köhler and Klüver: (1) using a stick to rake in an out-of-reach object; (2) using a shorter stick to rake in a longer stick, and then using the longer stick to rake in an out-of-reach object; (3) using a stick to push an object through a hollow brick, pushing it away from himself; (4) moving a box beneath a suspended out-of-reach object and climbing up on it; (5) using a long stick as a pole to climb up to get a suspended out-of-reach object; (6) climbing on the box with a stick and trying to climb the stick to reach the objective; (7) using a cloth as a rake thrown over an out-of-reach object. The animal was given a hollow stick and a thin stick; he looked inside the hollow stick, but did not insert the thin stick.

3. According to tests administered by Spinozzi and Poti' (1989), Carlotta had not achieved the sixth stage in the sensorimotor intelligence series at 31 months of age when she solved the double-stick task. They therefore concluded that the double-stick task is performed in Stage 5 of this series in the cebus. Comparable data on the development of this task in human children need to be collected.
4. Although such "innate motor patterns" as reflex scratching in frogs involve considerable variability and adaptative responsiveness to environmental cues, as well as hierarchical organization, these features are apparently designed to achieve "motor constancy" or "equifinality" such that "a system's movement is on the whole determined by initial conditions, [but] its final equilibrium position does not depend on these conditions, nor do the temporal perturbations during the movement" (Berkinblit et al., 1986).

## References

Alcock, J. (1972). The evolution of the use of tools by feeding animals. *Evolution*, *26*, 464–473.

Altmann, J. (1974). Observational study of behavior: Sampling methods. *Behaviour*, *49*, 213–266.

Antinucci, F., & Visalberghi, E. (1986). Tool use in *Cebus apella*: A case study. *International Journal of Primatology*, *7*(4), 349–362.

Barlow, G. (1968). Ethological units of behavior. In D. Ingle (Ed.), *The central nervous system and fish behavior* (p. 217). University of Chicago Press.

Barlow, G. (1977). Modal action patterns. In T. Sebeok (Ed.), *How animals communicate* (pp. 98–134). Bloomington: Indiana University Press.

Beck, B. (1975). Primate tool behavior. In R. Tuttle (Ed.), *Socioecology and psychology of primates* (pp. 413–447). The Hague; Mouton.

Beck, B. (1980). *Animal tool behavior*. New York: Garland Press.

Berkinblit, M. B., Feldman, A. G., & Fukson, O. I. (1986). Adaptability of innate patterns and motor control mechanisms. *Behavioral and Brain Sciences*, *9*, 585–638.

Case, R. (1985). *Intellectual development: Birth to adulthood*. New York: Academic Press.

Chevalier-Skolnikoff, S. (1989). Spontaneous tool use and sensorimotor intelligence in *Cebus* compared with other monkeys and apes. *Behavioral and Brain Sciences*, *12*(3), 561–627.

Fragaszy, D. (1987). Time budgets and foraging behavior in wedge-capped capuchins (*Cebus olivaceous*): Age and sex differences. In D. M. Taub & F. A. King (Eds.), *Current perspectives in primate social dynamics* (pp. 159–174). New York: Van Nostrand Reinhold.

Gibson, K. R. (1977). Brain structure and intelligence in macaques and human infants from a Piagetian perspective. In S. Chevalier-Skolnikoff & F. Poirier (Eds.), *Primate biosocial development* (pp. 113–158). New York: Garland Press.

Gibson, K. R. (1983). Comparative neurobehavioral ontogeny: The constructionist perspective in the evolution of object manipulation. In E. de Grolier (Ed.), *Glossogenetics* (pp. 41–66). New York: Harwood Academic Press.

Gibson, K. R. (1985). Has the evolution of intelligence stagnated since Neanderthal man? In E. Butterworth, J. Rutkouska, & M. Scaife (Eds.), *Evolution and development* (pp. 102–144). Brighton, England: Harvester Press.

Gibson, K. R. (1989a). The ontogeny and evolution of the brain, cognition and language. In A. Lock & C. Peters (Eds.), *Handbook of symbolic intelligence*. Oxford University Press.

Gibson, K. R. (1989b). Tool use in cebus monkeys: Moving from orthodox to neo-Piagetian approaches. Commentary on Chevalier-Skolnikoff (1989).

Gould, J., & Marler, P. (1984). Ethology and the natural history of learning. In P. Marler & H. Terrace (Eds.), *The biology of learning* (pp. 47–74). New York: Springer-Verlag.

Gould, J., & Marler, P. (1987). Learning by instinct. *Scientific American*, *256*(1), 74–85.

Hall, K. R. L. (1963). Tool using performance as indicators of behavioral adaptability. In P. C.

Jay (Ed.), *Primate behavior: Studies in adaptation and variability* (pp. 131–148). New York: Holt, Rinehart & Winston.

Harlow, H. (1951). Primate learning. Ch. 7 In C. P. Stone (Ed.), *Comparative psychology* (2nd ed., pp. 183–238). Englewood Cliffs, NJ: Prentice-Hall.

Hinde, R., & Stevenson-Hinde, J. (Eds.). (1983). *Constraints on learning*. New York: Academic Press.

Izawa, K. (1978). Frog-eating behavior of wild black-capped capuchins (*Cebus apella*). *Primates, 19*(4), 633–642.

Izawa, K. (1979). Foods and feeding behavior of wild black-capped capuchins (*Cebus apella*). *Primates, 20*, 57–76.

Izawa, K., & Mizuno, A. (1977). Palm-fruit cracking behavior of wild black-capped capuchin (*Cebus apella*). *Primates, 18*, 773–792.

Karmilov-Smith, A., & Inhelder, B. (1975). If you want to get ahead, get a theory. *Cognition, 3*, 195–212.

King, B. (1986). Extractive foraging and the evolution of primate intelligence. *Human Evolution, 1*(4), 361–372.

Klüver, H. (1933). *Behavioral mechanisms in monkeys*. University of Chicago Press.

Köhler, W. (1927). *The mentality of apes*. London: Routledge & Kegan Paul.

Kortlandt, A., & Kooij, F. (1962). Protohominid behavior in primates (preliminary communication). *Symposia of the Zoological Society of London, 10*, 61–68.

Lorenz, K. (1965). *The evolution and modification of behavior*. University of Chicago Press.

Natale, F. (1989). Causality II: The stick problem. In F. Antinucci (Ed.), *Cognitive structure and development in nonhuman primates* (pp. 121–133). Hillsdale, NJ: Erlbaum.

Natale, F., & Antinucci, F. (1989). Stage VI object-concept and representation. In F. Antinucci (Ed.), *Cognitive structure and development in nonhuman primates* (pp. 97–112). Hillsdale, NJ: Erlbaum.

Parker, S. T. (1973). *Piaget's sensorimotor series in an infant macaque: The organization of nonstereotyped behavior in the evolution of intelligence*. PhD dissertation, University of California, Berkeley.

Parker, S. T. (1977). Piaget's sensorimotor series in an infant macaque: A model for comparing unstereotyped behavior and intelligence in human and nonhuman primates. In S. Chevalier-Skolnikoff & F. Poirier (Eds.), *Primate biosocial development* (pp. 43–112). New York: Garland Press.

Parker, S. T. (1987). The origins of symbolic communication: An evolutionary cost/benefit model. In *Symbolism and knowledge* (pp. 7–29). Geneva: Cahiers de la Fondation Archives Jean Piaget.

Parker, S. T., & Gibson, K. R. (1977). Object manipulation, tool use and sensorimotor intelligence as feeding adaptations in cebus monkeys and great apes. *Journal of Human Evolution, 6*, 623–641.

Parker, S. T., & Gibson, K. R. (1987). King fails to distinguish among various kinds of object manipulation and various kinds of cognition (letter to the editor). *Human Evolution, 2*(4), 379–381.

Piaget, J. (1952). *The origins of intelligence in children*. New York: Norton.

Piaget, J. (1954). *The child's construction of reality*. New York: Basic Books.

Piaget, J. (1962). *Play, dreams and imitation in childhood*. New York: Norton.

Piaget, J. (1970). *Structuralism*. New York: Harper Colophon.

Romanes, G. (1986). *Animal intelligence* (4th ed.). London: Kegan Paul, Trench & Co.

Schiller, P. (1952). Innate constituents of complex responses in primates. *Psychological Review, 59*(3), 177–191.

Schiller, P. (1957). Innate motor action as a basic for learning. In C. Schiller (Ed.), *Instinctive behavior* (pp. 264–287). New York: International Universities Press.

Siegler, R. (1986). *Children's thinking*. Englewood Cliffs, NJ: Prentice-Hall.

Spinozzi, G. (1989). Early sensorimotor development in cebus. In F. Antinucci (Ed.), *Cognitive structure and development in nonhuman primates* (pp. 55–66). Hillsdale, NJ: Erlbaum.

Spinozzi, G., & Poti', P. (1989). Causality I: The support problem. In F. Antinucci (Ed.), *Cognitive structure and development in nonhuman primates* (pp. 113–119). Hillsdale, NJ: Erlbaum.

Terborgh, J. (1983). *Five New World primates*. Princeton University Press.

Thomas, R. K. (1985). The assessment of cognitive development in human and nonhuman primates. In T. C. Anand Kumar (Ed.), *Nonhuman primate models for human growth and development* (pp. 187–215). New York: Alan R. Liss.

Torigoe, T. (1985). Comparison of object manipulation among 74 species of non-human primates. *Primates*, *26*(2), 182–194.

van Lawick-Goodall, J. (1970). Tool using in primates and other vertebrates. In D. Lehrman, R. Hinde, & E. Shaw (Eds.), *Advances in the study of behavior*, Vol. 3. New York: Academic Press.

Visalberghi, E. (1988). Responsiveness to objects in two social groups of tufted capuchin monkeys (*Cebus apella*). *American Journal of Primatology*, *15*(4), 349–360.

Visalberghi, E., & Antinucci, F. (1986). Tool use in the exploitation of food resources in *Cebus apella*. In J. C. Else & P. C. Lee (Eds.), *Primate ecology and conservation* (Vol. 2, pp. 57–62). Cambridge University Press.

Visalberghi, E., & Riviello, C. (1987). The reintroduction in a social group of "Carlotta," a hand-reared capuchin monkey. *International Zoo Yearbook*.

Visalberghi, E., & Antinucci, F. (1986). Tool use in the exploitation of food resources in *Cebus apella*. In J. C. Else & P. C. Lee (Eds.), *Primate ecology and conservation* (Vol. 2, pp. 57–62). Cambridge University Press.

Warden, C. J., Koch, A. M., & Fjeld, H. A. (1940). Instrumentation in cebus and rhesus monkeys. *Journal of General Psychology*, *56*, 297–310.

# *Part III*

# Questions regarding imitation, "language," and cultural transmission in apes and monkeys

# 9 Do monkeys ape?

*Elisabetta Visalberghi and Dorothy Munkenbeck Fragaszy*

> It is proverbial that monkeys carry the principle of imitation . . . they are animals that imitate for the mere sake of imitating. – Romanes (1884, p. 477)

*Scimmiottare* in Italian, *singer* in French, *macaquear* in Portuguese, *nachäffen* in German, *majmuna* in Bulgarian, *obez'janstvovat'* in Russian, *majmol* in Hungarian, *matpowac'* in Polish, and "to ape" in English – each of these verbs is derived from a linguistic root that in that particular language labels primates. Across different languages and cultures, these verbs consistently mean "to imitate." The convergence across different cultures of the terms denoting monkeys and imitation reflects the common view that monkeys are excellent imitators. It is quite true that nonhuman and human primates share basic features of anatomy and movement. Because of their physical similarities to humans, nonhuman primates are the best candidates to perform actions resembling our own, and this may encourage the idea that they copy us. But are primates really proficient imitators?

The functional value of imitation as a means of producing similar behaviors among members of a group has been emphasized by many (e.g., Davis, 1973; Hauser, 1988; Meltzoff & Moore, 1983). Imitation is particularly useful as a means of learning from others when the observer is not proficient, when opportunities for practice are limited, when the costs of errors are high, and when learning by individual experience would be a slow process. Imitation would seem to have great potential value to nonhumans, for example, in learning ways to find, capture, and process foods, just as it does for humans. Thus, there are reasons to suppose that imitation, as a form of learning and as a means of producing behavioral coordination among members of a group, might occur in nonhumans.

Imitation has a long and illustrious history as a concept in classical European philosophy (Scherrer, 1982). Empirical studies of imitation in animals began at the dawn of experimental psychology more than one hundred years ago (Galef, 1988). Early psychologists considered imitation as an easy way to acquire behaviors, even though failures were reported when investigators tried to demonstrate learning in this way (e.g., Thorndike, 1911; Watson, 1908). Romanes (1884/1977) wrote that savages, monkeys, and retarded people were especially good imitators. Even now, some reputable scientists "take for granted that social mammals, such as lions, acquire hunting skills only after a lengthy period of watching, imitation, and practice" (Diamond, 1987).

On the other hand, contemporary psychologists argue that imitation (broadly defined to include producing familiar acts, producing novel acts, and acquiring rule-guided behaviors) reflects considerable cognitive complexity (Bandura, 1986; Bruner, 1972; Yando, Seitz, & Zigler, 1978). Conceptual advances made recently by developmental psychologists (Meltzoff & Moore, 1983) and comparative psychologists (Zentall & Galef, 1988) provide us with an opportunity to rethink the relationships among imitation, sensorimotor abilities, and cognition in animals. In addition to holding a more differentiated view of what imitation requires, contemporary investigators recognize that a variety of processes occurring in social situations (subsumed under the general label of social learning) can influence an observer's behavior. All these processes lead to similar behaviors in observer and observed. Jointly, these two lines of thinking (that human imitation requires particular cognitive operations, the nature of which is as yet poorly understood, and that other aspects of social learning can also produce similar behavior) cast doubt on the simple assertion that similar behaviors among individuals in a group are evidence of imitation.

We consider in this chapter, through reexamination of observations and experimental findings, whether or not monkeys can imitate (Tomasello, P&G10). In this effort, we draw upon the perspectives of comparative psychologists and zoologists and of developmental psychologists studying children. We focus especially on the question of learning to use tools through the observation of a skillful model. Tool use is not common in the repertoire of monkeys in nature, although it can be exhibited in captivity. We assume that spontaneous use of tools requires either unlikely environmental circumstances or a set of cognitive achievements that stand at the outer edges of simian abilities, or both.

Provided that appropriate controls eliminate alternative explanations (and this is critical, as discussed later), acquisition of tool-using behavior from a model is strong evidence of imitative ability. It requires further thought, however, to sort out precisely what must be learned for an individual to use a tool. We discuss this problem as well. If an individual can understand the rule relating action, object, and outcome from the behavior of a model using a tool, then its own attempts to solve that task with the tool can be effectively focused on its salient aspects. Can a monkey learn a rule observationally? Is this necessary to acquire a new tool-using skill? We take the position that careful empirical study of social learning and of cognitive abilities in nonsocial settings is necessary to unravel this problem and the problem of imitation in general. Descriptions of spontaneous behavior, however suggestive, are not acceptable substitutes for the empirical work.

## Toward an operational definition of imitation

Today there is increasing awareness that the evidence for imitation in nonhumans, including primates, is mainly anecdotal. Few properly designed experiments have been conducted that would allow discrimination of the different processes leading to the repetition of a behavior (Beck, 1980; Galef, 1988; Hall, 1963). Those who have successfully demonstrated imitation (in a strong sense, as described later) have reported their findings with birds, not primates (Dawson & Foss, 1965; Galef, Manzig, & Field, 1986; Palameta & Lefebvre, 1987). We stress that because more parsimonious explanations than imitation for behavioral matching are equally plausible, clear evidence eliminating these explanations will always be required before the hypothesis of matching through imitation is supported. Later we lay out operational definitions of imitation and related processes that distinguish among these processes on behavioral grounds. We attempt to minimize the theoretical content of the definitions, in favor of descriptive clarity.

"Social learning" has been proposed as a generic term to describe instances in which learning is affected by social influences, as distinct from instances in which interaction with others is not influential (Box, 1984). Most authors recognize at least three main types of social learning: (1) social facilitation, (2) local and stimulus enhancement, and (3) imitation (e.g., Bandura, 1986; Beck, 1980; Clayton, 1978; Galef, 1988; Thorpe, 1956). For a careful historical review of the confusing and often conflicting terminology in social learning, see Galef (1988).

1. *Social facilitation.* Social facilitation is the increased probability of performing a class of behaviors in the presence of a conspecific performing the same class of behaviors. The observer in this case performs an act already in its repertoire (Clayton, 1978). Social facilitation increases homogeneity of behavior among members of a group at a given time by enhancing motivational homogeneity (e.g., to eat, to investigate, to rest), rather than by influencing the performance of a specific motor act.

2. *Local and stimulus enhancement.* In both these processes, the animal's attention is drawn to where other animals are, or to what the other animals are interested in, leading to a greater probability of approaching or contacting the object or place than if the model had not been observed. For example, investigation of an object by an individual attracts the attention of others to the same object. As in social facilitation, the observer performs acts already in its repertoire.

Galef (1988) suggests the term *social enhancement* to cover all those aspects of social influence in which the presence and activity of another animal influences in a nonspecific way the behavior of an observer. Social enhancement encompasses both facilitation and local enhancement.

3. *Imitation.* Galef (1988) notes that at least 22 different terms have been used to name imitation or processes that have been considered equivalent to

it. The plethora of terms indicates conceptual confusion about this phenomenon. We have found Mitchell's definition (1987) a useful starting point for the development of an operational definition of imitation. His definition subdivides a complex phenomenon into logical parts:

imitation occurs when:
(1) something C (the copy [the behavior]) is produced by an organism;
(2) where C is similar to something else M (the model behavior);
(3) observation of M is necessary for the production of C;
(4) C is designed to be similar to M

(Mitchell, 1987, p. 198)

We immediately run into a problem in using Mitchell's definition to distinguish between social facilitation and imitation. Mitchell requires similarity between the behavior of the model and the behavior of the observer, but that also may occur in social facilitation. The fourth provision (that the act is designed to be similar to M, indicating a planned motor copy) sets imitation apart from social facilitation, where there is unknown, perhaps diffuse motivational influence. However, design is not an observable characteristic of the behavior.

One pragmatic way to distinguish between imitation and social facilitation is to require that the probability of producing the modeled behavior through chance alone, or through motivational influence, be acceptably small. Requiring production of a behavior that is novel for the observer is one way to achieve this. Producing a novel behavior as a result of social facilitation is less probable than performing a familiar behavior. Therefore, on operational grounds, we add another feature to Mitchell's definition for use in comparative studies:

(5) the behavior C must be a novel behavior, not already organized in that precise way in the organism's repertoire.

Restricting evidence for imitation to cases in which *something new is acquired* recognizes the unique function that sets apart imitation from other forms of social learning. That is, imitation serves as a means of learning novel behavior directly from others. This makes it less likely to be found, but it means that identification will not be ambiguous.

It must be emphasized that novelty of a behavior is not a yes-or-no thing; some parts of a behavior may be novel (e.g., its orientation in space), and others may not be. Often, the specific action is already in the animal's repertoire, but the location where it is performed can render its function completely new. An example from our work with monkeys learning to use a tool to pound open a nut can illustrate this point. By 1 year of age, tufted capuchins (*Cebus apella*) routinely pound any suitable object on a substrate (Visalberghi, 1987). To crack open a nut, the subject must pound the tool on the nut, achieving a specific orientation between the tool and the object to be modified by the tool. When naive animals pound the tool here and there in the cage, after seeing the model pound open a nut with the tool, they are not necessarily imitating. Although their acts of pounding may be similar in form to the model's pounding, we cannot clearly attribute a rise in the probability of pounding to an intention to copy the model. The same could result from enhancement of interest in the tool or in vigorous exploration of objects in general. Further work defining the probability of changes in the frequency and orientation of pounding in various situations is needed before the inference of imitation is justified.

The methodological problem we face is the determination of the degree and nature of behavioral match that is required for unambiguous differentiation of imitation from other aspects of social learning, particularly social facilitation. Meltzoff and Moore (1983) provide an excellent discussion of the problem and some ways to circumvent it. The operational definition of imitation that we have developed is this: A behavior is imitated (1) if it is sufficiently similar to the behavior of the model, with sufficiency defined a priori, (2) if observation of the model is necessary for its production, and (3) if the behavior is novel for the observer, with novelty required in a dimension in which it is usually absent. Thus, in our previous example, an individual's pounding an object following observation of a model does not satisfy the third criterion. Pounding is a common activity; novelty in the choice of object to pound is also common. Intention is implicated in this definition through the requirements for matching and novelty, but it is not a criterional variable that can be assessed directly.

These are very general requirements that must be translated into specific rules by the experimenter to distinguish between imitation and other phenomena in a particular situation. An example of a clear set of rules for identifying imitation in pigeons acquiring a novel food-finding behavior has been given by Lefebvre and Palameta:

> In order to show rigorously that "imitation" underlies the diffusion of a feeding innovation, five conditions must be met: (1) inexperienced animals that are not given demonstrations by a conspecific model should not be able to learn the innovation by themselves; (2) animals given only information about where to find food (local enhancement) should not learn, or should learn more slowly, than animals given full demonstrations of both how and where to find food; (3) animals given partial experience at feeding by themselves in conditions similar to social test conditions should also not learn or should learn more slowly than animals given full demonstrations; (4) animals having observed a demonstration should be able to perform the innovation in the absence of the model, i.e., should not need social facilitation by the feeding model; (5) animals should use specific features of the model's technique when they perform the innovative feeding behavior. (Lefebvre & Palameta, 1988, p. 148)

This definition captures all features of our definition and, in addition, requires delayed performance of the observed behavior (deferred imitation), although it is not clear whether or not the initial performance of the innovative behavior must also be performed without the model. Incidentally, Lefebvre and Palameta call the phenomenon observational learning, rather than imitation. It is self-evident that several control conditions are required for unambiguous demonstration of imitation.

### Theories of imitation and methods of studying imitation in developmental psychology

Developmental psychologists, as the group of researchers most concerned with imitation in humans, have built up a foundation of procedural expertise that can be of immense value to comparative researchers. Developmental psychologists have been interested in the origins of imitative abilities in humans, as well as age related changes in imitation, and therefore they have identified early forms of imitation. Even neonates produce motor acts they observe in others and can reproduce (Meltzoff & Moore, 1983), as well as repetitions of their own behaviors (Piaget, 1954). However, researchers studying human infants follow elaborate procedures to rule out alternative explanations for the production of familiar behaviors by their subjects, such as determining the rate of performing a familiar act under baseline conditions or determining the frequencies of performance of an action in two groups of subjects, each of which observes a model performing a distinctly different action (cf. Meltzoff & Moore, 1983). These procedures have not yet been applied by researchers studying imitation in monkeys or apes, although they have been used by Palameta and Lefebvre (1987) with pigeons.

Recent advances in comparative work have shown that there are many alternative approaches to the study of social learning that are productive in a variety of species, including pigeons and rats (Zentall & Galef, 1988). But there is no grand theory integrating research in imitation and social learning. When a comparative psychological theory of imitation does appear for nonhuman primates, we suggest it will be closely linked to theories of imitation in human infants and children. Early efforts in this direction involved the application of a Piagetian scheme of sensorimotor development to observational studies of nonhuman primates (e.g., Chevalier-Skolnikoff, 1977; Parker, 1977). Parker (1977), for example, carefully observed the behavior of an infant stump-tailed macaque (*Macaca arctoides*) and concluded that the monkey "did display a few contagious performances of behavior patterns from his own repertoire, but he displayed no purposeful matching of his behavior patterns to those of other animals, nor did he come to imitate behavior patterns outside his repertoire as human infants do" (Parker, 1977, p. 65). Analogous results were found when Japanese and crab-eating macaques (*Macaca fuscata* and *M. fascicularis*, respectively) and tufted capuchins (*Cebus apella*) were tested for imitation using facial expressions and closing the fist as eliciting movements (E. Visalberghi, unpublished data). Overall, researchers have found better achievements (e.g., performance attributed to higher levels) in other series of the Piagetian sensorimotor development scheme (object concept, means–ends relationship) than in the imitation series by nonhuman primates. These findings suggest that the Piagetian series of sensorimotor development are independent of each other (cf. Parker, 1977). Imitation must be studied in its own right, not as an element in a larger coordinated program of cognitive development.

Meltzoff and Moore (1983) reached a similar conclusion about the accuracy of Piaget's original formulation of imitation, from opposite findings (i.e., that infants are already able to imitate at birth). Furthermore, deferred imitations of familiar acts, and evidence of deferred imitation of a novel act, are seen by 14 months, well before the period in which symbolic representational skills are postulated to appear in the Piagetian scheme (Meltzoff, 1988b). Yando et al. (1978) noted the discrepancies between changes in other aspects of cognitive development (as cast into Piagetian stages) and imitative ability in older children as well (2 years old and older).

Meltzoff and Moore (1983) argued that the human infant's capacity to copy motor acts seen in others involves detection of motor equivalences and cross-modal processing of visual and motor information. These abilities are present at birth; they do not develop after birth, as originally postulated by Piaget (1962). Nonhuman primates may share some of these capacities with humans, as in categorical perception of species-typical vocalizations, for example. Perrett's recent work (Perrett, Mistlin, Harries, & Chitty, in press; D. I. Perrett, pers. commun.) has shown that in macaques, certain cortical neurons in the temporal lobe respond equivalently to the sight of actions on objects produced by an individual's own movements or by the movements of another (but not to the visual properties of the movement alone, e.g., moving edges). This capacity could be a neuronal component underlying detection of motor equivalence and hence could play a role in matching an observed act and a produced act. In the human neonate, subcortical areas probably are involved in the process (Gibson, 1981), but cortical areas could assume this function later.

Bandura (1986), Bruner (1972), and Yando et al. (1978) emphasize the role of motivational variables in the occurrence of imitation. Imitative behaviors in young children can be induced by external reinforcements, but more often in natural circumstances children imitate without external reinforcement, and in playful circumstances. Thus, the relationship between imitative activity and competence in solving a problem is indirect (Bruner, 1972). This raises a problem for the study of imitation in nonhumans, where we have less confidence about our knowledge of motivational characteristics, and a more difficult task in identifying novel behaviors. But it suggests that efforts to have a monkey imitate facial expressions performed by a human, for example, are unlikely to succeed. It also suggests that we should expect a variety of behaviors during initial experiences with problem-solving tasks that are unrelated to the model's behavior, even if imitation occurs also. In short, we must be able to pick out imitative elements in a stream of ongoing behavior.

Our definition of imitation leaves open the question of the process underlying behavioral matching of novel behaviors. This is a crucial question, however, for understanding species differences in imitative capacity and the relationship between imitative capacity and representational capacity. Distinguishing between "rote" copying of a motor act and learning a rule guiding another's behavior is important for this purpose. We have already noted that certain species of birds can copy motor acts, just as can human neonates. In addition to an ability to mimic movements, humans are proficient at learning rules underlying another's behavior (Bandura, 1986; Rosenthal & Zimmerman, 1978). Bandura (1986) uses the term "observational learning" to refer to this form, and we adopt this convention. It would seem that learning to use a tool, for example, from observation of another without learning the rules guiding successful performance should be characterized by careful copying, step by step, as we do when instructed in the operation of a new machine whose workings we do not understand. Efforts to match the motor pattern and monitor the model should be prominent. Alternatively, gradual approximations to the correct form would be observed, as when a parrot learns to produce a new word. Learning some or all of the rules guiding goal directed behavior, such as use of a tool, would produce a different set of behaviors on the part of the observer. Exploration of the tool and its relation to the object or apparatus to be exploited would be expected. This is consistent with the playful nature of imitation evident in human children.

It would seem that the most common function of imitation in increasing problem-solving competence in human children is through children's ability to learn, at least partially, the rules guiding problem-solving behavior (i.e., observational learning is the important process). Properties of objects and their uses may be partially grasped from observation of others using them, but fuller understanding of the properties follows upon individual explorations (Bruner, 1972). This implies, naturally, that the observer's competence and the model's performance cannot be too distant; otherwise the observer is unable to grasp any aspect of the rules guiding the model's behavior (Rogoff & Wertsch, 1984). Using even a simple tool involves multiple rules governing the orientation of the tool to the object, the movements necessary to use it, and so forth. Perhaps here is where cognitive capacities play a role in species differences in learning to use tools, through observation or through direct experience. We shall see that there is some evidence, although it is not strong, that nonhuman primates are able to learn observationally that some relationship exists between the tool and the task to be solved, but that the precise relationships governing successful use must be discovered through individual exploration.

### Evidence for imitation of innovative behaviors in monkeys

Kummer and Goodall define innovative behavior as "a solution to a novel problem, or a novel solution to an old one; . . . a new discovery such as a food item not previously part of the diet" (Kummer & Goodall, 1985, p. 205). The dissemination of innovative behaviors in a group does not, however, necessarily reflect imitation, or even social learning. For example, the spread of selection of a new food item may reflect changes in encounter rates, as a function of abundance, or, if socially mediated, may reflect social facilitation and enhancement processes, rather than imitation. In this section we assess the evidence for imitation of innovative instrumental behaviors in monkeys.[1]

### *Observations*

The best known examples of dissemination of new behaviors are those of food washing in Japanese macaques (*Macaca fuscata*); for a review, see Nishida (1986). Monkeys living along a coastal area of Japan were provisioned with sweet potatoes on a sandy beach. The monkeys were seen to take their potatoes to a stream, and later the ocean, and rinse them in water. This "washing" removed the sand, and possibly also added a pleasant salty taste to the potatoes. However, the original studies reported no direct observation of acquisition or of imitation of washing. The conclusion that Kawamura reached in 1954 (cited and confirmed by Itani & Nishimura, 1973), that potato washing was initiated by a single monkey and was imitated by others, was inferred from the fact that, over time, the number of monkeys washing food (potatoes) increased. The alternative hypothesis, that in the same environment several monkeys could have *independently* discovered the new behavior, has received little attention. Furthermore, social enhancement is likely to have favored the acquisition of washing behavior by others. The model and the "copier" could each have held a potato and each gone to the water, perhaps to play, or because infants and mothers remained near each other. This could have resulted in incidental washing in several individuals. We shall return to this theme in the next section and show that playful experience with water can facilitate the appearance of food washing.

Another plausible scenario for the independent discovery of washing is retrieval by a naive animal of food left in the water by another animal (see the next section for observations supporting this speculation). A similar hypothesis for the dissemination of a behavior by birds (*Parus* spp.) – opening milk bottles to drink the cream at the top – has been supported empirically. This behavior, first reported by Fisher and Hinde (1949; see also Hinde & Fisher, 1972), has often been cited as an example of a behavior transmitted through observation of others. Sherry and Galef (1984), however, noted that chickadees learned to open milk bottles equally well from encountering already opened bottles as from observing models opening them.

In addition to the existence of several plausible alternative hypotheses, we doubt that imitation was involved in the dissemination of washing behavior in the Japanese macaques on the grounds that dissemination was so slow. Learning a new behavior observationally should result in a faster dissemination than should individual independent acquisition. Food washing in Japanese macaques did not spread with the speed we would expect from imitation: The first observation of food washing was made in 1953; 3 years later, in 1956, only 11 of about 25 monkeys had adopted this practice (Kawai, 1965). If they learned to wash food from others, why did it take so long? Opportunities for observation of proficient models and for practice were frequent enough for the behavior to have spread more quickly.

Another innovative behavior that appeared in the same group of Japanese macaques was a variation on washing potatoes. The monkeys were provisioned with wheat, as well as potatoes. The wheat was thrown onto the shore, as were the potatoes. Eventually, the monkeys were seen to carry handfuls of sand mixed with grain to the water, dunking the mixture. Sometimes a handful of material would be released, and the grains of wheat would float on the surface, to be retreived for sand-free eating. This was called wheat washing, or placer mining. Nishida (1986) suggested that wheat washing might be more difficult to acquire than washing potatoes, because the monkey had to release the food, rather than dipping it up and down in the water. In any case, that behavior also spread very slowly. Six years after it was first observed, only 39% of the members of the group displayed wheat washing (Nishida, 1986). Galef (in press), after close scrutiny of the data on food washing and wheat washing, concluded that explanations simpler than imitation could fully account for the dissemination of these behaviors.

Our next example concerns another species of macaque, the rhesus (*Macaca mulatta*), and a more difficult behavior, opening coconuts. Pounding a coconut on a surface to crack it open demands more precise orientation between objects than does washing behavior. Whereas holding even a portion of a potato in the water results in at least some success at food washing (i.e., it removes some of the sand), holding the coconut on or by an anvil does not crack it. The rhesus macaques of Cayo Santiago are provisioned with commercial chow, and they feed on several other natural foods they find on the island. However, they do not exploit the coconuts, which are abundant on the island. Of course, monkeys like this rich and tasty food, as evidenced by the quick consumption of the meat when it is incidentally available. But a coconut is enclosed in a hard shell and must be pounded, or thrown, on rocks or wood anvils to be cracked open. The data recorded over 30 years by Berard and other investigators at Cayo Santiago show that only two individuals acquired this behavior: WK and his younger brother, known as 436 (J. Berard, pers. commun.). WK was born in 1963 to the dominant female in Group F; WK became the dominant male in that group as a young adult. When he was 9 years old he left the group and became a peripheral male. After leaving his natal group, it was observed that when he opened coconuts, frequently they were taken from him by more dominant animals, and he gradually stopped opening coconuts. WK has not been observed cracking open coconuts in the last 10 years. Sometimes his younger brother, when far from the other monkeys, still pounds coconuts on the ground to crack them open.

The most surprising part of this account is that no other macaque, among a population of more than a thousand on the island during that period, "imitated" WK and 436. Berard (pers. commun.) sometimes saw as many as five other males watch 436 as he opened coconuts, waiting to get scraps of the meat, but he never saw those males attempt to open coconuts on their own.

Clearly, motivation was not a limiting factor in this case. It seems likely that the requirements to obtain a particular orientation of nut to anvil, and to perform a specific act, were simply too complicated for the observers to learn this skill from watching WK and his brother.

Schonholzer (1958) reported observations made in the Zurich Zoo on hamadryas baboons (*Papio hamadryas*) that similarly documented the lack of dissemination of an advantageous behavior, although the value of the behavior seems less than that of opening coconuts or washing sandy food. Two hamadryas used their tails to drink: Suspended by its hands, a hamadryas would lower itself down the vertical wall of the moat, dip its tail into the water below, climb up again, and suck the water off its tail. All 13 other baboons watched closely, and licked up drops, but never learned to obtain water with their tails.

More recently, Hauser (1988) described an innovative behavior in a group of free-ranging vervet monkeys (*Cercopithecus aethiops pygerythrus*) shown during a period of drought. The behavior consisted of immersing a seedpod from an *Acacia tortilis* tree into exudate from the same kind of tree, followed by ingestion of the pod. That behavior may have provided the monkey with a more efficient means of obtaining exudate, which probably is an important source of water during dry periods, than would dipping in its fingers, and/or immersion may have softened the pod so that it could be ingested more easily. That behavior was first shown by an adult female and was seen during the next 22 days in 6 other individuals out of a group of 10. In his discussion of the processes involved in the dissemination of the innovation, Hauser acknowledged the impossibility of sorting out the possible contributions of social facilitation, enhancement, or imitation. Thus, although the rate of dissemination of this behavior was much higher than for the innovations described for Japanese macaques, and the data in general were more detailed, the results are no more compelling. The problem of demonstrating imitation in spontaneous situations is not adequate observation; the problem is exclusion of alternative hypotheses.

Although some other examples could be cited in which innovative behaviors have not spread, such instances are not often reported in the primatological literature. This does not necessarily mean that innovations typically *do* spread. Several factors would lead to an overestimate of the frequency with which novel behaviors disseminate in a social group, compared with cases in which novel behaviors occur in only one individual. First, if a new behavior is shown by several individuals, the human observer has a greater probability of witnessing it. Second, if the new behavior appears to have adaptive value, it is likely to be reported for that reason. New behaviors with less obvious functional value are not likely to generate the same interest, especially if they are performed by a single individual. There are a few exceptions to this second point, such as an odd grooming posture that had no evident function, but spread in a chimpanzee population (McGrew & Tutin, 1978). But, in general, the relative frequencies with which dissemination and lack of dissemination of spontaneous new behaviors are mentioned in published papers are not accurate measures of the frequencies with which they happen. Empirical studies are needed to determine the prevalence of dissemination, and the mechanisms involved, when it does occur.

### *Experimental studies*

The acquisition of food washing by monkeys has been studied recently in our laboratories using captive groups of tufted capuchins (*Cebus apella*) and crab-eating macaques (*Macaca fascicularis*) (Visalberghi & Fragaszy, unpublished data). In one group of capuchins ($N = 5$) the monkeys were first given access to a basin of water, with toys nearby or in the water, during five 1-hour periods on consecutive days. Next they were exposed to a basin filled with water and a basin containing sand and pieces of fruit that had been coated with sand. Eight sessions of varying lengths (10–30 min) were conducted in this phase. The first episode of washing occurred within a few minutes, and within 2 hours all members of the group washed food. Food washing involved patterns of behavior comparable to those exhibited in playful activities with water and nonedible toys. The monkeys usually washed fruit by dropping it and then retrieving it while it was sinking or after it had sunk to the bottom. (Recall that Nishida, 1986, considered wheat washing more difficult than washing potatoes precisely because release of the food was necessary to wash wheat.)

When younger monkeys first washed food during play, their understanding of the practical value of their behavior was poor. Sometimes they did not eat the washed fruit, and at other times they ate sandy fruit with facial expressions of disgust, even though they had previously washed fruit. The time required to develop purposeful food washing varied among individuals. This may reflect differences in understanding the consequences of the behavior (i.e., that washing resulted in more palatable fruit).

Food washing was acquired in a few hours by all the macaques ($N = 4$) in a manner similar to that observed in the capuchins, except for the macaques' obvious pleasure in the water. They entered the basin of water like children in a wading pool, splashing, running, and wetting themselves from head to feet. As they often carried fruit into the water with them, it was cleaned in the process. They were also observed to wash fruit in a more deliberate manner outside of the playful context.

A second experiment extended the findings of the first study with capuchins (Visalberghi & Fragaszy, unpublished data). Following presentations of a pan of water (only) and clean food and water, two groups of capuchins (about 15 animals per group) were presented with a pan of water and sandy slices of fruit, or grain mixed with sand. The monkeys expressed some interest in the pan of water when it was presented alone, but none washed fruit or grain when these items (not covered or mixed with sand) were presented in conjunction with the pan of water. When sandy fruits were presented, the monkeys expressed obvious disgust at initial tasting, and persistent attempts were made to clean the fruit by rubbing it on the wall or between the hands. Sandy grain elicited the same reaction, and monkeys attempted to pick grain out of the sand, or else to throw handfuls on the floor to better separate the grain from the sand. Some monkeys in each group, however, washed grain, and in one group, some monkeys washed fruit. Behaviors similar to those observed in the previous study were seen: Sometimes fruit and grain were released in the water and retrieved; sometimes they were immersed and stroked through the water, but not released.

As in previous studies in these groups, a few animals solved the problem, and others exploited their actions. Many others had little access to the center of activity. Exploiters did not become washers, in line with other observations on these groups (Fragaszy & Visalberghi, 1989); see Giraldeau and Lefebvre (1987) for similar findings with pigeons. Washers did sometimes retrieve food from the water prior to their first attempts at washing, but as exploiters also did this, and often more frequently, this is not strong evidence that the retriever recognized from this experience the utility of bringing sandy food to the water. Social enhancement of interest in the water was evident, but no individual tried to wash or first washed immediately after observing another doing the same.

These studies indicate that the rapid acquisition of food washing observed in the first study was not dependent on previous experience with toys and water, nor was acquisition clearly accelerated from observation of others washing. Taking food to water is a simple behavior learned readily enough through individual exploration. Observation apparently affected individuals' behavior in this situation by attracting their attention to the water, and sometimes the fruit or grain in it, but not by causing individuals to recognize that the food must be brought to that place.

We know of one other study, again with capuchins, in which an individual in a social group possessed useful knowledge about how to solve a problem that was not acquired by other members of its group, even after many hundreds of demonstrations. Adams-Curtis (1987) presented a sequential mechanical puzzle to a group of capuchin monkeys. One individual learned to solve the four-step puzzle within five 20-min sessions. A detailed analysis of the behavior of one of her frequent observers during several additional sessions indicated that the observer's contacts with the different parts of the puzzle were no different after he had just seen a solution than at other times. He continued to contact the parts of the puzzle in a way unrelated to the order of acts necessary to solve the puzzle.

## Imitation of tool use in monkeys

In this section we review studies on tool using in monkeys to identify possible instances of imitation. Much of the confusion in the area of social learning arises from the fact that social facilitation, local and stimulus enhancement, and imitation can all contribute to the spread of a novel behavior. In order to reduce the confusion, it is vital to identify acts that are unlikely to be acquired as a result of enhancement or facilitation. Using a tool requires a precise sequence of behaviors performed in the appropriate locations. In a task with these requirements, enhancement of nonspecific activity by the observer after observing the model is more easily distinguishable from imitation or from rule-guided exploratory behaviors. We note that these studies were designed to study the acquisition of tool using in a general sense, rather than to study imitation per se. None of them included control groups of individuals exposed to a model that expressed equivalent interest in the object as did the tool-using model, but that did not use the tool (the behavior to be imitated), or other appropriate controls. Hence, these studies should not be regarded as conclusive, but even with their limitations they serve to counteract the sloppy thinking that would attribute virtually any dissemination of novel behaviors to imitation. They also serve as a reference point for future studies tailored to assess imitation in nonhuman primates.

Several studies by Beck and one by Westergaard have investigated the ability of monkeys to learn from models the use of tools. Beck (1972, 1973a,b) carried out several experiments in which captive baboons were presented with an opportunity to use a tool to obtain food. Few subjects succeeded in using a stick to rake in food. Beck wrote that "despite prolonged opportunity to observe a skilled tool user, other baboons do not imitate the behavior and do not even acquire enough information to accelerate their own mastery of the task" (Beck, 1976, p. 301).

A similar experiment carried out with a group of seven pig-tailed macaques (*Macaca arctoides*) led Beck (1976) to hypothesize that macaques might be better imitators than baboons. In that study, one individual appeared to imitate the first solver, using the same technique of throwing the tool by flinging its arm, palm downward on the tool, toward the distant food pan, and subsequently raking in the pan. In support of his hypothesis that baboons are less able to learn novel behaviors from observation of others than are macaques, Beck noted that the majority of field observations on innovative behaviors in macaques have concerned more than one individual, whereas in baboons innovations have been reported for single individuals (e.g., van Lawick-Goodall, van Lawick, & Packer, 1973).

Since Beck published those studies, an additional report has appeared indicating the occurrence of some form of social learning of tool use in another species of macaque. Westergaard (1988) presented an apparatus containing sweet syrup to a group of nine lion-tailed macaques (*Macaca silenus*). At first, the monkeys could obtain the syrup by probing into the apparatus with their fingers, until the syrup got too low for that technique to work. Twenty-six days after the introduction of the apparatus, two monkeys were observed using sticks to obtain syrup. A third animal used sticks on day 30, and a fourth (a youngster) on day 70. These monkeys also were observed to manufacture their tools by breaking branches off browse. The older animals did this almost as soon as they used sticks as tools, and the youngster did so 3 weeks after he started to use tools.

A second experiment was conducted in which two of the original tool-using monkeys and one non–tool user from the some group were put in with a naive fourth animal. Again the apparatus was made available on a continuous basis. The solvers, of course, used the apparatus immediately. The naive animal used tools on the second day, and the familiar nonsolver, who had seen others use the apparatus for months previously, used tools on the sixth day. That rapidity of acquisition suggests that social influences were important, but no firm conclusions can be drawn about the process.

### *Experiments on tool use in capuchin monkeys*

Capuchins are the most versatile tool users among monkeys (Beck, 1980; Parker & Gibson, 1977; for a critical review, see Visalberghi, 1990). In the past few years we have carried out several experiments on tool use in tufted capuchins (*Cebus apella*) in five different groups. In all cases, one or more individuals in a group acquired (spontaneously) a novel tool-using behavior to get a desirable food. In the course of these experiments, we became convinced that capuchins did not acquire new tool-using behaviors from watching alone, although other processes of social learning, especially local enhancement, were implicated (Fragaszy & Visalberghi, 1989, 1990). Negative results cannot prove the absence of a process; nevertheless, we agree with Thomas (1987) about the value of publishing such results: They allow us to determine the outer bounds of a species' abilities. We can never know the boundaries if we report only their successes. With that aim in mind, we next present summarized findings from our recent studies.

### *Experiments 1 and 2*

One group ($N$ = 6; Experiment 1) and two groups ($N$ = 27 and $N$ = 15; Experiment 2) of tufted capuchin monkeys were tested for their abilities to use tools to crack open nuts (Antinucci & Visalberghi, 1986; Visalberghi, 1987). Stones or wooden blocks (potential tools) were provided, together with nuts; tools and nuts were provided in numbers about equal to the size of the group. One individual in each group solved the task repeatedly (in Experiment 2, 44 times in one group, and 80 times in the other). The cagemates eagerly observed the performances of tool users, but nut cracking did not spread to other individuals. It must be noted that the motivation to get the kernel of the nut was high among the observers, and there were numerous tools and nuts available. Although some low-ranking individuals may have had limited opportunities to try using the tools, the majority of individuals contacted them. However, in each case a single individual remained the only tool user in the group.

*Experiment 3*

Westergaard and Fragaszy (1987) presented a group ($N = 9$) of tufted capuchin monkeys with an apparatus requiring the use of a stick to probe for syrup and (separately) an apparatus containing fruit juice that could be sponged out with paper towels. Eventually six individuals used sticks to probe for syrup, and about as many used paper towels to sponge for juice. In the case of the probing task, the first monkey to use a detached tool of her own fabrication (obtained by breaking a stick from a branch) exhibited the innovative behavior immediately after observing a cagemate awkwardly stuff an attached branch into one of the holes of the apparatus. A second instance that suggested a role for observation in the acquisition of proficient probing concerned a mother (a proficient tool user) and her infant daughter of less than 1 year. The infant was often with her mother at the apparatus, and the mother frequently allowed her daughter to take sticks coated with syrup from her, and even to place her hand on the mother's hand during the probing action. In this case, observation and coaction were concurrent events. This infant was the first infant (out of three in the group) to acquire the tool-using behavior.

Although these are suggestive observations, they are not compelling evidence for imitative capacity. We are unable to rule out competing hypotheses, particularly the hypothesis that individual experience coupled with social enhancement processes can account for the observed behavior.

*Experiments 4 and 5*

To examine more closely the temporal relationship between the behaviors of observers and tool users, we conducted a set of experiments with two groups of tufted capuchins (Fragaszy & Visalberghi, 1989). In both studies, an apparatus containing food and one or more tethered tools was provided. In Experiment 4, we glued 24 walnuts in an apparatus so that the use of a tool was required to crack them open, and four metal tools were tethered with the apparatus. In Experiment 5, inserting a stick into a clear plastic apparatus and pushing a sliding mechanism allowed sunflower seeds to drop into a food cup.

By confining activity with the tools to the apparatus site, it was possible to monitor the behavior of tool users and all other animals present at the apparatus concurrently. This permitted us to examine the temporal relationships of behaviors among tool users (models) and observers.

In both studies, five individuals from the two groups became proficient at using the tools. But the activities of proficient models with the tools at the apparatus were not associated with higher rates of exploration or contact with the tools by the other capuchins. Some observers witnessed hundreds of solutions, but still did not use or even contact a tool themselves. When the groups were given the rod apparatus (Experiment 5), but without the most frequent tool user in each group, other individuals explored the apparatus more, and a few displayed behaviors approximating the correct solution. One individual who had minimal access to the rod task solved it in a manner suggestive of intentional solution on her first try. But this task had elements similar to those of a task that many of these individuals already knew how to solve: Experiment 3, involving using a stick to probe for food (Westergaard & Fragaszy, 1987). We suggest that, at most, our subjects may have learned that *some* relationship existed between tool and task in Experiment 5, but not precisely what the relationship was or what action was required.

In Experiment 4, animals who learned to use the tools to open nuts spent more time at the apparatus alone than did the others, and they performed more exploratory behaviors combining objects at the apparatus. They clearly did not coordinate these behaviors with the actions of others at the apparatus, whether or not the others were successful at the task. In short, imitation of a model was not evident in either experiment.

### *Experiment 6*

Visalberghi and Trinca (1989) presented a horizontal transparent tube, baited in the middle, to a group of capuchins ($N = 4$). The reward was accessible only by using one of the provided sticks to push it out of the tube. Three capuchins succeeded, but the fourth (a young adult female) did not, even after repeated sessions (totaling more than 3.5 hours) in which she was alone with the apparatus. Subsequently, this subject was again placed with her proficient cagemates, together with the apparatus. The subject's behavior was carefully observed while others' solutions occurred, and while the apparatus was unattended by others. Despite the fact that she witnessed 57 solutions, the subject did not learn to use a tool while present with the models, nor when tested alone in six 30-min sessions. Her behavior during the modeling sessions indicated that she was not selectively attentive to the model or the apparatus during solutions. Later on, however, she succeeded at solving a similar but slightly easier tool-using task, consisting of a vertical transparent tube filled with sugary liquid. The liquid could be obtained only by using a stick, several of which were provided. She solved this problem after less than 30 min. When this monkey was presented with the horizontal tube again, she solved it right away. This result suggests that individual experience in a similar task was a more powerful aid to the monkey than was the information she was able to acquire from the behavior of models. If this finding is confirmed in other experiments directly comparing the efficiency of learning in different ways, it will strengthen the conclusions from the studies reviewed earlier that, for these monkeys, imitation of a model is indeed a very limited way to learn to use a tool.

## Comments on imitation in chimpanzees

An early report on the imitative skills of an infant chimpanzee came from Kellogg and Kellogg (1933), who raised their infant son together with an infant chimpanzee of similar age. After a long period of careful comparisons between the behaviors of their infant and the chimpanzee, they concluded that "we are accustomed to regard the chimpanzee, as a splendid imitator, . . . yet the child is a more versatile and continuous imitator than the animal" (Kellogg & Kellogg, 1933, p. 230).

However, imitative behaviors in apes have been reported frequently, although generally on the basis of weak data. Parker and Gibson (1979) suggested the existence of deferred imitation in apes on the basis of anecdotal reports in the literature, such as the reports of home-reared chimpanzees exhibiting behaviors remarkably convergent with those of their human housemates (e.g., Hayes & Hayes, 1952). Apes using sign language have been reported to "imitate" the previous (familiar) utterances signed by their human partners (Terrace, Petitto, Sanders, & Bever, 1979; see Greenfield, 1984, for an opposing view), but rarely have *new* signs been acquired by imitation (Gardner & Gardner, 1969). More recently, Mignault (1985) observed deferred imitation (e.g., conventional use of objects) in chimpanzees during play. Rather stronger evidence for imitation comes from de Waal's descriptions of young chimpanzees playfully (arbitrarily and transiently) adopting the hunched posture or limping gait of another individual in the group, or the particular style of the intimidation displays of the current dominant male in the group (de Waal, 1982, in press). We note that the best evidence for imitation in apes comes from observations of play, not problem-solving tasks, in line with Bruner's analysis (1972) of imitation in human children. Nonetheless, observations of play are not satisfactory for clear interpretation on the question of imitation in nonhumans. The question is not, as it is for humans, *When* does imitation occur? The question is, *Does* it occur? Answering the latter question will require stronger evidence than answering the former.

Despite the strong anecdotal evidence for imitation of gestural and postural behaviors in chimpanzees, imitation of a novel tool-using behavior has never been witnessed in umabiguous circumstances, in captivity or in natural environments. "Fishing" for termites (probing into a termite mound with a stalk of grass, withdrawing the stalk, and eating the termites clinging to it) is perhaps the best known tool-use behavior in wild chimpanzees. Many reports suggest that social learning plays a role in the acquisition of this behavior by young chimpanzees. Immature chimpanzees watch attentively while their companions fish for termites, and they perform elements of the model's behavior in a crude way (such as picking up sticks) while the model is busy at the termite mound. Nevertheless, the full sequence of acts is not performed without a great deal of individual experience with the task (Goodall, 1986; McGrew, 1977). Young chimpanzees occasionally do experience a form of social assistance with the task that may facilitate acquisition of termiting skills. Mothers often let their infants participate in termiting activity, allowing the infants, for example, to get termites out of the mound while holding the mother's stick (McGrew, 1977). We suggested earlier that coaction in this form and other opportunities to participate in the mother's activities facilitated the infant's acquisition of dipping in a similar task in one mother–infant pair in capuchins, and the same points are relevant here. In coaction, the infant's attempts to use the tool are correct from the beginning, because the mother also holds the tool and maintains the correct orientation between tool and mound. The infant is obtaining direct motor guidance in this fashion.

Nut cracking is another form of tool using documented in wild chimpanzees in which social learning is implicated (Hannah & McGrew, 1987). However, nothing can be said as yet about the mechanisms of transmission. Sumita, Kitahara-Frisch, and Norikoshi (1985) argued that imitative copying was involved in the dissemination of nut cracking in their captive group of chimpanzees. Their data show, however, that the unsuccessful chimpanzee (Charlie) observed the demonstrator longer than did the successful learner (Deko). Deko, on the other hand, played with the tools (stones) more than twice as often as Charlie. The results do not justify the strong conclusion that the observer imitated the proficient model; individual experience with the tools and with the appropriate behavior are just as likely to have contributed to Deko's success.

A recent and very welcome experiment examined whether or not captive juvenile chimpanzees could learn to use a tool to rake in food (Tomasello, Davis-Dasilva, Camak, & Bard, 1987). Subjects participated in several phases of testing: a pretest in which the observer had food and tool, but the model was merely present; a phase in which the model had a tool and obtained food, but the observer had no tool; a phase in which both observer and model had tools and access to food; a final phase in which the pretest was repeated (the observer had the food and tool, whereas the model did not). All four younger experimental subjects learned to solve the task in its simplest form; one control subject did also. Those authors commented on the lack of imitation evident on the part of their young subjects: "none of the subjects demonstrated an ability to imitatively [*sic*] copy the demonstrators' precise behavioral strategies" (Tomasello et al., 1987, p. 175). Tomasello and associates suggested, however, that the juveniles *did* learn something about the tool from the observer – something along the lines of the tool having a function to solve the problem – although they did not learn precisely what the relationship was between movement of the rake and solution of the problem. We have weaker evidence for the same phenomenon in capuchins, as described earlier.

Overall, we lack clear experimental results showing that apes imitate the use of tools from proficient conspecifics or learn observationally how to solve a problem. However, the observations of probable imitation of postures and gestures in apes, coupled with their superior performances (as compared with monkeys) on many tasks involving representational skills (e.g., Natale & Antinucci, 1989), suggest that apes should be better imitators than monkeys. Apes probably *do* ape each other, at least in behaviors not involving tools.

This is still nearly a statement of faith, however. Experimental exploration of this ability is definitely warranted.

## Conclusions

Imitation plays an important role in the acquisition of novel skills in humans (Bandura, 1986; Bruner, 1972; Piaget, 1962). But the evidence we have reviewed in this chapter suggests that, optimistically, imitation plays a limited role in the acquisition of novel behaviors in monkeys. Hauser's report (1988) on vervet monkeys dipping leaves into exudate from a tree stands alone as a well-documented field report on monkeys in which imitation seems even remotely likely. The slow diffusion of food washing in Japanese macaques, cited even today as an example of a behavior learned by imitation (Ridley, 1986), can be explained more parsimoniously as socially enhanced learning. Other examples of innovative behaviors that have diffused in social groups of primates are equally lacking as strong evidence for imitation.

The lack of imitation in monkeys is as apparent in tool-using behaviors as it is in arbitrary behaviors such as postures, gestures, or problem-solving behaviors not involving tools. Beck (1976) noted one instance in which an individual macaque appeared to use a direct motor-copy method to solve a problem (to rake in a pan). Westergaard and Fragaszy (1987) noted one suggestive incident in which a capuchin monkey appeared to grasp the principle of insertion with a tool from observing another, and Fragaszy and Visalberghi (1989) reported similar suggestive episodes with a few other capuchins in a different tool-using task. But none of these provides strong evidence for imitation, as alternative explanations cannot be ruled out. Furthermore, many more individuals displayed no evidence of learning from observation. Clearly, imitation is a fragile phenomenon in monkeys, at best. Why?

The absence of imitation in monkeys under the many circumstances in which they have been observed may be due to contextual factors limiting their behavior. We have some ideas about why imitation should be unlikely in monkeys observed in groups, even if they possess the capacity to imitate.

First, individuals who solve a problem often tolerate certain other animals near them, so that more than one animal is sharing in the success of the solver. This was clearly apparent in our studies (e.g., Fragaszy & Visalberghi, 1989) (Figure 9.1). Giraldeau and Lefebvre (1987) documented the power of this phenomenon (termed scrounging or exploitation) to suppress exploration and acquisition of novel behavior in pigeons, both in flocks and in pairs, where the observer member of the pair was allowed to share the food obtained by the model. Differentiation of a population into type A (solvers) and type B (exploiters) is perhaps a different matter than a "failure to learn" on the part of type B individuals. D. Quiatt (pers. commun.) has suggested an interpretation of individual differences among rhesus macaques (*Macaca mulatta*) living in Cayo Santiago in dealing with a simple problem: Some macaques do not lift the lids of the feed bins, relying on others to perform that

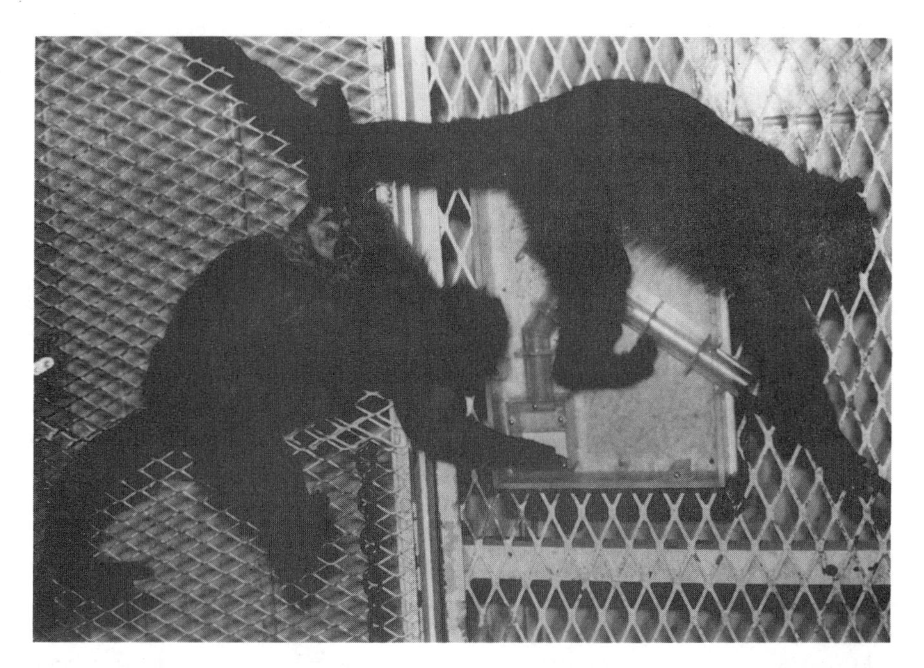

Figure 9.1. Capuchins at a vending machine: One individual pushes the tool into the apparatus; a second waits to obtain food from the food cup. Exploitation of tool users' actions by nonusers, like that shown in this picture, is common in groups of capuchins.

service for them. Quiatt has suggested that nonlifters may simply have alternative strategies for dealing with the concurrent demands of obtaining food and other aspects of life, such as maintaining social vigilance. Thus, failure to find imitation may reflect variation in strategies in a population (Clark & Ehlinger, 1987), rather than absence of the phenomenon.

Second, it is usually the case that one or more individuals in a group are constrained by social factors to limited access to desirable objects. These individuals have less opportunity to explore desirable objects, less opportunity to refine any special motor skills involved in manipulating those objects, and so forth. This apology for nonimitation is clearly more relevant to animals of low social status than to others.

Third, we suggest that coaction of a skilled model and a learner, in which the model allows the learner to participate intimately in its actions, is probably the most effective setting for learning a novel motor skill in nonhumans. This phenomenon occurs only when the learner poses no threat to the model, and it is probably restricted almost exclusively in normal circumstances to mother–infant pairs. Those instances among chimpanzees, for example, in which imitative learning seems plausible (such as using tools to dip for termites) involve behaviors in which infants can participate with their mothers. Coaction would not be "pure" imitation, as motor guidance is also involved, but it is a potentially powerful means of learning a novel skill.

But having said how social factors might constrain the appearance of imitation, we do not think that this is a satisfactory explanation for its general absence in monkeys. We think that the overwhelming lack of imitation reflects monkeys' cognitive constraints as well. These constraints probably hinge on representational capacity. Several authors have commented on the necessity of understanding the relations among actions and objects in order to grasp that imitation can be a means of solving a problem. Köhler, for example, wrote that imitation is the ability "to understand and intelligently grasp what the action of the other means" (Köhler, 1925/1976, p. 221). Bruner (1972) argued that visual imagery (rather than motor coding) is implicated in children's (delayed) imitation of another's acts, as the image of the other's acts is used to guide the observer's production of the same acts.

Our understanding of representational abilities in imitation and in tool-using behaviors in monkeys is limited. We have the most direct information and the greatest familiarity with these issues for capuchins. What we find is a mix of abilities. Capuchins are extremely successful tool users by nonhuman primate standards. Recent experiments, however, indicate that they are not reliably able to modify a tool in advance of its use, even after they have explored correct and incorrect tools many times (Visalberghi & Trinca, 1989). These findings indicate that they have an incomplete representation of the tool and/or its use. A limitation of representational capacity has also been found when capuchins have been tested on the Piagetian object permanence series (Natale & Antinucci, 1989). Capuchins, like macaques, cannot infer where an object has been hidden on the basis of hidden displacements, that is, when the monkeys view the hand moving the object, but not the object itself. A gorilla, in contrast, could solve this problem. Nor do capuchins display self-recognition when viewing mirror images of themselves (Anderson, in press; Visalberghi, Riviello, & Blasetti, 1988; R. C. Westergaard & D. M. Fragaszy, unpublished data). On the other hand, capuchins clearly do possess representation in some domains. Robinson (1986) has presented evidence that capuchins represent space and locations of objects (fruit trees) in their home ranges. D'Amato and Colombo (1988) and D'Amato, Salmon, Loukas, and Tomie (1985) have documented representation of serial order and transitivity of conditional relations in capuchins.

Our purpose here is not to review all the evidence relating to representation in capuchins, but merely to make the point that they possess some, though not all, of the representational abilities evident in humans. The relationship between representational abilities in these other domains and understanding the use of a tool, or imitating another using a tool (or performing some other novel behavior), must be better understood if we are to make sense of species differences in imitative capacity. Experimental studies are now planned or in progress in our laboratories and in others that should increase our understanding of the limitations of monkeys' representation of the properties of objects and our understanding of causality (including causal relationships among objects, and between the monkey's own action and its effect). Finally, careful experiments are also going forward to document the presence or absence of imitation in monkeys.

We turn now to the question of the significance of imitation and tool use in human evolution. Tool use and tool making often have been considered as having a central role in human evolution (Oakley, 1959; Washburn, 1960). Widespread and abundant fossil records of early tool use in hominids suggest that tool use was not confined to a limited number of individuals in a group. It is unlikely that widespread tool use within a group was the product of trial-and-error solutions reached independently by many individuals. On the contrary, social learning and, more specifically, imitation and apprenticeship probably were the usual ways for naive individuals to learn from skillful models, just as they are for contemporary humans (Yando et al., 1978).

Monkeys do not seem to be capable under common circumstances of learning tool use by imitation. The data for apes are scantier, but suggest similar, although less severe, limitations. Limitations in tool using are generally parallel to limitations in imitation. Passing from nonhuman primates to humans, we find an astronomical increase in tool using and a concomitant increase in imitation skills. What we want to argue here is that the two phenomena are not just proceeding together, linked by joint reliance on representational abilities or motor capacity. Rather, they amplify each other, producing an output that is not the sum but the product of the factors. Whereas in *Homo sapiens* (*Homo imitans*, as Meltzoff, 1988a, puts it), and to an unknown degree in ancestral hominids, any innovative behavior (including new ways to use a tool) can disseminate quickly, this phenomenon occurs only in circumscribed situations in apes, and it may be completely absent in monkeys. The cooccurrence of tool use and imitation has a snowball effect, and it is conspicuously absent in nonhumans.

The consequences of this amplification process in human evolution have not received adequate consideration, perhaps because of the widespread idea that human and nonhuman primates are all good imitators. Many people have argued that the intricacies of social behavior have been the most powerful factor supporting the evolution of intelligence in hominids (e.g., Byrne & Whiten, 1988). We suggest that technical intelligence and social intelligence are not independent of one another in the evolution of hominids (Parker & Gibson, 1979). Social learning, and particularly imitation, can link them.

## Acknowledgments

We are grateful to H. Kummer, F. de Waal, C. Castelfranchi, P. Renzi, E. Alleva, Paolo Barrosso, G. Jervis, and G. Giannoli for commenting on early drafts of this manuscript. We also thank the editors for their support. Research by the authors reported in this chapter was supported by grant 8503603 from the National Science Foundation and grant MHR01-41543 from the Public Health Service to D. Fragaszy and by a bilateral Italian/American grant to E. Visalberghi from the Consiglio Nazionale delle Ricerche. Preparation of the manuscript was supported by a Research Scientist Development Award from the Public Health Service to D. Fragaszy. This chapter was completed while the authors were visiting at the MRC Unit on the Development and Integration of Behaviour, Subdepartment of Animal Behaviour, University of Cambridge, Great Britain. We thank Robert Hinde, Barry Keverne, and other members of the subdepartment for their hospitality.

## Note

1. Numerous operant studies of observational learning using observer/model pairs have been published (e.g., Darby & Riopelle, 1959). This literature deserves to be revisited in light of current conceptions of social learning. Space does not permit adequate treatment of the topic here.

## References

Adams-Curtis, L. E. (1987). Social context of manipulative behavior in *Cebus apella*. *American Journal of Primatology*, *12*, 325.

Anderson, J. R. (in press). Responses of capuchin monkeys (*Cebus apella*) to different conditions of mirror stimulation. *Primates*.

Antinucci, F., Spinozzi, G., Visalberghi, E., & Volterra, V. (1982). Cognitive development in a Japanese macaque (*Macaca fuscata*). *Annali dell'Istituto Superiore di Sanita*, *18*, 177–184.

Antinucci, F., & Visalberghi, E. (1986). Tool-use in *Cebus apella*: A case study. *International Journal of Primatology*, 7, 349–361.

Bandura, A. (1986). *Social foundations of thought and action: A social cognitive theory*. Englewood Cliffs, NJ: Prentice-Hall.

Beck, B. B. (1972). Tool use in captive hamadryas baboons. *Primates*, *13*, 276–296.

Beck, B. B. (1973a). Observation learning of tool use by captive Guinea baboons (*Papio papio*). *American Journal of Physical Anthropology*, *38*, 579–582.

Beck, B. B. (1973b). Cooperative tool use by captive hamadryas baboons. *Science*, *182*, 594–597.

Beck, B. B. (1976). Tool use by captive pigtailed monkeys. *Primates*, *17*, 301–310.

Beck, B. B. (1980). *Animal tool behavior: The use and manufacture of tools by animals*. New York: Garland Press.

Box, H. O. (1984). *Primate behaviour and social ecology*. London: Chapman & Hall.

Bruner, J. S. (1972). Nature and uses of immaturity. *American Psychologist*, *27*, 687–708.

Byrne, R., & Whiten, A. (Eds.). (1988). *Machiavellian intelligence. Social expertise and the evolution of intellect in monkeys, apes, and humans*. Oxford: Clarendon Press.

Chevalier-Skolnikoff, S. (1977). A Piagetian model for describing and comparing socialization in monkey, ape, and human infants. In S. Chevalier-Skolnikoff & F. E. Poirier (Eds.), *Primate biosocial development: Biological, social, and ecological determinants* (pp. 159–188). New York: Garland Press.

Clark, A. B., & Ehlinger, T. J. (1987). Pattern and adaptation in individual behavioral differences. In P. P. G. Bateson & P. H. Klopfer (Eds.), *Perspectives in ethology* (pp. 1–48). New York: Plenum.

Clayton, D. A. (1978). Socially facilitated behavior. *Quarterly Review of Biology*, *53*, 373–391.

D'Amato, M., & Colombo, M. (1988). Representation of serial order in monkeys (*Cebus apella*). *Journal of Experimental Psychology*, *14*, 131–139.

D'Amato, M., Salmon, D., Loukas, E., & Tomie, A. (1985). Symmetry and transitivity of conditional relations in monkeys (*Cebus apella*) and pigeons (*Columba livia*). *Journal of the Experimental Analysis of Behavior*, *44*, 35–47.

Darby, C., & Riopelle, A. (1959). Observational learning in the rhesus macaque. *Journal of Comparative and Physiological Psychology*, *52*, 94–98.

Davis, J. M. (1973). Imitation: A review and critique. In P. P. G. Bateson & P. H. Klopfer (Eds.), *Perspectives in ethology* (pp. 43–72). New York: Plenum.

Dawson, B. V., & Foss, B. M. (1965). Observational learning in budgerigars. *Animal Behaviour*, *13*, 470–474.

de Waal, F. B. M. (1982). *Chimpanzee politics*. London: Cape.

de Waal, F. B. M. (in press). Behavioral contrasts between the two *Pan* species with special attention to tension regulation among captive bonobos. In P. Heltne (Ed.), *Understanding chimpanzees*. Chicago Academy of Sciences.

Diamond, J. M. (1987). Learned specialization of birds. *Nature*, *330*, 16–17.

Fisher, J., & Hinde, R. A. (1949). The opening of milk bottles by birds. *British Birds*, *42*, 347–357.

Fragaszy, D. M., & Visalberghi, E. (1989). Social influences on the acquisition and use of tools in tufted capuchin monkeys (*Cebus apella*). *Journal of Comparative Psychology*, *103*, 159–170.

Fragaszy, D. M., & Visalberghi, E. (1990). Social processes affecting the appearance of innovative behaviors in capuchin monkeys. *Folia Primatologica*, *54* (3–4).

Galef, B. G., Jr. (1976). Social transmission of acquired behavior: A discussion of tradition and social learning in vertebrates. In J. S. Rosenblatt, R. A. Hinde, E. Shaw, & C. Beer (Eds.), *Advances in the study of behavior* (pp. 70–100). New York: Academic Press.

Galef, B. G. (1988). Imitation in animals: History, definitions, and interpretation of data from the psychological laboratory. In T. Zentall & B. G. Galef (Eds.), *Psychological and biological perspectives* (pp. 1–28). Hillsdale, NJ: Erlbaum.

Galef, B. G. (in press). Tradition in animals: Field observations and laboratory analyses. In M. Bekoff & D. Jamieson (Eds.), *Methods, inference, interpretation and explanation in the study of behavior*. Boulder: Westview Press.

Galef, B. G., Manzig, L. A., & Field, R. M. (1986). Imitation learning in budgerigars: Dawson and Foss (1965) revisited. *Behavioral Processes*, *13*, 191–202.

Gardner, R. A., & Gardner, B. T. (1969). A standardized system of gestures provides a means of two-way communication with a chimpanzee. *Science*, *165*, 664–672.

Gibson, K. R. (1981). Comparative neuroontogeny, its implications for the development of human intelligence. In G. Butterworth (Ed.), *Infancy and epistemology* (pp. 52–82). Brighton, England: Harvester Press.

Giraldeau, L., & Lefebvre, L. (1987). Scrounging prevents cultural transmission of food-finding behavior in pigeons. *Animal Behaviour*, *35*, 387–394.

Goodall, J. (1986). *The chimpanzees of Gombe*. Cambridge, MA: Harvard University Press.

Greenfield, P. M. (1984). Perceived variability and symbol use: A common language–cognition interface in children and chimpanzees (*Pan troglodytes*). *Journal of Comparative Psychology*, *98*, 201–218.

Hall, K. R. L. (1963). Observational learning in monkeys and apes. *British Journal of Psychology*, *54*, 201–226.

Hannah, A., & McGrew, W. (1987). Chimpanzees using stones to crack open oil palm nuts in Liberia. *Primates*, *28*, 31–46.

Hauser, M. D. (1988). Invention and social transmission: New data from wild vervet monkeys. In R. Byrne & A. Whiten (Eds.), *Machiavellian intelligence* (pp. 327–343). Oxford: Clarendon Press.

Hayes, K. C., & Hayes, C. (1952). Imitation in a home-raised chimpanzee. *Journal of Comparative and Physiological Psychology*, *45*, 450–459.

Hinde, R. A., & Fisher, J. (1972). Some comments on the republication of two papers on the opening of milk bottles by birds. In P. H. Klopfer & J. P. Hailman (Eds.), *Function and evolution of behavior* (pp. 377–378). Reading, MA: Addison-Wesley.

Holloway, R. L. (1983). Human brain evolution: A search for units, models and synthesis. *Canadian Journal of Anthropology*, *3*, 215–230.

Itani, J., & Nishimura, A. (1973). The study of infrahuman culture in Japan. In E. W. Menzel (Ed.), *Precultural primate behavior* (pp. 26–50). Basel: Karger.

Kawai, M. (1965). Newly-acquired pre-cultural behavior of the natural troop of Japanese monkeys on Koshima Islet. *Primates, G*, 1–30.

Kellogg, W. N., & Kellogg, L. A. (1933). *The ape and the child*. New York: Whittlesey House.

Köhler, W. (1976). *The mentality of apes*. New York: Liveright. (Original work published 1925)

Kummer, H., & Goodall, J. (1985). Conditions for innovative behaviour in primates. *Philosophical Transactions of the Royal Society, London, B, 308*, 203–214.

Lefebvre, L., & Palameta, B. (1988). Mechanisms, ecology, and population diffusion of socially learned, food-finding behavior in feral pigeons. In T. R. Zentall & B. G. Galef (Eds.), *Social learning, psychological and biological perspectives* (pp. 141–164). Hillsdale, NJ: Erlbaum.

McGrew, W. C. (1977). Socialization and object manipulation by wild chimpanzees. In S. Chevalier-Skolnikoff & F. Poirier (Eds.), *Primate biosocial development* (pp. 261–288). New York: Garland Press.

McGrew, W. C., & Tutin, C. E. G. (1978). Evidence for a social custom in wild chimpanzees? *Man, 13*, 234–251.

Meltzoff, A. N. (1988a). *Homo imitans*. In T. R. Zentall & B. G. Galef (Eds.), *Social learning: Psychological and biological perspectives* (pp. 319–342). Hillsdale, NJ: Erlbaum.

Meltzoff, A. N. (1988b). Infant imitation after a 1-week delay: Long-term memory for novel acts and multiple stimuli. *Developmental Psychology, 24*, 470–476.

Meltzoff, A. N., & Moore, K. M. (1983). The origins of imitation in infancy: Paradigm, phenomena, and theories. In L. P. Lipsett (Ed.), *Advances in infancy research* (Vol. 2, pp. 265–301). Norwood, NJ: Ablex.

Mignault, C. (1985). Transition between sensorimotor and symbolic activities in nursery-reared chimpanzees (*Pan troglodytes*). *Journal of Human Evolution, 14*, 747–758.

Mitchell, R. W. (1987). A comparative-developmental approach to understanding imitation. In P. P. G. Klopfer & P. H. Bateson (Eds.), *Perspectives in ethology* (Vol. 7, pp. 183–215). New York: Plenum.

Natale, F., & Antinucci, F. (1989). Stage 6 object-concept and representation. In F. Antinucci (Ed.), *Cognitive structures and development of nonhuman primates* (pp. 97–112). Hillsdale, NJ: Erlbaum.

Nishida, T. (1986). Local traditions and cultural transmission. In B. B. Smuts, D. L. Cheney, R. M. Seyfarth, R. W. Wrangham, & T. T. Struhsaker (Eds.), *Primate societies* (pp. 462–474). University of Chicago Press.

Oakley, K. (1959). *Man the tool-maker*. University of Chicago Press.

Palameta, B., & Lefebvre, L. (1987). *Culturally transmitted foraging behavior in pigeons: Learning mechanisms and ecological correlates*. Paper presented at the International Ethological Congress, Madison, WI.

Parker, S. (1977). Piaget's sensorimotor series in an infant macaque: A model for comparing unstereotyped behavior and intelligence in human and nonhuman primates. In S. Chevalier-Skolnikoff & F. E. Poirier (Eds.), *Primate biosocial development* (pp. 43–112). New York: Garland Press.

Parker, S., & Gibson, K. R. (1977). Object manipulation, tool use, and sensorimotor intelligence in cebus monkeys and great apes. *Journal of Human Evolution, 6*, 623–641.

Parker, S., & Gibson, K. R. (1979). A developmental model for the evolution of language and intelligence in early hominids. *Behavioral and Brain Sciences, 2*, 367–408.

Perrett, D. I., Mistlin, A. J., Harries, M. H., & Chitty, A. J. (in press). Understanding the visual appearance and consequences of hand actions. In M. A. Goodale (Ed.), *Vision and action: The control of grasping*. Norwood, NJ: Ablex.

Piaget, J. (1954). *The construction of reality in the child*. New York: Ballantine Books.

Piaget, J. (1962). *Play, dreams and imitation in childhood*. New York: Norton.

Ridley, M. (1986). *Animal behaviour, a concise introduction*. Oxford: Blackwell Scientific Publications.

Robinson, J. G. (1986). Seasonal variation in use of time and space by wedge-capped capuchin monkey, *Cebus olivaceus*: Implications for foraging theory. *Smithsonian Contributions to Zoology*, No. 431.

Rogoff, B., & Wertsch, J. (1984). *Children's learning in the "zone of proximal development."* San Francisco: Jossey-Bass.

Romanes, G. J. (1884). *Mental evolution in animals*. New York: AMS Press.

Rosenthal, R., & Zimmerman, T. (1978). *Cognition and social learning*. New York: Academic Press.

Scherrer, E. (1982). Pre-evolutionary conceptions of imitation. In G. Eckardt, W. G. Bringmann, & L. Sprung (Eds.), *Contributions to a history of developmental psychology* (pp. 27–53). New York: Mouton.

Schonholzer, L. (1958). *Beobachtungen über das Trinkverhalten von Zootieren*. Unpublished doctoral dissertation, University of Zurich.

Sherry, D. F., & Galef, B. G. (1984). Cultural transmission without imitation: Milk bottle opening by birds. *Animal Behaviour*, *32*, 937–938.

Sumita, K., Kitahara-Frisch, J., & Norikoshi, K. (1985). The acquisition of stone-tool use in captive chimpanzees. *Primates*, *26*, 168–181.

Terrace, H. S., Petitto, R. J., Sanders, R. J., & Bever, T. J. (1979). Can an ape create a sentence? *Science*, *206*, 891–900.

Thomas, R. K. (1987). Overcoming contextual variables, negative results, and Macphail's null hypothesis. *Behavioral and Brain Sciences*, *10*, 680–681.

Thorndike, E. L. (1911). *Animal intelligence*. New York: Macmillan.

Thorpe, W. H. (1956). *Learning and instinct in animals*. London: Methuen.

Tomasello, M., Davis-Dasilva, M., Camak, L., & Bard, K. (1987). Observational learning of tool-use by young chimpanzees. *Journal of Human Evolution*, *2*, 175–183.

van Lawick-Goodall, J., van Lawick, H., & Packer, C. (1973). Tool-use in free-living baboons in the Gombe National Park, Tanzania. *Nature*, *241*, 212–213.

Visalberghi, E. (1987). The acquisition of nut-cracking behavior by 2 capuchin monkeys (*Cebus apella*). *Folia Primatologica*, *49*, 168–171.

Visalberghi, E. (1988). Responsiveness to objects in two social groups of tufted capuchin monkeys (*Cebus apella*). *American Journal of Primatology*, *18*, 349–360.

Visalberghi, E. (1990). Tool use in *Cebus*. *Folia Primatologica*, *54* (3–4).

Visalberghi, E., Riviello, M. C., & Blasetti, A. (1988). Mirror responses in tufted capuchin monkeys (*Cebus apella*). *Monitore Zoologico Italiano*, *22*, 555–556. (Abstract)

Visalberghi, E., & Trinca, L. (1989). Tool use in capuchin monkeys, or distinguish between performing and understanding. *Primates*, *30*, 511–521.

Washburn, S. L. (1960). Tools and human evolution. *Scientific American*, *203*, 62–75.

Watson, J. B. (1908). Imitation in monkeys. *Psychological Bulletin*, *5*(6), 169–178.

Westergaard, G. C. (1988). Lion-tailed macaques (*Macaca silenus*) manufacture and use tools. *Journal of Comparative Psychology*, *102*, 152–159.

Westergaard, G. C., & Fragaszy, D. M. (1987). The manufacture and use of tools by capuchin monkeys (*Cebus apella*). *Journal of Comparative Psychology*, *101*, 159–168.

Yando, R., Seitz, V., & Zigler, E. (1978). *Imitation: A developmental perspective*. Hillsdale, NJ: Erlbaum.

Zentall, T. R., & Galef, B. G. (1988). *Social learning: Psychological and biological perspectives*. Hillsdale, NJ: Erlbaum.

# 10 Cultural transmission in the tool use and communicatory signaling of chimpanzees?

*Michael Tomasello*

One of the most exciting findings to emerge from recent observations of free-ranging chimpanzees in equatorial Africa is that different populations behave differently: They eat different foods, use different tools, and communicate in different ways. These differences seem to persist across generations, and their geographical distribution (neighbors often differ more than do populations living a continent apart) makes it unlikely that they are due to genetic factors. Some researchers have therefore taken to speaking of chimpanzee "culture" (e.g., Goodall, 1973, 1986; McGrew, 1983; McGrew & Tutin, 1978; McGrew, Tutin, & Baldwin, 1979; Nishida, 1980; Nishida, Wrangham, Goodall, & Uehara, 1983; Sugiyama, 1985).

It is clear, however, that not all population differences are cultural in origin. Although the term *culture* is not easy to define, either when applied to human societies (e.g., White, 1959) or when applied to animal populations (e.g., Booner, 1980; Mainardi, 1980; Washburn & Benedict, 1979), at the very least culture would seem to require some form of social learning. Thus, no one would claim cultural transmission if all members of a population learned a particular behavior only because they each had been exposed to the same set of contingencies from the physical environment. For example, naturalistic observations have revealed that one population of wild rats dives into the Po River (in Italy) for mollusks, whereas another population of rats living on that same river does not. However, on the basis of a series of laboratory studies, Galef (1982) has concluded that the explanation for this is not social learning or cultural transmission. The explanation is that the river is low at some times of the year in the range of the first population, which encourages members of that population to dive down to the mud for food; they continue to dive as the river rises back to its normal level. Because the river does not change in this way in the range of the other population, this process of "environmental shaping" does not take place. The important point for current purposes is that such population differences in behavior may in some cases result solely from individual learning – shaped, as it were, by the different physical environments of the different populations. In these cases there is no cultural transmission, because there is no social learning.

Cultural transmission clearly is the appropriate concept, on the other hand, when adults intentionally teach youngsters or when youngsters imitate adults.

A variety of human behaviors seem to rely on just such explictly social learning processes, and in fact many human behavioral acquisitions simply would not be possible without them (Vygotsky, 1978). The human case provides the prototype for cultural transmission of course, because there is actual transmission of behavior – its topography is reproduced – from one generation to the next. And to anticipate a point to be argued later in this chapter, it is this reproduction or transmission of behavior that distinguishes cultural learning from other types of social learning more broadly defined.

The problem in the current context is that in many cases chimpanzee "culture" seems to involve a mixture of individual, social, and perhaps even cultural learning processes, and this makes straightforward interpretations difficult. Two very important cases in point are chimpanzee tool use and communicatory signaling. These behaviors are of special interest to students of human evolution because, in some theories at least, both were integral parts of the culturally transmitted behavioral complex involved in the process of hominidization (e.g., Isaac, 1983). Moreover, these two behaviors seem to involve analogous, if not homologous, cognitive structures: In both cases, the organism actively uses an outside element (tool or social signal) as an instrument to attain an otherwise unattainable goal (Bates, 1979; Hallinan, 1984). Some researchers have claimed cultural transmission for chimpanzees in both of these domains, but, as we shall see, the story is much more complex than that. In each of the two cases a variety of individual and social learning processes may underlie acquisition; moreover, these learning processes are different in the two cases because the outside element in the case of tool use is a physical object, whereas the outside element in the case of communication is itself a social behavior.

In this chapter, I review what is known about the tool use and communicatory signaling of common chimpanzees (*Pan troglodytes*), with a special focus on learning processes. In each case, I first review naturalistic observations in the wild. These are necessary for establishing the existence of particular behavioral traditions, but in most cases they are not sufficient to distinguish sensitively among the different types of learning that may be involved. I then review, for each case, observations of chimpanzees in captive or experimental settings – which in many cases allow for more detailed conclusions about learning processes. From these analyses, I derive a typology of individual, social, and cultural learning processes that includes both sensorimotor behaviors, such as tool use, and social-conventional behaviors, such as communicatory signaling.

## Tool use

### *Chimpanzees in the wild*

Most populations of wild chimpanzees exhibit some form of tool use,[1] and different types of tool use have been found associated with different populations. In some cases, one or some populations engage in a particular tool-use activity, whereas others do not; in other cases, there is variation in the way a particular tool is used. Only a few tool-use practices have been sufficiently studied to determine the extent of their use within and between populations – which is crucial for making inferences about potential transmission processes. The most complete data are available for those uses of tools designed to aid in the acquisition of food (and liquids).[2] In what follows, I relate all uses of tools for obtaining food of which I am aware, but focus on the use of tools in (1) capturing insects, (2) sponging liquids, and (3) cracking nuts.

*Capturing insects.* KASAKELA. The most extensively studied chimpanzee group (subspecies *Pan troglodytes schweinfurthii*) is the Kasakela community in the Gombe National Park of Tanzania (Goodall, 1964, 1968a, 1970, 1973, 1986). This group of individuals (ranging in size from 50 to 100 individuals over the years) has been studied at close range (through provisioning) for more than two decades. The chimpanzees at Gombe obtain much of their nourishment from termites, which are obtained almost exclusively through the use of tools. These termites (genus *Macrotermes*) build large mounds that are rock-hard during the summer dry season, but become less so in the wet season. The chimpanzees capture these insects, mostly in the wet season, by using twigs, grass, vines, bark, and fronds to "fish" down the passageways of the mounds. When the termites bite the foliage in a defensive reaction, the chimpanzees extract and eat them.

To be successful at termite fishing, a chimpanzee must

1. locate the passageways and scratch a hole to attain access,
2. choose a tool that is appropriately firm, yet supple,
3. modify the tool so that it is free of leaves and twigs,
4. navigate the tool into the winding passageways to a sufficient depth,
5. wiggle/vibrate the tool so that the termites are attracted to bite it, and
6. extract the tool without dislodging the termites (Teleki, 1974).

Over repeated trials, the end of the probe typically becomes frayed, at which time fishers bite off the frayed end and continue using the tool. When further repetitions result in a tool that is too short, they simply discard it and begin with another (they often bring several to a site before commencing). During their less frequent fishing bouts of the dry season, when the termites are farther below the surface, they use much longer sticks, and the technique becomes more difficult.

Young chimpanzees typically observe their mothers fishing, and play with their discarded tools, when they are 2–5 years old. Youngsters acquire adult-like competence at 5–6 years of age. The Gombe chimpanzees have been studied closely enough that we may say with impunity that virtually all members of the group fish for termites (Goodall, 1986).

Another insect-obtaining skill of the Gombe chimpanzees is called ant dipping (McGrew, 1974). Driver ants, which can inflict very painful bites, reside in underground nests, and each nest has a single opening at ground level. To capture these ants, the chimpanzee must

1. choose a long and fairly rigid stick,
2. poke it into the hole and stir the ants to action,
3. monitor their attack up the stick,
4. draw the stick out quickly when they are at the three-quarter mark,
5. "pull-through" the stick with the other hand, crumpling the ants into a ball and into the mouth in a single motion, and
6. chew ferociously so as not to be bitten inside the mouth.

Also, the chimpanzees must monitor the ants around their feet, and often they begin dipping or end up dipping from some sort of perch.

Ant dipping is quite widespread among members of the Gombe group, although they engage in it much less frequently than termite fishing. The ontogeny of ant dipping is somewhat delayed relative to termite fishing. Because of the fear of pain, infants keep their distance during the 2–5-year age period, and they therefore have less opportunity to practice, play, or observe. At 5–6 years old the juveniles begin attempting to dip, but they are not very skilled, and it is only a few years later that they attain adult competence (McGrew, 1977).

MAHALE K. Several other chimpanzee populations are located near Gombe, in the Mahale Mountains in the western part of Tanzania. Nishida and his colleagues at the Kasoje Chimpanzee Research Station have been studying these groups (of the same subspecies) for almost as long as Goodall has been studying the Gombe chimpanzees. The K group (20–30 members) has become habituated to humans and so may be studied at close range. The major insect-capturing tool-use technique of this group is fishing for wood-boring ants. This behavior has been extensively observed and documented (Nishida & Hiraiwa, 1982), and it has been established as a highly frequent and widespread behavior. Wood-boring ants (genus *Camponotus*) live inside trees with entranceways in the trunks or depressions where branches have broken off. The chimpanzees fish for these ants in the Gombe termite-fishing manner; that is, they gather and modify sticks, vines, leaves, and so forth, and push them into the ants' habitat. Because these ants can bite, however, the fisher must coordinate avoidance behavior with fishing. This often leads to some impressive acrobatics when the entranceways are in relatively inaccessible positions in a tree.

Members of the K group fish for ants year round, and virtually all of the individuals in the group have been observed to engage in this type of tool use. The ontogeny of ant fishing is somewhat delayed relative to termite fishing at Gombe, though perhaps not so delayed as ant dipping. The Gombe chimpanzees do not engage in this activity, despite the presence of this species of ant within their range. The Mahale K chimpanzees do not dip for driver ants, although driver ants are present within their range.

The K group at Mahale also fishes for termites, but not for the same genus of termite as the chimpanzees at Gombe – quite simply because these do not occur in their range – and not nearly as frequently as they fish for ants. Based mostly upon circumstantial evidence – finding tools on mounds the group members are known to frequent – Uehara (1982) documented exploitation of the termite *Pseudacanthotermes spiniger*. Throughout the year individuals break off and eat small pieces of the termite mound. This sometimes results in the ingestion of termites and helps them to determine the depth at which the termites are currently residing. At the beginning of the wet season, when the termites are at a moderate depth below the surface, the chimpanzees fish in the manner of the Gombe group. Near the end of the wet season, when the termites are mostly in the upper portions of the mound and the mound is soft, the chimpanzees simply destroy the mound with their hands and capture the termites manually. The distribution of these behaviors among group members is unknown.

MAHALE B. The nearby Mahale B chimpanzees, who have not been so closely observed because they are not habituated to humans, spend very little time fishing for ants. Nor do they eat the genus of termite eaten by the K group. Instead, they employ the termite-fishing technique of the Gombe chimpanzees on the same species of termite (McGrew & Collins, 1985). The dispersion of these behaviors among population members is not known, because most of the evidence for the behavior of these unhabituated animals is in the form of tools found on termite mounds within their range.

KASAKATI BASIN. Suzuki (1966) on several occasions observed several chimpanzees fishing for termites in the Gombe fashion in the Kasakati Basin of Tanzania.

WESTERN AFRICA. The populations of western Africa have been less extensively studied than those of the east. In 1974, Sabater Pi reported that in the Okorobiko Mountains, Rio Muni, chimpanzees (*Pan t. troglodytes*) capture termites of the same genus as those at Gombe, but in a different way. They use fairly large sticks to perforate the termite mound, after which they capture the termites with their hands. Because of the small number of observations made (eight direct observations), the extent of this behavior in the population is unknown. Sabater Pi found no evidence of this type of tool use 50 km away at Mount Alen, although he did at two other sites nearby (Jones & Sabater Pi, 1969). Sugiyama (1985), on the other hand, found tools of this larger type apparently used by a chimpanzee group in southwest Cameroon less than 100 km from Okorobiko (with perhaps a slight modification involving the fraying of the stick's end). McGrew et al. (1979) found evidence that a chimpanzee group (*Pan t. verus*) at Mount Assirik, Senegal (on the far west coast), fished for termites in the Gombe manner (with slight variations in the materials used and modifications made, especially their tendency to peel the bark off their tools). In the only documented case of its kind, McGrew and Rogers (1983) found evidence that a group in Gabon, West Africa, (*Pan t. troglodytes*) used both the Gombe and the Rio Muni techniques. Of the other western populations, neither the group at the Tai Forest (Ivory Coast) nor the group at Bossou (Guinea) has provided any evidence that they obtain termites in either of these ways, although *Macrotermes* are present in the range of the latter group.

*Sponging liquids.* When a chimpanzee at Gombe cannot reach some water that has collected in a tree hollow or the like, the chimpanzee often will pick some leaves, chew them into a "sponge," dip this into the water, and extract the sponge and drink the water. Goodall (1986) reported that in one 3-year period, 14 observations of this behavior were recorded, but it is unclear how many individuals were involved. Overall, it does not seem to be a frequent behavior. Goodall (1964, 1973) reported that leaf sponging occurred in its mature form at the surprisingly early age of 3–3.5 years, after a year or two of practice and observation.

Reports of this behavior in other populations are rare. Sugiyama and Koman (1979) reported one observation of a chimpanzee at Bossou dipping an unmodified leaf into water and then licking off the drops, and Brewer (1978) reported that several rehabilitants in Gambia apparently showed Gombe-like leaf sponging, with modification of the leaves.

*Cracking nuts.* Evidence that chimpanzees use tools to break open nuts has been found in a variety of western populations (*Pan t. verus*) in Liberia (Beatty, 1951; Anderson, Williamson, & Carter, 1983), Guinea (Sugiyama & Koman, 1979), Ivory Coast (Struhsaker & Hunkeler, 1971; Boesch & Boesch, 1983), and Sierra Leone (Whitesides, 1985). Only two of these reports were based on multiple direct sightings (as opposed to indirect sightings of tools and broken nuts found after the fact). In Bossou (Guinea), Sugiyama and Koman (1979) reported finding 29 cracking sites under oil-palm trees. These contained nut shells and almost always two stones: one for use as a tool and one for use as an anvil. Three direct sightings were made of chimpanzees placing nuts on an anvil (usually with a depression from previous use) and smashing them open with a tool.

Since 1979, Boesch and Boesch (1981, 1983, 1984) have made extensive observations of the chimpanzees of the Tai Forest (Ivory Coast) and have found a much more complex picture. Tai chimpanzees do not crack oil-palm nuts. Instead, they crack five other types of nut, most notably cola nuts and panda nuts, using three main techniques: cracking cola nuts on a ground anvil, cracking cola nuts in a tree, and cracking panda nuts on a ground anvil. In the first technique, the chimpanzee gathers some nuts from the ground and brings them to a place where they may be cracked – almost always a surface root that has been used previously for such a purpose. The chimpanzee brings a stick/club to the site, or else uses one already at the site. In the second method, the chimpanzee climbs the cola tree, with a club already in hand, and proceeds to pick nuts from the tree and crack them on a branch (sometimes steadying it with one hand and clubbing it with the other). This technique is quite complex, as the chimpanzee must hold some uncracked nuts, balance itself in the tree, keep the cracked nut from falling to the ground, and eat the nut (while retaining the club). Finally, panda nuts, which are too hard to be easily cracked with wooden clubs, are cracked almost exclusively with stones – mostly on root anvils. Because appropriate stones are scarce in the Tai Forest, chimpanzees often transport them hundreds of meters.

Although virtually all Tai population members engage in the behavior, nut cracking emerges relatively late in ontogeny. Boesch and Boesch (1984) reported that no chimpanzee younger than 5 years old succeeded in cracking a nut, though youngsters did play with nuts and tools, and there was some observation of adults during this period. Full adultlike behavior does not emerge until adolescence (9–10 years), with panda cracking lagging behind cola cracking.

None of the eastern populations of chimpanzees has been observed using tools to crack open fruits or nuts. This is true even though members of the Gombe group, for example, regularly eat oil-palm nuts whole and crack open other types of fruit by smashing it against a hard surface (without the use of a detachable tool of any sort).

### *Ecological interpretations*

Figure 10.1 shows the geographical distribution of the major tool-use techniques employed by wild chimpanzee groups to capture insects, sponge liquids, and crack nuts. (Note that only well-established tool-use traditions are included.[3]) The first thing to note is that genetic differences do not seem to provide a plausible explanation for these findings. Although nut cracking is confined to *Pan t. verus*, Gombe-like termite fishing has been observed in all three subspecies. Within a given subspecies there are, in many cases, very large population differences in tool use – especially among the eastern *Pan t. schweinfurthii* groups, who differ greatly despite living in comparative proximity.

To explain this geographical diversity, some researchers have attempted to examine ecological factors that might differ among populations. Such factors could potentially lead to population differences by providing different learning conditions for the individuals in the various populations, and that effect might be exacerbated by the fact that infants and youngsters accompany their mothers everywhere and are consequently exposed to many of the same learning conditions (cf. Galef, 1976, 1982). Where ecological differences among populations are found, therefore, researchers have remained skeptical about cultural transmission. In the absence of such differences, they have on several occasions invoked cultural transmission processes.

In a general analysis of chimpanzee termiting behaviors, McBeath and

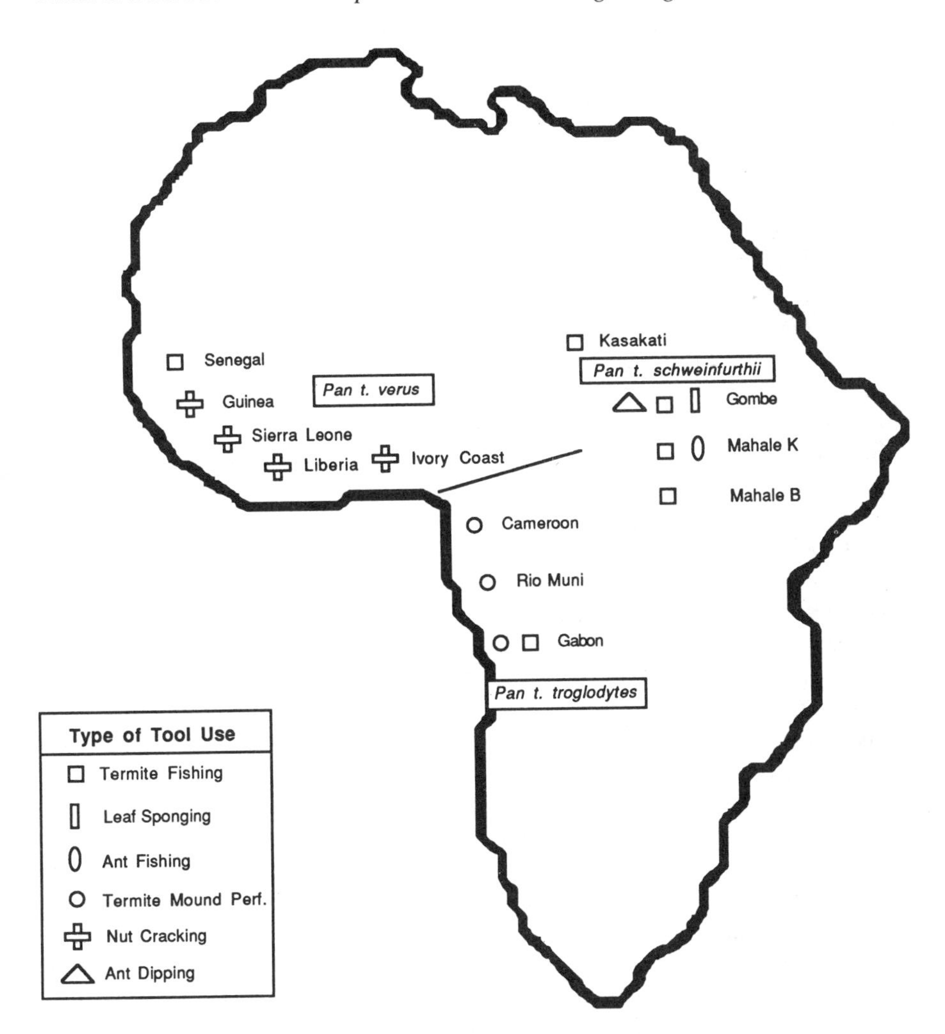


Figure 10.1. Geographic location of major types of chimpanzee tool-use activities.

McGrew (1982) concluded that the availability of suitable tool material was an important determinant of where and how different chimpanzee populations fished for termites. In a similar but more detailed comparison of the Mahale K and the Mahale B groups, Nishida and Uehara (1980) hypothesized that differences in the termite-capturing behaviors of these two groups were ultimately explained by differences of climate. In their view, climatic differences lead to differences in soil and vegetation, which lead to differences in termite availability and behavior and thus differences in the capturing techniques required. In another study of termiting behaviors, McGrew et al. (1979) reported some population differences that seemed to admit ecological interpretations of this sort, as well as some that did not. In comparing termiting techniques in three populations – Gombe, Mount Assirik, and Okorobiko (Rio Muni) – they concluded that the fishing technique of the Gombe and Mount Assirik groups and the perforating technique of the Okorobiko group were local adaptations to the differing demands of the two environments. On the other hand, these researchers could find no ecological reasons for differences between the two fishing groups in the choice of tool materials, for the way the tools were fashioned with respect to bark peeling, or for whether the group used one or both ends of the probe. They consequently attributed these differences to cultural tradition (see Goodall, 1986, for a similar claim).

The problem with inferring cultural tradition from such analyses is that relatively subtle ecological factors may be at work. In fact, it is impossible *in principle* for ecological analyses by themselves to answer definitively questions about learning processes. On the one hand, failure to find ecological differences between groups does not mean there are or were none. Perhaps bark peeling is advantageous for one group but not another, for example, because of some subtlety in the behavior of the particular termites on their range (they reside deeper, they grow larger, they have learned to be wary, their mounds are wetter inside, and so on ad infinitum), and this guides the individual learning of the members of the two groups. Even more subtly, perhaps a drought in the range of one group led to leaf sponging by encouraging the search for water in crevices (some of which also contained leaves), and that ecological condition no longer exists. In case such as these, cultural tradition is a plausible but erroneous conclusion.

On the other hand, the converse of this is also possible; that is, it is also possible that a population difference in behavior may be accompanied by an ecological difference that seems important to us, but in reality is irrelevant. For example, perhaps the group that leaf-sponges has more or better leaves and crevices than one that does not. This may lead researchers to suspect that ecological factors shaped individual learning, but it is nevertheless possible that the group that sponges passes this behavior to their progeny through some form of social or observational learning.

In all, it is clear that simple observation of animals in their natural habitats is not sufficient to determine the ways in which various behaviors may be acquired and transmitted (Galef, 1976, 1982, 1983). This is particularly true of chimpanzee behaviors such as tool use (and most others), for which we lack a history of individual learning. Although naturalistic observation is, and should be, the initial mode of investigating animal behavior, some of the questions rasied in these investigations can be answered only under more controlled conditions.

### *Chimpanzees in captivity*

Two classic studies of captive chimpanzees concluded quite different things about their abilities to learn from observation. Köhler (1927) repeatedly witnessed the complete failure of highly motivated chimpanzee subjects to benefit from observing conspecifics solving sensorimotor problems (e.g., obstacle and tool-use tasks). He argued that imitation is of limited usefulness in learning to solve complex problems, because an animal cannot imitate a solution unless it in some sense already understands the problem. Although he held out the possibility that observing another's solution could help a subject come to a better understanding of a particular problem situation, Köhler concluded that, overall, imitation is "a very rare occurrence [in the animal world] even among chimpanzees" (1927, p. 222). Hayes and Hayes (1952), on the other hand, reported that their human-reared chimpanzee Vicki (during the 1.5–3-year age period) solved several tool-use problems, after observing humans solve them, and that she imitated a variety of other behaviors as well. The discrepancy between these two reports could, of course, be due to any of a number of differences between the studies: Köhler's subjects were wild-born adult chimpanzees, whereas Vicki was reared by humans and tested in her youth; Köhler's subjects observed each other solving problems, whereas Vicki observed adult humans; and the precise problems involved were not identical in the two studies.

More recently, Mignault (1985) reported that young human-raised chimpanzees imitated humans using a variety of conventional objects (e.g., a hairbrush). Neither these observations nor any others involving human-reared chimpanzees (including the Hayes study), however, occurred in a study using experimental controls. This is important, because it is well known that chimpanzees can discover the use of many human objects and tools without the benefit of observation (e.g., Birch, 1945; Schiller, 1957). Without control conditions, therefore, it is not clear how a chimpanzee's performance compares to what it might have done without a demonstration. These considerations lead us to examine more carefully the ways in which one animal may learn to use tools by observing another, and which of these may properly be considered cultural transmission.

In Thorpe's classic formulation (1956) there are two important subtypes within the overall category of social or observational learning.[4] These are distinguished on the basis of whether or not the particular behaviors of one animal are actually reproduced in the behaviors of another. *Local enhancement* involves no reproduction and occurs when the behavior of a conspecific serves to draw an individual's attention to certain aspects of the situation – which then facilitates individual learning. *Imitation*, on the other hand, involves one animal reproducing in its behavior the behavior of a conspecific. Although Thorpe did not address the problem of culture directly, in the current interpretation only imitation may be considered cultural transmission, because only it involves the reproduction of behavior.

Thorpe's distinction is fine as far as it goes. The problem is that it does not do full justice to the complexities of enhancement effects and imitation in the tool-use situation. At least three different aspects of the tool-use situation may become salient ("enhanced") when an organism observes another using a tool: the tool, the goal, and the behavior of the demonstrator. Special attention to any one of these, or some combination of these, may lead to different learning processes. First, a focus on the demonstrator's association with the tool often will attract the observer to the tool, which sometimes leads to individual learning of its function – called by Beck (1980) and others *stimulus enhancement.* For example, when one animal observes another cracking nuts with a rock, it may become motivated to manipulate the rock – whose use it then discovers on its own, by accident, as it were. Second, a focus on the demonstrator's goal may lead the observer to be attracted to and seek to attain the goal. In some cases the animal may do this ignoring the tool, but in other cases both tool and goal become enhanced through observation. The observer may then attempt to "emulate" the demonstrator's behavior, that is, to reproduce the completed goal (e.g., an open nut) by whatever means it may devise, including using the tool in an unspecified manner. Clearly, this is highly intelligent problem solving, but it is not cultural transmission, because the observer is not attempting to reproduce the demonstrator's actual behavior. Wood (1988) calls this social learning process *emulation* and opposes it to the third learning process *impersonation* (the same as Thorpe's true *imitation*), in which the observer attends to and attempts to copy the behavior of the demonstrator.

What this means is that there are at least three different social learning processes that may produce the same result, especially when the behavior involved is familiar and simple. In order to answer questions about cultural transmission, therefore, true imitation (impersonation) must be distinguished from both stimulus enhancement and emulation. What is required is an experimental comparison of the observer's behavior following observation of a demonstration that merely enhances the salience of the tool and/or goal objects and its behavior following a demonstration of the actual use of the tool (Hall, 1963; Meltzoff, 1985). Although such a comparison has, so far as I know, never been made for chimpanzees, several recent studies of chimpanzees acquiring tool-use behaviors in at least partially controlled settings represent a step in this direction.

*Cracking nuts.* Sumita, Kitahara-Frisch, and Norikoshi (1985) trained three chimpanzees to use stone tools to crack nuts. They then released them back into a group of nine other captive individuals. Of these naive individuals, only one subject (a 3-year-old) acquired the new behavior during subsequent testing. Based on a careful analysis of all of this subject's behaviors with the tools and nuts, these authors concluded that she acquired the practice "through goal-directed trial-and-error" (p. 168), that is, without the influence of observation.

Hannah and McGrew (1987), on the other hand, found more substantial evidence for the role of observational learning. Sixteen wild-born chimpanzees, 5–20 years of age, were released from captivity onto a river island in Liberia. On the day of her release, one 9-year-old individual, who had been in captivity since she was 1 year old, began cracking oil-palm nuts with a concrete block on a concrete slab. On that same day, three other individuals attempted to crack nuts in a similar manner, though not in precisely the same manner, and not as successfully or skillfully as the inventor. Within 2 months, 13 of the 16 chimpanzees were cracking nuts. None had been observed to do this in captivity, and none of the 10 chimpanzees who had been released onto the island prior to the inventor had done so previously on the island. It should be noted that several of the individuals developed idiosyncratic methods of cracking the nuts.

*Sponging liquids.* Kitahara-Frisch and Norikoshi (1982) provided a 7-year-old chimpanzee with juice in a milk bottle, along with leafy branches for potential use as a sponge. During the second hour-long session, the subject began spontaneously to use leaves to sponge the liquid. During later sessions, she began to use the branch, ultimately discovering that a chewed end led to greater absorption. These authors argued that the ease of this discovery raises the possibility that "the sponge-making behavior observed in Gombe can most parsimoniously be interpreted as an incidental corollary of a highly variable and potentially meaningful expression of the chimpanzee's behavioral resourcefulness" (p. 41). Although Goodall (1973) reported that infants observed their mothers sponging in the wild, the only reported instances of immediate "imitation" all involved cases in which the infant simply used the leaves that the mother left in the crevice.

*Extending reach.* No laboratory studies have focused on chimpanzee termite fishing, ant fishing, or ant dipping. However, the tool-use behavior that has been most often studied resembles these techniques in that it involves the use of a stick to extend the animal's reach – usually through cage bars for food. Classic studies such as those of Köhler (1927), Birch (1945), Schiller (1957), and Menzel, Davenport, and Rogers (1970) demonstrate convincingly that chimpanzees *can* discover the use of such implements without the benefit of observation. In the only study of observational learning of these tasks, Menzel (1972) found that the use of "ladders" to provide access to otherwise inaccessible places (another type of extending reach) spread very quickly in his seminatural group of eight 6–7-year-olds. Immediately after an individual invented the practice, his two closest companions acquired it, and four of the five remaining individuals acquired it soon thereafter (after 5 years in which the practice had been nonexistent). The rapid acquisition of this behavior by nearly all group members implies that some form of observational learning was at work.

### *An experimental investigation*

Of the four studies that provided captive chimpanzees with the opportunity to learn a tool-use technique by observation of a more skilled conspecific, therefore, two suggested negative conclusions and two suggested positive conclusions about the role of observational learning. Although neither of the more optimistic studies had a control group, both reported rapid acquisition of a behavior by many group members after a long period in which it was absent. And though each of these types of tool use was acquired by some individuals without observation, the most plausible conclusion is that the behavior of the inventor had some effect on novices in both the Menzel (1972) and the Hannah and McGrew (1987) studies.

The nature of this effect is not clear, however. It is possible, perhaps even likely, that the learning of the novices in both cases was facilitated by the propitious arrangement of the elements of the tool-use situation: The nuts, the stone, and the anvil were all left in one place after the inventor was finished; the ladder was left leaning against the fence. This may have been sufficient for the novices to invent the behavior for themselves, as the inventor did. This interpretation is supported by the fact that both studies reported striking individual differences in the precise manner in which subjects performed the behavior, indicating at least some measure of individual learning. It is also possible, of course, that observers emulated the inventor's use of the stone or the cracking of the nut without copying her behavior, and it is also possible that the novices did actually imitate the behaviors of the inventor. In all, it is clear that whereas documenting the spread of a behavior among captive individuals provides valuable information, such studies are simply not designed to distinguish among learning processes at this level of detail.

Tomasello, Davis-Dasilva, Camak and Bard (1987) reported the only study of which I am aware that directly compared the performances of animals who observed a tool-use activity and those who did not. An adult female chimpanzee, the demonstrator, was trained to rake food items (fruits and nuts) into her cage with a metal T-bar. When a food item was in the center of the serving platform, she simply swept or raked the item to within reach. When the food was placed either along the side or against the back of the platform's raised edges, however, this strategy was seldom successful. Instead, she learned in each of these cases a more complex, two-step procedure. When the food was along the side, she pushed the tool beyond the food with one hand and reached with her other hand through the bars, grasped the tool 0.5 m up the shaft, and gently dragged the food along the edge to within reach. When the food was against the back edge, she gently tapped it off the edge, then switched her grip and raked it in.

The subjects were two groups of young chimpanzees, 4–6 and 8–9 years old, none of whom showed any tool-use behaviors in a pretest session. Subjects in the experimental condition were exposed to the adult demonstrator modeling all three of her strategies – for 5 trials prior to testing, and then continually during the next 15 trials of testing (over a 2-day period). Subjects in the control condition were exposed to the same demonstrator in an unoccupied state throughout. For the younger group, when presented with the opportunity to use the tool in the test trials, the experimental subjects ($n = 4$) learned its use (after only a few trials in most cases), whereas control subjects ($n = 3$) did not. (For unknown reasons, perhaps because of rearing history, the older subjects were not interested in the tool and for the most part showed no learning in either condition.) This pattern continued into a posttest in which there was no demonstration for either group. None of the experimental subjects, however, learned either of the demonstrator's more complex, two-step strategies – when the food was against the back or side edge – and this was despite the fact that the subjects were motivated, the demonstration was readily observable throughout their unsuccessful attempts, and the demonstrated behavior did not involve any movements the youngsters were incapable of making.

We interpreted these results as demonstrating that the young chimpanzees in this study benefited from observation in learning to use the tool. All animals were provided with the same tool and goal arranged in the same way, but only experimental subjects learned to use the tool. Because the experimental subjects did not learn to copy the demonstrator's more complex strategies, however, and because the raking strategy was most likely not a novel behavior, we concluded that the form of learning was not imitation. On the other hand, we also concluded that more than simple stimulus enhancement was involved in the learning of the experimental subjects. Further analyses demonstrated that whereas the two groups spent equal amounts of time manipulating the tool, implying that the tool was equally salient for both, control subjects simply played with the tool in an undirected manner and almost never used it in an attempt to get the food. We thus concluded that observation of the demonstrator served to draw the attention of the experimental subjects not just to the tool-object but also to the results that the demonstrator had produced. They then used their natural capacity for tool-use cognition (recall that they can learn this task on their own in some circumstances) to put these together in their subsequent acquisition of the skill. The youngsters were emulating the adult's use of the tool.

### *Cultural transmission of tool use?*

Close scrutiny of the laboratory studies thus shows that the case for chimpanzee imitation of tool use is not strong. None of the studies purporting to show its existence used the controls necessary to rule out other forms of social learning, and in the one experimental study in which there was ample opportunity to imitate, young chimpanzees only emulated the behavior of the demonstrator. Moreover, all three of the tool-use practices reviewed here have been invented by chimpanzees individually in the laboratory, without the benefit of observation, and both in the wild and in captivity (in studies where observation was possible and in those where it was not) there were distinct differences in the precise tool-use techniques used by individuals. The most plausible explanation of the laboratory findings, therefore, is that chimpanzees come to the task with a cognitive capacity for tool use and a social capacity for emulating the behavior of conspecifics. The opportunity to observe another animal using a tool thus helps them learn more about the task and what is required for success, but they do not imitate the other animal's behavior per se.

Given this background, the complex geographical patterns of tool use observed in wild chimpanzees may be explained most plausibly as a combination of individual learning and emulation, along with ecological factors and some cases of migration among groups. On this hypothesis, many of the tool-use activities practiced by particular groups must have first entered that group through spontaneous inventions by one or more individuals, as often happens in the laboratory. This would seem especially plausible in cases such as ant dipping, leaf sponging, and ant fishing that exist in one and only one population (see Burton & Bick, 1972, on founder effects). For other tools, original entry into a group may have taken place by the migration of chimpanzees from another group. This migration – along with at least one spontaneous invention somewhere – might help to explain, for example, that only groups within a certain area along the central western coast of Africa use tools to crack nuts and that only groups is the southwestern part of the continent use the mound-perforation technique for capturing termites. It is also possible that an animal from one group might observe and emulate those of another group – or even humans (Kortlandt, 1986) – without any animals actually migrating. Ecological explanations are, of course, also possible.

In cases such as the similar termite-fishing activities of the Gombe, Mahale B, Gabon, and Mount Assirik chimpanzees – living in entirely different parts of Africa – migration would seem to be a less plausible explanation. It is possible that individuals migrated across the continent, or that some "original" population had a particular behavior and after a fractionation only some of the new populations retained it, perhaps because the appropriate ecological conditions were differentially present (Parker & Gibson, 1979). Nevertheless, although all of the relevant ecological analyses have not been done, it would seem more likely in these cases that the similarities in tool-use practices among geographically isolated populations are due to parallel inventions of similar foraging strategies under similar ecological circumstances.

Once a behavior has entered a population, it is not likely that horizontal transmission among adults is of major importance in its dissemination. For one thing, older adult chimpanzees are not as likely as younger ones to adopt behavioral innovations (Tomasello et al., 1987; cf. Kawamura, 1959, on macaques), but more important, adult members of chimpanzee groups come into contact only irregularly, making it unlikely that they could observe each other in many of their tool-use activities. In contrast to their simplified counterparts in the laboratory, many of the tool-use activities in the wild would seem to be so complex – requiring the animal to locate the food and modify, transport, and manipulate the tool – that only one or a few observations of only one or a few aspects of the practice would not seem to be sufficient for learning. For all of these reasons, then, much of the burden for disseminating tool-use behaviors among the members of a chimpanzee group must fall to adult–child vertical transmission.

In each of the three forms of tool use reviewed, several years were required for youngsters to attain adultlike competence, usually during the 2–6-year age range, with more being required for the more difficult of these. During these years, youngsters mature motorically and socially, they develop cognitively, they play and practice their own tool-use activities, and they frequently observe their mothers using tools as well. The benefit a young chimpanzee derives from observation on any given occasion is dependent on its developmental level, as determined by all of these factors. Thus, for instance, a 2-year-old chimpanzee who knows nothing about termite fishing may observe its mother fishing and thus be attracted to the tool. After some play with sticks and some further cognitive developments (months or years later), it may notice that sometimes termites are on the stick when it is first picked up and that these may be eaten. At some later point during childhood, the youngster may observe and seek to emulate its mother's extraction of termites from within the mould's passageways. At each stage of this developmental process, the youngster's attention is drawn to the tool-use task, but what it takes away from the observation depends on what aspects of the problem situation it is capable of assimilating to its current knowledge (cf. Goodson & Greenfield, 1975, who found precisely this effect in preschool children's observational learning of a complex task). Much can be learned in this way, both individually (aided by the mother's propitious arrangement of tool and goal) and by emulation of the mother's result. It is certainly plausible that much, if not most, chimpanzee tool-use behavior is acquired in this way.

It would thus seem that while the social environment clearly exerts a powerful influence on the learning of young chimpanzees in the tool-use situation, chimpanzees do not as a matter of course transmit precise tool-use techniques from one generation to the next. The only evidence for such transmission comes from a few specific population differences that do not have plausible ecological explanations – such things as termite probes with frayed ends or peeled bark that are found in one population but not in another. But it is clear that ecological factors can never be ruled out entirely in such cases, especially because these particular cases were documented with indirect evidence only (finding the tools), and none of them involved observations across generations. It is also clear that child emulation of adults may produce similarities of behavior among members of a group without any imitative processes being employed. I thus conclude from all this that population differences in chimpanzee tool use are due not to cultural transmission but to the different learning conditions to which the youngsters in each population are exposed. In all cases the youngsters learn much of the task individually. In some cases they also learn some aspects socially. But in no case is there reproduction of behavior from one animal to another, and thus, by my definition, in no case is there cultural transmission.

### Communicatory signaling

Communicatory signaling behaviors have been systematically studied in three groups of wild and three groups of captive chimpanzees. I report first on observations of the wild groups together, classifying the specific communicatory signals very loosely according to their major contexts of use. I then report on each of the captive groups separately. In all cases, the focus is on signals that are unique to a particular group, although differences in the observational and definitional procedures of the different researchers sometimes make this determination difficult. I attempt to deal with these on a case-by-case basis, but in general I assume that minor variations in the descriptions of the "same" behavior result from methodological differences, not from fundamental differences of behavior. Only when a reported communicatory signal was distinctive in a group and was not reported for another group that was observed in similar contexts do I attribute a true population difference.

#### *Chimpanzees in the wild*

Researchers have investigated communicatory behaviors specifically in only three groups of chimpanzees in the wild. Those of the Kasakela group of Gombe were reported by Goodall (1968a,b, 1986), supplemented by Plooij's observations of infants (1978, 1980, 1984). The communicatory behaviors of the Mahale K group were reported by Nishida (1968, 1970, 1980). Those of a third group from the Budongo Forest, Uganda, in eastern Africa – observed, without provisioning, for 6 months – were reported by Sugiyama (1969). Random observations from a few other groups have been reported in some cases.

*Aggressive contexts.* Although the species-typical aggressive display of male chimpanzees is relatively stereotyped, every individual performance is in some ways different from every other. In general, the display involves some subset of the following: a visual glare; a bipedal swagger walk; a variety of waving and slapping movements of the arms; noise making by drumming on trees or the ground; throwing or brandishing objects, especially tree limbs; a sham attacking charge; and a variety of aggressive vocalizations. These are threatening behaviors only; an actual attack or fight will involve still other behaviors. Something resembling this display is seen in all three wild populations.

Goodall (1968a) also reported two sitting postures at Gombe that often preceded an aggressive display: a 'sitting hunch' and a 'quadrupedal hunch'. The other two researchers did not report these behaviors specifically, but because they are not particularly salient behaviors, this discrepancy is very likely due to differences in defining aggressive contexts, in observational method, and/or in the definitions of behavioral categories. However, Goodall also reported a distinctive 'head tip' and stylized 'arm raise' as threatening behaviors that were not reported by the other two researchers. In these cases, because of their distinctiveness, it seems less likely that these might have been overlooked or grouped under general categories. Another unique threat behavior was reported by Nishida (1970) for the Mahale K group: "standing on all fours with head down and with one hind leg often kept raised for some moments" (p. 49). This may be a variation on Goodall's 'quadrupedal hunch,' but it also seems likely that if Gombe chimpanzees had raise the hind leg in this distinctive way it would have been noticed and reported. The frequency and distribution of these threatening behaviors in their respective groups were not reported.

*Sexual contexts.* Like aggressive behavior, most sexual behavior has a large measure of species-typical patterns. The male courtship display involves many of the same behaviors as the aggressive display, although obviously in a different context. The male typically glares at the female, bipedal swaggers, engages in a sitting hunch, and/or waves a branch in her direction. These are almost always accompanied by some form of 'male invite,' which involves a display of the erect penis. The female, when she is soliciting, 'presents' her genitalia to the male. Again, this general pattern was reported for all three groups.

Goodall (1968a) reported that Gombe males engage in two other behaviors not reported for the other two groups:

1. 'tree leaping,' in which the male in a tree executes "a series of rhythmic swings through the branches, his body usually in an upright position while he faces in the general direction of the female" (p. 361), and
2. 'beckoning,' in which the male "in a bipedal posture, raises one arm level with his head or higher and then makes a swift 'sweeping toward himself' movement, his hand making an arc in the air" (p. 361).

These two behaviors accounted for 9% and 6%, respectively, of all male displays, although the number of individuals involved was not reported.

Nishida (1980) reported 23 instances of a 'leaf clipping' display in sexual contexts for the Mahale K group during a 1-year period. (The behavior was also seen about as often as a threat to humans and as a frustration reaction.) Usually, but not always, the male (19 of 23 cases), often without other sexual behaviors (13 cases), takes several leaves in its hand and noisily rips them apart with its teeth. This is often repeated and usually results in the female attending, approaching gradually, and copulating. The display was seen at least once in a sexual context for five of the seven males and three of the eight females observed. Nishida speculated that leaf clipping originated as a frustration response when an adolescent male (the group using the signal most frequently) could not court an estrous female because older males were around. The noise attracted the female's attention, she spied the adolescent's physiological signs of arousal, and approached the male and copulated. Nishida further speculated that this attention getter may be a "social custom" of the Mahale K group.

The leaf clipping display is almost surely not a part of the Gombe chimpanzees' courtship behaviors, nor was it observed at Budongo. Subsequent to the report from Mahale, however, leaf clipping was also reported at Bossou, Guinea (not one of the three groups systematically treated here, and located across the continent). Sugiyama (1981) reported that this behavior was used by almost all group members at Bossou, mostly in aggressive (toward humans) or frustration contexts. Only 3 of 44 instances observed occurred during sexual courtships, all of these by males.

*Submission, reassurance, appeasement, and greeting contexts.* In avoiding aggressive encounters, chimpanzees engage in a variety of behaviors, mostly involving a lowering of the body, a submissive facial expression, a whimpering vocalization, and cautious touching. Goodall (1968a) isolated the following: 'bowing, bobbing, and crouching,' 'presenting,' 'kissing,' 'touching,' 'mounting,' 'reaching hand to,' 'grooming,' 'holding hand,' 'patting,' and 'embracing.' Some subset of these is used in social contexts where affiliation and/or reassurance, as opposed to aggression, are the desired outcomes. Although not all of these specific behaviors were explicitly mentioned by the researchers at Mahale and Budongo, they agreed on these general types of behaviors; any discrepancies were most likely due to different methodologies and/or definitions.

Goodall (1968a), however, reported three behaviors so distinctive that it is unlikely that other researchers would have grouped them under another heading: 'wrist bending' and 'bending away' as submissive gestures, and 'rump turning' as a reassurance behavior. These were of a fairly low frequency at Gombe. Nishida (1970) mentioned in this context "inserting one hand under the arm" of a dominant individual as part of the appeasement process at Mahale, but its frequency was not reported.

*Feeding contexts.* All three groups of chimpanzees were seen using a 'hand beg' gesture in which food was solicited from a conspecific by holding the hand out, palm up, under the chin of the one who was eating. In the Gombe group this behavior was "frequent," at Bodongo it was observed only once, and at Mahale it was observed in almost all group members other than high-ranking males.

*Grooming contexts.* Often chimpanzees approach each other and simply sit and groom, with no apparent signals. Sometimes, however, an individual comes and sits in front of another individual in a more exaggerated way, presenting its back or head for grooming. 'Lip smacking' often accompanies the grooming, but has not been seen as a request for grooming. This is the general pattern for all three groups.

McGrew and Tutin (1978) reported what they considered a "social custom" used in the grooming interactions of the Mahale K group. In the grooming 'handclasp,' "each of the participants simultaneously extends an arm overhead and then either one clasps the other's wrist or hand, or both clasp each other's hand. Meanwhile, the other hand engages in social grooming of the other individual's underarm area revealed by the upraised limb" (p. 238). In 6 days of observation they saw 9 of the 17 adults and adolescents of the group engaged in this behavior.

The grooming 'handclasp' has not been observed in the Gombe chimpanzees nor in those of the Budongo Forest. Gombe chimpanzees do, however, accomplish the same end by a maneuver in which the participants hold on to an overhead branch with one hand. Recently, however, Ghiglieri (1984) reported the grooming 'handclasp' (termed 'A-frame') in one of the two groups of chimpanzees he studied in the Kibale Forest, Uganda, also in eastern Africa. It occurred in 38% of the sessions observed in one group, but was not observed in the other group only 10 km away. The function of this behavior – especially whether or not it serves as an invitation to a specific type of grooming – is unknown.

*Caregiving contexts.* Only the Gombe chimpanzees have been systematically observed in caregiving situations in the wild. Goodall (1968a) reported that mothers often touch their young infants or look over their shoulders "and make a beckoning movement" (p. 371) toward their older infants as signals to climb aboard for departure. Plooij (1978, 1984), in his study of the Gombe infants, reported three infant signals directed toward the mother (or other adult): The 'hand around head' gesture was used by infants sitting in adult laps to initiate play tickling, the 'arm high' gesture was used to initiate grooming, and the 'begging' gesture was used to beg food. It was not reported how many of the mothers and infants used these gestures.

*Play contexts.* Only the Gombe chimpanzees have been systematically observed in play contexts in the wild. Goodall (1968a, 1986) reported a 'gambol run,' a 'play walk,' a 'back present,' and a 'prodding' (all accompanied by a 'play face') as the main ways that play is initiated, mostly among youngsters. Plooij (1978) also reported 'leaf grooming' (holding and inspecting a leaf carefully) as a way to get others to approach, after which some other interaction such as grooming or play ensues. Also, he reported 'running away with an object' (while looking back at someone) as a play invitation. It was not reported how many individuals used any of these play gestures.

### *Chimpanzees in captivity*

Van Hoof (1973) studied a captive group of 25 chimpanzees in New Mexico for 200 hours over a 2-month period. Twenty-four of these were between 3 and 12 years of age; 11 were males, and 15 were females. The group had been together for 4 months when observations began. Van Hoof explicitly compared his findings to those of Goodall (1968b). His animals did not show 'tree leaping' (sexual context) or 'branching' (aggressive context), but there were no trees in the enclosure. They did not show 'arm raise' or 'head tip' in an aggressive context or 'play walk' in a play context or either of the distinctive submissive gestures of the Gombe group. Van Hoof observed a 'vertical head nod' in which the chimpanzee actively nods its head in an exaggerated fashion, but no functional context was given for this behavior. The two unique signals observed by van Hoof were a 'horizontal head shake' in play contexts and a 'vacuum thrust' from males as a sexual invitation to females. He did not report how many individuals used these two signals.

Berdecio and Nash (1981) studied two groups of five chimpanzees each, 1–7 years old, in Arizona for 150 hours over a 7-month period. Because these animals were all infants or juveniles, they did not show many of the sexual, aggressive, submissive-appeasement-greeting, caregiving, and grooming signals of wild chimpanzees. Of particular interest in these infants and juveniles were their play initiations, because these have not been systematically studied in the wild, except to a minor extent at Gombe. Eleven were distinguished: 'grab,' 'slap,' 'grab object from,' 'stamp,' 'hit out at,' 'reach out to,' 'jump upon,' 'kick out at,' 'wave object,' 'clap,' and 'jump.' Eight of these can be found in either Goodall's or Plooij's descriptions. The two dealing with objects were not reported in the wild in this form, though they seem similar to Plooij's 'running away with an object.' 'Clap' was not reported for the Gombe chimpanzees, and it seems certain that this would have been reported had it occurred there. These authors speculated that this might have been learned from humans.

Tomasello, George, Kruger, Farrar, and Evans (1985) observed a group of 14 chimpanzees at the Yerkes Primate Research Center field station in Atlanta, Georgia, for 250 hours over a 6-month period. The focus was on five infants, ages 1–4. Because the subjects were immature, many of the sexual, aggressive, and submissive-appeasement-greeting signals seen in the wild were not observed in this study. Of particular interest were the signals used by these infants in caregiving and play contexts. They used several of the signals reported in the wild: 'hand beg' (for food), 'back offer' (to solicit grooming), 'arms around head' (to solicit tickling), and 'look back' and 'head bob' (in play contexts). Four of the five, however, had distinctive 'touch side' behaviors to solicit nursing from their mothers and an 'arm on' behavior to solicit tandem walking; these have not been reported in the wild. Three of the infants either pointed or directed an adult's hand to that part of the body they wanted to be groomed or tickled; this behavior has not been reported in the wild, and possibly it is derived from interactions with humans. All of the infants slapped the ground as an attention getter to initiate play, and three of the five clapped their hands for the same end (usually accompanied by a play face and other play postures); these behaviors have not been reported in the wild. Play was also initiated by some youngsters with an 'arm raise' or a 'leg offer' gesture. Finally, four of the infants begged for food by means of 'lip lock' – a behavior in which the begger would grasp the eater's lower lip in its mouth, suck briefly, and then back off and look to the eater's face. This has not been reported for any other group, and it is unlikely that so distinctive a behavior could be overlooked.

*Learning processes*

Many of the signals reported for these chimpanzee groups may be species-specific actions and reactions in which learning plays a more or less minor role. If our interest is in cultural transmission, however, we may focus primarily on signals that occur in only one or a few populations. These presumably rely in large part on learning processes and thus presumably are capable, at least in principle, of participating in a cultural transmission process.

There are two basic ways that chimpanzees may acquire their learned communicatory signals. The first is conventionalization (Smith, 1977; Tomasello et al., 1985). In this process, social behaviors are transformed into intentionally produced communicatory signals. For example, a young chimpanzee might initiate play by directly jumping on, hitting, and wrestling with a peer, but this does not involve a communicatory signal per se because the initiator does not use one behavior to "stand for" or to represent something else in a semiotic relation. From the "sender's" point of view, there is no signal at all, only behavior. In contrast, if that same youngster learns to use a part of the behavioral sequence (e.g., 'arm raise') to stand for, and thus to elicit, a desired response from the peer – and that signal by itself serves to initiate the play – we can then speak of a semiotic relation and thus a communicatory signal.

The process of conventionalization does not take place all at once, but emerges from a series of social interactions. In general, this sequence is composed of four steps:

1. An individual behaves intentionally toward another (e.g., an infant who wants to nurse pulls away its mother's arm to gain access to the nipple).
2. The recipient reacts in a predictable way (e.g., the mother moves her arm to allow access).
3. On some subsequent occasion, the recipient anticipates this sequence on the basis of its first step (e.g., the mother withdraws her arm at the initial touch of the infant).
4. The initiator learns over repetitions of this sequence to shorten its behavior to just that initial step (e.g., 'touch side' as an intentional signal for eliciting the mother's receptivity to nursing).[5]

The important point for current purposes is that signals learned in this way are "shaped," as it were, by the social environment. The signal itself is not passed on from organism to organism.

The other learning process that may be involved in chimpanzees' acquisition of communicatory signals is, of course, imitation, of which there are two forms. The first is second-person imitation. For example, an individual who has been the recipient of a 'leg offer' gesture (as a solicit for play) may learn that gesture, without going through the entire conventionalization process, by imitating the initiator. This presumably requires some abilities of observational learning, although it is not clear at this point whether or not and in what ways it may differ from other forms of observational learning. Second-person imitation, it should be noted, is an important part of human infants' acquisition of linguistic conventions from a very early age (Snow, 1981). The other form of imitation is third-person imitation. In this case, the organism acquires a signal by observing and imitating other individuals signaling to each other, sometimes referred to as "eavesdropping." Although, again, the precise nature of the observational learning processes involved is not totally clear, such vicarious learning would seem to be cognitively more difficult than second-person imitation, and indeed, it is not employed in the language acquisition of human children, despite its obvious utility, until after they are well along in their acquisition of language (Snow, 1989).

These learning processes may produce either of two different types of learned chimpanzee signals. The first are called "intention movements": A behavior that is originally embedded in a social interaction is extracted and used to stand for the entire behavioral complex, as illustrated by the foregoing examples of 'touch side' and 'arm raise.' The second type are called "attention getters." These do not originate as parts of specific social interactions, and therefore their "meanings" are not tied to these. Instead, they are very general invitations for a conspecific to attend to the signaling organism, which often means attending to unlearned displays and physical characteristics of the signaling organism. For example, a chimpanzee in the mood to play might slap the ground in order to get a potential playmate to look at its (presumably unlearned) play face and posturing, which serve to initate the play.[6] Both of these signal types are important in chimpanzee communication, and it is possible that they are learned in different ways.

### *A longitudinal investigation*

Goodall (1967) concluded that during the chimpanzee's second year of life, "most, if not all, adult ritualized gestures and postures appear" (p. 342); Goodall (1968b) concluded that this occurs during the third year (p. 288). Although Goodall was mostly focused on ritualized (not conventionalized) signals, Tomasello et al. (1985) argued that this characterization was nevertheless misleading. It does not seem that the concept of young chimpanzees acquiring or developing adult gestures adequately characterizes the major developmental processes involved. Based on our data, we have argued that a more accurate chracterization of the process of development is one in terms of *ontogenetic adaptations*: At each developmental period the organism learns to communicate about the social functions that are important at that period, and these change during ontogeny.

Tomasello, Gust, and Frost (1989) provided further support for this perspective in a follow-up study of this same group of chimpanzees 4 years later, when they were 5–8 years old. We found even more evidence of a very close relationship between communicatory signals and the particular social contexts in which they were used:

1. Signals for social functions that were no longer important at the older age were no longer used (e.g., nursing),
2. Signals for functions that changed in important ways changed their form accordingly (e.g., gromming).
3. New signals were learned for newly emerged adultlike functions (e.g., appeasement).
4. Signals of functions important at both times were retained and augmented (e.g., play).

These observations thus support the hypothesis that much of the communication among chimpanzees is accomplished through learning and, moreover, that this learning is closely tied to the nature of the social interactions experienced at different developmental periods.

Longitudinal comparisons of chimpanzees at the two ages revealed little interindividual consistency or intraindividual stability. Overall, of the 28 gesture types used at either of the two ages, only 4 (14%) were used by all seven of the youngsters. Of the 74 individual signals used by the youngsters at the older age, only 27 (36%) were carryovers from the earlier age. These figures are inconsistent both with the idea that this group possessed a single repertoire of gestural signals that it passed along from one generation to the next and with the idea that each individual learned to communicate primarily by observing and imitating others. If all of the animals were imitating each other, much more uniformity and consistency would be expected.

Further support for this view comes from a detailed analysis of individual usage patterns; that is, Who used which signals with whom? Three usage patterns were distinguished: idiosyncratic, one-way, and reciprocal. First, 9 of the 28 signal types were used by one and only one youngster at either one or both time points (although each was used consistently and repeatedly by that one individual). Because there was no one else to imitate, these idiosyncratic signals, many of which were between mother and child, must have been learned by conventionalization, not imitation.

Second, 8 of 28 signal types were one-way signals; that is, they involved social functions that were solicited by youngsters from adults, but not vice versa. For example, adults never begged food or tickling from a youngster (nor did youngsters use these with each other). These signals therefore must have been learned by some means other than second-person imitation, because youngsters never were recipients of them. It is possible that they could have learned them by imitating other youngsters signaling adults (third-person imitation), but again the lack of uniformity argues against this. None of the 8 gesture types in this category was used by more than five of the seven youngsters, and most were used by less than half. Moreover, many of the interactions of which these signals were parts took place in the privacy between mothers and infants, more or less out of the direct view of other infants. Once again, therefore, it is most likely that these signals were learned by processes of conventionalization, not imitation.

Third, reciprocal signals were those that could be both produced and understood by at least two youngsters; for the most part, these were play gestures used among peers (e.g., 'arm raise,' 'leg offer,' 'ground slap,' 'throw stuff,' 'poke at,' etc.). Like the other signals, these could easily have been learned by individual conventionalization processes, with the similarity among individuals being due to the fact that these were common social experiences for all youngsters. However, an analysis of the recipients of these signals revealed that of the 31 individual reciprocal signals (representing nine types) acquired in the interim between the two studies, 25 were used by individuals who had themselves been recipients of that gesture at some time, thus providing an opportunity for second-person imitation. Whether or not this opportunity was utilized is unknown, but one further observation may shed light on this question.

Five months prior to the second set of observations, wood chips were introduced into the chimpanzees' compound as a ground cover. By the time of the second study, all seven juvenile chimpanzees threw wood chips at conspecifics as an attention getter, most often in association with play initiation; most of them generalized this to mud and dirt. Nothing like this was observed at the earlier time point, although mud, dirt, and hay were present. Although it is possible that each individual conventionalized the throwing of wood chips separately during the 5 months prior to the second study, the rapid and universal spread of this behavior would also be consistent with an influence of observation. Because presumably most individuals were recipients of this signal prior to their own production, we hypothesized that the learning process involved was second-person imitation. But it must be emphasized that this hypothesis is directly applicable only to this one signal, and only indirectly applicable to the other reciprocal signals; it could not possibly be true for most of the idiosyncratic and one-way signals. Moreover, the interpretation of precisely how this signal might have been learned is far from straightforward, as discussed later.

*Cultural transmission of signals?*

Table 10.1 lists those signals that are unique to each population, wild and captive, and that are fairly stable and widespread in that population.[7] In order to allow for methodological differences, I have been extremely conservative in positing a signal as unique to a group; I do so only when the signal is of a distinctive nature and authors from other groups have mentioned nothing remotely similar, despite an opportunity to have observed it. On the other

Table 10.1. *Unique communicatory signals for two wild and three captive groups of common chimpanzee* (Pan troglodytes)[a]

| Context | Gombe | Mahale K | van Hoof | Berdecio | Tomasello |
|---|---|---|---|---|---|
| Aggression | 'head tip'<br>'arm raise' | 'leg raise' | | | |
| Sex | 'beckon'<br>'tree leap' | 'leaf clip'[b] | 'vacuum thrust' | | |
| Submission, etc. | 'wrist bend'<br>'bend away'<br>'rump turn' | 'insert hand' | | | |
| Feeding | | | | | 'lip lock' |
| Grooming | | 'handclasp'[c] | | | 'point'<br>'direct-hand' |
| Caregiving | M: 'beckon'[d]<br>I: 'arm high'<br>'hand-head' | | | | 'touch side'<br>'arm-on' |
| Play | leaf groom | | 'head shake' | 'clap' | 'arm raise'<br>'ground slap'<br>'leg offer'<br>'throw stuff' |

[a] Budongo is not listed because it has no unique signals.
[b] Also at Bossou.
[c] Also at Kibale.
[d] M, mother; I, infant.

hand, so as not to miss any true population differences, I have been generous in including some signals that meet this criterion, but whose frequency in a particular population is unknown. Despite the obvious inadequacies of the data for current purposes, it is clear from Table 10.1 that at least some population differences in the communicatory signaling of chimpanzees do exist.

Based on the uneven distribution of signals within populations, both wild and captive, it would seem that conventionalization plays the major role in chimpanzees' acquisition of learned communicatory signals. As in the case of tool use, population differences in the use of conventionalized signals may arise quite naturally from the different learning conditions present in the different populations. As demonstrated in the two studies of Tomasello et al. (1985, 1989), different social functions lead to the conventionalization of different signals, and thus we might expect that any differences in group composition, organization, demographics, and behavior would naturally lead to differences in the specific signals conventionalized. Differences might also emanate from individual differences – a particular individual creates new signals in a way that others do not simply because of that animal's unique social persona (Goodall, 1986) – which may, of course, also lead to variation of signals within a population as well by virtue of differences among members within a group.

It must also be kept in mind, however, that lack of variation among individuals and/or populations in the use of particular signals does not mean that they were not conventionalized. Similarities in signal use may still result from conventionalization processes originating from similar social interactions. Thus, most chimpanzee infants probably will learn only one of a small set of intention movements to solicit nursing, because the underlying social interactive behaviors presumably are very similar for all. This is presumably the way many, but not all, human infants learn to request being picked up by raising their arms (Lock, 1978). In principle, it is at least theoretically possible that all learned chimpanzee signals are acquired by conventionalization and that all of the population differences observed among chimpanzee groups, both wild and captive, have been produced by differences in the social interactional "raw material" from which the different signals are conventionalized.

The role of imitation in chimpanzee communication is more uncertain. In general, if imitation were the prevalent way that chimpanzee signals are acquired, we should expect to see much more uniformity within a group's use of signals (and perhaps more systematic differences among groups). The fact is that many of the signals that would seem to be potentially useful to all group members are not learned by some individuals, and different group members commonly use different signals for the exact same function. The lack of uniformity within groups is an especially telling argument against third-person imitation, because for this learning process to take place, all that is required (given cognitive and motivational prerequisites) is visual access, which is certainly not a problem in captive groups nor for youngsters observing their mothers and siblings in the wild. In addition, if human children are any indication, third-person imitation is a very sophisticated cognitive process. Second-person imitation seems a much more likely possibility. If second-person imitation were the primary learning process in chimpanzee communicatory signaling, the heterogeneity within groups could be explained by positing irregular patterns of interaction among group members, and thus differential opportunity to learn.

Because of the rapid learning involved, imitation seems an especially attractive hypothesis for the 'throw stuff' signal in the Tomasello et al. (1989) study. But it is important to note that this signal was a general attention getter, not an intention movement. The precise way the movements of the throw were executed was not important: Some threw overhand, some threw underhand, and some threw backhand; this is not possible in the case of intention movements, such as 'leg offer,' for which the significance is intimately bound up with the precise movements involved. What was important for the 'throw stuff' signal, as for other attention getters, was reproduction of the *result*, namely, gaining the attention of a conspecific. It thus bears a close resemblance to the learning process of emulation as posited for tool use. Like many cases of tool use, therefore, it may be that young chimpanzees discover for themselves the precise way to gain the attention of others by producing an effect that they observe being made by others.

It is also important to note that in the case of attention getters, an animal that is not being directly addressed might in some cases have its attention drawn to the signaler and thus become, as it were, an unintentional addressee. (This would also be true of the other major group of attention getters for the Yerkes group involving the making of noise: handclap, ground slap, wall slap, etc.) For purposes of learning, this may be equivalent to the learning opportunities presented in cases of second-person imitation – the unintentional addressee has the same learning opportunity as the intended addressee – and thus it would seem that when attention getters are involved, the distinction between second-person imitation and third-person imitation is not always clear-cut. Based on these considerations, then, the current hypothesis is that the social learning of communicatory signals by young chimpanzees is operative only in cases of attention getters like 'throw stuff' and 'ground slap,' and even in these cases the behavior that comprises the signal is not impersonated or imitated, but only emulated.

These considerations also apply to the two most distinctive communicatory signals from the wild for which we have the most complete data: 'leaf clipping' and the grooming 'handclasp' (used by two populations each). Leaf clipping was reported for two groups living across the continent from one another, and so presumably in this case we must posit independent inventions. In the Mahale K group the signal is used primarily by males (directed to females) and primarily in sexual contexts. It thus would seem to be mostly a one-way signal – males direct it to females, not to each other – but note that because it is an attention getter, one male could easily be attracted to another's noise making. Note also that because it is an attention getter, the leaf clipping activity could easily be emulated (making noise with leaves) without any behaviors being copied per se, and in fact animals have been observed tearing up the leaves with their mouths, hands, and even feet. In the case of the Budongo group, the leaf clipping signal was observed mostly in aggressive contexts, illustrating the flexibility of attention getters, but presumably the possibilities presented for learning were the same as in the Mahale K group.

In the case of the grooming 'handclasp,' the two populations involved live in fairly close proximity, and so some form of migration is possible, although other groups living in closer proximity do not show the behavior, and thus independent invention is possible as well. Whatever its origin, this "signal" is clearly not an attention getter. The problem is that it does not seem to be an intention movement either. In fact, from the current perspective it may not even be a signal at all. The 'handclasp' is clearly not an instigation or signal for grooming; it is a part of the grooming process when grooming under arms is important. The Gombe chimpanzees, for example, accomplish the same end by hanging on to low-hanging limbs as they groom one another's underarms. It seems most likely to me, in any case, that this behavior is conventionalized in the same way that children learn to hold their heads a certain way to have their hair brushed, that is, as an anticipatory behavior, not an intentional communicatory signal. In conjunction with the fact that only half of the animals observed at Mahale engaged in this behavior, the most plausible view, in my opinion, is that the grooming 'handclasp' is not a good candidate for learning by imitation – certainly not as a communicatory signal.

If our interest is in cultural transmission, it is important to note one other fact about chimpanzee communication. Horizontal and vertical transmission (i.e., within and between generations, respectively) do not play the same role in the acquisition of communicatory signals that they do in the acquisition of tool use. Mother–child interaction is clearly not as important in communicatory signaling as it is in tool use; very few of the communicatory signals infants use with their mothers are of later communicative significance. In the Tomasello et al. (1989) longitudinal study, only 'hand beg' survived to the 5–8-year age range, the other infant-to-mother signals either disappearing with the disappearance of the function or being replaced by more mature forms. The child's interaction with nonparent adults, on the other hand, would seem to be more important than in the case of tool use, for example, in learning signals for appeasement and grooming. And, totally unlike the case of tool use, peer interaction is extremely important in the chimpanzee's acquisition of communicative competence. Youngsters learned many signals (especially those involving play) in interactions with each other, and these lasted well into adolescence. Although many of these ludic behaviors fade away as play becomes less important, others serve as the basis for the conventionalization of signals later used in more adultlike serious contexts, especially sex and aggression (e.g., Goodall, 1968a, reports 'arm raise' as an adult threat gesture).

My overall conclusion, then, is that the vast majority of chimpanzee communicatory signals are learned by conventionalization, and conventionalization is basically an individual learning process. Young chimpanzees interact socially and learn signals from this, but they learn them by having their own behavior "shaped" by the reactions of the interactant – in much the same way that the physical environment may sometimes shape behavior in the tool-use situation. Each signal is created anew from individual, though perhaps common, social interactions, with no attempt to reproduce the behavior of a conspecific. Youngsters learn very few of their signals from adults, and the possible cases of imitation – which are not numerous in any case – all involve attention getters whose results may be emulated without the actual copying of behavior. I thus conclude from all this that although there may be more suggestive evidence in the case of communication than in the case of tool use, overall imitation plays a minor role in chimpanzee communication, and thus there is little, if any, cultural transmission.

**Chimpanzee culture?**

Yes/no questions are seldom helpful when complex phenomena are at issue. I thus believe that the question of chimpanzee culture is best posed as a question of the nature of the learning processes that youngsters use to acquire mature forms of behavior. The learning processes that I have hypothesized

Table 10.2. *Learning processes in the tool use and communicatory signaling of chimpanzees*

| Type of learning | Tool use | Communication |
|---|---|---|
| Individual | Environmental shaping[a] | Conventionalization[a] |
| Social | Stimulus enhancement[a] | — |
| | Emulation[a] | Emulation of attention getters[a] |
| Cultural | Imitation[b] | Second person imitation[b] |
| | — | Third-person imitation[b] |
| | Instructed learning[c] | Instructed learning[c] |

[a] Strong evidence.
[b] Weak evidence.
[c] No evidence.

to be important in chimpanzee tool use and communicatory signaling are summarized in Table 10.2. The various processes have been classified as individual, social, or cultural, based on the discussions and analyses from throughout this chapter. In brief, where the environment, either physical or social, is responsible for shaping the behavior, individual learning is indicated. Where something about the task, but not a particular behavior, is learned socially from conspecifics, social learning is indicated. Where a particular behavior is reproduced or "transmitted" from one animal to another, the concept of cultural learning is invoked. I conclude this chapter by attempting to justify my contention that chimpanzee tool use and communicatory signaling involve individual and social, but not cultural, learning processes.

Some of the issues involved in the question of chimpanzee culture may be further highlighted by a brief description of cultural transmission in human beings. In all human societies, adults come prepared to teach youngsters (Mead, 1970), and youngsters come prepared to imitate adults (Meltzoff, 1988). Adults and children interact in a variety of ways and settings, some involving important cultural skills. In many of these, adults take an active interest in ensuring that children acquire the skills. In teaching children to use tools, for example, adults will typically draw children's attention to important parts of the task, encourage repeated attempts, help to identify errors and suggest correction techniques, and reinforce correct performance of the subskills involved in the task (Rogoff, 1986; Wertsch, 1985). On other occasions, of course, adults do not actively teach skills, but they simply engage in them, and interested children imitate their performance to the best of their abilities. On other occasions still, these two modes of learning are combined as adults actively incorporate into their teaching techniques the modeling of a skill or a part of a skill for the child, in conjunction with verbal instructions, for example.

Adults in most societies do not actively teach language and communication in this same way, but they do interact with children in ways that often achieve this same effect (e.g., Mannle & Tomasello, 1987). By attempting to communicate effectively with children, parents often provide implicit feedback about the conventionality of particular communicative devices (Demetras, Post, & Snow, 1986), and by conversing with them about a common topic they often provide them with salient and useful models to imitate (Snow, 1981; Tomasello & Farrar, 1986). The important role played by the child's imitative abilities in this process is well documented, and indeed it is hard to imagine how it could be otherwise, given that linguistic symbols are wholly conventional and arbitrary. In the domain of human communication, therefore, more of the acquisition burden is placed on imitation and less is placed on adult instruction than is the case in tool use. The important point for current purposes is this: Human adult–child interaction, in both tool use and communication, involves a large measure of adult instruction or child imitation (or some combination), and both of these serve to ensure to some degree the transmission of behavior from adult to child.

As the current review has shown, neither of these social learning processes seems to be of great importance for chimpanzees; this is almost certainly true in the domains of tool use and communicatory signaling and, most likely, in other domains not reviewed here. There have been a few reported instances of parental "teaching" in the wild, but all of these involved adults restraining youngsters from dangerous situations, such as eating unknown foods (Nishida et al., 1983). There have been no substantiated reports of intentional teaching in a more positive manner. In fact, in comparing a captive chimpanzee mother–child pair to a human mother–child pair in an identical situation (with objects), Bard and Vauclair (1984) concluded that the single most striking difference between the species was in the almost complete absence of teaching by the chimpanzee mother.

Imitation would also seem to be less important for chimpanzees than it is for humans. Kellogg and Kellogg (1933) reported that one of the most striking differences between the chimpanzee and the human child they were raising in tandem was the superior imitative abilities of the child (see Bates, 1979, for a brief review). What imitations there were involved mostly "mimicking," that is, reproducing a behavior outside any functional context – as parrots often mimic human speech. And as Kaye's studies (1982) of human infants have shown quite clearly, mimicking a behavior is not the same thing as imitating in a functional context, and, in fact, they may involve very different cognitive processes. And the Tomasello et al. (1987) experimental study of tool use certainly seemed to indicate that young chimpanzees, in captivity at least, do not as a matter of course turn their attention away from the task at hand to study the activities of conspecifics. If chimpanzees do imitate in some behavioral domains, they still do not seem to appreciate how widely this strategy might usefully be applied.

Skepticism about chimpanzee imitation may surprise those who have read about the spread of potato washing in Japanese macaques. But it must be pointed out that that behavior took almost 4 years to spread, and it spread to barely half the members of the group (Kawamura, 1959). The mechanism of transmission is thus not entirely clear, and many of the issues raised here with respect to chimpanzees could be raised in that case as well. Another seemingly contradictory case is provided by chimpanzee acquisition of human-like communication skills in the laboratory (e.g., Gardner & Gardner, 1984; Savage-Rumbaugh, 1986). In this case as well, however, the role of imitative processes has never been clearly established. There have been no controlled studies of the imitation of these animals, and all of the researchers have reported that individual shaping procedures often are necessary to teach the subjects production of particular gestural signals that involve more than simple pointing. This contrasts sharply with the ease with which human children imitate adult language, either spoken or signed.

For species whose ontogeny depends heavily on learning, behavioral continuity across generations may be maintained in any number of ways: environmental shaping, social learning, and cultural transmission, for example. My argument is that despite the minimal importance of cultural learning (instructed learning and imitation) in chimpanzee ontogeny, they achieve some of the same results by more general social learning processes, namely, emulation in their acquisition of tool use and conventionalization in their acquisition of communicatory signals. Whereas the social environment plays an absolutely indispensable role in both of these learning processes, they both rely primarily on individual learning and cognition, especially reasoning processes about the causal relations involved in object manipulation and social interaction (Parker & Gibson, 1979). Human children rely on these same cognitive capacities in these same domains, but they also benefit from adult instruction and imitation (Bates, 1979), and this supplies the underpinnings for human culture.

My overall conclusion is thus that chimpanzees have "culture" in a very different sense than do humans. Much of the continuity across generations in chimpanzee tool use and communication is maintained by the continuity of learning experiences available to individuals – which "shape" individual learning. Very little of their learning comes from observing and reproducing the behavior of conspecifics, which, in the current account, is a defining feature of cultural transmission. They thus have a very loose and approximate "culture" in which each generation struggles, as it were, to reach the skill level of its progenitors. In human societies, the more faithful transmission effected by intentional instruction and imitation preserves useful behavioral adaptations more faithfully across generations (Bullock, 1987; Parker, 1985), and it does this *even in the face of changing environments* (Boyd & Richerson, 1983), something that individual learning and environmental shaping cannot do. Cultural transmission of this sort thus lays the foundation for behavioral innovations by allowing youngsters in some cases to acquire skills at a younger age than their parents and thus to spend their adulthood attempting to improve upon them. This so-called ratchet effect is what constitutes cultural progress in human societies, and it simply does not seem to be a part of chimpanzee "culture." And a good argument could be made that it is precisely this progressive quality of human culture that underlies many of the unique aspects of human evolution, including exosomatic developments involving the use of tools, as well as the origins and evolution of the cultural conventions that compose human language.

Given the importance of these issues, we know surprisingly little about the social learning processes employed in the daily lives of chimpanzees, humans, or any other primate species. Although I have expressed doubts about the facile use of the term "culture" with respect to chimpanzees, in very few cases has the behavior of human children been examined with the skeptical scrutiny I have used in this analysis. I thus do not mean for my conclusions to be taken as definitively negative on this issue, but rather I mean them to stimulate a closer look at the underlying cognitive and social-cognitive processes involved in cultural transmission and related social learning phenomena. This will require especially the collection of more detailed information about dissemination patterns in the wild and more tightly controlled investigations of social learning processes in the laboratory – for the full range of primate species. Information from studies such as these will allow us to draw much more precise conclusions about the role of cultural transmission, both in the lives of our primate relatives and in the evolution of our hominid ancestors.

**Acknowledgments**

Portions of the research reported in this chapter were supported by NIH grant RR-00165 from the Division of Research Resources to the Yerkes Regional Primate Research Center. The Yerkes Center is fully accredited by the American Association for the Accreditation of Animal Care. Thanks to Sue Taylor Parker for valuable editorial assistance.

**Notes**

1. No tool use has been observed at the Kibale Forest (Ghiglieri, 1984) or in some populations of the Budongo Forest (Reynolds & Reynolds, 1965), despite lengthy observations in both cases.
2. The major classes of chimpanzee tool use that this excludes are (1) the use of leaves for personal hygiene and related activities, (2) the use of sticks and stones in agonistic contexts, (3) the use of sticks for probing/investigating holes and crevices (not fishing), and (4) the use of foliage in the building of nests and related activities.
3. Several ingenious foraging behaviors have been observed on a single occasion or a few occasions. For example, on one occasion the Bossou chimpanzees made protracted attempts to use appropriately modified branches to pull in an otherwise inaccessible fruit-bearing branch. Members of this same group have twice been observed to pound on a tree to rouse termites onto a stick placed in a hollow, and they have several times used sticks to obtain resin from inside a tree (Sugiyama & Koman, 1979). The Gombe chimpanzees have on several occasions been seen using sticks to roust insects, to break into nests, and to dip for honey. For whatever reason, these have not turned into populationwide behavioral traditions.
4. Thorpe also distinguished "social facilitation," in which an organism's pre-existing behavior is elicited by the performance of a similar behavior by a conspecific (e.g.,

contagious crying in human infants). By itself, this is not a functionally important process and should not be considered learning at all, because no new behaviors are acquired in this way. Also excluded in the current analysis are attempts to "mimic" the behavior of a demonstrator with no regard to its function – the way a parrot mimics human speech, for example.

5. It is important to note that for this process to take place, the organisms involved must possess very specific cognitive capacities. In particular, the recipient must be able to use one behavior as an index to anticipate another, and the sender must be able to analyze the causal relations among its behavior and the social result and to take advantage of this by being able to produce the cause (the piece of behavior that has become a signal) in order to elicit the desired result. Thus, while this conventionalization of "intention movements" such as 'touch side' and 'arm raise' may in some ways look like and produce results similar to those of phylogentically derived ritualizations, they are distinguishable, and they involve different ontogenetic processes.
6. It should also be mentioned that some signals may be composed entirely of unlearned behaviors that have become conventionalized; that is, the organism learns that certain of its behaviors are reliably reacted to in specific ways and so gains intentional control over them. For example, de Waal (1982) reported instances of chimpanzees intentionally displaying or hiding their unlearned 'fear grimace' for communicative purposes.
7. There have been many reported instances of spontaneous communicatory signals that were not repeated by an individual. Examples have been provided by de Waal (1982) and Menzel (1971), as well as by the researchers reported here. What would seem to be important for the stable establishment of a communicatory signal would seem to be its importance in a recurrent context, such as sex, aggression, play, etc.

## References


Anderson, J., Williamson, E., & Carter, J. (1983). Chimpanzees of Sapo Forest, Liberia: Density, nests, tools, and meat-eating. *Primates*, *24*, 594–601.

Bard, K., & Vauclair, J. (1984). The communicative context of object manipulation in ape and human adult–infant pairs. *Journal of Human Evolution*, *13*, 181–190.

Bates, E. (1979). *The emergence of symbols: Cognition and communication in infancy*. New York: Academic Press.

Beatty, H. (1951). A note on the behavior of the chimpanzee. *Journal of Mammalology*, *32*, 118.

Beck, B. (1980). *Animal tool behavior*. New York: Garland Press.

Berdecio, S., & Nash, A. (1981). *Chimpanzee visual communication: Facial, gestural, and postural expressive movements in young, captive chimpanzees*. Arizona State University Research Papers, No. 26.

Birch, C. (1945). The relation of previous experience to insightful problem solving. *Journal of Comparative and Physiological Psychology*, *38*, 367–383.

Boesch, C., & Boesch, H. (1981). Sex differences in the use of natural hammers by wild chimpanzees: A preliminary report. *Journal of Human Evolution*, *10*, 585–593.

Boesch, C., & Boesch, H. (1983). Optimisation of nut-cracking with natural hammers by wild chimpanzees. *Behaviour*, *83*, 265–286.

Boesch, C., & Boesch, H. (1984). Possible causes of sex differences in the use of natural hammers by wild chimpanzees. *Journal of Human Evolution*, *13*, 415–440.

Bonner, J. (1980). *The evolution of culture in animals*. Princeton University Press.

Boyd, R., & Richerson, P. (1983). *Culture and the evolutionary process*. New York: Wiley.

Brewer, S. (1978). *The forest dwellers*. London: Collins.

Bullock, D. (1987). Socializing the theory of intelligence. In M. Chapman & R. Dixon (Eds.), *Meaning and the growth of mind: Wittgenstein's significance for developmental psychology* (pp. 187–218). New York: Springer-Verlag.

Burton, F., & Bick, M. (1972). A drift in time can define a deme: The implications of tradition drift in primate societies for hominid evolution. *Journal of Human Evolution*, *1*, 53–59.

Demetras, M., Post, K., & Snow, C. (1986). Feedback to first language learners: The role of repetitions and clarification questions. *Journal of Child Language*, *13*, 275–292.
de Waal, F. (1982). *Chimpanzee politics.* New York: Harper & Row.
Evans, A., & Tomasello, M. (1986). Evidence for social referencing in young chimpanzees. *Folia Primatologica*, *47*, 49–54.
Galef, B. (1976). The social transmission of acquired behavior: A discussion of tradition and social learning in vertebrates. In J. Rosenblatt, R. Hinde, E. Shaw, & C. Beer (Eds.), *Advances in the study of behavior* (Vol. 6, pp. 123–148). New York: Academic Press.
Galef, B. (1982). Studies of social learning in Norway rats: A brief review. *Developmental Psychobiology*, *15*, 279–295.
Galef, B. (1988). Imitation in animals. In B. Galef & T. Zentall (Eds.), *Social learning: Psychological and biological perspectives* (pp. 1–34). Hillsdale, NJ: Erlbaum.
Gardner, R., & Gardner, B. (1984). A vocabulary test for chimpanzees. *Journal of Comparative Psychology*, *98*, 381–404.
Ghiglieri, M. (1984). *The chimpanzees of the Kibale Forest: A field study of ecology and social structure.* New York: Columbia University Press.
Goodall, J. (1964). Tool-using and aimed throwing in a community of free-living chimpanzees. *Nature*, *201*, 1264–1266.
Goodall, J. (1967). Mother–offspring relationships in chimpanzees. In D. Morris (Ed.), *Primate ethology* (pp. 287–346). London: Weidenfeld & Nicolson.
Goodall, J. (1968a). A preliminary report on expressive movements and communication in the Gombe Stream chimpanzees. In P. Jay (Ed.), *Primates: Studies in adaptation and variability* (pp. 313–374). New York: Holt, Rinehart & Winston.
Goodall, J. (1968b). Behaviour of free-living chimpanzees of the Gombe Stream Reserve. *Animal Behavior Monographs*, *1*, 163–311.
Goodall, J. (1970). Tool using in primates and other vertebrates. In D. Lehrman, R. Hinde, & E. Shaw (Eds.), *Advances in the study of behavior* (Vol. 3, pp. 195–249). New York: Academic Press.
Goodall, J. (1973). Cultural elements in a chimpanzee community. In E. Menzel (Eds.), *Pre-cultural primate behavior* (pp. 138–159). Basel: Karger.
Goodall, J. (1986). *The chimpanzees of Gombe.* Cambridge, MA: Harvard University Press.
Goodson, B., & Greenfield, P. (1975). The search for structural principles in children's manipulative play. *Child Development*, *46*, 734–746.
Hall, K. (1963). Observational learning in monkeys and apes. *British Journal of Psychology*, *54*, 201–226.
Hallinan, M. (1984). Biological evolution and the emergence of a cultural being. In T. Duster & K. Garrett (Eds.), *Cultural perspectives on biological knowledge.* Norwood, NJ: Ablex.
Hannah, A., & McGrew, W. (1987). Chimpanzees using stones to crack open oil palm nuts in Liberia. *Primates*, *28*, 31–46.
Hayes, K., & Hayes, C. (1952). Imitation in a home-raised chimpanzee. *Journal of Comparative and Physiological Psychology*, *45*, 450–459.
Isaac, G. (1983). Aspects of human evolution. In D. Bendall (Ed.), *Evolution from molecules to men* (pp. 324–352). Cambridge University Press.
Jones, C., & Sabater Pi, J. (1969). Sticks used by chimpanzees in Rio Muni, West Africa. *Nature*, *223*, 100–101.
Kawamura, S. (1959). The process of subcultural propagation among Japanese macaques. *Primates*, *2*, 43–60.
Kaye, K. (1982). *The mental and social life of babies.* New York: Halsted Press.
Kellogg, W., & Kellogg, L. (1933). *The ape and the child.* New York: McGraw-Hill.
Kitahara-Frisch, J., & Norikoshi, K. (1982). Spontaneous sponge-making in captive chimpanzees. *Journal of Human Evolution*, *11*, 41–47.
Köhler, W. (1927). *The mentality of apes.* London: Routledge & Kegan Paul.
Kortlandt, A. (1986). The use of stone tools by wild-living chimpanzees and earliest hominids. *Journal of Human Evolution*, *15*, 77–132.

Lock, A. (1978). On being picked up. In A. Lock (Ed.), *Action, gesture, and symbol: The emergence of language* (pp. 3–18). New York: Academic Press.
McBeath, N., & McGrew, W. (1982). Tools used by wild chimpanzees to obtain termites at Mt. Assirik, Senegal: The influence of habitat. *Journal of Human Evolution, 11*, 65–72.
McGrew, W. (1974). Tool use by wild chimpanzees in feeding upon driver ants. *Journal of Human Evolution, 3*, 501–508.
McGrew, W. (1977). Socialization and object manipulation of wild chimpanzees. In S. Chevalier-Skolnikoff & F. Poirier (Eds.), *Primate biosocial development* (pp. 261–288). New York: Garland Press.
McGrew, W. (1983). Animal foods in the diets of wild chimpanzees: Why cross-cultural variation? *Journal of Ethology, 1*, 46–61.
McGrew, W., & Collins, D. (1985). Tool use by wild chimpanzees to obtain termites in the Mahale Mountains, Tanzania. *American Journal of Primatology, 9*, 47–62.
McGrew, W., & Rogers, M. (1983). Chimpanzees, tools, and termites: New record from Gabon. *American Journal of Primatology, 5*, 171–174.
McGrew, W., & Tutin, C. (1978). Evidence for a social custom in wild chimpanzees? *Man, 13*, 234–251.
McGrew, W., Tutin, C., & Baldwin, P. (1979). Chimpanzees, tools, and termites: Cross-cultural comparisons of Senegal, Tanzania, and Rio Muni. *Man, 14*, 185–214.
Mainardi, D. (1980). Tradition and the social transmission of behavior in animals. In G. Barlow & J. Silverberg (Eds.), *Sociobiology: Beyond nature/nurture?* Boulder, CO: Westview.
Mannle, S., & Tomasello, M. (1987). Father, siblings, and the bridge hypothesis. In K. E. Nelson & A. van Kleeck (Eds.), *Children's language* (Vol. 6, pp. 23–41). Hillsdale, NJ: Erlbaum.
Mead, M. (1970). Our educational emphasis in primitive perspective. In J. Middleton (Ed.), *From child to adult: Studies in the anthropology of education* (pp. 120–141). New York: The Natural History Press.
Meltzoff, A. (1985). Immediate and deferred imitation in 14 and 24 month old infants. *Child Development, 56*, 43–57.
Meltzoff, A. (1988). Imitation, objects, tools, and the rudiments of language in human ontogeny. *Human Evolution, 3*, 45–64.
Menzel, E. (1971). Communication about the environment in a group of young chimpanzees. *Folia Primatologica, 15*, 220–232.
Menzel, E. (1972). Spontaneous invention of ladders in a group of young chimpanzees. *Folia Primatologica, 17*, 87–106.
Menzel, E., Davenport, R., & Rogers, C. (1970). The development of tool using in wild-born and restriction-reared chimpanzees. *Folia Primatologica, 12*, 273–283.
Mignault, C. (1985). Transition between sensorimotor and symbolic activities in nursery-reared chimpanzees. *Journal of Human Evolution, 14*, 747–758.
Nishida, T. (1968). The social group of wild chimpanzees in the Mahale Mountains. *Primates, 9*, 167–224.
Nishida, T. (1970). Social behavior and relationships among wild chimpanzees of the Mahale Mountains. *Primates, 11*, 47–87.
Nishida, T. (1980). The leaf-clipping display: A newly discovered expressive gesture in wild chimpanzees. *Journal of Human Evolution, 9*, 117–128.
Nishida, T., & Hiraiwa, M. (1982). Natural history of tool using behavior by wild chimpanzees in feeding upon wood-boring ants. *Journal of Human Evolution, 11*, 73–99.
Nishida, T., & Uehara, S. (1980). Chimpanzees, tools, and termites: Another example from Tanzania. *Current Anthropology, 21*, 671–672.
Nishida, T., Wrangham, R., Goodall, J., & Uehara, S. (1983). Local differences in plant feeding habits of chimpanzees between the Mahale Mountains and the Gombe National Park, Tanzania. *Journal of Human Evolution, 12*, 467–480.
Parker, S. (1985). A social-technological model for the evolution of language. *Current Anthropology, 26*, 617–639.

Parker, S., & Gibson, E. (1979). A developmental model for the evolution of language and intelligence in early hominids. *Behavioral and Brain Sciences*, *2*, 367–408.
Piaget, J. (1962). *Play, dreams, and imitation.* New York: Norton.
Plooij, F. (1978). Some basic traits of language in wild chimpanzees? In A. Lock (Ed.), *Action, gesture, and symbol: The emergence of language* (pp. 111–132). London: Academic Press.
Plooij, F. (1980). How wild chimpanzee babies trigger the onset of mother–infant play and what the mother makes of it. In M. Bullowa (Eds.), *Before speech: The beginning of interpersonal communication* (pp. 223–244). Cambridge University Press.
Plooij, F. (1984). *The behavioral development of free-living chimpanzee babies and infants.* Norwood, NJ: Ablex.
Reynolds, V., & Reynolds, F. (1965). Chimpanzees in the Budongo Forest. In I. DeVore (Ed.), *Primate behavior* (pp. 202–239). New York: Holt, Rinehart & Winston.
Rogoff, B. (1986). Adult assistance of children's learning. In T. Raphael (Ed.), *The contexts of school-based literacy* (pp. 73–91). New York: Random House.
Sabater Pi, J. (1974). An elementary industry of the chimpanzees in the Okoribiko Mountains, Rio Muni, West Africa. *Primates*, *15*, 351–364.
Savage-Rumbaugh, S. (1986). *Ape language: From conditioned responses to symbols.* New York: Columbia University Press.
Schiller, P. (1957). Innate motor action as a basis for learning. In P. Schiller (Ed.), *Instinctive behavior* (pp. 264–287). New York: International Universities Press.
Smith, J. (1977). *The behavior of communicating.* Cambridge, MA: Harvard University Press.
Snow, C. (1981). The uses of imitation. *Journal of Child Language*, *8*, 205–212.
Snow, C. (1989). Imitation: A trait or a skill. In G. Speidel & K. Nelson (Eds.), *The many faces of imitation* (pp. 314–333). Frankfurt: Springer-Verlag.
Struhsaker, T., & Hunkeler, P. (1971). Evidence of tool using by chimpanzees in the Ivory Coast. *Folia Primatologica*, *15*, 212–219.
Sugiyama, Y. (1969). Social behavior of chimpanzees in the Budongo Forest, Uganda. *Primates*, *10*, 197–225.
Sugiyama, Y. (1981). Observations on the population dynamics and behavior of wild chimpanzees at Bossou, Guinea, in 1979–1980. *Primates*, *22*, 435–444.
Sugiyama, Y. (1985). The brush-stick of chimpanzees found in southwest Cameroon and their cultural characteristics. *Primates*, *26*, 361–374.
Sugiyama, Y., & Koman, J. (1979). Tool-using and -making behavior in wild chimpanzees at Bossou, Guinea. *Primates*, *20*, 513–524.
Sumita, K., Kitahara-Frisch, J., & Norikoshi, K. (1985). The acquisition of stone tool use in captive chimpanzees. *Primates*, *26*, 168–181.
Suzuki, A. (1966). On the insect eating habits among wild chimpanzees living in the savanna woodland of western Tanzania. *Primates*, *7*, 481–487.
Teleki, G. (1974). Chimpanzee subsistence technology: Materials and skills. *Journal of Human Evolution*, *3*, 575–594.
Thorpe, W. (1956). *Learning and instinct in animals.* London: Methuen.
Tomasello, M., Davis-Dasilva, M., Camak, L., & Bard, K. (1987). Observational learning of tool use by young chimpanzees. *Human Evolution*, *2*, 175–183.
Tomasello, M., & Farrar, J. (1986). Joint attention and early language. *Child Development*, *57*, 1454–1463.
Tomasello, M., George, B., Kruger, A., Farrar, J., & Evans, E. (1985). The development of gestural communication in young chimpanzees. *Journal of Human Evolution*, *14*, 175–186.
Tomasello, M., Gust, D., & Forst, T. (1989). A longitudinal investigation of gestural communication in young chimpanzees. *Primates*, *30*, 35–50.
Uehara, S. (1982). Seasonal changes in the techniques employed by wild chimpanzees in the Mahale Mountains, Tanzania, to feed on termites. *Folia Primatologica*, *37*, 44–76.
van Hoof, J. (1973). A structural analysis of the social behavior of a semi-captive group of chimpanzees. In M. von Cranach & I. Vine (Eds.), *Social communication and movement* (pp. 49–91). London: Academic Press.

Vygotsky, L. (1978). *Mind in society: The development of higher psychological processes.* Cambridge, MA: Harvard University Press.
Washburn, S., & Benedict, B. (1979). Nonhuman primate culture. *Man, 14*, 163–164.
Wertsch, J. (1985). *Vygotsky and the social formation of mind.* Cambridge, MA: Harvard University Press.
White, L. (1959). The concept of culture. *American Anthropologist, 61*, 227–251.
Whitesides, G. (1985). Nut cracking by wild chimpanzees in Sierra Leone, West Africa. *Primates, 26*, 91–94.
Wood, D. (1988). *How children think and learn.* London: Basil Blackwell.

# 11 Primate cognition: From representation to language

*Jacques Vauclair*

After giving a brief definition of terms, I argue for the existence of basically similar ways of coding environmental stimuli in terms of cognitive organization throughout primate species, including humans.[1] Some examples are given in order to show that numerous species express behaviors involving various forms of schematization or mental representations of parts or wholes of their environment.[2]

I then examine communicative abilities in animals and in humans. Various characteristics of communicatory signals are discussed, namely, their disengagement from context and their intentionality, considered as possible criteria to assess their status as signs or symbols.

Next, I treat the question of language and debate the existence of linguistic competence in animals, particularly in primates. For that purpose, I present a critical review of the language studies in apes. A consideration of functional features of human language[3] (more precisely, of the notion of "radical arbitrariness" of the verbal sign) is applied to the performances of language-trained apes, with the aim of suggesting that these capacities do not qualify as truly linguistic.

In sum, I try to show that whereas most of the complex forms of cognitive activities of animals and humans can be viewed within a framework of biological continuity, language cannot be so interpreted. Rather, it is the result of simultaneous biological evolution and social evolution, which have led to a merging between communicational and representational systems.

## Definitions

I begin with definitions of the three main concepts:

> *Representation* is an individual phenomenon by which an organism structures its knowledge with regard to its environment. This knowledge can take two basic forms: either reference to internal substitutes (e.g., indexes or images) or use of external substitutes (e.g., symbols, signals, or words).
>
> *Communication* is a social phenomenon of exchanges between two or more conspecifics who use a code of specific signals usually serving to meet common adaptive challenges (reproduction, feeding, protection) and promote cohesiveness of the group.

> *Language* is conceived as a system that is both communicational and representational: It is grounded in a social convention that attributes to certain substitutes (called *signifiers*) the power to designate other substitutes (called *referents*).

With respect to the definition of representation, Piaget (1946) distinguished representation in its broad sense from representation in a more restricted sense. In the broad sense, representation equates with "all intelligence which no longer depends simply on perception and motions ... but rather on a system of concepts and mental schemes" (Piaget, 1946, p. 68). In its more limited sense, representation is conceived of as a function that relates an object (practical or abstract) to its substitute or representative. In contrast with the latter meaning, representation in the sense of adaptation or intelligence does not necessarily imply the presence of isolated substitutes. For the purpose of the present discussion, representation and communication are considered as separate elements. Although one can conceive of communicative events that do not possess representational elements (see the section on communication for an example), in most cases representational elements constitute necessary parts of the phenomenon of communication.

## Representation and mental organization in animals

It is now widely recognized that animals of different phyla exhibit behaviors that express an internal processing (or treatment) of external reality (Roitblatt, Bever, & Terrace, 1984). In comparative psychology, the concept of internal representation of objects and events was postulated by Tinklepaugh (1928) in monkeys, and later by Tolman (1948) in the notion of the *cognitive map*. Although representation has been invoked to explain the cognitive abilities of several species (insects, birds, rodents) (Straub, Seidenberg, Bever, & Terrace, 1979), the most prominent work has come from the primates. Among the clearest demonstrations was the work on cross-modal transfer in apes and monkeys (Davenport, Rogers, & Russel, 1975; Malone, Tolan, & Rogers, 1980), perceptual categorization (Sands, Lincoln, & Wright, 1982), and cognitive mapping (Menzel, 1973).

The studies on memory for visual patterns have also provided evidence regarding the existence of internal coding of stimuli by apes (Vauclair & Rollins, 1984; Vauclair, Rollins, & Nadler, 1983). In the study by Vauclair and Rollins, for example, the subjects (two juvenile chimpanzees) were presented with patterns composed of 5 lighted cells in a matrix of 25 cells. Patterns were reproduced immediately following their presentation by touching a sensor located in the center of each cell. Several patterns varying in complexity and symmetry were presented; transfer to novel patterns similar in structure but differing in orientation was also examined. The two main results of this work can be summarized as follows:

1. The complexity of a given pattern was related to the number of line segments that composed a pattern, not to the number and location of individual cells.

2. There was a remarkable degree of transfer from training patterns to new patterns. Such transfers, which involved unique response sequences to rotations of the same patterns used in training, imply that subjects attended to the structure of the patterns, not to individual cells.

Reproductive memory is quite different from recognition memory. Subjects in these experiments were required to reconstruct the pattern from memory, not just indicate that some previously seen pattern was familiar. Both chimpanzees performed this rather complex cognitive task with a high level of accuracy. The generalization of reproduction skills to novel patterns indicates that response production was not linked with a specific motor response and that chimpanzees had learned something about the perception and organization of the patterns themselves, or, to say it differently, they had acquired some form of internal representation of the figures to be reproduced.

All the studies mentioned earlier, as well as many others, imply a representation (use of an *internal substitute*) that takes different names, depending on authors. For example, it is a "code" for Terrace, a "concept of a stimulus" for Davenport. We must also consider other cases in which *external substitutes* can be used. More than 50 years ago, Wolfe (1936) showed that chimpanzees were able to manipulate a tool in order to bring grapes within reach. These apes continued to do this even if they were rewarded with tokens that they could later insert into a slot machine to obtain food. These tokens functioned as "abstract tools" that some individuals would keep or steal before using them to obtain food. Along this line, ape language studies have demonstrated the animal's capacity to use an object (piece of plastic, or a lexigram) as a substitute for its referent (Premack, 1972; Rumbaugh, 1977). I treat this point at some length later.

These data support the contention that "mental organizations exist in animals, in particular in nonhuman primates, as a necessary part of the perception of objects and their localisation and interrelationships in space and time" (Walker, 1983, p. 380). Given the genetic relation between apes and humans, it seems reasonable to infer evolutionary continuity in the processing of cognitive information. This inference of continuity has been accepted by most biologists, psychologists, and anthropologists.

This idea was suggested by Darwin in *The Descent of Man*: "The difference in mind between man and the higher animals, great as it is, certainly is one of degree and not of kind" (Darwin, 1871, p. 128). This idea is consistent with Piaget's theory, according to which there is a functional continuity from the primitive organisms to humans (Piaget, 1967), in that the same biological mechanisms are at work, namely, assimilation and accommodation. From this perspective, representational capacities of humans are conceived of as the end products of those general mechanisms. The data on object permanence across species (Etienne, 1984) and other studies on reasoning in monkeys (e.g., McGonigle & Chalmers, 1977) and in apes (e.g., Gillan, 1980) support this position vis-à-vis primates and some other mammals.

It is tempting to assume that similar processes act in the emergence of language; this assumption comes out of the definition one chooses for language. Piaget's conception of language, for example, is in obvious agreement with the theory of continuity. To him, language is not a specific entity but a particular aspect of the general capacity for representation:

> Language is only one aspect of the symbolic (or semiotic) function. This function is the ability to represent something by a sign or a symbol or another object. In addition to language the semiotic function includes gestures ... deferred imitation ... drawing, painting, modeling. It includes mental imagery which I have characterized elsewhere as internalized imitation. (Piaget, 1971, p. 45)

In addition to Piaget, one could mention other theorists of language origins (e.g., Hewes, 1973) according to whom human language derived from a system of vocalizations or gestures analogous to the systems of communication observed today in some animal species. For Lieberman (1984), the organization of language (e.g., syntax) is seen as a generalization of the mechanisms that control nonverbal behavior: "There is an evolutionary and morphological link between the neural bases of structured, automatized motor activity and syntax" (Lieberman, 1984, pp. 67–68).

## Communication

It is useful to start with what is commonly understood by the concept of communication. In the context of information theory (Miller, 1956), "communication" describes the phenomenon of transmission of information from one point to another point. This definition is very wide and corresponds to the definition of representation *sensu largo*, where every interaction, every behavioral sequence, can be described by an observer as an exchange of information and thus constitutes a communicatory act. When it is employed to describe, for example, information stored in DNA, interactions within an organism ("what the frog's eye tells the frog's brain," Maturana, 1959), communication takes on a quasi-metaphorical meaning. I consider this definition, wherein communication is applied to every sort of interaction, to be too general to allow the identification of events of communication; in that sense "eating" would also qualify as a communication.

I thus propose to restrict the scope of the concept of communication, and to differentiate it from representation, in order to retain its heuristic value. Moreover, in this restricted interpretation, this concept necessarily implies that communication occurs within the context of a social organization.[4]

In most cases, there is no problem in classifying interactions between animals as communication, such as the zigzag dance performed by sticklebacks during courtship or other acoustical, postural, or gestural behaviors expressed by insects, birds, and mammals in various social contexts. The problem arises, however, when we are to infer a criterion for intentionality or purposefulness in animal communication. I agree with von Glasersfeld's view that the "concept of purpose is essential for the definition of communication" and "that the purpose has to be on the side of the source or sender" (von Glasersfeld, 1977, p. 61).

For the sake of the present discussion, I consider that the philosophical concept of intentionality implies knowledge of the status of the message that is produced, or, in other words, a representation of the communicative behavior that is emitted. In fact, because this statement amounts to saying that these intentional behaviors represent functional equivalents of human language, intentionality is a relevant criterion for defining language, but is unsuitable for the definition of communication.[5] Of course, I am not trying to say that communicative behaviors in humans are limited to language. It appears that many communicative sequences in our species are bound to interpersonal relations in the immediate communication situation without automatically involving representational elements. An obvious example is provided by the displeasure vocalizations of babies that signal a state of discomfort to their mothers.

## Some features of communication

Criteria for comparing animal communication with human communication and language were proposed by Benveniste (1952) and then by Hockett (1960) and Thorpe (1972); they were based on lists of the most characteristic features of human language, to which different types of animal communication were compared. Although this approach can be biased by the traits one chooses to emphasize, it is worth mentioning some of them because of their direct relevance to the theoretical context of this chapter.

Hockett (1960) has suggested that one important design feature of language is "displacement," that is, the ability to "talk about things that are remote in space or time (or both) from where the talking goes on" (p. 90). Following this conception, a signal might become a symbol (equivalent to a verbal sign) when it can be used without direct connection to an experimental context.[6]

Von Glasersfeld (1977) argues that displacement fails to achieve exactly this transformation, because a mere delay (distance in time and space) does not change the one-to-one correspondence between the sign and the situation. With linguistic symbols,

> we can talk not only about things that are spatially or temporally remote, but also about things that have no location in space and never happen at all . . . in order to become a symbol, the sign must be detached from input. What the sign signifies, i.e., its meaning, has to be available, regardless of the contextual situation. (von Glasersfeld, 1977, p. 64)

The famous dance of honeybees that conveys information about the direction and the distance of a food source (von Frisch, 1950) is a good illustration of this process of symbolization. "Only a hypothetical bee which communicated about distance, direction, food sources, without actually coming from, or going to, a specific location could be said to use symbols" (von Glasersfeld, 1976, p. 222). Passingham said that "the bees certainly transmit information about the world, but they do not learn the basic code, and do not say anything else with it" (Passingham, 1982, p. 197; see also Benveniste, 1952). Another case in which animals appear to communicate directly about their environment ought to be mentioned. I refer to the alarm calls of wild vervet monkeys (Cheney & Seyfarth, 1982; Seyfarth, Cheney, & Marler, 1980). First of all, Seyfarth and coworkers showed that adult vervets gave a different call depending on which of three predators they had sighted. Second, playback experiments demonstrated that the calls were not functions of the length or the amplitude of the signal, nor were they dependent on the excitatory state and age of the sender. These results suggest that alarm calls have acquired a certain semantic value and hence that they are not totally tied to their context of production.

Both visual and gestural modalities offer cases of communicative displays that have been ritualized through evolution (e.g., the courtship behaviors among insects or birds). Other communicative behaviors have even taken a real status of conventionality. Some of the most sophisticated examples have been described for nonhuman primates – consider the presentation of the genitalia between baboons out of the context of reproduction (Wickler, 1972). There are also more advanced expressions of communicative signals, such as conventionalized[7] deception, that have been reported mainly for monkeys and apes (Quiatt, 1984). Because these deceptive behaviors raise the question of intentionality, they will be discussed at some length later.

The human observer cannot but be struck by the richness and variety of animal communicatory signals and by their versatile and modulated use. Because some interactive patterns (in both intraspecific and interspecific communication systems) demonstrate features such as decontextualization, semanticity, and even intentionality, one must wonder to what extent these patterns can be equated with symbolic communication and representation as they develop in human infants.

## Sensorimotor development in ape and human infants

There has been increasing evidence since the work of Bates, Benigni, Bretherton, Camaioni, and Volterra (1979), Bruner (1983), and Volterra (1987) that the use and mastery of linguistic signs are prepared during the sensorimotor period through the expression of deferred imitation and combinatorial and symbolic play, as well as through a wide range of interactive patterns between the child and the caregivers (mutual focus on inanimate objects, turn taking, giving, pointing, etc.). As far as ontogeny is concerned, observations of spontaneous object manipulations have shown a marked difference between human and chimpanzee infants. These differences can be best characterized by a lack of object–object relations, symbolic play, and conventional use of objects in the apes as compared with the behavioral repertoires of same-age human subjects (Mathieu & Bergeron, 1981; Mignault, 1985; Vauclair & Bard, 1983). It must be noted that the communicative style of adult primates in relation to object manipulation in infants is also different from the typical interactions seen in humans. For example, in contrast to humans, adult apes rarely act on objects with the apparent intent of engaging the attention of the infants (Bard & Vauclair, 1984).

What makes humans remarkable compared with apes is the early mutual exchange between adults and infants regarding inanimate objects. Bard and Vauclair have hypothesized that such interactive patterns might exist in apes for gestural and emotional exchanges, but not regarding physical objects other than food and, perhaps, nesting material (Bard & Vauclair, 1984). This feature and other aspects of human ontogeny (e.g., retarded locomotion) bring about a specific mode of contact with the external world (Spinozzi & Natale, 1986) that elsewhere (Vauclair, 1984) I have called a generalized ability to use intermediaries to control both social and physical objects. These cognitive and communicative instruments (referential communication, tool use, complex object manipulation) all refer to a sharing of significations or understandings (Bullowa, 1979) that structure the context in which language develops.

Of course, the studies just mentioned do not imply that some of the advanced skills that are already obvious in 1-year-old humans do not develop through observational learning or experimental training in juveniles or adult apes. In fact, all the "language studies" with chimpanzees have demonstrated, among other things, the abilities of apes to manipulate objects in complex ways and to attach arbitrary symbols to objects or actions.

### Language-trained apes

In order to introduce my discussion of human language and its specificity, I examine some of the performances of language-trained apes. In one experiment, when Sarah (the chimpanzee trained by Premack to name objects with plastic tokens) was shown an apple and asked to choose descriptive features of the apple (red vs. green, round vs. square, etc.), she correctly selected the features that characterized the apple. Then, after the apple was removed and replaced by the blue plastic triangle that stood for the object APPLE, Sarah was given a paired comparison test and was able to "assign the same features to the 'word' that she had earlier assigned to the object" (Premack, 1972, p. 98). The chimp clearly demonstrated her capacity not only to associate objects, action, or attributes of objects to specific tokens but also to associate one token to another token. Performances of the same level of complexity were reported for Sherman and Austin, the two chimpanzees trained by Savage-Rumbaugh and associates with arbitrary lexigrams assigned to the two object categories FOOD and TOOL. When shown an arbitrary pattern indicating, for example, BANANA, the chimps responded by pressing the key meaning FOOD, but if shown the lexigram for LEVER, they correctly pressed the TOOL key. In other words, the chimpanzees were able not merely to label objects, but to label substitutes (i.e., lexigrams) for those objects (Savage-Rumbaugh, Rumbaugh, Smith, & Lawson, 1980). Both Sarah's and Sherman and Austin's performances imply a form of mental organization in which artificial labels are related to each other rather than only directly related to representations of real objects.

A phenomenon even closer to naming has been reported recently (Savage-Rumbaugh, McDonald, Sevcik, Hopkins, & Rupert, 1986; Savage-Rumbaugh, Rumbaugh, & McDonald, 1985) in the behavior of Kanzi, a pygmy chimpanzee. Kanzi spontaneously presses lexigrams on the keyboard to request food, objects, and actions (e.g., tickling, chasing). It must be said, however, that Kanzi started pressing keys after a long observation of his foster mother's unsuccessful work on the keyboard. Kanzi is able to ask for things he wants and to tell the names of items in response to teacher's query. Though Kanzi's behavior is impressive because of its spontaneity, one needs to remember that his use of symbols appears mostly in a context of requests to "direct teacher's attention to places, things and activities" (Savage-Rumbaugh et al., 1985, p. 658). Kanzi is said to press keys that are "comments about surrounding events" (Savage-Rumbaugh et al., 1986, pp. 659–660). Unfortunately, the report provides no indication concerning the percentage of such activities. (See Greenfield & Savage-Rumbaugh, P&G20, for more on Kanzi's language abilities.)

To be sure, the special kind of training apes experience in language studies could enhance their abilities to solve various conceptual tasks. Specifically, Premack (1983) has suggested that apes exposed to language training are capable of solving problems involving an ability to operate on relations between relations (e.g., analogical reasoning), whereas non–language-trained animals appear to solve tasks implying only an imaginal representation (e.g., spatial problems).

The three examples I have cited all raise the same problem: Are the chimpanzees naming in the same way that children name with words? Several authors have already stressed the differences between an ape's use of artificial symbols and children's use of utterances. These differences can be summarized under two categories: disengagement from context and spontaneity (i.e., naming as an intent to communicate for its own sake, not solely the expression of a demand). Because of lack of space, I cannot comment further on the second functional difference, whereby children use language very early outside the context of request; on that topic, see Terrace (1985). Concerning the disengagement from context, it appears that the use of symbols by apes is closely tied to the achievement of immediate goals, because the referents occur in the context of behavior on their objects.

As Kaye pointed out, the repetitive pairing between an object and its substitute in training apes could yield to a wrong interpretation of the animal's "symbolic" activities, because object use by the chimpanzees may be "a human phenomenon due to the trainer's ability to construct an environment in which arbitrary chosen symbols (from his own point of view) function as indexical from the animal's point of view" (Kaye, 1982, p. 133). "To establish that an object (or action) is used symbolically in the absence of the item of which it refers, it must occur as an element related to other symbols" (Silverman, 1983, p. 133). Kanzi may be the exception,[8] but from these theoretical and methodological perspectives, there is no special reason to treat apes' performances differently from those of other beasts. I am convinced that apes display the most sophisticated form of representation in the animal kingdom (a kind of second power representation), but this phenomenon is insufficient in itself to qualify for linguistic status. To go beyond the one-to-one coorespondence between the sign and the actual perceptual situation, we need to introduce a third term. The relation between symbol and object is more than the simple correspondence between the two. Because the symbol is tied to a conception, we have a triangular connection among objects, symbols, and concepts: "It is the conceptions, not the things, that symbols directly mean" (Langer, quoted in von Glasersfeld, 1977).

### The concept of radical arbitrariness

I have already said that, for Piaget, language is one among several expressions of the semiotic function in humans, which mainly consists in establishing mental images. In symbolic play, for example, a mental image can be built from two objects (e.g., a truck, which is represented by a matchbox); from these two objects, a single representation has been constituted, with the consequence that they can be substituted for one another, and with the condition that the objects share some common characteristics. This notion that an object (called *referent*) can be represented by another object (called *signifier*) has been borrowed from the work of the Swiss linguist Ferdinand De Saussure (1916) and is still widely accepted and used by most psychologists and psycholinguists when dealing with language. The arbitrariness that characterizes verbal units lies in the relation between signifiers and referents: A linguistic sign is described as arbitrary, it bears no obvious resemblance to the object or event it represents.

The problem is that this notion of arbitrariness has been considerably simplified compared with its original meaning. According to De Saussure (Bronckart, 1977), two types of material reality have to be processed by the subject in order to establish a verbal sign: on the one hand, an acoustical substance, and on the other, a material substance that corresponds to the content to be expressed. In other words, a verbal sign is not simply a relation between material elements (sounds) and the content to which they refer (objects or actions), but rather the product of two representations: one representation built on the acoustical material and another representation built on the meaning (conceptual image). Thus, two mental images have been created in two distinct domains of reality. But these representations do not directly compose the ingredients of the sign. For De Saussure, an exclusive reference to an individual representation does not in any way explain why the two types of images correspond to each other. According to his theory, such representations have to be reorganized or reanalyzed through a social convention. Though the construction of conceptual and acoustical images is typically an individual activity, the basic operation of language, that is, designation or creation of signs, is performed through the social convention. Following this conception, a verbal sign is thus constituted by the relation between both types of images, where language has selected in a socially conventional and arbitrary manner a sequence of sounds to stand for a particular concept. This conventional and arbitrary relation between a signifier and a signified was characterized as "radical arbitrariness" by one of De Saussure's followers (De Mauro, 1969).

The basic features of linguistic signs, namely, the fact that they are not solely the products of individual mechanisms and that they change with time, have as consequences that the signs can be grasped only by opposition to other signs, as mentioned previously. Any sign takes its full meaning (or its *value*) by reference to the other signs of the language at a given time of its development.[9]

From an ontogenetic perspective, the emergence of referential naming in young children can be understood within a social and cultural framework: Access to naming (or designation) results from the participation of the child in a social convention. The outcome of this process is that each sign can be considered as a "value of representation" relative to the other signs of language. Moreover, these values can substitute for each other on a paradigmatic axis and thus become metaphors.

## Mental continuity

I return to the hypotheses underlying my definition of language considered as the result of representation and communication, and to the consequences of this definition for the question of mental continuity.

If we have to admit a gap between animals and humans, we have to look for it neither in the presentation per se nor in the communication per se but rather in the emergence in humans of verbal language that has, so to speak, mushroomed in both representative and communicative capacities.

The specificity of human language is above all of functional order. First, this system uses representative stimuli that allow the sender to know the status of the sent message, to control it, and to endow it with intentions. It should be noted that transmission of representations themselves is impossible. I cannot be sure that the meaning of a word I say is the same for the person to whom I direct it. Consequently, language works as a system of values, of reciprocal expectations. To say it differently, the processes of verbal communication always constitute a try, a hypothesis, and an intention from the sender to the receiver. Second, the substitutes that have been built so far on physical objects change their nature, because they result from a social convention that has determined the value of linguistic units.

This approach to language studies in apes implies that to be able to demonstrate that an ape uses signs that are equivalent to verbal signs, one must give evidence

1. that the ape possesses an individual representation of the substitute and of its content,
2. that a social convention has made possible an analysis of such representations, and
3. that those representations are relative values with respect to the stock of signs that the ape has at its disposal.

Clearly, such a demonstration has yet to be made regarding the so-called linguistic abilities of apes or other animals.

## Relations among representation, communication, and language

The remarks and criticisms I have made in this chapter came to my mind after consideration of the frequent absence of references to human psychology, and especially to developmental psychology, among those scientists who are studying complex cognitive behaviors in animals. This reference is sorely needed when we are concerned with language, a human acquisition par excellence. It must be said that if the first "language studies" satisfied themselves with more or less loose criteria to demonstrate the linguistic abilities of the apes they were testing, the literature published on these issues in recent years draws systematically on language development and other related fields; for an excellent demonstration, see Terrace (1985). This is, of course, not to say that the borrowing of methods and theories from human developmental psychology will bring an answer to each problem we face in animal psychology. Actually, several (sometimes contradictory) theories are flourishing in developmental psychology, but I believe that even conflicting approaches can serve to illuminate the questions we are concerned with in comparative psychology.

In this last section, I plan to reexamine the concepts of representation, communication, and language, with the goal of providing a heuristic framework with which to analyze abilities expressed by nonhuman primates.

We saw in the first section of this chapter that there is today a general tendency among researchers to admit that in the nervous systems of many (if not all) organisms there exist mechanisms to code external information. Such a view amounts to postulating that organisms do not react passively to stimuli and thus that they possess instruments to treat, filter, and select the flow of these stimuli. As a cognitive psychologist, I am inclined to designate such an activity under the heading of representation. In this sense, representation is just a different name for intelligence. But this broad interpretation of the concept is unsatisfactory if, on the one hand, we wish to grasp all the intricacies of the processing applied to physical and social objects and, on the other hand, we are interested in the possible hierarchy of phenomena as distantly removed from one another as innate releasing mechanisms (IRM) and language production.

A logical way to provide a more integrative framework for referring to the whole range of these diverse behavioral manifestations is to use some common key concepts. Keeping these key concepts in mind (e.g., substitutes, arbitrariness), I now reexamine representative and communicative performances of various primates.

Returning to representation, we saw earlier that this process functioned with substitutes, either internal or external. To start with internal substitutes, such substitutes can take different forms, from indexes to mental images. In the last 15 years or so, new findings about the precocious capacities of human babies have dramatically changed our conception regarding the early adaptation of humans. From a theoretical point of view, some authors (e.g., Mounoud & Vinter, 1981) have postulated that representation precedes symbolization. This idea is opposite to the position held by Piaget concerning the rather later emergence (end of the first year) of representations (evocation of an absent object) in the infant such as can be seen in deferred imitation or symbolic play. Following these novel approaches, neonates are already endowed with coding instruments; intermodal coordinations (Meltzoff & Borton, 1979) and imitation (Vinter, 1985) are examples of these precocious forms of coding. It can be noticed that such codes change through ontogeny: from sensory, at the beginning of life, they become perceptual at about 2 years of age, and later conceptual and then formal. Following this conception, activities as elementary as identification or recognition already imply representation.

If one applies the usual dichotomy between signifier and referent, one can easily conceive that a substitute is equivalent to a signifier, whereas the object or the situation for which it stands can be qualified as a referent.

Other forms of representations make use of external substitutes (see the earlier section on mental organization in animals). These kinds of substitutes can be either "motivated" (analogical) – that is, they bear some similarity to the referents they "re-present" – or conventional (e.g., as in the case of road signs). Thus, it appears that representation possesses some of the characteristics that are usually attributed to a linguistic sign. As is suggested by the definition of representation I proposed at the beginning of this chapter, however, different forms of representation function only for the benefit of the organism, in order to increase its knowledge of objects and situations. In contrast to representation, communication was described as a phenomenon that helps to regulate the interactions between two or more individuals: Its scope is thus not individual but collective (Bronckart, 1987). Accordingly, the concept of communication as a vector of information between a sender and a receiver through a channel (cf. its use in information theory) is too simplistic. The examples reported earlier have shown that in animal communication, the circulation of information depends on contextual factors, physical or/and social. An illustration is provided by the alarm calls ("call notes") emitted by vervet monkeys when they spot predators: Vervet monkeys produce different types of calls as a function of the nature of the predator they have encountered (Seyfarth et al., 1980; Struhsaker, 1967).

Between the expression of communicative signals that are the direct results of genetic evolution (such as ritualized behaviors) and the communicative aspect of human language, it is possible to see intermediate forms of communication. The conventional signals to be presented later can be understood as being of this intermediate form, because they are the results of social constraints acting on the group as well as of the control that an individual has of the situation. When the sender of information gets control over the message to be sent, we are a step closer to language, which implies a representation of the situation by the partners involved in the communicative exchange. In other words, such a process implies that the content of the message and its status (namely, its effect on the receiver) are known by the sender. Field studies with primates have yielded several instances of what can be called "premeditation": One can mention the case of infant stealing reported by Hrdy (quoted in Quiatt, 1984), which was preceded by an episode of grooming of the mother in order to render the infant more accessible. Other cases involve "cheating" and concealment of information, such as in the mating between young female hamadryas and subadult males that is performed out of sight of the dominant male; under these circumstances, the young females do not emit the typical vocalization they usually produce when they mate with the leader (Kummer, 1982).

In these cases, intentions can easily be inferred retrospectively by the human observer, but other, less cognitively oriented explanations could be proposed. In the case of infant stealing, for example, it is difficult to know if the goal has not been substituted by the communicator during the social exchange. In the case of the apparent deception of the dominant male by the female hamadryas, a more reductive explanation can be proposed, namely, aversive conditioning, because it is in the interest of the females (in order to avoid neck biting) to inhibit their vocalizations as much as they can when they are away from the dominant male.

A study (unique of its kind) to test experimentally chimpanzees' ability to convey and understand both accurate and misleading information concerning the location of hidden food showed that when the chimpanzee was competing with its trainer for access to food, it was capable (1) of conveying misleading information (e.g., by pointing to an unbaited container) and (2) of controverting the competing trainer's signals by avoiding the container toward which it oriented (Woodruff & Premack, 1979). For Woodruff and Premack, the chimpanzee's flexibility in dealing with the information it sends or receives demonstrates a real capacity for deceit. Is it, then, also a capacity for intentional communication? The answer to this question is complicated by the fact that "it is because intention is assumed to [underlie] deception that deception is so often presented, explicitly or implicitly, as evidence of intention" (Quiatt, 1984, pp. 33–34). Despite the critiques of some ethologists (Seyfarth et al., 1982), the Woodruff and Premack experiment represents an important contribution to an understanding of chimpanzees' communicative abilities, which under the conditions of long and systematic training can actively and appropriately mislead others.

It must be pointed out that, contrary to the active manifestation of deception obtained in the experimental situation, most expressions of deception in a natural setting consist of withholding or concealing information. It is obvious that this passive behavior can hardly be equivalent to human language. But, as Quiatt wrote, it certainly shows the ability of primates "to manipulate displays in complex and subtle ways" (Quiatt, 1984, p. 35). Quiatt added that "in intraspecific interactions in natural circumstances there is seldom any need for nonhuman primates to mask immediate intentions ... and it is unlikely that animals, including monkey and apes, can detach themselves sufficiently from displays which have been standarized by natural selection and adaptation to manipulate them like chess pieces" (Quiatt, 1984, p. 35).

The relations among representation, communication, and language in humans are far from clear. The task of interpreting an animal's communication and cognition with the former concepts may seem an elusive objective. Thus, caution is needed before drawing any definitive delineation between animal and human cognitive and communicative achievements. That notwithstanding, this chapter has attempted to examine some of the most elaborate behaviors expressed by animals in the light of corresponding realizations performed by humans. The main stance taken here concerns language, with its unique features of intentionality, conventionalization, and radical arbitrariness.

If we can consider that both animal communicative signals and human language are grounded in the biological makeup of the species and in their evolutionary history, only human language can be viewed as a predominantly social and personal system (actually, the true interfacing between the social environment and the individual). This double functional status of language (Bronckart, 1987) expresses the full complexity of the system, but it is at the same time at the origin of the possibilities this device opens: unity of the person, regulation of action, schema to translate other systems of signs and metalanguage.

**Acknowledgments**

I am grateful to K. Gibson and S. Parker for their helpful suggestions.

**Notes**

1. Certain portions of this work have been elaborated elsewhere (Bronckart, Parot, & Vauclair, 1987).

2. Although most of the examples provided are borrowed from the primatological literature, the basic approach and hypotheses should be valid for other animal species.
3. For convenience, most of my uses of the word "language" refer to spoken language, the latter being understood as the normal manifestation of the former in the vocal-auditory channel.
4. I do not intend to go into a discussion concerning the adaptive function of communication systems in animals. Recent views on this issue are available (Dawkins & Krebs, 1978; Parker, 1985, 1987).
5. The study of intentionality and thus of language requires a change of paradigm, that is, a move from questions of instantiation (Do apes have intentions?) to questions of attribution (Do apes think that others have intentions?). Such questions have only recently begun to be addressed (Premack & Woodruff, 1978). With respect to the criterion of intentionality, behavioral evidence, especially in primates, forces us to express some caution when we attribute intentionality only to human language. In fact, it appears that some communicative behaviors, such as those expressed in premeditation or deception, may involve intentionality in the sense defined in this chapter.
6. It is unfortunate that the use of concepts such as signal, sign, and symbol are inconsistent from one scholar to another. They seem to be tied to the linguistic and cultural traditions of those scientists who refer to them, but this fact may also reflect our disagreement or uncertainty concerning language definition. For example, in the United States (e.g., Peirce), the word "symbol" describes the highest forms of signals, namely, verbal signs, whereas in Europe (e.g., Piaget), "symbol" can designate an intermediary form between perceptual signals and linguistic signs.
7. For usages of the concepts of ritualization and conventionalization, I follow Smith (1977), for whom ritualized behaviors designate phylogenetically stereotyped behaviors, whereas conventionalized behaviors are based on individual experience (e.g., learning).
8. Seidenberg and Petitto (1987) have questioned the idea that Kanzi uses lexigrams as symbols, and they have proposed an alternative view whereby apes (in language studies) have learned the instrumental and pragmatic functions of lexigrams or signs, but not the concepts associated with them. On this controversial issue, see the replies by Nelson (1987) and Savage-Rumbaugh (1987).
9. The activity of translation can serve to illustrate the status of the signs' value and the difficulty of transposing a system of values from one language to another; for example, the sign "boeuf" in French possesses a value that corresponds neither to the English "ox" nor to "beef"; it is rather the sum of the two.

## References

Bard, K. A., & Vauclair, J. (1984). The communication context of object manipulation in ape and human adult–infant pairs. *Journal of Human Evolution*, *13*, 181–190.

Bates, E., Benigni, L., Bretherton, I., Camaioni, L., & Volterra, V. (1979). *The emergence of symbols*. New York: Academic Press.

Benveniste, E. (1952). Communication animale et langage humain. *Diogène, 1*, 1–7.

Bronckart, J.-P. (1977). *Théories du langage: Une introduction critique*. Bruxelles: Mardaga.

Bronckart, J.-P. (1987). Les fonctions de communication et de représentation chez l'enfant. In *La psychologie. Encylopédie de la Pléiade* (pp. 680–714). Paris: Gallimard.

Bronckart, J.-P., Parot, F., & Vauclair, J. (1987). Les fonctions de communication et de représentation chez l'animal. In *La psychologie Encyclopédie de la Pléiade* (pp. 92–122). Paris: Gallimard.

Bruner, J. (1983). *Child's talk*. New York: Norton.

Bullowa, M. (1979). Prelinguistic communication: A field for scientific research. In M. Bullowa (Ed.), *Before speech* (pp. 1–62). Cambridge University Press.

Cheney, D. L., & Seyfarth, R. M. (1982). How vervet monkeys perceive their grunts: Field playback experiments. *Animal Behavior*, *30*, 739–751.

Darwin, C. (1871). *The descent of man and selection in relation to sex*. London: John Murray.

Davenport, R. K., Rogers, C. M., & Russel, I. S. (1975). Cross-modal perception in apes: Altered visual cues and delay. *Neuropsychologia*, *13*, 229–235.

Dawkins, R., & Krebs, J. R. (1978). Animal signals: Information or manipulation? In J. R. Krebs & N. B. Davies (Eds.), *Behavioural ecology, and evolutionary approach* (pp. 282–309). Oxford: Blackwell Scientific Publications.

De Mauro, T. (1969). *Introduction à la semantique*. Paris: Payot.

De Saussure, F. (1916). *Cours de linguistique générale*. Paris: Payot.

Etienne, A. S. (1984). The meaning of object permanence at different zoological levels. *Human Development*, *27*, 309–320.

Gillan, D. J. (1980). Reasoning in the chimpanzee: II. Transitive inference. *Journal of Experimental Psychology: Animal Behavior Processes*, *7*(2), 150–164.

Hewes, G. W. (1973). Primate communication and the gestural origin of language. *Current Anthropoloty*, *14*, 5–24.

Hockett, C. F. (1960). The origin of speech. *Scientific American*, *203*(3), 88–96.

Kaye, K. (1982). *The mental and social life of babies*. Brighton, England: Harvester Press.

Kummer, H. (1982). Social knowledge in free-ranging primates. In D. Griffin (Ed.), *Animal mind – human mind* (pp. 113–130). Berlin: Springer-Verlag.

Lieberman, P. (1984). *The biology and evolution of language*. Cambridge, MA: Harvard University Press.

McGonigle, B. O., & Chalmers, M. (1977). Are monkeys logical? *Nature*, *267*, 694–695.

Malone, D. R., Tolan, J. C., & Rogers, C. M. (1980). Cross-modal matching of objects and photographs in the monkey. *Neuropsychologia*, *18*, 693–697.

Mathieu, M., & Bergeron, C. (1981). Piagetian assessment on cognitive development in chimpanzees (*Pan troglodytes*). In A. B. Chiarelli & R. S. Corruccini (Eds.), *Primate behavior and sociobiology* (pp. 142–147). Berlin: Springer-Verlag.

Maturana, H. (1959). What the frog's eye tells the frog's brain. *Proceedings of the IRE*, *47*(11), 1940–1951.

Meltzoff, A. N., & Borton, R. W. (1979). Cross modal matching by human neonates. *Nature*, *282*, 406.

Menzel, E. W. (1973). Chimpanzee spatial memory. *Science*, *182*, 943–945.

Mignault, C. (1985). Transition between sensorimotor and symbolic activities in nursery-reared chimpanzees (*Pan troglodytes*). *Journal of Human Evolution*, *14*, 747–758.

Miller, G. A. (1956). *Langage et communication*. Paris: PUF. Mounoud, P., & Vinter, A. (1981). Representation and sensorimotor development. In G. Butterworth (Ed.), *Infancy and epistemology* (pp. 200–235). Brighton, England: Harvester Press.

Nelson, K. (1987). What's in a name? Reply to Seidenberg and Petitto. *Journal of Experimental Psychology: General*. *116*(3), 293–296.

Parker, S. T. (1985). A social-technological model for the evolution of language. *Current Anthropology*, *26*, 617–639.

Parker, S. (1987). The origins of symbolic communication: An evolutionary cost–benefit model. In J. Montangero, A. Tryphon, & S. Dionnet (Eds.), *Symbolism and knowledge* (pp. 7–28). Geneva: Cahiers de la Fondation Archives Jean Piaget.

Passingham, R. (1982). *The human primate*. San Francisco: Freeman.

Piaget, J. (1946). *La formation du symbole chez l'enfant*. Paris: Dalachaux et Niestlé.

Piaget, J. (1967). *Biology and knowledge*. Edinburgh University Press.

Piaget, J. (1971). *Genetic epistemology*. New York: Columbia University Press.

Premack, D. (1972). Teaching language to an ape. *Scientific American*, *227*, 92–99.

Premack, D. (1983). The codes of man and beasts. *Behavioral and Brain Sciences*, *6*, 125–167.

Premack, D., & Woodruff, G. (1978). Does the chimpanzee have a theory of mind? *Behavioral and Brain Sciences*, *4*, 512–526.

Quiatt, D. (1984). Devious intentions of monkeys and apes? In R. Harré & V. Reynolds (Eds.), *The meaning of primate signals* (pp. 9–40). Cambridge University Press.

Roitblatt, H. L., Bever, T. G., & Terrace, H. S. (Eds.). (1984), *Animal cognition*. Hillsdale, NJ: Erlbaum.

Rumbaugh, D. M. (Ed.). (1977). *Language learning by a chimpanzee. The LANA Project*. New York: Academic Press.
Sands, F. J., Lincoln, C. E., & Wright, A. A. (1982). Pictorial similarity judgements and the organization of visual memory in the rhesus monkey. *Journal of Experimental Psychology: General*, *111*(4), 369–389.
Savage-Rumbaugh, S. (1987). Communication, symbolic communication and language: Reply to Seidenberg and Petitto. *Journal of experimental Psychology: General*, *116*(3), 288–292.
Savage-Rumbaugh, S., McDonald, K., Sevcik, R. A., Hopkins, W. D., & Rupert, E. (1986). Spontaneous symbol acquisition and communicative use by pygmy chimpanzees (*Pan paniscus*). *Journal of Experimental Pscyhology: General*, *115*(3), 211–235.
Savage-Rumbaugh, E. S., Rumbaugh, D. M., & McDonald, K. (1985). Language learning in two species of apes. *Neuroscience and Biobehavioral Review*, *9*, 653–665.
Savage-Rumbaugh, E. S., Rumbaugh, D. M., Smith, S. T., & Lawson, J. (1980). Reference: The linguistic essential. *Science*, *210*, 922–925.
Seidenberg, M. S., & Petitto, L. A. (1987). Communication, symbolic communication and language: Comment on Savage-Rumbaugh, McDonald, Sevcik, Hopkins, and Rupert (1986). *Journal of Experimental Psychology: General*, *116*(3), 279–287.
Seyfarth, R. M., Cheney, D. L., & Marler, P. (1980). Monkey responses to three different alarm calls: Evidence of predator classification and semantic communication. *Science*, *210*, 801–803.
Syefarth, R. M., Beer, C. G., Dennett, D., Gould J. C., Lindauer, M., Marler, P., Ristau, C., Savage-Rumbaugh, E. S., Solomon, R., & Terrell, H. (1982). Communication as evidence of thinking. In D. R. Griffin (Ed.), *Animal mind–human mind* (pp. 391–406). Berlin: Springer-Verlag.
Silverman, P. S. (1983). Attributing mind to animals: The role of intuition. *Journal of Social and Biological Structures*, *6*, 231–247.
Smith, W. J. (1977). *The behavior of communication, an ethological approach*. Cambridge, MA: Harvard University Press.
Spinozzi, G., & Natale, F. (1986). The interaction between prehension and locomotion in macaque, gorilla and child cognitive development. In J. G. Else & P. C. Lee (Ed.), *Primate ontogeny, cognition and social behaviour* (pp. 155–159). Cambridge University Press.
Straub, R. O., Seidenberg, M. S., Bever, T. G., & Terrace, H. S. (1979). Serial learning in the pigeon. *Journal of the Experimental Analysis of Behavior*, *32*, 137–148.
Struhsaker, T. (1967). Auditory communication among vervet monkeys (*Cercopithecus aethiops*). In S. Altman (Ed.), *Social communication in primates* (pp. 281–334). University of Chicago Press.
Terrace, H. S. (1985). In the beginning was the "name." *American Psychologist*, *40*(9), 1011–1028.
Thorpe, W. H. (1972). The comparison of vocal communication in primates and in man. In R. A. Hinde (Ed.), *Non-verbal communication* (pp. 27–47). Cambridge University Press.
Tinklepaugh, O. L. (1928). An experimental study of representative factors in monkeys. *Journal of Comparative Psychology*, *8*, 197–236.
Tolman, E. C. (1948). Cognitive maps in rats and men. *Psychological Review*, *55*, 189–208.
Vauclair, J. (1984). Phylogenetic approach to object manipulation in human and ape infants. *Human Development*, *27*, 321–328.
Vauclair, J., & Bard, K. A. (1983). Development of manipulations with objects in ape and human infants. *Journal of Human Evolution*, *12*, 631–645.
Vauclair, J., & Rollins, H. A. (1984). Reproductive memory for visual patterns in chimpanzees. *Primate Eye* [*Supplement*], *23*, 1–2.
Vauclair, J., Rollins, H. A., & Nadler, R. D. (1983). Reproductive memory for diagonal and nondiagonal patterns in chimpanzees. *Behavioural Processes*, *8*, 289–300.
Vinter, A. (1985). *L'imitation chez le nouveau-né*. Paris: Delachaux et Niestlé.
Volterra, V. (1987). From single communicative signal to linguistic combinations in hearing and

deaf children. In J. Montangero, A. Tryphon, & S. Dionnet (Eds.), *Symbolism and knowledge* (pp. 89–106). Geneva: Cahiers de la Fondation Archives Jean Piaget.

von Frisch, K. (1950). *Bees: Their vision, chemical senses, and language*. Oxford University Press.

von Glasersfeld, E. (1976). The development of lanuage as purposive behavior. In S. R. Harnard, H. D. Steklis, & J. Lancaster (Eds.), *Origins and evolution of language and speech* (pp. 212–226). New York Academy of Sciences.

von Glasersfeld, E. (1977). Linguistic communication: Theory and definition. In D. M. Rumbaugh (Ed.), *Language learning by a chimpanzee. The LANA Project* (pp. 55–71). New York: Academic Press.

Walker, S. (1983). *Animal thought*. London: Routledge & Kegan Paul.

Wickler, W. (1972). *The sexual code: The social behaviour of animals and men*. Garden City, NY: Doubleday.

Wolfe, J. B. (1936). Effectiveness of token rewards for chimpanzees. *Comparative Psychology Monographs*, *12*, 1–72.

Woodruff, G., & Premack, D. (1979). Intentional communication in the chimpanzee: The development of deception. *Cognition*, *7*, 333–362.

# *Part IV*

# Developmental perspectives on social intelligence and communication in great apes

# 12 The emergence of intentional communication as a problem-solving strategy in the gorilla

*Juan Carlos Gómez*

## Introduction

Since Wolfgang Köhler carried out his pioneering experiments on the intelligence of apes (Köhler, 1917), investigators have commonly tested the intelligence of anthropoids by confronting animals with situations in which a desirable goal is not directly available. The animal's task is to find an indirect way to get to the goal. In the simplest case, the solution consists of approaching the goal through a detour pathway; in more complex cases, the ape must use a tool to reach the goal (e.g., a box is placed under a goal hanging from the ceiling, or a stick is used to obtain an out-of-reach object) (Bingham, 1929; Guillaume & Meyerson, 1930; Köhler, 1917; Yerkes, 1927). Apes are able to solve these problems with remarkable efficiency, demonstrating, from a human Piagetian perspective, higher levels of sensorimotor or practical intelligence (Piaget, 1936).

Typically, in these experiments, the investigator keeps the animal isolated in the cage while the human observers watch from outside. Köhler's original experiments, however, did not actually conform to this arrangement. In many (if not most) experimental situations, Köhler himself and/or other humans were *inside* the animals' rooms, without any barrier separating them from their anthropoid subjects. This design allowed Köhler to observe what he considered "a most remarkable procedure" of problem solving. After failing to produce the orthodox solution (pushing a box) in the classical problem of the goal hanging from the ceiling, one of Köhler's chimpanzees developed the habit of approaching the human observer, taking him by the hand, and bringing him under the goal, where the animal would try to reach the goal by climbing on the observer.

This "use of the human as an instrument," as Köhler termed it, became a very popular procedure among his chimpanzees – so much so that eventually it had to be extinguished "in the interest of the experiments," because it had become an infallible and universal problem-solving strategy.

Unfortunately, Köhler did not carry out a systematic analysis of problem-solving strategies involving humans. What was their nature? Were they simple functional substitutes or generalizations from the use of boxes or other objects? Or were they *requests for help* addressed to the human? Köhler seemed to imply the first explanation, but sometimes he had the impression that the chimpanzees were requesting the help of humans.

This chapter addresses these questions, reproducing Köhler's original experimental arrangement (i.e., keeping humans in the company of apes during the solution of the problems). We compare the nature of strategies involving humans with the nature of those involving objects to discover if they are qualitatively different and, more specifically, if the strategies involving humans can be considered *acts of intentional communication* differentiated from the *acts of intentional manipulation* involved in the traditional strategies using objects.

## Subject and method

The subject was an infant female lowland gorilla (*Gorilla gorilla gorilla*) named Muni. The animal was born in the wild and arrived at the Madrid Zoo in the early summer of 1980. Her age was then estimated to be about 6 months.[1] She was hand reared in the nursery facilities of the zoo. Humans took care of her needs and acted as surrogate social companions during her first year in captivity. Although she interacted with other infant gorillas after her first year, hand rearing and therefore extensive contact with humans continued during the period considered in this chapter.

Researchers were integrated into the normal social environment of the infant gorilla. They spent about 3 hours per day in contact with the animal. During the first 4 months of the project, researchers spent 7 days per week with her; after that, about 5 days per week. Observations on different aspects of the subject's cognitive and social development were recorded by means of handwritten protocols. Observations of three kinds were recorded: (1) spontaneous behaviors, (2) occasional formal experiments, and, more frequently, (3) spontaneous experimental tasks designed to take advantage of the possibilities offered by the normal environment of the animal. Thus, a longitudinal corpus of observations on the behavioral and cognitive development of the gorilla was obtained, following a methodological approach very similar to that used in Piaget's studies of infant development (Inhelder & Matalon, 1960; Piaget, 1936, 1937).

In this chapter we draw on observations of the gorilla's behavior in problem situations involving an "out-of-reach, hanging" goal (recorded from July 1980 to May 1982). These situations included trying to reach a high platform, trying to open a door's latch, trying to reach suspended leaves of a tree or a piece of food, and so forth. When experimental interventions were carried out, humans used both food items, as in Köhler's experiments, and diverse objects of interest to the gorilla, including toys.

In all problem situations, at least one human remained within reach of the gorilla, to allow the possibility of solutions involving humans. Among the 22 months of records, the 352 observations that satisfied these conditions constitute the material that is analyzed in this study.

**Categorizing behaviors**

All 352 observations were considered to be *intentional problem-solving attempts* (i.e., the behaviors exhibited by the subject were interpreted as attempts to find an indirect way to reach an inaccessible goal). It should be noted that both successful solutions and failures were included in the corpus of 352 observations.

To qualify as an intentional problem-solving action, a behavior had to involve the following elements:

a *goal* (an environmental element whose attainment results in cessation of the intentional activity),
an *obstacle* or, more generally, conditions blocking the direct attainment of the goal, and
a *display of actions* designed to reach the goal.

*Goal directed behaviors* are characterized and identified by the *persistence* of the organism in the operations toward the goal, the *selection* of alternative routes or means, the introduction of *variations* in the operations as a function of feedback about the closeness of the goal, and the *cessation* of activity when the goal is reached (Bruner, 1982).

In our problem situations, the goal was highly variable (platforms, food items, swings, latches, etc.), but the obstacle was always the same: The vertical distance separating the gorilla from the goal was too great to allow her direct application of prehensive (in the case of goals to be taken) or locomotor (when the goal was a place to be reached) schemes.

All observations were initially classified as solutions involving objects or solutions involving humans. The behaviors in which the gorilla applied her schemes on inanimate elements of the environment (e.g., bringing a box under the goal and climbing on it) were considered to be *solutions involving objects* (SIO). All instances in which the schemes of the gorilla were applied on human beings (e.g., climbing on a human to reach an object hanging from a nearby tree) were considered to be *solutions involving humans* (SIH).

SIOs were classified according to the nature of the means scheme displayed by the gorilla. Those solutions involving an indirect displacement on inanimate environmental structures were called *roundabout strategies* (e.g., climbing up a tree to approach an object hanging from one of its boughs). Those involving the construction of an indirect way of changing the disposition of environmental elements were considered *instrumental strategies* (e.g., moving a box under the goal to climb on it).

The aim of this study, however, is to find out if there are qualitative differences between the SIOs and the SIHs, and, more specifically, if SIOs are manipulative in nature (i.e., the result of the gorilla's sensorimotor knowledge of the physical properties of objects and other elements of the inanimate environment), wheres SIHs are communicative in nature (i.e., the result of the gorilla's sensorimotor knowledge of the social properties of humans).

Thus, concerning strategies using humans, we are confronted with two

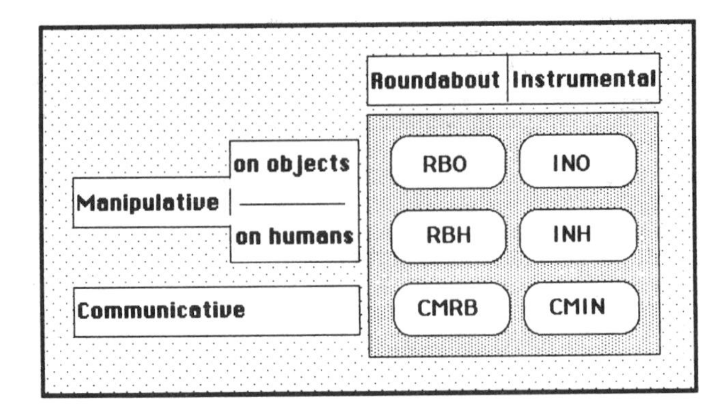


Figure 12.1. The six categories in which problem-solving strategies are classified. The categories are organized in a 3 × 2 table. There are three roundabout strategies (RBO, roundabout with object; RBH, roundabout with human; CMRB, communicative roundabout) and three instrumental strategies (INO, instrument with object; INH, instrument with human; CMIN, communicative instrumental strategy)

possibilities: First, suppose that SIHs do not take into account the social nature of humans, but are mere generalizations of SIO schemes applied to humans. In that case there will be roundabout strategies using humans (RBH) as a generalization of roundabouts with objects (RBO), and instrumental strategies using humans (INH), parallel to the instrumental use of objects (INO). The second possibility is that SIHs do take into account the social nature of humans and present the properties of intentional communication. In that case there will be communicative roundabouts (CMRB), in which communication with the human is used to carry out a detour approach to the goal, and instrumental communication (CMIN), in which the human is requested to change his or her position to help.

Thus, there are six possible ways to classify the gorilla's problem-solving strategies, as summarized in Figure 12.1.

To distinguish between manipulative and communicative behaviors we relied on studies of communicative development in prelinguistic human infants. Whether implicitly or explicitly, psychologists use two major criteria to characterize *intentional communication* in sensorimotor infants (Bates, 1976; Bates, Benigin, Bretherton, Camaioni, & Volterra, 1979; Bretherton & Bates, 1979; Bruner, 1975; Harding, 1981; Lock, 1980; Sugarman, 1984).

The first criterion is the *gestural* quality of behavioral schemes: Communicative actions are not designed to act as direct physical agents. Their intended effects do not occur as mechanical consequences of the addresser's action. That is because "gestures" involve transformations in the form of purposive behaviors that render them mechanically ineffective: increments and decrements, subtractions and additions to the morphological components of behavior, alterations in the pattern of display, and so forth (cf. Bretherton & Bates, 1979; Rivière & Coll, 1987).

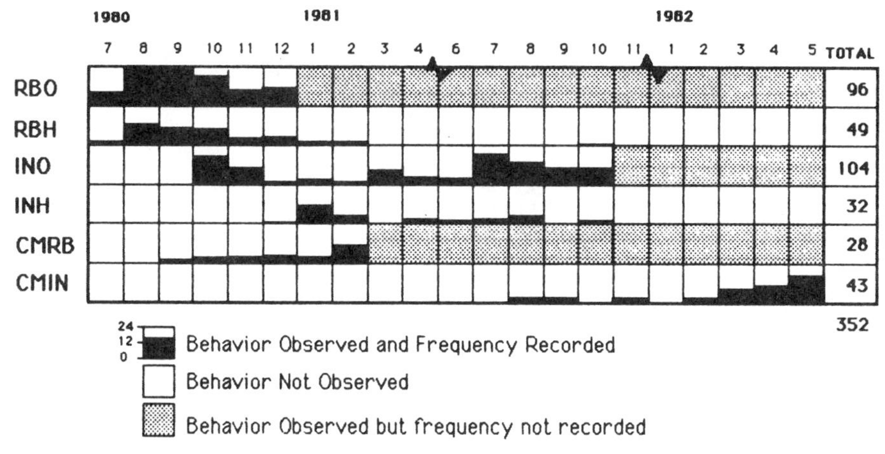


Figure 12.2. Number of observations corresponding to each category displayed as a function of time. On top appear the months of observation identified by their number in each year. Note that 5-81 and 12-81 are omitted because no observations were carried out in those months. On the left appear the six categories for problem-solving strategies. White cells indicate failure to observe the category in that month. Cells in black show the frequency of observation (from 1 to 24). Gray cells indicate that in those months the category continued to be observed, but its exact frequency was no longer recorded.

The second criterion for identifying intentional communication is visual behavior: A *look* must be directed at the human's eyes before, during, or after the performance of a communicative scheme. This criterion is explicitly mentioned by almost all authors involved in infant communication. Although often used as a decisive feature to define an action as communicative, visual behavior is very elusive, because it may be omitted by infants when the act of communication is habitual (Bates, 1976). It generally is reintroduced, however, when something unusual happens (e.g., an absence of response by the adult).

We considered both criteria – lack of mechanical functionality and looks at the eyes – essential for a scheme to qualify as communicative *for the first time*. Once recorded as such, however, instances of the scheme in which the visual feature was absent continued to be considered as communicative actions. Note that behaviors lacking only the mechanical functionality feature would not be classified as communicative if they were recorded before the first occurrence of the complete version of the scheme.

## Results

Figure 12.2 shows the number of observations recorded corresponding to each category. The observations are plotted against time to show their developmental course. Note that exact frequencies are available only for the emergence of each strategy. Once a strategy had become a stable and consistent procedure, only the presence of the strategy in the subject's repertoire was recorded. Three points emerge from Figure 12.2:

1. No category was void: All six strategies were used by the subject. Thus, to deal with humans, Muni relied on both manipulative and communicative strategies.
2. Different strategies appeared at different times. The first strategies involving humans were manipulative; communicative strategies always appeared after their manipulative counterparts.
3. Only two strategies were "transitory" (i.e., were used for a time and then disappeared from Muni's repertoire). These were precisely the two strategies described as application to humans of schemes developed to deal with objects. The other four remained in the behavioral repertoire of the gorilla.

In what follows, the six strategies are discussed in terms of their developmental courses.

## Roundabout strategies

### *RBOs and RBHs*

One of Muni's favorite playplaces was a concrete platform situated beside a lawn. Because the platform was too high for her to reach from the ground, many problems were generated by this situation.

*Obs. 1 (13.vii.80; RBO).*[2] The humans (H) who are accompanying the gorilla (G) are sitting on the platform. Two meters away from this point, a little tree stands just in front of the platform. The gorilla, who is exploring on the ground, suddenly becomes distressed and reaches toward the humans, trying to regain contact with them. After failing to climb directly on the platform, G turns to the tree, approaches it, and begins to climb it, looking back and forth between the tree and the platform. When she finally reaches the platform, G. approaches Hs and hugs them.

*Obs. 2 (8.vii.80; RBH).* H is sitting on the edge of the platform with her legs hanging down. G, who is on the ground, approaches, stands in front of H's legs, extends her arms, and grasps H's legs. Then she pulls herself up H's legs and climbs until she reaches the platform. Her visual behavior is exclusively directed at H's legs (normally those parts she is holding or is going to hold in the course of her climbing activity) and at the platform.

*Obs. 3 (13.viii.80; RBH).* G unsuccessfully tries to reach the platform from the ground. H, after watching her attempts, spreads himself on the ground perpendicular to the wall and puts his legs against the middle of the platform's wall. G approaches H from one side and tries to climb on H's lifted legs. This is, however, a difficult action, and she does not succeed. Suddenly, G gives up, turns 90° to her left, walks to H's trunk (which is lying on the ground), climbs on him, and easily climbs up the slope formed by H's thighs until she reaches the horizontal part of the legs. Once there, she has no problem reaching the platform. Visual behavior is again limited to looks at the different parts of H involved in her climbing attempts, intermingled with looks at the platform. Not a single look is directed at H's eyes.

An intelligent behavior is defined by Piaget (1936) as the coordination of two behavioral schemes so that one functions as the *means* that allows the second

– the *end* – to be carried out. When the scheme that functions as means is a displacement movement that involves no modification in the disposition of environmental elements, we are confronted with a *roundabout* intelligent behavior.

Observation 1 describes a typical roundabout procedure. When Muni fails to reach a goal (contact with humans, in this case) following a direct route, she moves away from the goal toward a structure (a tree) that allows her to overcome the obstacle (the distance between the ground and the platform) and then approach the goal.

Observations 2 and 3 demonstrate that Muni is able to use human bodies as roundabout means to reach the platform. Observation 2 shows a simple kind of roundabout approach without an initial move away from the goal. Observation 3 illustrates a more complex roundabout with humans. Note that the general features of Muni's behavior are exactly the same in the case of the tree (Obs. 1) and the body (Obs. 3): First, she tries to reach the goal directly; then she changes orientation, turning away from the goal and losing eye contact with it, and begins an approach based on the use of a new element as a bridge toward the goal. From the point of view of the behavior displayed by the gorilla, the only differences between the two cases lie in the different *physical* characteristics of a standing tree and a human body lying on the ground. In both cases, only *locomotor skills* – and the ability to organize them in an intelligent way – account for the behavior of the gorilla.

### *CMRBs*

*Obs. 4 (11.ix.80; CMRB).* A swing hanging from a tree, a favorite toy for Muni, is raised too high for her to reach it from the ground. G tries various approaches (all of them roundabout procedures through objects), but her efforts are unsuccessful. She gives up for a while. But as soon as H stands by the swing, G approaches him, lifts her arms toward H, and looks at his eyes. G "freezes" her posture without trying to climb up H. Only when H begins to take her in his arms does G actively collaborate in the lift. Once in H's arms, G turns to the swing and reaches it.

Note that in Observation 4 Muni has succeeded in covering the distance between the ground and the goal by taking advantage of a position spontaneously adopted by the human. Insofar as this performance involves using a scheme on a human to gain a position from which to obtain the goal, this roundabout procedure is comparable to those described in Observations 2 and 3. This time, however, Muni does not try to climb on the human applying a "climbing-on-something-like-a-pole" scheme, but adopts a posture – arms raised to the human while looking at his eyes – and keeps it until the human bows and takes her in his arms.

This is a *communicative scheme* in which the gorilla uses a gesture developed in a different context as a means for asking the human to take her in his arms (Gómez, 1989). This scheme meets the criteria for communication referred to earlier: It does not mechanically provoke its effect, and looks are directed at the human's eyes. Whereas in the previous cases the gorilla relied on her locomotor and climbing abilities to take advantage of the human's position, in this case she relies on a human's action to get the roundabout displacement to be carried out for her.

It must be noted that this request cannot be glossed as "raise G to the goal"; the only part encoded in the scheme displayed by the gorilla is "raise G." She requests to be taken in his arms, not to be raised to the swing. This request is used, however, as a means to the goal, and in this sense it forms part of a broader structure that also includes the swing.

This point is very important, because insofar as the goal is excluded in the arms-raising scheme, we are confronted with a communicative act with *no external referent*,[3] and therefore we remain at the level of simple interactive communication between two organisms without explicit encoding of external objectives.

During the first 4 months of observations, the only problem-solving strategies displayed by Muni were the roundabout procedures just discussed. This means that when the object or the human on which the detour approach had to be based did not occupy a suitable position, the gorilla was unable to reach the goal. The instrumental abilities required to modify an object's position or a human's position to render it useful did not appear until late 1980, when the animal was about 1 year old.

## Instrumental strategies

We consider any means–ends coordination of schemes that brings about a change in the spatial relationships between the goal and the intermediary to be an instrumental behavior. The "means" scheme is any behavior designed to modify the position of the intermediary object or human relative to the goal.

The first instrumental strategy to appear involved the use of inanimate objects (INO) (for a detailed account of the development of INOs, see Gómez, 1988). The following observations are examples of instrumental strategies with inanimate objects.

### *INOs*

*Obs. 5 (13.x.80; INO).* G's goal is to open a locked room, but the door's latch is too high. A pole is leaning against the wall about 1.5 m away. G approaches the pole, takes it with both hands, and holds it upright. Then she begins to move the pole toward the door; after a short displacement, she leans it again on the wall; a few seconds later, she holds the pole again and moves it a little nearer the door, leaving it against the wall. She repeats this action once more. When the pole is leaning against the wall 20 cm away from the door, she begins to climb up it. Because the inadequately secured pole accidentally slides to one side and remains fastened against the door's framework, she is able to reach the latch. G's visual behavior during her efforts consists of looking at different parts of the pole, at the upper zone of the wall, and at the door, mainly the latch area.

*Obs. 6 (iii.81; INO).* The problem is again to reach a door's latch. After an unsuccessful attempt to reach it directly, G approaches a toy tricycle (a favorite toy) placed several meters away and drags it without hesitations or pauses toward the door. As soon as the toy reaches the side of the door just opposite the latch, G stops the displacement and gets on the tricycle. She looks at the latch, but it is still too distant. Then she gets off the toy and pushes it until it is almost under the latch, looking alternately between the goal and the tricycle. She gets on the toy and reaches the latch.

The preceding observations reveal that instrumental strategies with objects in this problem-solving context consist of two parts: first, a displacement of an object toward the goal; second, the application of a roundabout scheme using the displaced object as an indirect means to the goal. Because the second part involves the same skills described in the previous section, it needs no further comment. Our analysis focuses on the first part, the instrumental action in the strict sense.

The instrumental action has two main components: an *interfacing contact* between the gorilla and the object, generally carried out with the hands, and a *force* or set of forces applied on the object through that interface. Each component allows the realization of movements organized according to a plan that determines their direction and extension. The crucial point is that *the movements of the object are brought about by mechanical causes*: The energetical and directional components of the vectors that cause the movement of the object are transmitted by the gorilla through the physical interface.

The precise form of the interface and the forces used by the gorilla vary according to the object used as intermediary and the displacement required by the situation. Different grasps and different forces are used to hold a stick upright and to drag a box along the floor. The nature of the actions shown in Observations 5 and 6, however, remains essentially the same. They are goal directed manipulations of objects adapted to the laws of mechanical causality; that is, the actions of the gorilla are designed to take into account factors such as weight, gravity, statics, texture, form, and so forth, in order to provide the necessary forces to cause a change in the object. The "physical knowledge" involved in these actions remains at the practical or sensorimotor level (Piaget, 1936).

We are now prepared to address the nature of instrumental problem-solving strategies involving humans.

## *INHs*

*Obs. 7 (8.i.81; INH).* In the course of mild rough-and-tumble play between G and H on the floor, H puts his feet against a wall so that his legs form a bridge to the higher part of the wall where there is a window with a sill. This is a favorite place for G. As soon as H adopts this posture, G climbs along his legs and, using each leg as a support for her feet, stretches to the sill. When G is about to reach her goal, H lets his legs slip to the ground. G then takes one of H's legs in both hands and pushes it up toward the goal. G uses all her strength, and her efforts are interrupted only if H himself keeps his leg against the wall. On such occasions, G tries again to climb up H's legs. As soon as H lets his legs fall again, however, G strives to lift them to the right position. G's initial lifting procedure consists of grasping H's legs from below with both hands and pushing up. From time to time, especially when his legs are already slightly lifted, G also pushes with her back, after slipping under his legs. G does not interrupt her efforts to allow H to lift his own legs, nor does she display a clear-cut expectation that the legs will stick against the wall. G's eyes are exclusively directed at the legs, the wall, and the window; she does not look at H's eyes.

*Obs. 8 (8.i.81; INH).* H is sitting on the ground just in front of the window area. Suddenly, G runs toward him and pushes against his chest (using head and shoulders) with enough force to make him fall backward to the wall. Then she climbs along his body to the windowsill. Her visual behavior is again exclusively concentrated on the wall and the parts of the human involved in her climbing.

*Obs. 9 (27.i.81; INH).* H is kneeling in front of a door. G, who is interested in the door's latch, grasps H by the neck and pulls, causing him bend down on all fours. G then gets on H's back, which now is a horizontal surface, stands on it, and reaches the latch. Once more, G's gaze is directed at the latch and different parts of H, but never at his eyes.

All the foregoing behaviors involve instrumental interventions designed to bring the human to a position suitable for use as a roundabout pathway toward the goal. From the point of view of the structure of the task, these behaviors and those in previous observations do not differ. In order to determine if the schemes used to manipulate persons are comparable to those used with objects, we analyze the behaviors along four dimensions: the form of the interface between the gorilla and the human, the strength displayed by the animal, the patterning of the strength, and the visual behavior of the gorilla.

The *interface* has a variety of forms: Muni uses her grasping hands to lift the legs of the human and to bend his body, her hands without grasping activity to push him on the chest, her shoulders and arms to push his legs up, and even her head to push against his chest.

The *strength* applied through the previously described actions is, more or less, what is required to move the body parts involved in each operation.

The *temporal patterning* of the action is approximately constant from the beginning to the completion of the action; that is, Muni does not allow any room for the appearance of the human's spontaneous activity.

The *gazing behavior* consists mainly of alternate looks between the goal and the parts of the human's body that are the focus of activities. Not a single look is directed at the human's eyes.

Thus, on the whole, no differences appear between these problem-solving activities involving humans and those involving objects analyzed in the previous section. They can be described as the application of instrumental schemes adapted to the properties of humans *as physical bodies*. Muni tries to provoke certain modifications in the human's position, applying the physical forces necessary to bring about the displacement mechanically. The only differences between INO and INH reflect the morphological peculiarities of the human body as an object (its specific form, weight, etc.). These, however, are "superficial" differences reflecting adaptations to particular "surface" properties. There is no evidence of adaptation to the "deep" properties that distinguish an object from a person.

To summarize, at this developmental point, and in the context of the problems we are considering, Muni treats humans by and large as she would treat a box or any other object to be manipulated. If we were to answer our main question (Does the gorilla treat humans as different from things?) on the basis of the foregoing evidence, the answer would be no.

## Communication as problem-solving strategy

### *CMINs*

*Obs. 10 (12.viii.81; CMIN).* G moves a box under the latch of a door, stands on it, and tries to open the door. Because it leads to a room disallowed to G, however, H prevents her from opening the door and withdraws the box. Soon afterward, G approaches H, takes him by the hand, and pulls him toward the door, which is about 4 m away. The initial pull is moderately strong; then it is weakened and becomes very soft, even nonexistent, despite the manual contact between G and H. During the walk to the door, G looks alternately at the door and at H's eyes. Once under the latch, G asks to be taken in his arms, and then reaches the latch.

*Obs. 11 (9.x.81; CMIN).* G has been moved to a new location, a cage with indoor and outdoor areas. She cannot open the new latches. Soon after her arrival in the cage, G approaches H and takes him by the hand to the door, using the same procedure described in Observation 10. Once there, she looks alternately at the latch and at H's eyes. H tells her he is not going to open the latch, and he leaves. Later, the same request is repeated, with identical result.

*Obs. 12 (22.iii.82; CMIN).* G and two Hs are inside an experimental room where some learning experiments are carried out with G. When the experimental session is over, G approaches H, who is sitting in a corner, takes him by the hand, and gently drives him to the door (same procedure as in Observation 10). Once in front of the door, G takes H's hand toward the latch zone; G keeps H's hand stretched midway to the latch, looking at it. Then she turns her head to H and looks fixedly at his eyes until she returns her look to the latch. H does not immediately open the door, and G asks to be taken in his arms. Once this is done, she again takes H's hand to the latch, although she could have opened it herself. This time H opens the door.

As in the previous section, these observations are instances of instrumental interventions by Muni using a human in order to solve a problem. Also as in the previous instrumental observations, two parts can be distinguished in these strategies: first, the displacement of the human toward a position where he can be of aid; second, attainment of the goal, taking advantage of the new position occupied by the human. Examining the displacement part using the same four dimensions as before, the following analysis can be made.

Muni uses a single form of *interface* in all cases: taking the human by the hand. It is a mild hand-in-hand grasping generally reciprocated by the human.

The *strength* displayed by the gorilla varies with the course of the displacement. Usually it consists of an initial pulling carried out with moderate strength, followed by a weakening of the grasp and of the pulling, which

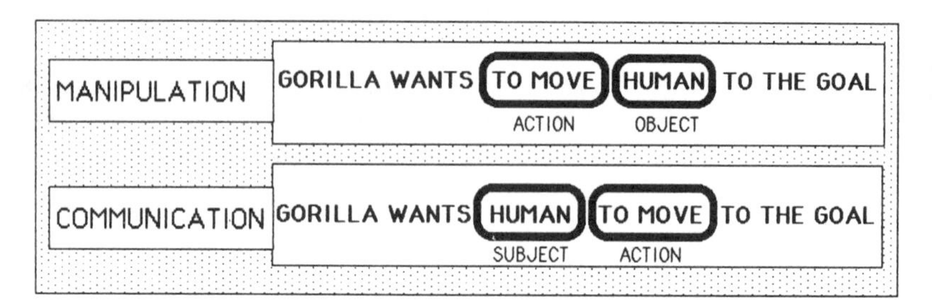


Figure 12.3. A comparison of the structure of instrumental problem-solving strategies used by Muni with humans. Note the different roles played by the human in the manipulative and the communicative versions of the scheme.

sometimes practically disappears. A brief reappearance of the initial strength occurs if there is a change of direction during the displacement or if reluctance is shown by the human. All in all, except on very rare occasions, the strength displayed by the gorilla even at those points was less than that required to displace a human body and, what is more important, much less than the force she actually showed when trying to displace huge objects.

*Visual behavior* is also substantially different. She continues to look at the goal, but she alternates with looks at the eyes of the human. This kind of gaze can hardly be explained as a simple act of checking a physical displacement, because for that purpose it would be more effective to look at the body as a whole or, if anything, at the feet and legs. These looks tend to appear in an alternation pattern, that is to say, forming a pattern in which a look at the goal is followed by a look at the eyes, or vice versa.

On the whole, this analysis makes it clear that the behavior displayed by Muni in Observations 10–12 was not intended to bring about the displacement of the human from one point to another mechanically. On the other hand, the direction of her soft pulling, her looks at the goal, and the contextual information available make it plain that her aim was to "get the human under the goal."

Now, if we assume that the gorilla wants the human under the goal, yet she does not try to displace him to it, the only alternative interpretation is that *the gorilla wants the human to move himself* to the desired position.

The critical difference between these two alternatives is illustrated in Figure 12.3. Whereas in the first case the human is included in the gorilla's scheme as an *object* to be moved, in the second he is included as the *subject* of the action of moving. In Observations 10–12 the human is no longer treated as an object to be manipulated, but as a subject, that is, a social being who "behaves" and has control of his own actions. I do not claim that the gorilla has reflective or conscious knowledge of the human as a subject. Muni remains at the sensorimotor level of knowledge. Rather, I claim that the gorilla's actions are adapted to the "subjective" characteristics of humans. So long as we can speak of a sensorimotor concept of "object" on the basis of behaviors adapted to the physical properties of objects, we can also speak of a sensorimotor concept of "subject" on the basis of behaviors adapted to the properties of persons as social beings.

The same conclusion emerges from analyses of the second part of the strategies – the attainment of the goal once the human is in the desired position. The behavior of taking the hand of the human toward the latch (Observation 12) presents the same characteristics as the action of taking the human by hand to the door; it is not intended as a mechanical means to produce the result, and it is accompanied by looks at the human's eyes.

In this case, the fact that the gorilla includes the human in her scheme as the subject of the action TO OPEN is still clearer. The mechanical version of the scheme GORILLA WANTS HUMAN TO OPEN LATCH would be GORILLA WANTS TO OPEN LATCH WITH HUMAN. In the latter, the human (or rather, the human's arm) would be included as a physical intermediary to manipulate the latch. However, if Muni were actually trying to move the latch using the human's arm as a prolongation, she would never take it by the hand, but rather by the other extreme, to handle it as a stick.

On the other hand, in the second part of this observation, when Muni takes the human's hand toward the latch from his arms, she does not even try to establish contact between the hand and the latch, let alone to transmit a specialized movement using the human's hand as a mechanical intermediary.

Thus, the only possible reason why the gorilla includes the human in her schemes is that she considers the human as a subject capable of carrying out by himself the action of opening the door's latch.

### The sensorimotor concept of subject

The properties of objects are shape, texture, weight, and so forth, and they are placed in a physical space that imposes certain constraints on their manipulation (Piaget, 1937). Subjects have two essential properties that even the most elementary sensorimotor concept should reflect: *agentivity*, or the ability to generate their own movements and actions without external mechanical causation, and *sensibility*, or the ability to perceive things and events – both physical and social – and react to them.

An essential part of this "subject concept" is the causal link between the subject's perceptive and agentive sides, that is, the possibility that the subject's behavior is influenced by what she or he perceives. Intentional communication exploits this property of subjects.

When an organism produces a behavior with the intention that another organism perceive it and, as a consequence, carry out a certain action, it is engaging in *intentional communication* in its most basic form.[4] This is what happens in Observations 10, 11, and 12.

As previously stated, the two main indicators of the communicative nature of actions are absence of intended mechanical functionality and gaze at the partner's eyes. The first of these can be interpreted as an adaptation to the agentive properties of subjects. We have already seen that when Muni takes the human by hand to the door (Obs. 10–12), when she takes his hand toward the latch (Obs. 12), or when she raises her arms and waits to be lifted (Obs. 4), the design of these actions reflects the gorilla's adaptation to the fact that the human is the agent of the behavior she is trying to provoke.

The precise meaning of the gorilla's gazing behavior, however, is not so straightforward. In intentional problem-solving strategies in which only objects are involved (e.g., Obs. 1, 5, and 6), visual behavior follows a characteristic pattern: Looks are directed at the relevant elements of the situation – mainly the goal, the means, and the obstacle. It seems reasonable to suppose that the function of these looks is to control the relevant parts of the situation involved in the execution of the intelligent action. In the strategies involving the manipulation of the human's body as a mean (Obs. 7–9), this pattern is reproduced: The gorilla looks at the human's body parts involved in her actions. In the strategies classified as communicative (Obs. 10–12), however, a new part of the human is included as a target of visual behavior: the human's eyes. The implication is that for this new kind of strategy the eyes have become a relevant part of the human, an element of the situation to be controlled by the gorilla. My suggestion is that the human's eyes are relevant because the gorilla is using a communicative strategy, and an essential condition for communication to occur is that the addressee perceive the addresser's gestures. Thus, looking at the human's eyes is a way of controlling an essential factor of communicative causality – as essential as the existence of a physical interface is for mechanical causality.

This is why gaze has intuitively become such an important criterion to identify intentional communication in human infants. The gestural quality of actions is just a recognition that there are external causal agents apart from one's own actions and that one's behavior may be somehow linked to these agents. Looking at the eyes, however, is a recognition of the way this link is established: through the other's perception of one's gestural actions. This would also explain why gaze disappears in communicative actions that have become habitual and why it reappears when the adult's response is not normal, that is, when the external agent's involvement in the request needs to be checked.

We know that in addition to perception, other processes can be attributed to human adults responding to a gorilla or human infant: intentional attribution, belief, altruistic motivation, and so forth. The question is to what extent the gorilla or the infant must know about these central processes in order to produce an intentional communicative act on the sensorimotor level, or, to put it differently, to what extent the gorilla and the infant need a "theory of mind" to communicate (Bretherton, McNew, & Beeghly-Smith, 1981; Leslie, 1987; Premack & Woodruff, 1978). Bretherton et al. (1981) have suggested that the early communicative behaviors of human infants are the first indicators of their emerging theory of mind. Leslie (1987), however, argues that the theory of mind does not develop until after the sensorimotor

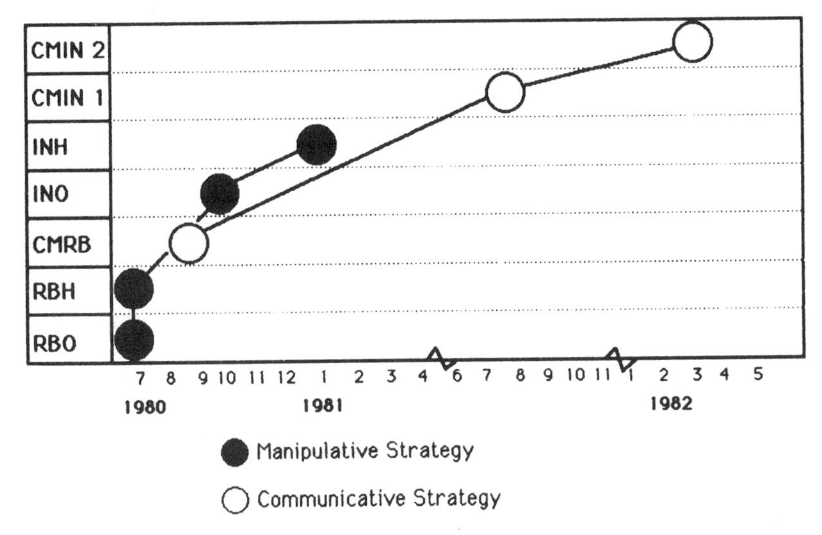


Figure 12.4. Onset moment for problem-solving strategies. Black circles represent manipulative strategies, and white circles represent communicative strategies. Note that CMIN has been divided into two different strategies: CMIN 1 to represent patterns of the kind exemplified in Observation 10 (displacement request plus communicative roundabout); CMIN 2 for patterns of the kind illustrated in Observation 12 (displacement request plus solution request).

period has been completed. My own suggestion is that the adaptations to the human adult's perceptual and cognitive processes evident in intentional prelinguistic communication constitute one of the sensorimotor foundations on which the theory of mind is constructed later.

## The developmental sequence of strategies

The emergence of intentional communication as a problem-solving strategy is compared with the development of the "traditional" manipulative strategies in Figure 12.4. Note that the latter include the manipulations of humans as objects and that the CMIN strategy has been separated into two different categories.

### *Roundabouts and instrumental manipulations*

All roundabout strategies, both communicative and manipulative, precede instrumental strategies developmentally. This is congruent with Köhler's suggestion that roundabout performances are easier than instrumental ones (Köhler, 1917). In fact, in the kind of problem considered in this chapter, an instrumental strategy consists of the intelligent coordination of a roundabout scheme and a manipulative scheme. Because the roundabout is contained as an element within the manipulative scheme, the structure of the instrumental procedure has to be more complex than the structure of the roundabout. Thus, the main achievement of the gorilla when she shows for the first time the INO strategy is the ability to intelligently coordinate two different action patterns into a single scheme.

Nonetheless, we cannot assume that the two components of the instrumental strategy are equally complex. It could be argued that the manipulative part of the instrumental procedure is inherently more complex because it involves changing the position of an external object relative to another external object (the goal), whereas in the roundabout the gorilla has only to change the position of her own body in relation to external objects.

### *Communication for interaction and communication about objects*

Communicative strategies always appear after strategies based on the use or manipulation of things. This is true, however, only if we consider roundabout and instrumental procedures separately, because communication appears two times as a problem-solving strategy: first, as the last step of roundabout procedures (CMRB), and second, as the culmination of instrumental strategies (CMIN). The first communications appear just before the first instrumental manipulations of objects (INO).

The two-step onset of intentional communication reflects the existence of two different kinds of communication: one that regulates simple interactions between two partners that involve no external object, and another that organizes interactions involving external objects. If we compare Observations 4 and 12, it becomes apparent that the arms-raising scheme involves only the gorilla and the human, whereas the two versions of the hand-taking procedure involve the coordination of the gorilla, the human, and an external object in the same scheme.

This distinction between communication for interaction and communication about external objects was noted by Sugarman (1973) in one of the first studies of infant sensorimotor communication and has recently been reformulated by Golinkoff (1983). Case (1984) analyzes it in terms of a general mechanism of sensorimotor development: the passage from *simple operational coordination* (actions organized around a single target to attain a goal) to *bifocal operational coordination* (actions organized around two targets to attain a goal). In the case of communicative development, bifocal coordinations involve an object and a person. I am going to retain the words "monofocal" and "bifocal" as descriptive terms, but without implying the conceptual distinctions suggested by Case.

### *Tool use and communication*

If we focus our attention on the sequences of development of INO, INH, and CMIN, it is clear that communication is a much later achievement than instrumental manipulation. The difference in timing between the first attempts to use things as instruments (INO) and the first attempts to use humans as instruments (INH) is perhaps because of the "hugeness" of humans as physical objects. Indeed, the extension of instrumentality to humans coincides in time with the beginning of the mastery of instrumental strategies and is contemporaneous with the generalization of INO patterns to a wider range of objects (Gómez, 1988).

The relationship between the ability to carry out instrumental actions with things and the ability to intentionally communicate with persons has been a major topic of research in developmental psychology (Bates, 1976; Bates et al., 1979; Harding, 1981; Sugarman, 1973, 1984). In human infants, the first, primitive manifestations of tool use appear at around the same time as intentional communication about external objects. This has been interpreted as evidence of a close cognitive relationship between these two abilities. The nature of this relationship, however, remains obscure. After reviewing the literature, Bates and Snyder (1987) concluded that "this relationship is correlational rather than sequential" (p. 187). Sarriá, Rivière, and Brioso (1988), however, found no linear correlation between intentional communication and the means–ends scale of the Uzgiris and Hunt instrument. Similarly, Martinsen and von Tetzchner (1988) were unable to find any clear-cut correlational trend when examining the developmental profiles of several infants observed every 2 weeks.

Because both abilities eventually appear in the same subject, our data on the gorilla can be seen as supporting the view that there is a general relationship between tool use and communication. This relationship, however, is not so close as in the human species, because in our subject there was a 10-month difference between the first manifestations of tool use (INO) and the appearance of intentional communication about external objects (CMIN)[5] – a temporal gap wider than any reported thus far in human infants. Similar "decalages" have been reported for nonhuman primates in other sensorimotor domains (Antinucci, Spinozzi, Visalberghi, & Volterra, 1982; Parker, 1977).

This result could be taken to support Harding and Golinkoff's view that the ability to use instruments is necessary but not sufficient to develop intentional communication about objects (Harding & Golinkoff, 1979). The wide gap shown by the gorilla could be explained as a species difference in the timing of the process that supplies the specifically communicative abilities that must be added to tool-using abilities for intentional communication to appear. One is tempted to conclude, then, that intentional communication is the result of combining the cognitive abilities underlying instrumental manipulations of objects (e.g., the bifocal mode of operation) with a set of specifically communicative abilities – what I have called the *subject concept*.

There emerges, however, a problem for this interpretation when one considers the results of this study closely. In September 1980, Muni had already demonstrated (with CMRB) that she was adapted to the human as a subject, that is, as a social being she could communicate with. In January 1981 she demonstrated (with INH) that she was able to coordinate the human and an external objective in a bifocal operation. Why, then, does "bifocal communication" appear only several months later?

For one thing, we cannot assume that the existence of simple interactive communication as revealed by CMRB patterns is evidence that the gorilla has acquired *the* subject concept. She merely demonstrates that her behavior is adapted to *certain* aspects of subjects. From a cognitive developmental point of view, we interpret such behavior as evidence that she possesses *a* subject concept (i.e., a practical understanding of subjects corresponding to a certain level of sensorimotor development). Similarly, when a human infant in the fourth stage of sensorimotor development who demonstrates the ability to uncover an object is confronted with the displacement of that object, then the limitations of the infant's object concept appear (Piaget, 1937). By analogy, one possibility is that the "subjective" sensorimotor knowledge required by CMIN strategies is more complex than that necessary to carry out CMRBs.

On the other hand, the existence of manipulative schemes involving "bifocal" coordinations does not imply that this level of operation is concurrently available for every task. The most abstract properties of sensorimotor schemes ("bifocality" in this case) may be content-dependent. If so, they will not transfer from one domain to another. In this case, the bifocal mode of operation, already available in the physical domain, will be reconstructed in the social domain by means of interactions with humans. This is another possible explanation for the "horizontal decalage" (Piaget, 1941) between CMRB and CMIN.

All in all, the results of this study point to the existence of a complex relationship, modulated by many different factors, between tool use and intentional communication. This is in agreement with the emerging view from human infant studies that emphasizes the nonsequential and nonlinear nature of the link between tool use and communication.

### *Origins of communicative schemes*

Given the developmental sequence RBO-RBH-CMRB and INO-INH-CMIN, in which the application of object schemes to humans always precedes the appearance of communicative patterns (Figure 12.4), it is tempting to conclude that communicative schemes are outcomes of a differentiation process that occurs when the gorilla wrongly applies physical schemes to humans (i.e., that communication arises out of physical operations carried out on human beings as a consequence of the special feedback provided by this special type of "object").

As a matter of fact, however, that is not the case with the two communicative problem-solving strategies considered here. Both the arms-raising scheme and the hand-taking patterns originally developed in a different context and only later were incorporated into the problem-solving situations (Gómez, 1989). The arm-raising gesture originated from attempts to gain the embrace position by means of climbing. So in this case the communicative scheme derived from an action intended to be functional by itself, but it originated in a context different from the problem tasks discussed here. Only after the scheme was established in the gorilla's repertoire did she begin to use it as a means to reach positions suitable for the solution of problems.

The hand-driving scheme originated in a completely different context: rough-and-tumble play with humans. Muni began to use it as a means to ask humans to play chase her. Later, the scheme was generalized to ask various displacements from the human, among them the displacement in problem situations.

Thus, neither CMRB nor CMIN resulted from a transformation of analogous schemes on objects, although the transformation of physically functional schemes into communicative patterns is one of the mechanisms of gesture formation in human and gorilla infants (Clark, 1978; Gómez, 1989).

### *Structure and function in communicative schemes*

From a *structural* point of view, the sequence of development of communicative strategies depicted in Figure 12.4 illustrates the transition from simple monofocal patterns to bifocal patterns and combinations of bifocal patterns. Thus, CMRB is a simple communicative act without external referents (monofocal). It is followed by the appearance of CMIN 1, which is a sequential combination of a bifocal pattern (taking a human by hand toward a place) and a monofocal pattern (a CMRB once the human is in the appropriate position). Finally, a combination of two bifocal patterns appears with the use of CMIN 2 strategies, in which two communicative acts with external referents are combined.

From a *functional* point of view, there is a transition from the simple request for help to the request for a solution to the problem. Thus, in CMRB Muni asks of the human a supportive action in a certain segment of the problem. In CMIN 1 she makes a more elaborate request, asking the human to move to a certain place in order that he can be of aid. Finally, in CMIN 2 she asks the human to solve the problem for her, using again an elaborate request that requires two successive steps: first, asking the human to move to the appropriate place and, second, asking him to carry out the solution. This illustrates how increasing structural complexity can be used in increasingly complex communicative functions.

## Summary and conclusion

When in 1917 Wolfgang Köhler published his "Intelligence Tests on Anthropoids" (such was the original title of what later would be translated into English as *The Mentality of Apes*), he demonstrated that apes are able to manipulate objects in an intelligent way to solve simple practical problems. Köhler's main interest was to show that apes possess a "practical understanding" of the physical world that allows them to organize their behavior into intelligent actions adapted to the requirements of the environment. The use of tools to obtain out-of-reach objects has become the emblematic behavior of Köhler's studies.

Tool use, however, was not the only type of strategy his apes relied on to solve the problems. Köhler briefly refers to a number of procedures spontaneously developed by his chimpanzees in which they tried to involve the human experimenters in the solutions of the problems. These were favorite procedures for the apes; they resorted to the strategy of seeking a human's help to solve many different problems. Because Köhler's concern was with manipulative intelligence, however, he paid little attention to procedures involving humans and dismissed them as an obstacle to the study of the manipulative intelligence of apes.

In this chapter I have reversed Köhler's interests, focusing on problem-solving strategies involving humans. Using longitudinal data from an infant gorilla for which both the use of objects and the use of humans to solve problems had been recorded, I have tried to analyze the nature of these strategies.

I have shown that strategies involving humans can be of two qualitatively different types: first, mere generalizations of the use and manipulation of objects in which the body of the human is treated just as one more physical object; second, strategies that present the characteristics of intentional communication, as defined in prelinguistic human infants, and that can be described as "requests for aid" addressed to the human to solve the problems.

I have argued that the second type of strategy is based on a practical understanding of the human as a subject, as opposed to a practical understanding of the human as an object on which the former strategies are based. This practical understanding of humans as subjects depends on the gorilla's adaptation to two basic properties of humans: their ability to act spontaneously (agentivity) and their ability to perceive things and react to these perceptions.

Gorillas (and human infants) develop behavioral schemes adapted to these properties (as shown by their intentional neglect of mechanical causality and the control of the adult's gaze), and as a consequence, they are said to be capable of intentional communication.

Comparative studies of communication in the sensorimotor period, like comparative studies of sensorimotor intelligence from a Piagetian point of view, must proceed in the future through a careful comparative analysis of the components and processes of communicative development in different species. Differences in timing, such as those found in this study, may reflect differences in the way each species adapts to the properties of the other as a subject, just as heterochronies and other peculiarities of sensorimotor development in nonhuman primates seem to reflect different ways of adaptation to the physical world within the general framework of sensorimotor intelligence.

**Acknowledgments**

The research reported in this chapter was funded in part by a grant from the Comisión Asesora de Investigación Científica y Técnica awarded to Dr. J. L. Linaza. We are grateful to the Madrid Zoo, especially to its director, Mr. Thomás Cerdán, for permission to work with the gorillas in the zoo facilities.

The corpus of data on which this chapter is based is the result of the joint efforts of the author and Paloma Fernández, who started up the project and shared all the tasks involved in it for many years. My thanks to José L. Linaza for his support and Mamen de Pablo for early contributions to the corpus of data. Special thanks to Sue Parker and Kathleen Gibson for their patient editorial work, especially difficult with a nonnative writer of English. Needless to say, they are not responsible for any shortcomings in this chapter.

**Notes**

1. Because age estimations leave open a considerable range of variability about the actual age of the animal, in this chapter emphasis is placed on the *dates* at which relevant behaviors appeared. According to the age estimate, Muni was about 8 months old at the time of the first observations reported here (July 1980) and about 2.5 years old at the time of the last observations (May 1982).
2. The date of the observation is placed in parentheses in this order: day, month, year, followed by the category in which it is included. Thus, Observation 1 was carried out July 13, 1980. It is an instance of roundabout behavior with object.
3. The term *referent* is used here in its most general sense to designate external objects included in prelinguistic communicative acts.
4. An organism can also exploit its knowledge about the properties of subjects to ensure that certain reactions of other organisms do not occur. Thus, an organism may learn to suppress a behavior, or to carry it out when the other cannot see it, in order to avoid a reaction not desired. This could be considered as intentional concealment or disguise, the first step toward intentional deception.
5. It must be noted that we are discussing communicative schemes used as problem-solving strategies, i.e., they are embedded in more complex behavioral structures. For example, in Observation 10, Muni asked the human to move to the door because she wanted to use him as a support to reach the latch and open the door, and she wanted to open the door because there was something inside the room that interested her. The communicative act is but the first step in a complex chain of goal directed behaviors. It was shown (Gómez, 1989) that Muni was able to do simple displacement requests – i.e., not embedded in problem-solving situations – a couple of months before their appearance as a problem-solving strategy. Thus, the actual gap between the beginnings of tool use and communication about external objects was a little narrower: only about 8 months.

**References**

Antinucci, F., Spinozzi, G., Visalberghi, E., & Volterra, V. (1982). Cognitive development in a Japanese macaque (*Macaca fuscata*). *Annali dell'Istituto Superiore di Sanita*, *18*(2), 177–184.

Bates, E. (1976). *Language and context: The acquisition of pragmatics*. New York: Academic Press.

Bates, E., Benigni, L., Bretherton, I., Camaioni, L., & Volterra, V. (1979). *The emergence of symbols: Cognition and communication in infancy*. New York: Academic Press.

Bates, E., & Snyder, L. (1987). The cognitive hypothesis in language development. In I. C. Uzgiris & J. M. Hunt (Eds.), *Infant performance and experience: New findings with the ordinal scales* (pp. 168–204). Chicago: University of Illinois Press.

Bingham, H. (1929). Chimpanzee translocation by means of boxes. *Comparative Psychology Monographs*, *5*(3), 1–91.
Bretherton, I., & Bates, E. (1979). The emergence of intentional communication. In I. C. Uzgiris (Ed.), *Social interaction and communication during infancy* (pp. 81–100). San Francisco: Jossey-Bass.
Bretherton, I., McNew, S., & Beeghly-Smith, M. (1981). Early person knowledge as expressed in gestural and verbal communication: When do infants acquire a "theory of mind"? In M. E. Lamb & L. R. Sherrod (Eds.), *Infant social cognition* (pp. 333–373). Hillsdale, NJ: Erlbaum.
Bruner, J. S. (1975). From communication to language: A psychological perspective. *Cognition*, *3*, 255–287.
Bruner, J. S. (1982). The organization of action and the nature of adult–infant transaction. In M. von Cranach & R. Harré (Eds.), *The analysis of action* (pp. 313–327). Cambridge University Press.
Case, R. (1984). *Intellectual development: Birth to adulthood*. Orlando, FL: Academic Press.
Clark, R. (1978). The transition from action to gesture. In A. Lock (Ed.), *Action, gesture and symbol: The emergence of language* (pp. 231–257). New York: Academic Press.
Golinkoff, R. M. (1983). Infant social cognition: Self, people and objects. In L. S. Liben (Ed.), *Piaget and the foundations of knowledge* (pp. 179–200). Hillsdale, NJ: Erlbaum.
Gómez, J. C. (1988, June). *The development of means–ends intelligent coordinations in an infant gorilla*. Paper presented at the Third European Conference on Developmental Psychology, Budapest.
Gómez, J. C. (1989). *El desarrollo de la comunicación intencional en el gorila*. Unpublished doctoral dissertation, Universidad Autónoma de Madrid.
Guillaume, P., & Meyerson, I. (1930). L'usage de l'instrument chez les singes I. *Journal de Psychologie*, *27*, 177–236.
Harding, C. G. (1981). *A longitudinal study of the development of the intention to communicate*. Doctoral dissertation, University of Delaware (UMI No. 8123796).
Harding, C. G., & Golinkoff, R. M. (1979). The origins of intentional vocalizations in prelinguistic infants. *Child Development*, *50*, 33–40.
Inhelder, B., & Matalon, B. (1960). The study of problem-solving and thinking. In P. H. Mussen (Ed.), *Handbook of research methods in child development* (pp. 421–455). New York: Wiley.
Köhler, W. (1917). Intelligenzprüfungen an Anthropoiden. *Abhandlungen der Preussische Akademie der Wissenschaften. Physikalische-Mathematische Klasse*. Nr. 1.
Leslie, A. (1987). Children's understanding of the mental world. In R. L. Gregory (Ed.), *The Oxford companion to the mind* (pp. 139–142). Oxford University Press.
Lock, A. (1980). *The guided reinvention of language*. New York: Academic Press.
Martinsen, H., & von Tetzchner, S. (1988, June). *The development of intended communication*. Paper presented at the Third European Conference on Developmental Psychology. Budapest.
Parker, S. T. (1977). Piaget's sensorimotor period series in an infant macaque: A model for comparing unstereotyped behavior and intelligence in human and nonhuman primates. In S. Chevalier-Skolnikoff & F. Poirier (Eds.), *Primate biosocial development* (pp. 43–112). New York: Garland Press.
Piaget, J. (1936). *La naissance de l'intelligence chez l'enfant*. Neuchâtel: Delachaux et Niestlé.
Piaget, J. (1937). *La construction du réel chez l'enfant*. Neuchâtel: Delachaux et Niestlé.
Piaget, J. (1941). Le méchanisme du développement mental. *Archives de Psychologie*, *28*, 215–285.
Premack, D., & Woodruff, G. (1978). Does the chimpanzee have a theory of mind? *Behavioral and Brain Sciences*, *4*, 515–526.
Rivière, A., & Coll, C. (1987). Individuation et interaction dans le sensorimoteur: Notes sur la construction génétique du sujet et de l'objet social. In M. Siguán (Ed.), *Comportement, cognition, conscience* (pp. 201–240). Paris: Presses Universitaires de France.

Sarriá, E., Rivière, A., & Brioso, A. (1988, June). *Observation of communicative intentions in infants*. Paper presented at the Third European Conference on Developmental Psychology. Budapest.

Sugarman, S. (1973). *A description of communicative development in the pre-language child*. Unpublished BA thesis, Hampshire College.

Sugarman, S. (1984). The development of preverbal communication. In R. L. Schiefelbusch & J. Pickar (Eds.), *The acquisition of communicative competence* (pp. 23–67). Baltimore: University Park Press.

Yerkes, R. (1927). The mind of a gorilla. I & II. *Genetic Psychology Monographs*, *2*, 1–193, 377–551.

# 13 "Social tool use" by free-ranging orangutans: A Piagetian and developmental perspective on the manipulation of an animate object

*Kim A. Bard*

## Introduction

Primates explore their world in large part through the manipulation of objects. Object manipulation is of particular interest to psychologists because it is often linked with practical intelligence. Intellectual structures develop through interaction of one's actions on the environment and the maturing of the nervous system (Piaget, 1952, 1954). In fact, cognitive development of children during the first 2 years of life, as described by Piaget, is highly correlated with patterns of maturation in cortical function (Gibson, 1977, 1981). The flexibility and complexity of an individual's actions on the environment may also be indicators of intelligence (Lethmate, 1982; Maple, 1980; Menzel, Davenport, & Rogers, 1970; Parker, 1974). Furthermore, "an evolutionary connection may be assumed between the rise of intelligence and extensive manipulativeness" (Lethmate, 1982, p. 59).

### *Application of Piagetian theory*

Piaget has shown that the manner in which young children interact with and explore their environment "can serve as a 'window' on cognitive development" (Belsky & Most, 1981, p. 630). The application of Piaget's theory to the study of nonhuman primates seems particularly appropriate for the following reasons:

1. Piaget provided an observational methodology that can be applied to nonhuman primates.
2. Piagetian theory is developmental and, as such, is applicable to describing development in other species.
3. Piagetian perspectives have contributed considerable organization to discrete data relevant to intelligence.
4. Because of the high degree of genetic relatedness (King & Wilson, 1975), the biological basis of the Piagetian perspective allows for meaningful comparisons of development and intelligence among the great apes, and between the great apes and humans.

The Piagetian framework provides a model for intellectual development that "has the advantage of providing a continuous series of forms and functions linking great ape levels of achievement with subsequent hominid levels of achievement" (Parker & Gibson, 1982, p. 27). Piagetian theory, moreover, has been applied to comparative studies of object manipulation and cognitive development in monkeys and apes (e.g., Antinucci, 1981; Chevalier-Skolnikoff, 1976; Hughes & Redshaw, 1974; Mathieu, Daudelin, Dagenais, & Decarie, 1980; Natale, Antinucci, Spinozzi, & Poti', 1986; Parker, 1977; Parker & Gibson, 1977, 1979; Vauclair & Bard, 1983).

Great apes appear to proceed along the same course of development as human infants in the following series of the sensorimotor period: sensorimotor intelligence, causality, the concept of the object (object permanence), and certain modes of imitation. Development proceeds in the same sequence (Antinucci, 1981; Chevalier-Skolnikoff, 1983; Hughes & Redshaw, 1974; Mathieu & Bergeron, 1981; Mathieu et al., 1980; Redshawm, 1978). The major differences between apes and human infants are in the rates of development, the frequencies of performance, and, of course, the highest developmental levels achieved; for example, some preoperational abilities (Mignault, 1985) and some concrete operational abilities have been documented in chimpanzees (Boesch & Boesch, 1984; Muncer, 1983; Woodruff, Premack, & Kennel, 1978).

Great ape infants appear to have the same underlying cognitive structures as human infants, but those structures are expressed in different forms of behavior (Antinucci, 1981; Mathieu, 1982). Human infants, in Piagetian sensorimotor Stage 5, for example, experiment with object–force relations, a form of tertiary circular reaction. The classic example is an infant throwing food down from the high chair, watching the path of the food while varying the height, position, and force with which it is thrown (Piaget, 1952). Ape infants express a similar capacity for object–force relations in social play (Antinucci, 1981; Kellogg & Kellogg, 1933; Mathieu, 1982; Redshaw, 1978; Vauclair & Bard, 1983). Young apes can observe the subsequent effects of their actions on the behavior of a social partner when they vary their proximity, running speed, and the force and placement of hits.

Tertiary circular reactions and an understanding of both means–ends relations and independent causal agents have been demonstrated in chimpanzees and orangutans. These conceptual abilities have been documented in the form of tool usage, other complex functional object–object relations, and object–force relations. It is argued that communicative gestures should be included in this list for the great apes, as they are for human infants.

The application of various psycholinguistic concepts has been helpful in linking studies of early language acquisition with studies of preverbal, or nonverbal, communication (e.g., Bates, Benigni, Bretherton, Camaioni, & Volterra, 1979; Bates, Camaioni & Volterra, 1975). Studies utilizing "speech acts theory" (e.g., Bates, O'Connell, & Shore, 1987; Greenfield & Smith, 1976) have postulated that the underlying cognitive organization is similar for both linguistic and nonlinguistic aspects of early communication (e.g., Bretherton, 1988). Thus, the communicative messages inherent in prelinguistic behavior have been identified: "Infants under 1 year of age can express communicative intent through directed gestures alone" (Bretherton, 1988, p. 226).

The definition of the term *gesture*, however, has caused some confusion in the literature (e.g., Volterra, 1987). The term has been used, for instance, to describe behaviors as diverse as direct reaching for food and holding a hand underneath a consumer's mouth while alternating eye gaze between the food and the consumer.

The confusion in terminology can be illustrated with a description of the development of begging behavior in feral chimpanzees. *Begging behavior*, which first appears at 9–10 months in feral chimpanzees (Plooij, 1978, 1984), consists of goal directed sequences, that is, a series of actions persistently directed toward obtaining food. This behavior of chimpanzees 9 months of age does not include gaze alternation or any other indication of coordination between the behavior that was directed at the mother with that directed at obtaining the food.

In contrast, the *begging gestures* that occur in adult chimpanzees consist of touching the consumer's hand or mouth while alternating eye gaze between the food (held in hand or mouth) and the possessor's eye (Plooij, 1984; van Lawick-Goodall, 1968). Gaze alternation is one of the most easily observed qualifiers for intentional communication. Thus, I can conclude that in adult chimpanzees, begging gestures serve a function of intentional communication.

Because both begging behavior and begging gestures have been labeled "gestures" in the literature (Plooij, 1984; Silk, 1978), it is unclear when communicative gestures first develop in feral chimpanzees. The age at which the behavioral pattern of communicative gestures, as defined here, first develops in feral chimpanzees is only approximately known. Early begging behavior is indicative of goal directed or intentional behavior. Differentiation of the two types of solicitation is necessary to describe ontogeny adequately.

It is important to distinguish gestures (characterized here as intentional communication) from goal directed sequences (i.e., intentional behavior). *Intentional behavior* consists of a series of actions, with at least one serving as a "means," and another action used to attain the goal. It is not necessary for actors to be able to account for or be conscious of the nature of their intentions (Bruner, 1982). Intentional behavior involving social means is evident in young chimpanzees (Boesch, 1986; de Waal, 1982; Goodall, 1986; Köhler, 1925; Plooij, 1978, 1984; Yerkes, 1943), bonobos (Jordan, 1982; Savage-Rumbaugh, 1984), gorillas (Gómez, 1986), orangutans (Bard, 1987; Laidler, 1978; MacKinnon, 1974; Rijksen, 1978), and humans.

Intentional behavior occurs developmentally prior to intentional communication (Bretherton, McNew, & Beeghly-Smith, 1981). In contrast to intentional behavior, *intentional communication* involves the following behavioral elements:

1. gaze alternation between the goal and partner (sometimes referred to as "visual referencing"),
2. use of signals that are varied as necessary to obtain the goal, and
3. "changes in the form of the signal toward abbreviated and/or exaggerated patterns that are appropriate only for achieving a communicative goal" (Bates et al., 1979, p. 36).

Specifically, intentional communication involves a complex goal directed sequence that involves the coordination of action involving external objects with social signaling (Sugarman, 1983).

Gestures such as those used in begging for food are intentional communication. Chimpanzees, orangutans, and humans at later ages exhibit gestures resulting from ritualization of actions and those with conventionalized meaning (e.g., Bates et al., 1979; Gardner & Gardner, 1969; Goodall, 1986; Laidler, 1978, 1980; Miles, 1986; Tomasello, George, Kruger, Farrar, & Evans, 1985; Tomasello, Gust, & Frost, 1989).

Human infants enlist the aid of an adult to obtain objects (food or nonfood items) by using a gesture (such as pointing and/or vocalizing) while gazing alternately between the adult and the object desired. "In form, these episodes exactly resemble the tool use characteristic of Stage V, except that now the infant's tools are gestures. Clearly, the infant's use of these developing tools is intentional" (Frye, 1981, p. 325; see also Bates et al., 1979).

The expression of *communicative gestures* also indicates an ability to recognize causal agents (Bates & Synder, 1987; Harding, 1984; Sexton, 1983; Wolf, 1982). The infant who is able to use the communicative gesture "recognizes the mother's role in the achievement of goals and the potential of using her as a means" (Harding, 1984, p. 131). Chimpanzees of about the same age may also possess this ability:

> With the onset of begging between the ages of 9 and [12.5] months, followed by the use of the other gestures initiating tickling (HOH), grooming (AOH), and approach (GOR), it may be assumed that the chimpanzee infant understands the role of the mother (and others) as an agent and that it possesses a true communicative ability, as was argued earlier [Plooij, 1978]. (Plooij, 1984, pp. 117–118)

Trevarthen and Hubley (1978) label true communicative ability *secondary intersubjectivity*. They concur that in humans it occurs during the same developmental period as complex functional object relations (i.e., Piagetian Stage 5), but they argue that it is an indicator of an exclusively human system of communication based on an innate "rule of sharing." The complexity of the human communication system is surely beyond the range of maximum ability of the apes (e.g., Greenfield & Savage-Rumbaugh, 1986; Seidenberg & Petitto, 1987), but recent studies investigating the cognitive correlates of prelinguistic communication have specified abilities that human and great ape infants have in common.

The abilities of orangutan, chimpanzee, gorilla, and human infants to use another individual or a communicative gesture as a means to obtain both social and nonsocial goals have been documented by a variety of investigators: orangutan (Bard, 1987; MacKinnon, 1974; Rijksen, 1978); chimpanzee (de

Waal, 1982; Goodall, 1986; Jordan, 1982; Köhler, 1925; Savage-Rumbaugh, 1984; Yerkes, 1943); gorilla (Gómez, 1986, 1988, P&G2). Moreover, tertiary circular reactions and an understanding of means–ends relations have been demonstrated in chimpanzees and orangutans in the form of tool usage, complex functional object–object relations, and complex object–force relations (Bard, 1987; Chevalier-Skolnikoff, Galdikas, & Skolnikoff, 1982; Mathieu & Bergeron, 1981).

Recent studies have documented the Piagetian cognitive abilities that either parallel or actually support prelinguistic communication in human infants (e.g., Harding, 1983, 1984). The current study describes a similar pattern of behavioral development for orangutan infants. In the 7–10-month-old orangutan infant, goal directed sequences focusing on obtaining mother dependent food items were successful. At approximately this age, the human infant attempts to "set mother's hand in motion" or direct behaviors to objects held by the mother without referring to the animate nature of the mother (Case, 1985; Piaget, 1952).

The 2-year-old orangutan looks at the actions of the mother, waits, and acts in synchrony with the activity of the mother. He shows evidence of coordinating social agent and object when he directs behavior at the mother as agent and not just at her appendages. Similar coordination by a 2-year-old while begging for food from the mother has been reported in feral chimpanzees (Ghiglieri, 1984, 1988). These interactions involve a new level of coordination initiated by the infant that parallels prelinguistic communications between human infants and their mothers: "Dyadic interaction is realized in a new way with the two partners being treated less like user and tool and more as two separate and equal partners" (Wolf, 1982, p. 314).

Informal observations of a 2-year-old orangutan, moreover, have provided evidence of exploration of object–force relations as he "plays" with moving a small tree through space by shifting his weight in different directions with different amounts of force. Free-ranging orangutans exhibit object–force relations as their predominant Stage 5 ability (Bard, 1987; Chevalier-Skolnikoff et al., 1982), whereas human infants exhibit a preponderance of object–object relations (Piaget, 1952). As is the case for human infants, it appears that these relational abilities in orangutan infants develop in "temporal synchrony" with developments in social agent–object coordination (Sugarman, 1984).

The differences between intentional behavior and intentional communication presented earlier parallel the differences between goal directed sequences and tool use. The following Piagetian definitions clarify this point. Goal directed sequences are coordinations of secondary circular reactions (i.e., Piagetian Stage 4), whereas tool use is evidence of a tertiary circular reaction (Piagetian Stage 5). Case (1985), with a neo-Piagetian perspective, actually relabeled Stage 4 as *bifocal coordinations*, indicating that at this stage the infant is able to coordinate actions on an object or to coordinate actions on a social agent. With the relabeling of Stage 5 as *elaborated coordinations*, the emphasis is placed on the infant's new ability to coordinate actions on objects with actions on social agents. This ability was alternately labeled a coordinated person–object sequence by Sugarman (1983, 1984).

The higher-level coordinations (fifth stage tertiary circular reactions) evident in the complex sequences of tool use and intentional communication both involve the elaboration of intermediary means. In tool use, an object is acted upon in the capacity of an intermediate means in order to obtain a goal. The communicative gesture is used as a substitute for previous direct manipulations of an animate object and hence can be considered a "social tool." The social agent is used to obtain the goal.

In other words, tertiary circular reactions can take the form of tool usage (complex functional object–object relations) and communicative gestures (complex functional social agent–object relations). These forms of behavior reflect the same level of cognitive complexity. At least in humans they probably reflect functioning of the same cognitive structures:

> Three kinds of tools develop during cognitive stage 5: the use of objects on objects, the use of [human] agents to obtain or operate on objects, and the use of the object itself to operate on [human] attention. (Bates et al., 1975, p. 219; brackets are added to terms to generalize from humans to the great apes)

### *The orangutan question*

Among the great apes, orangutans present a challenge to psychologists because they readily demonstrate such cognitive abilities as tool use in the laboratory or the zoo environment, yet rarely, if ever, have been observed to use objects as tools in the natural habitat (Galdikas, 1982; MacKinnon, 1974; Rijksen, 1978). Additionally, their opportunities for expression of similar cognitive skills within the social context (as mentioned earlier) are limited by their relatively solitary life-style.

This distinction highlights the need to differentiate between competence (as in Piagetian cognitive structures) and performance (observable forms of behaviors that are the results of interaction between cognitive structures and the environment). In Piagetian terms, tool use is the result of the discovery and utilization of another object as a means to an end. Other manipulative behaviors may reflect an equivalent level of cognitive functioning, for example, the "social tool use" exhibited through the use of communicative gestures (e.g., Bates et al., 1979; Frye, 1981).

The orangutan infant's first contact with food items is characterized by fortuitous grasping of the portions of food that the mother normally discards during food processing. With increasing development of hand–eye coordination and interest in acting on the world, the young orangutan makes contact with food items, gradually achieving the ability to obtain food through directed reaching and grasping. At some point in the first 2 years of life, these solicitations may become ritualized gestures oriented to the mother. Communicative gestures are important as intentional manipulations and are equivalent, in terms of means–end relations, to instrumentalization (Bates et al., 1979; Frye, 1981).

Although young orangutans in the wild have been observed obtaining some food items from their mothers (Galdikas & Teleki, 1981; Horr, 1977; MacKinnon, 1974; Rijksen, 1978), detailed observations of these exchanges have rarely been possible under the field conditions where orangutans are high in the trees and visibility is obscured by the leafy canopy. In order to overcome these limitations, I initiated a developmental study of the food-sharing process at the provisioning site of the Orangutan Research and Conservation Project, located within the Tanjung Puting Reserve, Indonesia, where unobstructed and close-range observations could be collected on videotape.

This constitutes the first study focused on the development of manipulative behavior in free-ranging infant orangutans. Because the setting for this study was the site of the longest continuous field study of feral orangutans, more accurate age estimates were available for this study than had been available in previous studies (Borner, 1979; Horr, 1977; MacKinnon, 1974; Rijksen, 1978; Rodman, 1979). Because orangutan mothers had been habituated in advance of this study, they allowed their infants to exhibit behaviors that ordinarily would not occur in the presence of observers.

The specific aims of this study were to document the different types of actions young orangutans use to obtain food, the maternal responses to these actions, and the abilities of orangutans of different ages to obtain food either from the mother or independently. The larger objective of this work was to document the form and complexity of manipulation involving animate objects among young orangutans. Object manipulation is important because, within a Piagetian perspective, it can provide comparative data on the development of sensorimotor intelligence in young orangutans.

## Method

### *Subjects*

Biological offspring of five free-ranging orangutan mothers were studied: Tom (male), 1–5 months of age; Riga (female), 1–10 months of age (broken down into two time periods, 1–6 months and 7–10 months); Arnold (male), 2 years; Mooch (male), $3\frac{1}{2}$ years; and Siswi (female), 5 years of age. The mother of the $3\frac{1}{2}$-year-old was feral; all other mothers were ex-captive and were fully rehabilitated to a free-ranging state prior to mating and rearing their offspring. Each of the offspring traveled with his or her mother and shared the same nest with her at night.

### *Procedure*

The subjects were videotaped during feeding time at the provisioning sites of the Orangutan Research and Conservation Project in the Tanjung Puting Reserve of central Indonesian Borneo, where a variety of cultivated fruits and vegetables were provided twice daily at 700 and 1700 hours. A total of

Table 13.1. *Ratio of mother dependent to independent food acquisition behaviors by young orangutan infants of different ages*

| | Age of subject | | | | | |
|---|---|---|---|---|---|---|
| Behavior | 1–5 mo | 1–6 mo | 7–10 mo | 2 yr | 3.5 yr | 5 yr |
| Mother dependent | 83.7 | 70.8 | 79.6 | 46.3 | 71.7 | 7.6 |
| Independent | 16.3 | 29.2 | 20.4 | 53.7 | 28.3 | 92.4 |
| ($n$) | (270) | (130) | (137) | (542) | (446) | (250) |

18 hours of videotape was collected over a 9-month period from September 1983 through May 1984.

The data were systematically collected from the videotape using a detailed coding system. Food acquisition behaviors were the main focus, that is, any infant-initiated action that brought the infant into closer proximity or contact with food. The following information was coded:

1. the type of action used by the infant,
2. whether or not the food that the infant attempted to get was in the mother's possession,
3. the maternal response to each of the infant's actions, and
4. whether or not the infant succeeded in getting food, which was noted at the end of each infant action–maternal response sequence.

## Results

The infant could direct behavior toward food that the mother had ("mother dependent") or toward food that was not in the mother's possession ("independent").

Table 13.1 shows that for the two orangutans younger than 1 year and for the $3\frac{1}{2}$-year-old, more than three-quarters of all behaviors were mother dependent. An important but brief aside: Because the mother of Mooch, the $3\frac{1}{2}$-year-old, was feral, she was cautious at the provisioning site, which at times was occupied by many other orangutans and some native people. It was my impression that this caused her to restrict her offspring's developed ability to forage more independently. Fewer than half of the behaviors for the 2-year-old (Arnold) were mother dependent, as compared with independent. Only a small portion of 5-year-old Siswi's food acquisition behaviors were mother dependent.

My major interest was the infant's ability to manipulate the mother; therefore, the following results (Tables 13.2–13.4) deal only with the behaviors that were mother dependent. The 5-year-old was excluded from these analyses because the actual number of mother dependent behaviors for this subject was small.

Table 13.2. *Proportions of kinds of mother dependent food acquisition behavior displayed by young orangutan infants of different ages*

| | Age of subject | | | | |
|---|---|---|---|---|---|
| Action type | 1–5 mo | 1–6 mo | 7–10 mo | 2 yr | 3.5 yr |
| Gesture | 0.0 | 0.0 | 0.0 | 0.8 | 17.3 |
| Pull | 16.4 | 13.5 | 27.6 | 27.7 | 30.4 |
| Grasp | 31.9 | 31.5 | 33.3 | 38.0 | 37.2 |
| Mouth | 15.0 | 16.8 | 24.8 | 16.9 | 7.4 |
| Reach | 28.8 | 27.0 | 6.7 | 5.8 | 6.1 |
| Look | 7.5 | 11.2 | 7.6 | 10.7 | 1.6 |
| ($n$) | (226) | (89) | (105) | (242) | (312) |

### *Infant actions*

The different kinds of food acquisition actions used by each subject are listed next in order from lowest to highest complexity. Only the highest-order action was recorded.

*Look* was defined as visual attention directed toward food. It was coded only when an additional action did not occur within 3 sec.
*Reach* was defined as a directing of the infant's hand or foot toward, but not touching, the mother or the food. Thus, reach was coded only when contact was *not* made.
*Mouth* was defined as movement of the lips or face that brought the infant into contact with or close proximity to the food.
*Grasp* was defined as contact of the infant's hand or foot with the mother or with food.
*Pull* was defined as an additional action, performed after grasping the mother, that consisted of drawing that part of the mother's body closer to the infant.
*Gesture* was defined as an action directed toward the mother that did not involve the physical manipulation of her body. Usually this took the form of an open cupped hand, palm up, held underneath, but not necessarily touching, the mother's chin.

Table 13.2 shows the proportions of mother dependent behaviors that were of each action type for each orangutan infant. Several points emerge. First, not all young orangutans used communicative gestures. Only a very small percentage of behaviors (less than 1%) by the 2-year-old were gestures. But almost a fifth of the behaviors of the $3\frac{1}{2}$-year-old were gestures.

Second, infants of all ages used the indirect means of obtaining food by pulling on the mother. The proportions of actions of this type were approximately the same in the two youngest subjects (16.4% and 13.5%). More than one-fourth of the actions of the infants 7–10 months and older were pulls. Gesturing and pulling, the two types of indirect means that involved the mother, together accounted for a quarter of the actions of the 7–10-

Table 13.3. *Proportions of types of maternal responses to the mother dependent food acquisition behaviors of young orangutan infants of different ages*

| | Age of subject | | | | |
|---|---|---|---|---|---|
| Maternal response | 1–5 mo | 1–6 mo | 7–10 mo | 2 yr | 3.5 yr |
| Give | 7.1 | 6.5 | 6.4 | 4.8 | 10.2 |
| Allow | 7.5 | 10.8 | 21.1 | 29.8 | 25.3 |
| Ignore | 36.3 | 64.1 | 46.8 | 50.2 | 40.0 |
| Reject | 47.3 | 18.5 | 24.8 | 12.7 | 21.5 |
| (*n*) | (226) | (92) | (109) | (251) | (320) |

month-old infant (Riga), more than a quarter of the actions of the 2-year-old (Arnold), and almost half of the actions of the $3\frac{1}{2}$-year-old orangutan (Mooch).

Third, the proportions of grasps remained approximately the same for all the orangutans. Finally, reaches accounted for more than a quarter of the behaviors of the 1–5-month-old (Tom) and the 1–6-month-old (Riga), but accounted for only small percentages of the behaviors of older infants (6.7%, 5.8%, and 6.1%, respectively). Note that the percentage of reaches dropped for Riga (by the age of 7–10 months) and that the early evidence of gestures occurred in Arnold (at the age of 2 years). Therefore, it appears that the communicative gesture does not develop directly from a reach.

### *Maternal responses*

Table 13.3 shows the proportions of four different types of maternal responses to the infants' food acquisition behaviors (the label for each column is the age of the subject): giving, allowing, ignoring, or rejecting. At present, the analysis describes only the relative proportions of different types of maternal responses summed over all the different kinds of mother dependent infant actions.

*Give* was defined as active transfer of food to the infant. All the mothers showed roughly the same proportions of gives (only a slightly larger ratio was found for the mother of Mooch, the $3\frac{1}{2}$-year-old infant).

*Allow* was defined as a relatively passive maternal reaction of letting her body be moved, the food be moved, or the food be taken by the infant. Small percentages of allow were found for the mothers of the two youngest infants (Tom and Riga). Approximately a quarter of the behaviors of the mothers of the older orangutans were allows.

*Ignore* was defined as no discernible response.

*Reject* was defined as an active repulsion of the infant's attempt to gain access or proximity to the food. The mother could reject the action by acting on the food, such as moving it out of the infant's reach, or by acting on the infant, such as repositioning, restraining, or biting at the infant.

Table 13.4. *Proportions of successful and unsuccessful mother dependent food acquisition behaviors displayed by young orangutan infants of different ages*

| Outcome | Age of subject | | | | |
|---|---|---|---|---|---|
| | 1–5 mo | 1–6 mo | 7–10 mo | 2 yr | 3.5 yr |
| Successful | 12.4 | 18.0 | 49.5 | 40.6 | 39.4 |
| Unsuccessful | 87.6 | 82.0 | 50.5 | 59.4 | 60.6 |
| (*n*) | (226) | (92) | (107) | (251) | (320) |

Rejections were exhibited most frequently by the mother of Tom, the youngest male; the mother of Riga, the youngest female, in contrast, appeared to use the more passive *ignore* rather than the active *reject* to refuse her offspring's solicitations. About a quarter of maternal responses to Riga, the 7–10-month-old, and Mooch, the $3\frac{1}{2}$-year-old, were rejections. A small proportion of maternal responses to Arnold, the 2-year-old infant, were rejections.

There are important points to note in Table 13.3. The first is that the percentage of active rejections by the mother of Tom (the 1–5-month-old infant) was high, up to three times higher than the percentages of rejections by mothers of older offspring. The second is that the proportion of behaviors resulting in potential food transfer, active gives and passive allows, seemed to show a positive relationship with age. The proportions of gives plus allows were close to 15% for the mothers of the two youngest infants, whereas for the mother of the 7–10-month-old the proportion was 25%, and it was higher still for the mothers of older offspring (approximately 35%).

### *Success*

Table 13.4 shows the ratios of successful to unsuccessful behaviors. The 1–5-month-old infant (Tom) obtained food only a small proportion of the times that he acted (and only ate about half of what he got, the rest being mouthed or chewed on, but not ingested). The 1–6-month-old (Riga) was also primarily unsuccessful in acquiring food items from the mother (and she ate less than a fifth of the food that she obtained). Riga at 7–10 months of age obtained food half of the times that she acted (and two-thirds of the time she ate what she got). The 2-year-old (Arnold) and the $3\frac{1}{2}$-year-old (Mooch) obtained food less than half the times that they directed behaviors toward getting food (but on over 90% of the occasions the food was ingested).

Shifting focus from mother dependent to independent behaviors (refer to Table 13.1), we can see that fewer than a third of the food acquiring behaviors of the orangutans younger than 1 year were independent, that is, did not involve food held by the mother. More than half of the actions of the 2-year-

Table 13.5. *Proportions of successful and unsuccessful independent food acquisition behaviors displayed by young orangutan infants of different ages*

| | Age of subject | | | | | |
|---|---|---|---|---|---|---|
| Outcome | 1–5 mo | 1–6 mo | 7–10 mo | 2 yr | 3.5 yr | 5 yr |
| Successful | 18.2 | 55.3 | 60.7 | 84.9 | 97.6 | 84.4 |
| Unsuccessful | 81.8 | 44.7 | 39.3 | 15.1 | 2.4 | 15.6 |
| (*n*) | (44) | (38) | (28) | (291) | (126) | (231) |

old were directed at independent food items. Like the younger infants, the $3\frac{1}{2}$-year-old infant directed about a quarter of his actions to independent food. The 5-year-old subject, in contrast, directed almost all her food acquiring behaviors to independent food items (and thus returns to consideration in the following analysis).

The growing ability of the young orangutans to gain access to food independently of their mothers is illustrated by the data presented in Table 13.5. For the most part (over 80%), the actions of the 1–5-month-old infant (Tom) aimed at getting food independently of the mother failed (and on only half of those infrequent occasions did he eat the food that he obtained). The 1–6-month-old (Riga) was able to obtain food independently of the mother on more than half of her attempts (but she ingested food on only a quarter of those occasions). The 7–10-month-old (Riga) was successful in more than half of her attempts to gain food (and on more than half of those occasions she ingested the food). The 2-year-old (Arnold), $3\frac{1}{2}$-year-old (Mooch), and 5-year-old (Siswi) subjects were highly successful in their attempts to get food independently. In addition to the proportionate data mentioned earlier, the total *numbers* of independent behaviors were much higher for the older orangutans (2 years and older) than for the infants less than 1 year old.

The following major patterns emerge from these preliminary analyses:

1. The young orangutan uses various actions to get food. The relative proportions of these actions are different for orangutans at different ages.
2. The ability of the infant to get food changes developmentally. This is true for success in getting the food that the mother has, as well as for success in getting food independently.
3. The ratio of behaviors involving the mother versus those not involving the mother changes with age.
4. The responses of the mother to the infant's food acquisition behavior also change with the infant's age.

## Discussion

The first description of age related changes in the form of infant actions within the context of food sharing is one outcome of this research. Although other investigators have described food sharing gestures in both orangutans

(MacKinnon, 1974; Rijksen, 1978) and chimpanzees (Kurods, 1980; Plooij, 1978; van Lawick-Goodall, 1968), they have not used a developmental approach. The following discussion focuses first on the development of communicative gestures used in solicitations ("begging") for food and then on developmental changes in success rates.

### *Infant actions*

The youngest subjects (Tom and Riga, at 1–6 months of age) reached for food and sometimes reached for the mother's mouth, but they did not use gestures. However, when the infant hit the mother's mouth area, whether intentionally or not, the mother often dropped food from her mouth. This behavioral interaction has also been observed in feral bonobo (pygmy chimpanzee) mother–infant pairs (Kuroda, 1980). Such interactions may become conventionalized over time, as has been reported for communicative gestures in young group-reared chimpanzees (Tomasello et al., 1989). Communicative gestures that included 'response waiting' were reported to occur only in chimpanzee offspring older than 2 years of age (Tomasello et al., 1985).

Rudimentary gestures were observed in Arnold, an orangutan 2 years of age. These goal directed actions took the form of 'hold out the hand' (Rijksen, 1978).

A more sophisticated form of gesture (directed at the mother for food items in her possession), seen in one orangutan in this study, consisted of an open, palm-up hand held underneath but not necessarily touching the mother's chin. This communicative gesture was observed to occur only in Mooch, the $3\frac{1}{2}$-year-old feral orangutan. Because these gestures involve 'response waiting' and variations in signals to obtain the goal, they qualify as *intentional communication*.

In contrast to chimpanzees, infant orangutans in the current study were not observed to gaze alternately between the mother's eyes and food items while begging. This may have been because the $3\frac{1}{2}$-year-old orangutan of the current study was usually on the back of his mother's neck, with the one arm that was used for gesturing draped over her shoulder, his face oriented in a parallel plane and pointed in the same direction as his mother's face. It is possible that tactile behavior serves the equivalent referencing function in orangutans, as suggested by laboratory work with bonobos, the other great ape that is primarily arboreal (Savage-Rumbaugh, 1984). Although this interpretation is feasible, given the close contact between mother and infant, it was not investigated in this study and remains an interesting topic for future research.

Owing to low visibility of the fine details of food sharing interactions, gaze alternation was not observed in orangutans in the field setting (Bard, 1987). At least by 2 years of age, orangutan infants appear to be capable of begging with gaze alternation (as demonstrated by one videotaped observation of a terrestrial solicitation at the provisioning site by the 2-year-old toward his older "sister," an unrelated individual adopted by his mother before his birth).

Although the oldest subject, Siswi, the 5-year-old, was not observed using communicative gestures with her mother (in fact, she rarely solicited from her mother), she did use intentional communication, including a hand-held-out behavior accompanied by vocalizations, gaze alternation, response waiting, and alternate behavior (when the desired goal was not obtained) with humans. This behavior, which was observed informally during other non-videotaped observation periods, often resulted in her acquiring food from human providers. When food was laid out for the orangutans to obtain on their own, during videotaped observation sessions, of all the infants, Siswi most often chose to obtain food independently from the pile and rarely attempted to obtain food from her mother.

The begging gesture in the $3\frac{1}{2}$-year-old orangutan seems to have been a clear instance of intentional communication, even though gaze alternation did not cooccur. It involved the following components: response waiting, alternate behavior if the desired outcome was not forthcoming, and cessation of the behavior when the desired goal was obtained (Bruner, 1973; Damon, 1981; Tomasello et al., 1985). Additionally, the form of the gesture, a hand held palm-up underneath the mother's chin, was not a direct attempt to reach for food, but was directed communicatively at the mother. The behavior of gesturing in orangutans, even without gaze alternation, therefore, appeared to serve a communicative function. Behaviors exhibited when gesturing was not successful consisted primarily of attempting to get off the mother's body in order to obtain food independently.

As mentioned previously, the mother of the $3\frac{1}{2}$-year-old appeared wary at the provisioning site and often prevented her offspring from foraging independently by repositioning or restraining the infant on her body. She appeared willing to share often and, perhaps, even to share the better parts of food items in order to minimize the infant's attempts to move off or far from her.

Food sharing communications fit the model that social objects support mutually intentional relations (Bruner, 1973; Damon, 1981). The complexity of communication within the food sharing context of orangutans can be seen, for example, in a series of interactions that involved turn-taking and vocal communication by the infant, which was followed by the only observed instance of a mother breaking off a piece of food with her hand and placing it in her offspring's mouth. The complexity of coordination during some food sharing interactions is consistent with the food sharing hypothesis of the evolution of language (e.g., Parker, 1985; Parker & Gibson, 1979).

This study of free-ranging orangutans suggests that orangutans as young as 7–10 months of age can manipulate their mothers in order to obtain food items that are in the mothers' possession. The behavior pattern of "pulling" can be considered intentional if Piagetian criteria of goal directed sequences of behavior are applied. The goal directed sequence includes manipulations of the mother, specifically, grasping and/or pulling on her hand or mouth as an intermediate means toward the goal of obtaining food in her possession. In addition, the behavior pattern of the gesture, an abbreviated and probably conventionalized pattern, was observed in an orangutan of $3\frac{1}{2}$ years of age, but not in younger orangutans. The one older orangutan studied did not exhibit gestures, but was observed engaging in intentional communication in attempts to acquire food from human providers.

### *Success*

The differential abilities of infants of different ages to manipulate their mothers in order to obtain food in the mothers' possession were measured. Both of the orangutans who were less than 6 months old were able to obtain food in fewer than one-sixth of their mother dependent attempts. One subject was more successful at obtaining food independently than from his mother, but the other subject was less successful in obtaining discards and other food *not* in the mother's possession.

The success of orangutans less than 6 months of age was comparable to the success that has been reported for young chimpanzees and bonobos. Overall measures of success reported for 0–6-month-old chimpanzees varied from 20% for naturally occurring food items to 50% for provisioned bananas (Silk, 1979). Similarly, young bonobos (of unspecified ages) were reported to be most successful at obtaining food that was neither difficult to obtain nor difficult to process (Kano, 1980). Gorillas are unique among the great apes in not exhibiting any food sharing, even among mother–infant pairs (Fossey, 1979, 1983; Watts, 1985).

Arnold, one orangutan who was 2 years old during the current study, was observed to direct almost as many solicitations to his mother as to food items out of her possession. He was twice as successful at obtaining food independently as he was at obtaining food from his mother. In contrast, chimpanzees 2 years of age have been reported to be less successful at obtaining provisioned food from their mothers (Silk, 1979).

The $3\frac{1}{2}$-year-old orangutan (Mooch) was more likely to solicit foods from his mother than he was to obtain food independently. Approximately 40% of his solicitations directed toward food in his mother's possession were successful. Almost all his independent food directed actions were successful. Comparable success rates have been found in plant food sharing for $3–3\frac{1}{2}$-year-old feral chimpanzees (Silk, 1979).

### *Summary of results*

The young orangutan subjects of the current study displayed differences in their abilities to manipulate their mothers and in their success at obtaining food items through solicitation. As was expected, older offspring displayed more sophisticated forms of solicitations and had greater success at obtaining food independently.

Because the number of subjects in this study was small, however, it is difficult to separate individual differences that are known to play major roles in behavioral interactions, such as food sharing in chimpanzees (Goodall, 1986), from age differences.

Dyadic differences were apparent, for instance, between the two youngest orangutan subjects and their mothers. Although the two received equal percentages of nonpositive maternal responses (i.e., negative = *reject*, and neutral = *ignore*), the proportions that were neutral, as opposed to negative, differed: Tom's mother more often actively rejected his solicitations, whereas Riga's mother primarily ignored her solicitations. The observed differences in maternal responses were undoubtedly influenced by such infant characteristics as persistence and activity. With few subjects, caution is advised in attributing variability to any single factor, such as age.

## Conclusions

In free-ranging orangutans the cognitive ability to use tools is demonstrated in social interactions and takes the form of communicative gestures that are used as tools, a social means to manipulate the mother in order to obtain food (Bard, 1987). The similarities between social tool use and traditional object–object tool use are discussed briefly in the final part of this section.

This study of the manipulation of an animate object suggested that social tool use was a relatively common occurrence in at least one young free-ranging orangutan. The form of the tool, however, differed from the traditional description of a detached object (Beck, 1980; Galdikas, 1982; Parker & Gibson, 1977; van Lawick-Goodall, 1968). Young orangutans use communicative gestures to manipulate their mothers to receive food items and to help in locomotion (Bard, 1987, in press). The Piagetian perspective of instrumentalization (i.e., the discovery and utilization of an object, apart from the self, as a means to an end) provides support for this interpretation of the phenomena of social tool use.

Young orangutans of different ages use different kinds of behaviors to obtain food. This is an important point, because it is commonly assumed that actions directed toward obtaining food take the form of gestures only. Examination of other behaviors used by young orangutans on an action-by-action basis may give us a clearer picture of the development of gestural communication and the other types of means, both direct and indirect, by which infants obtain food.

Orangutans less than 10 months of age engage in *begging behavior*. Infants 3–4 months old obtain some food items by grasping discarded pieces of food. Orangutans 7–10 months old can also pull on parts of the mother's body to get food. Arnold, a 2-year-old, twice exhibited rudimentary gestures. These types of goal directed sequences are evidence that young orangutans engage in *intentional behavior*, as defined by Bruner (1982) and Piaget (1952).

In contrast, orangutans $3\frac{1}{2}$ years and older exhibit a *begging gesture*, evidence of *intentional communication*. This complex behavioral sequence is built upon the foundation of goal directed sequences. Intentional communication was evident when an orangutan coordinated two types of behavioral sequences, in other words, when the orangutan obtained the goal of food through indirect manipulation of a social agent. This indirect manipulation consisted of a communicative gesture given to the mother. The elaboration of actions directed toward an intermediate means allows for the cognitive complexity evident in the use of a communicative gesture to be equated with the cognitive complexity evident in tool use. Hence the term "social tool use" is applied for the use of a begging gesture as defined here.

The use of the phrase "social tool use" is somewhat problematic. I use it here to emphasize that the cognitive complexity necessary for the use of communicative gestures is equal to the cognitive complexity necessary for the use of objects as tools. The question that is raised when gestures are used communicatively to obtain food from a social agent is, what is the tool? Is the mother the tool, because it is her behavior that brings the infant and the goal together? Or is the gesture the tool, because it is the elaborated and abbreviated means used by the infant to influence the behavior of the mother to obtain the goal?

It is the consensus of many researchers that the mother is utilized as an "instrument." It is her behavior that is manipulated by the infant in order to obtain an otherwise unattainable goal. The image invoked is usually one of direct action, for example, a "form of instrumentality in which children, by pushing, pulling and vocalizing, maneuver an adult into doing something they cannot accomplish for themselves" (Wolf, 1982, p. 302). Young orangutans, gorillas, and chimpanzees also use this direct approach, but it occurs at earlier ages than does the use of intentional communication.

The difficulty that I see in the notion of the mother as an instrument is in the lack of distinction between the infant acting directly on the mother with intentional behavior (as in pushing and pulling on the mother's body parts) and the more complex behavior of the infant acting indirectly on the mother with intentional communication (by using a gesture that serves to coordinate objects and social agents). The mother's behavior is not under the infant's control in the same manner of an inanimate object used as a tool. The intermediate means, discovered by the 14–20-month-old human infant and the 24–42-month-old orangutan, is the communicative gesture, which functions to coordinate the actions of the social agent with the desired actions on an inanimate object. The means are indirectly related to the goal: "It was when the children began to subordinate the use of one physical object to maneuver a second (sensorimotor stage V) that they showed a coordination of object pursuit and social focus in their interpersonal exchanges" (Sugarman, 1984, p. 43).

The gesture as the tool is the interpretation that I prefer, because it is the behavioral evidence that the infant is capable of intentional communication. Gestures have both instrumental and symbolic components. Gestures, as they are used in food sharing among orangutans and chimpanzees, and otherwise labeled "request" gestures (Acredolo & Goodwyn, 1988), "instrumental" gestures (Zinober & Martlew, 1985), and "proto-imperatives" (Bates et al., 1975) in human infants, have in common the intention to regulate another's behavior. Moreover, they are "symbolically representative of a desire rather than directly instrumental in attaining a goal" (Acredolo & Goodwyn, 1988, p. 453), as shown by the abbreviated nature of the action, useful only in communication, not in attaining the goal directly (Bates et al., 1979). In this sense, language also can be said to be a tool, although this greatly oversimplifies the nature and varied functions of language.

In a subsequent paper on development in food sharing interactions in orangutans (Bard, 1989), I document the relationships between the parts of each action event, that is, between infant actions and maternal responses and their consequences. Second, and perhaps more important, these subsequent analyses are used to document "bouts" or sequences of chained behaviors. Some food directed behaviors, especially those directed to the mother, occur in what is usually referred to as a solicitation bout. Analysis of bouts will elucidate the developmental roles played by different action types used to obtain food within the social context.

This study used Piagetian constructs to distinguish various kinds of manipulative behaviors. The structural basis for coding the complexity of manipulative behavior allows the following two forms of "tertiary circular reactions" to be equated: object–object relations (which include functional relations among objects, such as tool use) and communicative gestures (which serve as functional relations between objects and social agents and can be labeled social tool use). This study has documented the occurrence of communicative gestures in free-ranging orangutans and presented an argument for consideration. The cognitive ability to use tools in feral orangutans is expressed in the use of communicative gestures.

**Acknowledgments**

This study was made possible through a grant from the L. S. B. Leakey Foundation. Additional funding was provided by NIH grant HD-19060 to R. D. Nadler, by the Department of Psychology, Georgia State University, and by NIH grant RR-00165 to the Yerkes Regional Primate Research Center of Emory University. My Indonesian sponsor were Lembaga Ilmu Pengetahuan Indonesian (Indonesian Institute of Sciences) and Perlindungan dan Pengawetan Alam (The Nature Protection and Wildlife Management Department of the Forestry). Support in the form of generous loans of technological equipment was provided by Dr. Birutė Galdikas, director of the Orangutan Research and Conservation Project, the National Geographic Society, by Dr. Jim Dabbs of the Department of Psychology at Georgia State University, by Dr. Frederick King, Ms. Cathy Yarbrough, Mr. Victor Speck, and Dr. Kenneth Gould of the Yerkes Regional Primate Research Center, and by my friends Josh Schneider and Bill Hopkins.

I wish to acknowledge the words of wisdom and field advice given to me by Drs. Jeremy Dahl, Lyn Miles, Peter Rodman, Irwin Bernstein, Terry Maple, Gary Shapiro, Suzanne Chevalier-Skolnikoff, and especially Carey Yeager, Jim Murphy, Ellen Bradfield, Joel Volpi, and Neil Belman provided valuable assistance during the early phases of this project, and thanks are given to Jacques Vauclair, Lisa Jones, Gillian Webb-Gannon, Nellie Johns, Muriel Bard, Sharon Doyle, and Karen Gundaker for their assistance in the later phases. For the experience of a lifetime, I wish to extend my deepest appreciation to Dr. Biruté Galdikas. I wish to thank Lil Turner and Jim Young for spending many long hours coding videotapes. The expert typing skills and cheerful assistance of Ms. Peggy Plant and Ms. Ann Gore Ness facilitated the initiation, preliminary, and final presentations of this chapter.

I would like to thank the following people for their helpful comments on one or many versions of the written work: Mike Tomasello, Jeremy Dahl, Georgetta Cannon, Bill Hopkins, Duane Rumbaugh, Lauren Adamson, Ron Nadler, and especially Roger Bakeman. Special thanks are extended to the editors, Sue Taylor Parker and Kathleen Rita Gibson, who have made significant improvements in this chapter. I want to close by expressing my gratitude to all my subjects. Their "particularly orangutan" perspective on life allowed me to keep mine.

## References

Acredolo, L., & Goodwyn, S. (1988). Symbolic gesturing in normal infants. *Child Development, 59*, 450–466.

Antinucci, F. (1981, October). *Cognitive development in a comparative framework.* Paper presented at a symposium in honor of Jean Piaget, Rome.

Bard, K. A. (1987). *Behavioral development in young orangutans: Ontogeny of object manipulation, arboreal behavior, and food sharing.* PhD dissertation, Georgia State University.

Bard, K. A. (1989). *Intentional communication in young free-ranging orangutans.* Unpublished manuscript.

Bard, K. A. (in press). Cognitive competence underlying tool use in free-ranging orangutans. *Annales de la Fondation Fyssen.*

Bates, E., Benigni, L., Bretherton, I., Camaioni, L., & Volterra, V. (1979). *The emergence of symbols: Cognition and communication in infancy.* New York: Academic Press.

Bates, E., Camaioni, L., & Volterra, V. (1975). The acquisition of performatives prior to speech. *Merrill-Palmer Quarterly, 21*(3), 205–226.

Bates, E., O'Connell, B., & Shore, C. (1987). Language and communication in infancy. In J. Osofsky (Ed.), *Handbook of infant development* (2nd ed.). New York: Wiley.

Bates, E., & Synder, L. (1987). The cognitive hypothesis in language development. In I. C. Uzgiris & J. M. Hunt (Eds.), *Infant performance and experience: New findings with the ordinal scales* (pp. 168–204). Urbana: University of Illinois Press.

Beck, B. (1980). *Animal tool behavior: The use and manufacture of tools by animals.* New York: Garland Press.

Belsky, J., & Most, R. K. (1981). From exploration to play: A cross-sectional study of infant free play behavior. *Developmental Psychology, 17*(5), 630–639.

Boesch, C. (1986, July). *Cooperation in hunting in wild chimpanzees.* Paper presented at the presession "Evolutionary Perspectives on Cognitive Ontogeny" at the eleventh annual meeting of the International Primatological Society, Göttingen, West Germany.

Boesch, C., & Boesch, H. (1984). Mental map in wild chimpanzees: An analysis of hammer transports for nut cracking. *Primates, 25*(2), 160–170.

Borner, M. (1979). *Orang utan: Orphans of the forest.* London: W. H. Allen & Co.

Bretherton, I. (1988). How to do things with one word: The ontogenesis of intentional message making in infancy. In M. D. Smith & J. L. Locke (Eds.), *The emergent lexicon: The child's development of a linguistic vocabulary* (pp. 225–260). New York: Academic Press.

Bretherton, I., McNew, S., & Beeghly-Smith, M. (1981). Early person knowledge expressed in gestural and verbal communication: When do infants acquire a "theory of mind"? In M. E. Lamb & L. R. Sherrod (Eds.), *Infant social cognition: Empirical and theoretical considerations* (pp. 333–373). Hillsdale, NJ: Erlbaum.

Bruner, J. S. (1973). Organization of early skilled action. *Child Development, 44*, 1–11.

Bruner, J. S. (1982). The organization of action and the nature of the adult–infant transaction. In E. Z. Tronick (Ed.), *Social interchange in infancy* (pp. 25–35). Baltimore: University Park Press.

Case, R. (1985). *Intellectual development: Birth to adulthood*. New York: Academic Press.

Chevalier-Skolnikoff, S. (1976). The ontogeny of primate intelligence and its implications for communicative potential: A preliminary report. *Annals of the New York Academy of Sciences*, *280*, 173–211.

Chevalier-Skolnikoff, S. (1983). Sensorimotor development in orang-utans and other primates. *Journal of Human Evolution*, *12*, 545–561.

Chevalier-Skolnikoff, S., Galdikas, B. M. F., & Skolnikoff, A. (1982). The adaptive significance of higher intelligence in wild orangutans: A preliminary report. *Journal of Human Evolution*, *11*, 639–652.

Damon, W. (1981). Exploring children's social cognition on two fronts. In J. Flavell & L. Ross (Eds.), *Social cognitive development* (pp. 154–175). Cambridge University Press.

de Waal, F. (1982). *Chimpanzee politics: Power and sex among apes*. New York: Harper & Row.

Fossey, D. (1979). Development of the mountain gorilla (*Gorilla gorilla beringei*) through the first thirty-six months. In D. Hamburg & E. R. McCown (Eds.), *The great apes* (pp. 139–186). Menlo Park, CA: Benjamin-Cummings.

Fossey, D. (1983). *Gorillas in the mist*. Boston: Houghton Mifflin.

Frye, D. (1981). Developmental changes in strategies of social interaction. In M. Lamb & L. R. Sherrod (Eds.), *Infant social cognition: Empirical and theoretical considerations* (pp. 315–331). Hillsdale, NJ: Erlbaum.

Galdikas, B. M. F. (1982). Orang-utan tool use at Tanjung Puting Reserve, Central Indonesian Borneo (Kalimantan Tengah). *Journal of Human Evolution*, *11*, 19–33.

Galdikas, B. M. F., & Teleki, G. (1981). Variation in subsistence activities of female and male pongids. *Current Anthropology*, *22*(3), 241–256.

Gardner, R. A., & Gardner, B. T. (1969). Teaching sign language to a chimpanzee. *Science*, *165*, 664–672.

Ghiglieri, M. P. (1984). *The chimpanzees of the Kibale Forest: A field study of ecology and social structure*. New York: Columbia University Press.

Ghiglieri, M. P. (1988). *East of the mountains of the Moon: Chimpanzee society in the African rain forest*. New York: Free Press.

Gibson, K. R. (1977). Brain structure and intelligence in macaques and human infants from a Piagetian perspective. In S. Chevalier-Skolnikoff & F. E. Poirier (Eds.), *Primate biosocial development* (pp. 113–158). New York: Garland Press.

Gibson, K. R. (1981). Comparative neuro-ontogeny, its implications for the development of human intelligence. In G. Butterworth (Ed.), *Infancy and epistemology* (pp. 52–82). Brighton, England: Harvester Press.

Gómez, J. C. (1986). The development of intentional communication as a problem-solving strategy in the gorilla. *Primate Report*, *14*, 178.

Gómez, J. C. (1988). Tool-use and communication as alternative strategies of problem-solving in the gorilla. *Primate Report*, *19*, 25–28.

Goodall, J. (1986). *The chimpanzees of Gombe: Patterns of behavior*. Cambridge, MA: Harvard University Press.

Greenfield, P., Savage-Rumbaugh, E. S. (1986, July). *The rule-based use of combination by a pygmy chimpanzee*. Paper presented at the presession "Evolutionary Perspectives on Cognitive Ontogeny" at the eleventh annual meeting of the International Primatological Society, Göttingen, West Germany.

Greenfield, P., & Smith, J. (1976). *The structure of communication in early language development*. New York: Academic Press.

Harding, C. (1983). Setting the stage for language acquisition: Communicative development in the first year. In R. M. Golinkoff (Ed.), *The transition from prelinguistic to linguistic communication: Issues and implications* (pp. 93–115). Hillsdale, NJ: Erlbaum.

Harding, C. (1984). Acting with intention: A framework for examining the development of the intention to communicate. In L. Feagans, C. Garvey, & R. Golinkoff (Eds.), *Origins and gorwth of communication* (pp. 123–135). Norwood, NJ: Ablex.

Horr, D. (1977). Orangutan maturation: Growing up in a female world. In S. Chevalier-Skolnikoff & F. Poirier (Eds.), *Primate biosocial development* (pp. 289–321). New York: Garland Press.

Hughes, J., & Redshaw, M. (1974). Cognitive, manipulative, and social skills in gorillas. *Jersey Wildlife Preservation Trust, 11th Annual Report* [*Extract*], pp. 53–60.

Jordan, C. (1982). Object manipulation and tool-use in captive pygmy chimpanzees (*Pan paniscus*). *Journal of Human Evolution, 11*, 35–39.

Kano, T. (1980). Social behavior of wild pygmy chimpanzees (*Pan paniscus*) of Wamba: A preliminary report. *Journal of Human Evolution, 9*, 243–260.

Kellogg, W. N., & Kellogg, L. A. (1933). *The ape and the child*. New York: McGraw-Hill.

King, M. C., & Wilson, A. C. (1975). Evolution at two levels in humans and chimpanzees. *Science, 188*, 107–116.

Köhler, W. (1925). *The mentality of apes*. London: Routledge & Kegan Paul.

Kuroda, S. (1980). Social behavior of the pygmy chimpanzee. *Primates, 21*(2), 181–197.

Laidler, K. (1978). Language in the orang-utan. In A. Lock (Ed.), *Action, gesture, and symbol: The emergence of language* (pp. 133–135). New York: Academic Press.

Laidler, K. (1980). *The talking ape*. London: Collins.

Lethmate, J. (1982). Tool-using skills of orang-utans. *Journal of Human Evolution, 11*, 49–64.

MacKinnon, J. (1974). The behavior and ecology of wild orangutans (*Pongo pygmaeus*). *Animal Behaviour, 22*, 3–74.

Maple, T. (1980). *Orang-utan behavior*. New York: Van Nostrand Reinhold.

Mathieu, M. (1982, August). *Intelligence without language: Piagetian assessment on cognitive development in chimpanzee*. Paper presented at the joint meetings of the International Primatological Society and the American Society for Primatologists, Atlanta.

Mathieu, M., & Bergeron, G. (1981). Piagetian assessment on cognitive development of chimpanzees (*Pan troglodytes*). In A. B. Chiarelli & R. S. Corruccini (Eds.), *Primate behavior and sociobiology* (pp. 142–147). New York: Springer-Verlag.

Mathieu, M., Daudelin, N., Dagenais, Y., & Decarie, T. G. (1980). Piagetian causality in two house-reared chimpanzees (*Pan troglodytes*). *Canadian Journal of Psychology, 34*, 179–186.

Menzel, E. M., Davenport, R., & Rogers, C. (1970). The development of tool using in wild-born and restriction-reared chimpanzees. *Folia Primatologica, 12*, 273–283.

Mignault, C. (1985). Transition between sensorimotor and symbolic activities in nursery-reared chimpanzees (*Pan troglodytes*). *Journal of Human Evolution, 14*, 747–758.

Miles, H. L. (1986). Cognitive development in a signing orangutan. *Primate Report, 14*, 179–180.

Muncer, S. J. (1983). "Conservations" with a chimpanzee. *Developmental Psychobiology, 16*, 1–11.

Natale, F., Antinucci, F., Spinozzi, G., & Poti', P. (1986). Stage 6 object concept in nonhuman primate cognition: A comparison between gorilla (*Gorilla gorilla gorilla*) and Japanese macaque (*Macaca fuscata*). *Journal of Comparative Psychology, 100*(4), 335–339.

Parker, C. E. (1974). The antecedents of man the manipulator. *Journal of Human Evolution, 3*, 493–500.

Parker, S. T. (1977). Piaget's sensorimotor period series in an infant macaque: A model for comparing unstereotyped behavior and intelligence in human and nonhuman primates. In S. Chevalier-Skolnikoff & F. E. Poirier (Eds.), *Primate biosocial development* (pp. 43–112). New York: Garland Press.

Parker, S. T. (1985). A social-technological model for the evolution of language. *Current Anthropology. 26*(5), 617–639.

Parker, S. T., & Gibson, K. R. (1977). Object manipulation, tool use and sensorimotor intelligence as feeding adaptations in cebus monkeys and great apes. *Journal of Human Evolution, 6*, 623–641.

Parker, S. T., & Gibson, K. R. (1979). A developmental model for the evolution of language and intelligence in early hominids. *Behavioral and Brain Sciences, 2*, 367–407.

Parker, S. T., & Gibson, K. R. (1982). The importance of theory for reconstructing the evolution of language and intelligence in hominids. In A. B. Chiarelli & R. S. Corruccini (Eds.), *Advanced views in primate biology* (pp. 42–64). New York: Springer-Verlag.

Piaget, J. (1952). *The origins of intelligence in children*. New York: Norton.

Piaget, J. (1954). *The construction of reality in the child*. New York: Basic Books.

Plooij, F. X. (1978). Some basic traits of language in wild chimpanzees. In A. Lock (Ed.), *Action, gesture, and symbol: The emergence of language* (pp. 111–131). New York: Academic Press.

Plooij, F. X. (1984). *The behavioral development of free-living chimpanzee babies and infants*. Norwood, NJ: Ablex.

Redshaw, M. (1978). Cognitive development in human and gorilla infants. *Journal of Human Evolution*, *7*, 133–141.

Rijksen, H. D. (1978). *A field study on Sumatran orangutans* (*Pongo pygmaeus abelii, Lesson 1827*). Wageningan: H. Veenman & Zonen.

Rodman, P. (1979). Individual activity profiles and the solitary nature of orang-utans. In D. Hamburg & E. McCown (Eds.), *The great apes* (pp. 235–255). Menlo Park, CA: Benjamin-Cummings.

Savage-Rumbaugh, E. S. (1984). *Pan paniscus* and *Pan troglodytes*: Contrast in preverbal communicative competence. In R. L. Susman (Ed.), *The pygmy chimpanzee: Evolutionary biology and behavior* (pp. 395–413). New York: Plenum.

Seidenberg, M. A., & Petitto, L. A. (1987). Communication, symbolic communication, and language. Comment on Savage-Rumbaugh, McDonald, Sevcik, Hopkins, & Rupert (1986). *Journal of Experimental Psychology: General*, *116*(3), 279–287.

Sexton, M. E. (1983). The development of understanding of causality in infancy. *Infant Behavior and Development*, *6*, 201–210.

Silk, J. B. (1978). Patterns of food sharing among mother and infant chimpanzees at Gombe National Park. Tanzania. *Folia Primatologica*, *29*, 129–141.

Silk, J. B. (1979). Feeding, foraging and food sharing behavior of immature chimpanzees. *Folia Primatologica*, *31*, 123–142.

Sugarman, S. (1983). Discussion: Empirical versus logical issues in the transition from prelinguistic to linguistic communication. In R. M. Golinkoff (Ed.), *The transition from prelinguistic of linguistic communication* (pp. 133–145). Hillsdale, NJ: Erlbaum.

Sugarman, S. (1984). The development of preverbal communication: Its contribution and limits in promoting the development of language. In R. L. Schiefelbusch & J. Pickar (Eds.), *The acquisition of communicative competence* (pp. 23–67). Baltimore: University Park Press.

Tomasello, M., George, B., Kruger, A., Farrar, M. J., & Evans, A. (1985). The development of gestural communication in young chimpanzees. *Journal of Human Evolution*, *14*, 175–186.

Tomasello, M., Gust, D., & Frost, G. T. (1989). The development of gestural communication in young chimpanzee: A follow up. *Primates*, *30*, 35–50.

Trevarthen, C., & Hubley, P. (1978). Secondary intersubjectivity: Confidence, confiding, and acts of meaning in the first year. In A. Lock (Ed.), *Action, gesture, and symbol: The emergence of language* (pp. 183–229). New York: Academic Press.

van Lawick-Goodall, J. (1968). The behavior of free-living chimpanzees in the Gombe Stream area. *Animal Behavior Monographs*, *1*, 161–311.

Vauclair, J., & Bard, K. A. (1983). Development of manipulations with objects in ape and human infants. *Journal of Human Evolution*, *12*, 631–645.

Volterra, V. (1987). From single communicative signal to linguistic combination in hearing and deaf children. In J. Montangero, A. Tryphon, & S. Dionnet (Eds.), *Symbolism and knowledge* (pp. 89–106). Geneva: Cahiers Fondation Archives Jean Piaget.

Watts, D. (1985). Observations on the ontogeny of feeding behavior in mountain gorillas. *American Journal of Primatology*, *8*, 1–10.

Wolf, D. (1982). Understanding others: A longitudinal case study of the concept of independent agency. In G. E. Forman (Ed.), *Action and thought: From sensorimotor schemes to symbolic operations* (pp. 297–327). New York: Academic Press.

Woodruff, G., Premack, D., & Kennel, K. (1978). Conservation of liquid and solid quantity. *Science*, *202*, 991–994.

Yerkes, R. M. (1943). *Chimpanzees: A laboratory colony*. New Haven: Yale University Press.

Zinober, B., & Martlew, M. (1985). The development of communicative gestures. In M. Barrett (Ed.), *Children's single-word speech* (pp. 183–215). New York: Wiley.

# 14 The development of peer social interaction in infant chimpanzees: Comparative social, Piagetian, and brain perspectives

*Anne E. Russon*

## Introduction

Following Harlow's work (e.g., Harlow & Harlow, 1965), numerous studies have confirmed the fundamental importance of the peer affectional system in the social development of humans and nonhuman primates (e.g., Mason, 1963; Lewis & Rosenblum, 1975). Peer-oriented activities predominate during youth in most social primate species, including humans, and the formation of stable peer relationships early in life appears essential to the development of normal adult social behavior (Argyle, 1969; Harlow & Harlow, 1965; Lewis & Rosenblum, 1975; Loizos, 1967; Mason, 1963; Suomi & Harlow, 1975; van Lawick-Goodall, 1968). As a result of extensive research since the 1970s, we now recognize that peer contact can affect social development even during infancy. Distinctly structured peer relationships of considerable complexity can begin to develop as early as 1–2 months of age in both human and nonhuman primates. In hand-reared nonhuman primates, infant peer contact is able to offset loss of the mother (Fritz, 1984).

Although by now the published literature on early peer social development is substantial, it is deficient in two aspects that have guided the design of the current project. First, prior research on infant peer social development had focused either on humans or on a narrow range of monkey species, primarily rhesus macaques. I found anecdotal evidence but no published systematic studies on infant peer social development in the great apes. This is a serious deficit, as the great apes are the closest living genealogical relatives to humans. Second, there has been virtually no communication between the researchers working with nonhuman primate infant peers and those working with humans. The one attempt to introduce these two bodies of research, Lewis and Rosemblum's volume (1975) on peer relations, has had little success in integrating the field, despite the fact that a synthesis of these two bodies of research should have considerable potential value. The substantial similarities between nonhuman primates and humans during infancy are indisputable, especially in the areas of attachment and cognition (e.g., Chevalier-Skolnikoff, 1983; Harlow, 1962; Harlow & Harlow, 1965; Hinde, 1986; Hinde & Spencer-Booth, 1971; Mathieu, 1982; Mathieu & Bergeron, 1981; Mignault, 1985; Parker & Gibson, 1977; Redshaw, 1978).

To begin to bridge that gap, this chapter describes an observational study of infant chimpanzees and then compares its results to the published work on infant peer social development in humans and nonhuman primates. To assist these comparative analyses, material has been incorporated from comparative studies of Piagetian sensorimotor development and infant brain development.

**Infant peer social development in the apes**

The literature on apes contains only a few incidental references to interactions with age-mates for infant chimpanzees and gorillas. In part this may be because infant peer contact is likely to be infrequent in free-living apes. Unlike monkeys, apes do not normally live in large stable groups. Orangutans are primarily solitary, gorillas live in stable small groups, and chimpanzees experience shifting mosaics of social contacts. Groupings of chimpanzee mothers with dependent offspring are, however, common (Goodall, 1986; Sugiyama & Koman, 1979). Thus, chimpanzee mothers create opportunities for their infants to contact one another. When infant peer contact is possible, as in hand-reared chimpanzee infants, interaction is common.

Data on chimpanzees derive from several studies. Van Lawick-Goodall (1968) observed some infants in her pioneering field study; Tomilin and Yerkes (1935) studied rare fraternal twins born in captivity; Jacobsen, Jacobsen, and Yoshioka (1932) reported on one laboratory-reared infant, Alpha, who was introduced to another similarly aged infant at 10–12 months of age; and Savage, Temerlin, and Lemmon (1973) observed a contrived mother–infant group with several infants of various ages under 1 year.

In chimpanzees, the earliest contacts with any social partner consist of elicited responses that are reflexlike (Mason, 1965b). The earliest spontaneous infant–infant contacts described involve reaching toward another infant in attempts to touch (van Lawick-Goodall, 1968). These occur between 2 and 4 months of age. The onset of social play emerges between 5 and 6 months. Both rough-and-tumble and chase-tag forms of social play occur in chimpanzee infants, although the ages at which they commonly emerge have not been reported. The use of objects to initiate interaction has been reported twice: Once a 7.5-month-old play-offered food (the partner was not supposed to take the food) (Jacobsen et al., 1932), and once an infant of unreported age approached a peer with a toy, then bounced off with it while looking over its shoulder (van Lawick-Goodall, 1968). One "object possession" conflict was observed, over an 8-month-old's mother (Savage et al., 1973). There have been reports of security-seeking, protective, and prosocial behaviors in infants 8 months of age or older. These include running to a twin partner for refuge and touching a frightened twin to comfort her (Jacobsen et al., 1932), "baby-sitting," helping a 1-month-old to walk, and infant–infant food sharing (Savage et al., 1973).

Only one study was found on infant gorillas: Redshaw's study (1976) of social behavior in human-reared lowland gorillas. This study followed the development of play and social behavior between two infants (Redshaw, 1976) hand reared together from 10 weeks and 2 weeks of age. The two were not peers in the strict sense of "equals," and the study's main focus was development beyond the first year of life. However, a few observations on their interactive behavior during the first year (34–54 and 26–46 weeks of age) appear relevant. Redshaw first observed social exchange between the infants at 7 and 5 months of age. Around this age, when the older infant threw a tantrum (made faces and shrieked), the younger became tense and might pout and whimper. A month or so earlier, the infants would bite and take toys from one another. Interaction frequency increased near the end of the first year, when wrestling with a good deal of mouthing was a popular form. Chase games developed later. Play faces, laughs, and play gaits all occurred in play.

Although the data here are meager, they suggest patterned development in peer interaction in infant apes. I designed an empirical study to generate a more systematic, detailed description of these developmental patterns in chimpanzee infant peers.

## Methods

*Subjects.* The subjects were two infant chimpanzees (*Pan troglodytes*), a female, Sophie, and a male, Spock, born March 5 and March 9, 1976, respectively, in a captive colony of chimpanzees at the Institute for Primate Studies of the University of Oklahoma. The infants were half-siblings, born to different mothers. They were separated from their biological mothers immediately after birth and flown to Montreal on March 16, 1976. From that time they were raised together in a semifree nursery environment at the Laboratory for the Study of Comparative Development of the University of Montreal.

*Living conditions.* The nursery-laboratory was housed in a prefabricated mobile home, on the campus of the university, that had been modified for research purposes (e.g., one-way mirrors were installed to permit covert observation). The chimpanzees had use of a sleeping room and a large playroom equipped with furniture, caregiving materials, and a changing variety of toys appropriate for human infants of similar developmental levels.

Social conditions were planned to simulate essential features of infant chimpanzees' natural social environment. For primary attachment, each infant was provided with its own mother substitute, in the form of a female undergraduate student, from the date of arrival at the laboratory. Each mother substitute was responsible for her infant's physical and social well-being. Both spent 8 hours per day, 5 days per week with their respective infants until 6 months of age, then 4 hours per day, 6 days per week until 18 months of age. A stable eight-member research team well known to the infants provided additional caregiving via a rotating 24-hour shift schedule. The infants remained together as much as possible, as social isolation is particularly devastating to young chimpanzees (Daudelin, 1980; Mason, 1965a).

To accommodate the infants' needs and rhythms, structuring of the daily schedule was kept to a minimum. Beyond the basics of care, no instructions were given to the substitute mothers or the research staff about how or when to interact with the infants. The only constraints regularly imposed on the infants' activity were the scheduling of their participation in research studies and the prohibition of particular behaviors, such as biting, leaving the playroom unaccompanied, and using certain objects.

These conditions seemed reasonably effective in creating a nurturant environment, as Spock and Sophie manifested very few stereotyped abnormal behaviors (e.g., Spock adopted a security blanket; Spock and Sophie established a dependency relationship between themselves that was displayed when mothers were absent).

*Data collection.* Social interaction data were collected from videotaped samples of the infants' spontaneous activity. For each month the video sample consisted of 16 focal individual samples of 20 min duration each collected in two biweekly sessions (8 from each of two contexts, infants plus mothers and infant peers only, with 4 of each focused on Spock and the other 4 on Sophie). Videotaping was covert, via one-way mirrors. The time of day and ordering of social contexts were varied to control for order effects and activity cycle bias. The video samples covered 10–50 weeks of life for a total of 53.75 hours of raw data.

*Data coding.* Coding required two passes through the data, the first to identify interactions and describe them behaviorally, the second to identify and classify higher-level interactive units, sequences of interactive behavior showing unified thematic content.

For the first pass, a social interaction was defined as one or a series of exchanges between infant partners. An exchange consisted of a contingency between partners' behaviors, identified whenever one infant's behavior observably affected the other's. An interaction started at the onset of behavior eliciting the first exchange; it terminated at the end of whatever response failed to induce a further exchange within 10 sec (Bronson, 1981). The definition did not impose conditions on the social directedness of behaviors exchanged, meaningfulness of exchanges, or temporal organization. None of these characteristics are evident in the earliest peer exchanges in either humans or nonhuman primates (Mueller & Lucas, 1975; Plooij, 1984).

Behavioral descriptions of each of these peer interactions included component behavior patterns, actor, partner, time of occurrence, and qualitative/contextual information. Behavior patterns were distinguished morphologically with respect to the body part(s) involved, positions or postures assumed, and type and direction of any movement (e.g., turn toward, hit, touch, run away from, throw object at, sway stance). The inventory of behavior patterns was derived from the research literature on chimpanzees (Mason, 1965a,b; van Hooff, 1973; van Lawick-Goodall, 1968), human

Table 14.1. *Themes in infant peer interaction*

| Class | Description and examples |
|---|---|
| Reflexive | Reflexes or reflexive-quality behavior (e.g., flinch, cringe, freeze, orientation reflex, cling) |
| Exploratory | Curiosity-motivated behavior directed at acquisition of knowledge about the target (e.g., manipulations such as touching, poking, mouthing; intense visual inspection) |
| Playful | Nonserious, pleasure-motivated behavior indulged in for its own sake; actor- versus target-dominated behavior (e.g., almost any pattern performed exuberantly, loosely, and often with a play face; play chases; rough-and-tumble) |
| Assertive | Pursuit or persistence in achieving the actor's set goal in the face of obstruction or conflict; aggression treated as a subclass in which ritualized or actively destructive patterns are employed (e.g., take object, push in front of partner, pursue fleeing partner; hit, pull hair, bite, tantrum, hover-threat posture) |
| Submissive | Abandonment or modification of set goal in the face of obstruction or conflict; includes ritualized submissive patterns (e.g., run away from partner, cede object, presenting posture, take abandoned object) |
| Security | Fear- or discomfort-motivated behavior directed at eliciting protective behavior from another (e.g., cling, cling-walk, hoos and whimpers, excursions, sit near, body contact) |
| Prosocial | Behavior apparently directed at effecting positive social exchange between individuals, such as helping, sharing, and comforting (e.g., smile, vocalize, offer-show-exchange objects, "talk" to partner, groom, protect, or soothe partner) |
| Imitative | Direct and immediate reproduction of a partner's behavior, motivation unclear (variously exploratory or prosocial) (e.g., most commonly, various forms of object manipulation) |

infants and toddlers (see the human studies reviewed in the Discussion, as well as Blurton-Jones, 1974, and McGrew, 1972), and various other nonhuman primate species (see the studies reviewed in the Discussion, particularly Harlow, 1969). The inventory was left open to permit addition of novel elements while maintaining basic comparability with the source studies.

On the second pass, I coded these interactions for thematic content based on the behavioral descriptions. The catalog of content themes was based on the list of themes identified in published infant peer studies, modified to produce a set of mutually exclusive categories defined along one single dimension: interactants' motivation vis-à-vis the peer partner (Brenner & Mueller, 1982; Harlow, 1962, 1969; van Hooff, 1973; van Lawick-Goodall, 1968). The categories derived are summarized in Table 14.1.

In coding for thematic content I made no attempt to code whole interactions as units, as some earlier researchers had done (e.g., Harlow, 1969). Human infant peers do not integrate the content of their interactions – that is, the meaning that the acting infant intends to convey is not the meaning that the receiving infant perceives and responds to (Brenner & Mueller, 1982).

Pilot work on the infant chimpanzee interactions showed similar lack of integration between the two partners' contributions. This lack of intersubjectivity in infant peer exchanges contrasts sharply to the high levels of intercoordination and even ritualization seen in prelinguistic infant–adult exchanges in humans (e.g., Bard & Vauclair, 1984; Bruner, 1983; Stern, 1978; Trevarthen, 1979). Further, even each infant's interactive contribution did not represent one unified motivational theme. An infant often shifted themes within one interaction. For example, what started as a serious chase might turn into a play chase, or rough-and-tumble play into security seeking. Second, an infant would sometimes blend two themes in its behavior: Either it fused two motives, it was ambivalent and expressed two conflicting tendencies in its behavior, or its behavior was simply ambiguous to the observer. My solution was to let the themes define the units, by partitioning each infant's interactive contribution into a set of behavioral sequences that each represented a consistent theme. These thematically consistent individual sequences were the units classified. For blended sequences, I assigned a dual classification that was weighted accordingly in analyses.

For subsequent comparative analyses, published descriptions of peer interactions were reinterpreted using this derived catalog. Overall, this posed no serious difficulties. It did generate one terminological departure from the published literature that is important to mention. Object conflicts, a common class of peer interaction described in human infant peer studies, appear in my system as exchanges of assertive sequences (e.g., take object) with either assertive or submissive response sequences (e.g., guard object, cede object). Objects are described as central contextual features.

*Results*

Coding reliability was assessed via two procedures. Two coders worked to agreement on first-pass coding on one randomly chosen sample per month. Once high interrater agreement was achieved, one coder, the author, then completed the remainder of the first-pass coding for that month. Reliability assessments of second-pass coding were at the intrarater level, as only the author was involved, via comparisons of two separate codings of a second subset of randomly selected samples (one per month). The second coding was performed several months after the first to ensure relative independence. Intrarater reliability coefficients, calculated as the proportion of agreements to overall judgments made $[A/(A + D)]$, ranged from .70 to .81. The most common types of inconsistency involved identification of blends (one coding identified a sequence as a blend of two themes, and the other, as only one of the two themes) and the grouping versus splitting of sequences (one coding identified two sequences that the other grouped as one). On inspection, most such disagreements involved attentional components that had little impact on partner responses. Neither form of inconsistency seemed to seriously invalidate the coding system or the data as coded.

Table 14.2. *Quantitative measures of infant peer interaction*

| Month | Age | No. of interactions | Total (sec) | Duration $\overline{X}$ | Duration (*SD*) | Duration Percentage greater than median[a] |
|---|---|---|---|---|---|---|
| May | 2(5)–2(20) | 39 | 1,170 | 30.0 | (43.2) | 41.0 |
| June | 3(3)–3(17) | 19 | 375 | 19.7 | (16.2) | 42.0 |
| July | 4(7)–4(18) | 99 | 3,930 | 35.9 | (49.5) | 47.5 |
| August | 4(30)–5(15) | 64 | 1,714 | 26.8 | (24.3) | 48.0 |
| September | 6(11)–6(25) | 73 | 1,773 | 24.3 | (24.5) | 38.5 |
| October | 7(9)–7(24) | 87 | 2,301 | 26.4 | (25.4) | 55.5 |
| November | 8(5)–8(20) | 65 | 1,705 | 26.2 | (27.6) | 54.0 |
| December | 9(4)–9(18) | 86 | 3,141 | 36.5 | (39.6) | 59.5 |
| January | 10(0)–10(16) | 104 | 3,144 | 30.2 | (40.4) | 49.0 |
| February | 10(26)–11(12) | 55 | 1,995 | 36.3 | (35.2) | 56.5 |
| Total | | 1,691 | | | | |

[a] Overall median for duration = 18 sec.

Table 14.3. *Percentage of interactive sequences per class per month*

| Class | May (2.5) | June (3.5) | July (4.5) | Aug. (5.5) | Sept. (6.5) | Oct. (7.5) | Nov. (8.5) | Dec. (9.5) | Jan. (10.0) | Feb. (11.0) |
|---|---|---|---|---|---|---|---|---|---|---|
| Reflex | 30.3 | 27.8 | 16.9 | 10.5 | 5.1 | 13.5 | 6.0 | 11.7 | 6.2 | 8.6 |
| Exploration | 37.7 | 45.4 | 27.2 | 19.2 | 8.6 | 12.5 | 8.7 | 8.1 | 3.9 | 4.6 |
| Play | 5.3 | 12.4 | 9.6 | .9 | 1.6 | 3.4 | 3.1 | 12.0 | 7.4 | 12.6 |
| Assertion | 0.0 | 6.2 | 7.8 | 17.1 | 25.2 | 15.0 | 18.6 | 20.4 | 29.8 | 24.1 |
| Submission | 19.7 | 8.2 | 24.3 | 38.1 | 44.4 | 45.6 | 49.1 | 42.2 | 45.0 | 48.9 |
| Security | 7.0 | 0.0 | 13.0 | 12.9 | 13.3 | 8.1 | 12.4 | 3.9 | 5.5 | .6 |
| Prosocial | 0.0 | 0.0 | 1.2 | 1.3 | 1.8 | 1.9 | 2.4 | .9 | 2.1 | .6 |
| Imitation | 0.0 | 0.0 | 0.0 | * | 0.0 | 0.0 | * | .8 | .1 | 0.0 |

*Note*: Asterisk signifies one observation, percentage < .005.

Quantitative descriptive summaries of the 1,691 interactions identified over the 10 months are presented in Tables 14.2 and 14.3 and in Figures 14.1 and 14.2. To assess between-class developmental changes, monthly scores were calculated as the proportion of interactive units in each theme class to the total number of units identified. Blended sequences, those given a dual classification, were treated as contributing a frequency of .5 to each of the two themes involved. Within each class, month-to-month comparisons were made to identify continuing and changing patterns. These quantitative results are not discussed separately; rather, they are integrated into the discussions of qualitative results.

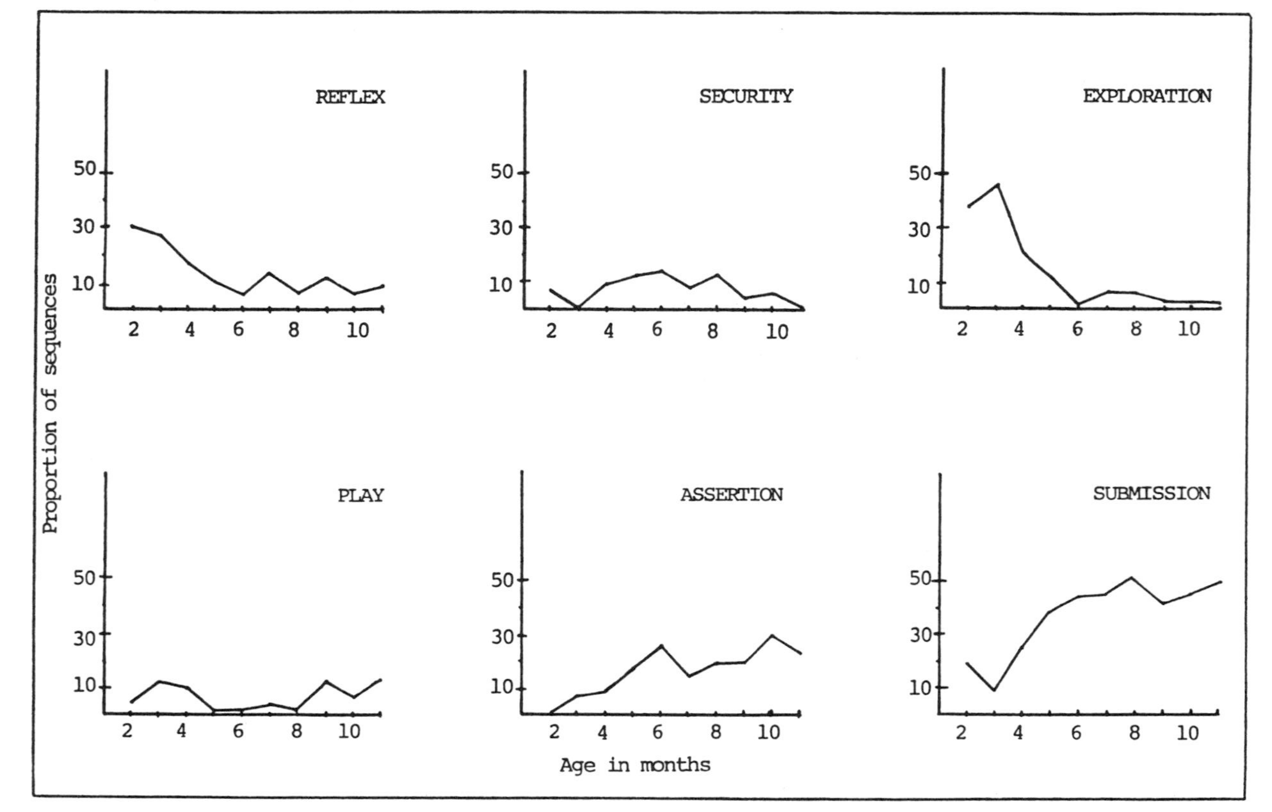


Figure 14.1. *Proportion of interactive sequences in each prominent interactive category, per monthly sampling period.*

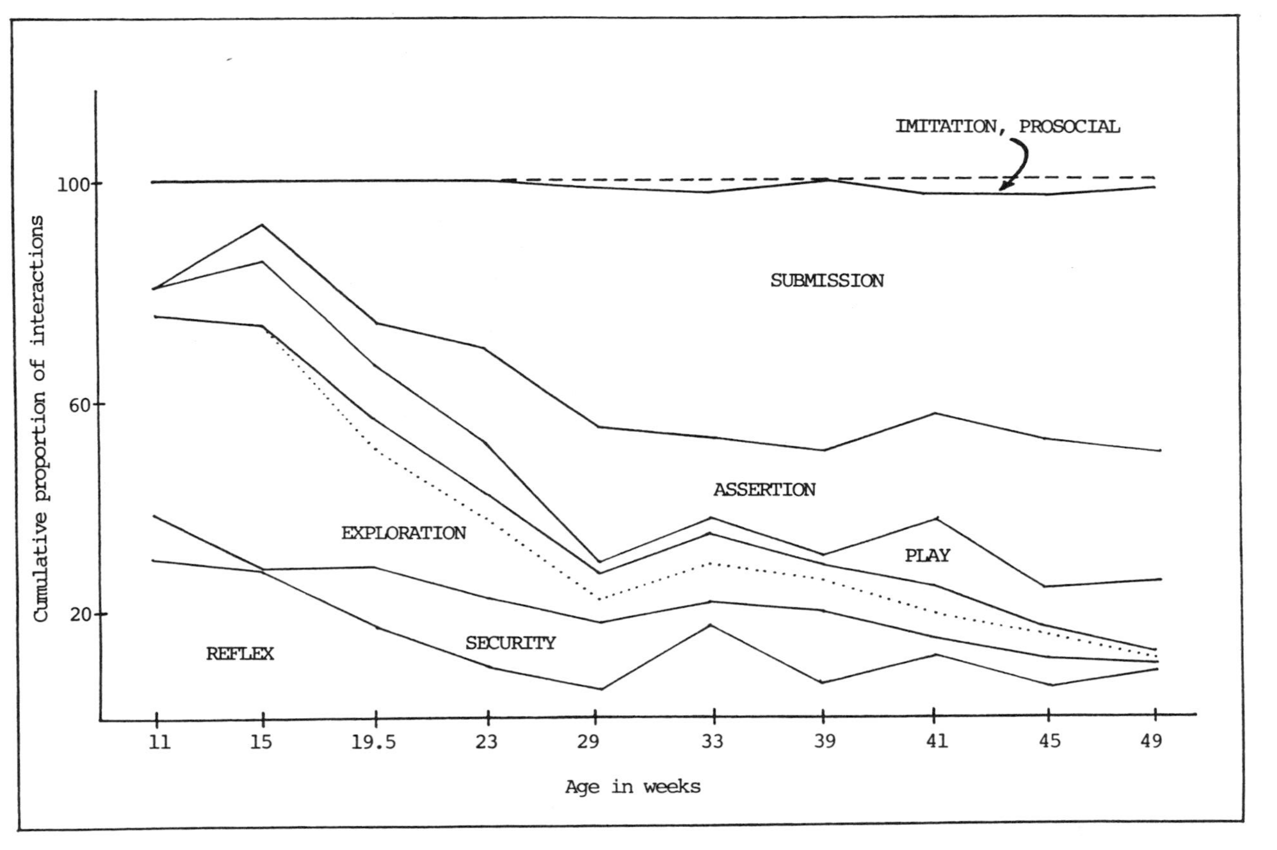


Figure 14.2. *Proportion of interactive sequences in each interactive category, per monthly sampling period: cumulative representation.*

The qualitative results consist of summaries of the between- and within-category developmental patterns in the themes in infant chimpanzees' social interactions. Systematically followed are age and order of appearance, periods of greatest prominence, and typical behavioral content. Individual incidents that appeared important are noted where they occurred. Piagetian sensorimotor phenomena were not followed systematically in this study (though they were elsewhere – see Bergeron, 1979, and Mathieu & Bergeron, 1981), but were noted where they were recognized. In the following discussions of the patterns identified, unless otherwise noted ages refer to Sophie, using the notation Y:MM(DD) (Y = years, MM = months, DD = days). Spock was 4 days younger.

*Reflexive sequences* were most prominent from 2 to 4 months, although already during the second month they were less common than exploratory sequences. Reflexive sequences found within these early peer interactions consisted of very short reactions elicited by movements, sounds, or coincidental contacts from the other infant (Obs. 1). Peer interaction did not tend to continue following the reflexive sequence.

*Obs. 1.* At 0:2(5), Spock and Sophie were lying close together in their crib. In trying to move, Spock fell and bumped Sophie. She immediately flinched and crouched, then turned her head to look in Spock's direction. Spock grimaced and squeezed his eyes closed at the bump, then began to grasp about in front of his body with both hands as though attempting to cling to something. He first managed to grab one of Sophie's legs, but continued to make grasping-clinging movements. This pulled him toward Sophie until his arms were around her shoulders. Sophie froze through Spock's clinging movements, making small "hoo" vocalizations after several seconds.

Initially, almost any peer stimulus could provoke reflexive responses. This corresponds with Plooij's finding (1984) that prior to 2–3 months, chimpanzee infants are responsive to stimulus intensity versus quality. In fact, these reflexive sequences were readily classed into four "types" – orientation/attention (visual orientation reflex, turning toward), self-protection (duck head, close eyes, freeze), withdrawal (pull/turn away, cringe), and security seeking (clinging and grasping, "hoo" vocalizations, whimpers) – corresponding to the intensity of the eliciting stimulus. Each type appeared to represent the roots of a voluntary interactive theme emerging later.

From 3 to 4 months, only unusual, unexpected, or very intense stimuli from the peer elicited reflexive responses. The responding infant was often able to continue interactive behavior with the peer once the offending stimulus was dealt with, commonly with voluntary forms of exploration, withdrawal, or security seeking. From 5 or 6 months, reflexive sequences occurred only as transient interruptions in ongoing peer interaction, following which the disrupted activity was rapidly resumed (Obs. 2). Reflexive responses did not altogether disappear by the end of the period studied.

*Obs. 2.* At 0:9(8), the chimpanzees were with their substitute mothers in the playroom. In engaging Sophie in play from the monkey bars, Spock swung and leaned toward her, bumped her, and then sat on her. On his approach Sophie abruptly pulled back one arm at risk of being bumped and looked at him; on being sat upon, she flinched and jerked her head away, with her mouth falling open. Within 3 sec she resumed playing with Spock.

Reflexive sequences were, then, of paramount importance to the emergence of infant peer interaction. Initially they provided the only venue for and content of peer exchange. Subsequently they served to focus attention on a peer partner as worthy of further involvement. Their importance in peer interaction decreased sharply as voluntary forms of peer related activity emerged.

*Exploratory sequences* were likewise most common in the early months, peaking in relative frequency during the third month. The earliest exploratory sequences seen during the second month consisted mainly of visual attention. This visual attention sometimes continued to inept and poorly directed mouthing, touching, and gross body contact. As sequences, these often consisted of voluntary continuations from initial reflexive responses to the peer. Some took the form of Piagetian circular reactions (see Sophie, Obs. 3). Sophie's leg flexion–extension cycle in Obs. 3 was identified as a primitive secondary circular reaction – primitive in the sense that the focus of the circular reaction, self or environment, was not clear (M. Mathieu, unpublished data, cited in Bergeron, 1979).

*Obs. 3.* At 0:2(7), Spock and Sophie were lying close to one another in their crib. Accidentally, while flexing her arms and legs, Sophie's feet bumped Spock, twice. Eight seconds later, Sophie repeated her leg extension in a more controlled manner, resuming the contact with Spock and this time maintaining it. She extended her legs even farther once she had made contact, which pushed Spock's hand slightly. Two minutes later Sophie again initiated contact with Spock by extending her legs brusquely, then repeated it 5 sec later and again maintained the contact. In response, Spock looked at her with a play face and grasped at her: She flinched and withdrew her foot. Thirty seconds later, she contacted Spock by extension of her legs, again brusquely. She repeated the contact 5 sec later and maintained it until Spock, looking at her with a play face, grasped her foot. She then withdrew her foot, and the two lost contact.

By 3–4 months, exploratory sequences involved better-formed patterns: orientations and manipulations, and approaches once the infants could walk (around 4 months of age). These were spontaneously versus reactively initiated and actively directed toward the peer partner. The exploratory repertoire continued to be elaborated and refined until 5 months of age. Sometimes mutual exploration was observed, although until 5–6 months of age such exchanges were "presocial" in the sense that inanimate objects and peers were not differentiated behaviorally (Harlow, 1962, 1969). From 5 months of age exploratory sequences decreased considerably in importance in peer social interactions, to the point of occurring only when some particularly novel aspect of the partner appeared (see Obs. 4) or when trying to access the partner's object.

*Obs. 4.* At 0:8(17), the infants were playing around a cognitive testing apparatus newly installed in the playroom, a 1-m-cube cage with wooden sides, open back, and wire grille front. Spock and Sophie approached one another, he from outside the cage and she from inside, so that they met face to face separated by the grille. Sophie sat down and looked at Spock through the screen, and he leaned close and looked at her face intently for 10 sec. After a short mutual pause, the two simultaneously leaned toward each other again, Sophie visually inspecting Spock for several seconds, then standing up and trying to touch him through the wire grille. He made small "hoo" vocalizations and leaned away, then Sophie turned away and left.

Between 4 and 5 months, manipulation appeared that was rougher, less cautious, and less attentive than other exploratory manipulations. For example, at 0:4(16) Sophie gradually shifted from gentle exploratory manipulation of Spock to butting him with her head and pushing hard against him, then seizing and pulling at him repeatedly with both arms. These rougher manipulations often appeared in later replications of peer-directed circular reactions, which began as exploratory and subsequently shifted to playful or aggressive. They constituted a low, stable percentage of interactive sequences (about 5%) through the remainder of the period studied.

*Play sequences* occurred throughout the developmental period studied. Play represented a relatively small proportion of each month's interactive sequences and showed no clear developmental trends quantitatively. The quantitative variation that did occur seemed to reflect environmental variation, as play commonly appeared only in relaxed, nonstressful situations (note in Figure 14.1 that the Play profile is almost the inverse of the Security profile). Early play, between 2 and 5 months, was clearly presocial in nature: One infant was in a playful mood that incidentally incorporated the other and resulted in social exchange, or else some aspect of an ongoing interaction was experienced as pleasurable and evoked playful behavior (see Spock's behavior in Obs. 3; a play face and active involvement with Sophie appeared only after several contacts). These play sequences contained most of the various actions the infants were capable of, except withdrawals and security patterns. In play sequences, the infants performed these actions in a loose, exuberant, nonstandard fashion with extensive repetition, accompanying them with inattention and play markers (play faces).

From 5 months, play sequences that were clearly social emerged. Social play faces appeared by 5 months, reciprocation of play by 6–7 months, and rough-and-tumble (wrestling) and approach–withdrawal (play chases) by 9–10 months. All of the play chases observed between Spock and Sophie appeared to be "pseudo" rather than real play chases (see Obs. 5), in that they were only partly social. In these pseudo–play chases, each infant was primarily engaged in repeating its own individual movements from an originally serious chase; it did, however, often look at the partner and intermittently and imperfectly synchronize its movements with the partner's. Similar pseudo–play chases were observed between human toddlers 15–18 months of age; their structure was described as two intercoordinated secondary circular reactions (Mueller & Lucas, 1975).

*Obs. 5.* At 0:9(18), the two were alone in the playroom. Spock had been loitering around a small mat under a table. An interaction started when Sophie approached and took the mat from him (stamped her foot on the mat, then butted him). Spock sidled away, looking back toward her as he left, and Sophie appeared to follow him. Some 20 sec later he again approached the mat from behind Sophie's back and touched it. After a few seconds Sophie spontaneously left the mat, but soon looked back to see Spock walking and stamping on it. She returned to displace him again, and he ran away. Sophie left spontaneously, and Spock returned to take the mat when she had left, but soon she returned to chase him off again – a set of exchanges that was repeated easily a dozen times, becoming increasingly boisterous (especially in Sophie), then overtly playful. For example, late in the running back and forth, Sophie lunged, jumped, and ran rather than walked; butted Spock; let herself fall down in front of him; flailed arms while running; and hovered. Spock once swayed while approaching; ran, lunged, or walked bipedally; fell or threw himself down; hovered; made play faces. The two often looked at each other at the ends of their respective trajectories. Several times during later repetitions Spock started to "flee" before Sophie started to chase him, resulting in his running toward rather than away from her, so that they met face to face; also, rather than following Spock, Sophie tended to move back and forth beside him so that she often actually passed him. The play paused at one point when it had become exceedingly boisterous, and they bumped into one another. Spock immediately clung tensely to Sophie, and the two sat together for about 5 sec. Sophie started the running game again, and Spock joined in. The sample finished before the interaction terminated.

The first few *assertive sequences* were identified at 3.5 months of age. These consisted of persistence with goal directed activity following its interruption. For example, at 0:3(16), Sophie tightened her grasp on Spock when he tried to pull away from her; he, consequently, pulled away harder. Assertive behavior commonly occurred in object-centered exchanges. In the earliest incidents, it was nonsocial and purely object-focused in nature (see Obs. 6); later, it had a clearly social focus (see Sophie, Obs. 7).

*Obs. 6.* At 0:4(30), Spock was sitting in a beanbag chair, and beside him, on the floor, Sophie was engaged in manipulating a toy rake. When she accidentally bumped him with it, he began to watch, then grabbed one end of the rake himself. For the next while, both infants manipulated different portions of the rake. Finally, Spock's attention and manipulation shifted to Sophie herself. On one occasion, after grabbing Sophie, he again grabbed the rake and, to facilitate manipulations, twisted it such that it was pulled from Sophie's hand (he "took" it from her). Sophie grabbed her end of the rake again, and the two again manipulated it in parallel activity.

*Obs. 7.* At 0:10(0), a small table was placed near the monkey bars, and the two infants could easily climb between the two. Spock came to a place X in the monkey bars beside the table. Sophie, on the table, proceeded to come over and take the X spot from him (she seized the bars near him, bumped him, and grasped one of his feet). Spock moved away and lost X, then tried to put his hand on the table. Sophie turned away from X and walked on the table: Spock removed his hand. He touched the table tentatively twice more, the first time glancing at Sophie and backing off quickly at her movements, the second time withdrawing his arm and climbing away when she walked directly toward him. About 10 sec later he came back toward X; Sophie guarded it. He reached toward the table; she took it. He swung close to X; Sophie took X again. He climbed onto the table; she pushed him and took the table back. The interaction ended when Sophie jumped on the floor.

From about 4.5 months of age the frequency of occurrence of assertive sequences increased dramatically. This trend continued in an approximately linear fashion through the remainder of the year, to the extent that assertive sequences ultimately represented one of the most important forms of peer-directed interactive behavior (Figure 14.2). This increase was associated with several factors: emerging individual differences, social relationships, and an increase in the frequency of peer conflict. Individual differences started to appear from 4–5 months, when Sophie, but not Spock, sharply increased assertive interactive behavior. Corresponding social roles emerged and were maintained within their dominance relationship once it was established (at 0:5) (Daudelin, 1980) – Sophie was dominant, Spock subordinate. Conflicts increased with object interest, as infants increasingly tussled over possession. From 5 to 6 months, over two-thirds of assertive sequences occurred in the context of object conflicts. Finally, pursuits appeared in the assertive repertoire around 9–10 months of age. These occurred within object conflicts, increasing their length and intensity, but at the same time apparently serving as the basis for play chases (see Obs. 5).

From 6 to 8 months of age, interactive behavior of a clearly aggressive nature appeared, although rarely. In Sophie, this appeared as assertive behavior intensified to the point of being physically hurtful, apparently intentionally so. For example, at 0:6(24), Spock tried to move away when Sophie seized him, so she grasped the hair on his back and pulled so hard that she lost her grip; at 0:8(20), when Spock refused to cede an object to her, she reached over, grabbed the hair on his back and pulled hard. Spock's aggressive peer behavior appeared a little later in the form of ritualized aggressive patterns, such as a hover/sway stance followed by a lunge at 0:7(6), stamps at 8–9 months, and throwing/shaking objects at 9–10 months. The finding of ritualized patterns in Spock, but not Sophie, suggests sex differences in the development of aggressive behavior; however, I found no mention of such in the literature for this age range.

The earliest *submissive sequences* at 2–3 months were a few voluntary withdrawals (e.g., controlled turning away of the head, pulling away of body parts, or recoiling/squirming at rather low intensity tactile stimulation from the peer). These were often accompanied by uneasiness or anxiety, as indicated by knit/oblique eyebrows, frequent visual monitoring and/or glancing at the partner, and rocking from side to side. The only changes to 4 months of age were increases in the intensity of contact necessary to elicit withdrawals, the range and specificity of body movements, and frequency of occurrence.

As with assertive sequences, three factors (individual differences, dominance, and objects) were associated with the increase in occurrence of submissive sequences. At 4–5 months, Spock sharply increased his frequency of withdrawals (90% of the monthly total). Spock maintained a subordinate role with respect to Sophie from 4 to 5 months of age. Finally, object conflicts increased, and in these, Spock consistently ceded (Obs. 5 and 7). Sometimes he lost objects as a side effect of his withdrawal from Sophie, as he tended to spontaneously abandon them in his withdrawal (Obs. 8 and 9).

*Obs. 8.* At 0:4(16), the infants were in the playroom with their substitute mothers. Sophie approached, and Spock lifted his head, saw her, and paused his mouthing of a table. Spock then abandoned his object and started to move away from Sophie toward his mother, although Sophie soon wandered off without making any attempts to take his object. As she passed, he glanced at her; following his withdrawal, he glanced at her again, and only returned to his table when she was gone.

*Obs. 9.* At 0:11(12), Spock was sitting on the floor with a "new" toy created by Sophie's mother, a plastic ring placed around the handle of a toy shovel. Sophie turned and looked toward Spock and the toy, then approached. As she approached, Spock looked at her and paused his manipulations, then immediately dropped the toy as though it were suddenly very hot, turned away almost reflexively with an expression of surprise/alarm on his face, and ran away as fast as he could. Sophie paused where Spock had been, but made no attempts to take his toy. Only when she had left did he come back and pick it up again.

Both object-centered activity and dominance increasingly characterized the patterning in submissive sequences for the rest of the year. Object conflicts, which Spock consistently lost by active submission, accounted for almost 50% of all interactive sequences by 8–9 months. In addition, the developmental pattern for submissive sequences was almost parallel to that for assertive sequences (Figure 14.1), reflecting the finding that the two were often exchanged in object conflicts. From at least 7–8 months, dominance was so pervasive that it permeated most other facets of their interaction. For example, mock withdrawals in play, all imitation, and most security solicitations were performed by Spock; protection was always provided by Sophie.

Only one ritualized submissive pattern was tentatively identified: a submissive presentation. At 0:7(23), Sophie had been rough with Spock; he started to move away from her, but then turned to come back. When near, he turned again so that his hindquarters were directed at her and paused in that position.

*Security oriented sequences* were observed from the first samples taken. These consisted of security seeking via various forms of physical contact (e.g., clinging/grasping movements, maintaining contact) and infantile distress vocalizations (commonly whimpers and hoos). Like submission, security seeking was commonly accompanied by indices of fear or uneasiness, but it was different in that it was elicited by more intense stimulation or instabilities in the infants' environment (e.g., moving to a new room, mother absent, unusual loud noises). Initially the nonspecificity of reactions was so extreme that the peer partner could simultaneously create the discomfort and provide comfort (see Obs. 1).

As with play, there were no identifiable age related trends quantitatively. Security oriented behavior accounted rather consistently for a small proportion of interactive sequences (5–13%). The little variation in frequency found seemed related to environmental instabilities rather than to developmental processes (see Obs. 10). A few qualitative developmental changes were identified: the degree of contact normally sufficient to calm a security-seeking infant diminished with age from the initial tense, close physical contact (preferably ventral) to periodic proximity (via excursions, common by 4–5 months) and then to mere visual contact (7–8 months). A few new elements appeared in the security oriented repertoire – excursions (4–5 months), cling or tandem walking (7–8 months), and protective behavior (offering a cling, Sophie only, 4–5 months). Protective behavior normally calmed the partner, who consequently lowered the intensity of security seeking and finally abandoned the contact spontaneously. In about half the incidents that elicited protective behavior by Sophie, she initially tried to avoid Spock's security-seeking contact but cooperated when his demands intensified (see end of Obs. 10).

*Obs. 10.* On 0:4(18), the chimpanzees spent their first day in the playroom and appeared uneasy. Each was mainly engaged in solitary activity, but stayed near the other. Once Spock hit a toy bear that had a bell in it; the noise frightened Sophie, and she clung to him. After a few seconds of sitting together in the beanbag chair, Spock made one small excursion, coming back to touch Sophie for a few seconds, then a second from which he returned to touch, lie on, and then rock back and forth in contact with her. The third time he went off on his own a loud external noise occurred, so he again returned to contact Sophie, rocking back and forth for about 10 sec. When he left Sophie a fourth time, she searched for him visually. This time he wandered too far away, got "lost," uttered a very loud distress vocalization (a series of hoos), and came running back to her. In a later excursion, when Spock went particularly far away, Sophie searched for him visually by getting up, lifting her head, and craning her neck in the direction in which he had disappeared. Finally, Spock left Sophie on an excursion (Sophie watched), coming back 10 sec later to cling and lie on her, rocking. Sophie attempted to pull away from Spock, but when she succeeded in breaking the contact with him, he hooed and reinitiated it (cling, lie down, touch, rock back and forth, pout expression). This time Sophie cooperated with his security seeking, allowing him to stay in contact with her and at most shifting her position.

The earliest gestures that could be considered *prosocial* occurred at 4–5 months. These consisted of providing protection or comfort for the security-seeking partner, as described earlier. Commonly, Spock sought such contact, and Sophie would drop or pause her other activities for several seconds to cling-walk or remain quietly in contact with him (see end of Obs. 10, Obs. 11).

*Obs 11.* At 0:6(11), Spock engaged Sophie in a cling-walk, she cooperated, and he finally let go. When she tried to move away, he grabbed her again, and she sat quietly with him for a few seconds. He again calmed and let go, and Sophie again started off on her own. Spock turned toward her, wiped his nose, ran after her, and again contacted her. At this, Sophie began to move away, but then spontaneously approached Spock and herself initiated a cling-walk. Spock readily accepted, only repositioning one arm to take his normal "initiator's" position.

Less common forms of prosocial behavior included cooperation with the partner's exploratory manipulations (between 3–4 and 11–12 months) and object "sharing" (8–9 months). As an example of the former, at 0:11(12) Sophie stopped her activity and became very still and attentive while Spock manipulated her foot. For her, this may have constituted presenting for grooming. In object sharing, as opposed to ceding, the actor actively accommodated partner-initiated parallel activity with the actor's object (8–9 months), but did not cede his or her own use of the object. Prosocial sequences were extremely rare, never accounting for more than 2.5% of the total per month. These sequences showed little identifiable developmental change, possibly because of this scarcity.

From 5 to 6 months of age, *imitation* was observed in peer interaction, but very sporadically (total 9 incidents). In the earliest example, at 0:5(1), each infant in turn replicated the other's behavior while watching the partner – Sophie hit a small desk that both were climbing, then Spock hit it, then Sophie again. Most imitative behavior occurred later (7 of 9 incidents after 9 months of age) and most was performed by Spock. Most occurred within interactive play at 9–10 months. A little occurred in object-focused interaction. For example, during a bout of rough-and-tumble play at 0:9(18) Sophie turned away to mouth a part of the monkey bars as Spock watched her, and then he leaned over and mouthed the same part of the bars; in an object-focused exchange at 0:10(0) Spock approached Sophie, who was visually inspecting his box, watched intently, and then explicitly leaned over to visually inspect it himself. All such imitative sequences were very short and unilateral. The actor replicated only one or two of the partner's patterns before returning to his or her prior activity, and the imitative behavior had little or no effect on the partner's behavior.

The first incident, hitting the desk, appeared to represent some form of constrained copying in that the model's behavior combined with the physical context merely triggered an appropriate known action. In the later incidents (9–10 months) it was more likely that the imitator intentionally accommodated to the partner's behavior, albeit only to known actions, because the imitator abruptly changed his or her ongoing activity to exactly copy the partner. In these later examples, imitation appeared related to dominant–subordinate roles, because in all cases the individual subordinate in the current context performed the imitation (e.g., Sophie was situationally subordinate if interested in an object that Spock currently possessed).

## Discussion

For richer interpretation, these empirically derived developmental patterns were compared with corresponding infant peer development in other primates and with comparable periods in Piagetian sensorimotor development and brain development.

### *Infant peer social development in monkeys and in humans*

Comparisons with parallel peer social development in other primates allow identification of general versus species-specific patterns. I reviewed studies concentrating on optimal ontogenetic processes while maintaining some degree of ecological validity. The following selection criteria were used:

1. Subjects were between the ages of 2 and 18 months for Western humans, and 1–6 months for monkeys. These ranges approximate the developmental period studied in chimpanzees: 2–12 months. Over most of this period, development proceeds about 50% faster in chimpanzees than in humans, and two to three times faster in the monkeys studied than in chimpanzees (Bergeron, 1979; Gibson, 1981; Mason, Davenport, & Menzel, 1968; Parker, 1977).
2. Maximum age differences between infant partners were 2–3 weeks. Age differences greater than this produce clear inequalities between partners in both humans and nonhuman primates (Baldwin, 1969; Maudry & Nekula, 1939).
3. The peer events described were interactions. Descriptions of individual social behavior (that simply directed at or simply affecting a partner) were excluded. Social interaction and social behavior are not interchangeable, because they show different developmental patterning (Bronson, 1981; Mueller & Brenner, 1977; Vandell, Wilson, & Buchanan, 1980). Only interactive measures reflect all participants as well as the skills necessary for effecting conjoint action (Bronson, 1981; Mueller & Brenner, 1977; Ross, Goldman, & Hay, 1979).
4. Infants were studied in familiar environments with familiar peer partners. Infants are highly reactive to novelty/familiarity in their physical and social environment (e.g., Jacobson, 1981; Uzgiris & Hunt, 1975). In my observational study (as normally in primates), early development occurs within a relatively constant environment (Rosenblum, Coe, & Bromley, 1975).
5. Adults, particularly caregivers, were normally involved with infants.
6. Various inanimate objects were available to infant subjects.

Because the bulk of the published studies had focused on the thematic content of individual interactions (as opposed to quantitative change, structure or "syntax," or relationships), the present chapter maintains that focus. Other facets of infant peer social development have been discussed elsewhere (Russon, 1985).

Patterns of peer social development in infant monkeys derive from three main types of studies:

1. studies of infants reared in laboratories under various conditions of social deprivation versus access (Deets, 1974; DeJonge, Dienske, van Luxemburg, & Ribbens, 1981; Hansen, 1966; Harlow, 1969; Harlow & Harlow, 1965; Suomi & Harlow, 1975),
2. studies of colonies living in seminaturalistic captive conditions such as zoos (Bolwig, 1980; Handen & Rodman, 1980; Harrington, 1978; Nakamichi, 1983; Rhine & Hendy-Neely, 1978; Rosenblum et al., 1975), and
3. studies of feral groups (Baldwin, 1969; Baldwin & Baldwin, 1978; Berman, 1982a,b; Chalmers, 1980; Cheney, 1978; Norikoshi, 1974; Owens, 1975).

The laboratory studies focused primarily on rhesus macaques; those of captive colonies and feral groups included several species of Old World monkeys (various macaques and baboons), two of New World monkeys (howlers, squirrels), and one prosimian (a lemur, included with monkeys because reported peer patterns were similar). Despite great variability in social behavior from one monkey species to another, these studies reported very uniform patterns in infant peer social development across species. The patterns identified are tentatively discussed as common to monkeys. They are detailed elsewhere (Russon, 1985) and are summarized in Appendix 14.A.

Descriptions of infant and toddler peer exchanges in humans derive from 14 studies (Becker, 1977; Brenner & Mueller, 1982; Bridges, 1933; Brownlee & Bakeman, 1981; Finkelstein, Dent, Gallagher, & Ramey, 1978; Holmberg, 1980; Jacobson, 1981; Mueller & Brenner, 1977; Mueller & DeStefano, 1973; Mueller & Lucas, 1975; Rubenstein & Howes, 1979; Shirley, 1933; Vandell et al., 1980; Vincze, 1971). These studies were conducted variously in the infants' homes, laboratory settings, a baby hospital, and play groups or day-care centers. Some were longitudinal in design, although none covered the complete age range of interest. The majority tended to focus on the second rather than the first year of life, and only three studies considered infants under 6 months of age. The patterns identified are detailed elsewhere (Russon, 1985) and are summarized in Appendix 14.B.

Table 14.4 compares these developmental patterns in monkey, chimpanzee, and human infant peer interactions via the interactive behaviors reported. Behaviors are organized into two groups: those strictly focused on peers versus those incorporating inanimate objects. Behaviors are placed at the earliest age observed; when ages were imprecise or not reported, chronological placement is suggested by a question mark. Vertical spacing for chronological age is adjusted for each species group, so that rows represent roughly comparable developmental levels. The letter in parentheses following each behavior entry indicates the theme it appears to represent.

### *Contributions of comparative studies on cognitive and brain development*

Over the past decade the Piagetian framework has been of great interest to comparative researchers because of Piaget's focus on the biological foundations of intelligence (Mathieu, Daudelin, Dagenais, & Decarie, 1980). Some empirical studies have been made of development through Piaget's stages of sensorimotor intelligence in the nonhuman primates (e.g., Chevalier-Skolnikoff, 1977, 1982; Parker, 1977; Parker & Gibson, 1977; Redshaw, 1978), including our chimpanzee dyad (Bergeron, 1979; Mathieu & Bergeron, 1981). The findings from these studies provide a second basis for interpreting patterns of infant peer social development across the primate groups.

Because Piaget's model of the development of intelligence deals mainly with physical causality and logicomathematical reasoning, the Piagetian studies further provide an opportunity to explore the relationship between social and physical cognition, an issue that is currently of considerable interest. Some researchers have suggested that early developments in social and physical understanding are closely linked, in that advances in social functioning derive from achievements in physical understanding (e.g., Mueller & Lucas, 1975). More recently, others have suggested that social functioning

Table 14.4. *Infant peer social development in monkeys, apes, and humans*

**Monkeys**

| Age[a] (mo) | Peer-focused | Theme | Object-focused | Theme |
|---|---|---|---|---|
| 0.5–1 | Attend to sound, movement | (R,E) | | |
| | Proximity, accidental contact | (R,E) | | |
| 1.5 | Touch, contact, look, mouth | (E) | | |
| ? | Excursions | (S) | Object-centered contacts | (?) |
| ? | Reciprocate play, play bites | (P) | | |
| 2–3 | R&T | (P) | | |
| ? | | | Friendly, mutual competitive exploration of objects | (P,A) |
| ? | Groom | (O) | | |

**Chimpanzees**

| Age (mo) | Peer-focused | Theme | Object-focused | Theme |
|---|---|---|---|---|
| ~2 | Attend to sound, movement | (R,E) | | |
| 2–3 | Contacts, cling | (E,S) | | |
| | Amusement, withdraw | (P,M) | | |
| 3–4 | Assertive behavior | (A) | | |
| | Protection | (S,O) | | |
| 4–5 | "Sorties," proximity | (S) | Nonsocial "takes" | (E) |
| 5–6 | Dominance | (A,M) | Social takes | (A) |
| | Early social play | (P) | Struggles | (M) |
| 7–7 | Aggression | (A) | | |
| 7–8 | Ritualized aggression | (A) | Play-offer food | (P,O) |
| 8–9 | "Baby-sit" | (O) | | |
| 9–10 | R&T | (P) | Share food | (O) |
| | PCE | (P) | Imitate behavior | (I) |
| | | | Throw, tease | (A) |
| 10–11 | JCR | (P) | | |

**Humans**

| Age (mo) | Peer-focused | Theme | Object-focused | Theme |
|---|---|---|---|---|
| 2–3 | Attend to sound, movement | (R,E) | | |
| 3–4 | | | Nonsocial "takes" | (E) |
| 4–5 | Mutual touch | (E) | | |
| | Pleasure | (P) | | |
| 5–6 | | | | |
| 6–7 | Mutual smile, vocalize, reach | (O,E) | | |
| 7–8 | Aggression, crawl over | (A,E) | Social takes, struggle | (A) |
| | Victim smile | (P) | | |
| 8–9 | | | | |
| 9–10 | | | Object-centered contacts | (?) |
| 10–11 | Group "games" | (P) | Offer, show | (O) |
| 11–12 | Imitate | (I) | | |
| 12–13 | | | Exchange | (O) |
| 13–14 | JCR | (P) | Imitate behavior to object | (I) |
| | Imitate laughs | (I,P) | | |
| 14–15 | PCE | (P) | | |
| | Imitative interaction | (I) | | |
| 15–16 | "Talk" to peer | (O) | | |

|  |  |  |  |  |  |  |  |  |  |  |
|---|---|---|---|---|---|---|---|---|---|---|
| ? | Chase, tag | (P) |  |  |  | 16–17 |  |  | Ballgames | (P) |
|  |  |  |  |  |  |  | Chase, tag, catch | (P) |  |  |
|  |  |  | 11–12 | Protection | (S,P) | 17–18 | Smile, laugh | (O) |  |  |
|  |  |  |  |  |  |  | R&T | (P) |  |  |

*Abbreviations:* R = reflex; E = exploration; P = play; A = assertion; M = submission; S = security; O = prosocial; I = imitation; ? = uncertain; JCR = joint circular reaction; PCE = play chase + errors; R&T = rough-and-tumble.

[a] Vertical spacing for age is adjusted so that rows represent roughly comparable developmental levels across species groups. Dotted lines marks ages when walking is commonly achieved. A question mark indicates that ages are imprecise or not reported; chronological placement is suggested.

Table 14.5. *Stages in Piagetian general sensorimotor development*

|  | Age (mo) |  |  |  |  |  |  |
|---|---|---|---|---|---|---|---|
|  |  | Apes[b] |  |  | Chimpanzees[c] |  |  |
| Stage | Monkeys[a] | Chimpanzee | Orangutan | Gorilla | Sophie | Spock | Humans[d] |
| 1. Reflex | 5–0.5 | 0–2 | 0–1 | 0–1 | 0–1 | 0–1 |  |
| 2. Primary reactions | 0.5–0.7 | 2–5 | 1–2.5 | 1–2.5 | 1–3 | 1–3.5 | 1–4 |
| 3. Secondary reactions | 0.7–2 or 2.5 | 5–6 | 2.5–5.5 | 3–7 | 3–6.5 | 3.5–7 | 4–8 |
| 4. Coordinations | 2 or 2.5–4 | 6–11.5 | 5.5–11 | 7–14 | 6.5–10 | 7–11 | 8–12 |
| 5. Tertiary reactions | 4–? | 11.5–? | 11 –48 | 14–? | 10–20 | 11–? | 12–14 |
| 6. Insight | ? | 18–? |  |  | 20–? | ?–? | 18–24 |

*Sources:* [a] Parker (1977) and Chevalier-Skolnikoff (1982) for stump-tailed macaques and Hanuman langurs; [b] Chevalier-Skolnikoff (1982); [c] Mathieu & Bergeron (1981) and Bergeron (1979) for Sophie and Spock (female showed Stage 6 behavior from 1 year, male beyond 24 months); [d] Summary from Chevalier-Skolnikoff (1982) and Mathieu & Bergeron (1981).

requires a set of concepts that are largely distinct from those appropriate for the physical environment (e.g. Byrne & Whiten, 1988; Hinde, Perret-Clermont, & Stevenson-Hinde, 1985) and that the two therefore develop independently. The availability of empirical data on both developmental patterns, especially on the same set of subjects, provides a rare opportunity to assess some of these relationships. Appendix 14.C provides brief outlines of Piaget's model of human intellectual development during infancy, called sensorimotor intelligence, and of this sensorimotor development across the primates. Table 14.5 compares progress in the general sensorimotor series in monkeys, apes, and humans, based on a recent review of the empirical studies (Dore & Dumas, 1987).

Some comparative research has investigated relationships between Piagetian sensorimotor development and brain development in the primates (e.g., Gibson, 1977, 1981; Fischer, 1987). Gibson's studies have identified remarkably close correlations between the cognitive and behavioral advances mapped in the Piagetian model and cortical maturation, as indexed by myelination. Fischer's study suggests similar correlations with cortical synaptogenesis. These findings are useful in suggesting the mechanisms underlying the developmental patterns identified in infant peer behavior and in suggesting which patterns in early development represent unified versus differentiated processes.

*General primate patterns in infant peer social interaction*

Three comparable periods could be identified in infant peer social development across all the primates studied. These were directly comparable to Harlow's first three "stages" in the monkey peer affectional system: reflexive, exploratory, and social (Harlow, 1969). These stages are defined in terms of infants' motivation vis-à-vis one another.

The reflexive stage of peer interaction could be identified prior to 5 weeks in monkeys, 2.5 months in chimpanzees, and 2–3 months in humans. In this stage in Harlow's rhesus infants, sounds and movements could elicit reactions, often reflexive orientation, and sometimes locomotion and exploration. Social exchange with peers occurred only by accident, if the source of these sounds and movements was a peer or if mutual interest in some other event accidentally resulted in peer contact. Behavior was described as presocial in quality. Peer interactive behaviors reported in apes and humans during this period are very similar in their qualities. The only difference noted between species is that because infant monkeys achieve independent locomotion during this period, they can independently make physical contact with peers. Because human and chimpanzee infants do not achieve independent locomotion until later, physical contact with peers is possible only if infants are placed together.

The exploratory stage spans 1–2 months in monkeys. Investigatory manipulations are its central features; infants directly orient toward other infants and investigate them by peering or gently mouthing or manipulating them (touching, clasping, limited body contact) (Harlow, 1969). Similar types of peer contact appeared at equivalent levels in both chimpanzees (by 2.5–4 months, touching, mouthing, body contact, and reaching were common) and humans (at 3–5 months, mutual touching, chewing a companion's finger, attention to others' movements). They were likewise prevalent over comparable periods: to 4–6 months in chimpanzees and 7–8 months in humans. These early exploratory contacts were uniformly described as presocial.

The social stage is also paralleled across these primates. In Harlow's model, peer encounters in this stage consisted initially of rough-and-tumble play and, later, chase play. The onset occurred at 2–3 months in rhesus as well as in other monkeys. These patterns emerged at equivalent developmental points in apes and humans: 4–6 months in chimpanzees, 5–7 months in gorillas, and 7–8 months in humans. Both forms of social play reported in monkeys (rough-and tumble and chase) also occurred in apes and humans. The ages at which both rough-and-tumble and "real" (versus "pseudo") play chases appeared roughly corresponded across the primate groups. The major developmental transition involved in all species was from peer behavior that was presocial to behavior that had social qualities.

These peer periods correspond to major features of both Piagetian sensorimotor development and brain development. The peer reflexive stage covers the first two stages in Piaget's model (reflex, and primary circular reactions). Very few data are available on the quality of peer interaction during this first stage, and virtually none from the Piagetian reflex period. Chimpanzee and human observations begin within Piaget's Stage 2 (Spock/Sophie, 2 months; human infants, 2–3 months). What peer behavior has been reported is nonetheless consistent in quality with that expected at Stage 2: The earliest peer behavior consists primarily of invariant responses to classes (here, intensity) of stimuli, motor patterns are undifferentiated, visual attention to peers (complex visual stimuli) is common at the end of Stage 2, and one example of rhythmic grasping about in the absence of something to grasp occurred at the end of Stage 2 (Spock, Obs. 1). Gibson's analyses (1977) of correlated cortical maturation suggest that behaviors characteristic of Piaget's Stage 1 reflect subcortical control, and those characterizing Stage 2 reflect initial functioning of premotor and sensory association areas. The qualities of peer behavior concur.

The peer exploratory stage corresponds closely to Stage 3 in Piaget's general sensorimotor series in onset, duration, and quality of behavior (Bergeron, 1979; Mathieu & Bergeron, 1981). Plooij (1984), studying wild chimpanzees, found a similar change in the quality of infant behavior around 2–3 months. Within the peer exploratory stage, as in Stage 3, behavior is characterized by simple actions; both semiintentionality and secondary circular reactions occur. The qualities of Stage 3 behavior indicate emergence of the infant's cognitive capacities to perceive sequential relations between its own actions and resulting visual and auditory events – that is, limited perception of causal/

temporal relations (Gibson, 1977; Parker, 1977). The infants' peer exploratory behavior did begin to show sequencing (e.g., look-approach-manipulate sequences appeared) and some awareness of the effects of their own actions on the recipient peer (e.g., rough or playful manipulations were not attempted until gentler ones proved acceptable). Following Gibson's analysis, these peer related abilities concur with the Piagetian capabilities in signaling initial functioning of sensory association areas of the cortex. The semiintentionality found in both realms correlates with incipient frontal association area functioning.

The onset of the peer social stage corresponds to the transition between Stages 3 and 4 of general sensorimotor intelligence and the onset of Stage 4 of both causality and object concept. Borrowing from the model of sensorimotor processes, this peer related change probably relates to (1) burgeoning interest in events external to the self and (2) the beginnings of understanding that forces external to the self are involved in the causes of events and that objects have independent, permanent existences. The critical abilities apparent in Stage 4 are those of coordinating schemata both sequentially and simultaneously in pursuit of a clear goal. Gibson's analyses (1977) of brain cognitive correlates suggests that this ability to coordinate (voluntarily combine, recombine, and inhibit a variety of schemata) plus goal-directedness indicate initial function of frontal association areas; the perception of the independent existence of objects and causes depends on more mature functioning of parietal and temporal association areas. These capabilities are clearly evident in both forms of social play: rough-and-tumble and play chases.

There is evidence from various studies on social development to suggest that some of these Stage 4 Piagetian achievements are initially social. Studies of sensorimotor development in Spock and Sophie showed that Stage 4 causality was achieved for persons around this point, but not for inanimate objects until 3 or 4 months later (Bergeron, 1979). Similarly, in human infants, "person permanence" may be attained at around this point and earlier than "object permanence" (Bell, 1970; Sroufe, 1979; Chevalier-Skolnikoff, 1982). This finding may hold for some monkeys, as by 12 weeks both bonnet and pig-tailed macaque infants can identify, via preference, their own mothers from other adult female conspecifics (Rosenblum & Alpert, 1974, 1979; Swartz, 1982). Achievement of person permanence around this point can be seen in peer development, because dyadic relationships begin to form, and these require recognition of individuals. A well-defined dominance relationship emerged between Spock at Sophie by 5 months (Daudelin, 1980). Peer dominance relationships have been identified between human infant peers as early as 6–12 months (Missakian, 1980; Strayer & Trudel, 1984), and between one pair of Japanese macaque twins by the end of the second month (Matsui, 1979; Nakamichi, 1983).

Further relevant to the timing and nature of these three peer stages is Fischer's speculation (1987) concerning links between cognitive advances and a second feature of brain development: concurrent cortical synaptogenesis.

Synaptogenesis consists of the formation of the synapses connecting neurons. The developmental pattern in synaptogenesis during infancy may be general across primates, as it is similar in human and stump-tailed macaque infants (Fischer, 1987; Rakic, Bourgeois, Eckenhoff, Zecevic & Goldman-Rakic, 1986). It may underlie cognitive advances by permitting greater coordination of cognitive functions. Synaptogenesis shows a spurt, whose onset (2–4 months in humans, 1 month in stump-tailed macaques) coincides with the emergence of the abilities to control single-dimension variability and to use simple actions to realize simple goals, or the transition to Piagetian sensorimotor Stage 3. The end of this spurt (7–8 months in humans, 2–3 months in stump-tailed macaques) occurs at the same time as the ability to coordinate systems of actions, the transition to Piaget's Stage 4. These two changes correspond directly in timing and in behavioral quality with the two stage transitions in infant primate peer development.

The peer findings suggest a basic primate pattern in early peer social development. Further, the parallels established with sensorimotor and brain development suggest that some early developmental patterns in social and physical world functioning constitute unified rather than independent processes and that cortical maturation constitutes one important underlying mechanism.

It should be noted, however, that there is a distinct possibility that some of these developmental parallels in social and physical world capabilities may be limited to the peer context. When infant–adult dyads are considered, for example, the parallels may weaken, or divergence may begin earlier (e.g., Bard & Vauclair, 1984). A first reason for this caution is that remarks in the peer literature suggest that infants are rather unclear as to the social versus inanimate status of peers. Mueller and DeStefano (1973) found that human infants were interested in peers only when they took on "object status," that is, when immobile, such as lying on the floor having a diaper changed. Infants are probably less clearly identifiable as social objects than are adults, because infants are less adept at voluntarily displaying signals that characterize social versus inanimate objects (e.g., the facial configuration, vocalizations, contingent responsivity). A second reason for the overlap is that infants may "try out" knowledge acquired in one domain on the other. I observed incidents in which infant behavior was applied to both realms: In one, a ritualized social behavior was explicitly directed at inanimate objects as well as to a social partner (Spock seemed to try out threat gestures on toys before directing them at Sophie). For these reasons, the parallels between physical and social cognitive development may be weaker in other social contexts.

It is unlikely, however, that all parallels would disappear. Some are based on Piagetian markers, such as circular reactions, which appear to be relatively domain independent. Circular reactions could be considered to represent a method used to explore or consolidate new understanding or knowledge; such methods are applicable across various domains. Parallels of this sort would be expected to maintain across different problem domains.

*Localized cross-species similarities and differences*

Various interactive characteristics were found that were not universally shared across these primate groups. Both human and chimpanzee, but not monkey, infants show an extensive focus on inanimate objects in their peer interactions, additional forms of primitive and complex play, prosocial behavior and imitation. Human infants differ from the nonhuman primates in the timing of the two advanced forms of social play. Only nonhuman primate infant peers demonstrate security oriented behavior; only human infant peers engage in vocal exchanges and throw–receive ballgames.

*The role of objects in infant peer interaction.* In both human and ape, but not monkey, infants, with age an increasing proportion of peer contacts involve inanimate objects (for data on humans, see Mueller & Brenner, 1977; Mueller & Lucas, 1975; Vandell et al., 1980; and Vincze, 1971). In both apes and humans, a few object "takes" have been reported at presocial levels, probably as accidental consequences of exploratory manipulations of the partner–object ensemble. An object-centered "stage" is readily identified when objects constitute the central focus of peer interaction. Typical are object possession struggles, which appear at comparable ages, 4–6 months in chimpanzees and gorillas, 6–8 months in humans. This stage persists over comparable periods, 4–6 to 7–8 months in chimpanzees and gorillas, 6–8 to 10–11 months in humans. Marking its end is a change in focus from inanimate objects to peers, signaled by the social use of objects. This occurs in human infants around 10–11 months of age, and in chimpanzees around 7–8 months (e.g., play offer of food, share food).

Several links are suggested with the Piagetian findings:

1. Although both human and ape infants incorporate objects into their peer interactions, reports consistently leave the impression that humans do so more extensively than apes. This corresponds with comments in comparative Piagetian studies that apes appear less interested in objects than humans. It has been suggested that this difference may relate to language (e.g., Vauclair, 1982).

2. The object-centered period emerges at the point that social and inanimate objects are differentiated, the transition between Stages 3 and 4 of general sensorimotor development. The preference for inanimate objects has been suggested to reflect the fact that objects are easier to predict than social partners.

3. The transition to peer- from object-centered behavior may correlate with the achievement of Stage 5 sensorimotor capabilities. This peer related change is marked by the social use of objects, that is, their use as tools to achieve social goals. Stage 5 is characterized by interest in object–object relationships, thus primitive tool use. In both humans and chimpanzees, the first effective social tool use, peer object exchanges, does coincide with Stage 5 onset. More primitive social object use (showing objects without actual exchange) predates formal acquisition of Stage 5; this suggests that, as with

person versus object permanence, Stage 5 understanding may evolve via social intermediaries.

4. Ape–human differences in preferred types of peer social object use correspond to Piagetian findings. Chimpanzees appear more concerned with assimilative and possessive uses of objects (e.g., takes, threat displays), whereas humans seem more interested in accommodating to the particular features of objects (e.g., ballgames). This agrees with the finding from comparative Piagetian studies that chimpanzees appear less interested than human children in the functional or constructive use of objects (Gibson, unpublished data, cited in Morell, 1985; Mathieu & Bergeron, 1981). Because the rearing conditions provided for the chimpanzees reported here encouraged human-like object use, this set of differences likely represents species characteristics.

Infrequent object involvement in infant peer encounters probably is characteristic of the monkey species studied. The Piagetian studies have consistently found that both apes and humans use objects differently and much more extensively than do those monkey species studied. Even feral chimpanzees are spontaneous and frequent tool users and makers; they further engage in other types of object manipulation, and even during infancy they incorporate objects into play (van Lawick-Goodall, 1968).

*Early play.* Some additional early forms of play have been described in humans and chimpanzees, but not monkeys. At comparable presocial levels (2.5–3 months in Spock and Sophie, 4–5 months in human twins studied by Shirley [1933]), infants may demonstrate "amusement" in the form of smiles or play faces in response to peer contact. Also in both, rudimentary forms of social play appear prior to rough-and-tumble and play chases, again at comparable ages. Typical examples are exchanges of smiles and laughs at 6–7 months in humans, and "social" play faces at 5 months and reciprocation of play at 6–7 months in chimpanzees. In monkeys, the only mentions of early play are that it sometimes occurs while infants are still on their mothers' backs and that it is gentler than later social play.

For the most part, these play differences involve the social use of a few facial expressions (play faces/smiles and/or laughs) during sensorimotor Stage 3. Chevalier-Skolnikoff's analysis (1982) of facial behavior in various primate species permits speculation concerning the source of these differences. She noted that human infants at this level voluntarily and repeatedly exchange smiles with partners, whereas monkey infants display such facial expressions only as involuntary reactions. Patterns were not clear in the apes she studied. Species differences may then exist in the voluntary and intentional use of facial expression as the main component of interaction, versus reactive and involuntary use as an accompaniment to interaction.

*Complex play.* One form of social play, which I termed a pseudo–play chase, occurred in both human and ape infant peer interactions. In pseudo–play chases, each infant runs back and forth repeatedly and intermittently coordinates its routine with the partner's (see Obs. 5). Mueller and Lucas (1975) characterized these as intercoordinated secondary circular reactions. They appear as expected during sensorimotor Stage 4, but near the end (9–10 months in chimpanzees, 14–15 months in humans). Given that the coordination involved is with a partner's behavior, and that social partners are seen as less predictable than inanimate objects, the finding that such social coordinations postdate those involving inanimate objects is understandable.

Such primitive play chases have not been described for monkeys, for which there are two possible explanations. The simpler is that available descriptions of monkey infant play chases are simply too general to capture these relatively subtle structural features, although they do in fact occur. The second, more interesting explanation again relates to differences in the patterning of sensorimotor development between primate groups. The monkey species commonly studied (Japanese and stump-tailed macaques) rarely, if at all, perform circular reactions (Antinucci, Spinozzi, Visalberghi, & Volterra, 1982; Chevalier-Skolnikoff, 1982; Parker, 1977). They perform other operations at the same Piagetian stage level without the repetition (Parker, 1977). As a circular reaction constitutes the basic structure of these pseudo–play chases, it is probable that they do not occur in infant monkey peer interactions. Greater detail in observation and study of a greater variety of monkey species are clearly suggested.

*Prosocial behavior.* Prosocial activity appears between human and chimpanzee infant peers, but not monkey infant peers. Prosocial behaviors are common between human infants near the end of the first year (e.g., offer, show, and exchange objects; "talk" to peer; encourage peer). These appear between infant chimpanzees at the equivalent point, 8–9 months of age (play offer and share food; protection), although they occur less frequently than in humans. The only potentially prosocial peer activity reported in other primate groups is grooming, observed between lemur twins at unreported ages (Harrington, 1978). The differences between nonhuman primates and humans may relate in part to the differences in social object use already discussed, as much of the reported prosocial behavior involved objects. Concerning the ape–human difference, studies of bonobos would be interesting, as they appear more affiliative than common chimpanzees (Savage-Rumbaugh, 1984).

*Imitation.* Imitation is widely described as a major theme in human peer interaction from about 12 months of age. Such peer imitation has been observed between chimpanzee infants, but only in the dyad reported here and very rarely. Other than parallel object activity in rhesus infants (Deets, 1974), peer imitation has not been reported in monkeys (Visalberghi & Fragaszy, P&G9; Gibson, P&G7; Tomasello, P&G10). As with prosocial behavior, although chimpanzees share the imitative theme with humans, it appears to have much less importance for them at these developmental levels. Also as with prosocial behavior, these differences likely represent species differences rather than methodological artifacts. Bonobo behavior would again be interesting. Second, although most of the peer imitation between Spock and Sophie appeared at 9–10 months, comparable to its emergence in human peers, one primitive incident was observed at 5 months, much earlier than any reported for human infants. My speculation is that primitive imitation does occur between younger human infants, albeit rarely, and that its identification requires extremely close monitoring. Human infants are capable of primitive forms of imitation from the first month of life (Meltzoff & Moore, 1983; Piaget, 1951). Infant peers may on occasion produce behavior that is worth copying, though probably by accident or because of developmental similarity, because as models, infants have little control over the stimuli they present to their age-mates.

*Developmental ordering in advanced social play.* Chimpanzee, gorilla, and monkey infant peers develop rough-and-tumble play prior to play chases. This has been explained in terms of the more advanced understanding, role taking, and synchronization required for chases (Harlow, 1969). In contrast, rough-and-tumble appears later than play chases in human infant peers. This may relate to species differences in rates of motor development. From my own observations, until about 18 months of age human children seem to be very disturbed by actions that endanger their rather precarious command of upright mobility. The lag in rough-and-tumble in humans may therefore reflect distaste, related to slower locomotor development, rather than a lag in interpersonal understanding. Alternatively, if only "real" play chases are considered, because pseudo–play chases may not occur in infant monkeys, the human pattern is more similar to that seen across the nonhuman primates. In humans, real play chases appear around the same point as rough-and-tumble: 16–18 months of age.

*Security exchanges.* Security exchanges occurred between infant peers in monkeys and chimpanzees, but not humans. These most clearly occur between peers raised in very close proximity: "twinned" infant monkeys (Deets, 1974), twin lemurs (Harrington, 1978), and the current chimpanzee dyad. Extensive security oriented behavior between infant peers probably is environmentally induced, by atypically close rearing conditions. Such mutual peer directed security-seeking behavior was the classic distinguishing feature of "together–together" rhesus infants who were peer raised and mother deprived (Chamove, 1973). It would then be expected in similarly raised human peers, especially twins; reports confirm this for older peer-raised children (Freud & Birmingham, 1944; Freud & Dann, 1951; Spiro, 1958), though none were found involving infants.

*Vocal exchanges and ballgames.* Vocal exchanges and throw–receive ballgames were observed in infant peer interactions in humans, but not in nonhuman primates. That vocal exchanges are exclusive to humans is commonly explained as related to language; why ballgames should occur only in humans is less evident.

The difference in occurrence of ballgames between humans and apes could be artifactual, because (1) incorporation of objects into infant peer play is common to both and (2) orangutans and gorillas have been observed to use balls around the end of the first year, at the Stage 5 level (Chevalier-Skolnikoff, 1982). The difference between humans and the nonhuman primates may be more important when the structures underlying ballgames are considered. The ballgames observed were sophisticated throw–receive exchanges appearing around 16 months of age (Mueller & Lucas, 1975). These require Stage 5 sensorimotor intelligence at a minimum, and possibly Stage 6, via the anticipation needed to catch or receive balls (Chevalier-Skolnikoff, 1982). Consequently, such ballgames are unlikely to appear in monkeys. Monkeys achieve only some Stage 5 capabilities during this period (Chevalier–Skolnikoff, 1982; Parker, 1977), and they do not appear to attain Stage 6 at all. In the ape infants studied, the level of functioning necessary is not fully achieved by the end of the first year. The appearance of such ballgames in apes might then simply be delayed, because of differential developmental rates. The chimpanzees I studied were at best entering Stage 6 by 12 months, and these abilities were not evident in their peer interactions. Because they did achieve many Stage 6 capabilities over the next few months, ballgames of this complexity were possibly imminent. A final, more likely possibility is that ballgames do not occur in nonhuman primates because of interest. As chimpanzee infants seem less interested than humans in functional properties or constructive uses of objects (e.g., Gibson, unpublished data, cited in Morell, 1985; Mathieu & Bergeron, 1981), it is possible that ballgames, which are based on the special properties of balls, might not develop simply because of lack of interest. Gibson tried to teach an older home-reared chimpanzee to play such ballgames; she did not succeed, despite the fact that her subject would have achieved Stage 6 (Morell, 1985).

## Conclusion

Although the current chimpanzee descriptions are based heavily on data from only one interactive dyad reared under atypical conditions, and although individual differences are highly important in chimpanzees, nonetheless the general patterns observed correspond both to the anecdotal reports from other infant chimpanzee peers and, in readily interpretable ways, to patterns reported in human infants and other ape and monkey infants. The chimpanzee patterns identified should then be sufficiently valid for the general level of discussion involved.

These findings do support the notion of a basic primate pattern in early peer social development. Overall correspondences reflect not only major themes enacted in peer interaction but also their order and relative rate of appearance. These correspondences also correlate with patterns in Piagetian sensorimotor development and cortical maturation. The most generalized parallels are found at the earliest levels (the presocial period, where infants do not differentiate peers from inanimate objects), leading to the suggestion that the initial development in both Piagetian and peer social domains involves a unified process for understanding the environment and interacting with it and that cortical maturation constitutes an important underlying mechanism.

Both human and ape infants appear to achieve some separation between peers and inanimate objects, because from the onset of the peer social stage there are observable behavioral distinctions between peer and inanimate object schemes. However, the separation apparently is not complete, as Piagetian–social parallels are found beyond this point. These later parallels are more localized than the earlier ones, consisting of specific interactive patterns versus globalized behavioral characteristics (e.g., social use of circular reactions, social uses of objects). These later parallels also relate to the structures rather than the thematic content of interactions (e.g., the structure underlying play chases or the complexity of interactive object use). Mueller and Lucas's study (1975) on human infants is the only study to date to focus on structural properties of peer interaction. The suggestion from our analyses is that this approach should be especially fruitful for exploring later links between social and physical cognitive development.

From the point where peer and inanimate object schemes can be distinguished, the issue of the relationship between social and physical cognitive development becomes relevant. Several of the parallels found indicate that physical understanding can influence social functioning. In species and at levels where objects are incorporated into peer interaction, the qualities of and limits on object related capabilities can constrain the level of ongoing peer interaction. This is particularly so when objects define the nature of the interaction, as in ballgames. In other cases, achievements in physical cognition (e.g., secondary circular reactions) form the basis for advances in social functioning (e.g., pseudo–play chases). Other parallels suggest the converse, that advances in social functioning precede and thus may promote advances in physical understanding (e.g., person and object permanence). In sum, the evidence from the infant peer context suggests bidirectional effects, in agreement with Case's contention (1985) that object manipulation skills and social behavior (and language) mature in synchronous and mutually facilitory fashion.

A number of cautions should be kept in mind as to the generality of these findings. The species similarities in infant peer interactions coexist with important species differences in adult–infant interactions (e.g., Bard & Vauclair, 1984). The social–physical cognition relationships in development may vary importantly between social contexts and between primate species. The nature and extent of these relationships suggested here may be specific to the peer context, as the peer–object distinction appears somewhat fuzzy to infants, although at least some of the parallels appear relatively domain independent (e.g., those based on circular reactions). Between primate species, the relative importance of inanimate objects should logically influence the nature of these relationships. To the nonhuman primates, inanimate objects are clearly less important than they are to humans; one would thus expect advances in social understanding to dominate intellectual development in the nonhuman primates. In humans, as inanimate objects are clearly of great interest, social processes may dominate development less, or they may in some areas be subordinate to physical understanding.

The species differences in infant peer social development indicate, among other things, increasing behavioral complexity from the monkeys through chimpanzees to humans. Based on the current data, chimpanzees appear to be closer to humans than to monkeys in the developmental patterning of their peer interactions; they are capable of more of the richness and complexity characteristic of human infants versus monkey infants, and they express a great proportion of the range of themes seen in humans. As in other domains, although similarities are clearly and widely present across primate species, and development is even accelerated in the nonhuman primates at the earlier stages, the major impression is of increasing divergence between species and of subsequent lag in the nonhuman primates at the upper developmental levels.

Some of the differences between the species groups studied are readily attributable to between-study variation in rearing conditions and methodologies (e.g., peer dependency). Others, notably social object use, forms of play, imitation, and prosocial behavior, appear to represent species differences that are reflected as well in Piagetian and brain patterns. All are least apparent in the monkeys, intermediate in chimpanzees, and prevalent in humans. As all three have been repeatedly proposed as crucial adaptative advantages in humans, they likely do represent legitimate species differences that are of particular importance to social development in humans. Tracing closely the functions of these three forms of social exchange could then be of particular value in investigating the unique qualities of early human social development.

These findings on infant peer social development highlight both basic patterns characterizing the primates as a family and factors differentiating species groups within the primates. They thus target specific features to be explored in identifying the common elements and unique nature of social development in each species group. The links indicated with concurrent Piagetian and cortical development suggest important biological foundations to the whole developmental package during infancy. They also indicate that during infancy, certain developmental changes of major importance are general rather than domain-specific and affect a wide range of behaviors. The points of both correspondence and difference identified may be useful in suggesting areas to explore in attempting to understand the links between social and physical world cognition.

## Appendix 14.A: Peer social development in monkeys

The studies of monkey infant peer social development, including one study of a set of prosimian twins, reported very uniform patterns. One probable reason for the uniformity is that most researchers have adopted the developmental model proposed by Harlow for the content of infant monkey peer social contacts (Harlow 1962, 1969; Harlow & Harlow, 1965). This model is based on observations of free activity in groups of laboratory-reared rhesus infants given cloth mother surrogates. It proposes four ordered stages for peer encounters characterized by participants' motivation with respect to one another. In the first, reflex stage, predominant between 2 and 4 weeks of age, sounds and movements from the environment may elicit reactions, often reflexive orientation, and, given their precocious motor development, sometimes locomotion and exploration. If the source of stimulation is an age-mate, a social exchange may ensue; if not, because same-age infants tend to respond to stimuli in very similar ways, mutual interest in a display may accidentally result in social contact. Harlow considered this stage to be a laboratory artifact, as normally infants of this age are in constant contact with their mothers. In the second, exploratory stage, peaking around 5–6 weeks of age, infants directly orient toward other infants and investigate them by peering and gently mouthing or manipulating them (touching, clasping, limited body contact). Harlow considered both of the first two stages to be presocial, in that inanimate objects and peers are not differentiated. The third stage, social play, is predominant between 10 and 14 weeks and marks the first truly social stage in peer social development. Peer encounters consist initially of contact play (rough-and-tumble or wrestling play) and, later, approach–withdrawal play (chasing back and forth, together with frequent changes of leader and follower roles). Harlow considered approach–withdrawal play to be the more advanced form, requiring a capability for coordination of roles and synchronization of behavior not needed for rough-and-tumble play. His fourth stage is aggressive play, characterized by rough-and-tumble play boisterous to the point of being hurtful. This stage appears near the end of the first year, beyond the period of interest.

The studies reviewed support Harlow's model. The only exceptions reported were a few additional forms of peer interaction, occurring primarily in infant monkeys experiencing unusually close living conditions (Deets, 1974; Harrington, 1978; Nakamichi, 1983; Rosenblum et al., 1975). First, some involvement of inanimate objects in monkey infant peer interaction was mentioned by both Deets (1974) and Rosenblum et al. (1975). Deets observed what he termed "imitation" – exploration of a target being explored by the partner, which is referred to as parallel activity in human research – in unrelated same-age rhesus infants laboratory reared as "twins" (two per rhesus mother). Rosenblum et al. (1975) noted friendly, mutual competitive exploration of inanimate objects among rhesus infants around the age that rudimentary contact play emerged. Second, joint excursions away from the mother and grooming were observed in two naturally occurring sets of twins: *Lemur macaco* (Harrington, 1978) and Japanese macaques (Nakamichi, 1983). Grooming might be considered as prosocial and/or security oriented (Mason, 1965a,b). Joint excursions appear to be security oriented. The infants' ages were not reported.

Overall, monkey infant peer development seems to show a narrow range of peer activity and clearly defined developmental patterning of a simple, linear nature.

## Appendix 14.B: Infant peer social development in humans

The earliest peer interactions in human infants are reported between 2 and 3 months of age. These consist of attentional responses to peers' movements or sounds (Bridges, 1933). By 4–5 months, peer exchanges include more extensive forms of contact, such as touching, followed contingently by mutual touches, mouthing, or other (undescribed) responses (Shirley, 1933; Vincze, 1971). There are some indications that peer contact may be pleasurable (one of the fraternal twins studied by Shirley showed amusement when his fingers were chewed as he reached toward his twin sister). Between 3 and 5 months, Vincze observed "object takes" on rare occasions, although these appeared to be accidental in nature and nonsocial in focus.

A major shift in the quality and focus of peer interaction appears around 6–8 months of age, as inanimate objects are differentiated from peers, and "social" qualities appear in peer exchanges (i.e., qualities reflecting response to communicative as opposed to merely physical aspects of stimulation). Infants 6–8 months old commonly establish contact with peers through activity with inanimate objects (Mueller & Lucas, 1975; Vandell et al., 1980; Vincze, 1971). They increasingly fight over toys, sometimes just because peers possess them. Also relatively common from this point until the end of the first year are exchanges of smiles, vocalizations, and laughs (Bridges, 1933; Vandell et al., 1980; Vincze, 1971), reaches, touches, gestures such as flapping arms and clapping, and some agonistic behaviors such as hitting and pushing (Bridges, 1933; Vandell et al., 1980).

Around the end of the first year, independent locomotion becomes possible (walking commonly develops around 10–12 months). This apparently leads to rougher contacts and more serious conflicts (Vincze, 1971), although overt aggression has rarely been reported. Positive affect in peer interaction also appears around this point and subsequently increases, although aggressive behavior and object possession struggles continue to occur (Brenner & Mueller, 1982; Bridges, 1933; Mueller & Lucas, 1975). Several forms of peer interaction emerge here, characterized by their social versus object focus and their positive affect. These include social play, prosocial behavior, and imitation, as described next.

Social play appears around 10–12 months. Vincze's infants first participated in group games at 10–11 months (but in a group with toddlers up to 6 months older); Shirley's twins laughed when they replicated a short exchange sequence verbatim at 11–12 months; Bridges's infants copied each other's laughs at 13–14 months. Three distinctive forms of social play appeared later: running chase from 15 months, throwing–receiving ballgames from 16 months, and rough-and-tumble play from 17 months (Brenner & Mueller, 1982; Mueller & Lucas, 1975). Some of the run-chase interactions took the form of what they termed "intercoordinated secondary circular reactions." In these, one child would repeatedly run away, and the second would follow, but the partners would makes errors in the synchronization or sequencing of their behaviors. In advanced chase and ballgames, properly synchronized and complementary behaviors were observed.

Prosocial contacts were first reported at 10–11 months by Vincze, who observed infants using toys as social devices by offering or showing them (but not giving them) to each other. By 12–13 months the first peaceful offer–receive exchanges of objects occurred (Brenner & Mueller, 1982; Vincze, 1971). These became increasingly common through 18 months of age. From as early as 15–16 months of age, toddlers sometimes "talked" to one another, and at 17–18 months they smiled and laughed to encourage their peer partners (Brenner & Mueller, 1982).

Imitation is mentioned from as early as 11–12 months and continues as a prominent theme in infant peer interaction through 18 months of age. Bridges's infants (1933) were observed at 11–12 months making similar vocal sounds and patting or rocking in imitation of one another, and at 13–14 months they were smiling and laughing in imitation. Mueller and Lucas (1975) observed "constrained" copying of a partner's behavior toward an object from as early as 13–14 months, and imitation of the partner's interactive behavior at 14–15 months. The earlier imitation was "constrained" in the sense that the copy appeared to represent activation of an already acquired scheme rather than deliberate accommodation to the model's actions. The occurrence of imitative exchanges from as early as 15–18 months of age has been reported in at least two other studies (Brenner & Mueller, 1982; Mueller & DeStefano, 1973). The descriptions of peer imitation suggest that it is of considerable importance to peer social activity during the second year of life.

By 18 months of age, the range of content observed in interactions includes prosocial behavior, imitation, run-chase and rough-and-tumble play, ballgames, vocal exchanges, object exchanges or struggles, and aggression (Brenner & Mueller, 1982).

Because objects consistently figure centrally in human infant peer interactions, a brief survey of their role in peer development is in order. Human infants have been reported to show interest in one another's toys from as early as 3–5 months of age (Vincze, 1971), but until peers are discriminated from inanimate objects around 6–8 months of age, interest appears to be aroused more by the peer–object ensemble than by the object alone (Mueller & Lucas, 1975). For several months thereafter, objects appear to be considered more interesting and important than peers: Object tussles are common (Vincze, 1971), and attention to peers decreases when objects are present (Vandell et al., 1980). Around 1 year of age, the relative importance of peers versus objects reverses. Peers seem to take on greater importance than objects, because objects may be used to initiate or facilitate social exchange (Mueller & Brenner, 1977; Vincze, 1971). Notwithstanding this increase in positive affect associated with objects, object possession struggles continue through 18 months and may become increasingly aggressive (Bridges observed "aggression" at 13–14 months, and bites and hits at 14–15 months).

## Appendix 14.C: Piagetian sensorimotor development in human and nonhuman primates

Piaget's sensorimotor intelligence, as he defined it for human infants, is the intelligence that exists prior to language; it is essentially practical and must be supported by perceptuomotor experience. In Piaget's model, sensorimotor intellectual development covers several domains. Those that have been studied in both humans and nonhuman primates are general sensorimotor intelligence (general sensory and motor adaptations to the environment), the object concept (the understanding that objects continue to exist despite alterations of their appearance or even disappearance), causality (understanding of what forces cause events to occur in the external world), and imitation (the sequence of increasingly more advanced forms of imitation). Piaget saw sensorimotor progress in these areas as proceeding via six ordered stages that appear over the first 2 years of life (in humans), at about the same rate for each area. Each stage reflects a move to more advanced, complex cognitive functioning. This progress is considered to be the result of the child's progressive construction of reality based on the child's own actions on the external world. The cognitive changes are reflected in a set of behavioral phenomena specific to each stage.

The most general of these series, the sensorimotor intelligence series, has been the one most commonly used in human and ape studies to mark and compare developmental progress. Its six stages and their behavioral indicators are as follows:

1. *Reflex* – exercise of reflexes; unlearned, involuntary stereotyped reactions to stimulation.
2. *Primary circular reactions* – first acquired adaptations (acts modified as a function of experience); self-oriented simple actions showing acquired adaptation to impinging stimuli, and sustained through repetition; thus, "primary" (self-oriented) "circular" (repeated) "reactions" (actions).
3. *Secondary circular reactions* – emergence of means–ends distinction, of environment oriented (i.e., secondary) simple actions that are semiintentional

(initial acts are not intentional, but subsequent acts are), and of intentional sustaining of initially accidental environmental effects via repetition (i.e., circularity).

4. *Coordinations* – true intentionality (means–ends coordination); coordination of several environment oriented actions with the intention of achieving a goal established at the outset.
5. *Tertiary circular reactions* – true experiments on objects (creation of new means to reach fixed goals); environment oriented actions that are repeated with systematic variation in experiments with relations among objects, space, and force (i.e., tertiary).
6. *Insight* – emergence of mental representation (search for new means to predefined goals by mental versus trial-and-error processes, perceptuomotor combinations); solving problems about object-force-space relationships without repeated trial-and-error physical experimentation.

General conclusions concerning the nonhuman primates must be cautiously drawn, because these empirical studies have been based on very few individuals or species, and variability is great between both individuals and species. It does appear that all the apes studied (chimpanzees, gorillas, and orangutans) show sensorimotor development parallel to that in humans, in that they progress through similar stages in the same order. Rates, however, vary, in that the developmental sequence is initially accelerated in the apes, but may begin to lag by Stage 6. Early acceleration appears closely related to earlier physical maturation.

## Acknowledgments

This research was supported by doctoral fellowships from the Social Sciences and Humanities Research Council of Canada, the Province of Québec Bourses de la Direction Générale des Etudes Supérieures, and the Canadian Association of University Teachers (J. H. Steward Reid Fellowship). I am greatly indebted to Dr. Mireille Mathieu for the opportunity to study these chimpanzees under her direction at her Laboratory of Comparative Development, as well as to Daniel Paquette, N. Daudelin, and the other members of the research group for their support and contributions to this work. The critical comments offered by S. Parker and K. Gibson on an earlier version of this manuscript were most welcome and valuable.

## References

Antinucci, G., Spinozzi, G., Visalberghi, E., & Volterra, V. (1982). Cognitive development in a Japanese macaque (*Macaca fuscata*). *Annal: dell'Istituto Superiore di Sanita*, *18*(2), 177–184.

Argyle, M. (1969). *Social interaction*. London: Methuen.

Baldwin, J. D. (1969). The ontogeny of social behavior of squirrel monkeys (*Saimiri sciureus*) in a semi-natural environment. *Folia Primatologica*, *11*, 35–79.

Baldwin, J. D., & Baldwin, J. I. (1978). Exploration and play in howler monkeys (*Alouatta palliata*). *Primates*, *19*(3), 411–422.

Bard, K. A., & Vauclair, J. (1984). The communicative context of object manipulation in ape and human adult–infant pairs. *Journal of Human Evolution*, *13*, 181–190.

Becker, J. M. T. (1977). A learning analysis of the development of peer-oriented behavior in 9-month-old infants. *Developmental Psychology*, *13*, 481–491.

Bell, S. M. (1970). The development of the concept of object as related to mother–infant attachment. *Child Development*, *14*, 291–311.

Bergeron, G. (1979). *Développement sensori-moteur du jeune chimpanzé (Pan troglodytes) en liberté restreinte*. Unpublished master's thesis, University of Montréal.

Berman, C. M. (1982a). The ontogeny of social relationships with group companions among free-ranging infant rhesus monkeys. I. Social networks and differentiation. *Animal Behavior*, *30*, 149–162.
Berman, C. M. (1982b). The ontogeny of social relationships with group companions among free-ranging rhesus monkeys. II. Differentation and attractiveness. *Animal Behavior*, *30*, 163–170.
Blurton-Jones, N. G. (1974). Ethology and early socialization. In M. P. M. Richards (Ed.), *The integration of a child into a social world* (pp. 263–294). Cambridge University Press.
Bolwig, N. (1980). Early social development and emancipation of *Macaca nemestrina* and species of *Papio*. *Primates*, *21*(3), 357–375.
Brenner, J., & Mueller, E. (1982). Shared meaning in boy toddlers' peer relations. *Child Development*, *53*(2), 380–391.
Bridges, K. B. M. (1933). A study of social development in early infancy. *Child Development*, *4*, 36–49.
Bronson, W. C. (1981). Toddlers' behaviors with agemates: Issues of interaction, cognition, and affect. *Monographs on Infancy*, *1*, 1–127.
Brownlee, J. R., & Bakeman, R. (1981). Hitting in toddler–peer interaction. *Child Development*, *52*(3), 1076–1079.
Bruner, J. (1983). *In search of mind: Essays in autobiography*. New York: Harper & Row.
Buhler, C. (1935). *From birth to maturity: An outline of the psychological development of the child*. London: Routledge & Kegan Paul.
Byrne, R., & Whiten, A. (Eds.). (1988). *Machiavellian intelligence: Social expertise and the evolution of intellect in monkeys, apes, and humans*. Oxford University Press.
Case, R. (1985). *Intellectual development: Birth to adulthood*. New York: Academic Press.
Chalmers, N. R. (1980). The ontogeny of play in feral olive baboons (*Papio anubis*). *Animal Behavior*, *28*, 570–585.
Chamove, A. (1973). Rearing infant rhesus together. *Behaviour*, *47*(1–2), 48–66.
Champoux, M., Schneider, M., & Suomi, S. (1987, June). *Environmental enrichment and affective responses in infant rhesus monkeys*. Paper presented at the annual meeting of the American Society of Primatologists, Madison, WI.
Cheney, D. L. (1978). The play partners of immature baboons. *Animal Behavior*, *26*, 1038–1050.
Chevalier-Skolnikoff, S. (1977). A Piagetian model for describing and comparing socialization in monkey, ape, and human infants. In S. Chevalier-Skolnikoff & F. Poirier (Eds.), *Primate bio-social development: Biological, social and ecological determinants* (pp. 159–187). New York: Garland Press.
Chevalier-Skolnikoff, S. (1982). A cognitive analysis of facial behavior in Old World monkeys, apes, and human beings. In C. T. Snowdon, C. N. Brown, & M. R. Petersen (Eds.), *Primate communication* (pp. 308–368). Cambridge University Press.
Chevalier-Skolnikoff, S. (1983). Sensorimotor development in orangutan and other primates. *Journal of Human Evolution*, *12*, 545–561.
Costello, M. (1987, June). *Tool use and manufacture in manipulanda-deprived capuchins* (*Cebus apella*). Paper presented at the annual meeting of the American Society of Primatologists, Madison, WI.
Daudelin, N. (1980). *Etablissement d'une relation de dominance entre deux chimpanzés (Pan troglodytes) durant leur première année de vie*. Unpublished doctoral dissertation, University of Montréal.
Deets, A. C. (1974). Age-mate or twin siblings: Effects on monkey age-mate interactions in infancy. *Developmental Psychology*, *10*(6), 913–928.
DeJonge, G., Dienske, H., van Luxemburg, E., & Ribbens, L. (1981). How rhesus monkey infants budget their time between mothers and peers. *Animal Behavior*, *29*, 598–609.
Doré, F. Y., & Dumas, C. (1987). Psychology of animal cognition: Piagetian studies. *Psychological Bulletin*, *102*(2), 219–233.
Finkelstein, N. W., Dent, C., Gallagher, K., & Ramey, C. T. (1978). Social behavior in infants and toddlers in a day-care environment. *Developmental Psychology*, *14*(3), 257–262.

Fischer, K. W. (1987). Relations between brain and cognitive development. *Child Development*, *58*, 623–632.
Freud, A., & Birmingham, D. (1944). *Infants without families*. New York: International Universities Press.
Freud, A., & Dann, S. (1951). An experiment in group upbringing. *Psychoanalytic Study of the Child*, *6*, 127–168.
Fritz, J. (1984). *Handrearing*. Unpublished manuscript.
Gibson, K. R. (1977). Brain structure and intelligence in macaques and human infants from a Piagetian perspective. In S. Chevalier-Skolnikoff & F. Poirier (Eds.), *Primate biosocial development* (pp. 113–157). New York: Garland Press.
Gibson, K. R. (1981). Comparative neuro-ontogeny, its implications for the development of human intelligence. In G. E. Butterworth (Ed.), *Infancy and epistemology* (pp. 52–82). Brighton, England: Harvester Press.
Goodall, J. (1986). *The chimpanzees of Gombe: Patterns of behavior*. Cambridge, MA: Harvard University Press.
Handen, C. E., & Rodman, P. S. (1980). Social development of bonnet macaques from six months to three years of age: A longitudinal study. *Primates*, *21*, 350–356.
Hansen, E. W. (1966). The development of maternal and infant behavior in the rhesus monkey. *Behaviour*, *27*, 107–149.
Harlow, H. F. (1962). The development of affectional patterns in infant monkeys. In B. M. Foss (Ed.), *Determinants of infant behavior* (pp. 75–88). New York: Wiley.
Harlow, H. F. (1969). Age-mate or peer affectional system. In D. Lehrmann, R. Hinde, & E. Shaw (Eds.), *Advances in the study of behavior* (Vol. 2, pp. 333–383). New York: Academic Press.
Harlow, H. F., & Harlow, M. K. (1965). The affectional systems. In A. Schrier, H. Harlow, & F. Stollnitz (Eds.), *Behavior of non-human primates* (Vol. 2, pp. 287–334). New York: Academic Press.
Harrington, J. E. (1978). Development of behavior in *Lemur macaco* in the first nineteen weeks. *Folia Primatologica*, *29*, 107–128.
Hinde, R. A. (1986). Can nonhuman primates help us understand human behavior? In B. B. Smuts, D. L. Cheney, R. M. Seyfarth, R. W. Wrangham, & T. T. Struhsaker (Eds.), *Primate societies* (pp. 413–420). University of Chicago Press.
Hinde, R. A., Perret-Clermont, A.-N., & Stevenson-Hinde, J. (Eds.). (1985). *Social relations and cognitive development*. Oxford University Press.
Hinde, R. A., & Spencer-Booth, Y. (1971). Effects of brief separation from mother on rhesus monkeys. *Science*, *173*, 111–118.
Holmberg, M. C. (1980). The development of social interchange patterns from 12 to 42 months. *Child Development*, *51*, 448–456.
Howes, C. (1980). Peer play scale as an index of complexity of peer interaction. *Developmental Psychology*, *16*(4), 371–372.
Jacobsen, C. F., Jacobsen, M. M., & Yoshioka, J. G. (1932). Development of an infant chimpanzee during her first year. *Comparative Psychology Monographs*, *9*, (No. 1, Serial No. 41).
Jacobson, J. L. (1981). The role of inanimate objects in early peer interaction. *Child Development*, *52*, 618–626.
Lewis, M., & Rosenblum, L. A. (1975). *Friendship and peer relations*. New York: Wiley.
Loizos, C. (1967). Play behaviour in higher primates: A review. In D. Morris (Ed.), *Primate Ethology*. London: Wiedenfeld & Nicholson.
McGrew, W. G. (1972). *An ethological study of children's behavior*. New York: Academic Press.
Mason, W. A. (1963). The effects of environmental restriction on the social development of infant monkeys. In C. A. Southwick (Ed.), *Primate social behavior*. New York: Van Nostrand.
Mason, W. A. (1965a). Determinants of social behavior in young chimpanzees. In A. M. Schrier,

H. F. Harlow, & F. Stollnitz (Eds.), *Behaviour of nonhuman primates* (Vol. 2, pp. 335–364). New York: Academic Press.

Mason, W. A. (1965b). The social development of monkeys and apes. In I. DeVore (Ed.), *Primate behaviour: Field studies in monkeys and apes* (pp. 514–543). New York: Holt, Rinehart & Winston.

Mason, W. A., Davenport, R. K., Jr., & Menzel, E. W., Jr. (1968). Early experience and the social development of rhesus monkeys and chimpanzees. In G. Newton & S. Levine (Eds.), *Early experience and behavior* (pp. 440–480). Springfield, IL: Thomas.

Mathieu, M. (1982, August). *Intelligence without language: Piagetian assessment of cognitive development in chimpanzee.* Paper presented at the Congress of the International Society of Primatology, Atlanta.

Mathieu, M., & Bergeron, G. (1981). Piagetian assessment on cognitive development in chimpanzee (*Pan troglodytes*). In A. B. Chiarelli & R. S. Corruccini (Eds.), *Primate behavior and sociobiology* (pp. 142–147). Berlin: Springer-Verlag.

Mathieu, M., Daudelin, N., Dagenais, Y., & Décarie, T. G. (1980). Piagetian causality in two house-reared chimpanzees (*Pan troglodytes*). *Canadian Journal of Psychology*, *34*(2), 179–186.

Matsui, T. (1979). Futagozaru wagenkini so datta (The twin Japanese monkeys survived). *Monkey*, *23*, 6–12.

Maudry, M., & Nekula, M. (1939). Social relations between children of the same age during the first two years of life. *Journal of Genetic Psychology*, *54*, 193–215.

Meltzoff, A. N., & Moore, M. K. (1983). The origins of imitation in infancy: Paradigm, phenomena, and theories. In L. P. Lipsitt (Ed.), *Advances in infancy research* (Vol. 2, pp. 265–301). Norwood, NJ: Ablex.

Mignault, C. (1985). Transition between sensorimotor and symbolic activities in nursery-reared chimpanzees (*Pan troglogytes*). *Journal of Human Evolution*, *14*, 747–758.

Missakian, E. (1980). Gender differences in agonistic behavior and dominance relations of Synanon communally reared children. In. D. R. Omark, F. F. Strayer, & G. Freedman (Eds.), *Dominance relations: An ethological view of human conflict and social interaction* (pp. 397–413). New York: Garland Press.

Morell, V. (1985). Challenging chimpanzees. *Equinox*, *22*, 17–18.

Mueller, E., & Brenner, J. (1977). The origins of social skills and interaction among playgroup toddlers. *Child Development*, *48*, 854–861.

Mueller, E., & DeStefano, C. (1973). *Sources of toddlers' peer interaction in a playgroup setting.* Unpublished manuscript, Boston University.

Mueller, E., & Lucas, T. (1975). A developmental analysis of peer interaction among toddlers. In M. Lewis & L. A. Rosenblum (Eds.), *Friendship and peer relations* (pp. 223–258). New York: Wiley.

Nakamichi, M. (1983). Development of infant twin Japanese monkeys (*Macaca fuscata*) in a free-ranging group. *Primates*, *24*(4), 576–583.

Norikoshi, K. (1974). The development of peer-mate relationships in Japanese macaque monkeys. *Primates*, *15*(1), 39–46.

Owens, N. W. (1975). Social play behaviour in free-living baboons, *Papio anubis*. *Animal Behavior*, *23*, 387–408.

Parker, S. (1977). Piaget's sensorimotor series in an infant macaque: A model for comparing unstereotyped behavior and intelligence in human and nonhuman primates. In S. Chevalier-Skolnikoff & F. E. Poirier (Eds.), *Primate bio-social development: Biological, social, and ecological determinants* (pp. 43–112). New York: Garland Press.

Parker, S. T., & Gibson, K. (1977). Object manipulation, tool use and sensorimotor intelligence as feeding adaptations in cebus monkey and great apes. *Journal of Human Evolution*, *6*, 623–641.

Piaget, J. (1951). *Play, dreams and imitation in childhood.* London: Routledge & Kegan Paul.

Plooij, F. X. (1984). *The behavioral development of free-living chimpanzee babies and infants.* Norwood, NJ: Ablex.

Rakic, P., Bourgeois, J. P., Eckenhoff, M. F., Zecevic, N., & Goldman-Rakic, P. (1986). Concurrent overproduction of synapses in diverse regions of the primate cerebral cortex. *Science*, *232*, 232–235.

Redshaw, M. (1976). *The development of play and social behavior in two lowland gorilla infants*. Jersey Wildlife Preservation Trust 13th annual report.

Redshaw, M. (1978). Cognitive development in human and gorilla infants. *Journal of Human Evolution*, *7*, 133–141.

Rhine, R. J., & Hendy-Neely, H. (1978). Social development of stumptail macaques (*Macaca arctoides*): Momentary touching, play, and other interactions with aunts and immatures during the infants' first 60 days of life. *Primates*, *19*(1), 115–123.

Rosenblum, L. A., & Alpert, S. (1974). Fear of strangers and specificity of attachment in monkeys. In M. Lewis & L. A. Rosenblum (Eds.), *The origins of fear* (pp. 165–193). New York: Wiley.

Rosenblum, L. A., & Alpert, S. (1979). Response to mother and stranger: A first step in socialization. In S. Chevalier-Skolnikoff & F. Poirier (Eds.), *Primate bio-social development* (pp. 463–478). New York: Garland Press.

Rosenblum, L. A., Coe, C. L., & Bromley, L. J. (1975). Peer relations in monkeys. The influence of social structure, gender, and familiarity. In M. Lewis & L. A. Rosenblum (Eds.), *Friendship and peer relations* (pp. 67–98). New York: Wiley.

Ross, H. S., & Goldman, B. M. (1977). Establishing new social relations in infancy. In T. Alloway, P. Pliner & L. Krames (Eds.), *Attachment behavior: Advances in the study of communication and affect*, Vol. 3, pp. 61–79. New York: Plenum.

Ross, H. S., Goldman, B. M., & Hay, D. F. (1979). Features and functions of infant games. In B. Sutton-Smith (Ed.), *Play and learning* (pp. 107–121). New York: Gardner Press.

Rubenstein, J. L., & Howes, C. (1979). Caregiving and infant behavior in day care and homes. *Developmental Psychology*, *15*, 1–24.

Russon, A. E. (1985). *The ontogeny of peer social interaction between infant chimpanzees: A description and comparative analysis*. Unpublished doctoral dissertation, University of Montreal.

Savage, E. S., Temerlin, J. W., & Lemmon, W. B. (1973). Group formation among captive mother–infant chimpanzees (*Pan troglodytes*). *Folia Primatologica*, *20*, 453–473.

Savage-Rumbaugh, E. S. (1984). *Pan paniscus* and *Pan troglodytes*: Contrasts in preverbal communicative competence. In R. Susman (Ed.), *The pygmy chimpanzee: Evolutionary history and behavior* (pp. 395–413). New York: Plenum.

Schaffer, H. R., & Emerson, P. E. (1964). The development of social attachments in infancy. *Monographs of the Society for Research in Child Development*, *29*(3), 1–77.

Shirley, M. (1933). *The first two years: A study of 25 babies* (2 vols.). Institute of Child Welfare Monograph Series, No. VII. Minneapolis: University of Minnesota Press.

Spiro, M. (1958). *Children of the kibbutz*. Cambridge, MA: Harvard University Press.

Sroufe, L. A. (1979). Socioemotional development. In J. D. Osofsky (Ed.), *Handbook of infant development* (pp. 462–516). New York: Wiley.

Stern, D. (1978). *The first relationship*. Cambridge, MA: Harvard University Press.

Strayer, F. F., & Trudel, M. (1984). Developmental changes in the nature and function of social dominance among young children. *Ethology and Sociobiology*, *5*, 279–295.

Sugiyama, Y., & Koman, J. (1979). Social structure and dynamics of wild chimpanzees at Bossou, Guinea. *Primates*, *20*(3), 323–339.

Suomi, S. J., & Harlow, H. F. (1975). The role and reason of peer relationships in rhesus monkeys. In M. Lewis & L. A. Rosenblum (Eds.), *Friendship and peer relations* (pp. 153–185). New York: Wiley.

Swartz, K. B. (1982). A comparative perspective on perceptual, cognitive, and social development. *Journal of Human Evolution*, *11*, 315–320.

Tomilin, M. I., & Yerkes, R. M. (1935). Chimpanzee twins: Behavioral relations and development. *Journal of Genetic Psychology*, *46*, 239–264.

Trevarthen, C. (1979). Communication and co-operation in early infancy. A description of primary intersubjectivity. In M. Bullowa (Ed.), *Before speech* (pp. 321–347). Cambridge University Press.

Uzgiris, I., & Hunt, J. M. (1975). *Assessment in infancy: Ordinal scales of psychological development.* Urbana: University of Illinois Press.

Vandell, D. L., Wilson, K. S., & Buchanan, N. R. (1980). Peer interaction in the first year of life: An examination of its structure, content, and sensitivity to toys. *Child Development, 51*, 481–488.

van Hooff, J. A. R. A. M. (1973). A structural analysis of the social behaviour of a semi-captive group of chimpanzees. In M. von Cranach & I. Vine (Eds.), *Social communication and movement: Studies of interaction and expression in man and chimpanzee* (pp. 75–162). New York: Academic Press.

van Lawick-Goodall, J. (1968). The behaviour of free-living chimpanzees in the Gombe Stream Reserve. *Animal Behaviour Monographs, 1*, 161–311.

Vauclair, J. (1982). Sensorimotor intelligence in human and non-human primates. *Journal of Human Evolution, 11*, 257–264.

Vincze, M. (1971). The social contacts of infants and young children reared together. *Early Child Development and Care, 1*, 99–109.

Westergaard, G. C. (1987, June). *Lion-tailed macaques manufacture and use tools.* Paper presented at the annual meeting of the American Society of Primatologists, Madison, WI.

# 15 Spatial expression of social relationships among captive *Pan paniscus*: Ontogenetic and phylogenetic implications

*Ben G. Blount*

## Introduction

An exploratory study of a captive group of bonobos (*Pan paniscus*) was undertaken to describe the expression of social relationships through the utilization of space. The research objectives were to describe how the animals distributed themselves in the cage space, to describe spatial aspects of their social interaction, and to demonstrate a relationship between those phenomena. Distinctive patterns of spatial distribution, in fact, were observed for two subgroups of the population, and those distributions were reflected in the patterns of social approaches that members of the group adopted. The importance of space utilization resides in its ability to reflect stable social relationships, and thus dominance, in the virtual absence of aggression and overt gestures and signals.

The presence of a juvenile female in the group revealed a sharp contrast to the adult social space utilization. Her behavior indicated that although she reflected the female pattern of occupation of cage space, she had not yet begun to show the patterns of interaction consistent with social use of space. Adult interactions with her showed two distinctive patterns, one based on mutual avoidance and one based on use of overt gestures. The differential behavior shown by the adults among themselves, in their interactions with the juvenile, and by the juvenile herself suggest cognitive-based decisions. Inferences about cognition in ontogeny and phylogeny are tentatively drawn from the observed phenomena.

The social relationships among the bonobos are seen as a consequence of stable expression of dominance. Dominance is widely viewed as a principal mechanism for structuring social interaction among members of social groups. Despite the central importance of the concept, research in recent years has indicated that the expression of dominance in nonhuman primate societies is sometimes difficult to ascertain. The commonly accepted definition of dominance as patterned directionality of aggression (Bernstein, 1981) has been expanded in recent years to include other behavioral phenomena. Because two of those phenomena, social use of space and use of gestures, are employed in the present study, a review of their place in dominance is necessary.

One source of expanded perspectives on the expression of dominance has been research on chimpanzees (*Pan troglodytes*). Dominance relationships among chimpanzees have proved difficult to recognize, and descriptions of dominance ranking among social groups of chimpanzees are especially problematic. Dominance does not easily generalize across chimpanzee age–sex categories and types of activities. In one study, Noe, de Waal, and van Hooff (1980) distinguished different hierarchies for males, based on agonism and bluff, and for females, based on competition for food. Moreover, being dominant and being influential did not necessarily mean the same thing for males, because dominance rank did not directly translate ino access to resources. Dominance appeared to be due to patterned relationships independent of competition for resources, in part.

Recently, de Waal (1986) recast dominance as reconciled hierarchy. By that he meant a formalization of real dominance relationships and a linkage to those of friendly coexistence, expressed through signaling. The resultant formal dominance is seen to overlap with, but to be different from, real dominance, which is essentially success in aggressive encounters. Formal dominance is viewed as a buffer to aggressive encounters, reducing their frequency and disruptiveness. Exchange of signals prevents, or at least inhibits, actual aggression.

De Waal's distinction between formal dominance and real dominance should prove to be highly useful. In some species, however, episodes of agonistic behavior are too infrequent to allow observers to identify rank hierarchies. A case in point is the patas monkey (*Erythrocebus patas*). The low salience of agonistic rank related behavior among captive and wild populations of patas monkeys led Rowell and Olson (1983) to question the adequacy of the "who does what to whom" (i.e., dominance-based) description of social organization. They presented as an alternative mechanism the visual monitoring of spatial distance among troop members. Behavioral maneuvers on the basis of activity, social bonding, and kinship regulated spatial patterns and expressed the social organization of the patas troops. In situations where the usual spatial maneuvers were more difficult to accomplish, displays to signal presence and behavior were more likely to occur.

Expression of social relationships through management of spatial distance and of formal dominance through gestures is a useful perspective in studying the social behavior of bonobos. Virtually all observers of bonobo behavior in naturalistic studies have noted the low levels and mild forms of aggression among them, as compared with chimpanzees (e.g., Badrian & Badrian, 1984; Kano & Mulavwa, 1984; Mori, 1984). Aggressive behavior among bonobos in captivity has also been reported to be exceptionally low (Jordan, 1977; Patterson, 1979; Savage-Rumbaugh & Wilkerson, 1978). The expression of dominance through patterned directionality of aggression is less useful as a

Table 15.1. *Yerkes sample population*

| Animal | Sex | Age | Years at Yerkes | Born |
|---|---|---|---|---|
| Bosondgo | M | 14 | 8 | Wild |
| Linda | F | 30 | 2 | Wild |
| Laura | F | 15 | 4 | Captivity |
| Lorel | F | 13 | 4 | Captivity |
| Lisa | F | $2\frac{1}{2}$ | $2\frac{1}{2}$ | Captivity |

descriptive or analytic concept than is monitoring of spatial proximity and formal dominance in the study of bonobos.

## Study population

The group of bonobos used in the study were at the Yerkes Regional Primate Research Center near Lawrenceville, Georgia. Like all captive populations of bonobos, the group was small, containing only five individuals: one adult male, three adult females, and one juvenile female. The name, sex, and approximate age of each animal are given in Table 15.1.

The females were all related to each other. Linda (Ln) was the mother of Laura (La) and Lorel (Lo), and Laura was the mother of the juvenile female, Lisa (Ls). The male, Bosondjo (Bo), was related to the females. The group had been together for approximately 2 years at the time of data collection.

## Procedures

### *Data collection*

Focal animal samples (Altmann, 1974) of 10-min duration were taken throughout the day from 9:00 A.M. to 4:00 P.M. during the period 12 October through 12 December 1983. A total of 225 focal samples was collected, 45 for each of the animals. After each third focal sample, a 10-min ad libitum sample was taken to enhance the probability of observing infrequently occurring behaviors. Seventy-two samples were taken. After each ad libitum sample, an instantaneous scan sample was made of each animal's location within the caged area. A total of 315 scan samples were taken.

### *Proximity by cage location*

The bonobos' quarters consisted of a single building containing indoor and outdoor areas. The outside portion was divided into two equal parts, separated by a solid concrete wall. The top, ends, and sides of each outdoor

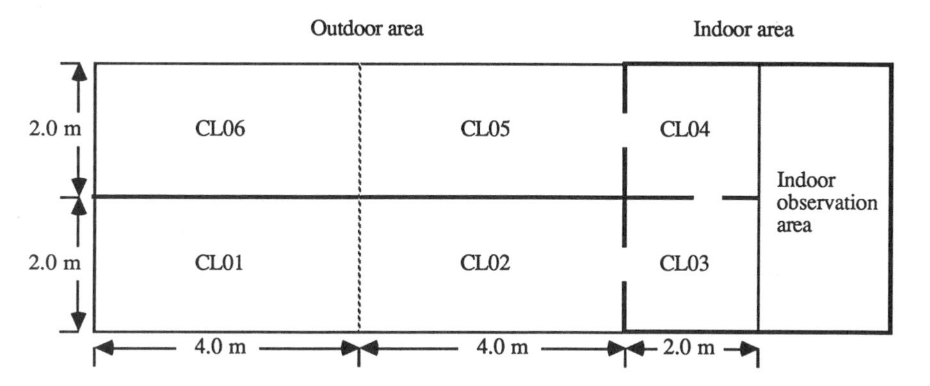


Figure 15.1. Diagram of bonobo enclosure.

area consisted of heavy mesh wire, as did the ceiling and the end of the indoor area. The indoor area was also divided into two sections, connected by a small doorway, and each indoor section was connected by a small doorway to the adjacent outside area. The structure thus contained four distinct areas, divided by interior walls and linked by doorways. A diagram of the building is shown in Figure 15.1.

Because the outdoor areas were considerably larger than the indoor areas, they were divided diagrammatically into two halves for purposes of the scan samples. A grid of six areas (cage locations 01–06) of approximately the same size was thereby available for location of each animal within the compound in any given scan. The totals for each animal across the scans allowed for cage location profiles to be established.

Two or more animals frequently were in the same grid during a scan. In order to obtain a measure of which animals were most likely to be found in the same grid area, a coefficient of association was calculated for each dyad. The formula for the calculation was

$$C = a + b/2ab$$

where $a$ and $b$ represent the two members of the dyad (Cole, 1949).

A negative measure of association was also obtained by counting the number of times that an animal appeared alone in a grid.

### *Proximity by dyads*

In order to obtain a measure of social proximity independent of cage location, observations from the focal animal and ad libitum data were used. For each possible dyad, proximity was described on two measures: 'nearby' (the members were within 1–2 m of each other) and 'near' (the members were within 1 m of each other but not in contact). The total occurrences allowed for a description of members' proximity to each other by dyad.

*Social interaction and dominance*

Although aggression was rare among the bonobos, a number of acts did occur. Those were distinguished as contact aggression and aggressive displays. Contact aggression included 'grab' and 'slap', and aggressive displays included 'demonstration' (a leaping kick against the wall or wire mesh) and 'slide object' (running behind and sliding an object, usually an automobile tire, on the floor of the compound in the general direction of another animal). The aggressive acts are summed by dyad.

The aggressive acts give only a suggestion, at best, of dominance relations among the group. Because dominance is considered to provide an animal some modicum of control over a resource, behavior that affects control of resources can be taken as expressive of dominance, however subtly or indirectly. One type of control concerns whether or not social interaction occurs and, if it does, what form it takes. Social spacing of animals in the cage is one component of such control.

Another, more direct, means of control of social interaction is a social approach, which is defined as the approach of one animal to another within range of touching. A count was made of the number of social approaches that each animal made to each of the others, and the total number for each dyad was calculated.

The social approach data were further analyzed to determine if one member of each dyad tended to initiate the social approaches significantly more than the other. That measure allows identification of the dyadic members who exercised comparatively more control over whether or not a social interaction occurred.

Displacement of one animal by another is commonly taken as evidence for expression of dominance. In order to obtain a measure of that phenomenon, the result of each social approach was distinguished as avoidance or as affiliation. Avoidance was defined as moving more than 1 m away from an approaching animal within 5 sec of the approach. If the animal who was approached did not avoid the approacher or interacted actively with him or her, the behavior was scored as affiliative. Avoidance and affiliation were summed by dyad.

*Gestures*

The bonobos used several gestures in their social interactions. Two were made by moving the head, one by nodding it back and forth rhythmically, and one by shaking it from side to side. They were labeled, respectively, 'head nod' and 'head shake'. Two gestures were made by using one of the hands. One was a sudden relaxation of the wrist, allowing the hand to flap downward – a 'wrist flap' – and the other was a tap or poke with a finger against the body of another animal – a 'tap/poke'. The numbers of occurrences were tabulated by dyad.

Table 15.2. *Occurrences of location in cage by grid*

| Cage location | Bo | La | Ls | Ln | Lo |
|---|---|---|---|---|---|
| 01 | 44 | 9 | 7 | 7 | 10 |
| 02 | 29 | 107 | 74 | 81 | 64 |
| 03 | 24 | 89 | 83 | 107 | 102 |
| 04 | 52 | 70 | 100 | 89 | 94 |
| 05 | 44 | 28 | 39 | 27 | 20 |
| 06 | 122 | 12 | 12 | 4 | 25 |
| $\chi^2$ | 120.8 | 166.5 | 143.6 | 194.0 | 150.9 |
| $p$ | <.001 | <.001 | <.001 | <.001 | .001 |

## Results

### *Proximity by cage location*

The number of times that each animal was scored present in each of the six grids is reported in Table 15.2. Assuming that all animals had equal probability of being in any one of the grids on any given scan sample, nonrandomness of the observed distribution can be tested with $\chi^2$, and $z$ tests can be applied to determine if presence in a particular cage location contributed significantly to a nonrandom distribution. In these and all subsequent statistical tests, the probability level of significance was predetermined to be equal to or less than .05.

Each animal showed a significant nonrandom distribution among the six areas of the cage. The $z$ tests gave the following results: Bosondjo was in the cage locations 01, 05, and 06 more often than expected, in area 04 at an expected rate, and in locations 02 and 03 less than expected; Laura, Lisa, Linda, and Lorel were all in areas 02, 03, and 04 more often than expected and in areas 01, 05, and 06 less than expected. The spacing pattern is clear. Bosondjo tended to be in the upper half of the outdoor cage area, or at the far outdoor end of the lower half. The females tended to be at the end of the cage area opposite from Bosondjo, indoors or in the outdoor area directly adjoining the lower indoor area. The male and the females thus tended to form separate clusters. Only in grid 04 were the rates as expected or greater than expected for all five of the animals.

The coefficients of association are given in Table 15.3. They show that the highest rate of association was between Laura and her juvenile daughter Lisa. Intermediate rates obtained among the adult females, and the rates between Bosondjo and all of the females were considerably lower.

The instances during the scan samples when an animal was alone in a grid area can be taken as a negative measure of association. A combination of the scan, focal animal, and ad libitum data give the following results: Bosondjo

Table 15.3. *Coefficients of association by dyad*

| | Bo | La | Ls | Ln | Lo |
|---|---|---|---|---|---|
| Bo | — | | | | |
| La | .057 | — | | | |
| Ls | .079 | .232 | — | | |
| Ln | .048 | .178 | .159 | — | |
| Lo | .070 | .159 | .133 | .184 | — |

342, Lorel 160, Linda 138, Lisa 115, and Laura 93. Bosondjo was alone three to six times more often than any of the other animals. Laura and Lisa were least likely to be alone, and the occurrences for Linda and Lorel were intermediate between Bosondjo's totals and those of Laura and Linda.

*Proximity by dyads*

The numbers of occurrences of dyadic 'nearby' and 'near' are given in Table 15.4. Assuming equi-probability of each dyad for each category, $\chi^2$ can be used to test for differences between expected and observed rates. The distribution for dyadic rates of 'nearby' and 'near' are nonrandom ($\chi^2 = 162.06$ and 393.94, respectively, $df = 9$).

A $z$ test was applied to determine the contribution of each dyad to the nonrandom distribution of 'nearby' and 'near'. For the category 'nearby', the dyads of Bosondjo and each of the females occurred at rates less than expected, the dyads of Lorel and each of the other females occurred at expected rates, and the dyads involving Laura, Lisa, and Linda all occurred at rates greater than expected. The category 'near' revealed, again, that Bosondjo and each of the females had rates lower than expected, whereas all of the female dyads occurred at rates greater than expected, except for Lisa–Linda and Lisa–Lorel, which were at expected rates.

Collectively, the dyadic proximity measures indicate that Bosondjo, on the one hand, and the females, on the other hand, tended to have different and distinctive patterns. Among the females, all except Lorel tended to be within 1–2 m of each other at rates greater than expected. Within closer proximity, 1 m or less, all of the females showed rates greater than expected, except for Lisa and the two females other than her mother. Stated otherwise, the three adult females tended to be within 1 m of each other, in dyads, at rates greater than expected, and the same was the case for Lisa and her mother, Laura.

*Social interaction and dominance*

The numbers of acts of contact aggression and of aggressive displays are shown in Table 15.5. Only three acts of contact aggression occurred, and 27

Table 15.4. *Occurrences of 'nearby' and 'near' by dyad*

| | Nearby[a] | | | | | Near[b] | | | | |
|---|---|---|---|---|---|---|---|---|---|---|
| | Bo | La | Ls | Ln | Lo | Bo | La | Ls | Ln | Lo |
| Bo | — | | | | | — | | | | |
| La | 42 | — | | | | 33 | — | | | |
| Ls | 55 | 144 | — | | | 30 | 194 | — | | |
| Ln | 44 | 99 | 142 | — | | 11 | 127 | 96 | — | |
| Lo | 44 | 72 | 92 | 78 | — | 33 | 105 | 96 | 169 | — |

[a] $\chi^2 = 162.06; p < .001$.
[b] $\chi^2 = 393.94; p < .001$.

Table 15.5. *Occurrences of contact aggression and aggressive displays by dyad*

| Actor | Recipient | Act | Occurrences |
|---|---|---|---|
| *Contact aggression* | | | |
| Bo | Lo | Slap | 1 |
| La | Lo | Grab | 1 |
| La | Ln | Grab | 1 |
| *Aggressive displays* | | | |
| Bo | La | Demonstrate | 11 |
| Bo | Ln | Slide object | 5 |
| Bo | La | Demonstrate | 5 |
| Bo | Lo | Demonstrate | 1 |
| Ln | Lo | Slide object | 4 |
| La | Bo | Slide object | 1 |

instances of aggressive displays were recorded. Of those, 22 were made by Bosondjo, and 16 of those were directed to Laura. All 16 were in cage location 06, and all five of the aggressive displays against Linda were in the same location.

Social approaches occurred among all possible dyads, but they were not randomly distributed ($\chi^2$ = 840.08, $df$ = 9). The distribution is reported in Table 15.6.

Analysis of the data in Table 15.6 ($z$ tests) shows that the dyads of Lisa–Bosondjo and Lisa–Laura occurred at rates greater than expected, those between Linda–Laura and Linda–Lorel occurred at expected rates, and all other dyads occurred at rates less than expected. In other words, all dyads involving Lisa and those involving Bosondjo were at rates less than expected, except with each other, and the rates among the adult females were at expected rates, except for Laura–Lorel.

Table 15.6. *Occurrences of social approaches by dyad*

| | | Within dyad | | | | | |
|---|---|---|---|---|---|---|---|
| Dyad | No. | Direction | No. | Direction | No. | $\chi^2$ | $p$ |
| Bo + La | 27 | Bo → La | 19 | La → Bo | 8 | 4.48 | <.05* |
| Bo + Ls | 98 | Bo → Ls | 52 | Ls → Bo | 46 | 0.37 | >.50 |
| Bo + Ln | 5 | Bo → Ln | 2 | Ln → Bo | 3 | 0.02 | >.75 |
| Bo + Lo | 37 | Bo → Lo | 18 | Lo → Bo | 19 | 0.02 | >.75 |
| La + Ls | 314 | La → Ls | 37 | Ls → La | 277 | 183.44 | <.01* |
| La + Ln | 83 | La → Ln | 24 | Ln → La | 59 | 14.76 | >.01* |
| La + Lo | 54 | La → Lo | 28 | Lo → La | 26 | 0.07 | >.75 |
| Ls + Ln | 45 | Ls → Ln | 39 | Ln → Ls | 6 | 24.20 | <.01* |
| Ls + Lo | 39 | Ls → Lo | 26 | Lo → Ls | 13 | 4.33 | <.01* |
| Ln + Lo | 91 | Ln → Lo | 71 | Lo → Ln | 20 | 28.58 | <.01* |

Table 15.7. *Outcomes of social approaches: Occurrences of avoidance and affiliation by recipient*

| | | | Outcome | | |
|---|---|---|---|---|---|
| Initiator | Instances | Recipient | Avoids | Affiliates | $p$ |
| Bo | 19 | La | 7 | 12 | .125 |
| Bo | 52 | Ls | 26 | 26 | .500 |
| Bo | 2 | Ln | 0 | 2 | .079 |
| Bo | 18 | Lo | 9 | 9 | .500 |
| La | 8 | Bo | 1 | 7 | .017* |
| La | 37 | Ls | 24 | 13 | .035* |
| La | 24 | Ln | 7 | 17 | .021* |
| La | 28 | Lo | 4 | 24 | <.001* |
| Ls | 46 | Bo | 20 | 26 | .189 |
| Ls | 277 | La | 144 | 133 | .255 |
| Ls | 39 | Ln | 33 | 6 | <.001* |
| Ls | 26 | Lo | 23 | 3 | <.001* |
| Ln | 3 | Bo | 1 | 2 | .281 |
| Ln | 59 | La | 22 | 37 | .026* |
| Ln | 6 | Ls | 6 | 0 | <.001* |
| Ln | 71 | Lo | 33 | 38 | .278 |
| Lo | 19 | Bo | 7 | 12 | .125 |
| Lo | 26 | La | 8 | 18 | .025* |
| Lo | 14 | Ls | 14 | 0 | <.001* |
| Lo | 20 | Ln | 4 | 16 | .004* |

The distribution of initiations of social approaches within each dyad is also reported in Table 15.6. Analysis of the distributions by $\chi^2$ indicates that six of them were nonrandom: Lisa and each of the three adult females, Linda and the two other adult females, and Bosondjo–Laura. In each of her dyads, Lisa

Table 15.8. *Occurrences of gestures by dyad*

| Dyad | Gesture | | | |
|---|---|---|---|---|
| | Head nod | Head shake | Wrist flap | Tap/poke |
| La → Ls | 28 | 2 | 7 | 30 |
| Bo → Ls | 9 | 10 | 3 | 2 |
| Ls → La | 0 | 20 | 0 | 1 |
| All others | 3 | 7 | 0 | 0 |

initiated social approaches disproportionately, and the same was the case for Linda in each of her dyads with the adult females. In the Bosondjo–Laura dyad, Bosondjo initiated approaches more often than did Laura. To state the results differently, Lisa tended to initiate social approaches significantly, except when she was interacting with Bosondjo, and Bosondjo tended to initiate social approaches significantly only if he was interacting with Laura. Linda initiated them significantly unless she was interacting with Lisa or Bosondjo.

The numbers of social approaches that led significantly to avoidance or to affiliation for each dyad are given in Table 15.7. The $z$ test (approximation to the binomial) was used to test for significance. The dyads in which avoidance occurred at rates greater than expected all involved the juvenile Lisa. Linda and Lorel avoided her when she approached them, and she avoided all of the adult females, including her mother, when they approached her. No other dyads showed significant rates of avoidance.

Several dyads showed significant rates of affiliation following social approaches. Laura's approaches to all of the other animals (except Lisa), Lorel's approaches to the adult females, and Linda's approaches to Laura were in that category. Stated otherwise, all of the adult females' approaches to each other showed rates of affiliation greater than expected, except for the dyad Linda–Lorel, and among the adult females only Laura's approaches to Bosondjo showed rates of affiliation higher than expected.

## *Gestures*

The numbers of occurrences of head nod, head shake, wrist flap, and tap/poke are given in Table 15.8. It is clear that the gestures were virtually never used by the adults in interaction with each other. When they occurred, they were almost always in interactions between Lisa and her mother Laura and between Lisa and Bosondjo. Except for head shake, in which approximately one-half of all occurrences were made by Lisa to her mother Laura, Lisa was the recipient in almost all cases. The two dyads involving Lisa that exceeded expected rates of occurrence were the ones in which the gestures were used.

## Discussion

The measures of spatial proximity on the basis of cage location – grid, coefficient of association, alone – all give similar results. Bosondjo tended to stay spatially separated from the females, who in turn tended to be clustered at the opposite end of the caged area. The dyadic measures of proximity independent of cage location reflect a similar pattern. Bosondjo tended to be more spatially distant from the females than they were from each other. The females, moreover, tended to be within 1 m of each other at rates of occurrence greater than expected. Collectively, the proximity measures show that this social group of bonobos constituted two subgroups in terms of space utilization: Bosondjo constituted one subgroup, and the females, including the juvenile Lisa, constituted the other.

The social approach data reflect the spatial patterning of the two subgroups for the adults. Bosondjo and the adult females approached each other at rates less than expected, and the females approached each other at expected rates, except for one dyad. The juvenile, Lisa, however, showed a pattern anomalous in comparison with the adult females. Although she was engaged in social approaches with her mother at rates greater than expected, she showed rates less than expected with the other females. She also interacted with Bosondjo at a rate that was a polar opposite to those for the other females. Not only did Lisa show levels of social approach distinctive from those for the other females, she also initiated the approaches significantly more than did any other animal.

Lisa's anomalous patterns of social interaction become even clearer when the outcomes of social approaches are considered. Lisa avoided, at significant rates, the social approaches of all of the females, including her mother, and the two adult females other than her mother avoided her when she approached them. No other member of the social group showed a significant rate of avoidance of social approaches. The adult females, in contrast, tended to show rates of affiliation with each other higher than expected following social approaches.

The proximity measures and the social approach data suggest that the use of space by adult members is reflected in their social interactions. The organization of spacing expresses the stable social relationships among them. Following Lehner (1978), the relationships of animals commonly reflect priorities of access to fixed space, and the priorities can be viewed as expressions of dominance patterns. Space control features appear to be characteristic of dominance relationships, and the tendency for Bosondjo to occupy the outdoor upper half of the cage and for the females to occupy the other end is an expression of space control. The spatial separation seems to be a buffer to true dominance (i.e., patterned directionality of aggression).

The control of space seemed to lie in large part in Bosondjo's realm. He engaged in direct forms of aggressive display only when the females were in the area of the cage that he frequented most. The females, however, did not react similarly when he was in their areas of the cage. The females, however, appeared to exercise relatively greater control over whether or not interaction occurred. They made social approaches to each other significantly more often than they did to Bosondjo. Bosondjo, it should be noted, did not tend to avoid the social approaches when the females made them. The patterns of spatial use were not simple products of Bosondjo's self-isolation from the adult females, because he did not avoid them when he was approached socially.

Lisa appeared not yet to be socialized into the spatial expression of formal dominance. She showed the same spatial patterns of cage use as the adult females, but she did not reflect the patterns in her social behavior. She especially did not reflect the adult patterns of avoidance and affiliation to social approaches. The lack of predictability in her behavior, as compared with the adult animals, may have been responsible for the overt use of gestures with her. The gestures may have been needed because her behavior did not reflect and thus could not have been taken to represent the spatial buffering evident in the adult behavior. Notably, the gestures were used in the dyads that exceeded the expected rates: those with Bosondjo and Laura. Linda and Lorel, on the other hand, tended not to initiate social approaches to Lisa, and they did not use gestures with her.

To be clear, the argument here is not that Laura and Bosondjo were using gestures to socialize Lisa. It is more likely that they used them to mitigate the unpredictability of Lisa's behavior, as compared with the stable patterns of the adults. Learning, on Lisa's part, may have been a consequence of their actions, but there is no evidence of intentional socialization.

A feature of particular interest in the adult patterns of behavior and in Lisa's departures from them was that the direct expression of formal dominance seemed to be either context-specific or individual-specific. As evidence of context specificity, Bosondjo's aggressive acts occurred only in his preferred area of the cage, and the females' high rates of social approaches tended to occur at their preferred end of the cage and exclusive of Bosondjo. Evidence for individual specificity was the almost exclusive use of gestures with Lisa by her mother and by Bosondjo.

## Summary

The spatial and interactional behaviors of the Yerkes bonobos suggested that the adults employed selected strategies based on cognitive assessment. The type of behavior in which they engaged depended on what part of the caged area they were in. Social relationships were expressed through the use of space and in interactions. The behavior of the juvenile Lisa, however, was instructive in understanding the bonobo space–interaction pattern. Although her patterns of cage use were the same as those of the adult females, her interaction patterns were different. What that suggests is that her spatial behavior in terms of cage use had not yet been linked to other social behavior.

Such linkage eventually must be on the basis of cognitive development. Whether the linkage is through mere associative learning or through a reinterpretation of spatial proximity to integrate it with other aspects of social behavior is not known. The strategic use of gestures to minimize unpredictability of behavior, however, reflects the capacity of the species to engage in flexible behavior that is based on learning and cognition. The gestures do not constitute language in the full sense of the word, but they do reflect intelligence.

## Acknowledgments

This investigation was supported in part by NIH grant RR-00165 from the Division of Research Resources to the Yerkes Regional Primate Research Center. The Yerkes facility is fully accredited by the American Association for Accreditation of Laboratory Animal Care.

## References

Altmann, J. (1974). Observational study of behavior: Sampling methods. *Behavior*, *49*, 227–267.

Badrian, N., & Badrian, A. (1984). Social organization of *Pan paniscus* in the Lomako Forest, Zaire. In R. Susman (Ed.), *The pygmy chimpanzee: Evolutionary biology and behavior* (pp. 325–346). New York: Plenum.

Bernstein, I. (1981). Dominance: the baby and the bathwater. *Behavior and Brain Sciences*, *4*, 419–457.

Cole, L. (1949). The measurement of interspecific association. *Ecology*, *30*, 411–424.

de Waal, F. (1986). The integration of dominance and social bonding in primates. *Quarterly Review of Biology*, *61*, 459–479.

Jordan, C. (1977). *Das verhalten zoolebender Zwergschimpansen.* PhD dissertation, Goethe University, Frankfurt.

Kano, T., & Mulavwa, M. (1984). Feeding ecology of the pygmy chimpanzee *Pan paniscus* at Wamba. In R. Susman (Ed.), *The pygmy chimpanzee: Evolutionary biology and behavior* (pp. 233–274). New York: Plenum.

Lehner, P. (1978). *Handbook of ethological methods.* New York: Garland Press.

Mori, A. (1984). An ethological study of pygmy chimpanzees in Wamba, Zaire: A comparison with chimpanzees. *Primates*, *25*, 255–278.

Noe, R., de Waal, F., & van Hooff, J. (1980). Types of dominance in a chimpanzee colony. *Folia Primatologica*, *34*, 90–110.

Patterson, T. (1979). The behavior of a group of captive pygmy chimpanzees (*Pan paniscus*). *Primates*, *20*, 341–354.

Rowell, T., & Olson, D. (1983). Alternative mechanisms of social organization in monkeys. *Behaviour*, *86*, 31–54.

Savage-Rumbaugh, S., & Wilkerson, B. (1978). Socio-sexual behavior in *Pan paniscus* and *Pan troglodytes*: A comparative study. *Journal of Human Evolution*, *7*, 327–344.

*Part V*

# Development of numerical and classificatory abilities in chimpanzees and other vertebrates

# 16 The development of numerical skills in the chimpanzee (*Pan troglodytes*)

*Sarah T. Boysen and Gary G. Berntson*

Estimation of quantity and/or related phenomena (e.g., estimation of time) have been demonstrated in a variety of species. Numerous hypotheses have been proposed to account for a range of behaviors observed in rats, raccoons, various species of birds, rhesus monkeys, and chimpanzees relative to the emergence of counting skills in children. An understanding of the cognitive processes that underlie numerical skills in nonhuman animals could provide information critical to clarifying the evolution of cognitive capabilities.

Capaldi and Miller (1988) suggested that despite opportunities to employ other strategies for task solution, animals readily use counting cues to solve a variety of instrumental problems. Capaldi and his colleagues further proposed that counting mechanisms may explain a wide range of learning situations in rats and other mammals.[1] In contrast, Davis and Memmott (1982) proposed that animals engage in counting as a last-resort strategy and questioned the evolutionary significance of such capacities, given that these behaviors often require intensive training in a highly structured laboratory situation (e.g., Davis, 1984; Ferster, 1964).

The degree to which counting behaviors in nonhuman species fit the criteria for true counting, as defined for children (Gelman & Gallistel, 1978), is a point of controversy (Davis & Perusse, 1988). Interestingly, discussion of related issues pervades the human developmental literature (e.g., Gelman & Gallistel, 1978; Piaget, 1952), including the age at which children are capable of demonstrating counting, as opposed to subitizing (Beckmann, 1924; Mandler & Shebo, 1981), that is, the proposed ability to estimate small numbers of items in an array through a direct perceptual apprehension mechanism (von Glasersfeld, 1982). Such estimation of numerosity, typically applied to arrays composed of one to five items, appears in preschool children and often is observed in adult humans (Beckwith & Restle, 1966).

Beckmann's data (1924) suggest that subitizing develops in children after they have already learned to apply counting rules for the estimation of arrays (Gelman & Tucker, 1975). Gelman and Gallistel (1978) concur, noting that the essential ingredients for counting are present in most 2-year-old children, and thus subitizing could be viewed as a perceptual grouping strategy that aids in the process of abstracting a numerical representation. These authors propose that subitizing is, in fact, a higher-level ability than counting and that rapid numerical judgments based on such a process should not be considered primitive. Nonetheless, subitizing continues to be seen as a more primitive level of number estimation (Cole & Scribner, 1974).

Despite such arguments, the results of numerous animal "counting" experiments have been explained through some process related to, or analogous to, subitizing as a lower-level mechanism (Davis & Bradford, 1987; Davis & Memmott, 1982). Several experiments that purported to address "counting" employed simultaneous presentation of relatively small numbers of items (Hicks, 1965; Pepperberg, 1987). As noted by Davis and Bradford (1987), such experiments do not preclude a simpler explanation than counting. In light of these considerations, Davis and his colleagues incorporated sequential events in a more naturalistic experiment that would encourage *protocounting*, a term used to describe data highly suggestive of a counting process that excludes subitizing as an alternative explanation (Davis & Perusse, 1988).

Because chimpanzees have demonstrated a remarkable facility for symbol manipulation and complex information processing (Gardner & Gardner, 1984; Premack, 1986; Savage-Rumbaugh, 1986), and because they share a considerable portion of their evolutionary history with humans, we are interested in the degree to which the acquisition of number concepts might parallel the developmental sequence observed in children.

**Preliminary training of numerical skills**

The Primate Cognition Project was established in 1983 with the arrival of two young male chimpanzees, on permanent loan from the collection of the Yerkes Regional Primate Research Center, Emory University, Atlanta, Georgia. A young female joined the project in February 1984, from the Columbus Zoological Gardens, Powell, Ohio. Prior to their work with numbers, these animals engaged in a variety of conceptual and information-processing tasks, including one-to-one correspondence, nonrepresentational imitation of body parts, cross-modal discriminations, drawing (Boysen, Berntson, & Prentice, 1987), the use of colors as attributes, and a vigilance task of sustained attention. They had been trained earlier on color and shape discriminations and slide recognition. Thus, all three animals had considerable experience in teaching and testing situations in the laboratory, with matching-to-sample the predominant teaching mode.[2] Although none of those studies were directly related to the work with numbers reported here (with the exception of the one-to-one correspondence task), those tasks provided the animals with a variety of conceptual experiences that contributed, in part, to their subsequent acquisition of number concepts.

Formal training for number concepts began with a new variation of the earlier one-to-one correspondence task. This involved the presentation of three round placards, with a black marker affixed to one, and the others left blank (Figure 16.1A). A single food item was presented, and the chimps were

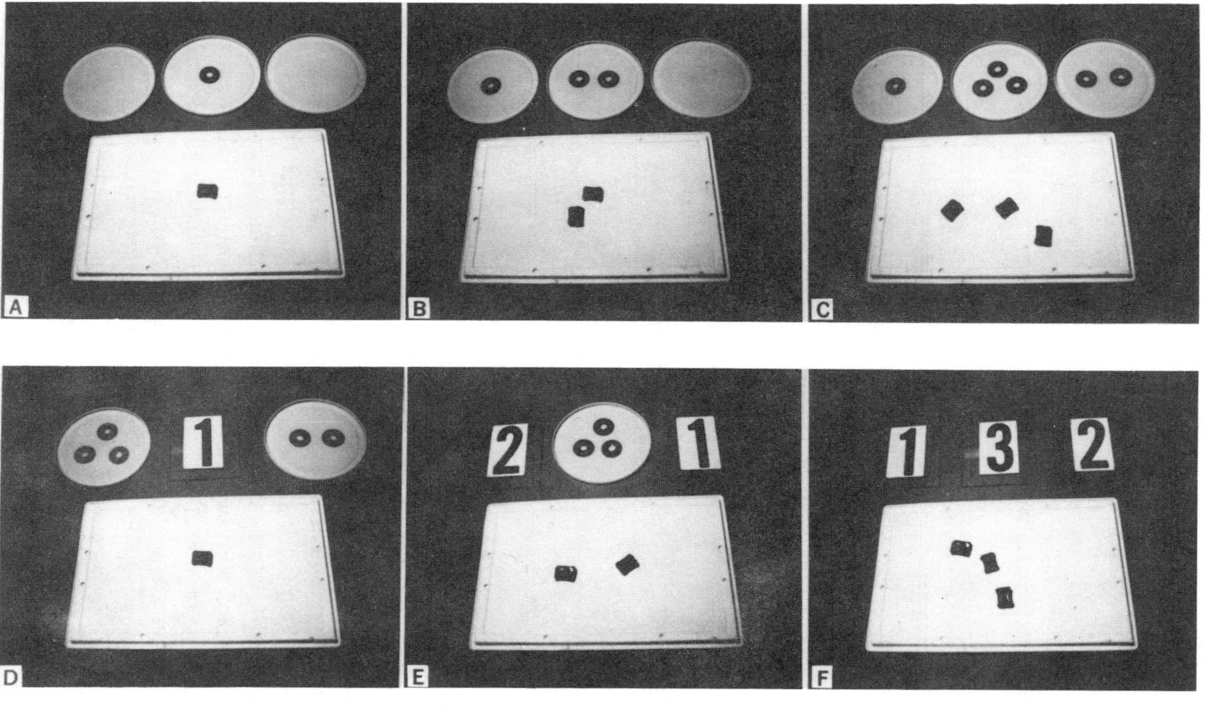


Figure 16.1. *One-to-one correspondence stimuli, in order of introduction (A–C); stimuli for introduction of arabic numbers (D–F).*

Table 16.1. *Introduction of Arabic numbers: Trials to criterion*

| Subject | Sessions | Trials to criterion | | | Total |
|---|---|---|---|---|---|
| | | No. 1 | Nos. 1, 2 | Nos. 1–3 | |
| Darrell | 24 | 25 | 150 | 50 | 325 |
| Kermit[a] | 18 | — | — | 375 | 375 |
| Sheba[a] | 14 | — | — | 325 | 325 |

[a] These two animals were introduced to all three Arabic numbers at once following one-to-one correspondence training. Note that total trials to criterion did not differ for the two procedures.

Table 16.2. *Introduction of receptive training with Arabic numbers 1–3*

| Subject | Trials to criterion | Sessions | Overall correct (%) |
|---|---|---|---|
| Darrell | 201 | 9 | 70 |
| Kermit | 315 | 16 | 72 |
| Sheba | 282 | 13 | 69 |

encouraged verbally to select the placard with one marker. If correct, the teacher responded verbally with "One," placed the food item directly on top of the marker, and permitted the chimp to eat the food. The position of the marked placard was varied randomly every few trials to control for position bias. Once the animals were selecting the correct placard reliably, a second placard with two markers was introduced (Figure 16.1B). At that point, two food items were always presented, and the chimps quickly learned to select only the placard with two markers. On subsequent trials, either one *or* two food items were presented, and the animals then had to make a specific discrimination, depending on the number of food items available on a given trial. Following reliable performances with such mixed trials of one or two foods, three markers were affixed to the final placard, and mixed trials of one, two, or three food items were introduced (Figure 16.1C). Reliable performance was achieved by all animals, with no significant decrement in performance during blind testing (Boysen & Berntson, 1986b).

*Introduction of Arabic numbers*

Arabic numbers were introduced by replacing the marked placards with placards bearing the corresponding Arabic numeral (Figure 16.1D–F). These stimuli consisted of 5- × 7.5-cm black numbers on a silver background affixed to clear Plexiglas placards. Number stimuli were introduced one at a time, as depicted in Figure 16.1D–F. Thus, the placard bearing one marker was the

Figure 16.2. Testing context and stimuli for receptive use of Arabic numbers (number comprehension).

first to be replaced with the Arabic number 1 (Figure 16.1D), and mixed trials of one, two, or three foods were completed until reliable performance was achieved. The Arabic number 2 then replaced the placard with two markers (Figure 16.1E), and performance was permitted to stabilize. The Arabic number 3 was then introduced (Figure 16.1F). For Kermit and Sheba, all three Arabic numbers were introduced at once, as depicted in Figure 16.1F, and mixed trials were begun immediately. As seen in Table 16.1, overall performances for the introduction of Arabic numbers did not differ significantly in trials to criterion for the three animals (Boysen & Berntson, 1986b).

### *Receptive Arabic number training*

At this point in their training, the animals were reliably associating the correct Arabic number symbol with arrays of discrete edibles, thus "producing" a label for collections of one to three food items. The productive use of symbol labels, however, did not necessarily ensure that the animals were capable of number comprehension. Consequently, specific training for receptive number skills was undertaken. In this phase, Arabic symbols were individually presented on a video monitor (Figure 16.2). The numbers to be decoded appeared individually and remained on the screen throughout a trial. The placards used in the one-to-one correspondence task served as choice stimuli (Figure 16.1). When a given Arabic number appeared on the screen, the chimpanzee was to select the placard bearing the corresponding number of markers. Mixed trials were again employed, with the number 1, 2, or 3 presented on the monitor, and the positions of the choice stimuli were varied randomly. Overall performance on the receptive (comprehension) task is depicted in Table 16.2 (Boysen & Berntson, 1989a).

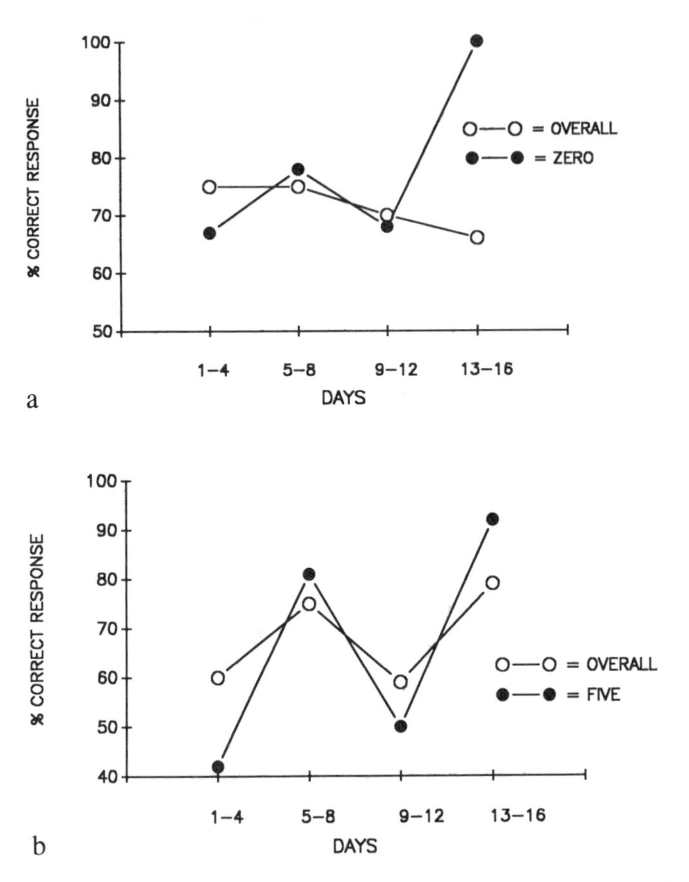


Figure 16.3. (a) Acquisition of number 0 and concurrent performance with previously acquired numbers 1–4. (b) Acquisition of number 5, and concurrent performance with numbers 0–4.

*Number labeling of object arrays*

To evaluate whether or not the animals could apply number labels to arrays of novel objects, common household objects (e.g., flashlight battery, spoon) were presented for counting in a double-blind, novel test. The chimps readily applied Arabic numbers to collections of one to three objects. Following each trial, the animals were given a matching number of food items as a reward for correct choices (Boysen & Berntson, 1986c).

*Introduction of numbers 4, 0, and 5*

The Arabic number 4 was introduced next into the productive numbers task described previously. Although the animals had no previous experience with the number 4, they were readily able to incorporate the use of this new number into their existing repertoire. Unlike numbers 1–3, the number 4 was

Figure 16.4. Testing context for functional counting task.

introduced directly as an Arabic number, rather than through the one-to-one correspondence task. The number 0 was introduced next by adding a number placard bearing the 0 symbol to the available alternatives, and presenting an empty food tray to the animals. The experimenter verbally and gesturally encouraged the chimps to attend to the empty tray. Performance on 0 trials is depicted in Figure 16.3A, as well as ongoing performance on all other numbers (1–4). Subsequently, the number 5 was introduced among mixed trials of 0–5. Acquisition of the number 5 is depicted in Figure 16.3B, with concurrent performances on all other numbers also shown (Boysen, 1987).

*Functional counting task*

Davis and Bradford (1987) suggested that most studies of countinglike behavior in animals had been derived from very structured, artificial laboratory settings. In an effort to move beyond the constrained setting of earlier numbers tasks and more directly address the issue of subitizing, a functional counting task was devised that incorporated a "foraging" feature (Boysen & Berntson, 1989a). Three food sites (approximately 6 ft apart) were chosen that formed a rectangular pathway in the laboratory (Figure 16.4). A work station was situated at Point A, which consisted of a wooden platform where tasks and testing were typically completed (Figure 16.4). The experimenter sat on the platform, with the number placards in ordinal sequence, and the chimp was free to move around the testing area. On most trials, the chimp

Table 16.3. *Functional counting task*

| | No. of sessions | No. of trials (correct/total) | Probability Correct | Probability Chance | $\chi^{2a}$ |
|---|---|---|---|---|---|
| Nos. 1–3 | 7 | 60/82 | .73 | .33 | 58.32 |
| Nos. 1–4 | 10 | 75/89 | .84 | .25 | 166.42 |
| Blind sessions | 4 | 28/38 | .74 | .25 | 49.83 |

[a] $p < .001$; $df = 1$.

moved directly from Point A to Site 1 (a tree stump), on to Site 2 (a food bin attached to an empty cage), next to Site 3 (a plastic dishpan), and then returned to the work station (Point A).

Sheba, who was then 5.8 years old, participated in the functional counting study (Boysen & Berntson, 1989a). Oranges were chosen to bait the food sites because they were not a preferred food yet were visually salient. One to three oranges were placed in any two of the three sites on each trial. For example, two oranges might be hidden at Site 1, and one orange at Site 3. Sheba was required to move from place to place, examine the arrays present, and return to the work station, where the number alternatives were available. Her task was to select the number that correctly represented the total number of oranges hidden among the sites. During the first three trials of the task, the experimenter walked with Sheba, to draw her attention to the hidden fruit. Sheba's movement was also monitored visually over several trials by the experimenter, who offered verbal encouragement as Sheba gained facility at moving independently from one site to another. It had been hypothesized that Sheba could be taught to (1) move from location to location, (2) attend to the separate arrays, and (3) learn to "sum" the arrays, in order to report the correct total number of food items observed. The initial results of the task, from Session 1, revealed that Sheba was able to perform this new task at a level significantly above chance (Table 16.3), and thus no explicit additional training procedures were necessary. Sheba readily adapted to the numerous novel demands of the functional counting task and continued to demonstrate statistically significant performance throughout subsequent testing. By Session 8, the number 4 was introduced as a response option, and combinations of oranges totaling from one to four (i.e., 2 & 2, 1 & 3, 2 & 0, 1 & 1, etc.), were employed. Blind tests, with the experimenter seated behind Sheba, revealed no appreciable decrement in performance (Figure 16.5, Table 16.3) (Boysen & Berntson, 1989a).

### *Symbolic counting task*

Sheba's performance with the functional counting task suggested that she had greater flexibility with number concepts than might have been predicted from

Figure 16.5. Blind testing context for functional and symbolic counting tasks.

individual skills acquired through prior training. One test of her capabilities would be to replace the arrays of oranges used in the functional task with their Arabic number representations and observe her performance. Again, we had hypothesized that Sheba might come to use number symbols in such a manner following extensive training. Because she had required little additional training for the functional counting task and had performed at significant levels from its inception, we elected to explore the possibility that representational addition with numbers might also be possible.

The symbolic counting task followed the same procedures described previously for the functional task and employed the same three food sites (Figure 16.4). Instead of encountering oranges at the various locations, Sheba found Arabic number placards. She was then required to return to the work station at Point A and select the number that represented the total number depicted by the numerals hidden at the two sites. As with the functional task, Sheba was unable to see the numbers at the sites from her position at the work station (Point A).

Sheba's initial performance on the symbolic counting task, conducted during a morning session, was significantly above chance. She was 80% correct on these exploratory trials, and therefore double-blind trials were completed during an afternoon session the same day. Sheba's overall performance, including blind sessions, is depicted in Table 16.4. Although it must be emphasized that these results represent the use of a limited repertoire

Table 16.4. *Symbolic counting task*

| | No. of sessions | No. of trials (correct/total) | Probability | | $\chi^{2a}$ |
|---|---|---|---|---|---|
| | | | Correct | Chance | |
| Nos. 1–4 | 3 | 16/21 | .76 | .25 | 28.94 |
| Blind sessions | 3 | 17/20 | .85 | .25 | 35.27 |

[a] $p < .001; df = 1$.

of numbers and their arithmetic combinations, they nonetheless suggest that such numerical skills may be available to a nonhuman primate who has had intensive training on the productive and receptive use of numbers.

Among the criteria specified for demonstrating the existence of an enumeration or counting process in preschool children, Gelman and Gallistel (1978) proposed three "how-to-count" principles: cardinality, ordinality, and the stable-order principle. Cardinality involves the application of the final number tag as a representation of the entire array. Sheba was able to apply this principle reliably when labeling groups of items and when specifying the sum of two arrays separated in time and space. Her appropriate application of Arabic numbers in all number related tasks also implies an understanding of the stable-order principle. Although ordinality has been evaluated for Sheba in a separate study (Boysen & Berntson, 1988), an appreciation of ordinality (that numbers represent an ordered series) is also implicit in this study. An understanding of ordinality is a necessary logical precursor of her demonstrated abilities with rudimentary addition. Sheba's ability to employ numbers to count novel objects demonstrates her understanding of the Gelman and Gallistel (1978) abstraction principle, which specifies that numbers can be used to count virtually anything. Sheba has demonstrated competence in counting foods or objects, starting with any item in the array. According to their final principle, the order-irrelevance principle, true counting entails an understanding that items in an array may be counted in any order (Gelman & Gallistel, 1978).

*Partitioning and tagging behavior*

In addition to her capabilities with a range of number concepts and demonstrated number reasoning abilities, Sheba also exhibited a range of "motor tagging" behaviors during counting (Boysen, 1988). These behaviors emerged spontaneously after approximately 1.5 years of training with numbers. They included touching food items or objects prior to selecting the number that represented the array, or moving each item apart from others in the array and then touching it before selecting the correct number. These behaviors are similar to those Gelman and Gallistel (1978) referred to as partitioning and tagging. Those investigators observed that young children pointed to or touched items to be counted prior to stating the number that represented the array. They proposed that such motor behaviors allowed the child to attend to items that had already been included in their count and those that remained to be counted. Typically, children stated each number aloud as they touched or pointed to each item. Later, as the children gained experience and additional practice with counting, these behaviors dropped out.

A study is currently under way to evaluate the functional significance of tagging behaviors in Sheba (i.e., to determine if they serve as an organizing strategy to facilitate counting). Anecdotal accounts for other species (Davis & Perusse, 1988) have described behaviors that may have served as motor markers, that is, may have functioned in a manner similar to those described by Gelman and Gallistel (1978) for children. In Sheba's case, tagging was particularly interesting because it was not specifically taught, encouraged, or reinforced in any way. Once the behaviors emerged, however, they were remarkably persistent and were exhibited during most number related tasks. Sheba was also observed to count, as evidenced by partitioning and tagging, during free play with objects, or when provided with snacks that were collections of edibles (i.e., grapes or peanuts).

In their discussion of numerical abilities in a 3.5-year-old home-reared chimpanzee, Viki, Hayes and Nissen (1971) described several number related tasks. These included number matching problems using cards with dots that varied in their spatial configuration, presented in a matching-to-sample format. Viki's performances on this and other number tasks were compared with those of children, who performed significantly better. Periodic testing on selected number tasks over the next several years did not reveal notable improvements in Viki's skills. In conclusion, they reported that Viki's numerical abilities resembled those of precounting children, though she did not perform as well as some bird species (Koehler, 1950). Hayes and Nissen suggested that chimpanzees and humans may differ phylogenetically in numerical abilities and that number conceptualization may be a capacity that is "peculiarly limited" in apes (Hayes & Nissen, 1971). It is interesting that they mention, relative to the success of Koehler's birds, that perhaps the techniques used for testing them were more meaningful than counting spots on a card was to a chimpanzee (Pepperberg, P&G18).

The breadth of numerical abilities demonstrated by the champanzees, particularly Sheba, indicates that number conceptualization is not as limited in this species as previously thought. A pragmatic approach to teaching numerical skills (i.e., counting gumdrops is likely more meaningful to a chimpanzee than counting spots on a card) may have contributed to the conceptual breakthrough. Clearly, the capacity for high-level manipulation of number symbols, including the generative application of number reasoning principles such as summing arrays, is possible for the chimpanzee when the necessary prerequisite concepts are acquired in an appropriate teaching/training context.

**Discussion**

In the series of studies reported here, numerical competence was established in a chimpanzee (*Pan troglodytes*). Sheba was able

1. to demonstrate an understanding of one-to-one correspondence,
2. to demonstrate the ability to apply the counting principles (cardinality, ordinality, and stable-order principle),
3. to demonstrate an appreciation for the abstraction and order-irrelevance principles,
4. to show transfer of training by counting novel objects using Arabic numbers, and
5. to show evidence of number reasoning principles not specifically taught, permitting the correct addition of physical objects or Arabic numbers presented as separate arrays for summing.

All these accomplishments support the proposal that Sheba has a concept of number (Davis & Perusse, 1988) and can flexibly apply her understanding of numbers in novel situations.

From a human developmental perspective, counting is an emergent skill of great interest and the focus of much debate among developmental psychologists, educational psychologists, and information-processing psychologists (Fusion, 1982; Gelman & Gallistel, 1978; Klahr & Wallace, 1976). Counting is one of the skills neo-Piagetians have used in order to demonstrate that children acquire a variety of logical capacities long before the age of concrete operations (Case, 1985). The development of counting has also been studied in relation to developments in other cognitive domains. Case (1985), for example, specifies four substages of cognitive development in counting and other domains during the so-called relational stage of development from 1 to 5 years: operational consolidation (Stage 0), operational unifocal coordination (Stage 1), bifocal coordination (Stage 2), and elaborated coordination (Stage 3).

During the initial substage of operational consolidation, which emerges between the ages of 1 and 1.5, children in Case's study solved a simple relational problem that required that they imitate the experimenter's action of placing a finger on each of a row of dolls, saying "Nice dolly" as each was touched. One of the control structures required for the task, one-to-one correspondence between each doll and the verbal enumeration, is a relational component from which counting emerges. Competence at one-to-one correspondence was first established in the chimps shortly after their introduction to the project, at age 3–3.5 for Kermit and Darrell, and age 2.5 for Sheba, and was readily acquired (Boysen & Berntson, 1986b).

During the next developmental substage, operational coordination, which emerges in children between 1.5 and 2 years, children in Case's study were able to touch two dolls in sequence. This task required that the child coordinate the motor action of designating one doll, then focus attention on the next doll and repeat the motor act of tagging. Though not formally tested initially, such sequential tagging emerged spontaneously (i.e., without specific shaping or reinforcement) in Sheba at age 4.5 and was likely acquired by imitation of the teacher's actions (Boysen & Berntson, 1989a). Additional studies are planned to clarify the role these motor actions play during counting in a chimpanzee. The conditions under which tagging emerged, however, are sufficiently interesting to explore as additional young animals are introduced to counting skills. That such tagging emerged concurrently with the introduction of larger numbers (4, 5) to the animals' repertoire suggests that tagging during the early phases of learning served as an organizing strategy to count arrays that were greater than three.

During the next substage, bifocal coordination, between 2 and 3.5 years of age, children were able to coordinate the act of tagging items and uttering number words simultaneously. At the beginning of the stage, children undergo a transition from counting two objects to coordinating tagging and repeating the correct number words for arrays of four or five (Case, 1985). Although Sheba could not verbalize, she was nevertheless, at age 5, able to coordinate the tagging of individual items in arrays of zero to five foods or objects and then designate the correct Arabic label for the array. Preliminary evaluation of her tagging behaviors at age 7, using numbers 0–5, revealed a correlation of .75 between the number of items in the array and the number of motor tags, a correlation of .85 between the number of items and her response, and a correlation of .67 between the number of tags made and the number Sheba selected as the cardinal label for the array (Boysen, 1988). This suggests that the component processes involved in simultaneously touching and stating the number by children in Substage 2, bifocal coordination, were comparable to those involved in Sheba's skills at tagging and labeling of arrays.

During the final substage, elaborated coordination, between the ages of 3.5 and 5, children were able to count items that were spread randomly and were able to count some subset of the array, such as only girl dolls, from among an array containing dolls of both genders (Case, 1985). The control sequence for the latter operation requires children to elaborate on prior structural components of the counting sequence through the addition of a decision loop specifying whether or not an individual item in the array (e.g., doll) belongs to the to-be-counted set (e.g., only girl dolls).

Such component skills were also observed in our chimpanzees. They were able to specify the number of apples in an array composed of apples and bananas (S. T. Boysen & G. G. Berntson, unpublished data). As each array of mixed fruit was presented, the experimenter held up an exemplar fruit that specified which subset of the two types of fruit was to be counted. Evidence for this skill in the chimpanzees was demonstrated from the first session that testing was initiated. At that time, Kermit and Darrell were approximately 8 years old, and Sheba was 6.5 years, although it is possible that the capacity might have been observed earlier had testing been attempted. Sheba and Darrell have also demonstrated the ability to label subsets of colored wooden shapes, presented in arrays of different colors and quantities. As in the procedure described for counting subsets of fruits, the experimenter presented arrays of one to five shapes (circles, triangles, squares, and pentagons) of two different colors and then presented a red, yellow, green, or blue placard that specified the color of the subset to be counted.

In children, it appears that the control structures for dealing with quantitative relations are similar to the structures underlying spatial, social, and linguistic relations (Case, 1985). In addition, both the substage and rate of passage through stages are similar across domains. Case's work provides documentation for early childhood developmental changes that had not previously been clearly specified by Piaget. One trend that is apparent is that as they approach the end of the stage, children utilize the component control structures as elements that can be coordinated to generate qualitative shifts in thinking (Case, 1985). Case cited one such shift observed in the area of logical or scientific reasoning (including such competencies as an appreciation of causality, or an understanding of class inclusion) that required the coordination of classification with counting. Case (1985) proposed that such coordinated, qualitatively enhanced abilities have a significant sociocultural consequence, in that children (and perhaps some chimpanzees) with such logical insights become ready for formal schooling.

Such qualitative shifts have also been observed in the chimpanzees, as described earlier for demonstrated competence in simple addition of physical items and number symbols (Boysen & Berntson, 1989a). Although prior training on labeling arrays did not entail the coordination of novel features that were introduced with the functional and symbolic counting tasks, Sheba was nevertheless able to demonstrate immediate competence with these tasks, representing a qualitative shift in her number reasoning abilities. This suggests that Sheba, like children, was able to integrate the component control structures gained through acquired skills at counting foods and objects and integrate them in two novel situations that were considerably more complex than any prior numerical task.

Although the numerical competencies demonstrated by Sheba are striking, many aspects of the number skills of the chimpanzee remain to be clarified, and comparative relationships need to be more fully elaborated and empirically tested. Continued exploration of the capacity for numerical competence in the chimpanzee may provide new insights into the cognitive structure of the apes, the phylogenetic origins and evolution of human cognition, and the ontogeny of number concepts in human children.

**Acknowledgments**

This research was supported in part through the Office of Research and Graduate Studies, The Ohio State University, and by grants from the National Institute of Mental Health (RO3 MH44022-01) and the National Science Foundation (BNS-8820027) to Sarah T. Boysen. Two of the chimpanzees were provided by Yerkes regional Primate Research Center, which is supported by NIH grant RR-00165, Division of Research Resources. The Yerkes Center and the Laboratory Animal

Center, The Ohio State University, are fully accredited by the American Association for Laboratory Animal Care. We would like to thank the Columbus Zoo for their cooperation and the contributions of Linda Brent, Gina Long, Carolyn Lake, Beth Radisek, Victoria Yang, and Mike Moore to the project. Correspondence should be addressed to Dr. Sarah T. Boysen, Primate Cognition Project, Room 48, Townshend Hall, Department of Psychology, 1885 Neil Avenue, The Ohio State University, Columbus, OH 43210.

## Notes

1. Church and his colleagues have contributed a wealth of empirical data supporting a relationship between the discrimination of number and time estimation in rats (Church & Meck, 1984). They have proposed that the internal mechanism used for counting is also used for timing in the rat. The proposed mechanism may be used in the "event" mode for counting sequential events, or in a "run" and "stop" mode for timing duration of an event. In an elegant series of experiments, the psychophysical functions for timing and number were shown to fit a scalar expectancy model with the same parameter values for each mode (Meck & Church, 1983). That suggested that rats used the same mechanism for both counting and timing, or had access to two separate mechanisms with the same sensitivity. In either case, they concluded that rats are able to differentially invoke the use of either mode, depending on the specific task demands. The ready access of such a sophisticated mechanism in rodents suggests that counting mechanisms may functionally subserve other cognitive processes in a wide range of mammalian species.
2. In addition, Sheba had participated in studies of the heart rate indices of recognition of humans and chimpanzees (Boysen & Berntson, 1986a; Boysen & Berntson, 1989b) and cardiac correlates of vigilance (Berntson & Boysen, 1987).

## References


Beckmann, H. (1924). Die Entwicklung der Zahlleistung bei 2–6 jahrigen Kindern. *Zeitschrift für Angewandte Psychologie*, *22*, 1–72.

Beckwith, M., & Restle, F. (1966). Process of enumeration. *Psychological Review*, *73*, 437–444.

Berntson, G. G., & Boysen, S. T. (1987). Cardiac reflections of attention and preparatory set in a chimpanzee (*Pan troglodytes*). *Psychobiology*, *15*, 87–92.

Boysen, S. T. (1987, May). *Scrutinizing subitizing: Further studies of numerical skills in the chimpanzee* (*Pan troglodytes*). Paper presented at the 59th meeting of the Midwestern Psychological Association, Chicago.

Boysen, S. T. (1988, August). *Numerical competence in animals: Functional or frivolous*? Paper presented at the annual meeting of the American Psychological Association, Atlanta.

Boysen, S. T., & Berntson, G. G. (1986a). Cardiac correlates of recognition in the chimpanzee (*Pan troglodytes*). *Journal of Comparative Psychology*, *100*, 321–324.

Boysen, S. T., & Berntson, G. G. (1986b, May). *Clever Kermit: Counting capabilities in a chimpanzee*. Paper presented at the 58th meeting of the Midwestern Psychological Association, Chicago.

Boysen, S. T., & Berntson, G. G. (1986c, July). *Pongid pedagogy*. Paper presented at the International Primatological Society Congress, Göttingen.

Boysen, S. T., & Berntson, G. G. (1988, April). *Four is more: Evidence for an understanding of ordinality in a chimpanzee* (*Pan troglodytes*). Paper presented at the 60th meeting of the Midwestern Psychological Association, Chicago.

Boysen, S. T., & Berntson, G. G. (1989a). Numerical competence in a chimpanzee (*Pan troglodytes*). *Journal of Comparative Psychology*, *103*, 23–31.

Boysen, S. T., & Berntson, G. G. (1989b). Conspecific recognition in the chimpanzee (*Pan troglodytes*): Cardiac responses to significant others. *Journal of Comparative Psychology*, *103*, 215–220.

Boysen, S. T., Berntson, G. G., & Prentice, J. (1987). Simian scribbles: A re-appraisal of drawing in the chimpanzee (*Pan troglodytes*). *Journal of Comparative Psychology, 101*, 82–89.

Capaldi, E. J., & Miller, D. J. (1988). Counting in rats: Its functional significance and the independent cognitive processes which constitute it. *Journal of Experimental Psychology: Animal Behavior Processes, 14*, 3–17.

Case, R. (1985). *Intellectual development: Birth to adulthood.* New York: Academic Press.

Church, R. H., & Meck, W. H. (1984). The numerical attributes of stimuli. In H. L. Roitblat, T. G. Bever, & H. S. Terrace (Eds.), *Animal cognition* (pp. 445–464). Hillsdale, NJ: Erlbaum.

Cole, M., & Scribner, S. (1974). *Culture and thought: A psychological introduction.* New York: Wiley.

Davis, H. (1984). Discrimination of the number "three" by a raccoon (*Procyon lotor*). *Animal Learning and Behavior, 12*, 409–413.

Davis, H., & Bradford, J. (1987). Simultaneous numerical discrimination by rats. *Bulletin of the Psychonomic Society, 25*, 113–116.

Davis, H., & Memmott, J. (1982). Counting behavior in animals: A critical review. *Psychological Bulletin, 92*, 547–571.

Davis, H., & Perusse, R. (1988). Numerical competence in animals: Definitional issues, current evidence and a new research agenda. *Behavioral and Brain Sciences, 11*, 561–579.

Ferster, C. B. (1964). Arithmetic behavior in chimpanzees. *Scientific American, 210*, 98–106.

Fuson, K. C. (1982). An analysis of the counting-on solution procedure in addition. In T. C. Carpenter, J. M. Moser, & T. A. Romberg (Eds.), *Addition and subtraction: A cognitive perspective* (pp. 67–81). Hillsdale, NJ: Erlbaum.

Gardner, R. A., & Gardner, B. T. (1984). A vocabulary test for chimpanzees (*Pan troglodytes*). *Journal of Comparative Psychology, 98*, 381–404.

Gelman, R., & Gallistel, R. (1978). *The child's understanding of number.* Cambridge, MA: Harvard University Press.

Gelman, R., & Tucker, M. F. (1975). Further investigations of the young child's conception of number. *Child Development, 46*, 167–175.

Hayes, K. J., & Nissen, C. H. (1971). Higher mental functions of a home-raised chimpanzee. In A. M. Schrier & F. Stollnitz (Eds.), *Behavior of non-human primates* (Vol. 4, pp. 59–115). New York: Academic Press.

Hicks, L. H. (1965). An analysis of number-concept formation in the rhesus monkey. *Journal of Comparative and Physiological Psychology, 49*, 212–218.

Klahr, D., & Wallace, J. (1976). *Cognitive development: An information-processing view.* Hillsdale, NJ: Erlbaum.

Koehler, O. (1950). The ability of birds to "count." *Bulletin of Animal Behavior, 9*, 41–45.

Mandler, G., & Shebo, B. J. (1981). Subitizing: An analysis of its component processes. *Journal of Experimental Psychology: General, 111*, 1–22.

Meck, W. H., & Church, R. M. (1983). A mode control model of counting and timing processes. *Journal of Experimental Psychology: Animal Behavior Processes, 9*, 320–334.

Pepperberg, I. M. (1987). Evidence for conceptual quantitative abilities in the African grey parrot: Labelling of cardinal sets. *Ethology, 75*, 37–61.

Piaget, J. (1952). *The child's conception of number.* New York: Norton.

Premack, D. (1986). *Gavagai.* Cambridge University Press.

Savage-Rumbaugh, E. S. (1986). *Ape language: From conditioned response to symbol.* New York: Columbia University Press.

von Glasersfeld, E. (1982). Subitizing: The role of figural patterns in the development of numerical concepts. *Archives de Psychologie, 50*, 191–218.

# 17 Spontaneous sorting in human and chimpanzee

*Tetsuro Matsuzawa*

## Introduction

Spontaneous sorting of objects was investigated in human and chimpanzee subjects. Recent advances in the study of animal cognition have revealed that some components of human cognitive abilities are shared by nonhuman primates, especially the great apes. Many studies have shown that the great apes (chimpanzees, pygmy chimpanzees, gorillas, and orangutans) are capable, to some extent, of both producing and comprehending "words" in natural and artificial language systems (Asano, Kojima, Matsuzawa, Kubota, & Murofushi, 1982; Gardner & Gardner, 1969; Patterson, 1978; Premack, 1971; Rumbaugh, 1977; Terrace, 1979). Apes can place one thing in symbolic correspondence to another. They can correlate one thing not only with another "thing" but also in terms of "color" (Matsuzawa, 1985a), "number" (Matsuzawa, Asano, Kubota, & Murofushi, 1986), and other attributes. However, the limits of apes' ability to acquire such a system is controversial. Can they (Terrace, 1979) organize the "words" into a "sentence"? The question seems to be too vague to be answered, because the terms themselves cannot be easily defined. The fundamental question is how the apes classify and organize the world. The application of cognitive skills to hierarchical organization is a main concern of this study.

In a recent study of languagelike skills, a chimpanzee named Ai showed spontaneous organization of her acquired "words" (Matsuzawa, 1985b). The chimpanzee was required to name the number, color, and object for 300 types of samples by pressing the corresponding "word" keys (Figure 17.1). Each key displayed a complex geometric figure that represented a "word." Although no particular sequence of describing samples was required, the chimpanzee favored two sequences (color/object/number, such as "red/pencil/five," and object/color/number, such as "pencil/red/five") among the six possible alternatives. In both sequences, numerical naming was always last. The chimpanzee spontaneously organized the "word order." The word order, once established, was used to name a new set of items and persisted for a long time (Matsuzawa, 1987b). The chimpanzee also learned to use the letters of the alphabet as names of individual humans and chimpanzees. Then the chimpanzee was required to name two individuals in free order by

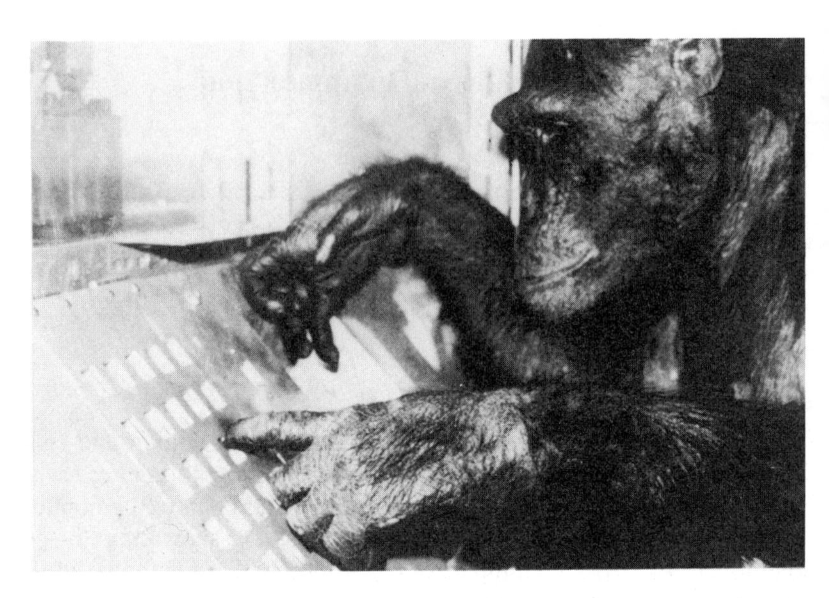

Figure 17.1. The chimpanzee Ai is naming object/color/number by pressing the corresponding "word" keys in the computer-controlled "language" system. The chimpanzee spontaneously developed "word order" reflecting the perceptual/cognitive world. The same kind of cognitive rules can be identified under the simpler test situations described in this study.

pressing a conjunction key "and" when two photographs were presented at once. She showed a strong tendency to name chimpanzees before humans (Matsuzawa, 1989).

Spontaneous organization can be found in other levels of cognitive skills. The chimpanzee learned to construct complex geometrical figures from elemental figures in a "constructive matching-to-sample" task (Matsuzawa, 1986). The samples were various kinds of complex figures consisting of nine elemental figures, such as circle, rectangle, oblique line, wave, and so on. When a sample appeared on a video screen, the chimpanzee was required to construct the same figure by choosing elemental figures provided on the keys. Although the order of construction was free, she showed a strong tendency to construct the complex figures starting from their outer contours (Matsuzawa, 1989).

The spontaneous organization the chimpanzee showed cannot be explained in terms of differential reinforcement, because no particular organization was required for reinforcement. It seems to reflect the internal perceptual/cognitive structure of the chimpanzee. Being allowed to freely behave to solve the problems, the chimpanzee spontaneously developed the solutions, governed by the cognition of the problems. However, the tests described earlier were so complicated and specific to the chimpanzee that they cannot be directly applied to human children. Although many researchers have devised tests to evaluate cognitive development in human children (Greenfield,

Nelson, & Saltzman, 1972; Sugarman, 1983), few tests have been applied to the apes. The study reported here was an attempt to devise a simple test exploring cognitive development in human and chimpanzee using exactly the same method. The test was a spontaneous sorting task. The subjects were given two plates and a number of objects. Then they were encouraged to put them on the plates. A chimpanzee and human children of various ages were compared in the simple sorting test.

## Method

### *Subject*

A 26-year-old female chimpanzee, Sarah, was the chimpanzee subject. She had engaged in various kinds of cognitive tasks, including the training of languagelike skills (Premack, 1976; Premack & Premack, 1983). The human children tested ranged from 1 to 5.5 years old.

### *General procedure*

The tester and the subject sat face to face. Two white plastic plates were symmetrically placed in front of the subject. The plates were 6.5 in. in diameter and were placed about 4 in. apart, edge to edge. Several objects were scattered randomly between the plates and the subject. Then the subject was encouraged to put them all on the plates when the tester tapped the center of each plate with an index finger. After the subject completed putting them on the plates, the subject was always praised verbally. In the case of the chimpanzee subject and some human children, they received food rewards in addition to verbal praise. It must be noted that the subject was always praised whenever he or she put all the objects on the plates, no matter in what assortment. If the subject did not complete putting all items, the tester encouraged him or her to do so. The tester recorded the spatial arrangement and the temporal order of object sorting onto the two plates. The objects were small household objects: various kinds of foods, metal, paper, plastic, and other materials.

In the following experiments, each trial employed a set of two kinds of objects, totally different in color, shape, and other features. For example, bells (round, silver, metal) and blocks (cubic, brown, wood) were used in one trial, red grapes and green plastic toys were used in the next, and so forth. Every trial employed trial-unique objects; no objects were repeated. In the following explanation, one kind of object is represented by the letter *A*, and the other by *B*. The number of each kind of object was varied to reveal the nature of sorting.

The following five variants were tested: *AAB*, *AABB*, *AAAB*, *AAABB*, and "modified matching-to-sample" (MMTS). For example, the *AAB* task meant that two *A*'s and one *B* were given to the subject. MMTS was a variant of the *AAB* task: Three items, *AAB*, were prepared, but in advance one *A* was put on one plate, and the *B* on the other plate. Then the subject was given the remaining *A* only and encouraged to put it on either plate. The subjects did not always participate in every task. A session of each task consisted of 10 trials, except for very young children around 1 year old. If necessary, several sessions were added to confirm the test results. A session lasted approximately 5–15 min, depending on the task and on the subject.

## Results

### *AAB sorting task*

There were three possible spatial arrangements in the task. First, two *A*'s were put together on one plate, and one *B* was put on the other. That solution was named *categorical sorting*, represented by *AA*:*B*. Second, two *A*'s were split onto two plates, and *B* was put on one of the two. That solution was named *symmetrical sorting*, represented by *AB*:*A*. The reason that solution was called symmetrical is explained in a later section. It must be noted that the temporal order of sorting was neglected in the definition. Third, all three items were put on one plate, and the other plate was left empty. That solution was named *one-side sorting*, represented by *AAB*:. Those three exhausted the possible spatial arrangements.

How can we evaluate the frequency of the three possible types of sorting? A random sorting model can provide the standard. If the subject picked up the items one by one, neglecting *A* or *B*, and put them equally often on the two plates, the expected probability value can be calculated from the binomial distribution. The probabilistic expectations of occurrence of the three possible solutions are as follows: *AA*:*B* = 25%, *AB*:*A* = 50%, and *AAB*: = 25%.

Figure 17.2 shows typical performances of symmetrical sorting and categorical sorting by human children and the typical performance of Sarah. The data for all subjects are summarized in Table 17.1. The subjects were classified into four groups. First, the youngest children (age less than 2 years) always showed one-side sorting. They always put all items on one plate and left the other plate empty. It is very difficult for children less than 1 year old to put the objects on the plates. Second, the children 2–2.5 years old began to put the items on both plates, but there was no systematic sorting rule. That can be named *both-sides sorting*. All types of sorting occurred in approximately chance proportions. Third, the children 2.5–3.5 years old almost always showed symmetrical sorting. Fourth, the children older than 3.5 years almost always showed categorical sorting.

The adult chimpanzee, Sarah, almost always showed categorical sorting. Her results were quite similar to those for the oldest group of human children.

Sorting can be described not only in terms of the final spatial arrangement but also in terms of the temporal order of putting the items on the plates. Suppose that the subject picks up and puts the three items on the plates one by one. There are three possible temporal orders: *A*-*A*-*B*, *B*-*A*-*A*, and

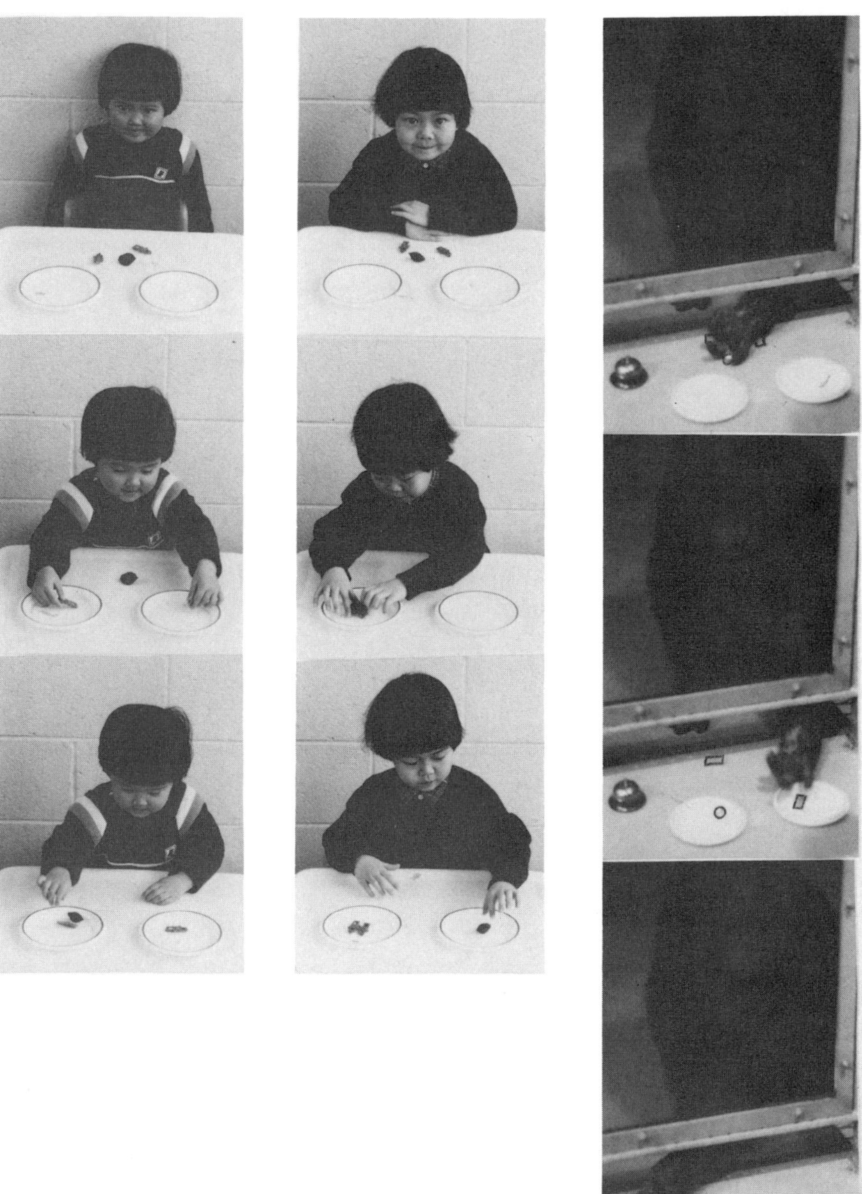

Figure 17.2. Spontaneous sorting in the *AAB* task: typical performances of symmetrical (left column) and categorical (central column) sorting by human children and the typical performance of Sarah (right column).

*A-B-A*. The first two sequences can be joined as categorical classification of the two objects with respect to temporal order. The probabilistic expectation of occurrence of categorical classification is 67%. Children indicated by asterisks in Table 17.1 were those whose temporal order of sorting was

Table 17.1. *Spontaneous sorting (AAB task) by human and chimpanzee subjects*

| | | | | Spatial arrangement | | | |
|---|---|---|---|---|---|---|---|
| Subject | Sex | Age (yr:mo) | Trials | Category (*AA*:*B*) | Symmetry (*AB*:*A*) | One-side (*AAB*:) | TO |
| Sarah | f | 26:00 | 40 | 82.5 | 17.5 | 0 | * |
| Yuko-S | f | 5:06 | 20 | 100 | 0 | 0 | * |
| Yoko | f | 4:03 | 40 | 90 | 10 | 0 | * |
| Ryota | m | 3:11 | 20 | 100 | 0 | 0 | * |
| Tanty | f | 3:06 | 10 | 0 | 100 | 0 | * |
| Luke | m | 3:05 | 10 | 0 | 100 | 0 | * |
| Amanda | f | 3:00 | 30 | 3.3 | 96.7 | 0 | * |
| Yuko-M | f | 2:06 | 40 | 0 | 100 | 0 | * |
| Yumi | f | 2:07 | 10 | 20 | 50 | 30 | * |
| Tadashi | m | 2:05 | 15 | 13.3 | 40 | 46.7 | * |
| Koichi | m | 2:04 | 10 | 0 | 0 | 100 | * |
| Sho | m | 2:04 | 10 | 0 | 0 | 100 | ns |
| Yuko-X | f | 2:00 | 10 | 0 | 10 | 90 | ns |
| Kyoko | f | 1:07 | 10 | 0 | 0 | 100 | ns |
| Mami | f | 1:02 | 5 | 0 | 0 | 100 | ns |
| Momoyo | f | 1:01 | 9 | 0 | 0 | 100 | ns |
| Kiyoshi | m | 1:01 | 2 | 0 | 0 | 100 | ns |
| Chance level | | | | 25 | 50 | 25 | |

*Notes:* The percentage frequency of occurrence of each sorting type is shown for each subject. The letter code of spatial arrangement indicates how the objects were divided onto the two plates. The temporal order (TO) is explained in the text. An asterisk indicates a significantly higher occurrence of temporal categorical sorting than the change level; "ns" indicates a nonsignificant difference. Sarah is the only chimpanzee.

categorical. The youngest group of human children did not show the tendency.

Both children whose spatial arrangements were categorical and those whose arrangements were symmetrical tended to sort categorically in terms of temporal order. All these subjects preferred an *A*-*A*-*B* to a *B*-*A*-*A* sequence and seldom used an *A*-*B*-*A* sequence. Children in the stage of symmetrical sorting by location typically put one *A* on one plate and another *A* on the other. In many cases a child would pick up the two *A*'s simultaneously in both hands and put one on each plate at once. Then the subject would pick up the *B* and hesitate to put it on a plate. In some instances a subject would complain of the lack of another *B* or demand it by gesture. In other instances a subject would put *B* on the center between the two plates or transfer *B* from one plate to the other repetitively. Children in the stage of categorical sorting first put two *A*'s together on one plate and then put *B* on the other. Two *A*'s were sometimes picked up simultaneously in both hands, but were still put on the same plate. They were stacked one on the other, or placed side by side in some instances. Sarah did not use both hands simultaneously, but the other temporal characteristics of her sorting behavior accorded with the typical categorical children.

Finally, it must be noted that children in the categorical stage can sort objects symmetrically, but those in the symmetrical stage never sort them categorically. This shows the irreversible developmental process of sorting skills.

### *AABB sorting task*

Two pairs of objects allowed four possible spatial arrangements. First, two *A*'s could be put together on one plate, and two *B*'s on the other. That solution was named categorical sorting, represented by *AA*:*BB*. Second, two *A*'s could be split onto two plates, and two *B*'s also split. That solution was named symmetrical sorting, represented by *AB*:*AB*. Third, two *A*'s could be put together on one plate, and two *B*'s split between the plates, resulting in two *A*'s and one *B* on one plate, and one *B* on the other. Notice that *A* and *B* are changeable in this notation. This solution is named mixed or miscellaneous sorting, represented by *AAB*:*B*. Fourth, all items could be put on one plate, and the other plate left empty. That solution was named one-side sorting, represented by *AABB*:.

The random sorting model can provide the probabilistic expectation of occurrence of each of the four sorting types as follows: *AA*:*BB* = 12.5%, *AB*:*AB* = 25%, *AAB*:*A* = 50%, and *AABB*: = 12.5%.

Figure 17.3 shows typical performances of categorical and symmetrical sorting for human children and the performance of Sarah. The data for all subjects are summarized in Table 17.2. The subjects were classified into four groups. First, the youngest children (less than 2 years old) always showed one-side sorting. They always put all items on one plate and left the other empty. Second, children 2–2.5 years old began to put the items on both plates, but unsystematically (both-sides sorting). Third, children 2.5–3.5 years old almost always showed symmetrical sorting. Fourth, children more than 3.5 years old preferred to do categorical sorting.

The chimpanzee Sarah also preferred categorical sorting. The results for Sarah were similar to those for the oldest group of human children.

There are three possible temporal orders of placing the two kinds of objects, ignoring left–right position of the plates: *A-A-B-B*, *A-B-B-A* and *A-B-A-B*. (Note that the sequence *B-A-A-B* is equivalent to *A-B-B-A*.) Only the first sequence can be called categorical classification with respect to temporal order. The probabilistic expectation of occurrence of the temporal categorical classification is 33%. The subject indicated in Figure 17.3 showed a significant level of the temporal categorical classification. The younger groups of children did not show this tendency.

As for *AABB* sorting, both children who sorted categorically and those who sorted symmetrically with respect to spatial arrangement clearly showed

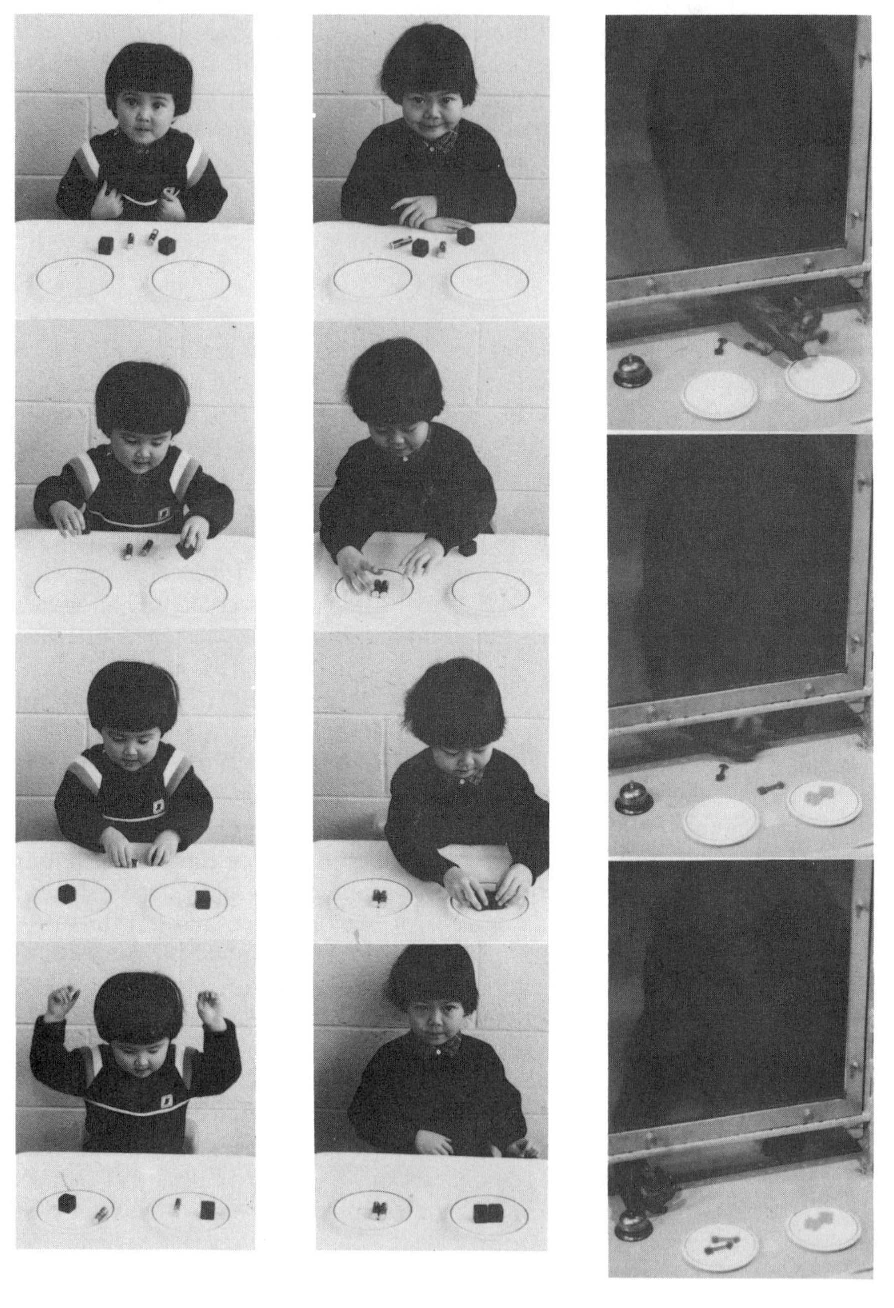

Figure 17.3. Spontaneous sorting in the *AABB* task: typical performances of symmetrical and categorical sorting by human children and the typical performance of Sarah.

Table 17.2. *Spontaneous sorting (AABB task) by human and chimpanzee subjects*

| Subject | Sex | Age (yr:mo) | Trials | Spatial arrangement: Category (*AA*:*BB*) | Symmetry (*AB*:*AB*) | Mixed (*AAB*:*B*) | One-side (*AABB*:) | TO |
|---|---|---|---|---|---|---|---|---|
| Sarah | f | 26:00 | 60 | 76.7 | 3.3 | 20 | 0 | * |
| Yuko-S | f | 5:06 | 10 | 100 | 0 | 0 | 0 | * |
| Yoko | f | 4:03 | 40 | 75 | 25 | 0 | 0 | * |
| Ryota | m | 3:11 | 10 | 100 | 0 | 0 | 0 | * |
| Tanty | f | 3:06 | 10 | 0 | 100 | 0 | 0 | * |
| Luke | m | 3:05 | 10 | 20 | 70 | 10 | 0 | * |
| Amanda | f | 3:00 | 20 | 0 | 100 | 0 | 0 | * |
| Yuko-M | f | 2:06 | 40 | 0 | 100 | 0 | 0 | * |
| Tadashi | m | 2:05 | 10 | 0 | 20 | 20 | 60 | ns |
| Koichi | m | 2:04 | 10 | 0 | 0 | 0 | 100 | * |
| Sho | m | 2:04 | 10 | 0 | 0 | 0 | 100 | ns |
| Kyoko | f | 1:07 | 2 | 0 | 0 | 0 | 100 | ns |
| Chance level | | | | 12.5 | 25 | 50 | 12.5 | |

*Notes*: The percentage frequency of occurrence of each sorting type is shown for each subject. The other notations are as in Table 17.1.

categorical classification with respect to temporal order. All these subjects first placed two *A*'s and then put two *B*'s on the plates. The two kinds of objects were strictly separated in the temporal order of sorting. Children in the stage of symmetrical sorting typically split two *A*'s, putting one on each plate, using both hands, and then split two *B*'s in the same fashion. They usually looked at both plates repetitively while holding one object of a pair in each hand. Children in the stage of categorical sorting sometimes put two *A*'s simultaneously on one plate, and then two *B*'s on the other. The way of placing two *A*'s corresponded to the way of placing two *B*'s in many cases. When one pair was stacked, the other pair was also stacked. When one pair was placed side by side, the other pair was similarly placed.

### *AAAB sorting task*

The symmetrical structure of the provided objects was totally destroyed in the *AAAB* task. There were four possible spatial arrangements in this task. First, three *A*'s could be put together on one plate, and one *B* put on the other. That solution was named categorical sorting, represented by *AAA*:*B*. Second, three *A*'s could be unevenly split between the two plates, with two *A*'s on one plate and one *A* on the other, and then *B* could be put on the second plate to equalize the number of objects. That solution was named

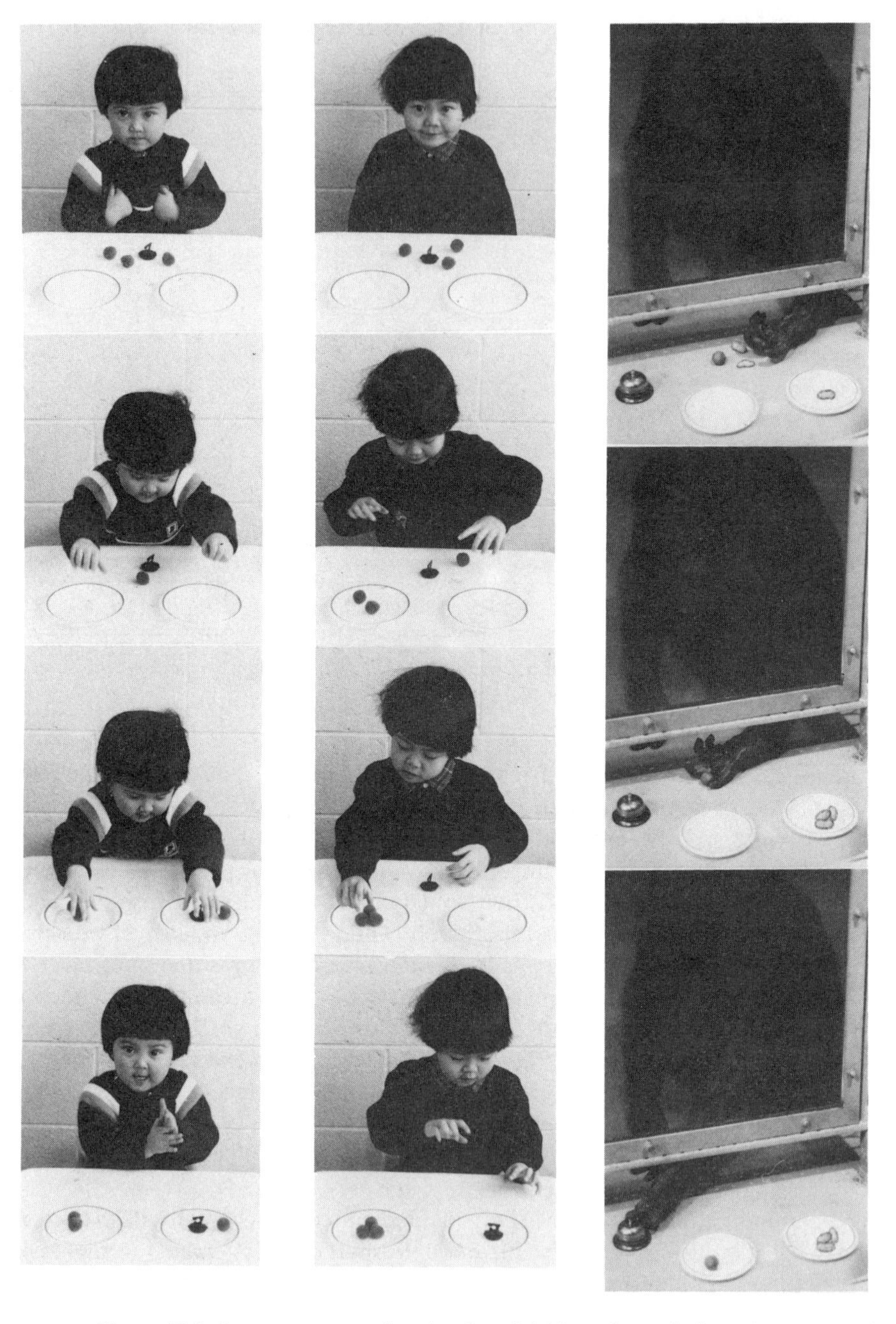

Figure 17.4. Spontaneous sorting in the *AAAB* task: typical performances of symmetrical and categorical sorting by human children and the typical performance of Sarah.

Table 17.3. *Spontaneous sorting (AAAB task) by human and chimpanzee subjects*

| | | | | Spatial arrangement | | | | |
|---|---|---|---|---|---|---|---|---|
| Subject | Sex | Age (yr:mo) | Trials | Category (*AAA*:*B*) | Symmetry (*AA*:*AB*) | Mixed (*AAB*:*A*) | One-side (*AABB*:) | TO |
| Sarah | f | 26:00 | 40 | 62.5 | 37.5 | 0 | 0 | * |
| Yuko-S | f | 5:06 | 10 | 100 | 0 | 0 | 0 | * |
| Yoko | f | 4:03 | 40 | 70 | 22.5 | 7.5 | 0 | * |
| Ryota | m | 3:11 | 10 | 100 | 0 | 0 | 0 | * |
| Tanty | f | 3:06 | 10 | 0 | 90 | 10 | 0 | * |
| Luke | m | 3:05 | 10 | 0 | 100 | 0 | 0 | * |
| Amanda | f | 3:00 | 10 | 0 | 90 | 10 | 0 | * |
| Yuko-M | f | 2:06 | 60 | 0 | 100 | 0 | 0 | * |
| Chance level | | | | 12.5 | 37.5 | 37.5 | 12.5 | |

*Notes:* The percentage frequency of occurrence of each sorting type is shown for each subject. The other notations are as in Table 17.1.

symmetrical sorting, represented by *AA*:*AB*. Third, three *A*'s could be split between two plates, with *B* joining the two *A*'s. Equality of number was not achieved in that solution. That was named mixed sorting, represented by *AAB*:*A*. Fourth, all four items could be put on one plate, and the other plate left empty. That solution was named one-side sorting, represented by *AABB*:.

The random sorting model provides the probabilistic expectation of occurrence of each of the four sorting types as follows: *AAA*:*B* = 12.5%, *AA*:*AB* = 37.5%, *AAB*:*A* = 37.5%, and *AAAB*: = 12.5%.

Figure 17.4 shows typical performances of symmetrical and categorical sorting for children and the performance of Sarah. The data for all subjects are summarized in Table 17.3. The chimpanzee Sarah preferred categorical sorting, as did the oldest group of human children.

There are four possible temporal orders of placing the two kinds of objects: *A-A-A-B*, *B-A-A-A*, *A-A-B-A*, and *A-B-A-A*. The first two sequences can be called categorical classification with respect to temporal order. Individuals showing this tendency are indicated by an asterisk in Table 17.3. The results were consistent with the previous two tasks.

### *AAABB sorting task*

The number of objects can be increased to give a more complicated task. There are six possible spatial arrangements in this task: categorical or *AAA*:*BB*, symmetrical or *AAB*:*AB*, one-side or *AAABB*:, and the remaining alterna-

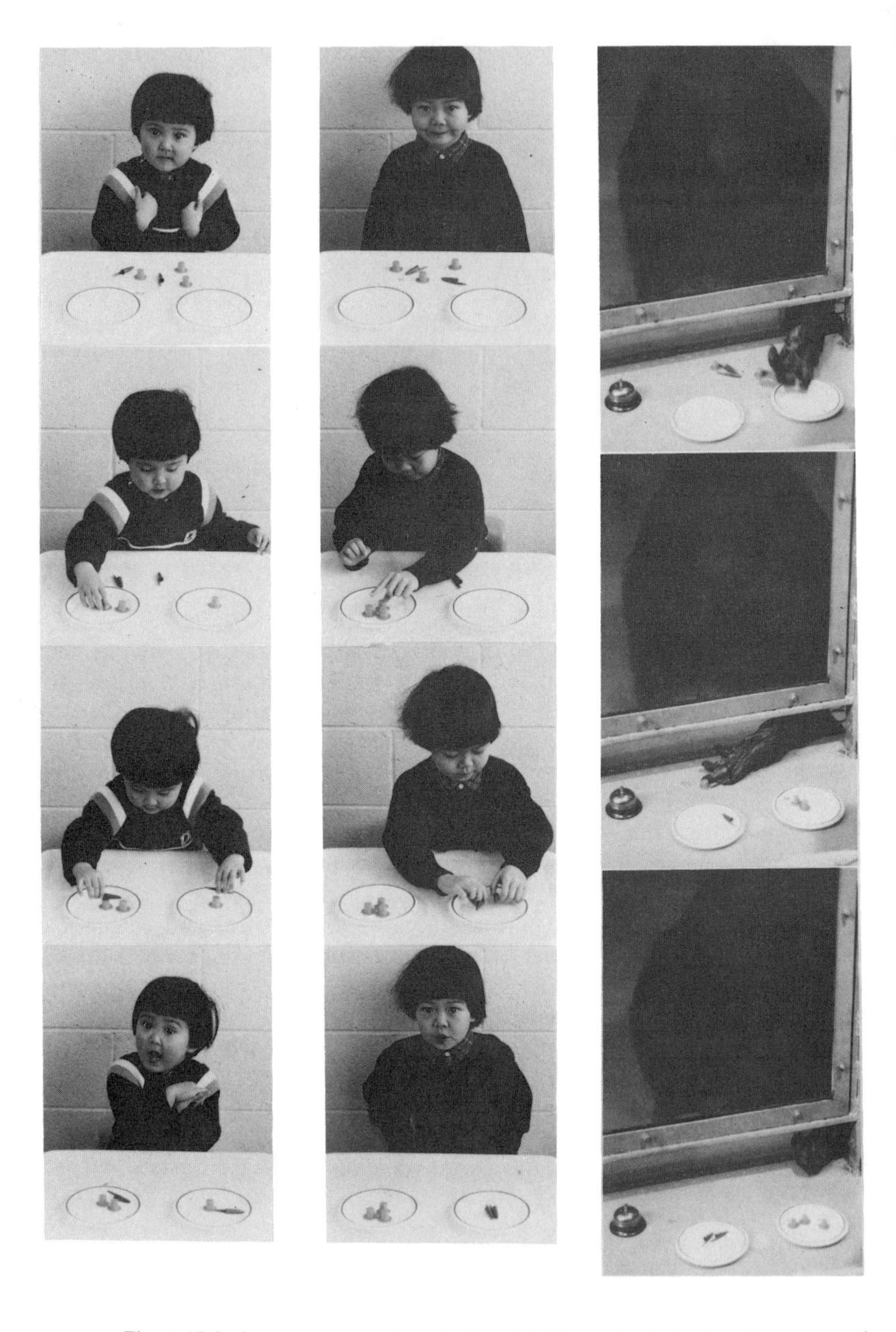

Figure 17.5. Spontaneous sorting in the *AAABB* task: typical performances of symmetrical and categorical sorting by human children and the typical performance of Sarah.

Table 17.4. *Spontaneous sorting (AAABB task) by human and chimpanzee subjects*

| | | | | Spatial arrangement | | | | |
|---|---|---|---|---|---|---|---|---|
| Subject | Sex | Age (yr:mo) | Trials | Category (*AAA*:*BB*) | Symmetry (*AAB*:*AB*) | Mixed (*AAB*:*B*) (*AABB*:*A*) (*AAAB*:*B*) | One-side (*AABB*:) | TO |
| Sarah | f | 26:00 | 20 | 80 | 5 | 15 | 0 | * |
| Yoko | f | 4:03 | 40 | 77.5 | 15 | 7.5 | 0 | * |
| Amanda | f | 3:02 | 10 | 0 | 90 | 10 | 0 | ns |
| Yuko-M | f | 2:06 | 40 | 0 | 95 | 5 | 0 | ns |
| Chance level | | | | 6.25 | 41.25 | 46.25 | 6.25 | |

*Notes:* The percentage frequency of occurrence of each sorting type is shown for each subject. The other notations are as in Table 17.1.

tives of mixed sorting, such as *ABB*:*AA*, *AAAB*:*B*, and *AABB*:*A* (Figure 17.5).

The random sorting model provides the expectation of each sorting type shown in Table 17.4. Results for temporal order were basically similar to those already described in the simpler tasks.

### *Modified matching-to-sample sorting task*

There are numerous possible variants of the sorting task as the numbers of the two kinds of objects are increased. How can we reduce the number of objects to reveal the nature of sorting behavior? The ultimate minimum task is "one-item sorting," in which only one item is given to the subject, and he or she is encouraged to put it on either plate. As a variant of the original *AAB* sorting task, this task can test which type of sorting the subject prefers between the two alternative solutions. For example, in the task named "modified matching-to-sample," an *A* was put on one plate, and a *B* on the other, in advance. The subject was given only one *A*. If the subject put the *A* with the other *A*, the final spatial arrangement was categorical. If the subject put it with *B*, then the spatial arrangement was symmetrical. The left–right positions of *A* and *B* objects placed in advance were counterbalanced within a session. This one-item sorting task allows unequivocal classification of sorting as either categorical or symmetrical.

Figure 17.6 shows typical performances of the two types and the performance of Sarah. Table 17.5 summarizes the results. The oldest group preferred categorical sorting, and the second oldest preferred symmetrical sorting. The younger children did not sort systematically in this situation.

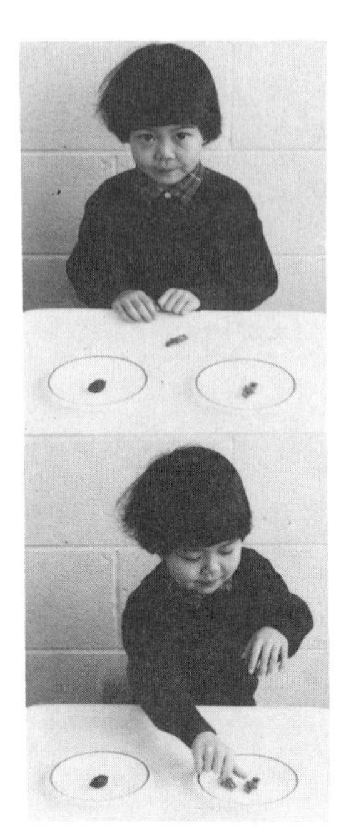

Figure 17.6. Spontaneous sorting in the modified matching-to-sample task: typical performances of symmetrical and categorical sorting by human children and the typical performance of Sarah.

Children around 3 years old almost always "mismatched" the objects. It has been hypothesized that human children naturally put the same things together in matching-to-sample tasks, but this result contradicts the view that the matching is intrinsic. Mismatching in this task was typical of children around 3 years old.

*Summary of the results*

The data are summarized in Table 17.6, in which the types of solutions are shown for each subject in each variant of the sorting task. Although there were various kinds of tasks, each subject consistently showed the same sorting tendency throughout the tasks. Four major stages of development can be discerned in the table. First, the youngest children, less than 2 years old, showed one-side sorting. Children in the second stage, from 2 to 2.5 years old, began to put the items on both plates, but sorting was neither symmetrical nor categorical, and was labeled both-sides sorting. Children in the third stage, 2.5–3.5 years old, showed typical symmetrical sorting. Finally,

Table 17.5. *Spontaneous sorting (modified matching-to-sample task) by human and chimpanzee subjects*

| Subject | Sex | Age (yr:mo) | Trials | Spatial arrangement Category (*AA*:*B*) | Spatial arrangement Symmetry (*AB*:*A*) |
|---|---|---|---|---|---|
| Sarah | f | 26:00 | 100 | 95 | 5 |
| Yuko-S | f | 5:06 | 10 | 100 | 0 |
| Yoko | f | 4:03 | 20 | 100 | 0 |
| Ryota | m | 3:11 | 10 | 100 | 0 |
| Luke | m | 3:05 | 10 | 80 | 20 |
| Tanty | f | 3:06 | 10 | 0 | 100 |
| Amanda | f | 3:00 | 20 | 15 | 85 |
| Yumi | f | 2:07 | 10 | 0 | 100 |
| Yuko-M | f | 2:06 | 20 | 0 | 100 |
| Yuko-X | f | 2:00 | 6 | 33.3 | 66.7 |
| Kyoko | f | 1:07 | 5 | 60 | 40 |
| Momoyo | f | 1:01 | 6 | 33.3 | 66.7 |
| Chance level | | | | 50 | 50 |

*Notes*: The percentage frequency of occurrence of each sorting type is shown for each subject. The other notations are as in Table 17.1.

children more than 3.5 years old and the adult chimpanzee Sarah showed categorical sorting.

## Discussion

The simple tests described earlier revealed four distinctive developmental stages of sorting behavior in human children. The adult female chimpanzee Sarah showed the same performance as the oldest group, those more than 3.5 years old. How can we explain the results? What is the nature of development in sorting behavior?

Classification of sorting types itself implicitly gives the explanation. Consider again the definition of each sorting type. One-side sorting is easily defined. Everything is put on one plate. Categorical sorting is also easily defined. There were always two kinds of objects in the task: One kind was put together on one plate, and the other kind was put on the other plate. What is symmetrical sorting? The reason we call the *AB*:*AB* arrangement symmetrical is obvious. However, why call the *AB*:*A*, *AA*:*AB*, *AAB*:*AB* arrangements symmetrical?

For children in the symmetrical sorting stage, the two plates placed symmetrically might result in the recognition of "a pair" of plates. A pair consists of two identical components. The two are physically the same. The children appeared to be trying to preserve that physical symmetry in their sorting.

Table 17.6. *Summary of results of the spontaneous sorting tests*

| | | | Sorting tasks | | | | |
|---|---|---|---|---|---|---|---|
| | | | *AAB* | *AABB* | *AAAB* | *AAABB* | MMTS |
| Subject | Sex | Age (yr:mo) | *AA*:*B*<br>*AB*:*A*<br>*AAB*: | *AA*:*BB*<br>*AB*:*AB*<br>*AABB*: | *AAA*:*B*<br>*AA*:*AB*<br>*AAAB*: | *AAA*:*BB*<br>*AAB*:*AB*<br>*AAABB*: | *AA*:*B* (C)<br>*AB*:*A* (S)<br>*AAB*: (O) |
| Sarah | f | 26:00 | C | C | C | C | C |
| Yuko-S | f | 5:06 | C | C | — | — | C |
| Yoko | f | 4:03 | C | C | C | C | C |
| Ryota | m | 3:11 | C | C | — | — | C |
| Tanty | f | 3:06 | S | S | S | — | S |
| Luke | m | 3:05 | S | S | S | — | C |
| Amanda | f | 3:00 | S | S | S | S | S |
| Yuko-M | f | 2:06 | S | S | S | S | S |
| Yumi | f | 2:07 | * | — | — | — | S |
| Tadashi | m | 2:05 | * | * | — | — | — |
| Koichi | m | 2:04 | O | O | — | — | — |
| Sho | m | 2:04 | O | O | — | — | — |
| Yuko-X | f | 2:00 | O | — | — | — | * |
| Kyoko | f | 1:07 | O | O | — | — | * |
| Mami | f | 1:02 | O | — | — | — | — |
| Momoyo | f | 1:01 | O | — | — | — | * |
| Kiyoshi | m | 1:01 | O | — | — | — | — |

*Notes*: Sorting preferences are indicated by C (categorical), S (symmetrical), O (one-side), and * (no preference). A dash means the subject did not receive the task.

The preservation of physical symmetry can be achieved by a process of one-to-one correspondence. Each member on one plate corresponds to a member on the other plate. It must be noted that correspondence of the members on the two plates is achieved on a strictly one-by-one basis. There is no correspondence as a group. Some tasks never permitted strictly symmetrical sorting because of the lack of corresponding members. For example, the four items of an *AABB* task can be divided on the two plates symmetrically, but the three items of an *AAB* task cannot. Subjects in the stage of symmetrical sorting have a strong tendency to solve each task by making the spatial arrangement as symmetrical as possible. When one *A* does not have a corresponding *A* to place on the other plate, the child is forced to approximate correspondence with the *B* alternative to equalize the *numbers* of objects on the two plates. If one *A* does not have even a *B* alternative, the child may simply place it on either plate, abandoning symmetry. The symmetrical solution of *AB*:*A* in the *AAB* task does not imply that *AB* on one plate corresponds to *A* on the other. The *AB*:*A* solution means that *A* on one plate corresponds to *A* on the other, and *B* is a kind of remainder.

This hypothesis can explain why children in the stage of symmetrical sorting did the curious mismatch in the modified matching-to-sample task. They put *A* on the plate with *B* rather than with *A* to make the whole arrangement as close as possible to the physical symmetry.

Although there are many possible variants of the sorting task, each task always has only a single one-side solution, a single categorical solution, and also a single symmetrical solution following this definition. Suppose that eight items of an *AABB* task can be divided on the two plates symmetrically, but is represented by *AAAABBBB*:. The symmetrical solution is *AABB*:*AABB*. The categorical solution is *AAAA*:*BBBB*. The other six possible solutions, such as *AAAB*:*ABBB* and *AAAABB*:*BB*, are classified as mixed or miscellaneous sorting, because it consists of diverse types excluding the three solutions defined earlier.

As the number of items increases, the probability of occurrence of the mixed sorting, following the random sorting model, dramatically increases. In other words, the probability of systematic sorting by chance decreases. However, human and chimpanzee subjects never preferred the mixed solution in any stage of development.

Children in the first stage neglected the nature of "a pair" of the plates. Putting all things on one plate was the only solution for children less than 2 years old. Children who had reached 2 years of age began to recognize "a pair" and also to classify two kinds of objects in the temporal order of their sorting. Behaviorally, they showed that *A* is *A* and not *B*. However, they had no way to make members of a pair correspond in a systematic arrangement on a pair of plates. Children about 2.5 years old can relate the members of a set of objects in one-to-one correspondence, as described earlier.

What kind of mechanism can we recognize in the most "advanced" stage of spontaneous sorting? One plausible explanation is the hierarchical organization of one-to-one correspondence. The things can be divided into two categories: *A* and *B*. A pair of plates can be used one for *A* and the other for *B*. In the categorized sorting, the category is treated as one "whole" to be put in correspondence with another "whole" category, as each individual member is made to correspond to the other in symmetrical sorting. *AA* on one plate and *BB* (or even *B* only) on the other can be seen as corresponding on the level of category. The ultimate basis for categorical sorting seems to be the cognition of "a pair" of plates and the correspondence of the two kinds of objects on the level of class or category. The result from the *AAAB* sorting task clearly shows that three *A*'s can correspond to one *B* for children in the categorical stage. Some additional tests revealed that human adults (9 of 9) did categorical sorting in the *AAB* task.

The chimpanzee Sarah showed categorical sorting in this study. Sarah adopted the same "rule" in a related task called "stacking blocks." Chimpanzees older than 4 years can stack blocks (1-in. cubes) up to 10, on the average (Matsuzawa, 1987). In a simple "free stacking" test, the subject received a bunch of blocks at once and was asked to use them up to build as many towers as she liked. Half of the blocks were painted white, and the other half were black. Sarah spontaneously preferred to put white blocks together to make a white tower, and black blocks together to build a black tower. The results clearly show that the cognitive rule, such as categorical sorting, is not task-specific but universal across the various tasks.

This study shows that a simple test such as sorting *AAB* can reveal qualitative differences in cognitive development for human children from age 1 to age 4 and above. The test can also provide a simple and unitary scale for comparing cognitive development between different species, because certain behaviors in sorting objects seem to be intrinsic not only to humans but also to other primates, such as the chimpanzee in this study. The test requires no previous training. Exactly the same method can be applied to subjects of different ages and different species.

**Acknowledgments**

The author acknowledges the helpful discussions and support of Dr. D. Premack. I also thank Dr. E. Segal for suggesting improvements to the original manuscript and Mrs. H. Takeshita and B. Dennis for collecting part of the data. This research was supported by an overseas research fellowship from the Ministry of Education, Science and Culture, Japan.

**References**

Asano, T., Kojima, T., Matsuzawa, T., Kubota, K., & Murofushi, K. (1982). Object and color naming in chimpanzees (*Pan troglodytes*). *Proceedings of the Japan Academy 58*(B), 118–122.

Gardner, R. A., & Gardner, B. T. (1969). Teaching sign language to a chimpanzee. *Science*, *165*, 664–672.

Greenfield, P. M., Nelson, K., & Saltzmann, E. (1972). The development of rule-bound strategies for manipulating seriated cups: A parallel between action and grammar. *Cognitive Psychology*, *3*, 291–310.

Matsuzawa, T. (1985a). Color naming and classification in a chimpanzee (*Pan troglodytes*). *Journal of Human Evolution*, *14*, 283–291.

Matsuzawa, T. (1985b). Use of numbers by a chimpanzee. *Nature*, *315*, 57–59.

Matsuzawa, T. (1986). Pattern construction by a chimpanzee. *Primate Report*, *14*, 225–226.

Matsuzawa, T. (1987). Stacking blocks by chimpanzees. *Primate Research*, *3*, 1–12.

Matsuzawa, T. (1989). Spontaneous pattern construction in a chimpanzee. In P. Heltne & L. Margardt (Eds.), *Understanding chimpanzees.* Cambridge, MA: Harvard University Press.

Matsuzawa, T., Asano, T., Kubota, K., & Murofushi, K. (1986). Acquisition and generalization of numerical labeling by a chimpanzee. In D. Taub & F. King (Eds.), *Current perspectives in primate social dynamics* (pp. 416–430). New York: Van Nostrand Reinhold.

Patterson, F. G. (1978). The gestures of a gorilla: Language acquisition in another pongid. *Brain and Language*, *5*, 72–97.

Premack, D. (1971). Language in chimpanzee? *Science*, *172*, 808–822.

Premack, D. (1976). *Intelligence in ape and man.* Hillsdale, NJ: Erlbaum.

Premack, D., & Premack, A. J. (1983). *The mind of an ape.* New York: Norton.

Rumbaugh, D. M. (1977). *Language learning by a chimpanzee.* New York: Academic Press.

Sugarman, S. (1983). *Children's early thought: Developments in classification.* Cambridge University Press.

Terrace, H. S. (1979). *Nim: A chimpanzee who learned sign language.* New York: Knopf.

Terrace, H. S., et al. (1979). Can apes create a sentence? *Science*, *206*, 891–895.

# 18 Conceptual abilities of some nonprimate species, with an emphasis on an African Grey parrot

*Irene Maxine Pepperberg*

## Introduction

Many researchers are actively engaged in cross-species comparisons of animal intelligence. These researchers make inferences about relative intelligence by analyzing animals' overt learning behaviors in a particular experimental paradigm (e.g., numerical competence, concepts of similarity and difference, foraging and memory), as reviewed by Kamil (1988a) and Macphail (1987). Such inferences are adequate provided that the behavioral analyses are complete in terms of cause, function, evolution, and development (Burghardt, 1984; Kamil, 1984; Snowdon, 1983). Unfortunately, the study of behavior still is often partitioned between those who, as a group, study cause and development and those who study evolution and function.[1] These two groups can be characterized, respectively, by the differing perspectives of comparative psychology and ethology (Marler & Terrace, 1984; Parker, P&G1). Comparative psychologists generally examine the cause and development of learning by comparing the mechanisms and processes in just a few animal species that are trained and tested under rigorously controlled laboratory conditions. Ethologists, in contrast, examine the function and evolution of learning by comparing natural behaviors in a variety of animals, from ants to elephants, often using closely related species to see if each species' performance depends on the particular aspects of the situation in which it is observed or tested (Kroodsma et al., 1984). Incomplete knowledge of each other's domains may cause both groups of researchers to perform experiments that are inadequate for the purpose of comparing intelligence.

In an attempt to integrate these two approaches, some investigators have adopted a cognitive perspective. A cognitive perspective assumes that each animal species is a "decision maker" that interacts with its environment in variable, flexible ways and develops its behavior accordingly (Kamil, 1984). Although most often associated with laboratory research on animal learning, cognitive approaches are also central to the work of many field ethologists (e.g., Gouzoules, Gouzoules, & Marler, 1985). The cognitive perspective not only has stimulated an interdisciplinary approach to research but also has generated interest in topics, methodologies, and animal species (especially nonprimates) that would have been ignored under other paradigms.

In this chapter I review some of these studies and thus provide background for the comparable chapters on primate abilities. Before proceeding, however, I briefly discuss some of the issues that may influence a researcher's choice of methodology, topic, and animal species.

### *Experimental strategies in the study of animal intelligence*

Researchers in some areas of animal behavior have become increasingly aware of how even subtle variations in experimental design may affect their results (Kamil, 1988b; Kroodsma, 1989; Kroodsma, Baker, Baptista, & Petrinovich, 1985; Menzel & Juno, 1985). Comparative studies, in particular, are subject to arguments concerning the effects of contextual variables on tests for general intelligence (Macphail, 1987). Some researchers, for example, have yet to recognize that the degree to which animals can be shown to exhibit complex cognitive abilities can be as much a function of the design of the training and testing procedures as of the inherent capacities of the animal subjects (see discussions in Cole, Hainsworth, Kamil, Mercier, & Wolf, 1982; Hayes & Nissen, 1956/1971; Mason & Hollis, 1962; Spetch & Edwards, 1986).

A separate but related issue involves the merits of using human frameworks for studying cognitive abilities of animals. Researchers such as Hodos (1982), for example, argue for an approach that evaluates an animal's performance with respect to the demands of its natural environment (Parker & Gibson, 1977; Snowdon, 1983). Hodos's point is that the abilities that are most often examined in the search for intelligent behavior may have little meaning in relation to the capacities that a nonhuman needs or exhibits in its natural environment. Researchers must therefore ask whether, by judging animals by human standards, they may be examining cognitive capacities that are relevant only for humans, rather than general, inherent abilities (Parker, P&G1).

The foregoing contention leads to yet another point central to this volume, that of the possible correlation between advanced cognitive abilities and complex communication systems – specifically language. As reviewed elsewhere (Pepperberg & Kozak, 1986), researchers continue to debate the extent to which (1) language influences concept formation, (2) conceptual capacities are prerequisites for language acquisition, and (3) concept formation and language development may initially be dissociated (Bates, Benigni, Bretherton, Camaioni, & Volterra, 1979; Ginsburg & Opper, 1969; Hall, Braggio, Buchanan, Nadler, Karan, & Sams, 1980; Piaget, 1952b, 1954; Premack, 1978, 1983; Rice, 1980; Steckol & Leonard, 1981; Steklis & Raleigh, 1979; Vygotsky, 1962). At one time, researchers expected to resolve this debate by comparing cognitive abilities in language-trained and non–language-trained animal subjects (e.g., Premack, 1983). The possibility existed, however, that the outcome of the comparative tests could be influenced by the protocols of the language training paradigm (Anderson,

1987) – a possibility that returns us to the first point raised in this section. In the next few paragraphs, therefore, I briefly discuss the problems that must be acknowledged and solved before animal capacities can be compared.

*Integrating techniques from field and laboratory.* The studies that have demonstrated complex, learned abilities in various animals have been those that have successfully integrated the rigorous experimental design of the laboratory with detailed knowledge of the behavioral ecology of the subject. Gill and Wolf (1975), for example, showed that some birds develop a strategy for optimal territoriality and do not, as once thought, rotely defend a defined area: A cost–benefit analysis of energy used in defense and energy obtained (in the form of nectar) from a specific area demonstrated that golden-winged sunbirds (*Nectarina reichenowi*) apparently vary their degree of territoriality in response to an evaluation of the average daily levels of food in their feeding range. Logan (1987) suggested that northern mockingbirds (*Mimus polyglottos*) behave somewhat similarly, basing their responses on competing demands for energy expenditure.

In a comparable manner, the investigation of learning in hummingbirds provided an instance in which researchers had to adapt their laboratory paradigm of account for their subjects' natural predispositions: Because many nectarivorous birds in the wild survive by foraging on a "win-shift" basis (i.e., a recently visited flower emptied of its nectar will subsequently not be a good food source), such birds perform poorly on experimental tasks that demonstrate intelligence by rewarding only "win-stay" behavior (Kamil, 1984). Without this knowledge of hummingbirds' normative field behavior, researchers might have drawn incorrect inferences from the birds' behavior in the laboratory. Such projects, though only a sample, demonstrate the importance of integrating techniques and perspectives of ethology and psychology in the comparative study of intelligence (Kamil, 1984, 1988a,b; Menzel & Juno, 1985; Snowdon, 1988).

*Use of human frames of reference for animal capacities.* Discussions on this issue generally center on an animal's "lack of need" for a particular human skill (e.g., sensitivity to number, concepts of same/different) (Seligman, 1970; Snowdon, 1988), but such arguments often lose their validity as our analyses of animal behavior become more sophisticated and we gain information about the types of tasks in which animals routinely engage in the wild (Griffin, 1976). Thus, although there is as yet no evidence that animals' concepts of number and same/different are identical with those of humans, recent studies, to be discussed later, have provided evidence for animal competencies in these areas at levels considerably more sophisticated than were once thought possible. Similarly, field researchers have now demonstrated a high degree of functional categorization of various calls in marmosets and tamarins, as reviewed by Snowdon (1987), and in the predator alarm calls of vervet monkeys (*Cercopithecus aethiops*) (e.g., Seyfarth, Cheney, & Marler,

1980; Struhsaker, 1967). The scientific community, however, initially questioned the "relevance" of laboratory experiments that demonstrated that nonhuman primates could engage in a form of referential labeling related to that of humans (e.g., Gardner & Gardner, 1969) when little evidence existed for any form of primate referential communication in the wild. These findings do not prove the value of using a human framework to judge animal capacities; however, discovery in the laboratory of an animal's aptitude for a task performed by humans often provides an impetus to search for an analogue in the wild.

*The relationship between language and cognition.* A wealth of conflicting evidence prevents investigators from making a clear choice among the three alternatives presented earlier concerning connections (or the existence of connections) between the development of language and cognition. Even so, some researchers choose to infer intelligence – or at least the existence of complex cognitive capacities – from evidence for certain natural or artificially induced communication patterns in their animal subjects (e.g., referentiality, information storage, rule-governed behaviors) (Schusterman & Gisiner, 1988). This choice of methodology implies a belief that the cognitive capacities involved in learning and using either a complex natural system or a human-based code indicate a high level of intelligence that can be employed in widely divergent situations. Although the data are as yet insufficient to prove that the underlying mechanisms for communication and other cognitive tasks (e.g., memory, information processing, categorization) are the same, or even that training in human-based communication codes enhances cognitive abilities (Macphail, 1987; Premack, 1983), various studies indicate that animals that use complex communication systems often evince other capacities as well (Parker & Gibson, 1979; Pepperberg, in press-b). Nevertheless, the assumption that sophistication in communication is anything more than an *indication* of complexity in other behaviors may be unwise (Rogoff, 1984). Alternatively, using a communication system as a *tool* to investigate levels of intelligence might be well advised (Pepperberg, 1987b). There is, for example, little disagreement that adult human cognitive abilities are most easily assessed through the use of human language. Various researchers have, in fact, used communication as a tool to assess cognitive competence in animals; some of their results are discussed later in this chapter and in other chapters (Greenfield & Savage-Rumbaugh, P&G20; Miles, P&G19).

### *The problem of species-specific differences in the study of animal intelligence*

The scientific community no longer views pigeons or rats as interchangeable models in which to discover general principles of learning nor expects that an animal's intelligence relates solely to its anatomical and neurophysiological similarity to humans (Dethier, 1984). Researchers who study comparative cognition, however, have yet to devise entirely satisfactory solutions to the problems caused by variation in observable behaviors across species. Kamil (1988a) has recently delineated the problems that occur when researchers either defend strict hierarchies with respect to learning that are indexed by the species' "closeness" to humans (Rumbaugh & Pate, 1984) or deny that species differences exist (e.g., Macphail, 1985, 1987). By defining animal intelligence as multidimensional rather than unidimensional (i.e., "It includes *all* processes that are involved in any situation where animals change their behavior on the basis of experience"), Kamil (1988a, p. 273) has suggested

1. that well-designed experiments enable a researcher to assess fairly the relative capacities of various species,
2. that relating the choice of species to the choice of problem under study can demonstrate patterns of strengths and weaknesses that are unique to each species, and
3. that such a perspective may encourage researchers to examine behaviors in species that are often poorly or inadequately studied.

In the spirit of Kamil's suggestions, I devote the rest of this chapter to describing some studies that employed methodologies and subjects that were somewhat out of the ordinary. Those studies are representative and by no means constitute an exhaustive review of this type of research. I chose the topics and, to a certain extent, the subjects because of their relationships to projects on primates described elsewhere in this volume.

## Nonprimate mammalian research: Label comprehension and rule-governed behaviors in marine mammals

Although studies of animal cognition most often focus on species with close phylogenic relationships to humans, such as monkeys and the great apes (Meador, Rumbaugh, Pate, & Bard, 1987), researchers have also investigated some capacities of marine mammals. Herman (1986, 1987) and Schusterman and Krieger (1984, 1986) have, respectively, examined dolphins (*Tursiops truncatus*) and sea lions (*Zalophus californianus*) for their capacities with respect to receptive language skills. These animal subjects appear capable of processing semantic and rule-governed features of simple artificial language systems: They understand imperative and interrogative strings of symbols and immediately transfer comprehension to novel strings.

The imperative strings direct the animals to take actions (e.g., RIGHT BALL TAIL-TOUCH) on labeled objects that are distinguished by sets of modifiers (e.g., right vs. left, surface vs. bottom). To perfom correctly, the animals must attend not just to the meaning of the symbols but also to the order of the symbols; for example, for HOOP BALL FETCH, the ball is taken to the hoop, but the hoop is taken to the ball for BALL HOOP FETCH (Herman, 1986; Schusterman & Krieger, 1984). One of the sea lions has also learned to respond to relational terms: She consistently chooses the larger of two objects even if the same object was the (relatively) smaller item on a previous test (Schusterman & Krieger, 1986).

The interrogative sentences are used to examine the animals' abilities to respond to questions about the existence of objects in their environment. One of the dolphins presses a paddle placed to its left (the NO paddle) to designate that an object is absent and a different style of paddle placed to its right (the YES paddle) if the object is present (Herman & Forestell, 1985). One sea lion responds with a nonreinforced "balk" when asked to perform an action (e.g., flipper-touch) on an absent object: The animal performs a visual search and then refuses to leave its station (Schusterman & Krieger, 1984). In this response to "absence," the behaviors of the marine mammals can be compared to the behaviors of chimpanzees (*Pan troglodytes*): Gardner and Gardner (1978) reported that their subjects used American Sign Language to comment upon the absence of a familiar object at a customary location, and Rumbaugh and Gill (1977) reported similar behavior in a chimpanzee trained to communicate through a computer-mediated system.

These projects demonstrate the attraction of studying species such as marine mammals whose cognitive capacities are generally not recognized. Each study also suggests that referential animal–human communication can be an important tool for assessing cognitive behavior. My own research illustrates further the importance of these points, as well as the value of an integrative approach. My work not only involves a nonmammalian subject – an African Grey parrot – and techniques of two-way, referential interspecies communication but also exemplifies the importance of integrating knowledge from the domains of psychology and ethology.

## Psittacine cognition: The study of an African Grey parrot

Over the past several years I have demonstrated that an African Grey parrot (*Psittacus erithacus*) named Alex possesses capacities once thought to be limited to humans and nonhuman primates (Premack, 1978). Specifically, Alex has used the sounds of English speech to identify, request, refuse, quantify, or comment upon more than 100 objects of various colors, shapes, and materials (Pepperberg, 1979, 1981, 1987a,b), has demonstrated a rudimentary capacity for categorization (Pepperberg, 1983, 1987c), and has displayed functional use of phrases such as "Come here," "No," "I want *X*," and "Wanna go *Y*," where *X* and *Y* are appropriate object or location labels (Pepperberg, 1981, 1988a,b,c, in press-b). This bird has also performed at the level of nonhuman primates (e.g., chimpanzees) on tests of object permanence (Pepperberg & Kozak, 1986).

Prior to my research, most studies on avian intelligence had been centered on pigeons (*Columba livia*), a species that has relatively little of the neural substrate now known to subserve avian intelligence (striatal areas) (Hodos, 1982; Stettner & Matyniak, 1968). The results of those studies suggested that pigeons (and, by extrapolation, other avian species), possibly because of their considerable anatomical, neurological, and behavioral differences from mammals, were unlikely to comprehend complex cognitive concepts (Premack, 1978). The scientific community had often ignored studies that demonstrated that certain other avian species (those with relatively large striatal areas, e.g., sturnids, corvids, and psittacids) readily learned complex behaviors comparable to those of primates on so-called intelligence tasks, such as insight detour problems and numerical discriminations (Gossette, 1967; Gossette, Gossette, & Riddell, 1966; Gramza, 1970; Kamil & Hunter, 1970; Koehler, 1943, 1950, 1953; Krushinskii, 1960; Lögler, 1959; Stettner & Matyniak, 1968). My findings provide further evidence that there is no a priori reason to reject birds as suitable subjects for cross-species examinations of cognitive abilities.

The two studies with Alex that I review in this chapter have been chosen because the results permit direct comparisons between and among various species, and some readers might be unaware of such research with an avian subject. I briefly discuss this parrot's ability to (1) identify quantity for collections of objects up to six (Pepperberg, 1987a) and (2) comprehend the relational concepts of "same" and "different" (Pepperberg, 1987c). For details of these experiments, the reader is referred to the original publications.

Because those studies used a system of vocal communication based on English, there exists the possibility that this special languagelike training has augmented Alex's cognitive abilities (Burdyn & Thomas, 1984; Premack, 1983). I consider the implications of these results for study of the relationship between language and cognition only briefly; see the Addendum at the end of this chapter (Pepperberg, 1987a–c, 1988c, in press-b; Pepperberg & Kozak, 1986). I emphasize, instead, how languagelike behaviors can be used to examine the *extent* to which an avian subject may possess mental capacities and comprehension comparable to those of humans and other primates. By studying the extent to which communicative and cognitive skills may be acquired and how such skills may function in a nonhuman, nonprimate, nonmammalian species, we may gain important insights into the mechanisms underlying cognitive and communicative capacities of all species. Such insight may help us formulate the appropriate questions for future comparative research on cognition and communication.

## *Methodology: A brief review*

Although I was not the first researcher to study avian–human communication, my students and I have been the first to achieve any measure of success. Previous researchers, well trained in the techniques of the psychology laboratory, but with little knowledge of avian behavior in the wild, had used operant conditioning techniques in their attempts to train referential, functional vocal communication in mimetic birds (e.g., Grosslight & Zaynor, 1967; Mowrer, 1950, 1954; review in Pepperberg, 1988b). Although the birds in those studies did learn to reproduce certain sounds, they were unable to attach meaning to their vocalizations. My students and I, however, by adapting various techniques to re-create, at least to some extent, the natural learning environment of our subject, have engendered referential, vocal learning in the laboratory.

Our experimental subject is an African Grey parrot, Alex, who was obtained from a pet store in the Chicago area in June 1977. At that time he was approximately 13 months old and had not yet received any formal vocal instruction (Pepperberg, 1981). To give this subject some control over his environment while maintaining experimental rigor, my students and I allow him free access, based on his vocal requests, to numerous areas in the laboratory room while we are present ~8 hours/day); in our absence he is confined to a cage (~62 × 62 × 73 cm) and the desk on which it rests. Water and a standard psittacine seed mix (sunflower seeds, dried corn, kibble, oats, etc.) are continuously available throughout the day; fresh fruits, vegetables, specialty nuts (cashews, pecans, almonds, walnuts), and toys are used in training and are provided at the bird's vocal requests.

*Training procedures: The model/rival approach, intrinsic rewards.* Our training program emphasizes the use of live, interacting tutors and intrinsic, rather than extrinsic, rewards (Pepperberg, 1981). Many previous programs designed to develop communicatory skills in both human and nonhuman subjects relied on noninteractive techniques and rewards that neither directly related to the skill being taught nor varied with respect to the task being targeted. Our procedures focus instead on demonstrating referential, contextual use of each targeted vocalization and on using as exemplars those objects and actions that themselves arouse the interest of the subject. These protocols provide the closest possible association of each object or action and the label to be learned. Because details of the training procedures for label acquisition and the rationale for their use have been described previously (Pepperberg, 1981, 1985, 1987b, 1988a,b), only a summary will be given here.

The primary technique, called the *model/rival* or M/R approach, employs humans to demonstrate to the parrot the types of targeted interactive responses. This procedure is based on a protocol developed by Todt (1975) for examining vocal learning in Grey parrots and on the social modeling theories of Mowrer (1950) and Bandura (1971, 1977). In the presence of the bird, one human acts as a trainer, and a second human acts as a trainee. The trainer presents objects, asks questions about these objects, gives praise and reward for correct answers, and shows disapproval for incorrect answers (errors similar to those being made by the bird at the time, e.g., "Wood" for "Green wood"). The trainee is both a *model* for the bird's responses and a *rival* for the trainer's attention. The roles of model/rival and trainer are frequently reversed to demonstrate the interactive nature of the system, and the parrot is given the opportunity to participate in these vocal exchanges.

At all times, Alex receives intrinsic rewards. At the start of the project, each correct identification was rewarded with the object that had been identified. Such a protocol demonstrated that referential nature of the communication code that was to be learned and was crucial for enabling the bird to acquire functional labeling ability (Greenfield, 1978; Pepperberg, 1978, 1981, 1983, 1988b). For the studies on numerical abilities and questions on same/different that involved collections of objects, the correct response was rewarded with *all* objects that were presented (e.g., four corks). So that Alex would work with objects in which he had little interest, we later modified the procedure to allow him to request alternative objects as his reward (Pepperberg, 1987a,b). He had already been trained to preface requests with the phrase "I want *X*" and to use object labels alone (e.g., "Blue wood") for identifications (Pepperberg, 1988a). A previous study (Pepperberg, 1987b, 1988a) showed that Alex usually (~75% of the time) did indeed want the objects he requested. Not only would he eat the walnut and use the key to scratch himself or the cork to clean his beak, but he would refuse substitutes, usually with the vocalization "No" and a repetition of the original request. Thus, during training and testing, the modified protocol, which separated requests from identifications, ensured that an inappropriate response was not a possible request for a preferred item: A vocalization such as "Cork" to a collection of keys was considered an error, whereas "I want cork" was taken as a valid (if interruptive) request. Only after Alex produced correct responses with respect to the targeted objects, and received and rejected them, were his requests for an alternative accommodated.

Although little has been published on natural psittacine behaviors, our procedures were consistent with the known constraints of both field and laboratory (Pepperberg, 1988b):

1. Communication among parrots in both wild and aviary settings appears to be primarily in the vocal mode (Busnel & Mebes, 1975; Dilger, 1960; Mebes, 1978; Power, 1966a,b; Serpell, 1981) and to be learned through social interaction (Nottebohm, 1970; Nottebohm & Nottebohm, 1969). Our training demands were thus closely related to naturally occurring tasks.
2. The conversational turn-taking we used in both our training and testing procedures resembles natural duetting behavior (Mebes, 1978; Wickler, 1976, 1980). Alex was therefore exposed to and required to participate in behaviors similar to those in which parrots engage in the wild; that is, his responses would be within his normal range of behaviors.
3. All training procedures involved the kinds of referential, contextually relevant situations that facilitate learning in both birds and mammals of behaviors that can otherwise be difficult to establish (Pepperberg, 1985, 1986).

*Test procedures.* A summary of the protocol for all tests is presented here; specific details can be found elsewhere (Pepperberg, 1981). Tests were designed to avoid trainer-induced cueing (Pepperberg, 1981). One precaution was to ensure that number trials were conducted by students who had never trained Alex on number labels and that same/different trials were conducted by students who had never trained same/different. A second precaution was a design that prevented both Alex and the principal trainer from predicting which questions or answers would appear on a given day. Tests were constructed as follows: On a previous day, all of the possible objects and topics to be tested would be listed by the principal trainer. A student not involved in testing would then set the question, form the pairs for same/different or collections for numerical questions, and randomly order all the questions, including those on topics other than number and same/different.[2] We thus would examine Alex's knowledge of numerical concepts or same/different by including, on each object identification test (tests of "What's this?") or during a training session on, for example, photograph recognition, one (or rarely two) number trials or questions in "What's same?" or "What's different?" Questions on either topic would be asked, on average, one to four times per week.

A student (secondary trainer) would present to the bird, in a variable but previously determined order, the objects to be identified. Alex was thus shown an exemplar or number of exemplars, asked "What's this?", "What color?", "What shape?", "How many?", "What's same?", or "What's different?" and was required to formulate a vocal English response from the 80 possible vocalizations in his repertoire (Pepperberg, 1987b). "What shape?" and "What color?" were used either for data collection of the type described earlier (Pepperberg, 1983) for categories or as a second question when Alex responded to "What's this?" by identifying correctly only the material of a colored or shaped object. Such "generic" answers were considered errors (Pepperberg, 1981). "How many?" was similarly used if he labeled only the material of a collection of objects (Pepperberg, 1987a).

The principal trainer was present; however, she sat in a corner of the room, did not look at the bird or the examiner during presentation of the test object(s), and did not know what was being presented. The principal trainer repeated out loud what she heard Alex say. (This repetition prevented the examiner from accepting an indistinct, incorrect vocalization that was similar to the expected, correct response, e.g., "Gree" for "Three.") If what the principal trainer heard was correct (e.g., the appropriate category label), Alex was rewarded by praise and the object(s). There were then no additional presentations of the same material during that test (i.e., there was only a single, first-trial response). If the identification was incorrect or indistinct, the examiner removed the object(s), turned her head (a momentary time-out), and emphatically said "No!" The examiner then implemented a correction procedure in that the misnamed object or collection was immediately re-presented until a correct identification was made; errors were recorded.[3] Alex thus found – and appeared to learn – that an incorrect identification (e.g., substitution of the name of a more desired object for the one presented) was fruitless; instead, a quick, correct identification allowed him to request the preferred item. Because immediate re-presentation of an object or collection of objects during a test occurred only when the response to the initial presentation was incorrect, the testing protocol penalized a "win-stay" strategy: Incorrect repetition of a previously correct response (e.g., the name of the previous exemplar) elicited no reward. The testing procedure thus provided a definite contrast to training protocols that rely on, and occasionally reinforce, repetitive behaviors. At all stages, the overall test score (results for "all trials") was obtained by dividing the total number of correct identifications (i.e., the predetermined number of objects or collections) by the total number of presentations required. First-trial results (percentage of first trials that were correct) are reported for comparison.

At the same time that we were conducting the studies on number concepts and same/different, we were training and testing additional labels (Pepperberg, 1987b,c, 1988a,b,c) training photograph recognition, and testing Alex on object permanence (Pepperberg & Kozak, 1986). Concurrent work on a variety of tasks is an important experimental protocol for this project: Alex becomes restless during sessions devoted to a single task. He ceases work, begins to preen, or interrupts with many successive requests for other items ("I want *X*") or changes of location ("Wanna go *Y*"). Similar "boredom" behavior has been observed in meerkats (*Suricata suricatta*) (Moran, Joch, & Sorenson, 1983), a raccoon (*Procyon lotor*) (Davis, 1984), chimpanzees (*Pan troglodytes*) (Putney, 1985), and rats (Davis & Bradford, 1986).

In addition to avoiding the boredom factor, intermingling different types of questions on tests or during training on other topics prevents "expectation cueing": In single-topic tests, contextual information (the homogeneous nature of questions that have a relatively restricted range of answers) could be responsible for a somewhat better performance than would otherwise be justified by a subject's actual knowledge of the topic. Alex, however, was never tested exclusively on questions of number or on same/different and, more important, was never tested successively in one session on similar questions ("What's same?") or questions that would have one particular correct response (e.g., "Three wood"). A question was repeated in a session only if his initial answer was incorrect (Pepperberg, 1981). Thus, even though the range of correct responses to questions of "What's same?" or "What's different?" was limited to three labels (color, shape, or matter), and responses on number to five labels, in any session Alex also had to choose from among many possible responses to other questions, such as "What's that?" or "What color?", in order to be correct (Pepperberg, 1983, 1987a,c).

## Results and a brief discussion of the importance of the findings for the comparative study of intelligence

### *Numerical abilities*[4]

*Background.* The ability to recognize and label different quantities of objects is a skill that all humans of normal intelligence eventually acquire, generally before the age of 8 years (Fuson, 1988). Although various studies have suggested that nonhumans have the capacity to process numerical information, the extent to which their abilities parallel those of humans is unclear (Davis & Memmott, 1982; Davis & Perusse, 1988; Pepperberg, 1987a). The continuing examination of numerical competencies across species, however, is an important aspect of comparative cognition, because some researchers (e.g., Gelman & Gallistel, 1986) treat the development of number concepts as a representative instance of the development of general cognitive capacities. The hypothesis is that advanced capacities that exist in one domain will exist in any other and can thus be used to index overall cognitive ability (e.g., Cheney & Seyfarth, 1986). Although the extent to which this hypothesis holds true remains to be fully tested (Rogoff, 1984), cognitive capacities are likely to extend beyond a training paradigm to a related situation. My initial investigation of numerical concepts was therefore designed as a test for the ability to generalize concepts of abstract categories: Could an African Grey parrot, previously trained to categorize objects on the basis of color and shape, learn to categorize collections of objects based on their quantity?

Results from other laboratories suggested that a Grey parrot would be a promising subject for a study on recognition and labeling of quantity. Koehler and his associates (Braun, 1952; Koehler, 1943, 1950; Lögler, 1959) had already performed systematic studies of numerical concepts in birds, demonstrating that ravens (*Corvus corax*), jackdaws (*Corvus monedula*), and Grey parrots were able to solve simultaneous match-to-sample problems for quantities up to and including eight. These birds would, for example, match quantities of disparate objects assembled in random configurations; that is, they would open a box lid that displayed eight dots if shown eight randomly sized pieces of plasticine as the sample. Koehler called this ability "thinking in unnamed numbers" or "non-numerical counting," because the birds could have made these distinctions on the basis of "same" versus "different" without necessarily counting in the human sense.

A detailed discussion of what operations constitute competence in counting is beyond the scope of this chapter. Analyses of the widely varying definitions can be found in Davis and Perusse (1988), Fuson (1988), Fuson and Hall (1983), Gelman and Gallistel (1986), Pepperberg (1988d), and Piaget (1941/1952a).[5] For the purposes of this chapter, it is sufficient to say that most researchers agree that the matching behavior of Koehler's birds is simpler and probably different from even the related task of verbal labeling of quantity. In humans, at least, the mechanisms involved in matching and labeling are likely to differ, because the two tasks appear to be mediated by separate areas of the brain (Geschwind, 1979). Given the complexity of these issues, my students and I were thus unlikely to determine whether or not an animal could "count" in the human sense. But because Alex had already demonstrated the capacity to acquire and use English vocalizations to identify, request, refuse, and categorize various objects, we could investigate this parrot's capacity to use vocal numerical labels to identify the quantities of collections of objects.

*Numerical concepts of an African Grey parrot: Comparisons with other animals.* The study was designed to answer the folllowing series of questions, each of which involves a more advanced stage of numerical competence:

1. Could a parrot use vocal numerical labels to distinguish cardinal sets simply as an additional attribute of the collection, much like color or shape? (See Matsuzawa, 1985, for comparable work with a chimpanzee.)
2. If the parrot succeeded in that task, could it then generalize that behavior to sets of novel objects, or had it merely learned to respond to a few particular groupings of familiar objects?
3. If successful, could the parrot then, without further training, respond correctly to objects placed in random arrays? That is, had the parrot merely associated each number label with a few particular display patterns, or had it abstracted a sense of quantity?
4. Could the parrot also abstract information about quantity in the face of irrelevant information, such as for subsets of heterogeneous collections? Demonstration of such abilities would not definitively answer the question whether or not the animal could count, but positive results would suggest that the parrot had acquired some concept of number (Davis, Albert, & Barron, 1985) that could be used to measure its abilities against those of other subjects.

The data, summarized here (see Pepperberg, 1987a, for details), show that Alex can use numerical labels to distinguish quantities of objects. He learned the vocal labels "Two" through "Six" and responded to questions about two to six keys, wooden sticks, clothespins corks, and pieces of paper with the labels for both the quantity and the objects (e.g., "Two cork") with an accuracy of 78.9% (183 correct responses to 232 questions). Our criterion for accuracy was strict: A correct response had to include correct identification of the material (e.g., wood vs. paper), which meant that Alex was also considered wrong even if he correctly identified the quantity but mislabeled the material. Moreover, approximately 60% of his errors were what we term "generic" – correct identification of the exemplar itself ("Key"), but failure to include the marker for quantity. Although generic errors were not incorrect in the sense of misidentification, such responses were counted wrong because they did not provide all of the requested information. His scores would have been higher had we evaluated only the numerical parts of his responses (i.e., 81.7%) or omitted his generic errors (91.7%).

We do not know the mechanisms that Alex used to produce the correct responses (see Pepperberg, 1987a, for discussion). It is, however, unlikely that he was responding to cues such as surface area, because the sizes and shapes of the exemplars (the pieces of paper, the collections of corks) varied considerably from trial to trial. Only the metal keys remained constant in size, and Alex performed no better with those objects than with any other. We used no food items, so that intensity of odor could not be a factor, and the various types of exemplars employed precluded factors such as brightness.

Because the capacity to transfer concepts from the learning situation to novel circumstances often indicates advanced cognitive abilities (e.g., Rozin, 1976), we needed to examine Alex's responses to two types of transfer trials:

1. how he performed when asked to produce the correct pairing of number and object labels on first presentation of familiar objects presented in novel multiple sets (e.g., the first presentation of two corks after training on two keys and two woods), and
2. his ability to label shapes he had never before seen, based (presumably) on their number of corners, such as his first response to two-corner wood (football shape) after seeing only "two wood."

He made no errors on these first-trial transfers on material labels and, with the exception of a few generic responses, immediately transferred use of quantity labels from training to testing exemplars (novel shapes as well as collections). The procedures employed in this first series of trials (Pepperberg, 1987a) were comparable to those employed for the chimpanzee Ai (Matsuzawa, 1985; Matsuzawa, Asano, Kubota, & Murofushi, 1986); note, however, that, unlike Alex, Ai did not immediately generalize to novel collections of familiar objects (see also Boysen & Berntson, P&G16).

The results, although interesting, did not entirely rule out the possibility that Alex was responding on some basis of familiarity. Although the arrangements of the particular objects in the particular arrays were novel (e.g., two corks), both the objects (corks) and the arrays themselves (as "two wood") were independently familiar. Thus, after Alex had acquired proficiency in labeling sets containing two to six familiar objects, we administered a more rigorous transfer test as part of the set of experiments described next: In contrast to Matsuzawa's questions to Ai (Matsuzawa, 1985; Matsuzawa et al., 1986), we queried Alex about various quantities for collections of entirely novel objects, and any quantity could be tested on any trial.

A separate but related question was whether or not Alex might be responding on the basis of similarity to either regular polygons or approximately linear patterns (i.e., to a set of separate perceptual units) rather than sensitivity to amount (e.g., "subitizing" or "clumping," as discussed later). Although trainers and examiners did vary the pattern of presentation (orientation and angular separation), and no two humans held the objects in precisely the same manner (see Matsuzawa, 1985, for similar hand-held presentations to chimpanzees), the number of ways to present six corks by hand is limited. For the shaped objects, size or surface area could not have been a cue, as the surface areas of the different batches of wood, paper, and rawhide shapes differed by about 10%, and exemplars from different batches were deliberately intermingled. Transfer between quantities of objects and shaped items (e.g., between five woods and a five-corner hide, or two keys and two-corner paper) suggested that this pattern recognition was not initially based on a single perceptual unit, but once the pentagonal shape had been related to the number 5, and the (two-corner) "football" to the number 2, subsequent shape recognition could have been by form. Alex's previously correct untutored identification[6] of shaped wood, paper, and rawhide objects could only suggest, but not prove, that he had transferred a concept of quantity from a collection of objects to a collection of corners arranged in particular

Figure 18.1. Objects presented to Alex in a random array. Six woolen pompoms, for example, were thrown onto the tray, and Alex was asked "What's this?" (from Pepperberg, 1987a).

patterns. Mandler and Shebo (1982), for example, found that canonical patterning (e.g., use of a triangular array for three, a square for four) simplified recognition of quantity for numerosities greater than three (von Glasersfeld, 1982). The next step, then, would be to learn if Alex could respond correctly, *without* further training, to subsequent collections of objects that would not allow him to employ mechanisms based on overall familiarity of the display or recognition of an oft-repeated pattern.

We therefore performed the following tests. We tested Alex on entirely novel collections whose objects he could not label. Objects were placed on a tray in linear arrays, but the spacing between objects in and among sets varied, so that the length of an array could not be a cue. We also performed additional trials with objects presented in random arrays. In these trials, exemplars were all somewhat familiar and were simply tossed onto the surface of the tray (Figure 18.1). Only if an object was rendered invisible by the placement of other objects was the arrangement altered in any way.

Alex correctly produced numerical labels in response to our queries for the two sets of tasks (Tables 18.1 and 18.2). His accuracy was 77–80%; if only first presentations are counted, his score was 75%.[7] Alex could therefore proficiently identify quantities of entirely new objects that varied considerably in size and shape from his original training exemplars or that were presented in random arrays; he had not simply learned to identify sets of familiar exemplars in familiar patterns. Because all but one set of transfer objects in the first set of tasks were entirely novel and all the results were

Table 18.1. *Responses to presentation of quantities of entirely novel objects*

| Object | Quantity | Score | Erroneous ID |
|---|---|---|---|
| Beads | 3 | + | |
| Antacid tablets | 5 | + | |
| Candies | 6 | + | |
| Candies | 2 | + | |
| Small bottles | 4 | + | |
| Small bottles | 2 | + | |
| Raisins | 5 | − | 6[a] |
| Beads | 6 | +[b] | |
| Antacid tablets | 4 | + | |
| Raisins | 3 | − | 2[a] |
| Toy cars | 4 | + | |
| Jelly worms[c] | 2 | + | |
| Spools | 5 | + | |
| Erasers | 3 | + | |
| Plastic hairclips | 6 | + | |
| Toy cars | 2 | − | key[a] |
| Erasers | 4 | − | box[a] |
| Spools | 6 | + | |
| Jelly worms | 5 | + | |
| Plastic hairclips | 3 | − | 4[a] |

*Notes:* Only one trial was administered per session; the order is that of the trials across sessions. Overall, Alex was correct 20/25 = 80%; for first trials only, his score was 15/20 = 75%. (Binomial test, $p < .0001$. Assume $p = 1/6$: 5 possible number labels + attempt at a material label. Note that this does not take into account the fact that the bird could produce any of the other labels in his repertoire.) Plus = correct response; minus = error.
[a] Alex was correct on second presentation.
[b] Alex responding "Six wood." The beads were wooden, and he had chewed one during the earlier trial.
[c] A gelatin-based candy shaped into wormlike form.
From Pepperberg (1987a).

based on Trial 1 transfers (Gardner & Gardner, 1978), these tests were significantly more stringent than those used in Matsuzawa's study (1985; Matsuzawa et al., 1986) on labeling of quantity by chimpanzees, as discussed later. Although this experiment did not enable us to determine the mechanisms Alex used, it appeared to eliminate simple pattern recognition in most cases (see Pepperberg, 1987a, for a detailed discussion).

*Numerical concepts in an African Grey parrot: Comparisons with children.* Another indication that Alex might have a concept of quantity, rather than just the capacity to form paired associations with particular exemplars, would be his ability not only to perform the labeling task with novel or randomly placed exemplars but also, *without* further training, to respond to questions concerning quantity for a particular targeted subset of objects within a *hetero-*

Table 18.2. *Responses to presentation of objects placed in random arrays*

| Object | Quantity | Score | Erroneous ID |
|---|---|---|---|
| Candies | 6 | +[a] | |
| Clips | 4 | + | |
| Pencils | 2 | +[b] | |
| Woolen pompoms | 6 | + | |
| Thimbles | 3 | – | 2[c] |
| Wooden cubes | 5 | + | |
| Pasta | 2 | + | |
| Beads | 4 | + | |
| Pennies | 5 | + | |
| Wood | 3 | – | 4[c] |
| Cork | 4 | + | |
| Scrapers[d] | 2 | + | |
| Pasta | 6 | + | |
| Chalk | 3 | + | |
| Paper[e] | 5 | – | 4,6 |
| Cork[e] | 6 | + | |
| Shower rings | 4 | + | |
| Rocks[e] | 2 | – | rocks[c] |
| Keys | 5 | + | |
| Hides[e] | 3 | – | 4-corner hide[f,c] |

*Notes:* Only one trial was administered per session; the order is that of the trials across sessions. For all trials, Alex was correct 20/26 = 77%; for first trials, Alex was correct 15/20 = 75% (Binomial test, $p < .0001$. Assume $p = 1/6$: 5 possible number labels + attempt at a material label. Note that this does not take into account all the possible vocalizations in his repertoire.) Plus = correct response; minus = error.

[a] In this one instance the candies were placed in two rows of three, but this was a pattern Alex had not previously experienced.

[b] The pencils crossed one another to form an X-like shape.

[c] Alex was correct on second presentation.

[d] Nail files.

[e] These objects were all of different sizes.

[f] One piece of rawhide was squarish.

From Pepperberg (1987a).

*geneous* display. Could he now, for example, answer queries about "How many wood?" for a collection of keys and wood? The overall pattern would then be irrelevant; to be correct, he would have to observe a multiobject array, decode the question to determine which objects were being targeted, and then produce vocally the appropriate label for the quantity. Note that Piaget (Inhelder & Piaget, 1958) suggested that children as old as 5–6 years might still have trouble quantifying groups versus subgroups: Given a collection of wooden beads of two colors, such children respond incorrectly to questions such as "Are there more wooden or brown beads?"

A previous study (Pepperberg, 1983) had demonstrated that Alex could

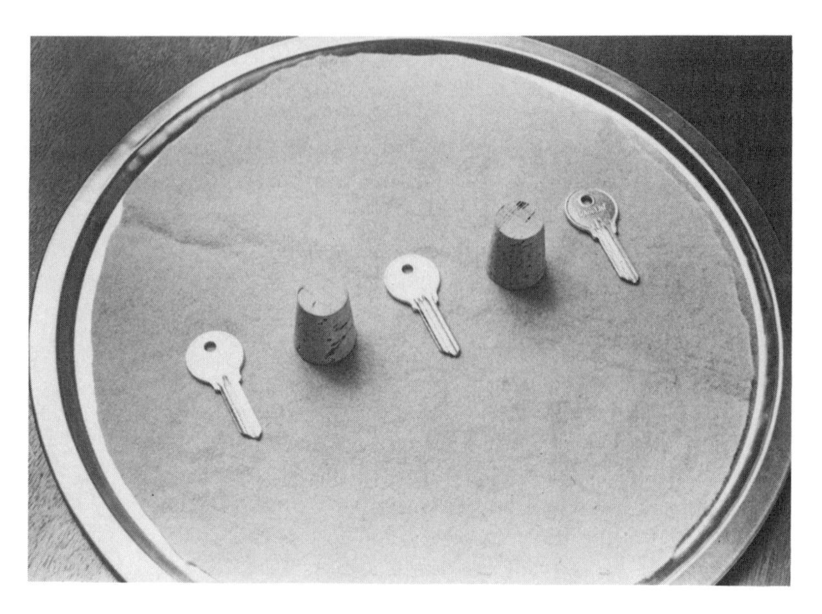

Figure 18.2. Heterogeneous sets of objects presented to Alex in an intermingled pattern (from Pepperberg, 1987a).

decode and reply vocally to queries about a subset of the total information available (e.g., the color *or* the shape of exemplars that simultaneously incorporated both variables). Success on a task involving subsets of objects seemed compatible with these earlier skills. Alex was thus presented with collections composed of two sets of objects and queried "How many *X*?", where *X* was the label for one of the two sets (e.g., $X$ = key). One of the collections contained nails (which he could not label) as the untargeted subset so that we could learn if "unlabelable" objects would be treated as part of the targeted subset. Because it was also possible that the set he could not label would be ignored, thus simplifying the task, only one such set was included in this series. (Additional tests on objects that could not be labeled are discussed later.) The objects were presented on the tray in two types of linear arrays. Objects of the two sets were either placed side by side in separate groups ("contiguous array") or intermingled as randomly as possible (Figure 18.2), but at no time were objects of the same set clustered: Distances between groups were never different from the distances between objects.

The nature of Alex's responses was of particular interest because he had never before been queried about heterogeneous sets. At no point had he ever been presented with a collection that could have represented anything resembling a choice of objects about which to comment (Pepperberg & Kozak, 1986). Nevertheless, in 7 of 10 first presentations (Table 18.3), Alex responded with the appropriate numerical label for the targeted set ($p$ = .0031, chance is conservatively 25%). Thus, he appeared to have some ability to look at a collection of items, target the appropriate set based on the ques-

Table 18.3. *Data on trials concerning heterogeneous groupings of objects*

| Object set | Array type | Targeted subset | Response | Correct |
|---|---|---|---|---|
| 2 wood, 3 key | C | keys | wood | − |
| | C | wood | 2 | + |
| | C | keys | key | − |
| | C | keys | 2 wood | − |
| | C | keys | 5 | − |
| | C | keys | 3 | + |
| 2 rock, 3 pennies | C | rock | 2 | + |
| 2 cork, 3 key | I | key | 3 | + |
| 2 wood, 3 clips[a] | I | wood | 2 | + |
| 2 nails, 4 jacks | C | jacks | 4 | + |
| 2 rose wool, 4 yellow wool | C | yellow wool | 4 | + |
| 2 cork, 4 pennies | I | cork | 2 | + |
| 2 key, 3 rock | I | rock | rocks | − |
| | I | rock | 3 | + |
| 2 wool, 5 wood[b] | C | wood | wool | − |
| | C | wood | 5 | + |

*Notes:* Alex was asked to produce the label for the number of the targeted subset. A complete transcript of Trial 1 is included. C refers to a contiguous grouping of objects in a subset, I to a randomly interspersed array. On initial questions, Alex was correct 7/10 = 70%, $p$ = .0031 (assume $p = 1/4$; most stringent case; assume no object labels, because those were given, and four possible number labels, because 6 was never involved, even though Alex could use it, as well as all his color and object labels); on all questions, 10/16 = 63%, $p$ = .0014. (Multiple entries denote multiple successive queries for that collection.)

[a] Clips are called "scraper."

[b] This was performed as a probe, to see if Alex would have trouble if the total number in the collection was more than he could label.

From Pepperberg (1987a).

tion of the trainer, and respond with the correct label for the quantity it represented, even when the sets of objects were interspersed.

It was also important to learn whether or not Alex had the capacity to produce the label representing the *overall* number of objects in a heterogeneous set. In experiments that initially focused exclusively on homogeneous sets, Gast (1957) found that children (ages 3–4) did not spontaneously transfer to collections containing two sets of objects, but would respond with only the numerical label for the larger set. Premack (1976) found similar limitations with chimpanzees. Gelman (1980) and Gelman and Gallistel (1986), however, found that children as young as $2\frac{1}{2}$ years had no problem with heterogeneous sets, and they suggested that the earlier findings were consequences of the chosen training procedures (cf. Siegel, 1982). Gast (1957) and Klahr and Wallace (1973) suggested that children might need a comprehensive categorical representation for all members of a heterogeneous set before they can judge the numerosity of the entire set (e.g., need a concept of the category "toy animals" in order to quantify a set of toy cats and dogs) (Greeno, Riley, & Gelman, 1984; Inhelder & Piaget, 1958;

Steffe, von Glasersfeld, Richards, & Cobb, 1983; von Glasersfeld, 1981a,b).

We did not know if Alex would need a comprehensive category label before he could label the quantity of a heterogeneous collection. We also did not know if he had general classifications for his various exemplars, as, for example, "food" versus "toy": Although he had demonstrated a concept of categorical class with respect to color and shape (Pepperberg, 1983) and had begun to comprehend a class based on material, he had had no formal training on labels for other categories. His only relevant experience was hearing students make comments such as "We're almost out of food" or "Let's work with the some toys." As we did not want to investigate his comprehension of such categories at the time, we developed the following protocol.

Trials consisted of presentation of different quantities of two different groups of objects. Each presentation included one subset of objects that Alex could not label. He was queried as to the quantity of objects in this group (e.g., "washers"). Thus, the questions could equally well be interpreted as "How many (specific) $X$?" or "How many (total) $X + Y$?" If Alex had been predisposed by the previous experiment to subdivide the collection, or if he were dividing the elements into two groups like Gast's subjects, then his response would have been the quantity for the objects he could not label: He had already shown that he would respond selectively to questions on heterogeneous sets of objects he *could* label, and the set of such objects was excluded by our questions. If, however, he were viewing the set as a collection, he would have responded with the label for the total quantity. Only five trials were presented, because more trials with repeated use of the word "toy" and the differences in our vocal queries (e.g., "washer" vs. "thimble") might have cued him that we were referring to the sets of unusual objects rather than to the collection as a whole.

A transcript of the interactions between the examiner and Alex for these five trials is reported in Table 18.4. When Alex was asked about the number of the objects that he could not label, he responded with the label for the overall quantity. Thus, despite the small number of trials, the data suggest that Alex, even with his training on homogeneous sets, was responding to the total number of objects in the heterogeneous collection. Only the results of the last trial suggest that he might have begun to respond with the quantity for the unfamiliar label.

The preceding experiments provide data for cross-species comparisons of numerical capacities (Matsuzawa, 1985; Matsuzawa et al., 1986; Boysen & Berntson, P&G16). The findings show that Alex can

1. respond appropriately to queries about different collections of objects with a vocalization chosen from a large repertoire (labels for colors, shapes, exemplar materials, categories, and requests),
2. generalize these labels to entirely novel sets and randomized patterns, and
3. demonstrate some facility with respect to heterogeneous collections.

The ability to label a quantity presented in these ways suggests that he has some ability both to conceptualize quantity and to code it abstractly. Whether

Table 18.4. *Partial transcript (responses of PI omitted) concerning heterogeneous groups of objects, some of which Alex cannot label*

1. 3 wood, 4 metal washers (popsicle sticks, contiguous groupings)
   *T:* How many washers?
   *A:* (Grabs a piece of wood and tears it apart. It is not replaced.)
   *T:* Alex, how many washers?
   *A:* Sih (six).
   *T:* That's the number of toys. How many washers?
   *A:* Sih, sih.
   *T:* That's how many *toys*. How many washers?
   *A:* Four (Alex takes washer momentarily, drops it, and requests pasta.)
2. 2 thimbles, 4 wood (plant stakes, contiguous groupings)
   *T:* Alex, look, how many wood?
   *A:* Four. (Alex is given a stake, which he chews. It is replaced, and the tray re-presented.)
   *T:* Alex, how many thimbles?
   *A:* Sih.
   *T:* That's how many *toys*. How many metal toys? ("Metal" is a label that he has not acquired, but on which he is being trained.)
   *A:* Two. (He is given a thimble, which he manipulates in his beak.)
3. 3 jacks, 2 bobbins (contiguous)
   *T:* Alex, how many bobbins?
   *A:* Five.
   *T:* Well, that's the number of toys. (Goes on to work with photos.)
4. 4 wood, 2 hearts (rubber heart-shaped 1-in. magnets, contiguous)
   *T:* Alex, how many hearts?
   *A:* Sih.
   *T:* That's the number of toys. How many hearts?
   *A:* Two.
   *T:* You're right. (Gives heart, which Alex immediately drops.)
5. 4 wool, 2 plastic hairclips (interspersed)
   *T:* How many plastics?
   *A:* Two.
   *T:* That's right, two. (Gives plastic pieces. Alex chews one apart.)

*Note*: *T*, trainer; *A*, Alex.
From Pepperberg (1987a).

or not Alex can ever be shown to "count" in the human sense is still unclear, but the experiments described here present significant evidence for certain, albeit limited, number-related abilities in a Grey parrot.

## *Concepts of same and different*[8]

There is significant debate concerning exactly what has been demonstrated in the various animal studies purporting to examine the concepts of same/different; see Premack (1983) and the commentaries therein, as well as Edwards, Jagielo, and Zentall (1983). Comprehension of the abstract relational concept of same/different is one of the cognitive capacities for which researchers continue to argue that species-specific differences exist (Premack, 1978, 1983); see the discussions in Pepperberg (1987c, 1988c, in press-b). Such relational concepts are particularly important in the study of relative intelligence, because they likely draw on more sophisticated capacities than do those involved in, for example, match-to-sample or oddity-from-sample (Case, 1985). Piaget (Inhelder & Piaget, 1958), for example, considered such ability for abstraction and relational concepts to mark the point at which children (generally around age 7) begin to reason at an adult level (see discussion in Premack, 1976).

According to Premack, comprehending same/different is more complex than comprehending match-to-sample and oddity-from-sample, because comprehension of same/different, unlike the other tasks, involves the ability to use arbitrary symbols to represent the relationships of sameness and difference between sets of objects. In generalized match-to-sample or oddity-from-sample, the subject demonstrates its understanding of the concept simply by showing a savings in the number of trials needed to respond to *B* and *B* as a "match" after learning to respond to *A* and *A* as a "match" (and, likewise, by showing a savings in trials involving *C* and *D* after learning to respond to *A* and *B* as "nonmatching"). In contrast, according to Premack, demonstrating comprehension of "same" requires that a subject recognize not only that two independent objects, $A_1$ and $A_2$, are blue but also that only a single attribute, the category color, is shared. Furthermore, the subject must realize that this attribute, or sameness, can be *immediately* extrapolated and *symbolically* represented not only for any two other blue items but also for two novel green items, $B_1$ and $B_2$, that have nothing in common with the original set of *A*'s. Likewise, the subject must demonstrate a concept of difference that can be extrapolated to two entirely novel objects. In other words, the subject is not just determining what constitutes categories (i.e., how the specific instances of categories are related), but how the relations between the instances of the categories are themselves related. Premack (1978, 1983) proposed that such abilities are likely to be limited to primates and, because of the requirement for symbolization, are most readily demonstrated by nonhuman primates that have undergone some form of language training (note also Inhelder & Piaget, 1958).

The results of early studies appeared to support Premack's claim. For example, some pigeons (*Columba livia*) trained on match-to-sample or oddity-from-sample tasks acquired a concept of "same" but not of "different," as discussed by Edwards et al. (1983) and Zentall, Edwards, Moore, and Hogan (1981). Later studies (e.g., Edwards et al., 1983; Zentall et al., 1981; Zentall, Hogan, & Edwards, 1984), however, have shown that pigeons can use symbols to demonstrate some degree of same/different concept transfer. Although the pigeons' performances were influenced by stimulus-specific associations, the birds did respond symbolically and needed fewer sessions to respond correctly to novel instances of same and different when the task remained the same than when it was reversed. Additional research (Santiago & Wright,

1984; Wright, Santiago, & Sands, 1984; Wright, Santiago, Urcuioli, & Sands, 1984) demonstrated that both monkeys (*Macaca mulatta*) and pigeons could use two-choice symbolic responses in same/different concept learning; the pigeons, however, showed a much lower level of transfer to novel items than did the monkeys. These experiments therefore suggested that an avian subject cannot use symbols for same and different in a manner fully comparable to that of language-trained chimpanzees or humans, or even appropriately trained monkeys.

Few other studies have investigated the concepts of same/different in avian subjects, even though categorization based on similarity or difference would seem to correlate with natural behaviors of individual recognition and vocal dueling and song matching (e.g., Beecher, Stoddard, & Loesche, 1985; Falls, 1985; Falls, Krebs, & McGregor, 1982; Kroodsma, 1979). Some birds can, for example, classify novel series of tones as "same as" or "different from" ascending or descending reference series, but only for sequences that lay within the range of frequencies used in training: starlings (*Sturnus vulgaris*), cowbirds (*Molothrus ater*), and mockingbirds (*Mimus polyglottos*) (Hulse & Cynx, 1985). In laboratory studies on the ability of budgerigars (*Melopsittacus undulatus*) to discriminate similarities and differences in calls of canaries (*Serinus canarius*), the budgerigars appeared to learn the unique characteristics of each individual canary call (Park & Dooling, 1985). Similar results were found in studies on song discriminating with cliff swallows (*Hirundo pyrrhonota*) and barn swallows (*Hirundo rustica*) (P. K. Stoddard & M. D. Beecher, pers. commun., 1986). Great tits (*Parus major*) can recognize test songs as similar to or different from one particular training song (Shy, McGregor, & Krebs, 1986) in a manner comparable to visual categorization by pigeons (e.g., Herrnstein, 1984), but none of these studies has shown actual labeling of the relation of sameness or difference, as, for example, demonstrated by transfer to entirely different exemplars (e.g., labeling which *aspects* are the same or different for calls or songs of other species). The study reported here was therefore undertaken to see if an avian subject, the aforementioned Alex, could use vocal labels to demonstrate symbolic comprehension of the concepts of same and different.

For the reasons discussed earlier, the task designed for Alex would have to be functionally equivalent to that used with Premack's chimpanzees. The task would have to ensure that

1. the symbolic concepts tested would be more abstract than those examined in standard conditional discrimination paradigms,
2. the subject would be given explicit, equal training on both the concepts of same and different,
3. the findings could not be dismissed as stimulus-specific associations, and
4. first-trial transfer test results could be examined for their significance.

To take into account these constraints and the history of the experimental subject with respect to use of vocal labels, I designed the following task. Alex was to be presented with pairs of objects that could differ with respect to three categories: color, shape, and material (e.g., a yellow rawhide pentagon and a gray wooden pentagon, a blue wooden square and a blue paper square). He would then be queried "What's same?" or "What's different?" The correct response would be the label of the appropriate *category*, not the specific color, shape, or material marker that represented the correct response (e.g., "color," not "yellow"). To be correct, therefore, Alex would have to

1. attend to multiple aspects of two different objects,
2. determine, from a vocal question, whether the response was to be on the basis of sameness or difference,
3. determine, based on the exemplars, what was the same or what was different (e.g., Were they both blue, or square, or made of wood?), and
4. produce, vocally, the label for this particular category.

Thus, the task required, at some level, that Alex perform a feature analysis of the two objects; the responses could not be made on the basis of total physical similarity or difference (Premack, 1983).

In comparison, most research on the same/different concept in animals uses

1. a two-choice design whereby the subject merely indicates whether pairs do or do not match,
2. topographically similar (and thus possibly easier to acquire) responses for both answers (e.g., lever pressing or key pecking) (Michael, Whitley, & Hesse, 1983), and
3. "same" pairs that are identical in all dimensions and "different" pairs that are different with respect to most, if not all, dimensions.

An equivalent task would be to ask Alex to view two objects and merely respond "Same" or "Different," the latter response occurring to anything that was different. Alex's task, however, would be considerably more difficult: He would have to respond with the vocal label for one of three dimensions that was the same or different for each pair, depending on the question he was asked. Even chimpanzees respond "Different" less reliably for pairs of objects differing on only one of three dimensions than for pairs differing on two dimensions (McClure & Helland, 1979).

Alex's responses on tests were unlikely to be based on absolute physical properties or on learning the answer to a given pair, because the number of possible permutations of question topic, correct response, and combination of exemplar attributes was very large.[9] Moreover, because Alex's response would be a *category* label rather than a specific object or attribute label, we could test Alex on novel objects whose labels were unknown. In sum, Alex would have to be able to transfer between like and unlike pairs of colors, like and unlike pairs of shapes, and like and unlike pairs of materials, all of which would vary from the training exemplars. That is, he would have to demonstrate transfer among stimulus domains as well as among various instances of each domain; see Premack (1976, pp. 354–355) for the importance of such transfers in determining that a behavior is not just stimulus generalization.

We also planned to determine if Alex was responding to the content of the questions (i.e., differentially processing "What's different?" vs. "What's same?"), not just to variations in the physical characteristics of the objects. It was possible that Alex, just by looking at the objects, might determine the *one* attribute that was the same or different, and simply respond on that basis. Thus, at random intervals, probes were to be administered in which he would be asked questions for which either of *two* category labels could be the correct response (i.e., he would be shown a yellow and a blue wooden triangle and asked "What's same?"). If he were ignoring the content of the question and answering on the basis of the attributes and his prior training, he would respond with the one wrong answer; if he were answering the question posed, he would have two possible correct responses. Note, too, that having two possible correct answers provided additional protection against expectation cueing. In sum, after training, Alex was to be given a series of questions involving, in random order, pairs of objects that were familiar but not used in training, pairs in which one or both exemplars were novel objects, and probes to determine if he were indeed processing the content of the questions.

The results of these experiments not only would provide information on the parrot's concepts of same and different but also would provide additional evidence for comprehension of concepts of categories, that is, the ability to recognize that two novel objects (e.g., pink and brown paper triangles) differed with respect to the category color even though the colors were untrained and the specific combination of color, shape, and material for each exemplar, as well as the combination for the pair, had never before been seen on a test. A correct response would suggest that Alex was not reacting to specific instances of color, shape, and material, but rather that he possessed an understanding of the categorical concepts themselves.

Alex's scores for tests on objects that were familiar but not used in training were 99/129 = 76.6% for all trials and 69/99 = 69.7% on first-trial performance ($p < .0001$ on binomial test, with a chance value of 1/3) (Figure 18.3). Note that the choice of 1/3 was conservative, in that it ignored the possibility that Alex could have said any number of vocalizations besides "Color," "Shape," and "Matter." In all cases, the first single word that Alex uttered was one of these category labels, although he produced other phrases that encoded requests for other objects or actions (e.g., "I want *X*") (Pepperberg, 1987a, 1988a). His scores for pairs consisting of objects that were no longer novel (additional presentations of exemplars previously presented as novel), but that contained a color, shape, or material he could not label (e.g., plastic), were 13/17 = 76.5% for all trials ($p = .0003$) and 10/13 = 76.9% for first trials ($p = .0014$).

Interestingly, Alex's scores on transfer tests containing novel objects were somewhat higher than his scores for familiar objects (i.e., 96/113 = 85% on all trials, 79/96 = 82.3% on first-trial performance, $p < .0001$, with chance value again of 1/3) (Figure 18.4). His scores for pairs containing one versus two totally novel objects (respectively 86% and 83% for first trials, $p <$

**WHAT'S SAME?**

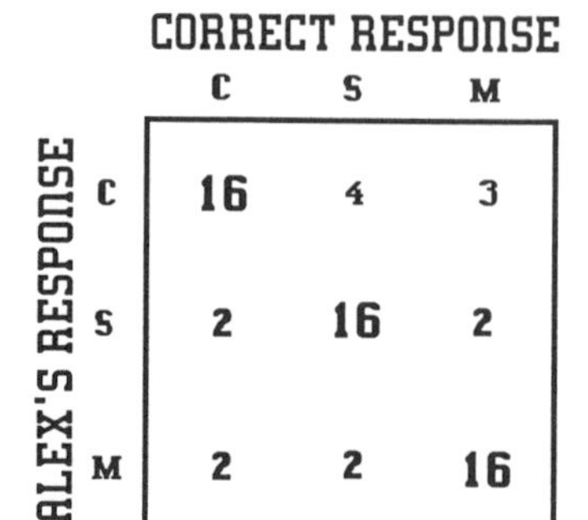

CORRECT RESPONSE

| ALEX'S RESPONSE | C | S | M |
|---|---|---|---|
| C | 16 | 4 | 3 |
| S | 2 | 16 | 2 |
| M | 2 | 2 | 16 |

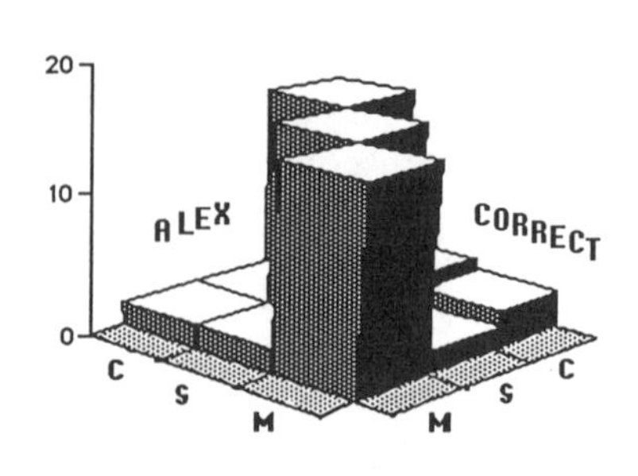


**WHAT'S DIFFERENT?**

CORRECT RESPONSE

| ALEX'S RESPONSE | C | S | M |
|---|---|---|---|
| C | 17 | 3 | 3 |
| S | 2 | 16 | 1 |
| M | 2 | 4 | 18 |

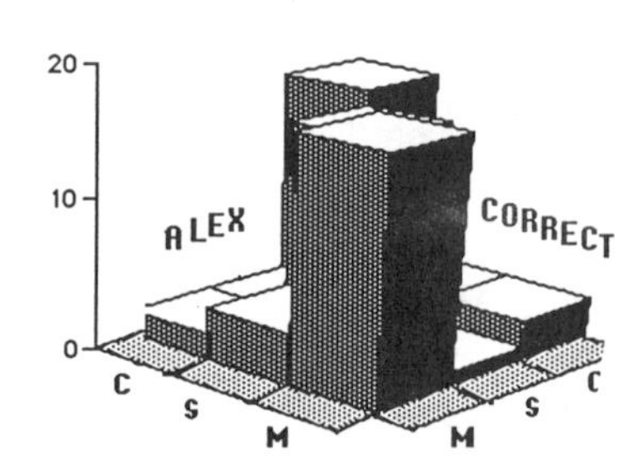


Figure 18.3. Alex's scores on familiar objects: C, color; S, shape; M, mah-mah (matter). From Pepperberg (1987c), in *Animal Learning & Behavior*, *15*, 423–432; reprinted by permission of Psychonomic Society, Inc. See also Pepperberg (in press-b).

.0001) differed little. Based on the test for differences in proportions at the .05 confidence level, however, his scores were not significantly better for questions involving all novel exemplars versus all familiar ones, when the results are examined for all trials, and were only just significantly better when the results are examined for first-trial performance only. Usually, subjects perform less well on transfer tests, but the results here were not surprising. Remember that Alex received the objects themselves as his primary reward and that there was therefore some inherent incentive to pay closer attention both to the objects and to the response when these reward objects were new items that were potentially interesting to chew apart, to try to eat, or to use for preening.

The results of the probes show that Alex was indeed processing the questions, as well as responding to the physical properties of the objects. His scores were 55/61 = 90.2% on all trials (binomial test, $p = .0001$, chance of 2/3) and 49/55 = 89.1% on first-trial performance ($p < .00001$) (Figure 18.5).

**WHAT'S SAME?**

| ALEX'S RESPONSE \ CORRECT RESPONSE | C | S | M |
|---|---|---|---|
| C | 17 | 3 | 3 |
| S | 1 | 16 | 0 |
| M | 2 | 1 | 17 |

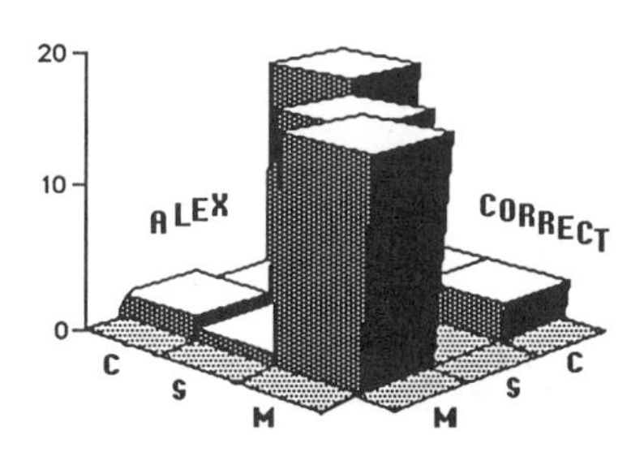


**WHAT'S DIFFERENT?**

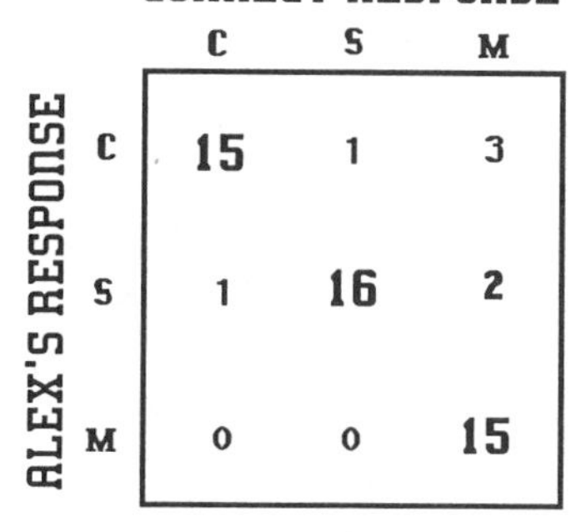

| ALEX'S RESPONSE \ CORRECT RESPONSE | C | S | M |
|---|---|---|---|
| C | 15 | 1 | 3 |
| S | 1 | 16 | 2 |
| M | 0 | 0 | 15 |

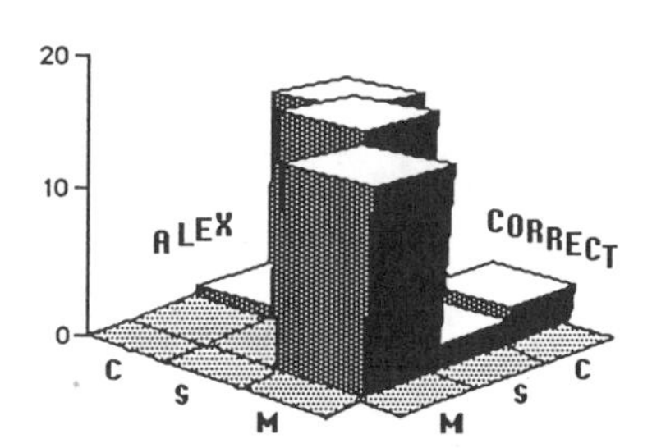


Figure 18.4. Alex's scores on novel objects: C, color; S, shape; M, mah-mah (matter). From Pepperberg (1987c), in *Animal Learning & Behavior*, *15*, 423–432; reprinted by permission of Psychonomic Society, Inc. See also Pepperberg (in press-b).

The data indicate that Alex showed symbolic comprehension of the concepts of same/different on a task comparable to that performed by primates. The test conditions, although not identical with those used by Premack (1976, 1983), were as rigorous as those used in his initial study with chimpanzees. Like the chimpanzees, Alex was presented the object pairs – including objects never before tested – simultaneously, rather than successively, so that, unlike animals that are trained on successive match-to-sample or oddity-from-sample, he was unlikely to be responding on the basis of "old" versus "new" or "familiar" versus "unfamiliar" (Premack, 1983, p. 127). Unlike the chimpanzees, Alex's task was one of symbolic *comprehension* of "same" and "different" (i.e., responding *to* questions of "What's same?" and "What's different?"); the task most commonly employed with the language-trained chimpanzees involved symbolic *production*, in which the animals responded *with* the labels "Same" and "Different." Nevertheless, Alex similarly had to respond symbolically on the basis of (1) the instances of the categorical

WHAT'S SAME?

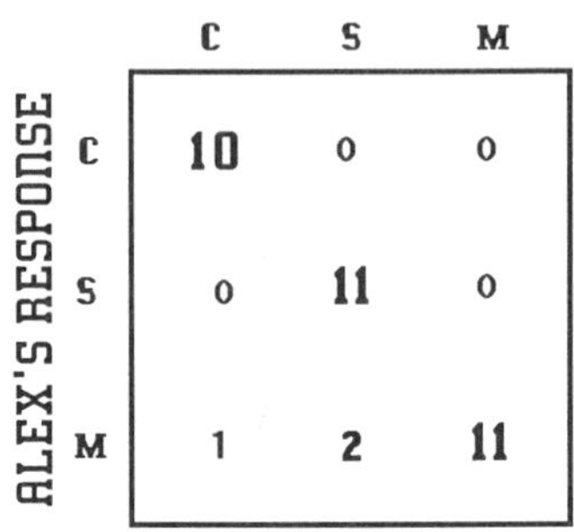

CORRECT RESPONSE

| ALEX'S RESPONSE | C | S | M |
|---|---|---|---|
| C | 10 | 0 | 0 |
| S | 0 | 11 | 0 |
| M | 1 | 2 | 11 |

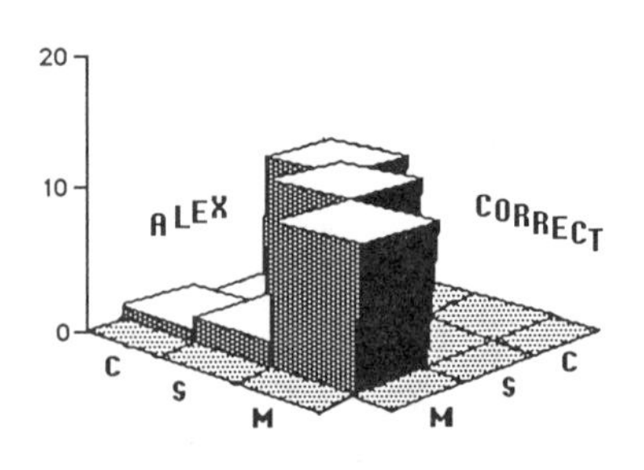


WHAT'S DIFFERENT?

CORRECT RESPONSE

| ALEX'S RESPONSE | C | S | M |
|---|---|---|---|
| C | 8 | 1 | 0 |
| S | 0 | 7 | 2 |
| M | 0 | 0 | 8 |

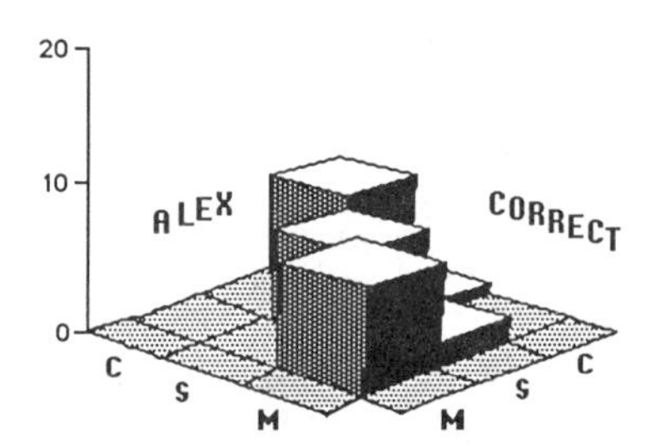


Figure 18.5. Alex's comprehension scores for the concept of same/different: C, color; S, shape; M, mah-mah (matter). From Pepperberg (1987c), in *Animal Learning & Behavior*, *15*, 423–432; reprinted by permission of Psychonomic Society, Inc. See also Pepperberg (in press-b).

classes the objects represented and (2) the symbolic relationship (the labels "Same" or "Different") requested by the experimenter – not on the basis of whether or not two items looked alike or were objects he had seen before. It is therefore unlikely that the task presented to Alex can be interpreted as a forced-choice conditional discrimination of the type generally used to determine nonhuman conceptual abilities.

Arguably, Alex did learn one form of conditional discrimination. He learned to respond to questions about two somewhat different objects phrased as "What's same?" and "What's different?" with categorical labels, whereas he maintained the ability to respond to questions about two objects phrased as "How many?" with the label "Two *X*" (Pepperberg, 1987a). Rather than detracting from the demonstrated abilities, comprehension and utilization of such a conditional constraint on the inherently conditional nature of the productive labeling task (Premack, 1976, pp. 140–142) more likely represent an additional cognitive achievement (e.g., Pepperberg, 1983; Premack, 1983, p. 129).

Although the findings presented here provide additional evidence for Alex's comprehension of concepts of categorical classes, our determination of the extent of his abilities is far from complete. He has, for example, just begun to respond appropriately to the questions "What's the color of the metal key?" and "What shape is the green wood?" for collections of differently colored and shaped exemplars of various materials (Essock, Gill, & Rumbaugh, 1977; Granier-Deferre & Kodratoff, 1986; Pepperberg, in press-a). It is also not yet clear if he can learn to respond directly to analogic "relations between relations," such as "Is the relation between $A_1$ and $A_2$ (same/different) as that between $B_1$ and $B_2$?" (Premack, 1983). Alex, however, has gone beyond match-to-sample or oddity-from-sample. He comprehends the vocal symbols "Same" and "Different," and he demonstrates a more abstract concept of category than in our previous study (Pepperberg, 1983). His responses in this study were not stimulus-specific, because he performed as well on entirely novel objects as he did on familiar ones, including those for which he had no labels. Thus, he appeared to learn the task in terms of the concepts of color, shape, and material.

## Concluding remarks

The foregoing results suggest that the cognitive capacities of our African Grey parrot can be compared favorably, at least in some domains, to those of primates such as chimpanzees. In some instances our data allow us to make comparisons with the capacities of human children. Our findings thus demonstrate the extent to which mammalian and nonmammalian abilities may be analogous, although not necessarily homologous.

The current data, however, cannot address questions concerning the evolutionary origins of comparable abilities in different species. Answers to those questions can come only when we learn more about these abilities in the wild: how they develop, why they may be needed, and how they are used. At present we can only speculate that the cognitive capacities that we see in the laboratory were indeed first developed under some set of evolutionary pressures. We are furthermore assuming that these capacities are the same as those used in the wild (i.e., that capacities that are adaptive in one instance can be applied in divergent situations). Such an assumption, although not without support (e.g., Cheney & Seyfarth, 1986; Rozin, 1976), has been challenged (Rogoff, 1984). In many cases, natural correlates have yet to be found for the advanced capacities exhibited in the laboratory.

Evidence for such natural correlates, however, often appears in the wild in alternative sensory modalities or contexts. Cross-modal transfer of abilities may thus be more common than is generally supposed (Meador et al., 1987). Significant transfer between auditory and visual modalities has been shown in the laboratory on some discrimination tasks for animals as disparate as prosimian bush babies (*Galago senegalensis*) (Ward, Yehle, & Doerflein, 1970), rabbits (*Oryctolagus cuniculus*) (Yehle & Ward, 1969), and rats (Church & Meck, 1984; Over & Mackintosh, 1969). Davis and Albert (1987), however, did not see such transfer for rats on a numerical task. For birds, there is evidence for acoustic correlates to the visual competence on numerical and same/different tasks that my students and I have demonstrated in the laboratory: Recent studies on bird-song processing indicate that birds may have a means of judging similarities and differences in acoustic patterns and their repetitions to an extent beyond "identical" or "nonidentical" and "more" versus "less." Birds may need to recognize particular sets of repetitions and subtle variations in the vocalizations of different neighbors to respond appropriately (Brown, 1985; Falls et al., 1982; Stoddard, Beecher, & Willis, 1988; Thompson, 1968; Wolfgramm & Todt, 1982). These data do not necessarily imply that a crow must "count" the number of caws in its neighbors' bursts, or that great tits necessarily pick out the same parameters as do scientists in distinguishing neighbor and stranger songs (Shy et al., 1986). The data do suggest that some sensitivity to features beyond those of "numerousness" or "identity" is likely to exist and that this sensitivity may be correlated with laboratory findings in other contexts.

In sum, I suggest

1. that successful examination of an animal's cognitive capacities requires integration of knowledge and techniques from both the field and the laboratory, and
2. that the techniques of interspecies communication provide a particularly powerful and efficient means by which to define and explore interspecies similarities and differences for various subjects (Griffin, 1976; Pepperberg, 1986, 1988a, in press-b).

My students and I have, through these methods, demonstrated that a nonprimate, nonmammalian subject has certain abilities that once were thought to be limited to primates, if not to humans (Premack, 1978, 1983). Our hope is that detailed examination of abilities in the laboratory will encourage field researchers to look for correlates (even in other modalities) in the wild and, of course, that laboratory studies will continue to provide additional information for the study of comparative intelligence.

**Addendum**

Alex's training and testing uses a code based on English speech, and some researchers would argue that his languagelike training may have prepared him for comprehension of the concepts of number and same/different beyond his "natural" capacities (Burdyn & Thomas, 1984; Pepperberg, 1987a,c, 1988c; Pepperberg & Kozak, 1986; Premack, 1983). Although some researchers (e.g., Rice, 1980) do not believe that linguistic input can teach a nonlinguistic concept for which a subject is unready, others (e.g., Premack, 1983) have suggested that the abilities of animals trained with language-related tasks significantly differ from the abilities of those without such training. Premack's contention is consistent with the writings of Vygotsky (1962), who suggested that even the earliest stages of language acquisition (from approximately 2 years in children) influence concept formation. The work of Lenneberg (1971) also supported Premack: Lenneberg proposed that development of most cognitive skills (both linguistic and nonlinguistic) requires the capacities to symbolize and comprehend relational concepts. Macphail (1987) took that argument further and claimed that language abilities are inextricably tied to the capacities that place humans at the pinnacle of cognitive abilities. Nevertheless, certain fairly complex numerical concepts and same/different relationships can be acquired by animals who have had no explicit training in "human language" (e.g., Davis & Bradford, 1986; Zentall et al., 1984). Such findings are in accord with those of Piaget (1952b): Although Piaget proposed that the emergence of a capacity for symbolization marks a major stage in cognitive development, he stressed such development in terms of the concept of internal representation rather than in terms of language. Language, in the Piagetian framework, may depend on representational ability, but representational ability need not be expressed solely through language.

**Acknowledgments**

This chapter is based on a paper presented to the Eleventh Congress of the International Primatological Society, Göttingen, West Germany, July 1986. The work described in this chapter was supported by grants from the Harry Frank Guggenheim Foundation and the National Science Foundation, BNS 7912945 and 8414483. Preparation of this chapter was supported in part by NSF grant BNS 8414483.

**Notes**

1. Many of the chapters in this volume, however, do attempt to integrate developmental and comparative (evolutionary) perspectives.
2. At the beginning of the project, each exemplar was assigned a number, and the test order was determined by a random number table. When the number of exemplars exceeded 10, a trainer not involved in testing picked the exemplars out of the toy box and wrote down the order of her choice. In the current studies, the test groups were listed on a page, the list itself was covered, and a trainer other than the examiner randomly ordered the list.
3. Occasionally it was the examiner who stood corrected: In approximately 1 in 20 trials, an examiner would err and scold Alex for a correct response. Alex would repeat his correct response, despite our procedures, which encourage a lose-shift strategy. The examiner would then recognize her error, and the bird would get his reward. Note that although this was not a formal blind test, it produced the same results. Also, other forms of blind tests have been performed: Alex has several idiosyncratic labels for objects (e.g., "banerry" for apple), and people unfamiliar with Alex and these labels have queried him about these objects; he has been correct on four of five trials.
4. Portions of this section have been excerpted from Pepperberg (1987a).
5. One problem in studying numerical competence, even in humans, is that there is no consistent use of terminology. Davis and Perusse (1988) have suggested a possible scheme, but at present the same terms represent very different levels of competence to different researchers. My use of the term *counting* is closely related to that of both Fuson (1988) and Davis and Perusse and is close to Piaget's concept (1941/1952a) of number: I use the term *counting* to refer to the ability (1) to produce a standard sequence of number words, (2) to apply to each individual term to be counted a unique number word, (3) to remember which items have already been counted, and (4) comprehend that the last number word said in counting tells how many objects are there (Pepperberg, 1987a).
6. These identifications do demonstrate an important ability unrelated to quantity: He had competently linked separate, appropriate vocalizations into novel combinations (e.g., "Two-corner wood") to describe novel objects. Whatever the mechanism, this segmentationlike behavior provides further evidence for a lexical sophistication not expected at the level of the parrot (e.g., Fromkin & Rodman, 1974).
7. Remember that Alex was given as many presentations as were necessary for him to

produce the correct response. He was, however, almost always correct on his second response if his first was incorrect (e.g., Table 18.2).

8. Portions of this section have been excerpted from Pepperberg (1987c [reprinted by permission of Psychonomic Society, Inc.], in press-b).
9. There were approximately 70 different possible objects (food items were rarely used), three possible correct responses, and two different questions for the same/different task alone. For example, if we asked "What's same?," desiring the response to be color, and chose a round green key as one of the exemplars, this key could be paired with 2-, 3-, 4-, 5-, or 6-cornered objects of paper, wood, or rawhide, 3-, 4-, or 6-cornered objects of plastic, plus objects such as green clothespins, wooden or plastic cubes or spheres, plastic boxes, cups, or barrels, etc. A similar set of permutations existed for responses of "shape," "matter," and the question "What's different?" In addition, Alex was currently being tested on numerical responses ("How many?") and additional labels ("What's this?", "What color?", etc.).

## References


Anderson, R. E. (1987). Intelligence and human language. *Behavioral and Brain Sciences, 10*, 657.

Bandura, A. (1971). Analysis of social modeling processes. In A. Bandura (Ed.), *Psychological modeling* (pp. 1–62). Chicago: Aldine-Atherton.

Bandura, A. (1977). *Social modeling theory*. Chicago: Aldine-Atherton.

Bates, E., Benigni, L., Bretherton, I., Camaioni, L., & Volterra, V. (1979). *The emergence of symbols: Cognition and communication in infancy*. New York: Academic Press.

Beecher, M. D., Stoddard, P. K., & Loesche, P. (1985). Recognition of parents' voices by young cliff swallows. *Auk, 102*, 600–605.

Braun, H. (1952). Ueber das Unterscheidungsvermögen unbenannter Anzahlen bei Papageien. *Zeitschrift für Tierpsychologie, 9*, 40–91.

Breland, K., & Breland, M. (1961). The misbehavior of organisms. *American Psycholgist, 16*, 681–684.

Brown, E. D. (1985). The role of song and vocal imitation among common crows (*Corvus brachyrhynchos*). *Zeitschrift für Tierpsychologie, 68*, 115–136.

Burdyn, L. E., Jr., & Thomas, R. K. (1984). Conditional discrimination with conceptual simultaneous and successive cues in the squirrel monkey (*Saimiri sciureus*). *Journal of Comparative Psychology, 98*, 405–413.

Burghardt, G. M. (1984). Ethology and operant psychology. *Behavioral and Brain Sciences, 7*, 683–684.

Busnel, R. G., & Mebes, H. D. (1975). Hearing and communication in birds: The cocktail party effect in intraspecific communication of *Agapornis roseicollis*. *Life Science, 17*, 1567–1569.

Case, R. (1985). *Intellectual development: Birth to adulthood*. New York: Academic Press.

Cheney, D. L., & Seyfarth, R. M. (1986). The recognition of social alliances by vervet monkeys. *Animal Behaviour, 34*, 1722–1731.

Church, R. M., & Meck, W. H. (1984). The numerical attribute of stimuli. In H. L. Roitblat, T. G. Bever, & H. S. Terrace (Eds.), *Animal cognition* (pp. 445–464). Hillsdale, NJ: Erlbaum.

Cole, S., Hainsworth, F. R., Kamil, A. C., Mercier, T., & Wolf, L. L. (1982). Spatial learning as an adaptation in hummingbirds. *Science, 217*, 655–657.

Crook, J. H. (1983). On attributing consciousness to animals. *Nature, 303*, 11–14.

Davis, H. (1984). Discrimination of the number three by a raccoon (*Procyon lotor*). *Animal Learning & Behavior, 12*, 409–413.

Davis, H., & Albert, M. (1987). Failure to transfer or train a numerical discrimination using sequential visual stimuli in rats. *Bulletin of the Psychonomic Society, 25*, 472–474.

Davis, H., Albert, M., & Barron, R. W. (1985). Detection of number or numerousness by human infants. *Science, 228*, 1222.

Davis, H., & Bradford, S. A. (1986). Counting behavior of rats in a simulated natural environment. *Ethology, 73*, 265–280.

Davis, H., & Memmott, J. (1982). Counting behavior in animals: A critical evaluation. *Psychological Bulletin, 92*, 547–571.
Davis, H., & Perusse, R. (1988). Numerical competence in animals: Definitional issues, current evidence, and a new research agenda. *Behavioral and Brain Research, 11*, 561–615.
Dethier, V. G. (1984). Levels of behavior and emergent mechanisms. In G. Greenberg & E. Tobach (Eds.), *Behavioral evolution and integrative levels* (pp. 121–132). Hillsdale, NJ: Erlbaum.
Dilger, W. C. (1960). The comparative ethology of the African parrot genus *Agapornis. Zeitschrift für Tierpsychologie, 17*, 649–685.
Edwards, C. A., Jagielo, J. A., & Zentall, T. R. (1983). Same/different symbol use by a pigeon. *Animal Learning & Behavior, 11*, 349–355.
Essock, S. M., Gill, T. V., & Rumbaugh, D. M. (1977). Language relevant object- and color-naming tasks. In D. M. Rumbaugh (Ed.), *Language learning by a chimpanzee* (pp. 193–206). New York: Academic Press.
Falls, J. B. (1985). Song matching in Western Meadowlarks. *Canadian Journal of Zoology, 63*, 2520–2524.
Falls, J. B., Krebs, J. R., & McGregor, P. K. (1982). Song matching in the great tit (*Parus major*): The effect of similarity and familiarity. *Animal Behaviour, 30*, 997–1009.
Fromkin, V., & Rodman, R. (1974). *An introduction to language*. New York: Holt, Rinehart & Winston.
Fuson, K. C. (1988). *Children's counting and concepts of number*. Berlin: Springer-Verlag.
Fuson, K. C., & Hall, J. W. (1983). The acquisition of early number word meanings: A conceptual analysis and review. In H. P. Ginsburg (Ed.), *The development of mathematical thinking* (pp. 49–107). New York: Academic Press.
Gardner, R. A., & Gardner, B. T. (1969). Teaching sign language to a chimpanzee. *Science, 165*, 664–672.
Gardner, R. A., & Gardner, B. T. (1978). Comparative psychology and language acquisition. In K. Salzimger & F. L. Denmark (Eds.), *Psychology: The state of the art*, Vol. 309 (pp. 37–76). New York Academy of Sciences.
Gast, H. (1957). Der Umgang mit Zahlen und Zahlgebilden in der frühen Kindheit. *Zeitschrift für Psychologie, 161*, 1–90.
Gelman, R. (1980). What children know about numbers. *Educational Psychologist, 15*, 54–68.
Gelman, R., & Gallistel, C. R. (1986). *The child's understanding of number* (2nd ed.). Cambridge, MA: Harvard University Press.
Geschwind, N. (1979). Specializations of the human brain. *Scientific American, 241*, 180–199.
Gill, F. B., & Wolf, L. L. (1975). Foraging strategies and energetics of east African sunbirds at mistletoe flowers. *American Naturalist, 109*, 491–510.
Ginsburg, H., & Opper, S. (1969). *Piaget's theory of intellectual development: An introduction*. Englewood Cliffs, NJ: Prentice-Hall.
Gossette, R. L. (1967). Successive discrimination reversal (SDR) performances of four avian species on a brightness discrimination task. *Psychonomic Science, 8*, 17–18.
Gossette, R. L., Gossette, M. F., & Riddell, W. (1966). Comparisons of successive discrimination reversal performances among closely and remotely related avian species. *Animal Behaviour, 14*, 560–564.
Gouzoules, S., Gouzoules, H., & Marler, P. (1985). External reference and affective signaling in mammalian vocal communication. In G. Zivin (Ed.), *The development of expressive behavior: Biology–environment interactions* (pp. 77–101). New York: Academic Press.
Gramza, A. F. (1970). Vocal mimicry in captive budgerigars (*Melopsittacus undulatus*). *Zeitschrift für Tierpsychologie, 27*, 971 –983.
Granier-Deferre, C., & Kodratoff, Y. (1986). Iterative and recursive behaviours in chimpanzees during problem solving: A new descriptive model inspired from the artificial intelligence approach. *Cahiers de Psychologie Cognitive, 6*, 483–500.
Greenfield, P. M. (1978). Developmental processes in the language learning of child and chimp. *Behavioral and Brain Sciences, 4*, 573–574.

Greeno, J. G., Riley, M. S., & Gelman, R. (1984). Conceptual competence and children's counting. *Cognitive Psychology*, *16*, 94–143.
Griffin, D. R. (1976). *The question of animal awareness.* New York: Rockefeller University Press.
Grosslight, J. H., & Zaynor, W. C. (1967). Verbal behavior in the mynah bird. In K. Salzinger & S. Salzinger (Eds.), *Research in verbal behavior and some neurophysiological implications* (pp. 5–9). New York: Academic Press.
Hall, A. D., Braggio, J. T., Buchanan, J. P., Nadler, R. D., Karan, D., & Sams, J. B. (1980). Multiple classification performance of juvenile chimpanzees, normal children, and retarded children. *International Journal of Primatology*, *1*, 345–359.
Hayes, K. J., & Nissen, C. H. (1971). Higher mental functions of a home-raised chimpanzee. In A. Schrier & F. Stollnitz (Eds.), *Behavior of nonhuman primates* (Vol. 4, pp. 59–115). New York: Academic Press. (Original work published 1956)
Herman, L. M. (1986). Cognition and language competencies of bottlenosed dolphins. In R. Schusterman, J. Thomas, & F. Wood (Eds.), *Dolphin behavior and cognition: Comparative and ecological aspects* (pp. 221–252). Hillsdale, NJ: Erlbaum.
Herman, L. M. (1987). Receptive competencies of language-trained animals. In J. S. Rosenblatt, C. Beer, M.-C. Busnel, & P. J. B. Slater (Eds.), *Advances in the study of behavior* (Vol. 17, pp. 1–60). New York: Academic Press.
Herman, L. M., & Forestell, P. H. (1985). Reporting presence or absence of named objects by a language-trained dolphin. *Neuroscience and Biobehavioral Reviews*, *9*, 667–681.
Herrnstein, R. J. (1984). Objects, categories, and discriminative stimuli. In H. L. Roitblat, T. G. Bever, & H. S. Terrace (Eds.), *Animal cognition* (pp. 233–261). Hillsdale, NJ: Erlbaum.
Hicks, L. H. (1956). An analysis of number concept formation in the rhesus monkey. *Journal of Comparative and Physiological Psychology*, *49*, 212–218.
Hodos, W. (1982). Some perspectives on the evolution of intelligence and the brain. In D. R. Griffin (Ed.), *Animal mind–human mind* (pp. 33–56). Berlin: Springer-Verlag.
Hulse, S. H., & Cynx, J. (1985). Relative pitch perception is constrained by absolute pitch in songbirds. (*Mimus*, *Molothrus*, and *Sturnus*). *Journal of Comparative Psychology*, *99*, 176–196.
Humphrey, N. (1979). Nature's psychologists. In B. Josephson & B. S. Ramchandra (Eds.), *Consciousness and the physical world* (pp. 57–75). New York: Pergamon.
Inhelder, B., & Piaget, J. (1958). *The growth of logical thinking.* New York: Basic Books.
Kamil, A. C. (1984). Adaptation and cognition: Knowing what comes naturally. In H. L. Roitblat, T. G. Bever, & H. S. Terrace (Eds.), *Animal cognition* (pp. 533–543). Hillsdale, NJ: Erlbaum.
Kamil, A. C. (1988a). A synthetic approach to the study of animal intelligence. In D. W. Leger (Ed.), *Nebraska symposium on motivation: Comparative perspectives in modern psychology* (Vol. 35, pp. 257–308). Lincoln: University of Nebraska Press.
Kamil, A. C. (1988b). Experimental design in ornithology. *Current Ornithology*, *5*, 313–346.
Kamil, A. C., & Hunter, M. W., III. (1970). Performance on object discrimination learning set by the greater hill mynah, *Gracula religiosa. Journal of Comparative and Physiological Psychology*, *13*, 68–73.
Klahr, D., & Wallace, J. G. (1973). The role of quantification operators in the development of conservation of quantity. *Cognitive Psychology*, *4*, 301–327.
Klahr, D., & Wallace, J. G. (1976). *Cognitive development: An information-processing view.* Hillsdale, NJ: Erlbaum.
Koehler, O. (1943). "Zähl"-Versuche an einem Kolkraben und Vergleichsversuche an Menschen. *Zeitschrift für Tierpsychologie*, *5*, 575–712.
Koehler, O. (1950). The ability of birds to "count." *Bulletin of Animal Behaviour*, *9*, 41–45.
Koehler, O. (1953). Thinking without words. In *Proceedings of the XIV International Congress of Zoology* (pp. 75–88). Copenhagen: Danish Science Press.
Kroodsma, D. E. (1979). Vocal dueling among male marsh wrens: Evidence for ritualized expressions of dominance/subordinance. *Auk*, *96*, 506–515.

Kroodsma, D. E. (1989). Suggested experimental designs for song playbacks. *Animal Behaviour, 37*, 600–609.
Kroodsma, D. E., Baker, M. C., Baptista, L. F., & Petrinovich, L. (1985). Vocal "dialects" in Nuttall's white-crowned sparrows. *Current Ornithology, 2*, 103–133.
Kroodsma, D. E., Bateson, P. P. G., Bischof, H.-J., Delius, J. D., Hearst, E., Hollis, K. L., Immelmann, K., Jenkins, H. J., Konishi, M., Lea, S. E. G., Marler, P., & Staddon, J. E. R. (1984). Biology of learning in nonmammalian vertebrates. In P. Marler & H. S. Terrace (Eds.), *The biology of learning* (pp. 399–418). Berlin: Springer-Verlag.
Krushinskii, L. V. (1960). *Animal behavior: Its normal and abnormal development.* New York: Consultants Bureau.
Lenneberg, E. H. (1971). Of language, knowledge, apes and brains. *Journal of Psycholinguistic Research, 1*, 1–29.
Logan, C. A. (1987). Fluctuations in fall and winter territory size in the northern mockingbird (*Mimus polyglottos*). *Journal of Field Ornithology, 58*, 297–305.
Lögler, P. (1959). Versuche zur Frage des "Zähl"-Vermögens an einen Graupapagein und Vergleichsversuche an Menschen. *Zeitschrift für Tierpsychologie, 16*, 179–217.
McClure, M. K., & Helland, J. (1979). A chimpanzee's use of dimensions in responding same and different. *Psychological Record, 29*, 371–378.
Macphail, E. M. (1982). *Brain and intelligence in vertebrates*. Oxford: Clarendon Press.
Macphail, E. M. (1985). Vertebrate intelligence: The null hypothesis. *Philosophical Transactions of the Royal Society, London, B, 308*, 37–51.
Macphail, E. M. (1987). The comparative psychology of intelligence. *Behavioral and Brain Sciences, 10*, 645–695.
Mandler, G., & Shebo, B. J. (1982). Subitizing: An analysis of its component processes. *Journal of Experimental Psychology, General, 111*, 1–22.
Marler, P., & Terrace, H. S. (Eds.). (1984). *The biology of learning*. Berlin: Springer-Verlag.
Mason, W. A., & Hollis, J. H. (1962). Communication between young rhesus monkeys. *Behaviour, 10*, 211–221.
Matsuzawa, T. (1985). Use of numbers by a chimpanzee. *Nature, 315*, 57–59.
Matsuzawa, T., Asano, T., Kubota, K., & Murofushi, K. (1986). Acquisition and generalization of numerical labeling by a chimpanzee. In D. M. Taub & F. A. King (Eds.), *Current perspectives in primate social dynamics* (pp. 416–430). New York: Van Nostrand Reinhold.
Meador, D. M., Rumbaugh, D. M., Pate, J. L., & Bard, K. A. (1987). Learning, problem solving, cognition, and intelligence. In G. Mitchell & J. Erwin (Eds.), *Comparative primate biology* (Vol. 2, Pt. B, pp. 17–83). New York: Alan R. Liss.
Mebes, H. D. (1978). Pair-specific duetting in the peach-faced lovebird. *Agapornis roseicollis. Naturwissenschaften, 65*, 66–67.
Menzel, E. W., Jr., & Juno, C. (1985). Social foraging in marmoset monkeys and the question of intelligence. *Philosophical Transactions of the Royal Society, London, B, 308*, 145–158.
Michael, J., Whitley, P., & Hesse, B. (1983). The pigeon parlance project. *VB News, 2*, 6–9.
Moran, G., Joch, E., & Sorenson, L. (1983, June). *The response of meerkats (Suricata suricatta) to changes in olfactory cues on established scent posts*. Paper presented at the annual meeting of the Animal Behavior Society, Lewisburg, PA.
Mowrer, O. H. (1950). *Learning theory and personality dynamics*. New York: Ronald Press.
Mowrer, O. H. (1952). The autism theory of speech development and some clinical applications. *Journal of Speech and Hearing Disorders, 17*, 263–268.
Mowrer, O. H. (1954). A psychologist looks at language. *American Psychologist, 9*, 660–694.
Nottebohm, F. (1970). Ontogeny of bird song. *Science, 167*, 950–956.
Nottebohm, F., & Nottebohm, M. (1969). The parrots of Bush Bush. *Animal Kingdom, 72*, 19–23.
Over, R., & Mackintosh, N. J. (1969). Cross-modal transfer of intensity discrimination by rats. *Nature, 224*, 918–919.
Owings, D. H., & Hennessy, D. F. (1984). The importance of variation in sciurid visual and vocal communication. In J. O. Murie & G. R. Michener (Eds.), *Biology of ground dwell-*

*ing squirrels: Annual cycles, behavioral ecology, and sociality* (pp. 169–200). Lincoln: University of Nebraska Press.

Owings, D. H., Hennessy, D. F., Leger, D. W., & Gladney, A. B. (1986). Different functions of "alarm" calling for different time scales: A preliminary report on ground squirrels. *Behaviour, 99*, 101–116.

Owings, D. H., & Leger, D. W. (1980). Chatter vocalizations of California ground squirrels: Predator- and social-role specificity. *Zeitschrift für Tierpsychologie, 54*, 164–184.

Park, T. J., & Dooling, R. J. (1985). Perception of species-specific contact calls by budgerigars (*Melopsittacus undulatus*). *Journal of Comparative Psychology, 99*, 391–402.

Parker, S. T., & Gibson, K. R. (1977). Object manipulation, tool use and sensorimotor intelligence as feeding adaptations in *Cebus* monkeys and great apes. *Journal of Human Evolution, 6*, 623–641.

Parker, S. T., & Gibson, K. R. (1979). A developmental model for the evolution of language and intelligence in early hominids. *Behavioral and Brain Sciences, 2*, 367–408.

Pepperberg, I. M. (1978, March). *Object identification by an African Grey parrot (Psittacus erithacus)*. Paper presented at a meeting of the Midwestern Animal Behavior Society, West Lafayette, IN.

Pepperberg, I. M. (1979, June). *Functional word use by an African Grey parrot*. Paper presented at the annual meeting of the Animal Behavior Society, New Orleans.

Pepperberg, I. M. (1981). Functional vocalizations by an African Grey parrot (*Psittacus erithacus*). *Zeitschrift für Tierpsychologie, 55*, 139–160.

Pepperberg, I. M. (1983). Cognition in the African Grey parrot: Preliminary evidence for auditory/vocal comprehension of the class concept. *Animal Learning & Behavior, 11*, 179–185.

Pepperberg, I. M. (1985). Social modeling theory: A possible framework for understanding avian vocal learning. *Auk, 102*, 854–864.

Pepperberg, I. M. (1986). Acquisition of anomalous communicatory systems: Implications for studies on interspecies communication. In R. Schusterman, J. Thomas, & F. Wood (Eds.), *Dolphin behavior and cognition: Comparative and ecological aspects* (pp. 289–302). Hillsdale, NJ: Erlbaum.

Pepperberg, I. M. (1987a). Evidence for conceptual quantitative abilities in the African Grey parrot: Labeling of cardinal sets. *Ethology, 75*, 37–61.

Pepperberg, I. M. (1987b). Interspecies communication; A tool for assessing conceptual abilities in the African Grey parrot (*Psittacus erithacus*). In G. Greenberg & E. Tobach (Eds.), *Language, cognition, consciousness: Integrative levels* (pp. 31–56). Hillsdale, NJ: Erlbaum.

Pepperberg, I. M. (1987c). Acquisition of the same/different concept by an African Grey parrot (*Psittacus erithacus*): Learning with respect to categories of color, shape, and material. *Animal Learning & Behavior, 15*, 423–432.

Pepperberg, I. M. (1988a). An interactive modeling technique for acquisition of communication skills: Separation "labelling" and "requesting" in a psittacine subject. *Applied Psycholinguistics, 9*, 59–76.

Pepperberg, I. M. (1988b). The importance of social interaction and observation in the acquisition of communicative competence: Possible parallels between avian and human learning. In T. R. Zentall & B. G. Galef (Eds.), *Social learning: Psychological and biological perspectives* (pp. 279–299). Hillsdale, NJ: Erlbaum.

Pepperberg, I. M. (1988c). Evidence for comprehension of "absence" by an African Grey parrot: Learning with respect to questions of same/different. *Journal of the Experimental Analysis of Behavior, 50*, 553–564.

Pepperberg, I. M. (1988d). Studying numerical competencies: A trip through linguistic wonderland? *Behavioral and Brain Sciences, 11*, 595–596.

Pepperberg, I. M. (in press-a). Cognition in the African Grey parrot (*Psittacus erithacus*). Further evidence for comprehension of categories and labels. *Journal of Comparative Psychology*.

Pepperberg, I. M. (in press-b). A communicative approach to animal cognition: A study of conceptual abilities of an African Grey parrot (*Psittacus erithacus*). In C. Ristau (Ed.), *Cognitive ethology: The minds of other animals*. Hillsdale, NJ: Erlbaum.

Pepperberg, I. M., & Kozak, F. A. (1986). Object permanence in the African Grey parrot (*Psittacus erithacus*). *Animal Learning & Behavior*, *14*, 322–330.

Piaget, J. (1952a). *The child's conception of number*. London: Routledge & Kegan Paul. (Original work published 1941)

Piaget, J. (1952b). *The origins of intelligence in children*. New York: International Universities Press.

Piaget, J. (1954). *The construction of reality in the child*. New York: Basic Books.

Power, D. M. (1966a). Agnostic behavior and vocalizations of orange-chinned parakeets in captivity. *Condor*, *6*, 562–581.

Power, D. M. (1966b). Antiphonal duetting and evidence for auditory reaction time in the orange-chinned parakeet. *Auk*, *83*, 314–319.

Premack, D. (1976). *Intelligence in ape and man*. Hillsdale, NJ: Erlbaum.

Premack, D. (1978). On the abstractness of human concepts: Why it would be difficult to talk to a pigeon. In S. H. Hulse, H. Fowler, & W. K. Honig (Eds.), *Cognitive processes in animal behavior* (pp. 421–451). Hillsdale, NJ: Erlbaum.

Premack, D. (1983). The codes of man and beast. *Behavioral and Brain Sciences*, *6*, 125ff.

Putney, R. T. (1985). Do willful apes know what they are aiming at? *Psychological Record*, *35*, 49–62.

Rice, M. (1980). *Cognition to language*. Baltimore: University Park Press.

Rogoff, B. (1984). Introduction: Thinking and learning in a social context. In B. Rogoff & J. Lave (Eds.), *Everyday cognition: Its development in social contexts* (pp. 1–8). Cambridge, MA: Harvard University Press.

Rozin, P. (1976). The evolution of intelligence and access to the cognitive unconscious. In J. M. Sprague & A. N. Epstein (Eds.), *Progress in psychobiology and physiological psychology* (Vol. 6, pp. 245–280). New York: Academic Press.

Rumbaugh, D. M., & Gill, T. V. (1977). Lana's acquisition of language skills. In D. M. Rumbaugh (Ed.), *Language-learning by a chimpanzee* (pp. 165–192). New York: Academic Press.

Rumbaugh, D. M., & Pate, J. L. (1984). Primates' learning by levels. In G. Greenberg & E. Tobach (Eds.), *Behavioral evolution and integrative levels* (pp. 221–240). Hillsdale, NJ: Erlbaum.

Santiago, H. C., & Wright, A. A. (1984). Pigeon memory: *Same/different* concept learning, serial probe recognition acquisition, and probe delay effects on the serial-position function. *Journal of Experimental Psychology: Animal Behavior Processes*, *10*, 498–512.

Schusterman, R. J., & Gisiner, R. (1988). Artificial language comprehension in dolphins and sea lions: The essential cognitive skills. *Psychological Record*, *38*, 311–348.

Schusterman, R. J., & Krieger, K. (1984). California sea lions are capable of semantic comprehension. *Psychological Record*, *34*, 3–23.

Schusterman, R. J., & Krieger, K. (1986). Artificial language comprehension and size transposition by a California sea lion (*Zalophus californianus*). *Journal of Comparative Psychology*, *100*, 348–355.

Seligman, M. E. P. (1970). On the generality of the laws of learning. *Psychological Review*, *77*, 406–418.

Serpell, J. (1981). Duets, greetings, and triumph ceremonies: Analogous displays in the parrot genus *trichoglossus*. *Zeitschrift für Tierpsychologie*, *55*, 268–283.

Seyfarth, R. M., Cheney, D. L., & Marler, P. (1980). Vervet monkey alarm calls: Semantic communication in a free-ranging primate. *Animal Behaviour*, *28*, 1070–1094.

Shy, E., McGregor, P. K., & Krebs, J. (1986). Discrimination of song types by male great tits. *Behavioural Processes*, *13*, 1–12.

Siegel, L. S. (1982). The development of quantity concepts: Perceptual and linguistic factors. In C. J. Brainerd (Ed.), *Children's logical and mathematical cognition* (pp. 123–155). Berlin: Springer-Verlag.

Snowdon, C. T. (1983). Ethology, comparative psychology, and animal behavior. *Annual Review of Psychology*, *34*, 63–94.

Snowdon, C. T. (1987). A naturalistic view of categorical perception. In S. Harnad (Ed.), *Categorical perception: The groundwork of cognition* (pp. 332–354). Cambridge University Press.

Snowdon, C. T. (1988). A comparative approach to vocal communication. In D. W. Leger (Ed.), *Nebraska symposium on motivation: Comparative perspectives in modern psychology* (Vol. 35, pp. 145–199). Lincoln: University of Nebraska Press.

Spetch, M. L., & Edwards, C. A. (1986). Spatial memory in pigeons (*Columba livia*) in an open-field feeding environment. *Journal of Comparative Psychology*, *100*, 266–278.

Steckol, K. F., & Leonard, L. B. (1981). Sensorimotor development and the use of prelinguistic performatives. *Journal of Speech and Hearing Research*, *24*, 262–268.

Steffe, L. P., von Glasersfeld, E., Richards, J., & Cobb, P. (1983). *Children's counting types: Philosophy, theory, and application*. New York: Praeger.

Steklis, H. D., & Raleigh, M. H. (1979). Requisites for language: interspecific and evolutionary aspects. In H. D. Steklis & M. H. Raleigh (Eds.), *Neurobiology of social communication in primates* (pp. 283–304). New York: Academic Press.

Stettner, L. J., & Matyniak, K. (1968). The brain of birds. *Scientific American*, *218*, 64–76.

Stoddard, P. K., Beecher, M. D., & Willis, M. S. (1988). Response of territorial male song sparrows to song types and variations. *Behavioral Ecology and Sociobiology*, *22*, 125–130.

Struhsaker, T. T. (1967). Auditory communication among vervet monkeys (*Cercopithecus aethiops*). In S. A. Altmann (Ed.), *Social communication among primates* (pp. 281–324). University of Chicago Press.

Thompson, N. S. (1968). Counting and communication in crows. *Communications in Behavioral Biology*, *2*, 223–225.

Todt, D. (1975). Social learning of vocal patterns and modes of their application in Grey parrots. *Zeitschrift für Tierpsychologie*, *39*, 178–188.

von Glasersfeld, E. (1981a). An attention model for the conceptual construction of units and number. *Journal of Research in Mathematical Education*, *12*, 83–94.

von Glasersfeld, E. (1981b). *Sensory-motor sources of numerosity*. Paper presented at the Eleventh Symposium of the Jean Piaget Society, Philadelphia.

von Glasersfeld, E. (1982). Subitizing: The role of figural patterns in the development of numerical concepts. *Archives de Psychologie*, *50*, 191–218.

Vygotsky, L. S. (1962). *Thought and language*. Cambridge, MA: M.I.T. Press.

Ward, J. P., Yehle, A. L., & Doerflein, R. S. (1970). Cross-modal transfer of a specific discrimination in the bush baby (*Galago senegalensis*). *Journal of Comparative and Physiological Psychology*, *73*, 74–77.

Warren, J. M. (1965). Primate learning in comparative perspective. In A. Schrier & F. Stollnitz (Eds.), *Behavior of nonhuman primates* (Vol. 1, pp. 249–281). New York: Academic Press.

Wickler, W. (1976). The ethological analysis of attachment. *Zeitschrift für Tierpsychologie*, *42*, 12–28.

Wickler, W. (1980). Vocal duetting and the pairbond. I. Coyness and the partner commitment. *Zeitschrift für Tierpsychologie*, *52*, 201–209.

Wolfgramm, J., & Todt, D. (1982). Pattern and time specificity in vocal responses of blackbirds *Turdus merula L. Behaviour*, *81*, 264–286.

Wright, A. A., Santiago, H. C., & Sands, S. F. (1984). Monkey memory: *Same/different* concept learning, serial probe acquisition, and probe delay effects. *Journal of Experimental Psychology: Animal Learning Processes*, *10*, 513–529.

Wright, A. A., Santiago, H. C., Urcuioli, P. J., & Sands, S. F. (1984). Monkey and pigeon acquisition of same/different concept using pictorial stimuli. In M. L. Commons, R. J. Herrnstein & A. R. Wagner (Eds.), *Quantitative analysis of behavior* (Vol. 4, pp. 295–317). Cambridge, MA: Ballinger.

Yehle, A. L., & Ward, J. P. (1969). Cross-modal transfer of a specific discrimination in the rabbit. *Psychonomic Science*, *16*, 162–164.

Zentall, T. R., Edwards, C. A., Moore, B. S., & Hogan, D. E. (1981). Identity: The basis for both matching and oddity learning in pigeons. *Journal of Experimental Psychology: Animal Behavior Processes*, *7*, 70–86.

Zentall, T. R., & Hogan, D. E. (1974). Abstract concept learning in the pigeon. *Journal of Experimental Psychology*, *102*, 393–398.

Zentall, T. R., Hogan, D. E., & Edwards C. A. (1984). Cognitive factors in conditional learning by pigeons. In H. L. Roitblat, T. G. Bever, & H. S. Terrace (Eds.), *Animal cognition* (pp. 389–405). Hillsdale, NJ: Erlbaum.

# *Part VI*

# Comparative developmental perspectives on ape "language"

# 19 The cognitive foundations for reference in a signing orangutan

*H. Lyn White Miles*

## Introduction

Beginning in the 1960s, Gardner and Gardner (1969), Premack (1972), and Rumbaugh, Gill, and von Glasersfeld (1973) first demonstrated that chimpanzees could represent words or ideas using a set of gestural signs, computer lexigrams, or plastic tokens. In subsequent research, Fouts (1973), Miles (1976, 1983), Patterson (1978), and Terrace, Petitto, Sanders, and Bever (1979) extended these language studies of the gorilla and orangutan and expanded the focus to such issues as ape-to-ape communication, discourse ability, and the relationship between language and other cognitive processes. Controversy arose over the degree to which an ape's use of these systems could be called "language," the extent of animal linguistic abilities, and whether or not other species, such as aquatic mammals and birds, could exhibit similar skills (Brown, 1973; Epstein, Lanza, & Skinner, 1980; Le May & Geschwind, 1975; Limber, 1977; Mounin, 1976; Pepperberg, P&G18); Sebeok & Umiker-Sebeok, 1980; Terrace et al., 1979).

Project Chantek is an attempt to advance our knowledge of animal intelligence and "language" ability through a developmental perspective. It consists of a longitudinal sign language study with an orangutan named Chantek. The orangutan is the only great ape from Asia (the chimpanzee and gorilla are found in Africa). The orangutan was thought by many to be less likely to develop language skills than chimpanzees, primarily because of beliefs that chimpanzees are more intelligent and evidence that the African apes had a more recent evolutionary separation from humans than did the orangutan. After several years of being immersed in a human community communicating with gestural signs, Chantek learned to use a vocabulary of 140 signs and invented additional signs of his own. How he learned to use signs and combine them to express new meanings became key issues in the effort to understand the stages of his linguistic and cognitive development and how this development compared with that of both normal and impaired human children. Two issues emerged as particularly important. The first was the process by which his signs became referential, that is, intentionally meaningful indicators related to the real world. The second was the developmental sequence of his cognitive abilities and how these related to stages of cognitive development in human children, as defined by Piaget and others. This chapter discusses the methods and results of Project Chantek and describes the development of referential abilities and the stages of cognitive development to which they were related.

It should be noted that the goal of this research was not to demonstrate whether or not Chantek had acquired "language." Linguists do not agree on the essential nature of language or on whether it is an ability totally distinct from animal communication or differs only in degree and not kind. Interest in early studies of ape language abilities focused on whether or not apes could learn adult human language. In contrast, the focus of Project Chantek is on a developmental perspective that seeks to identify the cognitive and communicative processes that might underlie language development. It is our conclusion that Chantek's communicative and cognitive skills allowed him to develop referential ability without extensive drills and that he has strong similarities to human children in the earliest stages of language development. Whether or not the early stages of child and ape linguistic development constitute language depends on the definition of language being used. Clearly, neither apes nor children are using complex forms of adult human language. The advantage of a developmental approach is that it views language acquisition as a *process* by which communication is established within a cognitive framework in a cultural context.

## Project Chantek

Chantek was born in captivity at the Yerkes Regional Primate Research Center on December 17, 1977. Project Chantek was established in the fall of 1978 with Chantek's transfer to the University of Tennessee at Chattanooga, where the majority of this research was conducted. In 1986, Chantek returned to the Yerkes Center, where the research was continued on a modified basis. Both of Chantek's parents, Datu and Kampong, were wild-caught orangutans living at the Yerkes Center. In prehistoric times, orangutans were found throughout Asia and perhaps other parts of the Old World. They are now found only in Southeast Asia, on the islands of Borneo and Sumatra, and are considered an endangered species. Generally believed to be genetically less closely related to humans than are chimpanzees and gorillas, orangutans have a surprising number of behavioral and biological similarities to humans, for example, in gestation period, brain hemispheric asymmetry, characteristics of dentition, copulatory behavior, and insightful style of cognition (Schwartz, 1987). One possibility is that humans and African apes may have diverged more recently, but orangutans may have retained the ancestral hominoid characteristics of populations that later gave rise to pongids (apes) and hominids, our ancestors. At any rate, it seems apparent that orangutan abilities have been underestimated in the past.

Chantek was briefly observed with his mother Datu at the Yerkes Center in the fall of 1978. At the age of 9 months he was brought to the University of

Tennessee campus and housed in a five-room 12 × 40-ft trailer. The trailer was situated in a courtyard enclosed by fencing and bounded on one side by a music conservatory. Chantek received care and sign training from a small staff of caregivers. The use of caregivers, not just trainers, reflected the developmental and anthropological approach of this study (Miles, 1983). Chantek was not just trained to use signs; he was immersed in a human cultural environment and learned the rules for behavior and interaction, a process anthropologists call *enculturation.* Learning these human cultural rules for interaction with others and the environment played an integral role in his language development. Because signing was established as the primary mode of communication, a realm of shared meanings formed a "culture of communication."

Chantek was taught to use the gestural signs of the American Sign Language for the deaf. His caregivers included both hearing companions and some deaf companions, who used gestural signs in predominantly English word order in a form often called pidgin Sign English. For the first several years, speech was not used with Chantek, although a variety of vocal sounds were made. Subsequently, Chantek was exposed to speech, and we commonly signed and spoke the word at the same time in a "telegraphic" simple grammar. This means that Chantek's modified gesture system possesses some gestural modulations of meaning that occur in American Sign Language, wherein they become grammatical devices similar to syntax in spoken English.

Chantek's sign training was broken down into four stages:

1. A social bond was created between Chantek and his companions.
2. Motivation, or a purpose to the communication, was established.
3. The rules for human communication and conversation were introduced to Chantek by encouraging him to pay attention to his caregivers and to participate in games and activities designed to allow him to practice attending, noticing, and taking turns.
4. He was taught specific gestural signs by molding his hands and later by imitation.

Signs were considered to be an active part of his vocabulary when he used them spontaneously and appropriately on half the days of a given month. Most signs achieved active acquired status. Finally, Chantek was encouraged to use his signs independently to transmit meaningful information about subjects of interest to him. Chantek was encouraged to take an interest in his natural environment and to communicate about daily events, such as feeding his pet squirrel, completing a puzzle, taking a walk, or visiting with the local mounted police who came to his yard. In turn, his signs were responded to as if they were meaningful communications, so that he learned to gain mastery over his environment through the shared use of symbols.

The role of communication in language acquisition is important (Bruner, 1975; Dore, 1974; Greenfield & Smith, 1976). A set of rules for noticing and sending and receiving messages must be established if two-way communication is going to develop. Motivation is a key factor. Chantek was encouraged to use signs when he seemed most attentive or involved with an activity or person. He was not pressured to learn new signs, to maintain a strict sign acquisition schedule, or to label objects repeatedly. The focus was on Chantek's meaningful and appropriate use of signs, rather than on his rate of sign acquisition. In fact, acquiring a large vocabulary was not a major goal; that chimpanzees and gorillas could acquire large vocabularies of signs or lexigrams seemed well established. Of greater significance was what these signs or lexigrams actually meant. Thus, attention was focused instead on the development of sign meanings and how these were used. The method was one anthropologists call *participant observation*, in which the anthropologist is both objective outside observer and participant in the effort to understand internal meanings. This process has also been called "the guided re-invention of language" (Lock, 1980). As human children learn language, they and their parents reinvent the language process by creating consistent meanings out of their shared experiences, making referential connections to the real world, and so forth. The goal of this approach was to encourage Chantek in spontaneous and initiated signing on topics of interest to him. Lack of attention to this process and directed pressure for apes to sign on command can result in an animal that will produce long strings of signs until he accidentally hits upon the "correct" one, in the observer's view. This procedure has led to false conclusions regarding ape language ability (Terrace et al., 1979).

Data were collected on Chantek's daily activities, social behavior, and interactions with his environment, as well as on the context or situation in which Chantek signed and the interactions between Chantek and his caregivers. These interactions were videotaped, and signing exchanges were translated into English glosses based on the procedures of Hoffmeister, Moores, and Ellenberger (1975) and Klima and Bellugi (1979). Also, experimental testing procedures were carried out to monitor Chantek's cognitive and linguistic development and to test specific abilities such as drawing, problem solving, and tool use.

### Chantek's use of signs

Chantek's first months were spent adjusting to his new environment. His sign training began in November 1979. In December 1979, after 1 month of enculturation, he produced his first signs: FOOD-EAT and DRINK. Chantek eventually learned to use approximately 140 signs, 127 of which met the acquisition criterion of spontaneous and appropriate usage on half the days of a given month. This criterion was intermediate between the Terrace et al. (1979) more liberal criterion of spontaneous usage on 5 consecutive days and the Gardner and Gardner (1969) more conservative criterion of spontaneous and appropriate usage on 15 consecutive days. Once Chantek acquired a sign, it usually remained an active part of his vocabulary or could be easily restored as long as there was an appropriate situation or referent.

The development of Chantek's vocabulary is shown in Figure 19.1. Chan-

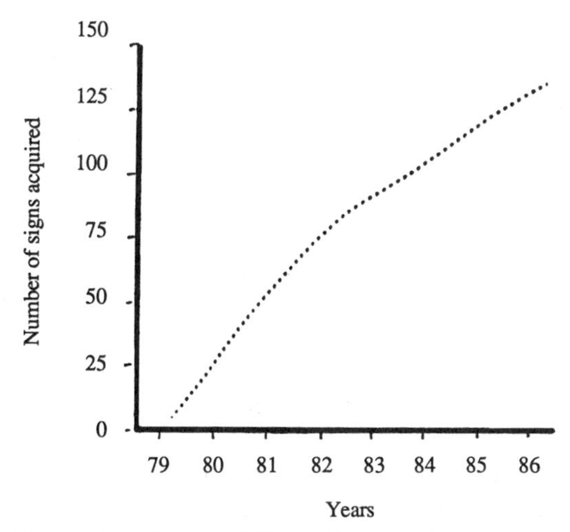


Figure 19.1. Growth of Chantek's vocabulary (January 1979–January 1986).

tek's vocabulary steadily increased throughout the primary study period at the University of Tennessee (despite an interruption in his training following his return to the Yerkes Center, he maintained his signing skills). During this study period, Chantek used approximately one-third to one-half of his vocabulary each day. As he acquired new signs, he incorporated them into his daily usage. During the first 22 months, Chantek acquired 44 signs. This is a rate slightly greater than that for other signing apes. In that same time period, Gardner and Gardner (1969) reported that the chimpanzee Washoe acquired 34 signs, and Terrace, Petitto, and Bever (1976) reported that the chimpanzee Nim learned 39 signs. In this initial period, Chantek learned three or four new signs each month. After this small vocabulary was in place, Chantek was able to request a variety of foods, drinks, and activities to meet his basic daily needs. Later, he added new signs to his vocabulary at a slower pace, in accordance with our requirement that the new vocabulary items be relevant to his interests and motivation. His vocabulary was not inflated with names for objects of little interest to him.

Motivation was a key factor in acquisition. Most of Chantek's signs were acquired within a couple of months of their introduction, although a few signs took longer to acquire. For example, Chantek was highly motivated to learn names for his favorite foods. He acquired the action sign GIVE and the food signs RAISIN, CHEESE, and BERRY in less than 20 days. On the other hand, he was uninterested in certain objects, such as shoes. He would use the sign regularly during a month, but it took over a year for it to meet the acquisition criterion. The locative DOWN created another problem. Chantek acquired the similar sign UP very early in his development and was highly motivated to use it. He signed UP initially as an action sign to ask his caregivers to pick him up and to carry him. Only much later did it become a decontextualized locative. But

Chantek almost never wished to be put DOWN, and thus he also took over a year to acquire this sign, although it was clear that he understood its meaning. For reasons of this sort, strict acquisition criteria provide a poor measure of the ape's understanding and use of a sign.

A second factor that affected acquisition was the degree of difficulty in executing the proper shape of the sign. In ape language studies, some signs are adapted in what Maestas y Moores (1980) terms "infant adapted" sign formation. For example, the sign KEY is made by placing the middle section of the index finger of one hand into the palm of the other hand. This sign was modified by keeping the index finger extended, with the fingertip touching the other palm. Because the manual dexterity of the orangutan is superior to that of the chimpanzee and gorilla (Parker, 1969), many of Chantek's later signs were not infant adapted, but taught in their adult form.

A third factor that affected acquisition was the conceptual difficulty of a sign. Table 19.1 lists the 11 semantic categories for the 127 signs that met the acquisition criterion (some signs fit more than one category). Although one criticism of ape language skills has been that apes are motivated to sign for food and drink only, these signs formed less than one-quarter of Chantek's vocabulary (22.7%). Table 19.1 shows that objects formed the largest sign category (33 signs), followed by actions (25 signs). Object and action signs composed almost half of his vocabulary. Other categories included 24 food signs, 12 name signs, 10 animal signs, 6 drink signs, 5 color signs, 5 locatives, 4 attributes, 3 emphasizers, 3 place names, and 2 pronouns.

Table 19.1 shows that semantic categories of signs formed roughly three groups: those that were easy, those of medium difficulty, and those that were difficult to acquire. Chantek was highly motivated to acquire signs for food, drinks, and emphasis or reoccurrence (such as HURRY and MORE); these signs he acquired in a short period. The middle group of signs included those for names, actions, colors, attributes such as BAD and HURT, places, objects, and animals. Those in the most difficult category were the most abstract – signs for locations and pronouns. Locative signs were actually as easy to acquire as the middle group. Their average acquisition period was distorted by the fact that it took over 2 years to acquire DOWN, because of Chantek's infrequent use of this sign. The pronouns YOU and ME were genuinely difficult to acquire. Interestingly, Chantek seemed to prefer to use the sign name of an individual rather than a pronoun, even when addressing that individual. Children also find pronouns confusing at first because the referent is flexible, depending on who is sending the message (de Villiers & de Villiers, 1978).

The ability of apes to engage in two-way exchanges resembling conversation has been a controversy in ape language research. Studies of child language have shown that an important shift in how children engage in discourse with their parents occurs during the second year of life (Brown, 1973; Caselli, 1983; Ervin-Tripp, 1970; Ferguson, Peizer, & Weeks, 1973; Kantor, 1982; Lock, 1980; Maestas y Moores, 1980; Sachs, Brown, & Salerno, 1976; Snow, 1972, 1977; Snow & Ferguson, 1977). At first, parents prompt children

Table 19.1. *Semantic categories of acquired signs (Jan. 1979–Jan. 1986)*

| | | | | | |
|---|---|---|---|---|---|
| *Objects (mean days to acquire = 197)* | | | | | |
| 1. brush[a] | 26 | 12. bag | 21 | 23. fence | 19 |
| 2. key | 192 | 13. toilet | 100 | 24. hair | 861 |
| 3. toothbrush | 143 | 14. oil | 75 | 25. house | 36 |
| 4. flower | 385 | 15. beard | 58 | 26. bike | 117 |
| 5. puppet | 181 | 16. mask | 110 | 27. eye-drink[b] | 72 |
| 6. hat | 713 | 17. ball | 162 | 28. cup | 39 |
| 7. jacket | 91 | 18. screwdriver | 122 | 29. box | 801 |
| 8. shoe | 915 | 19. Q-tip | 32 | 30. toothpaste | 46 |
| 9. straw | 35 | 20. car | 94 | 31. lipstick | 177 |
| 10. tree | 65 | 21. money | 95 | 32. barrel | 64 |
| 11. bed-sleep[a] | 136 | 22. ring | 464 | 33. earring | 57 |
| *Actions (mean days to acquire = 117)* | | | | | |
| 1. food-eat | 28 | 10. touch | 127 | 19. pick-groom | 54 |
| 2. drink[a] | 31 | 11. give | 19 | 20. keep | 195 |
| 3. come | 68 | 12. listen | 293 | 21. pull | 76 |
| 4. brush[a] | 26 | 13. share | 82 | 22. work | 183 |
| 5. tickle | 92 | 14. bed-sleep[a] | 136 | 23. kiss[c] | 1446 |
| 6. go | 28 | 15. piggy-game | 104 | 24. ride | 267 |
| 7. open | 63 | 16. smell | 58 | 25. no-teeth[b] | 400 |
| 8. chase | 29 | 17. bite | 38 | | |
| 9. hug | 267 | 18. break | 102 | | |
| *Foods (mean days to acquire = 53)* | | | | | |
| 1. food-eat[a] | 28 | 9. orange[a] | 52 | 17. butter | 130 |
| 2. nut | 53 | 10. cereal | 229 | 18. yogurt | 56 |
| 3. candy | 36 | 11. ice-cream | 36 | 19. grape[c] | 1399 |
| 4. banana | 20 | 12. berry | 18 | 20. carrot | 36 |
| 5. raisin | 17 | 13. peach | 24 | 21. cookie | 37 |
| 6. bread | 39 | 14. cheese | 15 | 22. chocolate | 25 |
| 7. apple | 33 | 15. meat | 49 | 23. jello | 115 |
| 8. cracker | 19 | 16. corn | 52 | 24. egg | 93 |
| *Names (mean days to acquire = 110)* | | | | | |
| 1. Chantek | 22 | 5. John | 223 | 9. Ray | 25 |
| 2. Ann | 74 | 6. Jackie | 189 | 10. Rick | 76 |
| 3. Lyn | 50 | 7. Richard | 31 | 11. Dave[b] | 301 |
| 4. Kim | 62 | 8. Michael | 25 | 12. Jeannie | 243 |
| *Animals (mean days to acquire = 208)* | | | | | |
| 1. dog | 160 | 5. alligator | 48 | 9. ape | 134 |
| 2. cat | 329 | 6. bug | 185 | 10. horse | 207 |
| 3. monkey | 24 | 7. squirrel | 208 | | |
| 4. bird | 697 | 8. elephant | 91 | | |
| *Drinks (mean days to acquire = 69)* | | | | | |
| 1. drink[a] | 31 | 3. Coke | 207 | 5. water | 42 |
| 2. milk | 21 | 4. coffee | 79 | 6. tea | 33 |
| *Colors (mean days to acquire = 120)* | | | | | |
| 1. orange[a] | 52 | 3. black | 55 | 5. green | 30 |
| 2. red | 435 | 4. white | 26 | | |

Table 19.1. *(cont.)*

| | | | | | |
|---|---|---|---|---|---|
| *Locatives (mean days to acquire = 297)* | | | | | |
| 1. up | 131 | 3. out | 67 | 5. down | 934 |
| 2. point | 29 | 4. in | 325 | | |
| *Other attributes (mean days to acquire = 136)* | | | | | |
| 1. dirty | 117 | 3. bad | 237 | | |
| 2. good | 38 | 4. hurt | 152 | | |
| *Emphasizers/reoccurrences (mean days to acquire = 69)* | | | | | |
| 1. more | 127 | 2. Hurry | 56 | 3. time | 25 |
| *Places (mean days to acquire = 175)* | | | | | |
| 1. Cadek-Hall | 40 | 2. Brock-Hall | 211 | 3. yard | 273 |
| *Pronouns (mean days to acquire = 565)* | | | | | |
| 1. me | 631 | 2. you | 502 | | |

[a] Sign belongs to more than one semantic category.
[b] Sign invented by Chantek.
[c] Sign not included in the mean because its teaching was interrupted.

through ritualized questions and other devices. Later, at approximately 15–16 months of age, children develop their own initiations.

Mandel and Fouts (1976), Patterson and Linden (1981), and Van Cantfort and Rimpau (1982) concluded that chimpanzees and gorillas can engage in meaningful sign exchanges with their caregivers and each other, much like human children. On the other hand, Terrace et al. (1979) concluded that their chimpanzee Nim had not mastered even the rudimentary skills of conversation. Nim interrupted and imitated his "teachers" and did not spontaneously initiate signing. This failure was seen as significant, because Terrace et al. (1979) stressed that spontaneous use of signs is an essential element of language development.

Chantek's two-way sign exchanges with his caregivers were compared with similar samples of Nim's conversations at the same age of 26 months, using precisely the same methods of data collection and analysis (Miles, 1983). The results showed that Chantek produced significantly less imitation and more spontaneous signing than did Nim. Whereas 38% of Nim's communications were imitations of his human companion's signing, Chantek's rate of imitation was only 3%. Conversely, whereas 4% of Nim's communication appeared to be spontaneously initiated by him, 37% of Chantek's communications were spontaneous. Following the study period, Chantek's rate of spontaneous signing increased to an average of 49% and ranged as high as 88% of his communications. Chantek continued to sign spontaneously even in special videotaped interactions in which his caregiver refrained from signing. During these interactions, Chantek continued to sign at the rate of about two communications per minute. Thus, mere imitation of the caregiver was ruled out as the source of Chantek's signing.

Chantek also interrupted approximately as often as he was himself interrupted by his caregiver. In the interaction comparison, 8% of Chantek's communications were interruptions of his caregiver, and 9% of his caregiver's communications interrupted him (based on criteria used with Nim). In comparison, Nim's caregivers interrupted him twice as often as Chantek was interrupted. This may have encouraged a different conversational model, resulting in Nim's high rate of 35% interruptions. Even this high rate may have been misunderstood, however, because overlap of signing occurs in human conversations (Baker, 1977), and this effect is not nearly as significant as it might be in speech (Van Cantfort & Rimpau, 1982). Nevertheless, the language model taught to Nim encouraged learning to interrupt.

In Chantek's second month of training he began spontaneously to combine signs into sequences, such as COME FOOD-EAT (Miles, 1983). Chantek came to use approximately 25 different combinations of signs each day. Most combinations were two or three signs in length. Some sign order regularities were observed. For example, Chantek was more likely to name an object and then sign GIVE if the object was present, and GIVE followed by the object name if it was not. This is particularly interesting because the form used with Chantek was always GIVE + (object name), and he did not model this variation on his caregivers' signing. The obvious presence of the object may have provoked Chantek to name it first, before he described the activity he desired to enact with it. Some children use a topic/comment form of grammar, and in American Sign Language the most important word of a sentence often is expressed first. It is important to keep in mind that the inflected nature of sign language does not require the order regularities found in English (Klima & Bellugi, 1979). However, Chantek's regularities in sign order suggest that Chantek developed a contextually dependent modulation of meaning analogous to the means by which a speaking child learns to use a certain word order (Greenfield & Savage-Rumbaugh, P&G20).

Tables 19.2 and 19.3 list Chantek's communications and their contexts for two 1-week periods at the University of Tennessee and the Yerkes Center in 1986. Tables 19.2 and 19.3 show that the majority of Chantek's communications were single signs or two-sign combinations, with an occasional three- or four-sign combination. The mean length of his communications taken in citation noninflected form remained approximately 2.0, characteristic of children in the earliest stage of language acquisition (approximately 2 years of age). Calculation showed that his mean length of utterance (MLU) based on gestural inflected modulations was slightly higher. That was in strong contrast with the long strings of up to 16 imitative signs produced by the chimpanzee Nim under different conditions of interaction (Terrace et al. 1979). Tables 19.2 and 19.3 show that typical sign combinations were COKE DRINK after finishing drinking his Coke, PULL BEARD while pulling a caregiver's hair through a fence, TIME HUG while locked in his cage as his caregiver looked at her watch, and RED BLACK POINT to a group of colored paint jars.

Table 19.2. *Chantek's signed communications (Feb. 10–16, 1986)*

| Communication | Context |
|---|---|
| *Food* | |
| apple[a] | at the kitchen |
| apple | for pineapple |
| banana | for banana |
| berry[a] | at the kitchen |
| berry ice-cream[a] | while locked in his cage |
| bread[a] | at the kitchen |
| butter nut[a] | at the kitchen |
| candy[a] | at the kitchen |
| carrot | for carrot |
| carrot point (refrigerator)[a] | at the kitchen |
| cereal | for box of oatmeal |
| cereal | for caregiver to put oatmeal in his hot chocolate |
| cereal coffee tea point (table)[b] | for box of oatmeal on the table while caregiver fixed a cup of tea |
| cereal point (refrigerator)[a] | while sitting on the platform before breakfast |
| cheese[a] | at the kitchen |
| cheese | for mozzarella cheese |
| chocolate[a] | while caregiver fixed him a cup of milk |
| cookie | for cookie |
| cookie cracker point (cookie) | at the kitchen while caregiver ate a cookie |
| cracker | for cookie |
| cracker[a] | at the kitchen |
| cracker point (floor)[a] | while locked in his cage |
| egg point (refrigerator)[a] | at the kitchen |
| food-eat[b] | for caregiver to give him lettuce |
| food-eat[b] | for caregiver to give him bell peppers |
| food-eat[b] | for caregiver to give him balloons |
| ice-cream | at the kitchen |
| jello | for cookie (form error) |
| nut butter cracker point (refrigerator)[a] | at the kitchen |
| nut cracker point (refrigerator)[a] | at the kitchen |
| nut point (refrigerator)[a] | at the kitchen |
| point (refrigerator) nut apple[a] | at the kitchen |
| *Drinks* | |
| cereal coffee tea point (table)[b] | for box of oatmeal on the table while caregiver fixed a cup of tea |
| milk | for milk |
| milk | caregiver asked what point (powdered milk)? |
| tea | while caregiver made tea |
| water | for bottle of water |
| water point (container)[c] | for empty container |
| *Objects* | |
| bag[a] | at the kitchen |
| bag | caregiver asked what point (bag)? |
| bed-sleep | while pulling caregiver into the hammock |
| car | as he passed caregiver's car on a walk |

Table 19.2. *(cont.)*

| Communication | Context |
|---|---|
| fence | caregiver asked what point (fence)? |
| jacket | for his jacket hanging in the kitchen |
| key[a] | at the locked partition gate |
| key | caregiver asked what (plastic spoon)? |
| mask | for Bugs Bunny mask |
| money | for token money in his cup |
| money point (token money) | for token money on the floor |
| nut | for small box of pistol caps |
| oil | for bottle of hand lotion |
| Q-tip | caregiver asked what (phone)? (form error) |
| ring | caregiver asked what point (ring)? |
| shoe | caregiver asked what point (shoe)? |
| time | when caregiver looked at her pocket watch |
| toothbrush[a] | at the kitchen (where toothbursh kept) |
| boothpaste[a] | at the kitchen (where toothpaste kept) |
| *Play* | |
| beard pull | during hair-pulling game |
| Chantek chase | at the kitchen |
| come Chantek | at the kitchen |
| Jeannie Chantek chase | at the kitchen with Kristine |
| keep come | at the kitchen for keep-away game |
| keep | to play keep-away with objects |
| key tickle | during tickling |
| kiss tickle[b] | during play |
| pull | to pull objects through the fence in the pull-game |
| pull beard | while pulling caregiver's hair through the fence |
| tickle | to start tickling |
| tickle no-teeth | during tickling |
| tickle chase | for caregiver to chase him along the fence |
| you pull | after tossing caregiver a balloon for the pull-game |
| *Nonplay actions* | |
| food-eat[b] | for caregiver to give him lettuce |
| food-eat[b] | for caregiver to give him bell peppers |
| food-eat[b] | for caregiver to give him balloons |
| hug | at the kitchen |
| kiss | during tickling |
| kiss tickle[b] | during play |
| no-teeth kiss | during play |
| open | at the locked partition gate |
| share | to caregiver with peanut butter cracker |
| time hug | while locked in his cage and caregiver looked at her watch |
| work | when he had no tokens and needed to work for money |

[a] Communication was displaced reference.
[b] Communication belongs in more than one category.

Table 19.3. *Chantek's signed communications (Feb. 24–Mar. 2, 1986)*

| Communication | Context |
|---|---|
| *Food* | |
| bread | for bean cake |
| bread[a] | while in his cage |
| bread point (bag)[a] | for garbage bag containing food |
| candy[a] | while in his cage |
| candy ice-cream | for fruit ice cream bar |
| candy point (front door)[a] | while in his cage |
| candy point (kitchen door)[a] | while caregiver was hosing his cage |
| candy point (kitchen door)[a] | while in his cage |
| cereal[a] | while in his cage before breakfast |
| cereal | for box of oatmeal |
| chocolate[a] | while in his cage before breakfast |
| Coke[a] | while in his cage |
| cookie[a] | while in his cage before breakfast |
| cracker[a] | while in his cage before breakfast |
| cracker[a] | while looking at bag of food |
| cracker[a] | for box of stone wheat crackers |
| ice-cream[a] | while in his cage |
| meat | for cheeseburger in wrapper |
| nut[a] | while looking at bookbag |
| nut point (bag)[a] | for paper bag containing food |
| orange | for orange |
| orange point (orange) | caregiver asked what want? |
| peach | for peach |
| raisin | for raisins |
| *Drinks* | |
| Coke drink | after finishing his Coke |
| cup | for cup of soda on the table |
| milk[a] | while in his cage before breakfast |
| milk drink[a] | while in his cage |
| time drink[a] | when caregiver looked at her watch |
| water | for cup of hot water |
| water point (glass) | to glass of water on the table |
| *Objects* | |
| bag | for plastic bag |
| bag | for roll of paper towels |
| car point (door)[a] | toward the closed door facing the street |
| cup point (cups) | for cups on the sink |
| key | caregiver asked what you need point (to lock on his cage)? |
| key point (keys)[a] | for caregiver's keys in her pocket when he wanted the door opened |
| point (cups) | to cups in his cage |
| time point (watch)[a] | for caregiver's pocket containing watch |
| toothbrush[a] | while in his cage in the morning |
| *Nonplay actions* | |
| bed-sleep | while in his cage |
| bed-sleep hug | while in his cage |

Table 19.3. *(cont.)*

| Communication | Context |
|---|---|
| candy food-eat[a,b] | while in his cage |
| candy point (candy) smell candy[b] | to visitor with a candy cane |
| cracker food-eat[a,b] | while in his cage |
| food-eat[a,b] | while in his cage before breakfast |
| food-eat good[a,b] | while in his cage before breakfast |
| give cracker[a,b] | to request more carrot chips |
| hug | when caregiver started to leave |
| hug | when Lyn arrived |
| hug cracker food-eat[a] | while in his cage |
| kiss | before kissing Lyn at bedtime |
| listen | when he heard gibbons vocalizing |
| listen point (radio) | to radio playing music |
| Lyn point (front of cage)[b] | for Lyn to come and rest with him |
| open | while looking at the closed front door |
| open | while in his cage |
| open point (front door) | while looking at closed front door |
| out | while in his cage |
| pick-groom | while resting in his cage with Lyn |
| point (door) give Ann[a,b] | while in his cage |
| smell | to smell the hand of a technician (Karen) |
| smell | to smell the hand of a supervisor (Jimmy) |
| smell you | for hand of visiting technician |
| smell point (front door) | while in his cage |
| *Attribute* | |
| red black point (paint jars) | to group of different colored paint jars (including red and black) |
| *Internal state* | |
| dirty | while in his cage before urinating |
| good | when caregiver returned after he misbehaved |
| *Specific person* | |
| Lyn | when Lyn arrived |
| Lyn point (front of cage)[b] | for Lyn to come and rest with him |
| point (door) give Ann[a,b] | while in his cage |
| *Animal* | |
| monkey point (front door)[a] | toward front door and grounds where monkeys were caged |

[a] Communication was displaced reference.
[b] Communication belongs in more than one category.

## The development of reference

The acquisition of representational skills is not a simple matter. Representational skills require an understanding that the sign or word stands for or represents a person, object, action, or idea. An ape's use of a sign may be simply a learned association between the gesture and the object (Savage-Rumbaugh, 1986). The ape may not form a concept of the referent, be able to generalize it, and understand that it can be classified into categories according to its attributes. It cannot be assumed that an ape knows that a sign refers to and symbolizes something in the real world in an abstract sense (Savage-Rumbaugh, Pate, Lawson, Smith & Rosenbaum, 1983). There may not be a natural progression from learned label to referential symbol. There may be stages or degrees of referential ability, or it may be compartmentalized so that some aspects are expressed while others are not. Even if the ape shows referential ability, this may develop easily if the necessary cognitive elements underlying referential ability are present, or it may emerge only after extensive drills, resulting in a learning-to-learn phenomenon.

Linguistic representation consists of at least three elements:

1. A sign must designate an element of the real world.
2. A shared cultural understanding about its meaning must exist.
3. The sign must be used intentionally to convey meaning.

There are several types of evidence that suggest that the subject may be using words or signs referentially: first, that signs can be used to indicate an object in the environment; second, that signs are not totally context dependent; third, that signs have relevant semantic domains or realms of meaning; fourth, that signs can be used to refer to objects or events that are not present.

Referential communication has several components and stages. In human children, the act of pointing to draw attention to an object, person, or event begins by 9–12 months of age and is followed by labeling, semantic overgeneralization or undergeneralization, and finally representational naming, as a word comes to have a more specific shared referential meaning (Bates, 1976; Lock, 1980). Savage-Rumbaugh et al. (1983) have argued that apes and children do not go through similar referential developmental stages. However, evidence from Chantek's use of signs suggests that Chantek exhibited each of the stages shown by children, beginning with what Bates (1976) called the *protodeclarative*.

The *protodeclarative* involves four substages. The first stage of exhibiting self occurs at around 9 months of age. The child voluntarily and intentionally repeats socially oriented actions. Chantek first engaged in similar behaviors, such as rattling objects and making raspberry noises to attract attention, at around the same age (10 months). The second stage involves showing objects. Chantek also advanced to this stage and frequently made eye contact while presenting an object, such as a blade of grass, for us to look at. The third stage consists of giving objects. Savage-Rumbaugh et al. (1983, p. 462) reported that this and later stages did not occur with the chimpanzees they trained to use computer lexigrams, or occurred only with extensive drills. However, Chantek often would pick up objects in his yard or on walks, make eye contact with his caregiver, and give an object. He was especially fond

Table 19.4. *Development of pointing (1979–1980)*

| Date | Age (mo) | First use |
|---|---|---|
| Jan. 1979 | 13 | ME (point handshape) |
| May 1979 | 17 | GO (meaning to move forward) |
| May 1979 | 17 | GO-THERE (meaning to move in a particular direction) |
| Sep. 1979 | 21 | (pointing to body parts in an imitation game) |
| Feb. 1980 | 26 | POINT (spontaneously pointed to indicate body parts; e.g., TICKLE POINT [nose]) |
| Mar. 1980 | 27 | POINT (spontaneously indicated food and drinks; e.g., POINT [bottle] DRINK) |
| Mar. 1980 | 27 | POINT (spontaneously indicated objects; e.g., OPEN POINT [door]) |
| May 1980 | 29 | POINT (spontaneously indicated locations; e.g., LYN GO POINT [location]) |

of giving rather wormy nuts from trees on campus that he would insist his caregiver taste first. Based on this spontaneous sharing, at 34 months of age Chantek learned the sign SHARE to express this giving and social exchange.

The final stage of the protodeclarative involves indicative pointing, with frequent eye contact with others. Table 19.4 shows the way in which Chantek developed indicative pointing. Several signs were introduced to Chantek at 13 months of age that utilized a point handshape, such as CANDY and LISTEN. Chantek also learned the sign ME (pointing to self), but it was not referential at first. It apparently functioned semantically to mean that he wanted to be the recipient of an action. At 17 months of age, Chantek learned the sign GO and first used this sign for his caregivers to move him from one location to another. Within a few days, and without any instruction from his caregivers, Chantek began to modulate the meaning of the sign by modifying the form of articulation. Chantek would articulate the sign in a directional orientation in order to control his caregiver's movements toward one location or another. At 21 months, a pointing game was introducted to Chantek in which his caregivers pointed to different body parts and asked WHERE NOSE?, WHERE EYES?, and so forth. Chantek responded by imitating the caregiver and pointing to his own nose, eyes, and other body parts. At 26 months of age, Chantek began spontaneously to point to where he wanted to be tickled. He also used the POINT sign to respond to questions such as WHERE YOU WANT TICKLE? At 27 months of age, Chantek first pointed to food and objects in an indicative manner, that is, to bring his caregiver's attention to the object or simply to acknowledge its presence. He also responded by pointing when asked WHERE HAT?, WHICH DIFFERENT?, and WHAT WANT? Finally, at 29 months of age, he began spontaneously to use the POINT sign in his requests to move about in combination with GO. For example, he would sign LYN GO POINT (to a location).

Thus, Chantek moved from exhibiting self and showing and giving objects at 13 months to distal indicating at 17 months, self-directed proximal pointing at 26 months, object oriented proximal pointing at 27 months, and distal pointing at 29 months. Finally, at 30 months of age he combined proximal and distal pointing in sign combinations. In their study of a deaf child, Hoffmeister and Moores (1973, p. 5) reported a similar pattern and argued that "within the early stages of sign language development . . . the proximal pointing indicates specific reference." Although this progression occurred later in Chantek than it normally occurs in human children, the stages were essentially the same, suggesting a developmentally delayed emergence of reference.

Because Chantek learned several signs before he acquired the POINT sign and specific reference, his early sign usage was likely associational. As he acquired more signs, however, our contextual data showed additional patterns similar to those in human children. Following preverbal pointing, a child goes on to learn labels for things and may semantically overextend or underextend the meanings of words and signs (Clark & Clark, 1977). For example, Braunwald (1978) conducted a study of the utterances of her child from 8 to 20 months and the contexts in which the utterances occurred. At first, the child used general sounds to refer to a variety of related contexts, such as spilled milk, milk in a bottle, and so forth. Later she could make more specific designations and comment on the attributes of features of her environment, such as "Cold milk." As Chantek's vocabulary increased, he began to create sign combinations such as RED BIRD and WHITE CHEESE FOOD-EAT, thus showing evidence of the ability to encode some attributes semantically. The context or situation in which a word or sign is used is also a clue as to the meaning of the word to a child. The meanings are not necessarily the same as adult meanings. For example, a child might say "Bow-wow" not only for dogs but also for cats, fur coats, and so forth (Bloom & Lahey, 1978; Clark & Clark, 1977; Nelson, 1973).

Several examples of such overextensions have been reported in the ape language literature (Gardner & Gardner, 1978; Patterson & Linden, 1981). For example, Patterson and Linden (1981) reported that the gorilla Koko saw a frog and signed RED FROG. They concluded that Koko was joking (because the frog was green). However, full contextual information might show that this was simply an error – Koko might frequently confuse names for different colors. Full contextual information would be necessary to support the conclusion that apes convey humor through signs. By collecting total contextual data for all of Chantek's signs and sign combinations, we are able to carry out analyses of errors, patterns of overextension and underextension, semantic processes, and so forth, and move beyond the level of anecdote.

The results show that most of Chantek's signs came to have adult meanings. For example, 96% of his uses of the sign LISTEN were appropriate. He first signed LISTEN GO when he heard musical instruments being played in the music building on campus. He later used LISTEN to refer to the sounds of birds, bells, buzzers, music boxs, sirens, automobiles, watches, thunder, and gunshot. However, a contextual analysis of Chantek's signs shows that an ape's use of signs does not always map adult usage and that the meanings of signs change over time. For example, Chantek increased his adult use of BANANA for bananas or banana-flavored foods from 61% in 1979 to 90% in 1981. In the same period, Chantek used CAT for cats, cat noises, and "meow" sounds made by cats and his caregivers. He also overextended CAT to include dogs, birds, and an opossum. The error pattern in his use of caregiver names is interesting. About half of his uses of LYN were correct; however, all but 4% of his errors were in reference to other caregivers. Thus, he had overextended LYN to refer to all caregivers, much as the chimpanzee Washoe overextended the name ROGER to refer to all humans. Interestingly, Chantek never used LYN or other caregiver names to refer to strangers. This use of LYN represented an overgeneralization or class error: LYN = Lyn + other caregivers.

Chantek also continued to make class errors regarding his color terms. For example, he mistakenly described strawberries as BLACK and ORANGE, although he usually called them RED. Color term accuracy is also not present in human children in the earliest stage of language acquisition. It is important to note that when Chantek was asked the color of strawberries, he never signed wildly inappropriate responses such as CAT or BAG. Some errors were also form errors – slight variations in how a sign was made that resulted in a very different meaning, much like a slip of the tongue. For example, Q-TIP and PHONE are made near the ear in similar ways. JELLO and COOKIE are also similar in form, with a movement present with the sign JELLO, but not COOKIE. These errors also indicate that these signs are processed and stored in cheremic form (similar to phonemes in speech).

An analysis of DIRTY reveals a shifting meaning pattern. The sign DIRTY was originally used for urine or feces, after the convention adopted by Gardner and Gardner (1969). By 1981, 70% of Chantek's communications using DIRTY were for urine or feces. However, he also overextended its meaning to refer to soiled objects as well as "bad" behavior when he was scolded. This represented an overlap in meaning. When the sign BAD was later introduced to describe disapproved behavior, Chantek decreased his use of DIRTY for these circumstances from 29% to 10% of his communications, and he began to use BAD instead. This resulted in a shifting semantic domain for DIRTY.

Chantek's use of LYN, BAD, and DIRTY represented simple errors or meaning shifts. More complex was his use of OPEN. Initially, Chantek signed OPEN and the first combination, OPEN FOOD-EAT, in front of closed doors, particularly the refrigerator and kitchen cabinets. He later generalized this signing to request the opening of jars and other objects. He used the sign twice to have his caregiver crack open a peanut shell. Superficially, it appeared that he understood the adult usage of OPEN. However, he also overextended the meaning of OPEN when he wanted a caregiver to move a large object or play apparatus such as the jungle gym. Without contextual data, there would have been a tendency to ignore this as a careless error and assume he had mastered adult usage. With contextual data it became clear that Chantek used OPEN in some cases to mean something like "move a large object." It also meant to make the object available for play by rocking it back and forth or moving it from one location to another. This is a broader meaning than to cause to be unclosed or uncovered.

Tables 19.2 and 19.3 show that Chantek continued to overextend some of his signs. For example, during a 2-week period, he overextended APPLE for pineapple, CRACKER for a cookie, NUT for small round pistol caps, BEARD for hair, NUT for peanut butter, MILK for hot chocolate, BREAD for bean cake, BAG for a roll of paper towels, and MEAT for a cheeseburger. In some cases Chantek learned more specific signs for these items.

Chantek also produced other overextensions of his signs. He has signed DOG for a sleeping black dog, barking dogs, a picture of a dog in his View-master, orangutans on television, barking noises on the radio, a bird, a horse, a tiger at the circus, a field of cows, a picture of a cheetah, and a noisy helicopter that presumably sounded like it was "barking." DOG apparently refers to dogs, pictures and sounds of dogs, and similar animals. He used BUG for crickets, roaches, a dead roach, a picture of a roach, a beetle, a slug, a small moth, a spider, a worm, a fly, a picture of a graph shaped like a butterfly, tiny brown pieces of cat food, and small bits of feces. He signed BREAK after he broke his toilet and also before he broke and shared pieces of a cracker. He called the hair on a caregiver's leg a BEARD, and he got into an oversized jumpsuit and called it a BAG. He signed BAD to himself before grabbing a cat, when he bit into a radish, and for a dead bird. These uses form a rich, varied, and often surprising corpus of contexts. However, what is significant is that in all of these examples, Chantek was attending to a variety of aspects, such as form, shape, color, texture, function, size, and so forth, in using his signs to describe, refer to, and categorize his environment.

The ability to refer to things or events not present is called displacement. An important transition in children's language development occurs when a child can communicate about other than the here and now and can make signs outside the specific context (Bates, 1979; MacNamara, 1972). Over the period of study, there were increasing numbers of occasions when it appeared that Chantek was signing about places or objects not present. For example, he signed CEREAL POINT (to the refrigerator) FOOD-EAT before breakfast, indicating that he knew his cereal was kept in the refrigerator. He frequently asked to go to places in his yard to look for animals, such as his tame squirrel, which served as a playmate. He also made requests for ICE-CREAM and signed CAR RIDE and pulled us toward the parking lot.

A developmental trend was clear in Chantek's displacements. In 1982, when he was 4 years old, 19% of Chantek's signing showed displacement. In 1986, when he was 8 years old, that rate had increased to 37% (Tables 19.2 and 19.3). Contextual factors were also important. After we introduced place names, such as CADEK-HALL and YARD, his displaced references increased. Seventy-six percent of his uses of the sign BROCK-HALL occurred while the building was not in view. In addition, after signing BROCK-HALL, he would point across campus in the direction of the building and pull his caregiver toward that area. He also correctly labeled slides of Brock Hall in double-blind vocabulary tests. Because he was more confined at the Yerkes Center than he was at the University of Tennessee, his rate of displaced references increased to 38%. When first transferred to Yerkes Center, he was understandably upset and concerned about separation from his caregivers. He gave the first clear evidence of displaced requests to see his caregivers at that point. For example, he signed GIVE ANN POINT (to the front door) and showed agitation as he watched the door for the different cars and individuals who passed by. Thus, although the major portion of Chantek's signing continued to be about the here and now, the growing references to people and things that were not present is further evidence that his signing moved beyond mere labeling.

Fouts (1973), Terrace (1979), and Patterson and Linden (1981) reported incidents in which they believed that apes might be using their signs to deceive. Deception is an important indicator of language abilities, because it requires a deliberate and intentional misrepresentation of reality. De Villiers and de Villiers (1978, p. 165) defined linguistic deceit as "the deliberate use of language as a tool to escape reprimand, with the content of the message unsupported by context, and designed to mislead the listener." In their view, deception is the ultimate achievement of discourse ability. Mitchell (1986) and Miles (1986) identified stages in the development of deception and pointed out that mature deception requires the ability to understand a construction of reality, how this is perceived by others, and how to negate this perception from the perspective of the other.

A study of the development of Chantek's ability to deceive showed that Chantek engaged in an average of three deceptive incidents per week (Miles, 1986). The majority of these incidents involved his use of the sign DIRTY in order to gain access to the bathroom, where he liked to play with the washer and dryer, soap dish, and so forth. On one occasion the caregiver became angry with Chantek because once in the bathroom he failed to use the toilet. Chantek sat on the toilet and pressed his penis, but failed to extract any drops of urine. He also used his signs to gain social advantage in games, to divert attention in social interactions, to avoid testing situations, and to avoid coming home after walks on campus.

Many examples of Chantek's deception show coordination and association of multiple behaviors with the proper signs. For example, on one occasion Chantek stole food from a caregiver's pocket while he simultaneously pulled her hand away in the opposite direction. On another occasion he stole the caregiver's pencil eraser, pretended to swallow it, and "supported" his case by opening his mouth and signing FOOD-EAT. However, he really held the eraser in his cheek, and later it was found in his bedroom, where he commonly hid objects. These examples show evidence of intentionality, premeditation, taking the perspective of the other, and displacement. These processes require that some form of mental image about the outcome of events be created. Desmond (1979) pointed out that deception requires the ability to model introspectively the actions of another. Meddin (1979) suggested that such "role playing" is evidence for symbolic functioning, and Vasek (1986) showed that similar perspective taking is essential to lying performed by humans.

Another example of the formation of some kind of mental image was found in Chantek's ability to respond to his caregiver's request that he improve the articulation of a sign. When his articulation became careless, a caregiver would ask him to SIGN BETTER. He then would sign in a slow emphatic way, and with one hand he would put the other hand in the proper sign shape. His changes provided important evidence that Chantek understood that the sign had to approximate an ideal shape and was not simply a trained motor response. This response required Chantek to have a mental image of the proper form of the sign. The fact that he would sign slowly and in a step-by-step manner indicated that he could also take the perspective of the other, as he gradually constructed the proper form of the sign. This interpretation was reinforced by the frequent eye contact he made with the caregiver as he offered the revised form of the sign.

Further evidence of a mental sign image came from an unusual aspect of Chantek's sign production – his execution of signs with his feet. Obviously, Chantek was not taught to sign with his feet, but the arboreal nature of orangutans may have predisposed him to regard his feet as another set of hands. At 13 months of age, Chantek first used his foot to sign FOOD-EAT. Later he executed seven other signs with his feet: DRINK, BRUSH, RAISIN, KEY, BANANA, and PUPPET. Chantek used one foot, both feet, and feet and hands in combination. Chantek had a strong left-hand preference in signing, but the data on foot preference have not yet been analyzed. It will be interesting if we discover that he also possesses "footedness." His foot signing was significant in that it indicated that Chantek understood the gestalt of a sign configuration and that a sign could be executed in a number of ways with different body parts. Chantek even began to use objects in relation to each other to form signs. For example, Chantek used the blades of scissors instead of his hands to make the sign for biting, applied to his body or to objects. For Chantek, then, a sign was not simply a motor skill that was associated with an object or action. By transferring the total shape of the sign, including configuration and movement, to another means of expression, he showed that he understood that the sign was an abstract representation in which the composite of elements stood for something else. That supports the evidence that his processing of signs was abstract and representational.

### Cognitive development and stages of representation

Language and cognition have been thought by some to be closely related, even homologous, in ontogeny and phylogeny (Bates, 1979). Developing

Table 19.5. *Chantek's performance on the Bayley Scales of Infant Development (Mental scale) (Nov. 1979–June 1986)*

| Date | Age (yr; mo) | Base–ceiling item | Mental age (mo.) | Raw score | Mental development index |
|---|---|---|---|---|---|
| Nov. 1979 | 1; 11 | 110–113 | 13.6–14.2 | 111 | — |
| Dec. 1979 | 2; 0 | 110–117 | 13.6–15.3 | 116 | 50 |
| Dec. 1979 | 2; 0 | 117–138 | 15.3–21.4 | 127 | 71 |
| Feb. 1980 | 2; 2 | 122–144 | 17.0–23.4 | 134 | 73 |
| Mar. 1980 | 2; 3 | 122–152 | 17.0–25.6 | 137 | 72 |
| Apr. 1980 | 2; 4 | 122–152 | 17.0–25.6 | 139 | 72 |
| Aug. 1982 | 4; 8 | 129–161 | 19.3–30+ | 146 | — |
| Jun. 1983 | 5; 6 | 134–161 | 20.0–30+ | 152 | — |
| Apr. 1984 | 6; 4 | 139–161 | 21.6–30+ | 154 | — |
| Jun. 1985 | 7; 6 | 139–162 | 21.6–30+ | 158 | — |

a cognitive framework within which to examine language development is essential. Piaget developed stages of cognitive development in human children, and these have been applied to studies of orangutans (Chevalier-Skolnikoff, 1983) and other primates (see Parker, P&G1, for a summary). Because a goal of Project Chantek was to study the development of Chantek's cognitive abilities, a program of both formal and informal cognitive testing was carried out with Chantek. Informal studies focused on his use of tools, symbolic play, ability to solve problems, puzzle completion, object sorting, painting, and drawing of geometric shapes. Chantek's own freestyle drawings resembled those of 3-year-old human children. He also learned to imitate drawing horizontal lines, vertical lines, and circles.

Formal cognitive testing was carried out with Chantek using standardized cognitive tests developed for human children that had not been normalized for an orangutan population. Three primary scales were administered: the Bayley Scales of Infant Development (Mental scale) (Bayley, 1969), the Uzgiris–Hunt Scales of Infant Development (Uzgiris & Hunt, 1975), and the Miller Composite Sensorimotor Scales (Miller, Chapman, Branston, & Reichle, 1979).

The Bayley scales were first administered when Chantek was 2 years old. The results of Bayley testing are shown in Table 19.5. At 2 years of age, Chantek scored a mental development index of 50, with an equivalent human mental age of 13.6 months (range 10–20 months). He rapidly advanced at 28 months to a human equivalent mental age of 17 months (range 12–24 months). Chantek reached an equivalent human mental age of 20.5 months (range 16–29 months) when he was $5\frac{1}{2}$ years old. On our last administration of the Bayley Mental scale, Chantek scored an equivalent human mental age of 21.6 months (range 19–30+ months). On the Bayley test, Chantek was able to point to specific pictures, build cube towers, fold paper, show an understanding of several prepositions, and put pegs in a board. He also was

Table 19.6. *Chantek's performances on the Uzgiris–Hunt Infancy Assessment Scales and the Miller Composite Sensorimotor Scales (June–Oct. 1982)*

| Date | Age (yr; mo) | Scale | Items | Mental age (mo) |
|---|---|---|---|---|
| *Uzgiris–Hunt Scales* | | | | |
| Jun. 1982 | 4; 6 | I | 1–7 | 0–1 |
| Jun. 1982 | 4; 6 | II | 1–10 | 1–4 |
| Jun. 1982 | 4; 6 | I | 8–15 | 0–1 |
| Jul. 1982 | 4; 7 | II | 11–12 | 1–4 |
| Jul. 1982 | 4; 7 | IIIb | 1–4 | 4–10 |
| Jul. 1982 | 4; 7 | IV | 1–3 | 10–12 |
| | | IV | 4–7 | 10–12 |
| | | V | 1–7 | 12–18 |
| | | V | 8–11 | 12–18 |
| | | VI | 1–3 | 18–24 |
| *Miller Composite Scales* | | | | |
| Oct. 1982 | 4; 10 | Object Permanence | 3–9 | 15–20 |
| | | Means–Ends | 3–8 | 18–21 |
| | | Causality | 3–8 | 18–21 |
| | | Space | 3–8 | 16 |
| Oct. 1982 | 4; 10 | Schemes in Relation to Objects | 0–7 | 15–20 |

able to understand the concept of "one" and complete several tasks requiring imitation of scribbles and completing a shape puzzle. Interestingly, Chantek failed items requiring extensive symbolic play and some culture-bound items, such as mending a doll exactly.

Chantek was also tested on the six scales of the Uzgiris–Hunt Infancy Assessment Scales, with the exception of the Vocal Imitation scale. The results of the Uzgiris–Hunt testing are shown in Table 19.6. All of the Uzgiris–Hunt items were appropriate for use with Chantek. He was tested six times over a 22-day period, and he successfully completed all of the items. This suggests that at $4\frac{1}{2}$ years (55 months), Chantek had achieved at least the cognitive performance level of a human child 18–24 months old.

Finally, Chantek was given the Miller Composite Sensorimotor Scales, which explore the relationship between the Piagetian sensorimotor stages and language development in children. The results are shown in Table 19.6. Chantek was given the set of five subscales and passed all subscale items except Item 8 on the Schemes in Relation to Objects subscale. Item 8 requires frequent use of one object to stand for another, pretend play, and the grouping and collecting of items. At the time of testing, Chantek gave evidence of these infrequently (Item 7). For example, he collected poker chips and sorted them into various piles and containers with good concentration, but he did not exhibit symbolic or pretend play during the sessions. Miller associated Level 7 with a mental age of 15–20 months in human children. Although not formally tested, in later years Chantek showed evidence of the ability to pass Level 8 through various forms of pretend play. Bates (1979) developed a number of criteria for studying symbolic play in children and animals, including partial and/or recombinative execution of behaviors, substitutability of one or more objects for objects usually involved in the execution of the scheme, signs of pleasure or lack of seriousness, and awareness of object substitution. Several of these criteria were present in Chantek's play pattern beginning in his second year. He engaged in chase games in which he would look over his shoulder as he darted about, although no one was chasing him. He also signed to his toys and offered them food and drink. Although none of these behaviors were as extensive as they would have been in a human child, they were present at low levels in several forms. The difference appears to be one of degree, not kind.

A second component of the Miller test is a series of language comprehension tasks, including asking a child to name objects and carry out action–object and agent–action–object routines. For example, a child would be told to have the "horsey kiss the ball." According to Miller, children in sensorimotor Stage 6 use two- and three-word combinations, but do not usually pass these language comprehension items. It is interesting to note that these items require a nonegocentric representation of the external world that provides a basis for symbolic reference. By 5 years of age, Chantek was able to pass these items.

Apart from formal testing, observational evidence also suggested that Chantek passed Piagetian sensorimotor Stages 5 and 6. Evidence of Stage 5 included experimentation in play and problem solving, such as vacuuming himself at the age of 17 months and feeding his toy animals at the age of 20 months. Evidence of Stage 6 included completion of tool-use sequences involving up to 22 problem-solving steps, delayed imitation such as attempting to "cook" his cereal in the kitchen, and self-recognition using mirrors, which began at 20 months of age, but was not fully consistent until much later, at 5 years of age.

There was also some evidence that Chantek showed cognitive skills associated with the next Piagetian stage, the preoperational stage, characteristic of 2–7-year-old human children. Chantek showed evidence of animism, a tendency to endow objects and events with the attributes of living things. He also showed evidence of the preconceptual substage characteristic of 2–4-year-old human children, in that he overgeneralized and showed perceptual constancy and representation through language and, to a lesser extent, symbolic play. He has not yet shown representation through drawing. In summary, the evidence suggests that Chantek completed the sensorimotor series and showed some aspects of the preoperational period at the cognitive level of a human child up to 4 years old.

The development of Chantek's signing and cognitive abilities can be organized into three basic stages: (1) instrumental performative, (2) subjective representation, and (3) nascent perspective taking. The instrumental perfor-

Table 19.7. *Instrumental performative stage of Chantek's development*

| Age (yr; mo) | Achievement |
|---|---|
| 1; 1 | Signs and sign combinations |
| 1; 2 | Behavioral deception (sneaking to the refrigerator) |
| 1; 3 | Imitation of signs |
| | Self-signing |
| 1; 5 | Piagetian sensorimotor Stage 5 |
| | Identification |
| | Directional pointing |
| | Modulations of signs (grammar) |
| 1; 8 | Active pretense and animism |
| 2; 0 | Vocabulary of 16 signs |

Table 19.8. *Subjective representation stage of Chantek's development*

| Age (yr; mo) | Achievement |
|---|---|
| 2; 1 | Displaced reference |
| 2; 2 | Self-pointing |
| 2; 3 | Proximal pointing |
| | Distal pointing |
| 2; 11 | Piagetian sensorimotor Stage 6 |
| 3; 2 | Planning through mental representations |
| | Self-signing (prior to acting toward a displaced object) |
| 3; 4 | Imitation of a two-dimensional representation (photograph) |
| 3; 9 | Signed deception (regarding a displaced object) |
| 4; 4 | Vocabulary of 89 signs |

mative stage extended through the age of 2 years (Table 19.7). In this first stage, he acquired 16 signs, which seemed (without additional evidence such as displacement – see the next stage) to be used primarily to manipulate or obtain people, objects, or actions, rather than to refer to them. His use of signs at this stage was most conservatively interpreted as instrumental, because he clearly understood that his signs had likely consequences, and he performed the signs to elicit those outcomes. Chantek completed sensorimotor Stage 5 and achieved an equivalent human mental age of about 21 months. During this stage, Chantek was able to imitate his caregiver's signs and behaviors, engage in behavioral deception by sneaking to the refrigerator, and sign to himself, his toys, and other animals. He modulated the meanings of his signs, displayed directional pointing, and showed rudimentary evidence of sympathetic identification, pretense, and animism.

The second stage of development, that of subjective representation, was differentiated based on evidence of reference by Chantek (Table 19.8). The subjective representation stage ranged from 2 years to almost $4\frac{1}{2}$ years of age (Table 19.8). In this stage, Chantek used his signs as symbolic representa-

Table 19.9. *Nascent perspective-taking stage of Chantek's development*

| Age (yr; mo) | Achievement |
|---|---|
| 4; 5 | Self-recognition (self-grooming in a mirror) |
| 4; 8 | Improved articulation of signs in response to requests to SIGN BETTER |
| 4; 10 | Perspective taking (manually directing caregiver's eye gaze before signing) |
| 5; 2 | Invention of signs |
| 6; 7 | Understanding of contrastive attributes |
| 6; 8 | Offered hands to be molded for new signs |

tions, but his perspective remained subjective. He gave the first evidence of displacement (signing about objects not present) and developed proximal pointing, which indicated that he had mental representations. Chantek completed sensorimotor Stage 6 and quadrupled his vocabulary to a total of 89 signs. He elaborated his deceptions and pretend play, identified and imitated two-dimensional representations such as a picture of a gorilla pointing to its nose, and combined proximal and distal pointing. He used objects in novel relations to one another to create new meanings. For example, he gave his caregiver two objects needed to prepare his milk formula and stared at the location of the remaining ingredient. He showed evidence of planning through mental representations and signed to himself about objects not present. He recognized himself in a mirror, although inconsistently. For the first time he also used signs in his deceptions. He used tools to explore his environment, and he used his caregivers as social tools to obtain attention.

The third stage, nascent perspective taking, ranged from about $4\frac{1}{2}$ years to over 8 years of age, during which his vocabulary increased to 140 signs, 127 of which met the acquisition criterion (Table 19.9). This stage appeared to be an elaboration of the previous stage. In the third stage, Chantek's representations became more objective and moved toward perspective taking, the ability to utilize the point of view of the other. The evidence indicated that Chantek mastered a nonegocentric point of view and was able to objectify his environment and others. He not only recognized himself in the mirror but also began to use it spontaneously for self-grooming. Most important, he was able to take the perspective of the other by getting the caregiver's attention and directing the caregiver's eye gaze before he began to sign.

It was at this point that he invented his first signs. By 8 years and 3 months of age he had invented five different signs: NO-TEETH (to indicate that he would not use his teeth during rough play), EYE-DRINK (for contact lens solution used by his caregivers), DAVE-MISSING-FINGER (a name for a favorite university employee who was missing a finger), VIEWMASTER, and BALLOON. He clearly understood that signs were representational labels, and he immediately offered his hands to be molded when he wanted to know the name of an object. He responded to his caregiver's requests to SIGN BETTER by altering the articulation of a sign. His cognitive development extended somewhat into the preoperational stage, which includes representation through symbolic play and mastery of rudimentary categorical conceptualizations. He developed an understanding of a number of contrasting concepts, such as hot/cold and same/different, and completed a number of complex sorting tasks that required an understanding of classification, classes, and categories.

By viewing the emergence of reference in Chantek's communications in perspective with his cognitive development, we can discern the specific processes involved in using signs symbolically, such as pointing and displacement. By the same token, a developmental approach reveals the history of the behavior. For example, our understanding of Chantek's improvement of the articulation of his signs on request was illuminated by his prior ability to transfer signs from his hands to his feet, showing that he conceived of the shape of a sign apart from its place of articulation and could separate and generalize its components. A developmental approach also shows that it is not enough to consider that apes "acquire" signs without analyzing the development of their meanings through semantic processes such as overextension and shifting semantic domains.

The prolonged period of intellectual and linguistic development that we discovered in Chantek underlines the importance of long-term developmental studies extending even into adulthood. Anecdotes can be useful, especially in the early stages of an investigation, but they fail to explain the meaning of what is being observed. This meaning can be gained only by understanding the development of a behavior, and its relationships to other behaviors, in context.

## Conclusion

The evidence suggests that Chantek acquired a sign vocabulary, invented new signs of his own, and used his signs in novel combinations to express various meanings. He developed referential ability, and that ability was elaborated and became more objective and less context dependent as his cognitive skills increased. Further, he developed that ability, viewed as an essential element of language, through an enculturation process similar to that in the social environment of human children, not as a result of extensive drills and training sessions that broke referential ability into small learning components. The development of reference was marked by the emergence of protodeclarative and indicative pointing, labeling, semantic overextension, and naming. But it was accompanied by other abilities that provided contextual support. During the study period, Chantek's cognitive development also progressed from Piaget's sensorimotor period through some aspects of the preoperational period. It appears as if this cognitive development supported and maintained Chantek's linguistic representational development and that this development formed three distinct stages, moving from performatives to subjective representations and finally representations showing perspective taking. The evidence also indicates that Chantek took the perspective of the other, understood that signs themselves were abstract representations, engaged in behavioral and linguistic deception, made reference to things not present, and formed mental representations of the real world.

The approach of this research can be helpful in understanding the development of intelligence in animals and the phylogenetic origins of human intelligence (Miles, 1975, 1990). The development of Chantek's abilities suggests a starting point for such an understanding. Chantek's skills appear to be high in memory related operations, combinatorial skills, semantic relations, deception, iconicity, and reference. They appear to be medium in the areas of imitation, symbolic complexity, displacement, and productivity. Finally, they appear to be low in elaboration of new information, symbolic play, and structural elements such as modulations and rule-following behaviors. At any rate, it suggests that the orangutan is a viable animal model for understanding language and intelligence and that within the lesser-known "red ape" lies not only a brain but also a mind.

## Acknowledgments

This research was supported by National Institutes of Health grant 1-R23-HD-14918, National Science Foundation grant BNS-8022260, and grants from the University of Chattanooga Foundation. The loan of Chantek was provided by the Yerkes Regional Primate Research Center, supported by NIH grant 00165. Thanks are extended to the editors of this volume, S. Parker and K. Gibson, and also A. Southcombe, J. Sachs, the members of Project Chantek, and especially R. Mitchell, for their assistance in this research.

## References

Baker, C. (1977). Regulators and turn-taking in ASL discourse. In L. Friedman (Ed.), *On the other hand* (pp. 215–236). New York: Academic Press.

Bates, E. (1976). *Language and context: The acquisition of pragmatics.* New York: Academic Press.

Bates, E. (1979). *The emergence of symbols: Cognition and communication in infancy.* New York: Academic Press.

Bayley, N. (1969). *Bayley scales of infant development.* New York: Psychological Corporation.

Bloom, L., & Lahey, M. (1978). *Language development and language disorders.* New York: Wiley.

Braunwald, S. R. (1978). Context, word and meaning: Towards a communicational analysis of lexical acquisition. In A. Lock (Ed.), *Action, gesture and symbol: The emergence of language* (pp. 485–527). London: Academic Press.

Brown, R. (1973). *A first language: The early stages.* Cambridge, MA: M.I.T. Press.

Bruner, J. S. (1975). The ontogenesis of speech acts. *Journal of Child Language*, *2*, 1–19.

Caselli, M. C. (1983). Communication to language: Deaf children's and hearing children's development compared. *Sign Language Studies*, *39*, 113–114.

Chevalier-Skolnikoff, S. (1983). Sensorimotor development in orang-utans and other primates. *Journal of Human Evolution*, *12*, 545–561.

Clark, H., & Clark, E. (1977). *Psychology and language: An introduction to psycholinguistics.* New York: Harcourt Brace Jovanovich.

Desmond, A. (1979). *The ape's reflexion.* New York: Dial Press.

de Villiers, J., & de Villiers, P. (1978). *Language acquisition.* Cambridge, MA: Harvard University Press.

Dore, J. (1979). A pragmatic description of early language development. *Journal of Psycholinguistic Research, 3,* 343–350.

Epstein, R., Lanza, R., & Skinner, B. F. (1980). Symbolic communication between two pigeons (*Columbia livia domestica*). *Science, 207,* 543–545.

Ervin-Tripp, S. (1970). Discourse agreement: How children answer questions. In J. Hayes (Ed.), *Cognition and the development of language* (pp. 79–107). New York: Wiley.

Ferguson, C., Peizer, D., & Weeks, T. (1973). Model-and-replica phonological grammar of a child's first words. *Lingua, 31,* 35–65.

Fouts, R. S. (1973). Acquisition and testing of gestural signs in four young chimpanzees. *Science, 180,* 978–980.

Gardner, R. A., & Gardner, B. T. (1969). Teaching sign language to a chimpanzee. *Science, 165,* 664–672.

Gardner, B. T., & Gardner, R. A. (1978). Comparative psychology and language acquisition. *Annals of the New York Academy of Sciences, 309,* 37–76.

Greenfield, P., & Smith, J. (1976). *The structure of communication in early language development.* New York: Academic Press.

Hoffmeister, R. J., & Moores, D. F. (1973, August). *The acquisition of specific reference in the linguistic system of a deaf child of deaf parents.* Research Report No. 53, Project No. 332189, U.S. Office of Education.

Hoffmeister, R. J., Moores, D. F., & Ellenberger, R. L. (1975). Some procedural guidelines for the study of the acquisition of sign language. *Sign Language Studies, 7,* 121.

Kantor, R. (1982). Communication interaction: Mother modification and child acquisition of American Sign Language. *Sign Language Studies, 36,* 233–282.

Klima, E., & Bellugi, U. (1979). *The signs of language.* Cambridge, MA: Harvard University Press.

Le May, M., & Geschwind, N. (1975). Hemispheric differences in the brains of great apes. *Brain, Behavior and Evolution, 11,* 48–52.

Limber, J. (1977). Language in child and chimp? *American Psychologist, 32,* 280–295.

Lock, A. (1980). *The guided reinvention of language.* New York: Academic Press.

MacNamara, J. (1972). The cognitive basis of language learning in infants. *Psychological Review, 79,* 1–13.

Maestas y Moores, J. (1980). Early linguistic environment: Interactions of deaf parents with their infants. *Sign Language Studies, 26,* 1–13.

Mandel, B., & Fouts, R. S. (1976, August). *Human–chimpanzee conversation in a social setting.* Paper presented at the annual meeting of the American Sociological Association.

Meddin, J. (1979). Chimpanzees, symbols, and the reflective self. *Social Psychology Quarterly, 42,* 99–109.

Miles, H. L. (1975, April). *Tool use and language in early hominids.* Paper presented at the annual meeting of the Northeastern Anthropological Association, Potsdam, NY.

Miles, H. L. (1976). The communicative competence of child and chimpanzee. In S. R. Harnard, H. D. Horst, & L. Lancaster (Eds.), *Origins of evolution of language and speech* (pp. 592–597). New York Academy of Sciences.

Miles, H. L. (1983). Apes and language: The search for communicative competence. In J. de Luce & H. T. Wilder (Ed.), *Language in primates: Implications for linguistics, anthropology, psychology and philosophy* (pp. 43–61). New York: Springer-Verlag.

Miles, H. L. (1986). How can I tell a lie?: Apes, language and the problem of deception. In R. W. Mitchell & N. S. Thompson (Eds.), *Deception: Perspectives on human and nonhuman deceit* (pp. 245–266). Albany: State University of New York Press.

Miles, H. L. (1990). The development of symbolic communication in apes and early hominids. In W. von Raffler-Engel (Ed.), *Studies in language origins,* Vol. 2. Menlo Park, CA: Benjamin.

Miller, J., Chapman, R., Branston, M., & Reichle, J. (1979). *Language comprehension in sensorimotor stages V and VI.* Unpublished manuscript.

Mitchell, R. W. (1986). A framework for discussing deception. In R. W. Mitchell & N. S. Thompson (Eds.), *Deception: Perspectives on human and nonhuman deceit* (pp. 3–40). Albany: State University of New York Press.
Mounin, G. (1976). Language, communication, chimpanzees. *Current Anthropology*, *17*, 1–7.
Nelson, K. (1973). Structure and strategy in learning to talk. *Monographs of the Society for Research in Child Development*, *38*, 1–2.
Parker, C. (1969). Responsiveness, manipulation and implementation behavior in chimpanzees, gorillas and orangutans. In C. R. Carpenter (Ed.), *Proceedings of the Second International Congress of Primatology* (Vol. 1, pp. 160–166). New York: S. Karger.
Patterson, F. G. (1978). Linguistic capabilities of a lowland gorilla. In F. C. C. Peng (Ed.), *Sign language and language acquisition in man and ape: New dimensions in comparative pedolinguistics* (pp. 161–201). Boulder, CO: Westview Press.
Patterson, F. G., & Linden, E. (1981). *The education of Koko*. New York: Holt, Rinehart & Winston.
Premack, D. (1972). Language in chimpanzees. *Science*, *172*, 808–822.
Rumbaugh, D., Gill, T., & von Glasersfeld, E., (1973). Reading and sentence completion by a chimpanzee (*Pan*). *Science*, *182*, 731–733.
Sachs, J., Brown, R., & Salerno, R. (1976). Adults' speech to children. In W. van Raffler Engel & Y. LeBrun (Eds.), *Baby talk and infant speech (neurolinguistics)* (pp. 240–245). Amsterdam: Swets & Zeitlinger.
Savage-Rumbaugh, E. S. (1986). *Ape language: From conditioned response to symbol*. New York: Columbia University Press.
Savage-Rumbaugh, E. S., Pate, J., Lawson, J., Smith, S., & Rosenbaum, S. (1983). Can a chimpanzee make a statement? *Journal of Experimental Psychology: General*, *112*, 457–492.
Schwartz, J. H. (1987). *The red ape: Orang-utans and human origins*. Boston: Houghton Mifflin.
Sebeok, T. A., & Umiker-Sebeok, J. (Eds.). (1980). *Speaking of apes: A critical anthology of two-way communication with man*. New York: Plenum.
Snow, C. (1972). Mothers' speech to children learning language. *Child Development*, *43*, 549–565.
Snow, C. (1977). The development of conversation between mothers and babies. *Journal of Child Language*, *4*, 1–22.
Snow, C., & Ferguson, C. (Eds.). (1977). *Talking to children: Language input and acquisition*. Cambridge University Press.
Terrace, H. S. (1979). *Nim: A chimpanzee who learned sign language*. New York: Knopf.
Terrace, H. S., Petitto, L., & Bever, T. (1976). *Project Nim progress reports I and II*. Unpublished manuscript.
Terrace, H. S., Petitto, L., Sanders, R., & Bever, T. (1979). Can an ape create a sentence? *Science*, *206*, 809–902.
Uzgiris, I. C., & Hunt, J. (1975). *Assessment in infancy*. Urbana: University of Illinois Press.
Van Cantfort, T., & Rimpau, J. (1982). Sign language studies with children and chimpanzees. *Sign Language Studies*, *34*, 15–72.
Vasek, M. E. (1986). Lying as a skill: The development of deception in children. In R. W. Mitchell & N. S. Thompson (Eds.), *Deception: Perspectives on human and nonhuman deceit* (pp. 271–292). Albany: State University of New York Press.

# 20 Grammatical combination in *Pan paniscus*: Processes of learning and invention in the evolution and development of language

*Patricia Marks Greenfield and E. Sue Savage-Rumbaugh*

In 1979, Terrace, Petitto, Sanders, and Bever asked "Can an Ape Create a Sentence?" Looking at the evidence from a syntactic, semantic, and conversational point of view, their answer was no. Their conclusion was based on evidence from their own research with a common chimp, Nim Chimpsky, as well as on their analysis of data from other studies of ape language (Gardner & Gardner, 1973; *Nova*, 1976; Premack, 1976; Rumbaugh, 1977). Our goal in this chapter is to demonstrate that with a different species, the bonobo or pygmy chimpanzee, under a different set of conditions, the answer can be yes.

The study of ape language is important in establishing the evolutionary roots of human language. This is a subject on which there has been tremendous controversy (Bronowski & Bellugi, 1970; Limber, 1977; Petitto & Seidenberg, 1979; Seidenberg & Petitto, 1979; Terrace et al., 1979; Terrace, Petitto, Sanders, & Bever, 1980; Terrace, 1983). Is human language unique? Is it in any sense discontinuous from all that has preceded it in evolution? (See Vauclair, P&G11.) Or can we find the evolutionary roots of human language in the linguistic capacities of the great apes?

The human being is a unique species, but so is each species. In addition to being unique, we have an evolutionary history. Our physical characteristics and our *behavior* have evolved over long periods of time.

Evolution is conservative: It modifies the material that already exists and builds on it. The conservatism of evolution is exemplified in the fact that 99% of our genetic material is held in common with the chimpanzees (King & Wilson, 1975), both the common chimps (*Pan troglodytes*) and the pygmy chimps or bonobos (*Pan paniscus*), who are the subjects of this chapter.

Evolutionary approaches to language contrast with an influential position on language: that of the linguist Noam Chomsky. Chomsky looks on language in general, and grammar in particular, as a sudden mutation in the human species. The anthropologist H. B. Sarles has called Chomsky the creationists' grammarian (Fouts & Couch, 1976; Sarles, 1972).

## Grammar

Grammar, the combinatorial aspect of symbolic communication, is often considered the sine qua non of human language (Chomsky, 1965, 1980). Apes have been shown to acquire vocabularies of meaningful symbolic elements in humanly devised languages analogous to single words in human language (Hill, 1978; Savage-Rumbaugh, Rumbaugh, Smith, & Lawson, 1980; Terrace et al., 1979). They have also spontaneously combined two or more elements in nonrandom ways (Fouts, 1974a; Fouts & Couch, 1976; Gardner & Gardner, 1969, 1971; Patterson, 1980).

From a linguistic point of view, however, this is not sufficient to demonstrate that they have grammar. A grammar is a set of formal rules for marking relations between categories of semiotic elements. At the minimum, a grammatical rule should satisfy five criteria. These criteria are implicit in modern linguistics; research in child language and ape language has caused them to become increasingly explicit. Different subsets of criteria have been emphasized by different investigators, as the reference citations indicate. Our view is that any subset of criteria is not adequate. All five of the following criteria are in fact necessary for a grammatical rule:

1. Each component of a combination must have independent symbolic status (Brown, 1973).
2. The relationship between the symbols must be reliable and meaningful (semantic) (Terrace et al., 1979).
3. A rule must specify relations between *categories* of symbols across combinations, not merely a relation between individual symbols (Bronowski & Bellugi, 1970; Terrace et al., 1979).
4. Some formal device, such as statistically reliable order (Braine, 1976; Goldin-Meadow & Mylander, 1984), must be used to relate symbol categories across combinations.
5. The rule must be productive (Hill, 1978; Savage-Rumbaugh, Sevcik, Rumbaugh, & Rubert, 1985; Terrace et al., 1979): A wide variety of spontaneous combinations must be generated (Petitto & Seidenberg, 1979; Terrace et al., 1979, 1980).

Because grammatical rules are creative and productive (Criterion 5), they allow a relatively small number of individual symbols to be combined into a large number of new sentences. Terrace et al. (1979, 1980) claimed that it was in this respect that chimpanzee language failed to meet the criterion of grammatical competence.

For example, Lana (studied by Rumbaugh and colleagues), Sarah (studied by Premack), and Nim (studied by Terrace and colleagues) learned grammatical rules but did not use those rules productively to generate spontaneous new combinations. In the case of Nim, Terrace and colleagues concluded that a large number of Nim's combinations were imitations of the preceding utterances of the trainer, and Nim's remaining combinations appeared to be based on rules for combining specific vocabulary items. There was a dearth of evidence for general rules governing whole classes of words, such as the rules for combining the class "noun" with the class "verb" to form a subject–predicate relation in human language.

In the case of Lana and Sarah, the experimenters imposed the formal device of symbol order on them; that predetermined symbol order was requisite to reward or reinforcement. Although Lana and Sarah learned the required symbol orders, the creative aspect of human grammar was lacking.

Matsuzawa (1985) reported spontaneous use of an ordering rule in the combinations of a common chimpanzee. However, because reinforcement was contingent on a very particular combination of symbolic elements (albeit in any order), the *selection* of semiotic elements was not spontaneous, and in that sense the rule could not be said to be a completely creative one.

In addition, rules for ordering two symbols often have been idiosyncratic to particular signs, rather than combinations between members of two symbol categories (Fouts & Couch, 1976; Terrace et al. 1979, 1980). However, because this limitation also exists at the two-word stage of child language (Braine, 1976), it is not a linguistic criterion that differentiates the chimp from the 2-year-old child.

Indeed, no previous study of ape language has demonstrated that apes satisfy all of the foregoing criteria. The grammatical evidence of other researchers has sometimes been compromised because they have not reported data completely or have not systematically eliminated immediate imitation from the analyses (Fouts, 1974a,b, 1975; Gardner & Gardner, 1969, 1971, 1974, 1978; Patterson, 1978, 1980; Terrace et al., 1979, 1980).

Delayed imitation is a hallmark of representational development at the end of human infancy. Comparative study of sensorimotor development among primates indicates that both the great apes and cebus monkeys are capable of delayed imitation (Chevalier-Skolnikoff, 1976, 1989; Parker & Gibson, 1979). Yet cebus has not shown symbolic capacities. Clearly, representation in the form of delayed imitation is not sufficient to assure symbolic communication, let alone productive grammar. (See Vauclair, P&G11, for a discussion of the disjunction between representation, symbols, and communication, both in human development and in cross-species comparison.) With respect to grammar, imitation leaves open the possibility that combinations may reflect productive use of grammar by human teachers rather than ape learners (Terrace et al., 1979).[1]

Grammatical rules should exist independently of highly structured training settings and imitation, and they should, at least in part, be determined by the ape as well as by its models. Early protohumans invented language; they did not merely learn it. To shed light on the evolution of grammar, it is necessary to demonstrate that apes have some capacity to invent grammatical rules. Although claims of innovative compound words abound in the language literature (Fouts, 1974a,b, 1975; Patterson, 1980), these are by their nature ambiguous one-time occurrences (Petitto & Seidenberg, 1979; Seidenberg & Petitto, 1979; Terrace, 1983; Terrace et al., 1979, 1980); in any case, such examples belong to the lexicon rather than to the grammar.

In contrast, we will demonstrate in the present chapter that a pygmy chimpanzee, member of a species virtually unstudied from the point of view of language, has not only *learned* but also *invented* productive protogrammatical rules. We use the term "protogrammar" to indicate the very simple nature of the rules; however, as the reader will see, the rules may well be as complex as those used by human 2-year-olds. If chimpanzees have but a protogrammar, so, it seems, do 2-year-old children.

Because so many semiotic and symbolic capacities have been found in the great apes in the last 20 years (Fouts, 1973; 1974a,b; 1975; Fouts & Couch, 1976; Gardner & Gardner, 1969, 1971, 1974, 1978, 1980; Greenfield & Savage-Rumbaugh, 1984; Jordan & Jordan, 1977; Miles, 1983; Patterson, 1978, 1980; Premack, 1970, 1971, 1976; Rumbaugh, 1977; Rumbaugh & Gill, 1976a,b; Rumbaugh, Gill, & von Glasersfeld, 1973; Savage-Rumbaugh, 1986; Savage-Rumbaugh, Rumbaugh, & Boysen, 1978a,b; Savage-Rumbaugh et al., 1980, 1985), Chomsky's emphasis on the uniqueness of human grammar has become the last bastion of the discontinuity theorists. Thus, the discontinuity position has come to hinge on the question whether or not language-trained apes have linguistic grammar above and beyond their imitation of human models (Terrace, 1983; Terrace et al., 1979, 1980).

## The evolution of an innate capacity

Chomsky nevertheless emphasizes one point with which we would strongly agree: that human language has an important genetic basis. Ultimately this point works *against* Chomsky's creationist position vis-à-vis language. If 99% of our genes are held in common with contemporaneous chimpanzees (King & Wilson, 1975), and if language is an extremely multigenic function (Studdert-Kennedy, 1988), then it is likely that much of the genetic basis of language is shared with present-day chimpanzees and with our common primate ancestors.

The logic of an evolutionary approach to behavior is as follows: If we find capacities in common between two related descendant species of a common ancestor, it is possible that, in both species, the capacity was inherited in some form from the common ancestor species. If the same behavioral capacity is found in not just two, but *all* the species stemming from a common ancestor, the presence of the genetic basis for the behavioral trait in the common ancestor becomes quite certain (Parker, P&G1).

Two points are often neglected in a theoretical consideration of ape language and its evolutionary implications. One is that human language was, at its origins, and continues to be (Parker, 1985) invented – not merely learned from a previous generation. This suggests that the *invention* of grammatical rules by apes would be stronger evidence of evolutionary continuity than mere learning of rules that have been demonstrated or taught by others in the environment. Our findings with Kanzi, a pygmy chimpanzee (*Pan paniscus*), provide evidence that bonobos can learn a simple grammar, but more interesting and more important, they can invent new protogrammatical rules –

that is, rules never demonstrated by any human or animal in the chimpanzees' social environment.

The second point is that the search for grammatical competence among apes has been very anthropocentric. Not only have ape researchers looked for human grammar in a general sense, but they have also assumed that the grammatical development of apes, if it occurs, will ressemble that of young human children (especially American children!) down to the very details. It may be that apes can develop grammatical rules, but that, at least in part, the nature and developmental order of their grammar derive from their species-specific and individual way of life. If that is the case, we would expect the *details*, if not the overall structural pattern, of chimpanzee grammatical development to diverge in some respects from that observed in human children; a similar point was made by Wiener (1984) and McNeill (1974). Drawing on data from Kanzi, a pygmy chimpanzee, we will demonstrate in the present chapter that his way of life results in a grammar that in some respects differs from the initial grammars of young children.

## A developmental approach to language evolution

A developmental approach has a particular role in the logic of evolutionary reconstruction. Evolutionary change modifies and builds on what is already there in a particular organism. A species' most recently evolved features *tend* to appear relatively late in ontogenetic development, although that is not an absolute law (Studdert-Kennedy, 1988). Nevertheless, the tendency can be explained as follows: The later a mutation or gene rearrangement (King & Wilson, 1975) occurs in ontogenetic development, the less likely it is to interfere with other, previously evolved, aspects of development. Consequently, the later a genetic change arises in development, the greater the chances of the altered organism's survival to reproduce, and consequently the greater the chances of the mutation's survival in the gene pool. Therefore, two related species generally are more divergent later in ontogenetic development, and more similar during earlier ontogenetic stages; von Baer's law (Gould, 1977) provides a slightly different rationale for this same conclusion. Applying this concept to language, one would expect chimpanzees' communicative systems to be more similar to the communicative systems of children during the earliest stages of language acquisition, less similar during later stages, and least similar during the adult stage. A developmental approach should therefore be more fruitful than the simple binary question "Do apes have language?"

Language does not appear full-blown in the human child. It develops quite gradually. Would we want to say that a 1-year-old has language? A 2-year-old? A 5-year-old? A 10-year-old? Clearly this would be a rather fruitless discussion, although, as Hill (1978) pointed out, there have been attempts to draw such a developmental line – for example, Limber (1977) proposed that human language begins at age 3. It is similarly fruitless to ask if apes taught humanly devised symbol systems do or do not have language. More profitable in the case of both child and ape is to ask what elements of language are present and what elements are absent at a particular point in development.

Human grammar becomes more complex with development. Simpler structures are components of the more complex ones. Therefore, each major step in the development of grammatical structure is logically, as well as ontogenetically, dependent on the preceding ones. As Parker and Gibson (1979) pointed out, the logical dependence characteristic of the ontogeny of language must also characterize its phylogenetic evolution.

With respect to major grammatical structures (e.g., the progression from one-word to two-word utterances, the progression from single-clause sentences to multiclause sentences), we would agree with Parker and Gibson that human language begins at age 3. It is similarly fruitless to ask if apes which it develops" (Parker & Gibson, 1979). Given that apes have not evolved humanlike language since diverging from the hominid line of evolution, they would at best be expected to have potentials for the primitive forms of linguistic grammar that may have been potential, or possibly even present, in our common ancestor. These historically early phylogenetic stages of grammatical structure, prerequisite to the more complex forms of human grammar today, could resemble developmentally early ontogenetic stages, similarly prerequisite to the complex structures of adult human grammar.

Therefore, we need to look for parallels between ape language and human language in the *earliest stage of development* and, having established these, see how far the apes can travel on the path toward human language. As a consequence, we searched for parallels to the earliest stages of children's grammar in the study of bonobo grammar to be reported here.

Our chief point of reference in the human literature will be the work of Goldin-Meadow and colleagues on sign language acquisition among deaf children of hearing parents (Feldman, Goldin-Meadow, & Gleitman, 1978; Goldin-Meadow, 1979; Goldin-Meadow & Feldman, 1977; Goldin-Meadow & Mylander, 1984). Like Kanzi, these children participate in the creation of their own language. Their caregivers, unfamiliar with sign language, do not provide a full-blown language model to be acquired. Without in any way implying that these deaf children are similar to apes, we believe that this parallel in their language learning conditions makes them the most interesting subjects of comparison for our study of grammar in *Pan paniscus*.

## Overview of the study of Kanzi

The pygmy chimpanzee or bonobo (*Pan paniscus*) is of special interest because the sociosexual behavior of the species is more like that of humans than is that of the common chimpanzee (*Pan troglodytes*) (Savage-Rumbaugh & Wilkerson, 1978). The question that inspired the ongoing study of Kanzi and other bonobos was whether or not bonobo communicative behavior would also develop more as human language does. Although genetic studies in-

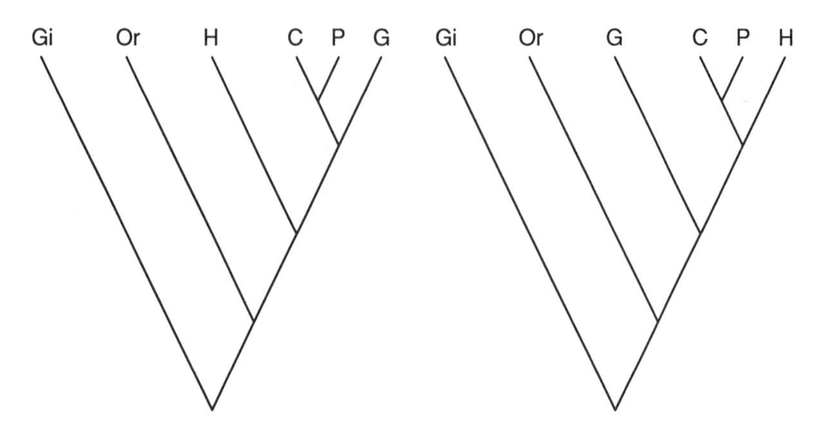


Figure 20.1. Two most likely family trees for apes and humans, based on genetic evidence: Gi = gibbon; Or = orangutan; H = human; C = chimpanzee; P = pygmy chimpanzee, bonobo; G = gorilla (from Weiss, 1987).

dicate that *Pan paniscus* and *Pan troglodytes* are equally close to the human species (Figure 20.1 shows the two most likely family trees), the small size and other morphological characteristics of the bonobo make it likely that *Pan paniscus* more closely resembles our common ancestor than does modern-day *Pan troglodytes* or *Homo sapiens* (Zihiman, 1979; Zihiman, Cronin, Cramer, & Sarich, 1978; Parker, P&G1).

The language learning environment of Kanzi and his half sister Mulika involved learning language in the ongoing course of communication, without formal training. Consequently, unlike Lana, Sherman, Austin, and other common chimpanzees previously trained by Rumbaugh, Savage-Rumbaugh, and colleagues (Greenfield & Savage-Rumbaugh, 1984; Rumbaugh, 1977; Rumbaugh & Gill, 1976a,b; Rumbaugh et al., 1973; Savage-Rumbaugh, 1986; Savage-Rumbaugh et al., 1978a,b, 1980), Kanzi and Mulika were not taught the names of objects by repeated practice that required them to associate the exemplar with its symbol. Although Kanzi and Mulika were exposed to the same basic system of lexigrams (geometric designs devised by humans to stand for things), unlike the common chimps mentioned earlier they were not restricted to the lexigram system.

Their human companions provided them with communicative models and encouraged communication. English was used freely around them, and as caregivers spoke they also pointed to the appropriate lexigram on the keyboard. Other informal gestures also served as a natural adjunct to speech. In addition, a number of American Sign Language (ASL) gestures were used by the caregivers, none of whom were fluent in ASL.

Comprehension of English words typically preceded the onset of lexigram usage (Savage-Rumbaugh, McDonald, Sevcik, Hopkins, & Rubert, 1986). That is, Kanzi and Mulika generally gave evidence of understanding a spoken word as it was used by their caregivers before they began to use its lexigram themselves. Comprehension of the spoken word, which extends to synthesized speech, has been documented in detail by Savage-Rumbaugh et al. (1986).

Although other studies of great ape communication in captivity had provided a communication environment similar to that experienced by human children (Gardner & Gardner, 1971, 1978, 1985; Miles, 1983; Patterson, 1978, 1980), this study of Kanzi and Mulika was the first to combine a human environment vis-à-vis communication with a physical environment and a structure of daily activities that were approximations in important respects of what the species would experience in the wild.

These pygmy chimpanzees were reared in 55 acres of forest outside of Atlanta, Georgia, where they encountered a variety of natural plants and wildlife. Many of their communications revolved around foraging activities as they traveled that area each day to find their food. No attempt was made to structure their activities during the day, except for when tests of symbol competency were given. At all other times, communication revolved around whatever happened to be of immediate interest to them.

In the previous studies of symbol use in *Pan troglodytes* by Rumbaugh, Savage-Rumbaugh, and colleagues, the lexigram system was attached to a computer keyboard and electronic display screen for ease and accuracy of recording communicative output. That was necessary because the vagueness of their pointing gestures made it ambiguous to merely have them point to the symbol they wanted to use. Kanzi and Mulika had the use of a similar system (with the addition of synthesized speech output) when they were indoors. However, that system was not portable enough to go with the chimps as they traveled about their 55 acres. Although portable computer systems were tried, none proved rugged enough to survive the chimpanzee way of life. The fact that these pygmy chimpanzees produced clear pointing gestures (Savage-Rumbaugh, 1984) made it possible to use a portable system (a thin laminated board containing an array of lexigram symbols). Figure 20.2 shows such a board. Both caregivers and chimpanzees could then point to the desired symbol when they were traveling about outdoors.

In the woods, lexigram usage was recorded by hand and entered into the computer at the end of the day. To make sure that this method was accurate, an analysis was done of 4.5 hours of videotape in which real-time coding was checked against coding of the videotape. The scoring was done independently by two different observers, with one observer scoring the behavior in real time and the other scoring the tape. At the time the scoring was done, the real-time observer did not know that the data would be used in the future for a reliability check, and thus the real-time scoring was not altered by the knowledge that this was a reliability check. Thirty-seven utterances were noted by both observers; nine utterances were noted on the videotape that were not seen by the real-time observer, suggesting that a number of lexigram usages were not noted during the busy flow of social interaction. Typically, when that happened, Kanzi was able to gain someone's attention by repeat-

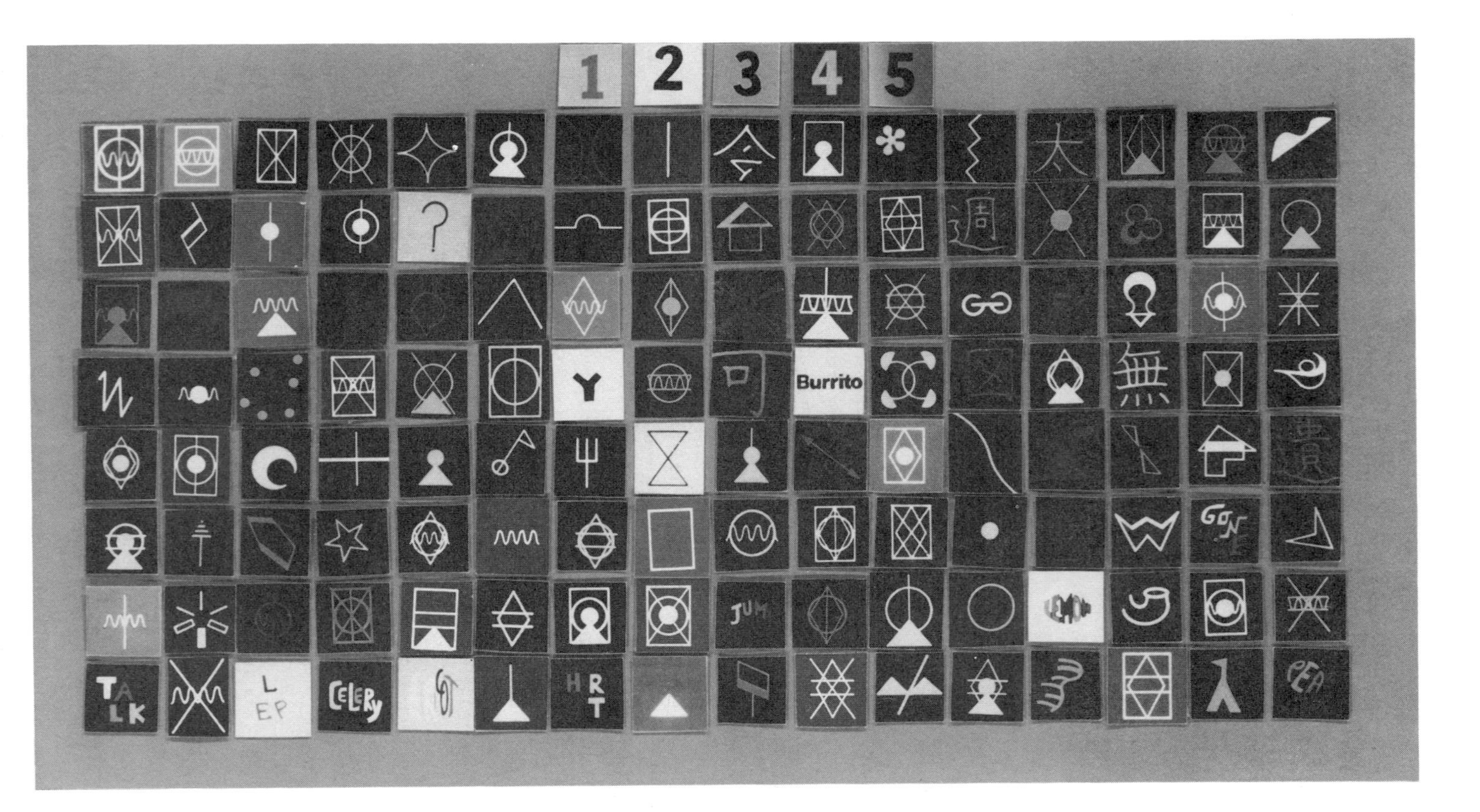


Figure 20.2. Kenzi's lexigram board. (The left side is shown above; the right side is shown below.)

FIGHT
sugar
SALT
Perrier
YOGURT
JELLO
Kool-Aid
TOMORROW
PLEASE
SUGAR CANE
QUIET
PLASTIC BAG
EASY
SLOW
BIG

ing himself. The two observers agreed 100% with regard to which lexigrams Kanzi used and whether or not they were used correctly in context. Thus, though the real-time observation may have lacked some of the redundancy of the chimps' natural conversation, it was very accurate and otherwise quite complete. Other details of the reliability check are available elsewhere (Savage-Rumbaugh et al., 1986).

Like their human caregivers, Kanzi and Mulika were not limited to lexigrams in their interspecies communication. The chimpanzees used gestures, as well as vocalizations. Although neither gestures nor vocalization had been formally taught, Kanzi and Mulika used both for communicative purposes. Human caregivers responded to the extent that they could appropriately interpret the chimpanzees' gestures and vocalizations.

Because Mulika, Kanzi's half sister, was younger and was not providing enough semiotic combinations for quantitative analysis, the data for the grammar study are based only on Kanzi, age $5\frac{1}{2}$ years.

Kanzi was first exposed to the use of lexigrams, gestures, and human speech while in the care of his mother, who was being given language training. Beginning at the age of 18 months, he began to show some interest in the lexigrams. Kanzi did not display regular intentional symbol use until $2\frac{1}{2}$ years of age, when he was separated from his mother for 4 months while she was housed at the Yerkes Field Station for breeding. The research team has kept a complete record of Kanzi's semiotic productions since that time. On the return of his newly pregnant mother, Kanzi chose to stay in the company of his human caregivers most of the time.

### Data and analysis

The data base for our analysis consists of all of Kanzi's two-element combinations (consisting of two lexigrams or of a lexigram plus a gesture) produced over a 5-month period (April–August 1986) when Kanzi was $5\frac{1}{2}$ years old. We also did a less intensive analysis of three-element utterance (three lexigrams or mixtures of lexigrams and gestures) during that same time period. The results of that analysis are reported at the end of this chapter.

Ours was not primarily a developmental analysis; rather, the goal was to ascertain whether or not Kanzi was using grammatical rules at his current, and therefore presumably his most advanced, level of functioning.

Normally, Kanzi had someone with him recording data 9 hours per day, 7 days per week. During that period, Kanzi produced a total of 13,691 utterances; 10.39% or 1,422 of those were combinations of two or more elements.

This is a much larger data base than any child studies have used to investigate grammatical development. Compared with our daily observations approximately 9 hours in length, longitudinal child language observations have averaged about 2 hours per month (e.g., Bloom, 1970; Bowerman, 1973a,b; Brown, 1973; Goldin-Meadow & Mylander, 1984). Even estimating that two observers were present to record context only half the time, we still are left with an average of about 4½ hours of observation per day.

Because, as Bloom (1970) discovered, the researcher cannot make inferences about grammatical relations without having information about meaning, we analyzed only the subset of the combinations for which sufficient situational context was recorded to make basic meaning relations clear. (That conformed to the practice used by Goldin-Meadow and Mylander, 1984, our major point of comparison for this analysis.) Context could be recorded only when two researchers were present with Kanzi (approximately half the time).

With reference to Criterion 5 – spontaneous productivity – we excluded partial or complete imitations (2.67% of total combinations) and utterances that were solicited in any way. In the latter category we omitted (1) Kanzi's responses to "test questions," where a human conversational partner already knew the "right" answer, and (2) Kanzi's responses to a human caregiver who withheld something contingent upon lexigram production.

The final corpus consisted of 723 spontaneous combinatorial two-symbol utterances, representing 51% of Kanzi's total recorded combinatorial output (some of which were longer utterances) during that period. The overwhelming majority of the eliminated utterances were not used simply because a second observer had not been present to record context.

This corpus of 723 two-element combinations is much more extensive than the usual child language corpus for a similar stage of development. Comparing our corpus specifically with those of Goldin-Meadow & Mylander (1984), we see that their data on semantic relations ranged from 20 to 437 utterances for the 10 subjects involved in the studies. Kanzi's sample of 723 utterances is clearly much larger than that for even their most productive child.

An important focus of our study was the source of any grammatical rules we might find. Could they have originated in a model provided by a caregiver, or were they Kanzi's own inventions? To get at the question of the grammatical models provided by humans, we examined video transcripts to ascertain the input provided by others in Kanzi's environment. In addition to this contemporaneous input, we examined video transcripts to analyze the communicative input provided 6 months earlier. These tapes, about 5½ hours in all, involved input from six caregivers, including all of the major ones. In general, the grammatical structure of the input was stable, and so we combined data from the two periods for the analyses that follow.

## Grammar and symbols

A grammar is based on combining semiotic elements, that is, elements of meaning. A semiotic element is a representational device, called a *signifier*, that stands for something else, called a *referent* or signified (Vauclair, P&G11).

According to the philosopher of language Peirce (1931), three hierarchical levels define how a semiotic element or sign is, in a particular usage, related to the thing it stands for:

1. An *icon* resembles the thing it stands for, its referent. A drawing is an example of something that generally has an iconic relationship to its referent.
2. An *index* derives its meaning from being connected to its referent in a particular situation. Pointing may be considered part of the situation of touching something to which attention is to be drawn. Pointing can therefore be considered to have an indexical origin. It also has an iconic element, for it physically resembles the act of touching.
3. A true symbol must be a concept in the sense that it is understood to stand for a whole category of referents and is not limited by its indexical origins in a particular situation; in short, it is decontextualized (Volterra, 1987). A symbol may have, but does not need to have, a relationship of similarity to its referents.

A symbol that does not resemble its referent in form is said to be arbitrary. Arbitrariness is not, however, criterial for a true symbol in Peirce's scheme, as it is in de Saussure's. De Saussure's classification has been used developmentally by Piaget (1962) and phylogenetically by Vauclair (P&G11). Apart from definitional differences, there is a terminology difference: Peirce's *icon* is de Saussure/Piaget's *symbol*, Peirce's *symbol* is closest to Piaget's *sign*. Arbitrariness is the criterion for Piaget's highest symbolic level, the sign. Peirce's criterion for the highest level is, in contrast, the categorical nature of the signifier.

Although it is sometimes claimed that words are true symbols, early words often have an indexical characteristic, as Piaget (1962) pointed out, for they may be restricted in usage to the particular situations in which they were first associated with a referent.

In the field of child language, the study of grammatical combinations often has been separate from the study of the symbolic status of individual words. A word may be a meaningful component in a semiotic combination without attaining the highest level of symbolic development. Thus, an early sentence in child language may, for example, consist of a combination of indices rather than true symbols. Many researchers have neveretheless examined the rules by which the combinations have been formed and have called such rules "grammar."

Because we have included gesture–lexigram combinations in our investigation of grammar, this point is of crucial importance. Whereas lexigrams do not have an iconic element, gestures do. Whereas lexigrams can transcend their indexical origins in a specific situation, the critical gesture of pointing cannot. Its meaning is always derived from its relationship to a referent in a particular situation. Hence, gestures may be on a lower semiotic level than lexigrams.

What is to count as a semiotic element? First of all, Roger Brown (1973) first advanced the idea that for a combination to be meaningful, each element must have some life independent of the combination (Criterion 1). Otherwise, it is always possible that the combination is not truly a combination, but an amalgam whose elements are psychologically not separable.

Relevant to this criterion, an individual lexigram was classified as a distinct semiotic element in Kanzi's productive vocabulary if it could function in isolation in the following way:

1. It first occurred spontaneously on 9 of 10 occasions.
2. Kanzi demonstrated behavioral concordance on 9 of 10 of the subsequent occasions.

As an example of concordance, if Kanzi requested a banana using the BANANA lexigram, he would be presented with a number of favorite foods. Concordance would be scored only if he subsequently selected the banana. This criterion embodies the philosophical point (Greenfield, 1980) that, contrary to behavioristic assumptions and the usual practice in operational definition, semantic intention must be defined proactively, in terms of behavior that occurs *after* rather than *before* or *with* the utterance.

Eighty percent of the lexigram vocabulary used in the corpus of combinations to be analyzed met these criteria for meaningful, spontaneous production of individual lexigrams. In adddition, 86% of the lexigram vocabulary Kanzi used in combinations was tested by asking him to listen to an English word and choose (from among three alternatives) the appropriate lexigram (Savage-Rumbaugh, 1988). During those tests, the researcher was unaware of the location of the correct lexigram. Of the tested lexigrams Kanzi used in combinations, he passed 73% of items on 100% of the trials and 16% of items on 75% of the trials. He responded to a small minority (12%) of the lexigrams used in combinations at a chance level. Hence, the precondition for the existence of a grammar – that it involve combinations of individually meaningful elements – was generally fulfilled. Note that this sort of rigorous testing of individual lexical items has never been carried out in connection with studies of children's grammar.

Must a semiotic element be a word (i.e., a linguistic symbol) to be considered part of a grammatical system? Lexigrams are formally like words; they are devices that are arbitrary in the sense that they do not resemble their referents. In this way, they can potentially be Piagetian signs. They can also function as symbols in the Peircian sense that each has a potential meaning apart from the situation in which it is used. Hence, lexigrams are formally suited to be part of a grammatical system.

Our study also considers the grammar of gesture combined with lexigram. What status can such combinations have in a grammar? Gestures are not formally like words. They are often iconic – the gesture resembling its referent in its form. Some gestures are also indexical: They derive their meaning from being part of the total situation in which they are used. For example, when Kanzi made the gesture of touching a person to denote an agent, that was an abbreviation of the full action of pulling the person into position for a desired action. As another example, a particular pointing gesture derives its specific referent from being part of the situation in which it is used. Such gestures do not have the same symbolic status as lexigrams or most words. It can be argued, however, that words such as pronouns are also indexical and still function in human grammar.

It has also been argued among researchers (e.g., Goldin-Meadow & Mylander, 1984; Petitto, 1987) studying the development of visual language in humans (i.e., the development of sign language in deaf children) that gesture, even the pointing gesture, does function linguistically in sign language more than it does in spoken language. With respect to pointing, the case has been made that pointing at something should be given at least the linguistic status of a demonstrative pronoun such as "this" or "that." Volterra (1987) made the argument that gesture becomes symbolic through the developmental process of decontextualization. However, Volterra did not find the symbolic development of gesture to be unique to children acquiring sign language. She found that the gestures of hearing children acquiring spoken Italian also decontextualized into "true" symbols with age.

**A comparative framework: Deaf children of hearing parents**

Most relevant to the comparison we shall make, Goldin-Meadow and her colleagues (Feldman, Goldin-Meadow, & Gleitman, 1978; Goldin-Meadow, 1979; Goldin-Meadow & Feldman, 1977; Goldin-Meadow & Mylander, 1984) have intensively studied the communication systems of deaf children of hearing parents who do not expose their children to sign language because they believe in the oral method of deaf education. The systems of gestural communication these children develop under these conditions have been widely accepted as showing that human children invent grammar without a model. If this is true, it is clearly relevant to the evolution of human language.

However, it is interesting to note that none of the grammatical rules created by Goldin-Meadow and colleagues' subjects consisted exclusively of pure symbols; the reader is referred especially to the monograph by Goldin-Meadow and Mylander (1984), the most comprehensive report on that research. Instead, those combinations typically consisted of an index – a pointing gesture – used to indicate some entity (e.g., a jar), plus an iconic sign to represent some action (e.g., a turning motion for "open"). There was no indication in Goldin-Meadow's examples whether or not the iconic sign had the categorical meaning required to qualify as a true symbol (e.g., Could the same sign be used to denote "open" in the context "open door"?).

Despite the fact that these deaf children may not have produced any purely symbolic combinations, the regularities of their combinatorial system have been widely accepted as showing the presence of a grammatical system. The studies by Goldin-Meadow and colleagues have provided a comparative rationale for considering Kanzi's systematic lexigram–gesture and gesture–gesture combinations as potentially constituting a part of grammar. Indeed, the nature of the rules for combining semiotic elements is, in principle, independent of the nature of the elements themselves. Noting that the form of Kanzi's lexigram–lexigram combinations may have been on a higher symbolic level than any of those used by Goldin-Meadow's deaf children because of the arbitrary nature of lexigrams, we shall leave the semiotic level of individual signifiers aside and deal with our main focus: the rules for combining them.

We have adopted the methodology and criteria used by Goldin-Meadow and Mylander (1984) with their deaf children for identifying types of combinations and for establishing grammatical rules in Kanzi's semiotic productions. Indeed, the similarities in language learning conditions facing these children and Kanzi made it particularly appropriate to use Goldin-Meadow's methodology (and even to compare the results, which we shall do later in this chapter).

More specifically, there were three similarities in the language learning conditions and semiotic productions of Kanzi and the deaf subjects of Goldin-Meadow and colleagues:

1. Neither was learning a completely conventionalized language. Although the deaf children did communicate gesturally with their parents, they had no preestablished model to follow. Although Kanzi did have a preestablished model in individual lexigrams spontaneously used in English word order, that model was incomplete, for his caregivers did not have preset gestures or combinatorial rules for combining gestures with lexigrams. In addition, one of Kanzi's invented rules, to be discussed later, was imitated by his caregivers.

2. Although most of the communicative input for both Goldin-Meadow's subjects and Kanzi consisted of speech, neither was capable of speech output. Kanzi understood some English speech (Savage-Rumbaugh et al., 1986), and the deaf children also had limited access to English speech through lipreading training (Goldin-Meadow & Mylander, 1984).

3. Pointing gestures were important parts of the semiotic combinations of both the deaf children and Kanzi.

## Grammar and semantic relations

Grammars can be based on two different types of formal combinatoral devices: (1) morphological inflections (e.g., *s* in English for the plural) and (2) syntax (i.e., rule-governed ordering of semiotic elements). The early stages of human grammar, whether the learners be hearing or deaf, and whether the language be inflected or not, involve word order rather than inflectional rules (Slobin, 1973).

Given the nature of the lexigram system, morphological inflection is excluded as a possible basis for grammatical rules. There was no evidence for morphological modification in Kanzi's gesture repertoire. His caregivers, not being fluent signers, presented almost no sign inflections to Kanzi in their input. We therefore concentrate exclusively on ordering rules.

Because we have already shown that data from Kanzi meet Criterion 1 (independent symbolic status of lexigrams) for a grammar, we now concentrate on Criteria 2–5 in presenting data. Our focus is on rules that generate sufficiently large numbers of combinations to yield statistically significant evidence of ordering. We treat data on lexigram–lexigram combinations separately from data on lexigram–gesture combinations. The body of gesture–gesture combinations in the period under consideration was not large enough for analysis.

The rules to be presented involve the ordering of categories to create productive semantic relationships (Criteria 2–5 for a grammatical rule). We have preferred to label a category according to the role of its referents in the ongoing situation (e.g., entity, action) rather than to use parts-of-speech labels such as noun and verb. In the first place, these latter get their meaning from the existence of a grammatical framework. We cannot assume the presence of such a framework. In the second place, the categories of noun and verb are more general than anything observed in our data. These same arguments were used by Bowerman (1973b) to promote the use of semantic rather than part-of-speech labels in early child language.

The semantic roles that were used to categorize Kanzi's two-element combinations were similar to the category systems used in child language by Bloom (1971), Brown (1973), and Goldin-Meadow and Mylander (1984). They are shown in Table 20.1. We present this table to give a complete summary of the entire corpus used in the analysis. The corpus itself will appear in Greenfield and Savage-Rumbaugh (in press). The table demonstrates that the range of relations used by Kanzi included many of the relations described for children at the two-word stage (Bowerman, 1973a; Brown, 1973; Goldin-Meadow & Mylander, 1984; Schlesinger, 1971). The 10 major relations (conjoined actions, agent–action, action–object, agent–object, entity–demonstrative, goal–action, conjoined entities, conjoined locations, location–entity, and entity–attribute) accounted for 93.5% of Kanzi's two-element lexigram–lexigram and lexigram–gesture output. In the classic child language data of Brown (1973), eight major relations accounted for 70% of the output of most children. Most interesting from a comparative standpoint, seven of those eight major relations (agent–action, action–object, agent–object, entity–demonstrative, goal–action [termed action and locative by Brown], location–entity, and entity–attribute) were found in Kanzi's data. Kanzi did not produce one relation frequently found in children: possession. Kanzi did produce one relation not frequently found (or perhaps not found at all) in children at the two-word stage: conjoined action relations, to be discussed in detail later.

As can be seen from Table 20.1, Kanzi rarely paraphrased or repeated himself or formed combinations that were semantically unrelated. In addition, details presented in the notes to Table 20.1 indicate high interobserver reliability for categorizing semantic relations. In other words, Kanzi's productions satisfied Criterion 2, a prerequisite for grammatical rules: They expressed reliable and meaningful relations.

Table 20.1. *Distribution of two-element semantic relations in Kanzi's corpus*

| Relation | | Example (of dominant order) |
|---|---|---|
| Conjoined actions | 92 | TICKLE BITE, then positions himself for researcher/caregiver to tickle and bite him. |
| Action–agent | 119 | CARRY *person* (*gesture*), gesturing to Phil, who agrees to carry Kanzi. |
| Agent–action | 13 | |
| Action–object | 39 | KEEPAWAY BALLOON, wanting to tease Bill with a balloon and start a fight. |
| Object–action | 15 | |
| Object–agent | 7 | BALLOON *person* (*gesture*), Kanzi gestures to Liz; Liz gives Kanzi a balloon. |
| Agent–object | 1 | |
| Entity–demonstrative | 182 | PEANUT (*demonstrative gesture*), points to peanuts in cooler. |
| Demonstrative–entity | 67 | |
| Goal–action | 46 | COKE CHASE; then researcher chases Kanzi to place in woods where Coke is kept. |
| Action–Goal | 10 | |
| (The above relations are analyzed for their ordering regularities in the tables and text that follow. The relations below either lacked ordering structure or were too infrequent to be subject to such an analysis.) | | |
| Conjoined entities | 25 | M&M GRAPE; caregiver/researcher: "You want both of these foods?" Kanzi vocalizes and holds out his hand. |
| Conjoined locations | 7 | SUE'S-OFFICE CHILDSIDE; wanted to go to those two places. |
| Location–entity | 19 | PLAYYARD AUSTIN; wants to visit Austin in the playyard. |
| Entity–location | 12 | |
| Entity–attribute | 12 | FOOD BLACKBERRY, after eating blackberries, to request more. |
| Attribute–entity | 10 | |
| Miscellaneous | 37 | These included low-frequency relations (less than seven) such as attribute of action, attribute of location, affirmation, negation, and relations involving an instrument |
| Two-mode paraphrase[a] | 4 | CHASE *chase* (*gesture*), trying to get staff member to chase him in the lobby. |
| No direct relation[b] | 6 | POTATO OIL; Kanzi commented after researcher had put oil on him as he was eating a potato. |
| Total | 723 | |

[a] There were no purely repetitious two-symbol utterances in the two-symbol corpus. This low-frequency category contains the closest phenomenon to a repetition.

[b] Although the "No direct relation" category was similar to the "Conjoined entities" category in consisting of two entities, the entities were linked by a common agent and action in the latter case, but not the former. In general, the system of semantic relations was based on those used for child language by researchers such as Schlesinger (1971), Brown (1973), and Goldin-Meadow and Mylander (1984).

*Note*: Two symbols were considered a combination if there was no other attentional focus intervening between the production of the two. Working from contextual notes, the first author categorized all utterances according to semantic relation. In order to test for interrater reliability, the second author then coded a randomly selected subset of 32 utterances, utilizing the same contextual notes. There was agreement on 29.5 of 32, or 92%, of all judgments. A reliability check also indicated that imitation could be judged as reliably from real-time coding as from videotape (Savage-Rumbaugh, McDonald, Sevcik, Hopkins, & Rubert, 1986).

Table 20.2. *Kanzi's two-element lexigram–lexigram combinations: Relations between action and object (animate and inanimate)*

*Examples (lexigrams are in small capitals)*

| | Action | Object | |
|---|---|---|---|
| Inanimate object | HIDE | PEANUT | Kanzi then hides some peanuts. |
| Animate object | GRAB | KANZI | Kanzi stuck his foot out to her, to show that she was to grab him. |

| | Action–object | Object–action | |
|---|---|---|---|
| *Development of Kanzi's lexigram order* | | | |
| Early (4/10–4/26/86) | 3 | 7 | |
| Late (4/29–8/30/86) | 31 | 6 | ($p$ <.00000)[a] |
| *Kanzi's human caregivers' lexigram order* | | | |
| | 51 | 7 | |

*Comprehension of action–object lexigram relation*
Correct: 17
Incorrect: 0

*Example of correct understanding*
Caregiver/researcher: PLAY HAT KEEPAWAY. Kanzi grabs the hat and shakes it at caregiver/researcher. (Note that this is not the obvious thing to do with a hat!)

*Example of meaningful misunderstanding (Nov. 1985)*
Caregiver/researcher: ICE (commenting on a big block of ice on TV). *Someone is HIDEing in the ICE.* Kanzi starts searching under the blankets. He apparently has understood the action–object relation, HIDE ICE, and is looking for the ice!

*Complete corpus of examples from the late (rule-bound) period*

| Action–object order | | Object–action order |
|---|---|---|
| BITE BALL (3) | HIDE AUSTIN (1) | BALL SLAP (4) |
| BITE ORANGEDRINK (2) | HIDE PEANUT (1) | SURPRISE HIDE (1) |
| BITE CHERRY (1) | HUG BALL (1) | WATER HIDE (1) |
| BITE COKE (1) | KEEPAWAY CLAY (4) | |
| BITE FOOD (1) | KEEPAWAY BALLOON (1) | |
| BITE TOMATO (2) | SLAP BALL (7) | |
| CARRY BALL (1) | TICKLE BALL (1) | |
| GRAB AUSTIN (1) | | |
| GRAB KANZI (2) | | |
| GRAB MATATA (1) | | |

*Note:* The variety of lexically distinct combinations indicates that position preferences of individual lexical items do not account for the action–object ordering rule of the late period. In addition, the fact that no one combination dominated provides even more conclusive evidence: The most frequent action–object combination, SLAP BALL, accounted for but a minority of the total instances of the late-period rule (7 of 31).

Not included in this table (but present in the totals for Table 20.1) are a small number (7) of action–object relations in which the object was expressed as a demonstrative gesture (e.g., GRAB

In the majority of frequent combinations, Kanzi tended toward a particular symbol order. We present our results by focusing on the category combinations that showed rule-bound regularity and that were productive enough to achieve statistical significance.

In terms of Criterion 5, productivity, we have satisfied the criterion of demanding a degree of spontaneity: All partial and complete imitations of the caregiver's prior utterances have been eliminated from the data to be presented in this chapter. Furthermore, in light of the claim by Terrace et al. (1979, 1980) that most of the combinations produced by Nim and other *Pan troglodytes* chimpanzees were imitated, it was significant that Kanzi rarely used imitation: Only 2.67% of the combinations in the period under study were full or partial imitations of the preceding utterances. That is at the low end of the range for normal children from age 2 to 3 years (Bloom, Rocissano, & Hood, 1976) and for deaf children without a sign model from about 1 to 4 years of age (Goldin-Meadow & Mylander, 1984 see also Miles, P&G19).

**Development of a rule learned from environmental models: Action precedes object**

Table 20.2 illustrates Kanzi's expression of action–object relations. During the first month of this study, Kanzi used no systematic order in these combinations. During the last 4 months, however, Kanzi employed, to a statistically significant degree, the symbol order used by his caregivers: action preceding object. This developmental trend from random ordering to an ordering preference was also found for human children at the two-word stage (Braine, 1976). Because his human caregivers generally overlaid lexigrams upon an English sentence, it was not surprising that his lexigram models followed English word order.

Unlike Nim (Terrace et al., 1979), Kanzi's preference for the action–object order was undisturbed by countertrends of individual lexical items. He thus fulfilled Criterion 4 for a grammatical rule: the presence of a formal device (order) relating the two semantic categories.

The variety of symbols he used in different utterances satisfied Criteria 3 and 5, a *productive* relation between *categories* of symbols.

Notes to Table 20.2 *(cont.)*

[demonstrative gesture]). All but one of these examples appeared in the last month of observation, and action–object order was dominant in this period (5 of 7 cases). However, the sample size was too small for the ordering trend to achieve statistical significance. We could consider this phenomenon of using a demonstrative gesture rather than the object lexigrams used earlier as an analogue of pronominalization development. (Note that the lexigram–gesture ordering rule presented in Table 20.3 may have overdetermined the symbol ordering pattern of these examples.)

[a] One-tailed significance test.

To determine if that semantic relationship was actually understood by Kanzi at the time he produced the combinations we are analyzing, we tallied naturally occurring examples of comprehension and miscomprehension of action–object lexigram relations expressed by human caregivers during the same period of time. The results (last figures in Table 20.2) show that Kanzi not only used but also understood that relationship. As in child language, errors (such as the last example in Table 20.2) were especially revealing: They showed a mental construction that could not have been environmentally cued.

**Invented rules**

*Place gesture after lexigram*

Not all of Kanzi's rules were learned from environmental models. Kanzi's human caregivers exposed him to the English model of word order: agent before action ("Human caregivers' order" in Table 20.3). His own lexigram–lexigram combinations (second row of figures in Table 20.3) showed signs of following this rule, but they were too infrequent to be statistically reliable. Kanzi made up his own rule, however, for combining agent gesture with action lexigram: His highly significant ordering rule, "place lexigram first," used the *opposite* ordering strategy from that of his caregivers' English-based rule (first row of figures in Table 20.3).

Data on Kanzi's comprehension from the same period reveal that he understood that symbols could represent relations in which agents carried out actions ("Comprehension of agent–action relations" in Table 20.3).

Kanzi's rule, "place lexigram first," had considerable generality as well as originality. The remainder of Table 20.3 shows how that rule was manifest in statistically significant symbol order for three other semantic relations (Criteria 4 and 5 of a grammatical rule). Most important, in none of the relations was there a human model for the rule. This is strong evidence for creative productivity (Criterion 5).

Although three of the four relations involved a demonstrative gesture, the fourth, goal–action, involved combining a lexigram with one of several other action gestures. Thus, to a limited extent, that rule involved relations between two categories (Criterion 3 of a grammatical rule): a larger category of lexigrams and a smaller category of gestures. In addition, the gestures conformed to Volterra's criteria (1987) for decontextualized symbols: According to her criteria, Kanzi's gestures would be considered symbolic signs, as found in deaf sign language, not mere gestures.

Note the regularity of the rule, even though the opposite order was equally possible and would be meaningful (indeed, it is used in English). This finding parallels the highly conventionalized (as opposed to random or idiosyncratic) use of pointing gestures observed by Goldin-Meadow and colleagues (Goldin-Meadow & Mylander, 1984) in the deaf children they studied.

Table 20.3. *Kanzi's ordering rule: Gesture follows lexigram*

*Relations between agents and actions*

Lexigram–gesture (10/10/85): CHASE (*gesturing to Mary*). Kanzi pulls on Mary. He is asked what he wants and communicates the foregoing. But Mary is playing with Mulika and denies his request until later. Later he produces the following:

Lexigram–lexigram (10/10/85): MARY CHASE. Mary agrees and chases after Kanzi.

| | Agent–action | Action–agent |
|---|---|---|
| *Kanzi's order* | | |
| Lexigram action–gesture agent | 7 | 116 ($p = .00000$)[a] |
| Lexigram–lexigram | 6 | 3 |
| *Human caregivers' order* | | |
| Lexigram action–gesture agent | 14 | 0 |
| Lexigram–Lexigram | 14 | 0 |

*Comprehension of agent–action relations*
Correct: 6
Incorrect: 0

*Example of correct understanding*
Caregiver/researcher: KELLY *and* ROSE CHASE. Kanzi looks at Rose, then Kelly. He then touches Kelly and pushes her arm toward Rose, signing CHASE. (Essentially, Kanzi has translated a lexigram sentence into a gestural sentence.)

*Complete corpus of examples of agent–action relations*

*Lexigram–gesture combinations*

| Order: Lexigram action–gesture agent (examples of rule) | Gesture agent–lexigram action (counterexamples to rule) |
|---|---|
| CHASE (dem. gest. to dog) (1) | (dem. gest. to person) CHASE (5) |
| CHASE (dem. gest. to person) (53) | |
| BITE (dem. gest. to person) (18) | |
| SLAP (dem. gest. to person) (10) | |
| CARRY (dem. gest. to person) (9) | (dem. gest. to person) CARRY (2) |
| TICKLE (dem. gest. to person) (8) | |
| HIDE (dem. gest. to person) (7) | |
| HUG (dem. gest. to person) (4) | |
| GRAB (dem. gest. to person) (4) | |
| KEEPAWAY (dem. gest. to person) (2) | |

*Lexigram–lexigram combinations* (too few for statistically significant rule)

| Order: Lexigram action–lexigram agent | Lexigram agent–lexigram action[b] |
|---|---|
| HIDE AUSTIN (2) | MATATA BITE (1) |
| CHASE DOG (1) | MATATA CHASE (1) |
| | MULIKA CHASE (1) |
| | MULIKA BITE (1) |
| | PENNY TICKLE (1) |
| | LIZ HIDE (1) |

Table 20.3 *(cont.)*

*Further examples of Kanzi's rule: Place gesture after lexigram*

| | Entity | Demonstrative |
|---|---|---|
| Kanzi | FOOD | (dem. gest.) |

He requests food from cooler by pushing FOOD key and then pointing to cooler.

| | Demonstrative gesture 1st | Demonstrative gesture 2nd |
|---|---|---|
| Kanzi | 67 | 182 ($p = .00000$)[a] |
| Human model | 3 | 2 |

*Partial corpus of examples of demonstrative–entity relations*

| Order: Demonstrative gesture 2nd (examples of rule) | Demonstrative gesture 1st (counterexamples to rule) |
|---|---|
| APPLE (dem. gest.) (1) | (dem. gest.) APPLE (3) |
| AUSTIN (dem. gest.) (1) | |
| BALL (dem. gest.) (2) | |
| BALLOON (dem. gest.) (2) | |
| BANANA (dem. gest.) (8) | (dem. gest.) BANANA (5) |
| BLACKBERRY (dem. gest.) (4) | (dem. gest.) BLACKBERRY (1) |
| BLUEBERRY (dem. gest.) (7) | (dem. gest.) BLUEBERRY (1) |
| BREAD (dem. gest.) (1) | (dem. gest.) BREAD (1) |
| BURRITO (dem. gest.) (3) | |
| BUTTER (dem. gest.) (1) | |
| CARROT (dem. gest.) (4) | (dem. gest.) CARROT (2) |
| CHERRY (dem. gest.) (2) | (dem. gest.) CHEESE (1) |
| COKE (dem. gest.) (9) | (dem. gest.) COKE (1) |
| EGG (dem. gest.) (4) | |
| FOOD (dem. gest.) (8) | (dem. gest.) FOOD (6) |
| GRAPE (dem. gest.) (10) | (dem. gest.) GRAPE (2) |
| HAMBURGER (dem. gest.) (2) | (dem. gest.) HAMBURGER (1) |
| | (dem. gest.) HOTDOG (1) |
| ICE (dem. gest.) (1) | (dem. gest.) ICE (3) |
| JELLY (dem. gest.) (6) | (dem. gest.) JELLY (2) |
| JUICE (dem. gest.) (10) | (dem. gest.) JUICE (4) |
| KEY (dem. gest.) (1) | |

| | Goal | Action |
|---|---|---|
| Kanzi | DOG | (go gesture) |

He then led to the dogs' pen.

| | Action gesture 1st | Action gesture 2nd |
|---|---|---|
| Kanzi | 0 | 30 ($p = .00000$)[a] |
| Human model | 0 | 0 |

Table 20.3 *(cont.)*

*Complete corpus of examples of goal–action relations expressed as lexigram–gesture combinations*

| Order: action gesture 2nd (examples of rule) | | Action gesture 1st (counterexamples to rule) |
|---|---|---|
| AUSTIN (go gesture) (8) | ORANGE (open gesture)[c] | None |
| AUSTIN (come gesture) (1) | PEANUT (go gesture) (2) | |
| BALL (go gesture) (1) | POTATO (go gesture) (1) | |
| BALL (chase gesture) (1) | STRAWBERRY (go gesture) (1) | |
| BLUEBERRY (come gesture) (1) | SURPRISE (come gesture) (2) | |
| CHILDSIDE (go gesture) (1) | SURPRISE (go gesture) (1) | |
| CLOVER (go gesture) (1) | SWEET-POTATO (go gesture) (1) | |
| DOG (go gesture) (1) | TOOLROOM (come gesture) (1) | |
| ICE (go gesture) (1) | WATER (come gesture) (1) | |
| MELON (go gesture) (1) | WATER (go gesture) (1) | |
| M&M (go gesture) (1) | | |

| | Object | Agent |
|---|---|---|
| Kanzi | BALLOON | (dem. gesture to person) |

Kanzi gestures to researcher; she gives Kanzi a balloon.

| | Agent gesture 1st | Agent gesture 2nd |
|---|---|---|
| Kanzi | 1 | 7 ($p < .03$)[a] |
| Human model | 0 | 0 |

*Complete corpus of examples of agent–object relations*

| Order: agent gesture 2nd (examples of rule) | Agent gesture 1st (counterexamples to rule) |
|---|---|
| BALL (dem. gest. to person) (1) | |
| BALLOON (dem. gest. to person) (2) | |
| JUICE (dem. gest. to person) (1) | |
| PEACH (dem. gest. to person) (1) | |
| PLAYYARD (dem. gest. to person) (1) | |
| SURPRISE (dem. gest.) (1) | (dem. gest. to person) SURPRISE (1) |

*Note:* Kanzi's ordering rule seems to have considerable generality, involving a large class of lexigrams and a small class of gestures (demonstrative and action). This small class of gestures resembles the "pivots" in children's two-word constructions. The rule is also quite complex, for it involves changing the position of the action symbol, depending on whether action is expressed by a lexigram (action lexigram–agent gesture) or by a gesture (goal lexigram–action gesture).

The alternative demonstrative–entity order, while in the minority, may represent the use of the demonstrative to indicate, while the other order is used to locate. This distinction has been reported by the Gardners (Gardner & Gardner, 1978) for *Pan troglodytes* and by Brown (1973) and Braine (1976) for human children. We would need to do further analysis of video data to know if Kanzi also makes this distinction.

*(cont.)*

Perhaps most interesting from a theoretical perspective is the seemingly arbitrary nature of this invented rule. The rule that "gesture follows lexigram" does not seem to have any basis in functional convenience. At one point Kanzi was observed to move away from a person he later would indicate as agent, go to the board (where he indicated an action lexigram), and then return to the person (using a gesture to designate her as agent). In that situation, the rule Kanzi had invented demanded extra motor steps and therefore seemed purely arbitrary.

*Ordering conjoined actions*

We continue by describing a second rule that Kanzi invented for himself: a rule for combining sequences of two action lexigrams. It was a rule that clearly manifested the interests and life-style of a pygmy chimpanzee, rather than a human. Table 20.4 presents all of the types of conjoined action combinations. Although at first it seemed that these were simply lists of actions, without any structure, we checked the privileges of occurrence in first or second position, drawing on an established method of linguistic fieldwork. We wanted to know if certain vocabulary items could appear only in the first position in a two-element combination, whereas others were constrained to the second position. This analysis yielded evidence of structure (Table 20.4). Certain action lexigrams (CHASE, TICKLE) had a statistically significant tendency to appear in the first position; others (HIDE, SLAP, BITE) tended to appear in the second position (Criterion 4). Still others (GRAB, HUG) showed no position preference. When we placed the first-positon actions in one category and the second-position actions in another, we got the categorical groupings shown in Table 20.4.

Notes to Table 20.3 *(cont.)*

Kanzi also produced 26 lexigram–lexigram combinations involving a relationship between action and goal. These, however, showed no ordering regularity (16 goal–action vs. 10 action–goal). There were no lexigram–lexigram combinations produced for the agent–object relation; lexigram–lexigram combinations were not possible for the entity–demonstrative relation because there was no demonstrative lexigram on the symbol board.

Note that the gestures for "go," "come," "bite," "tickle," "chase," "yes," "open," and "bad" entailed distinct topographies. Demonstrative gestures (dem. gest. in the table) indicating either people or things generally were produced by either pointing to or touching the person or the object of reference. This list of gestures is exhaustive for the two-element corpus under study.

[a] Test for significance of a proportion (two-tailed) (Bruning & Kintz, 1977).

[b] This is the English word order modeled by researchers.

[c] Kanzi wanted to open the cooler to get the orange.

Table 20.4. *Conjoined action lexigram combinations*

| | No. times 1st | No. times 2nd |
|---|---|---|
| *Prefers in 1st position* | | |
| CHASE | 19 | 8 ($p < .04$)[a] |
| TICKLE | 29 | 15 ($p < .04$)[a] |
| *Prefers in 2nd position* | | |
| HIDE | 2 | 9 ($p < .04$)[a] |
| SLAP | 1 | 6 ($p < .06$)[a] |
| BITE | 21 | 38 ($p < .04$)[a] |
| *No position preference* | | |
| GRAB | 5 | 4 |
| HUG | 7 | 5 |

*Examples of preferred orders (2/5/87)*

Kanzi: CHASE HIDE. After producing this lexigram combination, Kanzi gestures to the door to indicate that he wants to go out of the room to do this. He and Sue go out, and he runs away to be chased. He goes around to another room, where he tries to run in and hide behind the door. When Sue approaches, he rushes and hides behind the other door.

Kanzi: CHASE BITE. He runs away to be chased. Instead, Sue tries to bite him first. He will not let her. Only after she has chased him does he get in position to be bitten.

*Human model*: In 6 hours of videotape for November 1985, when Kanzi was already producing frequent action–action combinations, there was only one example of a caregiver combining the action words listed above, and that one example was a direct imitation of Kanzi. In April 1986, in about 2 hours of videotape, there were 10 examples, but the caregiver was imitating Kanzi in 9 of the 10 cases.

*Complete corpus of examples of conjoined action relations*

| Conform to above rules | | Counter to above rules | | Conflicting rules apply | |
|---|---|---|---|---|---|
| CHASE HIDE | (7) | HIDE CHASE | (2) | BITE HIDE | (1) |
| CHASE BITE | (6) | BITE CHASE | (2) | BITE SLAP | (2) |
| CHASE HUG | (4) | HUG CHASE | (1) | TICKLE CHASE | (3) |
| TICKLE HIDE | (1) | | | CHASE TICKLE | (2) |
| TICKLE SLAP | (3) | | | | |
| TICKLE BITE | (21) | BITE TICKLE | (13) | | |
| TICKLE GRAB | (1) | | | | |
| GRAB BITE | (4) | BITE GRAB | (2) | | |
| GRAB SLAP | (1) | SLAP GRAB | (1) | | |
| HUG BITE | (6) | BITE HUG | (1) | | |
| Totals | 54 | | 22 | | 8 |

[a] Test for significance of a proportion (two-tailed) (Bruning & Kintz, 1977).

*Note:* KEEPAWAY was used in only one conjoined action lexigram combination and was therefore omitted from the analysis. In addition, seven conjoined action combinations consisted of a lexigram plus a gesture and were therefore not included in the lexigram–lexigram analysis. All included a gesture representing "come" or "go"; in 5 of 7 cases the ordering followed the "gesture last" rule discussed earlier.

Clearly, there is an association between certain types of relations and their modes of expression. Whereas Kanzi, for example, favored two lexigrams to express conjoined action and action–object relations, he favored lexigram plus gesture for the goal–action and action–agent relationships. (The data in this table are slightly revised from those appearing in Savage-Rumbaugh 1988.)

Before concluding that these ordering tendencies constituted a rule, it was necessary to show that there was semantic reason to group TICKLE and CHASE in one category and BITE, SLAP, and HIDE in another (Criterion 1). Kuroda's formulation of the first-position lexigrams is that they function as invitations to play, whereas the second-position lexigram represents the play content that follows (S. Kuroda, pers. commun., 1987). In short, Kanzi's grammatical ordering reflected his action ordering. Grounded in action, this rule is fairly concrete. Note that the rules of human action would not have yielded such a syntactic rule, and indeed conjoined action combinations generally are rare in the speech of human children. In effect, Kanzi's rule of syntactic order corresponded to Kanzi's own rules of behavioral order and, indeed, to those of pygmy chimpanzees in general in the wild (S. Kuroda, pers. commun.).

Unlike the rules discussed up to now, these conjoined action order preferences lacked the minimum requirements of a proposition: one predicate and one argument. Conjoined action combinations simply chain two predicates. Lest we conclude that these structures were unrelated to human language, however, we must point out that many human languages have serial verbs where "conjoined action" word-order rules apply. Even in English, it is correct to say "go get," but not "get go." Although the particular conjoined action sequences were specific to Kanzi, and pygmy chimps more particularly, the concept of a rule-governed verb order certainly belongs in the ballpark of human language. Children at the two-word stage also formed conjoined structures (Brown, 1973), often expressing two arguments with no predicate.

Thus far, we have demonstrated a meaningful semantic relationship between the elements (Criterion 2) and consistent ordering to the point of statistical significance (Criterion 4). We have also demonstrated that the grammatical rule involved relations between categories of symbols (Criterion 3), for both categories in this invented rule involved at least two symbols each (although the categories were smaller than in the case of the previously discussed rules).

In addition to involving spontaneous, rather than imitative, combinations, the productivity criterion also requires a wide variety of combinations embodying the rule (Criterion 5). Of the 16 predicted combinations that would follow the rule, 10 were found in this period of data collection (Table 20.4, left column of "Complete corpus").

An even more stringent standard of productivity is novelty. The individual combinations were novel in the sense that they had not been modeled in Kanzi's environment; even more interesting, the rule itself was a novel creation ("Human model" in Table 20.4).

This phenomenon of creativity was similar to that found by Goldin-Meadow and Mylander (1984) for deaf children of hearing parents raised without sign language input. Most interesting, like the deaf children of hearing parents, Kanzi provided a model of this invented rule that caregivers ultimately imitated and, potentially, learned from him.

**Difference in symbol order signals difference in meaning**

Kanzi showed an incipient ability to use difference in symbol order to signal difference in meaning. When animate beings functioned as agents in Kanzi's lexigram–lexigram combinations, he tended to place them first, although there were too few for the trend to be statistically significant. When animate beings functioned as objects of action, Kanzi tended to produce them last, a trend that was significant at the .008 level (two-tailed binomial test). A $\chi^2$ test showed the difference between the orders used to signal the two different meanings to be significant at the .05 level. As an example, Kanzi produced GRAB MATATA, when Matata was grabbed, but MATATA BITE when Matata functioned as an agent. This is the beginning of autonomous syntax, in which symbol order signals meaning relations without the help of a disambiguating context.

**Three-symbol utterances**

Kanzi differed from Nim in that he produced *nonredundant* three-element combinations in which a pair of two-element combinations would be linked to add new information (Savage-Rumbaugh et al., 1986).

Only one three-element pattern reached sufficient productivity in the period under study to be analyzed quantitatively: the action–action–agent (demonstrative gesture) pattern. These combinations combined and preserved the ordering rules of their constituent two-element combinations perfectly (7 of 8 cases, $p$ = .0000, two-tailed significance-of-proportion test) (Bruning & Kintz, 1977), as do children's early multiword utterances (Braine, 1976). An example of this pattern is CHASE BITE person (demonstrative gesture). Here the actions chase and bite are ordered in accord with the conjoined action rule (Table 20.4), and the gesturally specified agent also conforms to the rule "place gesture last" (Table 20.3).

**Limitations of the grammar**

Kanzi showed some important differences from children in grammatical development. First, Kanzi's development was much slower. He took about 3 years from his first symbol to make the grammatical progress that children attain in about a year. Even at the point in the study just described, Kanzi was producing a much smaller proportion of combinations than a child would normally produce after 3 years of speech. Finally, in terms of pragmatic content, Kanzi had a much smaller proportion of indicatives or statements (4%), in comparison with requests (96%), than would be normal for a human child. This has also been observed in *Pan troglodytes* and may relate to a lesser proclivity for symbolization per se (Greenfield, 1978a; Terrace, 1985). However, part of the difference may reflect a bias stemming from the fact that in captivity, a chimpanzee's behavior and environment are under the control of humans, from whom he must request activities or objects. In the wild, a given animal might, for example, *state* his planned activity, rather than *requesting* it. This difference is quantitative rather than qualitative and may reflect a tendency on the part of the researchers to code chimpanzee statements as requests in the interests of conservative interpretation. The basic capacity to make a statement is present; evolutionary change could well have expanded on it.

Although Kanzi had been combining lexigrams for several years at the time of the grammar study, most combinations still were of only two elements, and most utterances still were single symbols. That was the same length limitation Terrace et al. (1979, 1980) found for Nim. In Kanzi's case, short symbol combinations may also have reflected a modality difference. Although his caregivers spoke in normal English sentences, they most frequently inserted only one or two lexigrams per sentence, reflecting the mechanical difficulty of the lexigram mode in generating longer utterances.

## Discussion: Implications for the evolution of language

### *The role of action*

Kanzi's self-created symbol-ordering rules seemed action-based:

1. In the conjoined action rule, symbol ordering followed the ordering of play action sequences.
2. In the "gesture follows lexigram" rule, symbol ordering arbitrarily sequenced two modalities of symbolic action.

The first invented rule seems to provide specificity and substance to the notion that language evolved as an instrument to plan coordinated action among human beings. Certainly, Kanzi's conjoined action utterances function as a means for him to plan his sequences of social play. A potential evolutionary implication is that grammatical combination arose partly as a tool to convey to others the planning of complex sequences of socially coordinated activity. Kanzi's conjoined action rule is pragmatically motivated, rather than arbitrary or abstract, but similar rules could have served as an evolutionary foundation for later, more abstract grammatical forms.

The second invented rule is, by contrast, arbitrary. Its function for Kanzi could be to enable him to order two symbolic modalities – lexigram and gesture – automatically, thereby minimizing physical awkwardness and facilitating the more rapid production of cross-modal symbol combinations. This interpretation rests on Lieberman's notion (1984) that syntax evolved to automaticize speech production, thereby enabling rapid speech output.

In sum, the nature of these particular rules suggests that the evolutionary origin of grammar lies in rules for sequencing actions. Insofar as the early stages of human ontogenetic development reflect our ape origins more strongly than do later stages, Greenfield's findings of the development of grammars of action in human infancy lend further support to this view. In the research of Greenfield and colleagues (Greenfield, 1978b; Greenfield, Nelson, & Saltzman, 1972), the parallels between action grammars and linguistic grammars are closest for the earliest stages of language. This suggests that linguistic grammars evolve, both phylogenetically and ontogenetically, out of rules for ordering action sequences, but that they transcend these origins in later stages of phylogenetic and ontogenetic development.

*Comparison with children at the two-word stage*

While creativity is the rule in normal language acquisition, the degree of independence from a model of Kanzi's invented combinatorial rules is probably matched only by that of young deaf children of hearing parents studied by Goldin-Meadow and colleagues (Goldin-Meadow, 1979; Goldin-Meadow & Feldman, 1977; Goldin-Meadow & Mylander, 1984). In addition, Kanzi's way of using gestural indication to specify the agent in an agent–action relation was paralleled by these same deaf children (Goldin-Meadow, 1979; Goldin-Meadow & Mylander, 1984).

With respect to hearing children, Kanzi's development of a generative action–object symbol-ordering rule replicates a pervasive tendency in normal language acquisition to induce word-order rules from syntactic models presented in the environment (Baker & Greenfield, 1988; Greenfield, Reilly, Leaper, & Baker, 1985; Slobin, 1973).

Another important source of similarity to human children was the range of semantic relations, displayed in Table 20.1, which overlaps so much with Brown's analysis (1973) of children acquiring a variety of human languages at the two-word stage. Not only do the high-frequency relations overlap, as mentioned earlier, but also the low-frequency relations overlap. For example, the conjoined entities and conjoined locations relations in Table 20.1 are equivalent to Brown's low-frequency relation of conjunction. Conjoined actions constitute a form of conjunction also, but Brown (1973) does not mention any conjoined actions in his analysis.

Although a "true" grammatical rule must use some formal device such as word order to mark the semantic relation, children do not always do this at the two-word stage (Bowerman, 1973b; Brown, 1973; Goldin-Meadow & Mylander, 1984). Some children use word order more consistently than others, but even relatively consistent children show variability when analyses such as that shown in Table 20.1 are done (Bowerman, 1973b; Brown, 1973; Goldin-Meadow & Mylander, 1984). Consequently, the fact that only a subset of Kanzi's semantic relations demonstrated statistically reliable order was in line with the findings from child language.

Indeed, out of six relations tested for ordering patterns by Goldin-Meadow and Mylander, the modal deaf child in their study did not show *any* statistically significant ordering patterns for two-sign sentences, and the mean was .9 (less than one ordering pattern per child). However, it must be remembered that the lack of ordering patterns, where it occurred in the deaf children, was correlated with an extremely small corpus of data. Nevertheless, Kanzi's data provided more evidence for the use of the formal device of symbol order in two-symbol utterances than did the data of these deaf children, widely considered to have created language.

In addition, the deaf children of hearing parents were limited to semantically based ordering rules; they did not create purely formal syntactic rules (Goldin-Meadow, 1979). Kanzi, in contrast, not only invented a semantically based rule (for symbolic ordering of serial actions) but also invented a purely formal rule of symbol order (place gesture after lexigram).

*The creation of ergativity by children and chimpanzees*

An ergative language is distinguished by the fact that objects of transitive verbs receive grammatical marking identical to subjects of intransitive verbs, whereas transitive subjects receive a distinctive marking. Ergative languages stand in contrast to another class of language, termed accusative. In an accusative language (such as English), both transitive and intransitive subjects receive identical grammatical marking, in contrast to the different marking of transitive objects. Ergative and accusative languages implicitly categorize grammatical roles in different ways.

Goldin-Meadow (1979) found that her deaf subjects of hearing parents provided evidence of an incipient ergative system. Kanzi also created an incipient ergative system: He placed both intransitive agents and transitive objects after the action symbol.

Whereas the deaf children initiated ergative structure in the absence of a grammatical model in their environment, Kanzi produced the rudiments of an *ergative* system in the face of the *accusative* model presented by his English-speaking caregivers. Ergativity was, therefore, a creative invention on Kanzi's part, just as it was for the deaf children.

Kanzi's ergative system was incomplete because he produced too few three-term transitive utterances (with agent, action, and object) to permit an analysis of the ordering of transitive agents. Except in the case of one subject, the deaf children's data were also incomplete. Another point to note about Kanzi's ergative system is that although his action–agent utterances had an intransitive surface structure, they usually were interpreted by his caregivers as being partial realizations of underlying transitive relations.

Kanzi's specific syntactic ordering – placing intransitive agent and transitive object *after* the action symbol – differed from that of the one deaf child with more complete data, who placed intransitive agent and transitive object *before* the action element. However, the important point is that Kanzi, like the deaf child, marked his intransitive agents and transitive objects in an identical fashion. In so doing, he implicitly created an ergative categorization of basic grammatical rules.

### *Pan paniscus, Pan troglodytes, and human children*

Brown (1970, 1973) and Gardner and Gardner (1971) have compared Washoe's two-sign utterances to those of children. Based on a set of six semantic relations that Brown used in his 1970 paper, the Gardners reported that 78% fell into those six categories (equivalent to our entity–attribute, goal–action, agent–action, action–object, and agent–object, plus possession, which we did not observe). Although we do not know how many of the 294 utterance types (any word pair could appear only once in the corpus) were imitated, and how many were spontaneous, the similarity to Kanzi's semantic relations, as well as to those of children, is striking.

Because the Gardners (1971) reported that they did not record sign order for the two-sign utterances, we do not know if any of the reported relations were marked by consistent ordering patterns. Nor did they report, with respect to their data, whether or not relations were marked with the inflections that are more important in American Sign Language than sign order. They did report one three-sign pattern that employed a consistent order pattern. Whereas Kanzi's three-symbol rule was an invented rule, Washoe's was modeled by her caregivers (Gardner & Gardner, 1971). Given this modeling effect, it would be particularly important to know to what extent these patterned three-sign utterances were direct imitations and to what extent they were spontaneous. Terrace and associates claimed that Washoe generally imitated her caregivers. But their conclusions were based on analyzing two commercial films, rather than on complete data. Whether or not Washoe's semantic relations were spontaneous to a significant degree, the data indicate that Kanzi's level of syntactic invention was greater than Washoe's both at the two-symbol level and at the three-symbol level.

### *The problem of a double standard*

Comparative developmental psycholinguistics has been plagued by a double standard. Because children ultimately develop language, their early stages are interpreted as having greater linguistic significance than the same stages in primates (de Villiers, 1984; Nelson, 1986). When children make up novel words on a one-shot basis, it is called lexical innovation (e.g., Clark, 1983). When chimpanzees do the same thing (e.g., Washoe's famous water-bird) (Fouts, 1974a), it is termed ambiguous (Terrace et al., 1979, 1980). One possible conclusion is that methodological standards in child language should be more rigorous. Another, more important conclusion is that we should compare developing behaviors across species objectively, without being influenced by the nature of later stages in either species. This methodological stricture is important because without it, it becomes impossible to compare developments in species whose early stages may be more similar than their developmental endpoints. Yet it is just such comparisons that may yield the most interesting behavioral data on the evolution of language.

*Language development, language history, and language variability: The role of context*

From this perspective, let us consider the controversy about rich interpretation. The fact that it is necessary to interpret the nonverbal context to assign meaning relations to children's two-word utterances has caused critics to say that the structure may be in the observer rather than in the child. If we move from the ontogeny of the child to the historical development of the language, Rolfe (1988) points out that the characteristic of early stages of language evolution is that the structures are not explicit, but require inference on the part of the listener. We may say the same for the child. Rich interpretation is required exactly *because* explicit syntactic and semantic marking is the product of development; it is not a characteristic of the early stages. Just as the early stages in language evolution are characterized by the necessity to infer structure from context, so are the early stages of child language. It is of evolutionary interest that bonobo language, like early human languages and like the language of young children, requires the listener to infer structure from context.

Moreover, even modern human languages, spoken by mature speakers, vary considerably in their reliance on inference from context, versus explicit grammatical marking (Comrey, 1987). Where relatively little information is grammatically marked, languages are termed "pragmatic." For example, as Duff (1989) points out, Chinese is considered to be just such a pragmatic language, in which much information must be recovered from context (e.g., agents, time reference), just as is required by Kanzi's productions and those of deaf children of hearing parents (Goldin-Meadow, 1979).

*Concluding comments*

A number of qualitative similarities between the two species, bonobo and human, have potential evolutionary significance:

1. The capacity for grammatical rules (including arbitrary ones) in Kanzi's semiotic productions shows grammar as an area of evolutionary continuity. Here we might prefer to speak of protogrammar rather than grammar. However, as stated earlier, the comparative data are such that if we speak of bonobo rules as protogrammar, we should apply the same term to the 2-year-old child.

2. The existence of action-based rules shows that the ordering of action sequences might be one of the evolutionary roots of grammar, even though this connection does not exist in normal human adults. Here it is interesting to note that when linguistic grammar breaks down in the agrammatism of Broca's aphasia, the grammar of manual action, as measured by a hierarchical construction task (Greenfield & Schneider, 1977), also breaks down (Grossman, 1980).

3. The finding of rule creation has implications for the evolutionary con-

tinuity of language. At various points in the history of our species, grammatical rules for human language were created, not merely learned. Although the rules just described were simple, they nevertheless were created de novo by Kanzi. This suggests that some rudiments of the ability to create a grammar have an ancient evolutionary history in a common ancestor of our chimpanzee and human species.

4. The particular grammatical structures invented by Kanzi, while distinct from those modeled by his caregivers, resemble those used by human beings:

(a) *Kanzi's conjoined action rule.* Ordered conjoined action sequences occur widely in human languages, where they are called serial verbs. Serial verbs are particularly important features of certain languages, figuring prominently in many West African languages, for example (Lord, 1989).

(b) *Combining gesture and lexigram.* The creative combination of semiotic elements from different symbolic levels was done by deaf children of hearing parents, who combined indexical points with iconic signs (Goldin-Meadow, 1979; Goldin-Meadow & Mylander, 1984). Kanzi's invented rule for combining lexigram with gesture (place gesture last) also involved different symbolic levels: He combined indexical gestures with arbitrary lexigrams.

(c) *Ergativity.* Kanzi invented a primitive version of an ergative grammatical system, despite the fact that an accusative system was modeled for him. In so doing, a bonobo chimpanzee spontaneously created one of the two logically possible grammatical groundplans utilized by all human languages (A. Duranti, pers. commun., November 1989).

Ideally, an evolutionary reconstruction is based on finding the same trait in all members of a genus. However, the more sophisticated grammatical qualities described in this chapter – notably protogrammatical rule invention – have not, so far, been either claimed or established in *Pan troglodytes*. At this point, we do not know if that is because of a true species difference or merely because of a difference in research methodology. An ongoing study involving the rearing of a pygmy chimpanzee with a common chimpanzee by Savage-Rumbaugh should begin to answer this question.

Simple recapitulationism is not a possible model for relating the evolution of human language to modern-day primate species, who have themselves evolved since the branching of the evolutionary tree. However, our evidence bears out the idea that structural dependencies govern development and therefore evolution (Parker & Gibson, 1979). Kanzi, like a human child, first acquired individual lexical items. At a later stage, he combined those elements into two-term relations. Still later, Kanzi systematically combined two-term relations into three-element utterances. Although the specific semantic content and formal nature of some of Kanzi's rules seemed unique to his species and symbol system, his global development of grammatical structure and his array of semantic relations mirrored that of the human child. Evolution could not have reversed this transspecific developmental sequence from simple to more complex structures.

In addition, Kanzi's array of semantic relations and the presence of rule-bound sequencing resembled the early language productions of human children. Because of the generally greater similarities in related species at the earlier points in ontogenetic development, these resemblances do not seem like an evolutionary coincidence. Finally, Kanzi's capacity to invent simple grammatical or protogrammatical rules provides clues as to the evolutionary origins of grammar, as well as a mechanism for historical language change in the absence of genetic evolution. This inventive capacity suggests that the ancestor of the pygmy chimpanzee may have had the cognitive prerequisites to invent a protogrammar. This protogrammar could then have provided an evolutionary foundation for the later development of full-blown grammar, just as the two-word stage of child language forms a developmental foundation for the more complex and abstract adult grammar that follows.

**Note**

1. For dissenting views on cebus imitation see Gibson, P&G7; Parker & Poti', P&G8; Visalberghi & Fragaszy, P&G9.

**Acknowledgments**

We would like to thank Elizabeth Rubert, Jeannine Murphy, Phillip Shaw, Rose Sevcik, and Kelly McDonald for assistance in all aspects of the data collection. Thanks also to Penny Nelson for help with data analysis and to Duane Rumbaugh and Terrence Deacon for stimulating discussion and to Laura Weiss for manuscript preparation. Alessandro Duranti provided helpful information concerning comparative linguistic structure. We very much appreciated the helpful comments on the first draft by Sue T. Parker and Susan Goldin-Meadow and the bibliographic assistance provided by Sue T. Parker.

The work described in this chapter and its preparation were supported by National Institutes of Health grant NICHD-06016, which supports the Language Research Center, cooperatively operated by Georgia State University and the Yerkes Regional Primate Research Center of Emory University. In addition, the research is supported in part by RR-00165 to the Yerkes Regional Primate Research Center of Emory University. PMG was supported by the Bunting Institute, Radcliffe College, by a grant from the Office of Naval Research to the Bunting Institute, Radcliffe College, by an award from the UCLA College Institute, and by the UCLA Gold Shield Faculty Prize.

**References**

Baker, N. D., & Greenfield, P. M. (1988). The development of new and old information in young children's early language. *Language Sciences*, *10*, 3–34.

Bloom, L. (1970). *Language development: Form and function in emerging grammars*. Cambridge, MA: M.I.T. Press.

Bloom, L. (1971). Why not pivot grammar? *Journal of Speech and Hearing Disorders*, *36*, 40–50.

Bloom, L., Rocissano, L., & Hood, L. (1976). Adult–child discourse: Developmental interaction between information-processing and linguistic knowledge. *Cognitive Psychology*, *8*, 521–552.

Bowerman, M. (1973a). *Early syntactic development: A cross-linguistic study with special reference to Finnish*. Cambridge University Press.

Bowerman, M. (1973b). Structural relationships in children's utterances: Syntactic or semantic?

In T. E. Moore (Ed.), *Cognitive development and the acquisition of language* (pp. 197–213). New York; Academic Press.
Braine, M. D. S. (1976). Children's first word combinations. *Monographs of the Society for Research in Child Development, 41* (1, Serial No. 164), 1–96.
Bronowski, J., & Bellugi, U. (1970). Language, name, and concept. *Science, 168*, 669–673.
Brown, R. (1970). The first sentences of child and chimpanzee. In R. Brown (Ed.), *Psycholinguistics* (pp. 208–231). New York: Free Press.
Brown, R. (1973). *A first language: The early stages.* Cambridge, MA: Harvard University Press.
Bruning, J. L., & Kintz, B. L. (1977). *Computational handbook of statistics*. Glenview, IL: Scott, Foresman.
Chevalier-Skolnikoff, S. (1976). The ontogeny of primate intelligence and its implications for communicative potential: A preliminary report. *Annals of the New York Academy of Sciences, 280,* 173–216.
Chevalier-Skolnikoff, S. (1989). Spontaneous tool use and sensorimotor intelligence in *Cebus* compared with other monkeys and apes. *Behavioral and Brain Sciences, 12*(3), 561–588.
Chomsky, N. (1965). *Aspects of the theory of syntax.* Cambridge, MA: MIT Press.
Chomsky, N. (1980). Human language and other semiotic systems. *Semiotica, 25*, 31–44.
Clark, E. (1983). Meaning and concepts. In P. Mussen (Ed.), *Handbook of child psychology: Cognitive development* (Vol. 3, pp. 787–840). New York: Wiley.
Comrey, B. (1987). Paper presented at the international Pragmatics Association, Antwerp.
de Villiers, J. (1984). Limited input? Limited structure. Commentary on Goldin-Meadow & Mylander. *Monographs of the Society for Research in Child Development, 49*, 122–129.
Duff, P. (1989, March). *"Protogrammar" in bonobos and deaf children (without ASL input).* Unpublished paper, Applied Linguistics Program, UCLA.
Feldman, H., Goldin-Meadow, S., & Gleitman, L. (1978). Beyond Herodotus: The creation of language by linguistically deprived deaf children. In A. Lock (Ed.), *Action, symbol, and gesture: The emergence of language* (pp. 351–414). New York: Academic Press.
Fouts, R. S. (1973). Acquisition and testing of gestural signs in four young chimpanzees. *Science, 180*, 978–980.
Fouts, R. S. (1974a). Language: Origins, definitions and chimpanzees, *Journal of Human Evolution, 3*, 475–482.
Fouts, R. S. (1974b). Capacities for language in great apes. In R. H. Tuttle (Ed.), *Socioecology and psychology of primates* (pp. 371–390). The Hague: Mouton.
Fouts, R. S. (1975). Communication with chimpanzees. In I. Eibl-Eibesfeld & G. Kurth (Eds.), *Hominisation und Verhalten* (pp. 137–158). Stuttgart: Gustav Fisher.
Fouts, R. S., & Couch, J. B. (1976). Cultural evolution of learned language in chimpanzees. In E. Simmel & M. Hahn (Eds.), *Communicative behavior and evolution* (pp. 141–161). New York: Academic Press.
Gardner, B. T., & Gardner, R. A. (1971). Two-way communication with an infant chimpanzee. In A. Schrier & F. Stollnitz (Eds.), *Behavior of nonhuman primates* (pp. 117–183). New York: Academic Press.
Gardner, B. T., & Gardner, R. A. (1974). Comparing the early utterances of child and chimpanzee. In A. Pick (Ed.), *Minnesota symposium on child language* (pp. 3–23). Minneapolis: University of Minnesota Press.
Gardner, B. T., & Gardner, R. A. (1978). Comparative psychology and language acquisition. *Annals of the New York Academy of Sciences, 309*, 37–76.
Gardner, B. T., & Gardner, R. A. (1980). Two comparative psychologists look at language acquisition. In K. Nelson (Ed.), *Children's language* (Vol 2, pp. 331–369). New York: Gardner Press.
Gardner, B. T., & Gardner, R. A. (1985). Signs of intelligence in cross-fostered chimpanzees. *Philosophical Transactions of the Royal Society, B, 308*, 159–176.
Gardner, R. A., & Gardner, B. T. (1969). Teaching sign language to a chimpanzee. *Science, 165*, 654–672.

Gardner, R. A., & Gardner, B. T. (producers). (1973). Teaching sign language to the chimpanzee (film). University Park, PA: Psychological Cinema Register.

Goldin-Meadow, S., & Feldman, H. (1977). The development of language-like communication without a language model. *Science*, *197*, 401–403.

Goldin-Meadow, S. (1979). Structure in a manual communication system developed without a conventional language model: Language without a helping hand. In H. A. Whitaker (Ed.), *Studies in neurolinguistics* (Vol. 4, pp. 125–209). New York: Academic Press.

Goldin-Meadow, S., & Mylander, C. (1984). Gestural communication in deaf children: The effects and noneffects of parental input on early language development. *Monographs of the Society for Research in Child Development*, *49*, (3–4, Serial No. 207), 1–120.

Gould, S. J. (1977). *Ontogeny and phylogeny*. Cambridge, MA: Harvard University Press.

Greenfield, P. M. (1978a). Commentary on Developmental processes in the language learning of child and chimp, by Savage-Rumbaugh and Rumbaugh. *Behavioral and Brain Sciences*, *4*, 573–574.

Greenfield, P. M. (1978b). Structural parallels between language and action in development. In A. Lock (Ed.), *Action, gesture, and symbol: The emergence of language* (pp. 415–445). London: Academic Press.

Greenfield, P. M. (1980). Towards an operational and logical analysis of intentionality: The use of discourse in early child language. In D. Olson (Ed.), *The social foundations of language and thought: Essays in honor of J. S. Bruner* (pp. 254–279). New York: Norton.

Greenfield, P. M., Nelson, K., & Saltzman, E. (1972). The development of rulebound strategies for manipulating seriated cups: A parallel between action and grammar. *Cognitive Psychology*, *3*, 291–310.

Greenfield, P., Reilly, J., Leaper, C., & Baker, N. (1985). The structural and functional status of single-word utterances and their relationship to early multi-word speech. In M. Barrett (Ed.), *Children's single-word speech* (pp. 233–267). London: Wiley.

Greenfield, P. M., & Savage-Rumbaugh, E. S. (1984). Perceived variability and symbol use: A common language–cognition interface in children and chimpanzees. *Journal of Comparative Psychology*, *98*, 201–218.

Greenfield, P. M., & Savage-Rumbaugh, E. S. (in press). Imitation, grammatical development, and the invention of protogrammar. In N. Krasnegor, D. M. Rumbaugh, R. Schiefelbusch, & M. Studdert-Kennedy (Eds.), *Biobehavioral foundations of language development*. Hillsdale, NJ: Erlbaum.

Greenfield, P. M., & Schneider, L. (1977). Building a tree structure: The development of hierarchical complexity and interrupted strategies in children's construction activity. *Developmental Psychology*, *3*, 299–313.

Greenfield, P. M., & Smith, J. H. (1976). *The structure of communication in early language development*. New York: Academic Press.

Grossman, M. A. (1980). A central processor for hierarchically-structured material: Evidence for Broca's aphasia. *Neuropsychologia*, *18*, 299–308.

Hill, J. H. (1978). Apes and language. *Annual Review of Anthropology*, *7*, 89–112.

Jordan, C., & Jordan, H. (1977). Versuche zur Symbol-Ereignis-Verknupfung bei einer Zwergschimpansen (*Pan paniscus* Schwarz, 1929). *Primates*, *18*, 515–529.

King, M. C., & Wilson, A. C. (1975). Evolution at two levels in humans and chimpanzees. *Science*, *188*, 107–116.

Lieberman, P. (1984). *The biology and evolution of language*. Cambridge, MA: Harvard University Press.

Limber, J. (1977). Language in child and chimp? *American Psychologist*, *32*, 280–293.

Lord, C. (1989). *Historical change in serial verb constructions in languages of West Africa*. Unpublished doctoral dissertation, University of California at Los Angeles.

McNeill, D. (1974). Sentence structure in chimpanzee communication. In K. Connolly & J. Bruner (Eds.), *The growth of competence* (pp. 75–94). New York: Academic Press.

Matsuzawa, T. (1985). Colour naming and classification in a chimpanzee (*Pan troglodytes*). *Nature*, *315*, 57–59.

Miles, H. L. (1983). Apes and language: The search for communicative competence. In J. De Luce & H. T. Wilder (Eds.), *Language in primates* (pp. 43–61). New York: Springer-Verlag.
Nelson, K. (1987). Reply to Seidenberg and Petitto. *Journal of Experimental Psychology: General, 116*(3), 293–296.
*Nova* (1976). The first signs of Washoe (film). New York: Time-Life Films.
Parker, S. T. (1985). A social technological model for the evolution of language. *Current Anthropology, 26*, 617–626.
Parker, S. T., & Gibson, K. R. (1979). A developmental model for the evolution of language and intelligence in early hominids. *Behavioral and Brain Sciences, 2*, 367–408.
Patterson, F. G. (1978). The gestures of a gorilla: Language acquisition in another pongid. *Brain and Language, 5*, 72–97.
Patterson, F. G. (1980). Innovative uses of language by a gorilla: A case study. In K. Nelson (Ed.), *Children's language* (Vol. 2, pp. 497–561). New York: Gardner Press.
Peirce, C. S. (1931). Division of signs. In C. Hartshorne & P. Weiss (Eds.), *The collected papers of Charles Sanders Peirce* (pp. 134–155). Cambridge, MA: Harvard University Press.
Petitto, L. A. (1987). On the autonomy of language and gesture: Evidence from the acquisition of personal pronouns in American Sign Language. *Cognition, 27*(1), 1–52.
Petitto, L. A., & Seidenberg, M. S. (1979). On the evidence for linguistic abilities in signing apes. *Brain Language, 8*, 162–183.
Piaget, J. (1962). *Play, dreams, and imitation*. New York: Norton.
Premack, D. (1970). A functional analysis of language. *Journal of the Experimental Analysis of Behavior, 14*, 107–125.
Premack, D. (1971). Language in chimpanzee. *Science, 172*, 808–822.
Premack, D. (1976). Language and intelligence in ape and man. *American Scientist, 64*, 674–683.
Rolfe, L. H. (1988, July). *Pragmatics and the evolution of syntax*. Paper presented at the NATO Advanced Workshop on the Evolution of Language, Cortona, Italy.
Rumbaugh, D. M. (Ed.). (1977). *Language learning by a chimpanzee: The Lana Project*. New York: Academic Press.
Rumbaugh, D. M., & Gill, T. V. (1976a). Lana's mastery of language skills. *Annals of the New York Academy of Sciences, 280*, 562–578.
Rumbaugh, D. M., & Gill, T. V. (1976b). Language and the acquisition of language-type skills by a chimpanzee (*Pan*). *Annals of the New York Academy of Sciences, 270*, 90–124.
Rumbaugh, D. M., Gill, T. V., & von Glasersfeld, E. C. (1973). Reading and sentence completion by a chimpanzee. *Science, 182*, 731–733.
Sarles, H. B. (1972, June). *The search for comparative variables in human speech*. Paper presented at the Animal Behavior Society meetings, Reno, NV.
Savage-Rumbaugh, E. S. (1984). *Pan paniscus* and *Pan troglodytes*: Contrasts in preverbal communicative competence. In R. L. Susman (Ed.), *The pygmy chimpanzee: Evolutionary biology and behavior* (pp. 395–413). New York: Plenum.
Savage-Rumbaugh, E. S. (1986). *Ape language: From conditioned response to symbol*. New York: Columbia University Press.
Savage-Rumbaugh, E. S. (1988). A new look at ape language-comprehension of vocal speech. In D. Leger (Ed.), *Comparative perspectives in modern psychology. Nebraska Symposium on Motivation* (Vol. 35, pp. 201–256). Lincoln: University of Nebraska Press.
Savage-Rumbaugh, E. S., McDonald, K., Sevcik, R. A., Hopkins, W., & Rubert, E. (1986). Spontaneous symbol acquisition and communicative use by pygmy chimpanzees (*Pan paniscus*). *Journal of Experimental Psychology: General, 115*, 211–235.
Savage-Rumbaugh, E. S., & Rumbaugh, D. M., & Boysen, S. (1978a). Linguistically mediated tool use and exchange by chimpanzee (*Pan troglodytes*). *Behavioral and Brain Sciences, 1*, (1–28), 539–554.
Savage-Rumbaugh, E. S., Rumbaugh, D. M., & Boysen, S. (1978b). Symbolic communications between two chimpanzees (*Pan troglodytes*). *Science, 201*, 641–644.
Savage-Rumbaugh, E. S., Rumbaugh, D. M., Smith, S. T., & Lawson, J. (1980). Reference: The linguistic essential. *Science, 210*, 922–925.

Savage-Rumbaugh, E. S., Sevcik, R. A., Brakke, K. E., Rumbaugh, D. M., & Greenfield, P. M. (in press). Symbols: Their communicative use, combination, and comprehension by bonobos (*Pan paniscus*). In L. P. Lipsitt & C. Rovee-Collier (Eds.), *Advances in infancy research*, Vol. 7. Norwood, NJ: Ablex.

Savage-Rumbaugh, E. S., Sevcik, R. A., Rumbaugh, D. M., & Rubert, E. (1985). The capacity of animals to acquire language: Do species differences have anything to say to us? *Philosophical Transactions of the Royal Society*, *B 308*, 177–185.

Savage-Rumbaugh, S., & Wilkerson, B. J. (1978). Socio-sexual behavior in *Pan paniscus* and *Pan troglodytes*: A comparative study. *Journal of Human Evolution*, *7*, 327–344.

Schlesinger, I. M. (1971). Production of utterances and language acquisition. In D. Slobin (Ed.), *The ontogenesis of grammar: A theoretical symposium* (pp. 63–101). New York: Academic Press.

Seidenberg, M. S., & Petitto, L. A. (1979). Signing behavior in apes: A critical review. *Cognition*, *7*, 177–215.

Slobin, D. I. (1973). Cognitive prerequisites for the development of grammar. In C. A. Ferguson & D. I. Slobin (Eds.), *Studies of child language development* (pp. 175–208). New York: Holt, Rinehart & Winston.

Studdert-Kennedy, M. (1988, June). *Introduction: Language development from an evolutionary perspective.* Paper presented at a conference on the biobehavioral foundation of language development. Leesburg, VA.

Terrace, H. S. (1984). Apes who "talk": Language or projection of language by their teaching? In J. De Luce & H. T. Wilder (Eds.), *Language in primates* (pp. 19–42). New York: Springer-Verlag.

Terrace, H. S. (1985). In the beginning was the "name." *American Psychologist*, *40*, 1011–1028.

Terrace, H. S., Petitto, L. A., Sanders, F. J., & Bever, T. G. (1979). Can an ape create a sentence? *Science*, *206*, 891–900.

Terrace, H. S., Petitto, L. A., Sanders, F. J., & Bever, T. G. (1980). On the grammatical capacity of apes. In K. Nelson (Ed.), *Children's language* (Vol. 2, pp. 371–496). New York: Gardner Press.

Volterra, V. (1987). From single communicative signal to linguistic combinations in hearing and deaf children. In J. Montanger, A. Tryphon, & S. Dionnet (Eds.), *Symbolism and knowledge* (pp. 89–106). Geneva: Jean Piaget Archives Foundation.

Weiss, M. L. (1987). Nucleic acid evidence bearing on hominoid relationships. *Yearbook of Physical Anthropology*, *30*, 41–73.

Wiener, L. F. (1984). The evolution of language: A primate perspective. *Word*, *35*, 255–269.

Zihlman, A. L. (1979). Pygmy chimpanzee morphology and the interpretation of early hominids. *South African Journal of Science*, *75*, 165–168.

Zihlman, A. L., Cronin, J. E., Cramer, D. L., & Sarich, V. M. (1978). Pygmy chimpanzee as a possible prototype for the common ancestor of humans, chimpanzees, and gorillas. *Nature*, *275*, 744–746.

# Index